BOSONIZATION

BOSONIZATION

Editor

Michael Stone

University of Illinois
Urbana–Champaign

World Scientific
Singapore • New Jersey • London • Hong Kong

Published by

World Scientific Publishing Co. Pte. Ltd.

P O Box 128, Farrer Road, Singapore 9128

USA office: Suite 1B, 1060 Main Street, River Edge, NJ 07661

UK office: 57 Shelton Street, London, WC2H 9HE

While every effort has been made to contact the publishers of reprinted papers prior to publication,we have not been successful in a few cases. Where we could not contact the publishers, we have acknowledged the source of the material. Proper credit will be given to these publishers in future editions of this work after permission is granted.

The editor and publisher would like to thank the following publishers of the various journals and books for their assistance and permission to include the selected reprints found in this volume:

American Institute of Physics
American Physical Society
Elsevier Science Publishers
Institute of Physics
Pitagora
Progress of Theoretical Physics
Springer-Verlag
The Royal Society

Library of Congress Cataloging-in-Publication
Bosonization / editor, Michael Stone.
p. cm.
ISBN 9810218478 -- ISBN 9810218486 (pbk)
1. Bosons. 2. Solitons. 3. Quantum field theory
4. Mathematical physics. I. Stone, Michael, Ph.D.
QC793.5.B62B67 1994
539.7'21--dc20 94-38551
CIP

Printed in Singapore.

PREFACE

The best way to learn any subject is to go to the masters and read their works. To this end part II of this book contains reprints of many of the "classic" papers relating to the method of bosonization. It should be of considerable use to students of low dimensional field theories and their applications in condensed matter and particle physics.

Sometimes it helps to have a parallel pedagogical text which can apply the benefit of hindsight to simplify the discussion. The four chapters of Part I are an attempt at such a text. I have tried to write at a level accessible to a graduate student who has had a first course in field theory. The first two chapters expound fairly standard material. The third reflects some personal bias as to what are interesting applications. The last chapter contains material that is perhaps less familiar, but deserves to be more widely known.

This book was written while I was at the Institute for Theoretical Physics in Santa Barbara. I would like to thank many of my colleagues at the ITP for their assistance. Joe Polchinski and Matthew Fisher have shared their insights into bosonization, and several times rescued me from the confusion generated by too many minus signs. Shyamoli Chaudhuri, Hartmut Monien, Ajit Srivastava, and Tony Zee have all helped me organize my thoughts. The efficient ITP staff reduced the burden of my duties and made it possible for me to have the time to write. I must thank Ping Ao, Daniel Boyanovsky, Eduardo Fradkin, Frank Gaitan, Juan Martinez, Mohit Randeria, and Sivaji Sondhi for asking questions and often providing answers. I must also thank Phylli Sladek for her friendship and for keeping me in contact with reality.

Michael Stone
Santa Barbara
Summer 1994

CONTENTS

INTRODUCTION

In 1975 the particle physics community was startled by a short paper written by Sidney Coleman [rep. 8] in which he demonstrated that the quantum Sine-Gordon model, a theory of interacting bosonic "mesons", was equivalent to the Thirring model of interacting massive fermions. Coleman further conjectured that the Sine-Gordon soliton was the fundamental fermion of the Thirring model. This assertion, that a "lump" in a Bose field might be a fermion, seems to have surprised even Coleman himself. A similar claim had been made by Skyrme in 1958 [rep. 2] but, like Skyrme's now familiar soliton model of the nucleon [1], the idea seemed too strange to be true. The new conjecture was more convincingly argued, but in the acknowledgments section of [rep. 8] Coleman thanks a colleague for reassuring him of his sanity. I myself remember being a second-year graduate student listening to Coleman begin a seminar in Cambridge (England) with the words "I may be crazy". He went on, of course, to give a characteristically sane and lucid exposition of his work.

Coleman had *not* been driven mad by spending too many years in too few dimensions. Nonetheless, the idea that Bose and Fermi models could be different facets of one underlying theory might have generated less surprise had we all been more aware of the long history of Bose-Fermi equivalences. In the condensed matter literature the tale goes back to 1934 when Felix Bloch was studying the rate of energy loss by charged particles traversing a metal [2]. Bloch claimed that the excitation spectrum of a degenerate Fermi gas was exhausted by its density fluctuations. Because of this, he argued, the properties of the many-fermion system could be obtained by keeping track of a far smaller number of (bosonic) plasmons. The converse notion was introduced by Pascual Jordan in a long series of papers published between 1935 and 1937 [3]. In these he advanced the idea that the photon might be composed of neutrinos. While the neutrino theory of light has faded from memory, mostly because it could only be made to work in one dimension, Jordan had correctly calculated the commutators of the one-dimensional fermion currents, showed how they were modified by the presence of an infinitely deep Dirac sea, and how this lead to the addition of what later became known as Schwinger terms [4]. Jordan used his result to represent the fermion currents in terms of a free boson field. He also discussed the commutators for currents on a finite interval, and in his work one can see the now familiar Kac-Moody central extension of the algebra generated by the Fourier components of the currents.

These early results were cited in 1950 by Sin-itiro Tomonaga as the motivation for his work on one-dimensional fermions [rep. 1]. Tomonaga knew that Bloch's claims were only approximate for three dimensions, but he realized that they were essentially exact for particles moving on a line. The formalism developed by Jordan allowed Tomonaga to present an extensive analysis of the properties of a one-dimensional Fermi gas using only the bosonic density operators.

In his paper Tomonaga regarded the Schwinger-term modification of the commutation relations to be a convenient approximation, but in 1965 Daniel Mattis and Elliot Lieb [rep. 3], correcting a subtle error in an earlier work by Luttinger [5], argued that the modification was an inevitable consequence of there being an infinite number of degrees of freedom, and made explicit the connection with the 1959 work of Schwinger [4]. Soon after Alba

Theumann [6] used Mattis and Lieb's boson fields to calculate the fermion propagator and so obtained the anomalous dimension of the interacting Fermi field. By 1974 Alan Luther, in collaboration with V. Emery and I. Peschel, had also demonstrated the Sine-Gordon Thirring model equivalence, except that being condensed-matter physicists they referred to the massive Thirring model as the backscattering Luttinger model [rep. 6 and rep. 7]. This work was published in condensed-matter journals and so was unknown to Coleman — although, as he explains in [rep. 8], he and Luther occupied offices in the same corridor.

Coleman's 1975 paper addressed only the equivalences between boson operators and operators bilinear in the Fermi field. Stanley Mandlestam's beautiful construction [rep. 9] of the individual Fermi fields as "kink" creation and annihilation operators completed the picture and confirmed that the Sine-Gordon soliton was indeed the Thirring fermion. Mandlestam's operator also has its antecedents in the condensed-matter literature, but I think it fair to say none of these prior works contained any hint of Mandlestam's simple and direct picture of the fermion operator creating a soliton.

A complicated boson current construction of the Fermi fields had appeared in the 1974 paper of Antonio, Frishman and Zwanziger [7]. Simpler vertex-operator forms for the Fermi fields were used by Schotte and Schotte, Mattis [rep. 5], and in the work of Luther and Peschel referred to above [rep. 6]. The Schotte and Schotte 1969 paper [rep. 4] is the earliest I have found that contains a recognizable vertex operator construction. These authors were studying the X-ray edge problem. Here a photon has just sufficient energy to promote a deeply bound core electron to the conduction band. The exact shape of the absorption threshold depends on a delicate interplay between the newly elevated conduction electron and the distortion of the extended states produced by the charge left on the ionized atom (Anderson's "orthogonality catastrophe" [8]). Schotte and Schotte represented the S-wave conduction-band electrons by means of Luttinger-Tomonaga oscillators and, needing to represent the creation of the new electron in the same manner, showed that one can use an exponential of the oscillator coordinates. They also showed that the non-canonical powers of t in the electron Green function were due to the anomalous dimension induced by the necessary normal ordering. The Schotte and Schotte construction was for a free fermion at a single location. A representation of interacting Fermi fields in arbitrary positions, and with the correct statistics, was achieved in the 1974 papers of Mattis [rep. 5], and Luther and Peschel [rep. 6]. Mattis' and Luther & Peschel's operator is identical to Mandlestam's, but being written in momentum space, its soliton-creating character is not apparent.

Subsequently the abelian bosonization schemes were generalized to non-abelian symmetries. This process was begun by the discovery by Marty Halpern of a representation for an $SU(N)$ current algebra [rep. 10]. In this representation the boson fields are in correspondence with a Cartan algebra (a maximally commuting set of generators) and the ladder operators are represented by vertex operators. Because of the non-uniform treatment of the generators, the group symmetry is not manifest in this type of construction. A manifestly covariant formalism was introduced by Ed Witten in 1984 [rep. 14].

Bosonization has also been extended to field theory on Riemann surfaces [rep. 18–24]. The motivation for doing this was provided by string theory, but I believe that the most interesting aspect of this work is the light it has cast on the algebraic geometry of such surfaces. In particular the coincidence of Bose and Fermi Green functions on a Riemann surface is equivalent to Fay's trisecant identity, a deep relation between theta functions that holds for period matrices deriving from algebraic curves, but not for general period matrices [9].

Another valuable by-product of the bosonization identities is the τ-function formalism of Sato, Sato, Miwa, Jimbo, Date and Kashiwara for families of integrable differential equations [1 and rep. 27].

References to papers not reprinted in Part II

[1] T. H. R. Skyrme, *Proc. Roy. Soc. London* **A247** (1958) 260; **A252** (1959) 236; **A260** (1961) 127.
[2] F. Bloch, *Z. Phys.* **81** (1933) 363; *Helv. Phys. Acta* **7** (1934) 385.
[3] P. Jordan, *Z. Phys.* **93** (1935) 464; **98** (1936) 759; **99** (1936) 109; **102** (1936) 243; **105** (1937) 114; **105** (1937) 229.
[4] J. Schwinger, *Phys. Rev. Lett.* **3** (1959) 296.
[5] J. M. Luttinger, *J. Math. Phys.* **4** (1963) 1154.
[6] A. Theumann, *J. Math. Phys.* **8** (1967) 2460.
[7] G. F. Antonio, Y. Frishman. D. Zwanziger, *Phys. Rev.* **D6** (1972) 988.
[8] P. W. Anderson, *Phys. Rev. Lett.* **18** (1967) 1049.
[9] D. Mumford, *Tata Lectures on Theta II* (Birkhäuser, Boston, 1984).
[10] M. Sato and Y. Sato, in *Lect. Notes in Num. Anal.* **5** (1982) 259.

Chapter 1
BOSONIZATION IN SPACE-TIME

1.1. Introduction

This chapter describes bosonization as an equality between Green functions. The euclidean space correlators of exponentials of free boson fields are found to miraculously coincide with the correlators of free fermions. This leads us to set up a table of correspondences between boson and fermion operators. The identity which makes all this work is easily proved, but like many identities seems to be pulled out of a hat. It is nonetheless the simplest manifestation of some deep mathematics which can only be hinted at in this introduction. For further details the reader must explore the reprinted articles.

1.2. Free Fermi Fields

In two euclidean dimensions the Dirac operator takes the form

$$D = \sigma_1 \partial_x + \sigma_2 \partial_y = \begin{bmatrix} 0 & 2\partial_z \\ 2\partial_{\bar{z}} & 0 \end{bmatrix}. \tag{1.2.1}$$

Spinor fields satisfying $D\psi = 0$ have upper components $\psi_R(z)$ which are holomorphic, *i.e.* $\partial_{\bar{z}}\psi_R = 0$, while their lower components, $\psi_L(\bar{z})$, are anti-holomorphic, $\partial_z \psi_L = 0$ (see Appendix 1A for notes on complex coordinates). In Minkowski space these chiral components would correspond to waves traveling to the right or to the left respectively. I will choose the time direction to be "1", so that $\gamma^1 = \sigma_1$, plays the role of "γ^0" in the definition $\overline{\psi} = \psi^\dagger \gamma^0$. The two-dimensional analogue of "γ^5" will be σ_3. With these conventions $\overline{\psi}_L = \psi_R^\dagger$ etc.

Because the propagator is equal to the Green function of the field equation, we can read off, with the help of (A.7), the free propagator

$$\langle \psi_R(z_1) \overline{\psi}_L(z_2) \rangle = \frac{1}{2\pi(z_1 - z_2)} \tag{1.2.2}$$

Multi-field correlators are built up from these with the aid of Wick's theorem, e.g.

$$\langle \psi_R(z_1) \cdots \psi_R(z_N) \overline{\psi}_L(\zeta_1) \cdots \overline{\psi}_L(\zeta_N) \rangle = \frac{1}{(2\pi)^N} \det \left| \frac{1}{z_i - \zeta_j} \right| \tag{1.2.3}$$

Correlators with both $\psi_R(z)$ and $\psi_L(\bar{z})$'s factor into a holomorphic part and an anti-holomorphic part.

$$\langle \psi_R(z_1) \psi_L(\bar{z}_2) \overline{\psi}_L(\zeta_1) \overline{\psi}_R(\bar{\zeta}_2) \rangle = \frac{1}{(2\pi)^2} \frac{1}{z_1 - \zeta_1} \frac{1}{\bar{z}_2 - \bar{\zeta}_2} \tag{1.2.4}$$

1.3. Free Bosons

The Green function for two-dimensional scalar fields must satisfy $-\nabla^2 G(\mathbf{r}) = \delta^2(\mathbf{r})$. Now, strictly speaking, there is no such $G(\mathbf{r})$, because the momentum space integral giving $G(\mathbf{r})$,

$$G(\mathbf{r}) = \int \frac{d^2\mathbf{k}}{(2\pi)^2} \frac{e^{i\mathbf{k}\cdot\mathbf{r}}}{\mathbf{k}^2}, \tag{1.3.1}$$

is logarithmically divergent at small $\mathbf{k}$. However, by introducing a small mass μ to serve as an infrared cut-off, we can approximate $G(\mathbf{r}) = -\frac{1}{2\pi}\ln\mu|\mathbf{r}| + const + O(\mu|\mathbf{r}|)$. The constant can be absorbed into a renormalization condition, so we will actually use the simpler expression $G(\mathbf{r}) = -\frac{1}{2\pi}\ln\mu|\mathbf{r}|$. All quantities of interest will be infrared finite, and we will always take the limit $\mu \to 0$ at the end of the calculation.

We now use $G = -\frac{1}{2\pi}\ln\mu|\mathbf{r}|$ to evaluate the generating functional for vacuum correlators of a real scalar field

$$\mathcal{Z}(J) = \int d[\varphi] e^{-\int d^2\mathbf{r}\{\frac{1}{2}(\partial\varphi)^2 + iJ(\mathbf{r})\varphi(\mathbf{r})\}} = \mathrm{Det}^{-\frac{1}{2}}(\nabla^2) e^{-\frac{1}{2}\int J(\mathbf{r})G(\mathbf{r}-\mathbf{r}')J(\mathbf{r}')d^2\mathbf{r}d^2\mathbf{r}'}. \tag{1.3.2}$$

Setting $J(\mathbf{r}) = \alpha\delta^2(\mathbf{r} - r_1) - \alpha\delta^2(\mathbf{r} - \mathbf{r}_2)$, we find

$$\langle e^{i\alpha\varphi(\mathbf{r}_1)} e^{-i\alpha\varphi(\mathbf{r}_2)} \rangle \propto e^{-\alpha^2 \frac{1}{2\pi}\ln|\mathbf{r}_1 - \mathbf{r}_2|} = \frac{1}{|\mathbf{r}_1 - \mathbf{r}_2|^{\alpha^2/2\pi}}. \tag{1.3.3}$$

Two conventions need to be explained regarding this formula. Firstly, in order to get a finite answer, I did not include the divergent "self energy" factors coming from the terms in the exponent where the Green function is to be evaluated at 0. This means that I have really evaluated correlators of the "normal-ordered" exponentials: $\exp\{\pm i\alpha\varphi\}$:. The omission of these divergences accounts for the dimensional mismatch between the right and left hand sides of (1.3.3). The normal-ordered exponential, unlike its un-ordered counterpart, is not dimensionless. Secondly, I have dropped reference to μ. If I *had* included a self-energy Green function in the form $\frac{1}{2\pi}\ln\mu a$, with a a short distance cut-off, the infrared cut-off μ would have canceled between the UV finite and the UV divergent terms leaving only a power, $|a|^{\alpha^2/\pi}$, in the numerator of the right hand side of (1.3.3). When present, this power serves to make the overall expression dimensionless. Normal-ordering provides a multiplicative renormalization of the operator which removes it.

Provided the total "charge", $\sum\alpha_i$, is zero, there is no μ-dependence in any correlator composed of operators of the form $\exp i\alpha_i\varphi$. For a correlator where this condition is not satisfied, the μ's give rise to an overall factor

$$C = \mu^{(\sum\alpha_i)^2/4\pi}, \tag{1.3.4}$$

which has a positive exponent. This tends to zero as the cut-off is removed, meaning that "non-neutral" correlators vanish.

From now on I will follow the precedents established in (1.3.3) by omitting all reference to μ, and, except where it would cause confusion, writing $\exp i\alpha\varphi$ for $: \exp i\alpha\varphi :$.

By taking $\alpha = 2\sqrt{\pi}$ we find

$$\langle e^{i2\sqrt{\pi}\varphi(\mathbf{r}_1)} e^{-i2\sqrt{\pi}\varphi(\mathbf{r}_2)} \rangle = \frac{1}{|\mathbf{r}_1 - \mathbf{r}_2|^2} = \frac{1}{z_1 - z_2} \frac{1}{\overline{z}_1 - \overline{z}_2}. \quad (1.3.5)$$

If we could write $\varphi(z, \overline{z})$ as the sum, $\varphi_R(z) + \varphi_L(\overline{z})$, of two independent fields we would be able to factorize the left hand side of (1.3.5) and assert that

$$\langle e^{i2\sqrt{\pi}\varphi_R(z_1)} e^{-i2\sqrt{\pi}\varphi_R(z_2)} \rangle = \frac{1}{z_1 - z_2}. \quad (1.3.6)$$

In the hamiltonian operator language there is no difficulty in such a decomposition: the quantum field φ obeys the equation of motion $\partial_z \partial_{\overline{z}} \varphi(z, \overline{z}) = 0$ whose solution is precisely the sum of two commuting fields, $\varphi_R(z)$ and $\varphi_L(\overline{z})$. In the path-integral formalism, fields which are being summed over do not in general satisfy the equation of motion, and a different interpretation is necessary. We can always write

$$\varphi(z, \overline{z}) = \int_{-\infty}^{z} \partial_z \varphi dz(z, \overline{z}) + \int_{-\infty}^{\overline{z}} \partial_{\overline{z}} \varphi(z, \overline{z}) d\overline{z}, \quad (1.3.7)$$

where $-\infty$ is some distant point where we can regard φ as vanishing. If φ were a sum of a holomorphic and an anti-holomorphic part, the derivatives and integrals would project out these two components. We therefore define

$$\varphi_R(z) = \int_{-\infty}^{z} \partial_z \varphi(z, \overline{z}) dz. \quad (1.3.8)$$

Despite the innocence of the notation, $\varphi_R(z)$ is an inherently non-local operator. Taking the integral along the x-axis, (1.3.8) becomes

$$\varphi_R(z) = \frac{1}{2}\varphi(x, y) + \frac{i}{2} \int_{-\infty}^{x} \partial_y \varphi(x, y) dx, \quad (1.3.9)$$

so

$$e^{i2\sqrt{\pi}\varphi_R(z)} = \exp\left\{ i\sqrt{\pi}\varphi - \sqrt{\pi} \int_{-\infty}^{x} \partial_y \varphi(x, y) dx \right\}. \quad (1.3.10)$$

Equation (1.3.10) is the euclidean version of the Mandlestam "Kink" operator for non-interacting fields [rep. 9]. By construction, the correlator of this operator and its conjugate has the same form as that of the free fermions (1.2.2).

Now we evaluate the correlator of a general product.

$$\langle e^{i2\sqrt{\pi}\varphi_R(z_1)} e^{i2\sqrt{\pi}\varphi_R(z_2)} \ldots e^{i2\sqrt{\pi}\varphi_R(z_N)} e^{-i2\sqrt{\pi}\varphi_R(\zeta_1)} \ldots e^{-i2\sqrt{\pi}\varphi_R(\zeta_N)} \rangle$$

$$= \frac{\prod_{i<j}(z_i - z_j)(\zeta_i - \zeta_j)}{\prod i, j(z_i - \zeta_j)}. \quad (1.3.11)$$

At first sight this product expression does not look much like the determinant in the multi-fermion correlator (1.2.3) — but there is a simple identity, attributed to Cauchy, asserting that they are identical! This is the coincidence that makes everything work.

Cauchy's Lemma [1]

$$\frac{\prod_{i<j}(z_i - z_j)(\zeta_i - \zeta_j)}{\prod i,j(z_i - \zeta_j)} = \det\left|\frac{1}{z_i - \zeta_j}\right| \tag{1.3.12}$$

is immediately plausible since both sides have the same poles and zeros. A formal proof is obtained by subtracting the ith row from the first and then doing the same to the columns. In this way one gradually extracts all the factors. Later we will see this identity again playing a vital role when we describe the operator basis for the Bose-Fermi equivalence.

A word about phases is needed. When we factor the correlator of the full exponentials to get the holomorphic part (1.3.11), a choice of signs is tacitly made. Minus signs could have been included in both the holomorphic and the anti-holomorphic factors without affecting the product. This ambiguity is part of the origin of the Fermi statistics in the $\exp 2i\sqrt{\pi}\varphi_R(z)$ operator. Since the path integral is over commuting fields one would not think that the order in which they appear in the correlator can matter, but a glance at (1.3.11) shows that the interchange of the first two exponentials is equivalent to re-labeling z_1 as z_2, and this interchange reverses the sign of the right hand side. This sign reversal is exactly what one expects for fermion correlators which come from integrals over anti-commuting Grassmann variables. The sign choice is equivalent to selecting a routing of the "branch cuts" that trail from the non-local $\varphi_R(z)$. Adiabatically interchanging the arguments of two of these operators results in a rearrangement of the cuts, and the induced Fermi-statistics minus sign comes from one now being on a different "sheet".

1.4. The Bosonization Rules

We have established a basic correspondence:

$$\psi_R(z) = \frac{1}{\sqrt{2\pi}} : e^{i2\sqrt{\pi}\varphi_R(z)} :, \tag{1.4.1}$$

where the "=" sign is to be interpreted as equality of arbitrary free correlators of the bosonic right hand side theory with the corresponding correlators in the free fermion theory. (In String Theory applications it is common to rescale φ to remove the factor of $2\sqrt{\pi}$. I will retain the present normalization in this chapter however)

When we construct a boson correspondence for $\psi_L(\bar{z})$ we could choose either $e^{\pm i2\sqrt{\pi}\varphi_L(\bar{z})}$, since either would give the correct correlators. It is convenient to select

$$\psi_L(\bar{z}) = \frac{1}{\sqrt{2\pi}} : e^{-i2\sqrt{\pi}\varphi_L(\bar{z})} : \tag{1.4.2}$$

so that in the bilinear combination $\psi_L^\dagger \psi_R$ the two parts of the boson field reconstruct the original $\varphi(\mathbf{r})$. We then have

$$\begin{aligned} \psi_L^\dagger(\bar{z})\psi_R(z) &= \bar{\psi}\frac{(1+\sigma_3)}{2}\psi = \frac{1}{2\pi} : e^{i2\sqrt{\pi}\varphi(\mathbf{r})} : \\ \psi_R^\dagger(z)\psi_L(\bar{z}) &= \bar{\psi}\frac{(1-\sigma_3)}{2}\psi = \frac{1}{2\pi} : e^{-i2\sqrt{\pi}\varphi(\mathbf{r})} : . \end{aligned} \tag{1.4.3}$$

or, equivalently,

$$\begin{aligned}\overline{\psi}\psi &= \frac{1}{\pi} : \cos 2\sqrt{\pi}\varphi : \\ \overline{\psi}\sigma_3\psi &= \frac{i}{\pi} : \sin 2\sqrt{\pi}\varphi : .\end{aligned} \tag{1.4.4}$$

There is, as yet, nothing built into the boson operators to make ψ_R and ψ_L anticommute with one another. This can be ensured by introducing Jordan-Wigner-like factors, or *cocycles*, but we will ignore this problem for now since it is not essential for subsequent developments (see [rep. 11]).

When we come to find the boson equivalents for the fermion currents we need to be more careful. A naive application of our correspondences to $\overline{\psi}\gamma^0\psi = \overline{\psi}\sigma_1\psi = \psi_R^\dagger\psi_R + \psi_L^\dagger\psi_L$ leads us to assert that

$$\psi_R^\dagger(z)\psi_R(z) \propto e^{2i\sqrt{\pi}(-\varphi_R(z)+\varphi_R(z))} \stackrel{?}{=} 1,$$

which is patently incorrect. What is wrong is that normal-ordered exponentials do not obey the simple "add the exponent" rule on multiplication. The correct form, taking into account which contractions are omitted when the exponentials are normal-ordered, is

$$: e^{i\alpha\varphi_R(z)} :: e^{i\beta\varphi_R(z')} := (z - z')^{\alpha\beta/4\pi} : e^{i\alpha\varphi_R(z)+i\beta\varphi_R(z')} : . \tag{1.4.5}$$

The normal-ordering symbol on the right hand side means "omit contractions between z and z'".

With this new rule in mind, we try again with a point-split current:

$$\begin{aligned}\psi_R^\dagger(z)\psi_R(z') &= \frac{1}{2\pi}(z - z')^{-1} : e^{2i\sqrt{\pi}(-\varphi_R(z)+\varphi_R(z'))} : \\ &= \frac{1}{2\pi(z - z')} - \frac{1}{2\pi}2i\sqrt{\pi}\partial_z\varphi_R(z) + O(z - z').\end{aligned} \tag{1.4.6}$$

Omitting the infinite term proportional to the identity operator is equivalent to normal-ordering the fermion current. Putting the parts together we get

$$\begin{aligned}j^1 &= : \overline{\psi}\sigma_1\psi := -\frac{i}{\sqrt{\pi}}(\partial_z\varphi_R(z) - \partial_{\bar z}\varphi_L(\bar z)) \\ &= -\frac{i}{\sqrt{\pi}}((\partial_z - \partial_{\bar z})\varphi_R + (\partial_z - \partial_{\bar z})\varphi_L) \\ &= -\frac{1}{\sqrt{\pi}}\partial_y\varphi(x, y)\end{aligned} \tag{1.4.7}$$

and similarly

$$j^2 =: \overline{\psi}\sigma_2\psi := +\frac{1}{\sqrt{\pi}}\partial_x\varphi(x, y) \tag{1.4.8}$$

Cross-differentiating (1.4.7) and (1.4.8) shows that the vector current is automatically conserved, $\partial_1 j^1 + \partial_2 j^2 = 0$, but conservation of the chiral current requires the equation of motion for the φ field.

We now have a table of correspondences between the Dirac bilinears and expressions containing the complete φ field:

$$\begin{aligned} \bar{\psi}\psi &\longleftrightarrow \frac{1}{2\pi} :\cos 2\sqrt{\pi}\varphi: \\ \bar{\psi}\gamma^5\psi &\longleftrightarrow \frac{i}{2\pi} :\sin 2\sqrt{\pi}\varphi: \\ :\bar{\psi}\gamma^\mu\psi: &\longleftrightarrow -\epsilon^{\mu\nu}\frac{1}{\sqrt{\pi}}\partial_\nu\varphi \end{aligned} \tag{1.4.9}$$

1.5. A Quantum Pythagoras Theorem

An amusing consistency check on our formalism is given by a quantum version of the familiar identity $\cos^2\theta + \sin^2\theta = 1$.

A two-dimensional Fierz transformation gives the relation

$$(\bar{\psi}\gamma^1\psi)^2 + (\bar{\psi}\gamma^2\psi)^2 = -\left((\bar{\psi}\psi)^2 - (\bar{\psi}\gamma^5\psi)^2\right), \tag{1.5.1}$$

both sides being equal to $4\psi_R^\dagger\psi_R\psi_L^\dagger\psi_L$. The minus sign comes from the need to interchange two of the anticommuting Fermi fields. Using the table of equivalences in (1.4.9), this implies the gnomic identity

$$\cos^2 2\sqrt{\pi}\varphi + \sin^2 2\sqrt{\pi}\varphi = -\frac{1}{\pi}(\partial\varphi)^2. \tag{1.5.2}$$

How is this to be understood?

The mystery is resolved when we remember that we should really have written

$$(:\cos 2\sqrt{\pi}\varphi:)^2 + (:\sin 2\sqrt{\pi}\varphi:)^2$$

on the left hand side, and $(:\cos 2\sqrt{\pi}\varphi:)^2$ is by no means equal to $:\cos^2 2\sqrt{\pi}\varphi:$. We try to find the correct equivalent of the product

$$\bar{\psi}(1+\sigma_3)\psi\bar{\psi}(1-\sigma_3)\psi = (\bar{\psi}\psi)^2 - (\bar{\psi}\gamma^5\psi)^2, \tag{1.5.3}$$

by point-splitting the two boson operators symmetrically about $\mathbf{r}$, and using (1.4.5) and its conjugate. We thus want the coincidence limit of

$$\frac{4}{4\pi^2} :e^{2i\sqrt{\pi}\varphi(\mathbf{r}+\mathbf{a}/2)}::e^{-2i\sqrt{\pi}\varphi(\mathbf{r}-\mathbf{a}/2)}: = \frac{1}{\pi^2}|a|^{-2} :e^{2i\sqrt{\pi}(\varphi(\mathbf{r}+\mathbf{a}/2)-\varphi(\mathbf{r}-\mathbf{a}/2))}:. \tag{1.5.4}$$

The symmetric difference in the exponent is, to $O(|a|^2)$, equal to $a^\mu\partial_\mu\varphi(\mathbf{r})$. Expanding the exponential to $O(|a|^2)$ gives

$$\frac{1}{\pi^2|a|^2}\left(1 + 2i\sqrt{\pi}a^\mu\partial_\mu\varphi(\mathbf{r}) - 2\pi a^\mu a^\nu\partial_\mu\varphi(\mathbf{r})\partial_\nu\varphi(\mathbf{r}) + O(|a|^3)\right). \tag{1.5.5}$$

We must average over the directions of point-splitting since we need a rotationally invariant product. This makes

$$a^\mu a^\nu/|a|^2 \to \frac{1}{2}g^{\mu\nu}. \tag{1.5.6}$$

The final operator product is

$$\overline{\psi}(1+\sigma_3)\psi\overline{\psi}(1-\sigma_3)\psi \longleftrightarrow \frac{1}{\pi^2}\frac{1}{|\mathbf{a}|^2} - \frac{1}{\pi}(\partial\varphi(\mathbf{r}))^2 + O(|a|). \tag{1.5.7}$$

Discarding the c-number divergence gives the right hand side of (1.5.2), including the minus sign from the Fermi interchange.

A precise restatement of (1.5.2) would be

$$:\left((:\cos 2\sqrt{\pi}\varphi :)^2 + (:\sin 2\sqrt{\pi}\varphi :)^2\right) := -\frac{1}{\pi}(\partial\varphi)^2. \tag{1.5.2a}$$

While everything worked out correctly in this instance, the sequence of manipulations necessary to confirm the simple Fierz identity should serve as a warning not to expect similar results to remain true for interacting theories. The operator product in (1.5.3) is singular when the two factors approach one another, and the degree of singularity will be altered by interactions. In general, composite operators are be defined by divergent operator products. They will not exhibit the nice cancellation of divergences between numerator and denominator that occurred in (1.5.7), and the remaining singular powers need to be absorbed into multiplicative renormalization constants. They will also typically mix with other operators having the same symmetry and the same or lower scaling dimension. An equality between composite operators needs therefore to be regarded warily. It almost certainly depends on choices of regularization procedures and renormalization conventions.

Appendix 1A. Complex Coordinates

In two-dimensional euclidean space it is convenient to introduce complex variables $z = x + iy$, $\overline{z} = x - iy$. Derivatives with respect to z, $\overline{z}$ are defined as

$$\partial_z = \frac{1}{2}(\partial_x - i\partial_y) \qquad \partial_{\overline{z}} = \frac{1}{2}(\partial_x + i\partial_y) \tag{1A.1}$$

so that

$$df = \partial_z f dz + \partial_{\overline{z}} f d\overline{z}. \tag{1A.2}$$

With these coordinates, even in euclidean space, a distinction must be made between raised and lowered indices. The metric is

$$ds^2 = g_{z\overline{z}}dzd\overline{z} + g_{\overline{z}z}d\overline{z}dz = dzd\overline{z} \tag{1A.3}$$

implying that $g_{z\overline{z}} = g_{\overline{z}z} = 1/2$ and $g^{z\overline{z}} = g^{\overline{z}z} = 2$. A euclidean vector A^μ gives rise to $z, \overline{z}$ covariant components $A_z = \frac{1}{2}(A_1 - iA_2)$, $A_{\overline{z}} = \frac{1}{2}(A_1 + iA_2)$ and contravariant components $A^z = (A^1 + iA^2)$, $A^{\overline{z}} = (A^1 - iA^2)$

On smooth functions (so that the order of differentiation can be interchanged) the laplacian,

$$\nabla^2 = g^{z\bar{z}}\partial_{z\bar{z}} + g^{\bar{z}z}\partial_{\bar{z}z}, \tag{1A.4}$$

can be written as $\nabla^2 = 4\partial_{\bar{z}z}$. The caveat about smoothness is not merely a polite nod to our more rigorous colleagues. It becomes significant when we come to seek the Green function for the laplacian. Use Stokes' theorem to write

$$2\pi i = \oint_{\partial\Omega} \frac{dz}{z} = \iint_{\Omega} d\left(\frac{1}{z}dz\right) = \iint_{\Omega} \partial_{\bar{z}}\left(\frac{1}{z}\right) d\bar{z}dz = \iint \partial_{\bar{z}}\left(\frac{1}{z}\right) 2idxdy \tag{1A.5}$$

for any region Ω containing the origin. From this we read off the notationally surprising, but useful, identity

$$\partial_{\bar{z}}\left(\frac{1}{z}\right) = \pi\delta^2(x,y). \tag{1A.6}$$

Thus

$$\partial_{\bar{z}z} \ln z = \partial_{\bar{z}}\left(\frac{1}{z}\right) = \pi\delta^2(x,y), \tag{1A.7}$$

while

$$\partial_{z\bar{z}} \ln z = 0. \tag{1A.8}$$

The Green function for the real operator ∇^2 is $\frac{1}{2\pi}\text{Re}\ \ln z = \frac{1}{2\pi}\ln|z|$, and we would have been wrong by a factor of two if we had naively interchanged the derivatives as after (1A.4). (The real and imaginary parts of equations (1A.7) and (1A.8) are $\nabla^2 \ln r = 2\pi\delta^2(x,y)$ and $(\partial_y\partial_x - \partial_x\partial_y)\theta = 2\pi\delta^2(x,y)$.)

Appendix 1B. Conformal Symmetry

The study of two-dimensional field theories has become almost synonymous with the study of conformal symmetry. This is by now a vast subject and it is presumptuous to even begin to review it in an appendix, but for pedagogical completeness I must describe the very basic ideas.

Under very general conditions (basically the existence of a local energy-momentum tensor), a massless two-dimensional field theory is symmetric under the group of conformal maps $z \to \zeta = f(z)$ [2]. The correlation functions of the theory are not invariant under these mappings, but rather they transform *covariantly*. This is best explained by using the correlation functions found in this chapter as illustrations.

The Green function for the Laplace operator on the plane $\mathbf{R}^2$ is

$$G(\mathbf{r},\mathbf{r}') = -\frac{1}{2\pi}\ln\mu|\mathbf{r}-\mathbf{r}'| = -\frac{1}{2\pi}\text{Re}\ \ln\mu(z-z'). \tag{1B.1}$$

On the infinite cylinder $\Omega = \mathbf{R}\times\mathbf{S}^1$ the corresponding Green function is

$$G_\Omega(\mathbf{r},\mathbf{r}') = -\frac{1}{2\pi}\ln\mu|\sin(\zeta-\zeta')/2| \tag{1B.2}$$

where $\mathbf{r} = (\tau, \theta)$, $i\zeta = \tau + i\theta$ and θ is an angle, *i.e.* $\theta + 2\pi n$ are the same point. We use this Green function to evaluate the two-point function on the cylinder as

$$\langle e^{i\alpha\varphi(\mathbf{r}_1)} e^{-i\alpha\varphi(\mathbf{r}_2)} \rangle_\Omega = \frac{1}{|\sin(\zeta - \zeta')/2|^{\alpha^2/2\pi}} \tag{1B.3}$$

Let us focus on the holomorphic part

$$\langle e^{i\alpha\varphi_R(\zeta)} e^{-i\alpha\varphi_R(\zeta')} \rangle_\Omega = \frac{1}{(\sin(\zeta - \zeta')/2)^{\alpha^2/4\pi}}. \tag{1B.4}$$

Now $z = e^{i\zeta}$ maps $\mathbf{R}^2$ conformally onto the cylinder, but the correlation function (1B.4) is not obtained from the correlation function in the plane, $1/(z - z')^{\alpha^2/4\pi}$, by simply substituting $z = e^{i\zeta}$. Instead, noting that $\frac{\partial z}{\partial \zeta} = ie^{i\zeta}$, we can write

$$\frac{1}{(2\sin(\zeta - \zeta')/2)^{\alpha^2/4\pi}} = \left(\frac{\partial z}{\partial \zeta}\right)^{\alpha^2/8\pi} \left(\frac{\partial z}{\partial \zeta'}\right)^{\alpha^2/8\pi} \frac{1}{(e^{i\zeta} - e^{i\zeta'})^{\alpha^2/4\pi}} \tag{1B.5}$$

We can generalize this result to a transformation $z = z(\zeta)$ mapping a scalar field theory defined in the region Ω_1 in the z-plane to a theory defined the region Ω_2 in the ζ-plane. The fields should satisfy corresponding boundary conditions (say Dirichlet) on the boundaries of the regions. Under such a map, a general correlator of operators $O_1, O_2, \ldots$, of exponential type transforms as

$$\langle O_1(\zeta_1) O_2(\zeta_2) \cdots \rangle_{\Omega_2} = \left(\frac{\partial z}{\partial \zeta_1}\right)^{\eta_1} \left(\frac{\partial z}{\partial \zeta_2}\right)^{\eta_2} \cdots \langle O_1(z(\zeta_1)) O_2(z(\zeta_2)) \cdots \rangle_{\Omega_1} \tag{1B.6}$$

where the η_i are the scaling dimensions, $\alpha_i^2/8\pi$, of the fields. Since there is a derivative factor for each operator, we can regard this as a transformation law for the operators themselves, and write

$$O_{\Omega_2}(\zeta) = \left(\frac{\partial z}{\partial \zeta}\right)^{\eta} O_{\Omega_1}(z(\zeta)) \tag{1B.7}$$

The physical reason for the extra derivative factors is the underlying cut-off. We can imagine defining the theory in region Ω_1 on a rectangular lattice coinciding with the squares of the coordinate grid with sides 10^{-9} m. The conformal image of this grid in Ω_2 is no longer comprised of squares but rather of distorted images of squares. If we used this distorted lattice to define the theory in Ω_2, the correlators would be the same as those in Ω_1 — we would merely be relabeling the points by ζ instead of z. What we ought to be calculating in Ω_2 is the correlator of a theory on a 10^{-9} m square lattice just as in Ω_1. This involves a local rescaling of the cut-off, and the consequent change in its wavefunction renormalization constant provides a scaling factor for each operator.

Not all operators transform in this simple multiplicative manner. For example if $\varphi(z)$ transforms in this way then $\partial\varphi/\partial z$ transforms inhomogeneously. Operators transforming as (1B.7) are called *primary fields*. The general primary field will normally have both holomorphic and antiholomorphic parts, with scaling dimensions η and $\overline{\eta}$ respectively. The overall scaling dimensions of the field will be $\eta+\overline{\eta}$. Since rotations are conformal transformations, we can use (1B.7) to see that the spin of the field is $\eta-\overline{\eta}$.

If a scalar primary field has correlator

$$\langle O(z)O(z')\rangle = \frac{1}{|z-z'|^{2\eta}} \tag{1B.8}$$

on $\mathbf{R}^2$, then it decays exponentially at long distance on the cylinder

$$\langle O(z)O(z')\rangle_\Omega = \frac{1}{|\sin(z-z')/2|^{2\eta}} \approx e^{-\eta|\tau-\tau'|}. \tag{1B.9}$$

The scaling dimension η of the operator O is therefore equal to the gap between the ground state and the first excited state of the quantum system on the circle, $\mathbf{S}^1$, to which O couples [3]. This is a useful result for numerical work.

An important example of an operator which transforms inhomogeneously is the energy-momentum or stress tensor. In a conformally invariant theory this symmetric tensor is traceless so the $T_{z\bar{z}} = T_{\bar{z}z}$ components vanish. In this case the conservation law $\partial^\mu T_{\mu\nu} = 0$ reads

$$\partial_{\bar{z}} T_{zz} = 0 \qquad \partial_z T_{\overline{zz}} = 0 \tag{1B.10}$$

showing that $T_{zz} \equiv T(z)$ is holomorphic and $T_{\overline{zz}} \equiv \overline{T}(\bar{z})$ is antiholomorphic.

Under a conformal transformation $T(z)$ transforms as

$$T_{\Omega_2}(\zeta) = \left(\frac{dz}{d\zeta}\right)^2 T(z(\zeta))_{\Omega_1} + \frac{c}{12}\{z,\zeta\}, \tag{1B.11}$$

where the constant c is called the *central charge* of the field theory, and

$$\{z,\zeta\} = \frac{z'''}{z'} - \frac{3}{2}\left(\frac{z''}{z'}\right)^2 \tag{1B.12}$$

is the *schwartzian*. Here $z' = \frac{dz}{d\zeta}$.

The unlikely combination of derivatives that comprise the schwartzian arise naturally in the theory of second order differential equations. Given an equation

$$\frac{d^2y}{dx^2} + W(x)y = 0, \tag{1B.13}$$

make the substitution $x = x(z)$ and find the differential equation obeyed by $y(z)$. Initially one gets an equation with a first-order $\frac{dy}{dz}$ term in addition to the second-order derivative. We absorb the first-order term and recover an equation of the original form by defining

$$y(z) = \left(\frac{dx}{dz}\right)^{\frac{1}{2}} \tilde{y}(z). \tag{1B.14}$$

We find that $\tilde{y}$ obeys

$$\frac{d^2\tilde{y}}{dz^2} + \left[\left(\frac{dx}{dz}\right)^2 W(x(z)) + \frac{1}{2}\{x,z\}\right]\tilde{y} = 0. \tag{1B.15}$$

Thus $W(z)$ transforms in a manner similar to T_{zz}

$$W(z) \to \left(\frac{dx}{dz}\right)^2 W(x(z)) + \frac{1}{2}\{x,z\}. \tag{1B.16}$$

By composing the maps $x \to z \to w$ we obtain Cayley's identity

$$\left(\frac{dz}{dw}\right)^2 \{x,z\} + \{z,w\} = \{x,w\} \tag{1B.17}$$

which would be tedious and unenlightening to derive directly from the definition of $\{x,z\}$. This identity shows that the transformation rule (1B.11) is functorially consistent.

I will not attempt to derive (1B.11) here, but to make it sound at least plausible, I will instead remark that a consequence of (1B.15) is the transformation formula for *one-dimensional* path-integrals under diffeomorphisms of the time variable. If one reparameterizes the time coordinate $t \to s = s(t)$ in the path integral

$$G(x_2, x_1, E) = \int_0^\infty dt \int_{x_1}^{x_2} d[x] e^{-\int_0^t dt\left\{\frac{1}{2}\dot{x}^2 + (V(x) - E)\right\}}, \tag{1B.18}$$

and compensates for the measure factor in the s integration by changing integration variables to $z = z(x)$ determined by

$$\left(\frac{dz}{dx}\right)^2 = \frac{ds}{dt}, \tag{1B.19}$$

then the kinetic term maintains the original form. Classically the potential would transform multiplicatively. In the path integral it acquires an inhomogeneous addition proportional to the schwartzian

$$G(x_2, x_1, E) = \sqrt{x'_1 x'_2} \int_0^\infty ds \int_{z(x_1)}^{z(x_2)} d[z] e^{-\int_0^s ds\left\{\frac{1}{2}\dot{z}^2 + (x')^2(V(x(z) - E) - \frac{1}{4}\{x,z\}\right\}}. \tag{1B.20}$$

In (1B.20) integral dots indicate derivatives with respect to s, and primes derivatives with respect to z. Both path integrals, (1B.18) and (1B.20), are defined as limits of equal "time" (t in one case, s in the other) discretizations. (1B.20) is most easily established by writing the Green function as the resolvant of the corresponding Schrödinger equation. Direct, but more complicated, derivations can be found in [4,5].

References

[1] H. Weyl, *The Classical Groups* (Princeton University Press, 1946), 202.
[2] A. A. Belavin, A. M. Polyakov and A. B. Zamolodchikov, *Nuc. Phys.* **B241** (1984) 333.
[3] J. L. Cardy *J. Phys.* **A17** (1984) L385.
[4] N. K. Pak and I. Sokmen, *Phys. Rev.* **A30** (1984) 1629.
[5] D. C. Khandekar and S. V. Lawande, *Phys. Rep.* **137** (1986) 115.

Chapter 2

TWO SOLVABLE MODELS

2.1. Introduction

The results of the previous chapter may be extended straightforwardly to include interactions. Using a perturbation expansion we express the correlator in the interacting fermion theory as a sum of correlators in the free theory. The free-theory bosonization of the last chapter then lets us write these as Bose correlators. Finally we interpret the new perturbation series as that of a boson theory whose interaction term is the bosonized equivalent of the fermion interaction.

In this chapter we will follow the above procedure for two simple systems: the massless Thirring model and the massless Schwinger model. In both these cases it results in a free field theory, effectively solving the model. Alternative solutions to these models are given in the two appendices.

2.2. Dirac Equation for Metals

Although the two models we will discuss in this chapter are relativistic field theories, both of them have important condensed matter physics interpretations. It will be helpful to have these interpretations in mind before we discuss the solutions.

To make the connection between the relativistic models and their condensed matter applications, consider free non-relativistic electrons moving in one dimension. The second-quantized hamiltonian for this system is*

$$\hat{H} = \int dx \psi^\dagger \left(-\frac{1}{2m} \partial_x^2 \right) \psi. \tag{2.2.1}$$

Here

$$\psi(x) = \sum_k \hat{a}_k e^{ikx} \tag{2.2.2}$$

is a Fermi field with $\{\psi^\dagger(x), \psi(x')\} = \delta(x - x')$.

Near the Fermi surface we can approximate the ψ field by

$$\psi(x) = e^{ik_f x} \psi_R(x) + e^{-ik_f x} \psi_L(x) \tag{2.2.3}$$

where for example

$$\psi_R(x) = \sum_{k=-\Lambda}^{\Lambda} \hat{a}_{k+k_f} e^{ikx} \tag{2.2.4}$$

contains only operators annihilating states within a momentum shell $[-\Lambda, \Lambda]$ about the right-hand Fermi surface. The cut-off Λ must be chosen small enough that the electron

*A deeper account is given in [rep. 12].

velocity is essentially constant within the interval, yet large enough that no process is able to change the occupation number of states outside the momentum shell. (If the physics does not allow this we must use a more sophisticated analysis that takes into account the curvature in the dispersion relation.)

Introducing a two-component field $\Psi = (\psi_R, \psi_L)$ and the Fermi velocity $v_f = k_f/m$, we can approximate

$$\hat{H} \approx H_D = \int dx \Psi^\dagger(x)(-iv_f\sigma_3\partial_x)\Psi(x). \tag{2.2.5}$$

which is a one-dimensional "relativistic" Dirac hamiltonian with the Pauli σ-matrix σ_3 acting on the left/right indices. The role of the speed of light is played by v_f.

The fermion number-density and current may similarly be approximated as

$$\hat{\rho}(x) = \hat{\rho}_R(x) + \hat{\rho}_L(x) = \psi_R^\dagger\psi_R(x) + \psi_L^\dagger\psi_L(x) = \Psi^\dagger\Psi(x) \tag{2.2.6}$$

$$\hat{j}(x) = v_f(\hat{\rho}_R(x) - \hat{\rho}_L(x)) = v_f(\psi_R^\dagger\psi_R(x) - \psi_L^\dagger\psi_L(x)) = v_f\Psi^\dagger\sigma_3\Psi(x) \tag{2.2.7}$$

The commutator of these operators contains the usual Schwinger term

$$[\hat{\rho}_R(x), \hat{\rho}_R(x')] = -[\hat{\rho}_L(x), \hat{\rho}_L(x')] = -\frac{i}{2\pi}\partial_x\delta(x - x'), \tag{2.2.8}$$

whose origin is easy to explain in this context. We revert to the full non-relativistic expressions for the density and current

$$\hat{\rho} = \psi^\dagger\psi \qquad \hat{j} = \frac{-i}{2m}\left(\psi^\dagger\partial_x\psi - (\partial_x\psi^\dagger)\psi\right). \tag{2.2.9}$$

For non-relativistic fermions the canonical commutation relations can be used without difficulty, and they give

$$[\hat{\rho}(x), \hat{j}(x')] = \frac{-i}{m}\left(\partial_x\delta(x - x')\psi^\dagger\psi + \delta(x - x')\partial_x\psi^\dagger\psi\right). \tag{2.2.10}$$

Because of the derivative appearing in the non-relativistic current operator there is no surprise at finding the derivative of a delta function in this result.

The relativistic approximation (2.2.5) becomes exact in the "Dirac Limit": $m \to \infty$, $k_f \to \infty$, with $v_f = k_f/m$ fixed. In this limit the curvature of the $E = k^2/2m$ dispersion curve vanishes near the points of interest, $k = \pm k_f$. Equally significantly the m in the denominator causes all fluctuations of the operator $\psi^\dagger\psi/m$ to be suppressed and it may be replaced by its expectation value $\langle\psi^\dagger\psi\rangle/m = k_f/\pi m = v_f/\pi$. Thus

$$[(\hat{\rho}_R + \hat{\rho}_L), v_f(\hat{\rho}_R - \hat{\rho}_L)] \to \frac{-iv_f}{\pi}\partial_x\delta(x - x'). \tag{2.2.11}$$

Arguing that $\hat{\rho}_R$ should commute with $\hat{\rho}_L$ then allows us to deduce (2.2.8).

2.3. Thirring Model

The massless Thirring model (see [1]) has a current-current interaction term $L_{int} = \frac{g}{2}\overline{\psi}\gamma_\mu\psi\overline{\psi}\gamma_\mu\psi$. (see 1.5.1 for a Fierz transformed version of L_{int}). Using the bosonization table (1.4.9), this interaction is equivalent to

$$\frac{g}{2}\overline{\psi}\gamma_\mu\psi\overline{\psi}\gamma_\mu\psi \longleftrightarrow \frac{g}{2\pi}(\partial_\mu\varphi)^2. \tag{2.3.1}$$

Adding the bosonized interaction term to the free action gives us

$$S = \int d^2x\frac{1}{2}\left(1+\frac{g}{\pi}\right)(\partial_\mu\varphi^2). \tag{2.3.2}$$

The net effect of the interaction is thus merely to rescale the field.

This rescaling *does* have consequences. We can write the exponential operators $\exp \pm i2\sqrt{\pi}\varphi$ as $\exp \pm i\beta\tilde{\varphi}$ where $\beta^2/4\pi = 1/(1+g/\pi)$ [rep. 8], and $\tilde{\varphi}$ is a field with the canonical normalization. The factor of β changes the critical exponents. For example the correlator

$$\left\langle\overline{\psi}(z)\frac{1}{2}(1+\gamma_5)\psi(z)\overline{\psi}(z')\frac{1}{2}(1+\gamma_5)\psi(z')\right\rangle = \langle e^{i2\sqrt{\pi}\varphi(z)}e^{-i2\sqrt{\pi}\varphi(z')}\rangle \tag{2.3.3}$$

is found to be

$$\left\langle\overline{\psi}(z)\frac{1}{2}(1+\gamma_5)\psi(z)\overline{\psi}(z')\frac{1}{2}(1+\gamma_5)\psi(z')\right\rangle = \frac{1}{|z-z'|^{2/(1+g/\pi)}}. \tag{2.3.4}$$

Clearly the correlators of all the fermion bilinears are as easy to evaluate as in the free theory. It is a little harder to obtain the correlators of the fermion fields themselves. This is because the exponentials contain time derivatives of the field. One must convince oneself that these must be written in terms of the field $\Pi(x)$ conjugate to the φ field, and the rescaling done in the phase-space path integral before the $\Pi(x)$'s are integrated out. The fermion correlators will be evaluated by hamiltonian methods in the next section. They are also obtained by a different approach in Appendix 2A.

An amusing consequence of the bosonization solution is that we can calculate the many-body ground-state wavefunction for the particles in the interacting Dirac sea. Suppose we have N fermions living on a circle of circumference 2π. (N should be large, so that the sea is deep and it does not matter whether we think of the problem as being intrinsically relativistic, or an approximation to a non-relativistic system.) We will find $\Psi(\theta_1, \theta_2, \ldots, \theta_N)$.

We will use a slightly modified form of the methods in [2]. We begin by finding the ground-state wavefunction Φ for the bosonized version.

Using Coleman's β, the bosonized action is

$$S = \frac{4\pi}{\beta^2}\int\frac{1}{2}(\partial\varphi)^2 d\tau d\theta \tag{2.3.5}$$

In a Schrödinger representation the wavefunction $\Phi(\varphi_c)$ is a functional of the boson field configuration. To compute it we take a path integral over φ's defined on the half-cylinder

$\Omega = [-\infty, 0] \times S^1$. The argument of the wavefunction, φ_c, appears as the boundary condition $\varphi_c(\theta) = \varphi(0, \theta)$ imposed on the circle at $\tau = 0$. The long euclidean time interval between $-\infty$ and 0 projects out the ground state, so the unnormalized wavefunction is given by

$$\Phi(\varphi_c) = \langle vac|\varphi_c\rangle = \int_0^{\varphi_c} d[\varphi] e^{-\frac{4\pi}{\beta^2}\int_\Omega \frac{1}{2}(\partial\varphi)^2 d\tau d\theta}. \tag{2.3.6}$$

One can evaluate the path integral by various routes. The following is a method that does not require any manipulation of divergent Fourier transforms.

Being quadratic, the path integral is, up to a φ_c independent factor, found by replacing the φ in the integrand with the solution of the equation of motion, in this case Laplace's equation with the given φ_c boundary values. Using the standard formula for the solution to the Dirichlet problem,

$$\varphi(\tau', \theta') = \oint d\theta \varphi_c(\theta) \partial_\tau G_\Omega(\tau, \theta; \tau'\theta')|_{\tau=0}, \tag{2.3.7}$$

and integrating by parts, we find that the exponent

$$E = \int_\Omega d\tau d\theta \frac{1}{2}(\partial\varphi)^2 \tag{2.3.8}$$

can be written in terms of the boundary data,

$$E = \frac{1}{2}\oint d\theta \varphi_c(0,\theta)\partial_\tau\varphi(0,\theta) = -\frac{1}{2}\int_{S^1}\int_{S^1} d\theta d\theta' \varphi_c(\theta)\varphi_c(\theta')\partial_\tau\partial_{\tau'} G_\Omega(\tau, \theta, \tau', \theta'). \tag{2.3.9}$$

In these formulae G_Ω is the Dirichlet Green-function on the half cylinder, i.e.,

$$\nabla^2_{\mathbf{r}} G_\Omega(\mathbf{r}, \mathbf{r}') = \delta^2(\mathbf{r} - \mathbf{r}') \tag{2.3.10}$$

and $G_\Omega(\mathbf{r}, \mathbf{r}') = 0$ if $\mathbf{r}$ is on the boundary circle.

The Green function G for the infinite cylinder is obtained by a conformal transformation of the $\mathbf{R}^2$ Green function, $G_0(\mathbf{r}, \mathbf{r}') = -\frac{1}{2\pi}\ln|\mathbf{r} - \mathbf{r}'|$, as

$$G(\mathbf{r}, \mathbf{r}') = -\frac{1}{2\pi}\text{Re}\,\ln(e^{iz} - e^{iz'}) = -\frac{1}{2\pi}\ln|\sin(z - z')/2|, \tag{2.3.11}$$

where $iz = \tau + i\theta$. The half-cylinder Green function, G_Ω, is then found from G by the method of images:

$$G_\Omega(\tau, \theta, \tau', \theta') = G(\tau - \tau', \theta - \theta') - G(\tau + \tau', \theta - \theta'). \tag{2.3.12}$$

We now use the property that G satisfies Laplace's equation, and the form of the arguments of G, to trade the partial derivatives with respect to τ for partials with respect to θ

$$\partial_\tau\partial_{\tau'} G_\Omega(\tau, \theta, \tau'\theta') = -\partial_\theta\partial_{\theta'} G(\tau - \tau', \theta - \theta') - \partial_\theta\partial_{\theta'} G(\tau + \tau', \theta - \theta'). \tag{2.3.13}$$

We need this expression only on $\tau = \tau' = 0$ where it is equal to $-2\partial_\theta \partial_{\theta'} G(0, \theta - \theta')$.

After a final integration by parts

$$E = \frac{1}{2} \iint d\theta d\theta' \partial_\theta \varphi_c(\theta) \partial_{\theta'} \varphi_c(\theta') \frac{1}{\pi} \ln |\sin(\theta - \theta')/2|. \tag{2.3.14}$$

The exponential of this $\Phi = \exp(-4\pi E/\beta^2)$ is the harmonic oscillator like ground state of the Bose field.

We convert Φ to a many-body Fermi wavefunction Ψ expressed in terms of the particle locations by using the bosonization rule $\rho = \frac{1}{\sqrt{\pi}} \partial_\theta \varphi$, and replacing the density with its first-quantized form $\rho(\theta) = \sum_{i=1}^{N} \delta(\theta - \theta_i)$. We find that

$$\Psi(\theta_1, \theta_2, \dots, \theta_N) = \left| e^{-i(N-1)\sum \theta_i} \prod_{i<j} (e^{i\theta_i} - e^{i\theta_j}) \right|^{4\pi/\beta^2} \tag{2.3.15}$$

Since we only know that there is some particle at the location θ_i, and not which particular particle it is, this expression is to be used for the standard ordering $\theta_1 < \theta_2 \cdots < \theta_N$ only. The values of Ψ for other orderings of the arguments are found by imposing the antisymmetry required by Fermi statistics.

This wavefunction (2.3.15) coincides with that of the Sutherland Model [3], a model with long-range $1/r^2$ interactions. That the Sutherland model has the same critical exponents as the Thirring model was shown in [4].

2.4. Hamiltonian Solution of the Luttinger-Thirring Model

For finding the fermion correlators in the Thirring model it is easiest to use hamiltonian methods. In this case we refer to the model as the Luttinger model. The Luttinger model on an interval of period 2π is defined by the hamiltonian

$$\mathcal{H} = 2\pi \int dx \left\{ \frac{1}{2} J_R^2 + \frac{1}{2} J_L^2 + \frac{g}{\pi} J_R J_L \right\}. \tag{2.4.1}$$

Here J_R, J_L are the currents for the left and right going fermions. They obey

$$[J_R(x), J_R(x')] = -[J_L(x), J_L(x')] = -\frac{i}{2\pi} \partial_x \delta(x - x'). \tag{2.4.2}$$

The first two terms in the hamiltonian are the Sugawara form for the free fermion action. That this is equivalent to the usual form follows from interpreting the free boson hamiltonian as the product of the Fermi currents. See [rep. 3] for more details. The second term coincides formally with (2.3.1)

The interaction may be decoupled by introducing a new set of currents $\tilde{J}_L, \tilde{J}_R$ defined by

$$\begin{aligned} J_R &= \cosh \alpha \tilde{J}_R + \sinh \alpha \tilde{J}_L \\ J_L &= \sinh \alpha \tilde{J}_R + \cosh \alpha \tilde{J}_L. \end{aligned} \tag{2.4.3}$$

If we set $-\tanh 2\alpha = g/\pi$ and express $\mathcal{H}$ in terms of $J_{L,R}$ the cross term disappears and

$$\mathcal{H} = \operatorname{sech} 2\alpha \int dx \left\{ \frac{1}{2}\tilde{J}_R^2 + \frac{1}{2}\tilde{J}_L^2 \right\}. \tag{2.4.4}$$

Both sets of currents obey the same commutation relations and are formal conjugates of each other — i.e., we can write down a formal expression for a unitary operator U such that $J_{R,L} = U^\dagger \tilde{J}_{R,L} U$. As usual, U is a proper unitary transformation only in a theory with a cut-off.

From (2.4.4) we see immediately that the speed of light has been renormalized. This cannot happen if we were to maintain manifest Lorentz invariance throughout, but this is the price we pay for the simplicity of the hamiltonian route. Another price is that we cannot simply identify the coupling constant g in (2.4.1) with that in section (2.3), despite their formal equivalence. The relation between the two coupling constants has to be found by calculating some quantity in both approaches.

We can write the currents as derivatives of two independent chiral boson fields

$$J_R = \frac{1}{2\pi}\partial_x \varphi_R \qquad J_L = \frac{1}{2\pi}\partial_x \varphi_L. \tag{2.4.5}$$

These fields obey

$$[\varphi_R(x), \varphi_R(x')] = -[\varphi_L(x), \varphi_L(x')] = i\pi \operatorname{sgn}(x - x'), \tag{2.4.6}$$

and then bosonized expressions for fundamental fermions in the system are

$$\psi_R =: e^{i\varphi_R} : \qquad \psi_L =: e^{-i\varphi_L} : . \tag{2.4.7}$$

(Notice that in this section we are using a different normalization for the scalar field. This is to make contact with condensed matter literature, for example [rep. 12] and [5].).

We calculate the fermion correlators by introducing new $\tilde{\varphi}_{L,R}$ in the same manner as $\tilde{J}_{L,R}$

$$\begin{aligned} \varphi_R &= \cosh\alpha \tilde{\varphi}_R + \sinh\alpha \tilde{\varphi}_L \\ \varphi_L &= \sinh\alpha \tilde{\varphi}_R + \cosh\alpha \tilde{\varphi}_L. \end{aligned} \tag{2.4.8}$$

The $\tilde{\varphi}_{L,R}$ are independent *free* fields so substituting (2.4.8) in (2.4.7) allows us to compute correlators. For example,

$$\langle \psi_R^\dagger \psi_L(z) \psi_L^\dagger \psi_R(z') \rangle = \frac{1}{|z - z'|^{2e^{2\alpha}}} \tag{2.4.9}$$

This coincides with (2.22) after one identifies $\beta^2/4\pi$ with $e^{2\alpha}$. This identification is confirmed by the computation of other Luttinger-Thirring correlators. To lowest order then, the g in (2.4.1) coincides with that in (2.3.1), but not to higher orders.

Following Haldane [rep. 12] and [5] we define two new fields

$$\begin{aligned} \theta(x) &= \frac{1}{2}(\varphi_R(x) + \varphi_L(x)) \\ \varphi(x) &= \frac{1}{2}(\varphi_R(x) - \varphi_L(x)), \end{aligned} \tag{2.4.10}$$

which obey

$$[\varphi(x), \varphi(x')] = [\theta(x), \theta(x')] = 0$$
$$[\varphi(x), \theta(x')] = \frac{i}{2}\pi \,\mathrm{sgn}(x - x'). \tag{2.4.11}$$

Thus

$$\rho = \frac{1}{\pi}\partial_x \theta \quad j = \frac{1}{\pi}\partial_x \varphi \tag{2.4.12}$$

and φ and ρ are pair of canonically conjugate fields.

$$[\varphi(x), \rho(x')] = i\delta(x - x') \tag{2.4.13}$$

One may give θ, φ physical interpretations: If the forces between the fermions are repulsive ($\beta^2 > 1$), the particles try to form a charge density wave — but fail because a continuous symmetry cannot be spontaneously broken in a strictly one-dimensional system. The quantity θ is proportional to the displacement of the particles from their would-be equilibrium position. Although its fluctuations always prevent the symmetry breaking, they become smaller as the repulsion grows strong. When the forces are attractive ($\beta^2 < 1$) the system tries to form a superconducting state (but again fails because the system is one-dimensional) and φ is the phase of the would-be superconducting order parameter.

Haldane [rep. 12] defines a set of higher spin operators by

$$\Phi_m(x) =: e^{i\varphi(x) + im\theta(x)} : \tag{2.4.14}$$

Clearly $\Phi_1(x) = \psi_R(x)$ and $\Phi_{-1}(x) = \psi_L(x)$. From (2.4.14) we see that

$$\Phi_m(x)\Phi_m(x') = e^{im\pi \,\mathrm{sgn}(x-x')}\Phi_m(x')\Phi_m(x), \tag{2.4.15}$$

so the Φ_m have Fermi statistics if m is an odd integer and Bose statistics if m is even. The correlator of a pair of these fields is

$$\langle \Phi_m(x,t)\Phi_m^\dagger(0)\rangle = \frac{1}{(x - ct')^\zeta}\frac{1}{(x + ct)^\eta} \tag{2.4.16}$$

where $c = \mathrm{sech}\, 2\alpha$ is the renormalized "speed of light" and

$$\zeta = \frac{1}{4}(me^\alpha + e^{-\alpha})$$
$$\eta = \frac{1}{4}(me^\alpha - e^{-\alpha}). \tag{2.4.17}$$

The quantity $\zeta - \eta$ is equal to m showing that these operators have spin $m/2$.

For the case $m = 1$ we find the fundamental Fermi correlator

$$\langle \psi_R^\dagger(x,t)\psi_R(0)\rangle = \frac{1}{(x - ct)}\frac{1}{|x^2 - c^2t^2|^{\sinh^2 \alpha}} \tag{2.4.18}$$

which may be compared with the euclidean space expressions given in Appendix 2A. After a little labour, one will see that the exponents agree when we take $\beta^2/4\pi = \exp 2\alpha$ as before.

In condensed matter applications Lorentz invariance is an artifact created by the approximation of restricting one's attention to states near the Fermi surface. Once the Lorentz symmetry is violated by the curvature of the band theory dispersion relation there is nothing to stop the higher spin Φ_m operators mixing with the fundamental fermion. In particular Haldane showed in [5] that one must write

$$\psi(x) = \sum_{m\,\mathrm{odd}} \Phi_m(x) e^{imk_f x} \tag{2.4.19}$$

to take into account the discreteness of the point particles. The various terms in (2.4.19) will renormalize with non-universal coefficients.

2.5. Schwinger Model

The Schwinger model [6–8] is a theory of one-dimensional fermions interacting with a gauge field. The action is

$$S = \int d^2x \left(-\frac{1}{2} F_{12}^2 + \overline{\psi}(\partial_\mu - ieA_\mu)\psi \right). \tag{2.5.1}$$

where $F_{\mu\nu} = \partial_\mu A_\nu - \partial_\nu A_\mu$. To solve the model one first integrates out the A_μ gauge field to get

$$S' = \int \overline{\psi}(\partial_\mu - ieA_\mu)\psi d^2x - \frac{e^2}{2} \int d^2x d^2x' \overline{\psi}\gamma_\mu\psi(x) G_{\mu\nu}(x-x')\overline{\psi}\gamma_\mu\psi(x'). \tag{2.5.2}$$

The Green function $G_{\mu\nu}(x-x')$ is the euclidean gauge field propagator in whatever gauge we choose. It is convenient to us a Coulomb gauge where the interaction is (apparently) instantaneous

$$G_{\mu\nu}(x-x') = \frac{1}{2}\delta_{\mu 0}\delta_{\nu 0}|x-x'|\delta(t-t'). \tag{2.5.3}$$

The bosonized version of the model is then

$$S' = \int \frac{1}{2}(\partial_\mu\varphi)^2 d^2x - \frac{e^2}{2\pi} \int dt dx dx' \frac{1}{2} \partial_x\varphi(x)|x-x'|\partial_{x'}\varphi(x'). \tag{2.5.4}$$

(we have reverted to the normalization of Chapter 1 for the boson field). We can simplify the double integral by integrating by parts to get

$$S' = \int \left\{ \frac{1}{2}(\partial_\mu\varphi)^2 + \frac{e^2}{2\pi}(\varphi - \varphi_\infty)^2 \right\} d^2x. \tag{2.5.5}$$

The quantity φ_∞ is the boundary value of the field which has to be the same at $\pm\infty$ since the system must be charge-neutral to have a finite energy.

We see that the final theory is essentially that of a free massive boson. The dynamical mass generation seems slightly mysterious in the relativistic field theory interpretation as it is due to the anomalous non-conservation of the axial current. The Schwinger model is often used as a solvable example of both the anomaly and the associated mass generation mechanism. In QCD, a similar mechanism due to the anomalous non-conservation of the flavour-singlet "ninth axial current" is believed to be responsible for the mass of the η' pseudoscalar meson [9–11]. The appearance of the difference $(\varphi(x) - \varphi_\infty)^2$ in the boson mass term allows the system to have spontaneously broken chiral symmetry (a global chiral transformation of the vacuum corresponds to $\varphi \to \varphi + const.$) while having no associated Goldstone boson.

The condensed matter interpretation of the system is much more straightforward. The axial charge $\psi^\dagger \gamma_5 \psi = \rho_R - \rho_L$ is the difference between the density of right moving particles and the density of left moving particles. The anomalous conservation law

$$\partial_\mu j_{5\mu} = e\frac{1}{2\pi}\epsilon_{\mu\nu}F_{\mu\nu} \tag{2.5.6}$$

reads, for a uniform charge distribution,

$$\partial_t(\rho_R - \rho_L) = \frac{1}{\pi}eE, \tag{2.5.7}$$

and simply expresses the acceleration of the center of mass of the charged system by the electric field. In this context the massive bosons are "plasmons" — phonons whose energy is pushed up by the coulomb energy. This interpretation is confirmed by comparing the induced (mass)^2, that is e^2/π, with the usual plasma frequency $\omega_p^2 = ne^2/m$ where n is the density of mass m particles in a neutralising background. For the one-dimensional spinless Fermi gas this density is k_f/π. Since we have set $v_f \equiv c = 1$, and $k_f = mv_f$, we find $n/m = 1/\pi$, and the mass-gap is exactly ω_p. The global symmetry $\varphi \to \varphi + const.$ reflects the fact that, although local charge fluctuations cost a lot of energy, the entire gas can be translated at no cost.

Another way of interpreting the mass generation is as Debye screening. In a degenerate plasma the potential due to an external charge distribution ρ_{ext} obeys

$$\left(\nabla^2 - e^2\frac{\partial n}{\partial \mu}\right)\phi = -\rho_{ext} \tag{2.5.8}$$

where $\partial n/\partial\mu$ is the density of states at the Fermi surface. For one-dimensional spinless fermions the density of states is $1/\pi$. External charges are therefore screened over a distance ξ, where $\xi^{-2} = e^2 \partial n/\partial\mu$.

We can include an external charge distribution in the Schwinger model by setting

$$\rho_{ext} = -e\frac{1}{\sqrt{\pi}}\partial_x\varphi_{ext}, \tag{2.5.9}$$

whence the Bose action becomes

$$\int\left\{\frac{1}{2}(\partial_\mu\varphi)^2 + \frac{e^2}{2\pi}(\varphi - \varphi_\infty + \varphi_{ext})^2\right\}d^2x. \tag{2.5.10}$$

The minimum energy configuration for smooth charge distributions is $\varphi = \varphi_\infty - \varphi_{ext}$. For point charges (*i.e.*, sudden jumps in φ_{ext}) the gradient terms become important, and φ will relax to its minimum energy value over a screening distance which is exactly ξ as given by (2.5.8).

To make the model more interesting we can also consider the case with N "flavours" of otherwise identical massless fermions. The action is

$$S = \int d^2x \left(-\frac{1}{2}F_{12}^2 + \sum_{\alpha=1}^{N} (\bar{\psi}_\alpha(\partial_\mu - ieA_\mu)\psi_\alpha) \right). \tag{2.5.11}$$

To bosonize we introduce one field φ_α for each fermion. The bosonized version of this model has action

$$S = \int d^2x \left(\frac{1}{2}\sum_{\alpha=1}^{N} (\partial_\mu\varphi_\alpha)^2 + \frac{e^2}{\pi}(\varphi_1 + \ldots \varphi_N)^2 \right) \tag{2.5.12}$$

I leave it as an exercise to the reader to calculate the various correlation functions and compare them with those exhibited in Appendix 2B.

Appendix 2A. Thirring Model via Chiral Rotations

This appendix provides an alternative solution of the Thirring model by using a chiral transformation to decouple the interaction. It is interesting to see how much harder one has to work when using this method.

The euclidean path integral for the Thirring model partition function is

$$\begin{aligned} \mathcal{Z} &= \int d[\psi]d[\bar{\psi}] \exp \int d^2x \left\{ \bar{\psi}\gamma^\mu\partial_\mu\psi - \frac{g}{2}\bar{\psi}\gamma^\mu\psi\bar{\psi}\gamma_\mu\psi \right\} \\ &= \int d[\psi]d[\bar{\psi}] \exp \int d^2x \left\{ \psi^\dagger\partial_1\psi - \psi^\dagger(-i\sigma_3\partial_2)\psi - \frac{g}{2}J_\mu J_\mu \right\}. \end{aligned} \tag{2A.1}$$

As before I am taking the "1" direction as time and setting $\gamma^1 = \sigma_1, \gamma^2 = \sigma_2, \gamma^5 = \sigma_3 = -i\gamma^1\gamma^2$. The action can then be written as $\psi^\dagger\partial_t\psi - H$, with $H = \psi^\dagger(-i\sigma_3\partial_2)\psi + \frac{g}{2}J_\mu J_\mu$ the Minkowski space hamiltonian.

We reduce the action to a quadratic in the Fermi fields by introducing a vector field A_μ:

$$\mathcal{Z} = \int d[\psi]d[\bar{\psi}]d[A_\mu] \exp \int d^2x \left\{ \bar{\psi}\gamma^\mu(\partial_\mu + iA_\mu)\psi - \frac{1}{2g}(A^\mu)^2 \right\}. \tag{2A.2}$$

Despite the minimal coupling, this is not a gauge-invariant action.

We next decompose A_μ into its solenoidal and curl-free parts,

$$A_\mu = \epsilon_{\mu\nu}\partial_\nu\theta_1 + \partial_\mu\theta_2, \tag{2A.3}$$

and introduce rotated fields $\psi_0, \bar{\psi}_0$ by

$$\begin{aligned} \psi &= e^{-\sigma_3\theta_1 - i\theta_2}\psi_0 \\ \bar{\psi} &= \bar{\psi}_0 e^{-\sigma_3\theta_1 + i\theta_2}. \end{aligned} \tag{2A.4}$$

Notice that there is no i in the chiral transformation.

When the action is written in terms of the new variables the various fields decouple. The partition function is

$$\mathcal{Z} = \int d[\psi_0]d[\bar{\psi}_0]d[\theta_1]d[\theta_2]e^J \exp \int d^2x \left\{ \bar{\psi}_0 \gamma^\mu \partial_\mu \psi - \frac{1}{2g}((\partial\theta_1)^2 + (\partial\theta_2)^2) \right\}. \qquad (2A.5)$$

The factor e^J is a jacobian which comes from the change in the Grassmann measure. We require the measure to be invariant under vector gauge transformations so there is no factor from the θ_2 rotation, but the chiral θ_1 transformation does not leave the measure invariant [12].

We find the change of measure by making the rotation in infinitesimal steps

$$\begin{aligned} d[\psi(x)]d[\overline{\psi}(x)] &\to d[e^{-\sigma_3 \delta\theta_1(x)}\psi(x)]d[\bar{\psi}e^{-\sigma_3\delta\theta_1}] \\ &= d[\psi(x)]d[\overline{\psi}(x)] \left(1 + \int d^2x \frac{\delta J}{\delta\theta_1(x)}(\delta\theta_1(x))\right). \end{aligned} \qquad (2A.6)$$

We now demand that in the path integral

$$\mathcal{Z}(A) = \int d[\psi]d[\bar{\psi}] \exp \int d^2x \left\{\bar{\psi}\gamma^\mu(\partial_\mu + iA_\mu)\psi\right\}, \qquad (2A.7)$$

we recover the effects of the chiral anomaly. Making the infinitesimal change $\delta\theta_1$ in the exponent of (2A.7) brings down the divergence of the axial current. The combined change, in both action and measure, is a change of variables and cannot change the overall partition function. Writing the null effect of this change of variables as an expectation gives

$$0 = \left\langle \int d^2x \left\{ \frac{\delta J}{\delta\theta_1(x)} + (-i)\partial_\nu(\bar{\psi}\gamma^\mu\psi\epsilon_{\mu\nu}) \right\} \delta\theta_1(x) \right\rangle. \qquad (2A.8)$$

We compare (2A.8) with the two-dimensional axial anomaly $\partial_\mu j_5^\mu = \frac{1}{2\pi}\epsilon_{\mu\nu}F^{\mu\nu}$ and find that

$$\frac{\partial J}{\partial\theta_1(x)} = -\frac{1}{\pi}F_{12} \qquad (2A.9)$$

As we gradually rotate away the gauge coupling, the original field strength $F_{12} = -\partial^2\theta_1$ decreases, so the total J is $\int d^2x \frac{1}{2\pi}(\partial\theta_1)^2$.

The partition function has now been written in terms of decoupled free fields.

$$\mathcal{Z} = \int d[\psi_0]d[\bar{\psi}_0]d[\theta_1]d[\theta_2] \exp \int d^2x \left\{ \bar{\psi}_0\gamma^\mu\partial_\mu\psi - \frac{1}{2}\left(\frac{1}{\pi} + \frac{1}{g}\right)(\partial\theta_1)^2 - \frac{1}{2g}(\partial\theta_2)^2 \right\}. \qquad (2A.10)$$

The operators whose correlators we wish to calculate are the original $\psi, \bar{\psi}$ fields which are related to the new fields via (2A.4).

For computational convenience set

$$\left(\frac{1}{\pi}+\frac{1}{g}\right)=\frac{1}{\kappa^2} \tag{2A.11}$$

and rescale $\theta_1 \to \kappa\theta_1$, $\theta_2 = \sqrt{g}\theta_2$. Then

$$\mathcal{Z}=\int d[\psi_0]d[\bar{\psi}_0]d[\theta_1]d[\theta_2]\exp\int d^2x\left\{\bar{\psi}_0\gamma^\mu\partial_\mu\psi-\frac{1}{2}(\partial\theta_1)^2-\frac{1}{2}(\partial\theta^2)^2\right\}, \tag{2A.12}$$

and

$$\begin{aligned}\psi &= e^{-\sigma_3\kappa\theta_1-i\sqrt{g}\theta_2}\psi_0\\ \bar{\psi} &= \bar{\psi}_0 e^{-\sigma_3\kappa\theta_1+i\sqrt{g}\theta_2}.\end{aligned} \tag{2A.13}$$

The chiral operators

$$\begin{aligned}\bar{\psi}\frac{(1+\sigma_3)}{2}\psi &= \bar{\psi}_{u0}\psi_{u0}e^{-2\kappa\theta_1}\\ \bar{\psi}\frac{(1-\sigma_3)}{2}\psi &= \bar{\psi}_{l0}\psi_{l0}e^{+2\kappa\theta_1},\end{aligned} \tag{2A.14}$$

are invariant under θ_2 vector rotations, and so have dependence only on θ_1. Their two-point correlators are

$$\left\langle\bar{\psi}\frac{(1+\sigma_3)}{2}\psi\bar{\psi}\frac{(1-\sigma_3)}{2}\psi\right\rangle=\langle\bar{\psi}_{u0}\psi_{u0}\bar{\psi}_{l0}\psi_{l0}\rangle\langle e^{-2\kappa\theta_1}e^{+2\kappa\theta_1}\rangle=\frac{1}{|z|^2}\cdot|z|^{4\kappa^2/2\pi} \tag{2A.15}$$

or

$$\left\langle\bar{\psi}\frac{(1+\sigma_3)}{2}\psi\bar{\psi}\frac{(1-\sigma_3)}{2}\psi\right\rangle=\frac{1}{|z|^x} \tag{2A.16}$$

with

$$x=2-\frac{4\kappa^2}{2\pi}=\frac{2}{\left(1+\frac{g}{\pi}\right)}=\frac{\beta^2}{4\pi}. \tag{2A.17}$$

The last equality is the definition of β^2.

The Fermi fields themselves partake of the θ_2 variables and, for example

$$\begin{aligned}\langle\psi_u^\dagger(z_1)\psi_u(z_2)\rangle &= \langle\psi_{u0}^\dagger(z_1)\psi_{u0}(z_2)\rangle\langle e^{-\kappa\theta_1(z_1)}e^{+\kappa\theta_1(z_2)}\rangle\langle e^{i\sqrt{g}\theta_2(z_1)}e^{-i\sqrt{g}\theta_2(z_2)}\rangle\\ &= \frac{1}{z}\cdot|z|^{\kappa^2/2\pi}\cdot\frac{1}{|z|^{g^2/2\pi}},\end{aligned} \tag{2A.18}$$

so

$$\langle\psi_u^\dagger(z_1)\psi_u(z_2)\rangle=\frac{1}{z}\frac{1}{|z|^y} \tag{2A.19}$$

with

$$y=\frac{g}{2\pi}-\frac{1}{2\pi}\frac{1}{\left(\frac{1}{g}+\frac{1}{\pi}\right)}=\frac{g^2\beta^2}{8\pi^3}=\frac{1}{2\pi^2}\frac{g^2}{1+\frac{g}{\pi}} \tag{2A.20}$$

The Fermi fields are no longer purely holomorphic since the interaction mixes left and right movers. The difference between the powers of z and $\bar{z}$ in their two point function is still unity. This is a consequence of their being spin-$\frac{1}{2}$ operators.

Appendix 2B. Solving the Schwinger Model by Chiral Rotation

Here we apply the same method to solve the Schwinger model. The euclidean path integral for the Schwinger model partition function is

$$\mathcal{Z} = \int d[\psi]d[\bar{\psi}]d[A_\mu] \exp \int d^2x \left\{ \bar{\psi}\gamma^\mu(\partial_\mu + iA_\mu)\psi - \frac{1}{2e^2}(F_{12})^2 \right\}. \tag{2B.1}$$

To make contact with the main text we will assume that there are N flavours of otherwise identical fermions.

We proceed exactly as with the Thirring model by decomposing A_μ into its solenoidal and curl-free parts,

$$A_\mu = \epsilon_{\mu\nu}\partial_\nu\theta_1 + \partial_\mu\theta_2, \tag{2B.2}$$

and introducing rotated fields $\psi_0, \bar{\psi}_0$ by

$$\begin{aligned} \psi_i =& e^{-\sigma_3\theta_1 - i\theta_2}\psi_{0i} \\ \bar{\psi}_i =& \bar{\psi}_{0i}e^{-\sigma_3\theta_1 + i\theta_2}, \end{aligned} \tag{2B.3}$$

one for each flavour i.

After including the jacobian, which is N times the single flavour Jacobian, the partition function becomes

$$\mathcal{Z} = \int d[\psi_0]d[\bar{\psi}_0]d[\theta_1]d[\theta_2] \exp \int d^2x \left\{ \bar{\psi}_0\gamma^\mu\partial_\mu\psi - \frac{N}{2\pi}(\partial\theta_1)^2 - \frac{1}{2e^2}(\partial^2\theta_1)^2 \right\}, \tag{2B.4}$$

Because the action is gauge invariant, the field θ_2 does not appear.

For computational convenience rescale $\theta_1 \to e\theta_1$, then

$$\mathcal{Z} = \int d[\psi_0]d[\bar{\psi}_0]d[\theta_1]d[\theta_2] \exp \int d^2x \left\{ \bar{\psi}_0\gamma^\mu\partial_\mu\psi - \frac{1}{2}\left(\frac{e^2 N}{\pi}(\partial\theta_1)^2 + (\partial^2\theta_1)^2 \right) \right\}, \tag{2B.5}$$

and the chiral operators become

$$\begin{aligned} \Sigma^+ &= \bar{\psi}\frac{(1+\sigma_3)}{2}\psi = \bar{\psi}_{u0}\psi_{u0}e^{-2e\theta_1} \\ \Sigma^- &= \bar{\psi}\frac{(1-\sigma_3)}{2}\psi = \bar{\psi}_{l0}\psi_{l0}e^{+2e\theta_1}, \end{aligned} \tag{2B.6}$$

There will be one of these operators for each flavour.

We will need the propagator for θ_1. It is

$$\mathcal{G}(m^2, \mathbf{r}) = \int \frac{d^2k}{(2\pi)^2} \frac{e^{i\mathbf{k}\cdot\mathbf{r}}}{k^2(m^2 + k^2)} = \frac{1}{m^2}\left(\int \frac{d^2k}{(2\pi)^2}\frac{e^{i\mathbf{k}\cdot\mathbf{r}}}{k^2} - \int \frac{d^2k}{(2\pi)^2}\frac{e^{i\mathbf{k}\cdot\mathbf{r}}}{k^2+m^2} \right) \tag{2B.7}$$

where $m^2 = Ne^2/\pi$. We will use the notation

$$G(m^2, \mathbf{r}) = \int \frac{d^2k}{(2\pi)^2} \frac{e^{i\mathbf{k}\cdot\mathbf{r}}}{k^2 + m^2}$$

for the conventional propagator with mass m. At large distance $G \approx e^{-mr}$.

Using the θ_1 propagator we find

$$\langle \Sigma^+(z)\Sigma^-(0)\rangle = \langle \bar{\psi}_{u0}\psi_{u0}\bar{\psi}_{l0}\psi_{l0}\rangle \langle e^{-2\kappa\theta_1} e^{+2\kappa\theta_1}\rangle$$
$$\propto \frac{1}{|z|^2} \cdot \exp \frac{4\pi}{N}\left(\frac{1}{2\pi}\ln|z| + G(m^2, |z|)\right) \tag{2B.8}$$

or

$$\langle \Sigma^+(z)\Sigma^-(0)\rangle \propto \frac{1}{|z|^{2-2/N}} \exp \frac{4\pi}{N} G(m^2, |z|) \tag{2B.9}$$

We can also look at the two point correlator of the operators

$$\chi_n^{\pm} = \prod_{i=1}^{n} \Sigma_i^{\pm} \tag{2B.10}$$

which are made by multiplying over the first n of the flavour indices. We find that these correlators

$$\langle \chi_n^+ \chi_n^- \rangle \propto \frac{1}{|z|^{2n-2n^2/N}} \exp \frac{4n^2}{N} G(m^2, |z|) \tag{2B.11}$$

have the feature that if $N = n$ the prefactor power is zero. Because the massive Green function $G(m^2, |z|)$ tends to zero at large r, the correlator tends to a constant. This failure of clustering is a sign of chiral symmetry breaking. It is not seen in the $n < N$ operators because zero modes in the fermion determinant suppress the field configurations contributing to the symmetry breaking. When $n = N$ these these zero modes are compensated by zeros in the denominators of the fermion propagators.

References

[1] B. Klaiber, in *Lectures in Theoretical Physics, Boulder 1968* (Gordon and Breach, New York 1968), 141.

[2] E. Fradkin, E. Moreno and F. A. Schaposnik, *Nucl. Phys.* **B392** (1993) 667.

[3] B. Sutherland, *J. Math. Phys.* **12** (1971) 246; see also *Phys. Rev.* **A4** (1971) 2019; **A5** (1972) 1372.

[4] N. Kawakami and S-K. Yang, *Phys. Rev. Lett.* **67** (1991) 2493.

[5] F. D. M. Haldane, *Phys. Rev. Lett.* **47** (1981) 1840.

[6] J. Schwinger, *Phys. Rev.* **128** (1962) 2425.

[7] J. Lowenstein and J. Swieca, *Ann. Phys.* (NY) **68** (1971) 175.

[8] A. Casher, J. Kogut and L. Susskind, *Phys. Rev.* **D10** (1974) 732.

[9] J. Kogut and L. Susskind, *Phys. Rev.* **D10** (1974) 3468.

[10] G. t'Hooft, *Phys. Rev.* **D14** (1976) 3432.

[11] E. Witten, *Nuc. Phys.* **B156** (1979) 269.

[12] K. Fujikawa, *Phys. Rev. Lett.* **42** (1979) 1195; *Phys. Rev.* **D21** (1980) 2848.

Chapter 3

SOME PHYSICAL APPLICATIONS

3.1. Introduction

In the last chapter we dealt only with massless fermions. Adding a mass takes us out of the domain of easily solvable systems, but this does not mean that bosonization is useless. It can be exploited quite profitably to calculate the leading terms in the effective action for systems with broken symmetries. There are many examples of this, but I will discuss two that led to my first learning some condensed matter physics. We will look first at some of the theory of charge-density waves (CDWs), and then at BCS superfluids.

3.2. Charge-Density Waves

Charge-density waves mostly occur in quasi-one-dimensional conductors. Examples are polymer chains such as polyacetylene, $(CH)_n$, or highly anisotropic crystals such as $NbSe_3$ where electrons hop predominantly along one crystal axis. The one-dimensional, non-relativistic hamiltonian (2.2.1) can be used as a simple model for these systems, provided that we substitute the band-theory effective mass, m_{eff}, in place of the electron mass.

Now suppose we add a periodic potential

$$V(x,t) = A\cos(2k_f x - \theta), \tag{3.2.1}$$

to the free theory hamiltonian (2.2.1). Here k_f is the Fermi momentum and θ a parameter that allows us to adjust the location of the potential. A term of this form will appear if there is a periodic deformation of the atomic lattice with wave-number $2k_f$. The reason for chosing this wave-number is that the potential will scatter particles from one Fermi point to the other by changing their momentum by $2k_f$. In the "relativistic" approximation to the hamiltonian, this process is described by

$$H_{mass} = \Delta\psi^\dagger\sigma_1 e^{i\sigma_3\theta}\psi, \tag{3.2.2}$$

which is just a chirally rotated mass term. By lifting the $k_f \leftrightarrow -k_f$ degeneracy, such a mass term opens up a 2Δ gap at the Fermi surface. The occupied states below the gap are thereby lowered in energy, and at low enough temperatures the energy released can offset the strain energy needed to distort the lattice. When this occurs the distortion and mass term will be spontaneously generated. We can think of the periodic distortion as forming a condensate of $2k_f$ phonons. Many one-dimensional conductors undergo such a *Peierls transition* to a CDW state. The CDW ground state reduces the lattice translational symmetry and under appropriate circumstances the *phason* field $\theta(x,t)$ is the corresponding Goldstone boson of broken translational invariance.

The bosonized lagrangian for the electrons, including the CDW mass gap, is

$$L_{electrons} = \frac{1}{2}(\partial_\mu\varphi)^2 + \Delta : \cos(2\sqrt{\pi}\varphi - \theta(x,t)) : . \tag{3.2.3}$$

We have set $v_f = 1$. Because we will mostly be treating these lagrangians semi-classically I will drop the normal ordering symbols for the rest of this chapter.

We must add to (3.2.3) a lagrangian for the θ field. In most cases this will be of the form

$$L_{phasons}(\theta) = \frac{1}{2}K'(\partial_t\theta)^2 - \frac{1}{2}K(\partial_x\theta)^2. \tag{3.2.4}$$

Here K and K' are parameters describing the elasticity and inertia of the phason field (phasons will not in general propagate at the Fermi velocity, so the action need not be Lorentz invariant).

When the mass gap is large enough the electrons will have little overlap with the states above the gap, and become adiabatically slaved to the motion of the phonon condensate. In the boson description this is reflected by φ sitting close to a minimum of the non-linear cosine potential. We can then approximate $2\sqrt{\pi}\varphi \approx \theta$. Substituting this back in the lagrangian we find that the effect of the electrons is to add a term

$$L_{Born-Oppenheimer} = \frac{1}{8\pi}(\partial_\mu\theta)^2 \tag{3.2.5}$$

to the lagrangian for θ, and so renormalize the phason motion.

Any deviation in the θ field from uniformity will compel φ to follow it, and so induce an excess local charge of

$$\delta\rho = -\frac{1}{\sqrt{\pi}}\partial_x\varphi \approx -\frac{1}{2\pi}\partial_x\theta. \tag{3.2.6}$$

Similarly a time-dependent θ induces a current

$$j = \frac{1}{2\pi}\partial_t\theta. \tag{3.2.7}$$

If the wavelength of the CDW is incommensurate with the periodicity of the one-dimensional conductor there will be no prefered location of the periodic CDW distortion relative to the underlying atomic lattice. The CDW distortion will then be able to "slide" freely with respect to the lattice. In this case the phason is gapless and is the Goldstone boson associated with the spontaneously broken translation invariance. A sliding CDW carries a current, even though a material with a gap is normally an insulator. In particular, sliding the potential (3.2.1) along at velocity v produces a current $j = \dot{\theta}/(2\pi) = vk_f/\pi$. Since the electrons have average density $n = k_f/\pi$, we can rewrite this as nv. We see that the *entire Fermi sea* is carried along by the sliding wave.

The principal force that the electrons feel is from the fixed lattice potential. We have accounted for this by taking m to be the effective band theory mass, m_{eff}. The periodic potential induced by the CDW distortion is only a small fraction of the lattice potential and it seems remarkable that such a small perturbation can drag all the electrons along with it. That the small potential triumphs over the large is a consequence of the extreme sensitivity of the electron gas to perturbations with wavevectors that span the Fermi surface, and so lift the $k_f \leftrightarrow -k_f$ degeneracy.

The Rebirth of the Schwinger Mechanism

So far we have ignored the electromagnetic interactions of the electrons. Usually CDW materials come as three-dimensional bundles of parallel one-dimensional chains. The θ phase on adjacent chains is usually correlated. If it is coherent across all the chains, the charge density induced by θ gradients will be independent of the transverse coordinate. The electrostatics of these charge sheets realizes the one-dimensional electrodynamics of the Schwinger model — but a Schwinger model now containing a fermion mass term. This mass term prevents a simple solution.

The bosonized form of the massive Schwinger model has lagrangian

$$L = \frac{1}{2}(\partial_\mu \varphi)^2 + \frac{e^2}{2\pi}(\varphi - \varphi_\infty + \varphi_{ext})^2 + \Delta \cos 2\sqrt{\pi}\varphi. \tag{3.2.8}$$

The presence of the cosine term prevents the φ field from freely tracking φ_{ext}. The quadratic term then provides an energy cost per unit length between well-separated test charges. This reflects the fact that a mass gap makes the electron gas an insulator, and insulators do not completely screen.

When we include a massless phason mode (3.2.8) becomes

$$\begin{aligned} L = &\frac{1}{2}(\partial_\mu \varphi)^2 + \frac{e^2}{2\pi}(\varphi - \varphi_\infty + \varphi_{ext})^2 \\ &+ \Delta \cos(2\sqrt{\pi}\varphi - \theta) + \frac{K'}{2}(\partial_t \theta)^2 - \frac{K}{2}(\partial_x \theta)^2. \end{aligned} \tag{3.2.9}$$

Now, by simultaneously varying θ, φ *can* adapt itself to φ_{ext}, and so completely screen a test charge. The system has become a conductor again — but the charge carrier is the sliding CDW.

The cross-cultural interplay between the axial anomaly that generates the Schwinger model's e^2/π photon mass and the sliding CDW was described independently in [1] and [2]. I remember being rather intrigued when these papers came out because a diagram-based discussion of this born-again Schwinger mass-generation mechanism was one of my thesis topics [3]. As a narrowly educated particle-physics graduate student, I knew very little about condensed matter physics, but I had, in the course of almost aimless playing with Feynman diagrams, discovered this conspiratorial cancellation between poles that let even massive fermions generate a Schwinger mass for the photon. I had no idea that there was a real physical model for the phenomenon.

A similar comment applies to the next topic:

Vacuum Decay

Not all CDW's have gapless phasons. In many cases the CDW has a wavelength that is commensurate with the lattice periodicity and is pinned in some preferred relation to the lattice. Consequently the θ field also has preferred values. This phase locking can be described by the lagrangian

$$L = \frac{1}{2}(\partial_\mu\varphi)^2 + \frac{e^2}{2\pi}(\varphi - \varphi_\infty + \varphi_{ext})^2 + \Delta\cos(2\sqrt{\pi}\varphi - \theta)$$
$$+ \frac{K'}{2}(\partial_t\theta)^2 - \frac{K}{2}(\partial_x\theta)^2 + \Delta'\cos Q\theta, \quad (3.2.10)$$

where Q is a parameter, often an integer, determined by the commensurability [4]. The lagrangian (3.2.10) gives rise to equations of motion with soliton solutions. In these θ passes from one minimum of $\cos Q\theta$ to another, increasing or decreasing by $2\pi/Q$ as it does so. Because φ will tend to track θ, these solitons carry charge e/Q [5]. In polyacetylene, for example, $Q = 2$, which would lead to solitons with charge 1/2 [6] were the charge not doubled by the two components of spin [7]. (In reality polyacetylene is not well described by the present model: we are explicitly assuming that the gap does not vary significantly, but in polyacetylene the gap vanishes in the soliton core.)

In the absence of solitons the pinned CDW system should be an insulator. We can still ask, however, what will happen if we apply an electric field E. Again assume that Δ is large enough to allow us to eliminate φ and obtain an action containing only θ. After rescaling θ and the "speed of light" we will end up with an effective lagrangian of the form

$$L = \frac{1}{2}(\partial_\mu\theta)^2 + M\cos\beta\theta. \quad (3.2.11)$$

To include the uniform external electric field we add a potential interaction

$$V(x)\rho(x) = \frac{Ex}{2\pi}\partial_x\theta. \quad (3.2.12)$$

This can be integrated by parts to give

$$L = \frac{1}{2}(\partial_\mu\theta) + M\cos\beta\theta + \frac{E}{2\pi}\theta. \quad (3.2.13)$$

The effect of the electric field is to tilt the cosine and produce a "washboard" potential. If $E < 2\pi M\beta$ the potential wells remain classically stable and the θ field will sit in one of the many local minima of the potential until it manages to tunnel through the barrier to the next lowest. There it will remain until it tunnels again. Since the average value of $\dot{\theta}$ will be non-zero due to the tunneling, a current will flow for arbitrarily small E.

Finding the tunneling rate is another problem I had fun with as a graduate student. Motivated by the work of Bender and Wu on high-order perturbation theory [8] I thought it might be interesting to investigate how a metastable vacuum state would decay in a scalar quantum field theory. The experts I consulted for advice on how to proceed all assured me that the action for getting a scalar field over a tunneling barrier was proportional to the volume of the system, and so decay could never occur when the volume was infinite. One day, while sitting in a lecture by J. Frölich at the 1975 Erice summer school, it occurred to me that one could "fermionize" and map the θ tunneling problem into QED in an external electric field. Obviously in this situation there will be pair-creation of fermions. Translating this back into the boson language reveals that the field theory is able to overcome the infinite barrier problem by tunneling only locally, producing a soliton-antisoliton pair between which is an ever-expanding region of lower energy "vacuum". In the present case the pair-produced fermions are not the original electrons but the fractionally charged solitons of [5].

For most values of β in (3.2.13) the fermionic solitons are interacting, but when $\beta^2 = 4\pi$ they are free and one can do the calculation exactly (see Appendix 3A). The result [9] is a pair creation rate per unit volume of

$$\Gamma = \frac{e|E|}{2\pi} \sum_{n=1}^{\infty} \frac{1}{n} e^{-n\pi m^2/e|E|} \tag{3.2.14}$$

where m is the soliton mass and e the soliton charge. One can use an instanton method to establish that this result remains true to leading order even when $\beta^2 \neq 4\pi$ [10].

Some experts still did not believe that the vacuum could decay in field theory, and [10] was initially rejected on the grounds that an unstable vacuum did not satisfy the Wightman axioms, and therefore could not exist! Fortunately (or unfortunately) Coleman's famous *Fate of the false vacuum* paper [11] then appeared as a preprint and I was able to cite it as my authority. Of course Coleman's analysis was much more comprehensive than mine, so few people read my brief note anyway. Such are the joys of being a student! I also found out that the high-order perturbation theory problem had been solved some months earlier by Lipatov [12], while the instanton method had been used much earlier by Langer in his work on metastability and nucleation in statistical mechanics [13].

The soliton pair creation mechanism was described explicitly in the context of CDW's by K. Maki [14]. Reading this paper provided my introduction to condensed matter physics. Unfortunately in real CDW systems the soliton pair production picture is considerably complicated by additional pinning due to impurities.

3.3. Zero Temperature Effective Action for BCS Superconductors

Bosonization also has its uses in two and three dimensions. The earliest discussion of higher dimensions to have seems to have been by Luther [15]. More recently the technique has been pursued by Haldane [16] and others [17,18] interested in the possible breakdown of Fermi-liquid theory in two dimensional strongly correlated systems [19].

A simpler application is the calculation of an effective action for Fermi superfluids. The Landau-Ginzburg effective action for a BCS superfluid near T_c is well known. This action is a functional of the BCS order parameter. Near $T = 0$ the situation is not so clear cut and there is some confusion in the literature over issues such as galilean invariance*. Surely for a simple, neutral, S-wave BCS superfluid it should be possible to derive a simple galilean invariant effective action that describes the gross superfluid dynamics of the condensate?

*The classic reference is Abrahams and Tsuneto [20] and, although the presentation may look different, I believe that nothing I am going to say here is inconsistent with their conclusions.

In this section I will use Fermi-surface bosonization to find such an effective action. The use of bosonization is not essential, but it makes the calculation simpler by obviating the need to evaluate diagrams. The resultant effective action leads to a Schrödinger-like equation of motion for the condensate, but the "wavefunction" in this equation is *not* the BCS order parameter.

We begin with a quick review of the field-theory basics of neutral BCS superfluids.

We write the partition function of a gas of non-relativistic spin-$\frac{1}{2}$ fermions interacting via a short-range potential as a Grassmann path integral

$$\begin{aligned}\mathcal{Z} &= \mathrm{Tr}(e^{-\beta H}) \\ &= \int d[\psi]d[\psi^\dagger]\exp - \int_0^\beta d^3x d\tau \left\{ \sum_{\alpha=1}^{2} \psi_\alpha^\dagger \left(\partial_\tau - \frac{1}{2m}\nabla^2 - \mu \right) \psi_\alpha - g\psi_1^\dagger \psi_2^\dagger \psi_2 \psi_1 \right\}.\end{aligned} \tag{3.3.1}$$

The indices $\alpha = 1, 2$ refer to the two components of spin. As is usual in the Matsubara formalism the Grassmann-valued Fermi fields are to be taken antiperiodic under the shift $\tau \to \tau + \beta$.

A positive value for g corresponds to an attractive potential. Given such an interaction, and a low enough temperature, the system should be unstable to the onset of superconductivity. To detect this instability we introduce an ancillary complex scalar field Δ which will become the superconducting order parameter. We use it to decouple the interaction

$$\begin{aligned}\mathcal{Z} = \int d[\psi]d[\psi^\dagger]d[\Delta]d[\Delta^*]\exp - \int_0^\beta d^3x d\tau \Big\{ &\sum_{\alpha=1}^{2} \psi_\alpha^\dagger \left(\partial_\tau - \frac{1}{2m}\nabla^2 - \mu \right) \psi_\alpha \\ &- \Delta^* \psi_2 \psi_1 - \Delta \psi_1^\dagger \psi_2^\dagger + \frac{1}{g}|\Delta|^2 \Big\}.\end{aligned} \tag{3.3.2}$$

The equation of motion for Δ shows us that $\Delta \equiv g\psi_2\psi_1$.

We may now integrate out the fermions to find an effective action for the Δ field. From this point on we will set the temperature, β^{-1}, to zero.

Taking note of the anticommutativity of the Grassmann fields, the quadratic form in the exponent can be arranged as a matrix

$$S = \int d^3x d\tau (\psi_1^\dagger, \psi_2) \begin{pmatrix} \partial_\tau - \frac{\nabla^2}{2m} - \mu & \Delta \\ \Delta^* & \partial_\tau + \frac{\nabla^2}{2m} + \mu \end{pmatrix} \begin{pmatrix} \psi_1 \\ \psi_2^\dagger \end{pmatrix}. \tag{3.3.3}$$

The fermion contribution to the effective action is the logarithm of the Fredholm determinant of this matrix of differential operators

$$S_F = -\ln \mathrm{Det} \begin{pmatrix} \partial_\tau - \frac{\nabla^2}{2m} - \mu & \Delta \\ \Delta^* & \partial_\tau + \frac{\nabla^2}{2m} + \mu \end{pmatrix} \tag{3.3.4}$$

We begin by first assuming that Δ is a constant. Under these circumstances S_F is given by

$$S_F = \int d^3x d\tau \left\{ - \int \frac{d^3k}{(2\pi)^3} \frac{d\omega}{2\pi} \mathrm{tr} \ln \begin{pmatrix} i\omega + \frac{k^2}{2m} - \mu & \Delta \\ \Delta^* & i\omega - \frac{k^2}{2m} + \mu \end{pmatrix} \right\} \tag{3.3.5}$$

It is convenient to introduce the notation $\epsilon = k^2/2m - \mu$, whence the momentum integral becomes

$$I = \int \frac{d^3k}{(2\pi)^3}\frac{d\omega}{2\pi}\ln(\omega^2 + |\Delta|^2 + \epsilon^2). \tag{3.3.6}$$

Since everything is spherically symmetric we can replace the integration over k by an integration over ϵ

$$I = \int \rho(\epsilon)d\epsilon\frac{d\omega}{2\pi}\ln(\omega^2 + |\Delta|^2 + \epsilon^2), \tag{3.3.7}$$

at the expense of introducing the density of states $\rho(\epsilon)$. We evaluate this by differentiating with respect to Δ

$$\frac{dI}{d\Delta} = \int d\epsilon\frac{d\omega}{2\pi}\rho(0)\frac{\Delta^*}{\omega^2 + |\Delta|^2 + \epsilon^2} \tag{3.3.8}$$

where, after noticing that the integral is peaked near the Fermi surface $\epsilon = 0$, we have approximated $\rho(\epsilon)$ by $\rho(0)$. After performing the frequency integral we find

$$\frac{dI}{d\Delta} = \frac{\rho(0)}{2}\int_{-\epsilon_D}^{\epsilon_D}\frac{\Delta^* d\epsilon}{\sqrt{\epsilon^2 + |\Delta|^2}} = -\frac{\rho(0)}{2}\Delta^*\ln\frac{|\Delta|}{\epsilon_D}. \tag{3.3.9}$$

To obtain a finite answer we have, as is customary, introduced a physically motivated cutoff at the energy ϵ_D, corresponding to the Debye frequency.

Putting this together with the $|\Delta|^2$ part of the exponent we find that the effective potential (the action per unit space-time volume) for Δ is minimized when

$$\frac{dV_{eff}}{d\Delta} = \left(\frac{\Delta^*}{g} + \frac{\rho(0)}{2}\Delta^*\ln\frac{|\Delta|}{\epsilon_D}\right) = 0 \tag{3.3.10}$$

or when

$$|\Delta| = |\Delta_0| = \epsilon_D\exp\left\{-\frac{2}{g\rho(0)}\right\}. \tag{3.3.11}$$

The effective potential itself is

$$V_{eff} = \frac{1}{g}|\Delta|^2 + \frac{\rho(0)}{4}\left(|\Delta|^2\ln\frac{|\Delta|^2}{\epsilon_D^2} - |\Delta|^2\right). \tag{3.3.12}$$

It will play no further role in our discussions.

We now wish to investigate the terms in the effective action $S_F(\Delta)$ that involve space-time gradients of Δ. Since Δ plays the role of a mass-gap we would expect to able to find an expansion in increasing orders and powers of derivatives of Δ, each extra derivative being accompanied by a factor of $|\Delta_0|^{-2}$. A diagrammatic evaluation of such an expansion is complicated, but we can get considerable insight into the problem by using some simple tricks, including bosonization.

Let us examine the kind of terms we expect to find by first considering the effect of uniform twists. Suppose that the phase of the order parameter varies linearly with position *i.e* $\Delta(x) = e^{2ik_s x}\Delta_0$. This should correspond to a uniform flow with superfluid velocity $v_s = k_s/m$.

We need to find the Fredholm determinant of

$$K = \begin{pmatrix} \partial_\tau - \frac{\nabla^2}{2m} - \mu & \Delta_0 e^{2ik_s x} \\ \Delta_0^* e^{-2ik_s x} & \partial_\tau + \frac{\nabla^2}{2m} + \mu \end{pmatrix}. \tag{3.3.13}$$

Now if

$$U = \begin{pmatrix} e^{ik_s x} & 0 \\ 0 & e^{-ik_s x} \end{pmatrix} \tag{3.3.14}$$

then

$$U^{-1} K U = \begin{pmatrix} \partial_\tau - \frac{(\nabla + ik_s)^2}{2m} - \mu & \Delta_0 \\ \Delta_0^* & \partial_\tau + \frac{(\nabla - ik_s)^2}{2m} + \mu \end{pmatrix}. \tag{3.3.15}$$

The determinant is not affected by such a unitary transformation* and we can now evaluate $\ln \operatorname{Det} K$ by Fourier transforming.

$$\begin{aligned} \ln \operatorname{Det} K &= \int \frac{d^3k}{(2\pi)^3} \frac{d\omega}{2\pi} \operatorname{tr} \ln \begin{pmatrix} i\omega + \frac{(k+k_s)^2}{2m} - \mu & \Delta_0 \\ \Delta_0^* & i\omega - \frac{(k-k_s)^2}{2m} + \mu \end{pmatrix} \\ &= \int \frac{d^3k}{(2\pi)^3} \frac{d\omega}{2\pi} \operatorname{tr} \ln \begin{pmatrix} (i\omega + v_s k) + \frac{k^2}{2m} - \left(\mu - \frac{k_s^2}{2m}\right) & \Delta_0 \\ \Delta_0^* & (i\omega + v_s k) - \frac{k^2}{2m} + \left(\mu - \frac{k_s^2}{2m}\right) \end{pmatrix} \end{aligned} \tag{3.3.16}$$

The net effect has been to shift $i\omega \to i\omega + v_s \cdot k$ and $\mu \to \mu - k_s^2/2m$.

If we momentarily forget about the shift in the ω variable, we find that, for small k_s,

$$S_F(\Delta_0 e^{2ik_s x}) = S_F(\Delta_0) - \frac{k_s^2}{2m} \frac{\partial S_F}{\partial \mu}. \tag{3.3.17}$$

But $-\partial S_F / \partial \mu$ is the number density of the system, so

$$S_F(\Delta_0 e^{2ik_s x}) = S_F(\Delta_0) + \rho \frac{k_s^2}{2m}. \tag{3.3.18}$$

This is what we would expect for a uniform flow of the entire fluid.

Let us now ask for the consequences of the sideways translation of the ω contour. In evaluating the density in (3.3.18) it is the ω contour integral that determines the occupation numbers of the various quasiparticle modes. These modes manifest themselves as poles in the $\langle \psi^\dagger \psi \rangle$ Green function at $\omega = \pm\sqrt{\epsilon^2 + |\Delta|^2}$, and the contour integral is to be closed in such a manner that the negative energy states are encircled and thus counted as occupied. These negative energy states are particle-like for $k \ll k_f$ and hole-like for $k \gg k_f$. If v_s is large enough, poles that were within the contour before the $iv_s \cdot k$ shift may no longer

*This is not true for relativistic systems where the determinant *is* altered by such chiral transformation because of anomalies. There are no anomalies for non-relativistic systems.

be encircled, and conversely, previously unoccupied states may be occupied. The physical reason is that, when seen from the rest frame, the quasiparticle energies in the moving fluid are doppler shifted from $\omega(k)$ to $\omega(k) + v_s \cdot k$. The use of a chemical potential μ to impose the constraint on the total number of particles implies equilibrium with the stationary walls of the container, and it is the negative energy states as seen from this rest frame that are occupied. There is a range of v_s for which the occupation numbers are unchanged, and consequently a critical value of v_s below which no "normal" fluid will be created. Above v_{crit} we need to replace the ρ in (3.3.18) by an effective ρ_s. Since the alteration of occupation numbers tends to reduce the momentum in the system we have $\rho_s < \rho$. Only the superfluid fraction is flowing. The normal component stays at rest.

We now consider the time variation $\Delta = e^{2i\Omega t}\Delta_0$. This shifts $\mu \to \mu - i\Omega$ so, again for small Ω,

$$S_F(e^{2i\Omega t}\Delta_0) = S_F(\Delta_0) + i\rho\Omega. \tag{3.3.19}$$

Thus we know that the euclidean action S_F will contain the terms

$$S_F(\Delta) \approx V_{eff}(\Delta) + i\rho\partial_\tau\phi/2 + \frac{\rho}{2m}(\partial_x\phi/2)^2, \tag{3.3.20}$$

where $\Delta = e^{i\phi(x,t)}|\Delta|$.

Notice that if we work in real time (as opposed to the Matsubara imaginary time we have been using) then, at least for $v_s < v_{crit}$, S_F is invariant under the combined phase transformation $\Delta \to (e^{imv_s x - i\frac{1}{2}mv_s^2 t})^2\Delta$. This corresponds to a simultaneous galilean transformation acting on all the particles in the fluid. When the fluid and its container are accelerated together the occupation numbers should remain unchanged even if $v_s > v_{crit}$. In this circumstance the contour should be arranged so that the $v_s \cdot k$ translation has no effect.

We will now use bosonization to directly calculate the effective action for slowly varying $\Delta(x,t)$. The discussion is based on methods introduced in [21].

Let us warm up with a one-dimensional system where we write as usual

$$\begin{aligned}\psi_i &= e^{ik_f x}\psi_{i,R} + e^{-ik_f x}\psi_{i,L}\\ \psi_i^\dagger &= e^{-ik_f x}\psi_{i,R}^\dagger + e^{ik_f x}\psi_{i,L}^\dagger.\end{aligned} \tag{3.3.21}$$

We introduce two two-component fields

$$\Psi_1 = \begin{pmatrix}\psi_{1,R}\\ \psi_{2,L}^\dagger\end{pmatrix} \quad \Psi_2^\dagger = \begin{pmatrix}\psi_{2,R}\\ \psi_{1,L}^\dagger\end{pmatrix}, \tag{3.3.22}$$

and their Hermitian conjugates. These are *not* the same combinations that we used for CDW systems. In terms of (3.3.32) the fermion number and current become

$$j = v_f \sum_{i=1,2}(\psi_{i,R}^\dagger\psi_{i,R} - \psi_{i,L}^\dagger\psi_{i,L}) = v_f \sum_{i=1,2}\Psi_i^\dagger\Psi_i \tag{3.3.23}$$

and

$$\rho = \sum_{i=1,2}(\psi_{i,R}^\dagger\psi_{i,R} + \psi_{i,L}^\dagger\psi_{i,L}) = \sum_{i=1,2}\Psi_i^\dagger\sigma_3\Psi_i \tag{3.3.24}$$

The quadratic form for the Fermi fields is approximated by linearizing about the Fermi surface and replacing $-\partial_x^2/2m - \mu$ by $-iv_f\partial_x$. We then have

$$L = \Psi_1^\dagger \begin{pmatrix} \partial_\tau - iv_f\partial_x & \Delta \\ \Delta^* & \partial_\tau + iv_f\partial_x \end{pmatrix} \Psi_1 + \Psi_2^\dagger \begin{pmatrix} \partial_\tau - iv_f\partial_x & -\Delta \\ -\Delta^* & \partial_\tau + iv_f\partial_x \end{pmatrix} \Psi_2. \quad (3.3.25)$$

The bosonization identities, with the speed of light set to unity, tell us that a Fermi action with a chiral mass term

$$L = \psi^\dagger(\gamma_\mu\partial_\mu + Me^{i\gamma_5\phi})\psi \quad (3.3.26)$$

is equivalent to the bosonic action

$$L' = \frac{1}{2}(\partial_\mu\varphi^2) + M\cos(2\sqrt{\pi}\varphi - \phi). \quad (3.3.27)$$

Provided the mass-gap is large enough, the cosine term suppresses fluctuations of the φ field, and so $2\sqrt{\pi}\varphi \approx \phi$. Using this relation the effective action becomes

$$L_{eff} = \frac{1}{8\pi}(\partial_\mu\phi^2). \quad (3.3.28)$$

Corrections to L_{eff} due to fluctuations in φ will contain higher gradients of ϕ. On dimensional grounds, each ∂_x will be accompanied by a factor of M^{-1}. Gradients of M will be accompanied by a factor of M^{-2}.

Using these results and restoring v_f we find that the real time action becomes

$$S_0 = (2)\int dxdt\left\{\frac{1}{8\pi v_f}\left(\frac{\partial\phi}{\partial t}\right)^2 - \frac{v_f}{8\pi}\left(\frac{\partial\phi}{\partial x}\right)^2\right\}. \quad (3.3.29)$$

where $\Delta = |\Delta|e^{i\phi}$. The factor of (2) outside the integral is from the two spin components. At the level of approximation we are using here (we have assumed that the gap is both small compared to the Fermi energy, and that it varies slowly in space and time) amplitude fluctuations of the gap do not contribute.

How can we can extend (3.3.29) to three dimensions? Let us first assume Δ varies only in the x direction. The quadratic form then couples momenta only in the x direction, so at each point on the Fermi surface we have a one-dimensional problem to solve. Summing bosonized lagrangians of the form (3.3.29) over each point of the Fermi surface gives us

$$S = (2)\int\frac{d^2k_\perp}{(2\pi)^2}\int dxdt\left\{\frac{1}{8\pi v_f\cos\theta}\left(\frac{\partial\phi}{\partial t}\right)^2 - \frac{v_f\cos\theta}{8\pi}\left(\frac{\partial\phi}{\partial x}\right)^2\right\}. \quad (3.3.30)$$

The angle θ is the angle between the normal to the (spherical) Fermi surface and the x direction.

We can write $d^2k_\perp$ in terms of the angle θ as

$$\int\frac{d^2k_\perp}{(2\pi)^2}\cdots = \frac{k_f^2}{(2\pi)}\int_0^1 \sin\theta d(\sin\theta)\ldots \quad (3.3.31)$$

Thus

$$\int \frac{d^2k_\perp}{(2\pi)^2}\cos\theta = \frac{1}{3}\frac{k_f^2}{2\pi}, \tag{3.3.32}$$

and

$$\int \frac{d^2k_\perp}{(2\pi)^2}\frac{1}{\cos\theta} = \frac{k_f^2}{2\pi}. \tag{3.3.33}$$

The three-dimensional equivalent of (3.3.29) is therefore

$$S = (2)\frac{k_f^2}{2\pi}\int dxdt\left\{\frac{1}{8\pi v_f}\left(\frac{\partial\phi}{\partial t}\right)^2 - \frac{v_f}{3}\frac{1}{8\pi}\left(\frac{\partial\phi}{\partial x}\right)^2\right\}. \tag{3.3.34}$$

As a check, notice that $\phi = 2(\delta k)x$ corresponds to giving every electron an extra δk of momentum. In this case the second term in the integrand gives

$$(2)\cdot\frac{1}{(2\pi)^3}\cdot\frac{4}{3}\pi k_f^3\cdot\frac{1}{2m}(\delta k)^2 \tag{3.3.35}$$

and, after identifying the product of the first three factors as the (both spin component) number density ρ_0, we get the kinetic energy.

The effect of the factor of 1/3 in the second term of (3.3.35) is to give a wave velocity of $c_s = v_f/\sqrt{3}$. This is the usual *hydrodynamic* sound velocity for a three dimensional Fermi gas. The principal effect of the gap on the dynamics of the system is thus to replace zero-sound by density waves which propagate at a velocity determined only by the mean density and the bulk modulus.

A similar argument, but one that maintains manifest rotational invariance, is to argue that the only important effect of spatial variations in ϕ is in the coupling of antipodal points of the Fermi surface. We can thus again decompose that action into a sum of actions over the Fermi surface. To get the correct measure, we write

$$\frac{d^3k}{(2\pi)^3} = k^2dk_\perp\frac{d^2\Omega}{(2\pi)^3} \approx \frac{k_f^2}{\pi}\frac{dk_\perp}{2\pi}\frac{d^2\Omega}{4\pi}. \tag{3.3.36}$$

which shows us how to express integrals over momentum space as integrals over the solid angle $d^2\Omega$ on the Fermi surface, together with an integral over $dk_\perp$ perpendicular to the surface.

We now write the three-dimensional bosonized action in this way

$$S = \frac{1}{2}(2)\frac{k_f^2}{\pi}\int\frac{d^2\Omega}{4\pi}\int d^3xdt\left\{\frac{1}{8\pi v_f}\left(\frac{\partial\phi}{\partial t}\right)^2 - \frac{v_f}{8\pi}(\mathbf{n}\cdot\nabla\phi)^2\right\}. \tag{3.3.37}$$

The symbol $\mathbf{n}$ denotes a unit vector in real space directed along the current k direction. The initial $\frac{1}{2}$ arises because each $\mathbf{n}$ accounts for a pair of antipodal points on the Fermi sphere and we do not want to over-count.

Using

$$\int \frac{d^2\Omega}{4\pi} n_i n_j = \frac{1}{3}\delta_{ij}, \tag{3.3.38}$$

we again find

$$S = (2)\frac{k_f^3}{24\pi^2}\frac{1}{m}\int d^3x dt \left\{ \frac{1}{2}\frac{3}{v_f^2}\left(\frac{\partial\phi}{\partial t}\right)^2 - \frac{1}{2}(\nabla\phi)^2 \right\}. \tag{3.3.39}$$

This action has appeared in the literature (e.g. [22]) but it is not galilean invariant. It does not, without further assumptions, reduce to conventional superfluid dynamics, and does not give rise to the expected Magnus force on vortices. To see what has gone wrong let us couple an external gauge field as a probe.

We require gauge invariance and use our knowledge of the transformation properties of Δ to argue that the effective action must be of the form

$$S = \frac{\rho_0}{4m}\int d^3x dt \left\{ \frac{1}{2}\frac{3}{v_f^2}\left(\frac{\partial\phi}{\partial t} - 2eA_0\right)^2 - \frac{1}{2}(\partial_\mu\phi - 2eA_\mu)^2 \right\}. \tag{3.3.40}$$

Differentiating with respect to the gauge field gives us the number current,

$$j_\mu = \frac{\rho_0}{2m}(\partial_\mu\phi - 2eA_\mu), \tag{3.3.41}$$

and number density,

$$\rho \stackrel{?}{=} -\frac{\rho_0}{2mc_s^2}\left(\frac{\partial\phi}{\partial t} - 2eA_0\right). \tag{3.3.42}$$

The first of these and these expressions is the expected form of the current. The second is consistent with the number conservation law which follows from the equation of motion for ϕ, but it does not seem quite right. We expect the equilibrium density, ρ_0, of fluid to be present even when ϕ is time independent. This "vacuum" charge has been omitted in (3.3.42). This is not surprising because the bosonization formulae give bosonic expressions for normal-ordered currents.

To get the correct number density requires us to add a $\rho_0 A_0$ term. Maintaining gauge invariance further requires us to add a time derivative to complete the covariant derivative. Thus we need a term

$$S_1 = \int d^3x dt \frac{1}{2}\rho_0\left(2eA_0 - \frac{\partial\phi}{\partial t}\right). \tag{3.3.43}$$

That the time derivative is necessary is also indicated by (3.3.19). Being a total derivative it does not affect the equations of motion but it does have some significant consequences. In the Matsubara formalism we sum over periodic configurations but $\partial_t\phi$ does not necessarily integrate to zero. Only the order parameter itself is required to be periodic; its phase can be multivalued. When there are vortices in the system, every point that is encircled by a vortex trajectory will have its value of ϕ incremented by 2π. The time derivative in (3.3.43) thus contributes a phase proportional to the area enclosed by the trajectory. This phase produces the Magnus force on the vortex [23]. That this is so should be clear by analogy with a particle moving in a magnetic field, where the action also accumulates a phase proportional the the area enclosed by the particle trajectory. The Magnus force is the direct analog of the Lorentz force on the particle.

The total action is now

$$S=\int d^3xdt\left\{\frac{1}{2}\frac{\rho_0}{mc_s^2}\left(\frac{\partial\phi/2}{\partial t}-eA_0\right)^2-\frac{\rho_0}{2m}\left(\nabla\phi/2-e\mathbf{A}\right)^2-\rho_o\left(\frac{\partial\phi/2}{\partial t}-eA_0\right)\right\}. \quad (3.3.44)$$

We could rest content with (3.3.44) and see whether we can extract superfluid mechanics from it. A more illuminating approach is to first derive a Schrödinger-like equation of motion. To do this we must promote ρ to the status of a dynamical variable. If we write

$$S=\int d^3xdt\left\{-\rho\left(\frac{\partial\phi/2}{\partial t}-eA_0\right)-\frac{\rho_0}{2m}\left(\nabla\phi/2-e\mathbf{A}\right)^2-\frac{1}{2}\frac{c_s^2m}{\rho_0}(\rho-\rho_0)^2\right\}, \quad (3.3.45)$$

then integrating out the ρ field, or eliminating it by using its equation of motion gives (3.3.44). To obtain a galilean invariant expression we must also replace the ρ_0 in front of the second term in the integrand of (3.3.45) by the full density, ρ. This step takes us beyond the approximations we have been using, but is justified because we know that the exact result must be consistent with (3.3.18), and that equation requires the coefficient to be ρ, and not ρ_0.

Now *define* the field $\Psi=\sqrt{\frac{\rho}{2}}e^{i\phi}$. For mnemonic purposes Ψ can be thought of as a "wavefunction" of the Cooper pairs. (It is *not* the BCS order parameter. There is no simple relation between $|\Delta|$ and ρ.) If we define the Cooper pair charge to be $e^*=2e$, its mass to be $m^*=2m$ and the equilibrium pair density to be $\rho_0^*=\rho_0/2$, we can write (3.3.45) in an appealingly simple form. Up to higher-order gradients of ρ^*, it is equivalent to

$$S=\int d^xdt\left\{i\Psi^\dagger(\partial_t-ie^*A_0)\Psi-\frac{1}{2m^*}|(\nabla-ie^*\mathbf{A})\Psi|^2-\frac{\lambda}{2}(|\Psi|^2-\rho_0)^2\right\}, \quad (3.3.46)$$

where $\lambda=\frac{c_s^2m^*}{\rho_0^*}$.

Varying this last action gives rise to the galilean invariant nonlinear Schrödinger equation for Ψ

$$i(\partial_t-ie^*A_0)\Psi=-\frac{1}{2m^*}(\nabla-ie^*\mathbf{A})^2\Psi+\lambda(|\Psi|^2-\rho_0^*)\Psi. \quad (3.3.47)$$

In this context the nonlinear Schrödinger equation is usually named after Gross and Pitaevskii [24–26]. It can be rewritten as the Euler equation for a compressible fluid (Appendix 3B).

I must stress that Ψ is neither the BCS order parameter, nor a genuine Cooper-pair wavefunction. It is simply a mathematical construct that permits us to write the low-energy effective action (3.3.45) in an easily digestible form. In particular (3.3.47) should not be thought of as a "time-dependent Landau-Ginzburg equation". A true Landau-Ginzburg equation would contain information about the dynamics of the order parameter amplitude. This amplitude mode has a frequency gap, so it does not appear in our low-energy description.

Appendix 3A. Pair Creation in an External Field

In this appendix we will calculate the pair creation rate for charged particles in an electric field. This result was originally obtained (in 3+1 dimensions) by Schwinger [27].

The pair creation rate per unit volume is given by twice the imaginary part of the ground-state energy-density, $\mathcal{E}$. This can be found from the partition function

$$\mathcal{Z} = \exp(-LT\mathcal{E}) = \mathrm{Det}(\gamma^\mu \nabla_\mu + m) \tag{3A.1}$$

of $1+1$ dimensional fermions in a uniform electric field. The fermions occupy a Euclidean spacetime region of duration T and length L .

Conjugating with $\gamma_5 \equiv \sigma_3$ does not alter the value of the determinant, so

$$\ln \mathrm{Det}(\gamma^\mu \nabla_\mu + m) = \frac{1}{2} \ln \mathrm{Det}\{(\gamma^\mu \nabla_\mu + m)(-\gamma^\nu \nabla_\nu + m)\}. \tag{3A.2}$$

Using $\{\gamma^\mu, \gamma^\nu\} = 2g^{\mu\nu}$ and $[\nabla_\mu, \nabla_\nu] = iF_{\mu\nu}$, this can be simplified to

$$\ln \mathrm{Det}(\gamma^\mu \nabla_\mu + m) = \frac{1}{2} \ln \mathrm{Det}(-\nabla^2 + m^2 + \sigma_3 e F_{12}). \tag{3A.3}$$

The logarithm of the determinant of an operator is equal to the sum over its eigenvalues, λ_n, of $\ln \lambda_n$. In the case of a uniform field $F_{12} = B$ the eigenstates $-\nabla^2 + m^2$ are Landau levels with $\lambda_n = eB(2n+1)$ and a common degeneracy $eBLT/2\pi$.

We use

$$\ln \lambda - \ln \mu = -\int_0^\infty \frac{ds}{s}(e^{-\lambda s} - e^{-\mu s}) \tag{3A.4}$$

so

$$\frac{1}{2} \ln \mathrm{Det}(-\nabla^2 + m^2 + \sigma_3 eB) = -\frac{eBLT}{4\pi} \sum_n \int_0^\infty \frac{ds}{s}$$
$$(e^{-2neBs} + e^{-2(n+1)eBs})e^{-m^2 s} + const. \tag{3A.5}$$

The two terms in the integrand correspond to the two eigenvalues of σ_3.

The energy density is thus

$$\mathcal{E} = \frac{eB}{4\pi} \int_0^\infty \frac{ds}{s} \coth(eBs) e^{-m^2 s}. \tag{3A.6}$$

This $\mathcal{E}$ is divergent. The divergence of the integral at small s, due to the quadratically divergent energy of the Dirac sea, and to the logarithmically divergent vacuum polarization, is of no interest to us here. We wish only to find the imaginary part, which is finite. We notice that a real Minkowski electric field, E, corresponds to taking a pure *imaginary* value for the Euclidean B field. When $B = iE$ (E real), the poles of $\coth(eBs)$ at $s = \pi n i/eB$ swing down onto the real axis. The contour has to be indented to avoid them, and the half circle indents give rise to imaginary contributions of πi times the residues of the integrand.

$$\mathrm{Im}\, \mathcal{E} = -\frac{eE}{4\pi} \sum_{n=1}^\infty \frac{1}{n} e^{-n\pi m^2/eE}. \tag{3A.7}$$

The imaginary part of the energy density implies a rate for the vacuum to decay to fermion-antifermion pairs of

$$\Gamma = \frac{eE}{2\pi} \sum_{n=1}^{\infty} \frac{1}{n} e^{-n\pi m^2/eE} \tag{3A.8}$$

per unit volume.

This calculation of the Fredholm determinant has a certain air of sophistication, so it is chastening to discover that the same result can be derived directly using only simple one-particle quantum mechanics.

We regard the vacuum as an infinitely deep sea with all the negative energy levels filled. Consider the evolution a single-particle state with momentum k under the influence of the electric field. The electric field has the effect of shifting $k \to k + eEt$. The Schrödinger equation for such a single particle state is

$$i\frac{\partial t}{\partial t}\begin{pmatrix} \psi_R \\ \psi_L \end{pmatrix} = \begin{pmatrix} k + eEt & m \\ m & -(k + eEt) \end{pmatrix}\begin{pmatrix} \psi_R(t) \\ \psi_L(t) \end{pmatrix} \tag{3A.9}$$

The time-dependent eigenvalues of the hamiltonian matrix are $E_\pm = \pm\sqrt{(k + eEt)^2 + m^2}$. If the state starts with negative k as a right-going particle deep in the Dirac sea and evolves adiabatically it will stay on the same branch of the square root and end up as a left-going particle diving deeper into the sea. When we solve the time dependent Schrödinger evolution exactly, however, the state has a finite probability p of crossing the $2m$ gap and ending up on the other branch of the square root. This Landau-Zener tunneling produces particle-hole pairs and its probability can be calculated exactly from the theory of hypergeometric functions. This probability is

$$p = e^{-\pi m^2/eE}. \tag{3A.10}$$

Now the number of filled sea states that attempt to cross the gap each second is equal to the density of states in k space, $L/2\pi$, times the "velocity" eE of these states. The crossing attempts by the different states are independent, so the probability $P(T)$ that no pair creation occurs during the interval $[0, T]$ is the product of the failure-to-cross probabilities, $1 - p$, for each state

$$P(T) = (1 - e^{-\pi m^2/eE})^{eELT/2\pi} = \exp\left\{-LT\left(-\frac{eE}{2\pi}\ln(1 - e^{-\pi m^2/eE})\right)\right\} \tag{3A.11}$$

The ground-state persistence probability, $P(T)$, decays exponentially with time, showing the pair creation rate per unit volume to be

$$\Gamma = -\frac{eE}{2\pi}\ln(1 - e^{-\pi m^2/eE}) = \frac{eE}{2\pi}\sum_{n=1}^{\infty}\frac{1}{n}e^{-nm^2\pi/eE} \tag{3A.12}$$

which is identical to the expression we obtained from the field theory formalism.

Appendix 3B. Madelung Fluids

Given a time-dependent nonlinear Schrödinger equation of the form

$$i(\partial_t - ieA_0)\Psi = -\frac{1}{2m}\sum_{a=1}^{3}(\partial_a - ieA_a)^2\Psi + \lambda(|\Psi|^2 - \rho_0)\Psi, \tag{3B.1}$$

we can recast it as the equation of motion of a charged compressible fluid. This observation was originally made by Madelung (although without the nonlinear term) very soon after the discovery of the Schrödinger equation [28].

We set $\Psi = \sqrt{\rho}e^{i\theta}$ and define a velocity field $\mathbf{v}$ in such a way that the current

$$\mathbf{j} = \frac{1}{2mi}\left(\Psi^*(\nabla - ie\mathbf{A})\Psi - ((\nabla + ie\mathbf{A})\Psi^*)\Psi\right) \tag{3B.2}$$

may be written $\mathbf{j} = \rho\mathbf{v}$. This leads to

$$\mathbf{v} = \frac{1}{m}(\nabla\theta - e\mathbf{A}). \tag{3B.3}$$

In the absence of vortex singularities in Ψ the vorticity, $\omega = \nabla \wedge \mathbf{v}$, is completely determined by the gauge field to be $\omega = -\frac{1}{m}\nabla \wedge \mathbf{A} = -\frac{\mathbf{B}}{m}$, i.e.,

$$m\omega + e\mathbf{B} = 0. \tag{3B.4}$$

When the gauge field is dynamical, and not just an external probe, this equation is responsible for the Meissner effect. A penetrating $\mathbf{B}$ field implies a uniform vorticity which would lead, in a sphere of radius R, to a kinetic energy that grows as R^5, i.e., faster than extensive.

With the definition (3B.3) the imaginary and real parts of (3B.1) become respectively the continuity equation

$$\partial_t\rho + \nabla \cdot \rho\mathbf{v} = 0, \tag{3B.5}$$

and the Euler equation governing the flow of a barotropic fluid

$$m(\partial_t\mathbf{v} + \mathbf{v}\cdot\nabla\mathbf{v}) = e(\mathbf{E} + \mathbf{v}\wedge\mathbf{B}) - \nabla\mu. \tag{3B.6}$$

The word *barotropic* refers to the simplifying property that the pressure term $\frac{1}{\rho}\nabla P$ which occurs on the right hand side of the conventional Euler equation is here combined into the gradient of a potential

$$\mu = \lambda(\rho - \rho_0) - \frac{1}{2m}\frac{\nabla^2\rho}{\rho} \tag{3B.7}$$

The potential contains the expected compressibility pressure, depending on the deviation from the equilibrium density, plus a correction depending on gradients of ρ. This correction is called the *quantum pressure*. If we had included $\hbar$ in (3B.1–3), it would not appear explicitly in (3B.4–6), but the last term of (3B.7) would be proportional to $\hbar^2$. For small density variations, and those are the only ones for which our derivation of the Gross-Pitaevskii equation is valid, the quantum pressure term is unimportant.

The Euler equation (3B.6) is most directly derived in the equivalent Bernoulli form

$$m(\partial_t \mathbf{v} - \mathbf{v} \wedge \omega) = e(\mathbf{E} + \mathbf{v} \wedge \mathbf{B}) - \nabla \left(\frac{1}{2} m \mathbf{v}^2 + \mu \right), \tag{3B.8}$$

where a cancellation of the $m\mathbf{v} \wedge \omega$ term against the $e\mathbf{v} \wedge \mathbf{B}$ term is evident on use of (3B.4). It is after this cancellation, and so without reference to $\mathbf{B}$ or ω, that the hydrodynamic picture of superconductivity is conventionally displayed [29]. I prefer to keep ω and B in (3B.8) and rewrite it as (3B.6). Then one can see that the only difference between the superfluid dynamics of the condensate and ordinary fluid dynamics lies in the constraint (3B.4).

References

[1] J. K. Krive and A. S. Rozhavkii, *Phys. Lett.* **113A** (1985) 313.

[2] Z-B. Su and B. Sakita, *Phys. Rev. Lett.* **56** (1986) 780.

[3] M. Stone, *Phys. Rev.* **D15** (1976) 1150.

[4] P. A. Lee, T. M. Rice and P. W. Anderson, *Solid State Comm.* **14** (1974) 703.

[5] M. J. Rice, A. R. Bishop, J. A. Krumhansl and S. E. Trullinger, *Phys. Rev. Lett.* **36** (1976) 432.

[6] R. Jackiw and J. R. Schrieffer, *Nuc. Phys.* **B190**(FSI) (1981) 253.

[7] W. P. Su, J. R. Schrieffer and A. J. Heeger, *Phys. Rev. Lett.* **42** (1979) 1698.

[8] C. M. Bender and T. T. Wu, *Phys. Rev.* **D7** (1993) 1620.

[9] M. Stone, *Phys. Rev.* **D14** (1976) 3568.

[10] M. Stone, *Phys. Lett.* **67B** (1977) 186.

[11] S. Coleman, *Phys. Rev.* **D15** (1977) 2929; C. G. Callan and S. Coleman, *Phys. Rev.* **D16** (1977) 1762.

[12] L. N. Lipatov, *Zh. Eks. Teor. Fiz. Pisma* **25** (1977) 116; *Zh. Eks. Teor. Fiz.* **72** (1977) 411.

[13] J. S. Langer, *Phys. Rev. Lett.* **21** (1968) 973; *Ann. Phys.* **54** (1969) 258.

[14] K. Maki, *Phys. Rev. Lett.* **39** (1977) 46.

[15] A. Luther, *Phys. Rev.* **B19** (1979) 320.

[16] F. D. M. Haldane, unpublished.

[17] A. Houghton and J. B. Marston, *Phys. Rev.* **B48** (1993) 7790.

[18] A. H. Castro Neto and E. Fradkin, *Phys. Rev. Lett.* **72** (1994) 1393.

[19] P. W. Anderson, *Phys. Rev. Lett.* **64** (1990) 1839.

[20] E. Abrahams and T. Tsuneto, *Phys. Rev.* **152** (1966) 152.

[21] F. Gaitan and M. Stone, *Annals of Phys.* (NY) **178** (1987) 89.

[22] N. R. Werthamer, *The Ginzburg Landau equations and their extensions*, in *Superconductivity*, R. D. Parks ed. (Marcel Dekker 1969).

[23] P. Ao and D. J. Thouless, *Phys. Rev. Lett.* **70** (1993) 2158.

[24] E. P. Gross, *Nuovo Cimento* **20** (1961) 454; *J. Math Phys.* **4** (1963) 195.

[25] L. P. Pitaevskii, *Zh. Eksp. Teor. Fiz.* **40** (1961) 646, translated in *Sov. Phys. JETP* **13** (1961) 451.

[26] V. L. Ginzburg and L. P. Pitaevskii, *Zh. Eksp. Teor. Fiz.* **34** (1958) 1240, translated in *Sov. Phys. JETP* **7** (1958) 858.
[27] J. Schwinger, *Phys. Rev.* **82** (1951) 664.
[28] E. Madelung, *Z. Phys.* **40** (1927) 322.
[29] W. F. Vinen, "A comparison of the properties of superconductors and superfluid helium," in *Superconductivity*, R. D. Parks ed., (Marcel Decker 1969).

Chapter 4

BOSONIZATION IN HILBERT SPACE

4.1. Introduction

In the previous chapters we described how the bosonization correspondences work for Green functions. Everything revolves around the identity (1.3.12) and the recognition that its two sides happen to be interpretable as Bose and Fermi n-point functions. Does this mean that bosonization a chance consequence of an obscure identity, or is there some deeper structure to be found?

In this chapter we will seek the deeper structure by examining how bosonization works in Hilbert space. We will find an interesting, and exact, bosonization map when the vacuum state is an infinitely deep Dirac sea. The fermionic Fock space built on this vacuum state turns out to be isomorphic to the bosonic Fock space for the oscillators in a scalar field. Surprisingly, the key to establishing the isomorphism is again the identity (1.3.12), but now seen as part of some classical mathematics linking the theory of symmetric polynomials with group representation theory.

The machinery described here provides the background to the coherent state path integral approach to bosonization. This method combines the geometric quantization of compact groups [rep. 28] with coherent-state path integrals for fermion systems [rep. 26] to derive the chiral boson action equivalent to a chiral fermion [rep. 29]. It is also the key to understanding the Kyoto school's approach [rep. 27] to families of integrable soliton equations (for a review see [1].) The machinery is also alluded to in some of the reprinted papers on Riemann surfaces [rep. 22,24].

4.2. Slater Determinants and Schur Functions

Consider right-moving relativistic fermions living on a circle of circumference 2π. A basis of plane-wave, single-particle states is given by the set of functions $\psi_n(\theta) = e^{in\theta}$. We will write these as z^n where z lies on the unit circle. If we were to take $H = -i\partial_\theta + 1/2$ as our hamiltonian, the ground state will be a Dirac sea with all the states having $n < 0$ occupied and those with $n \geq 0$ empty. Now an infinitely deep Dirac sea can be treacherous. It is safer to begin with a finite number of particles $1, \dots, N$, filling a shallow sea. For the moment assume that the sea bed corresponds to the state $z^0 = 1$, and its surface to z^{N-1}. Later we can take $N \to \infty$ and redefine the momenta so that the surface is at $n = 0$.

The Hilbert space of N-body wavefunctions is now spanned by occupation-number eigenstates which are $N \times N$ Slater determinants of the $z^n, n \geq 0$. The ground state is the Vandermonde determinant

$$\Psi_0(z_1, z_2, \dots, z_N) = \begin{vmatrix} z_1^{N-1} & z_1^{N-2} & \dots & 1 \\ z_2^{N-1} & z_2^{N-2} & \dots & 1 \\ \vdots & \vdots & \ddots & \vdots \\ z_N^{N-1} & z_N^{N-2} & \dots & 1 \end{vmatrix}. \tag{4.2.1}$$

We will also denote this determinant by the symbol $D(z)$.

The general Slater determinant can be written

$$\Psi_{\{\lambda\}}(z_1, z_2, \dots, z_N) = \begin{vmatrix} z_1^{\lambda_1+N-1} & z_1^{\lambda_2+N-2} & \dots & z_1^{\lambda_N} \\ z_2^{\lambda_1+N-1} & z_2^{\lambda_2+N-2} & \dots & z_2^{\lambda_N} \\ \vdots & \vdots & \ddots & \vdots \\ z_N^{\lambda_1+N-1} & z_N^{\lambda_2+N-2} & \dots & z_N^{\lambda_N} \end{vmatrix} \tag{4.2.2}$$

or, in compact but, I hope, self-explanatory notation

$$\Psi_{\{\lambda\}}(z) = \det |z_s^{\lambda_t+n-t}|. \tag{4.2.3}$$

Because of the identity of the particles we can, and will, arrange the columns of the determinant so that $\lambda_1 \geq \lambda_2 \geq \lambda_3$ *etc.* This wavefunction $\Psi_{\{\lambda\}}$ is still an N electron state, so it is a charge-neutral excitation of the ground state. If we expand out the determinant, each term has $\lambda_1 + \lambda_2 + \ldots + \lambda_n \equiv |\{\lambda\}| = M$ extra powers of z, so $\Psi_{\{\lambda\}}$ is an energy eigenstate with energy $E = M$ above the ground state. We can think of $\Psi_{\{\lambda\}}$ being created by sliding the ith electron in the sea up through λ_i steps in energy.

The labels $\{\lambda\}$ are conveniently displayed as *Young diagrams* — arrays with λ_i boxes in the ith row. For example

□□□□
□□□
□□□
□□

represents $\{4332\}$, or as it sometimes convenient to write $\{43^2 2\}$.

This pictorial representation of the neutral excitations has nice properties under particle-hole interchange: If we were, for example, to take the topmost electron and move it up four steps, we would represent the resulting state label by the Young diagram $\{4\}$:

□□□□

The particle-hole conjugate of this state has a hole four steps down in the sea. This state is made by raising the top four electrons each by one step, and its label is represented by $\{1111\}$, or equivalently $\{1^4\}$

□
□
□
□

i.e., as the *conjugate diagram* obtained by interchanging the rows and columns of the Young diagram. This is true in general — interchanging particles with holes interchanges rows and columns.

The set of distinct λ_i with $\sum \lambda_i = M$ are in one-to-one correspondence with the partitions of M. It is easy to see that generating function of the number, $p(M)$, of distinct partitions of M is given by

$$\mathcal{Z} = \prod_{n>0} \frac{1}{(1 - x^n)} = \sum_M p(M) x^M. \tag{4.2.4}$$

The generating function $\mathcal{Z}(x)$ is known in number theory as the "partition function". Now if we put $x = e^{-\beta}$ we obtain the familiar statistical mechanics partition function for the charge neutral excitations, and (4.2.4) becomes

$$\mathcal{Z} = \sum_M p(M)e^{-\beta M} = \frac{1}{\prod_{n>0}(1 - e^{-\beta n})}. \tag{4.2.5}$$

The right hand side of (4.2.5) is immediately recognizable as the partition function of an infinite ensemble $n = 1, 2, \ldots, \infty$ of bosonic oscillators with frequency n, and so already hints at bosonic description of the states. In Appendix 4B we extend this result to charged excitations and obtain a combinatoric proof of the Jacobi triple product identity.

To proceed further we must make a connection with the theory of symmetric functions (see [2] or [3]). We first observe that the Vandermonde determinant, $D(z) \equiv \Psi_0 \equiv \Psi_{\{\emptyset\}}$, is a factor of all the $\Psi_{\{\lambda\}}(z)$. This means that the quotient

$$\Phi_{\{\lambda\}}(z) = \Psi_{\{\lambda\}}(z)/\Psi_0(z) \tag{4.2.6}$$

is a *symmetric polynomial* in the z_i. Clearly all charge-neutral states $\Psi_{\{\lambda\}}$ are obtained by multiplying Ψ_0 by one of these symmetric polynomials, so the Hilbert space of charge-neutral excitations is isomorphic to the linear space spanned by them.

The polynomial $\Phi_{\{\lambda\}}(z)$ is the *Schur function* associated with the partition $\{\lambda\}$ of M. These functions are best known in physics as the characters of the groups $GL(n)$, $U(n)$, or $SU(N)$ [4]. The familar diagrammatic recipe (known to mathematicians as the Littlewood-Richardson rule [5]) for combining Young diagrams so as to find the representations occurring in the Clebsh-Gordon decomposition of direct products is nothing but the algorithm for decomposing products of the $\Phi_{\{\lambda\}}(z)$ into sums of $\Phi_{\{\lambda'\}}(z)$'s. Our interest, however, lies principally in the role they play in the theory of symmetric polynomials.

4.3. Symmetric Polynomials

In this section we will review those parts of the theory of symmetric polynomials that we need in the sequel. The material is standard and can be found in many books, e.g. [2,3].

Given a set of symbols α_i, $i = 1, \ldots, N$ (for example complex variables like our z_i), form the ring $\mathcal{S}(\alpha)$ of polynomials in the α_i which are invariant under arbitrary permutations of the α_i. Elements of this ring include: the *elementary symmetric functions* a_n, $n = 1, \ldots, N$ defined by

$$\prod(1 - \alpha_i x) = 1 - a_1 x + a_2 x^2 + \cdots \pm a_n x^n, \tag{4.3.1}$$

and the *homogeneous product sums* h_i, $i = 1, \ldots, \infty$, defined by

$$\frac{1}{\prod(1 - \alpha_i x)} = 1 + h_1 x + h_2 x^2 + h_3 x^3 + \ldots. \tag{4.3.2}$$

Inverting the power series definitions, we can express the a_n as polynomials in the h_n and *vice versa*. Remarkably the coefficients are integers in both directions. Sums of products of either the a_n or the h_n generate the ring $\mathcal{S}(\alpha)$.

The *power sums* S_n, $n = 1, \dots, \infty$ are defined as

$$S_n(\alpha) = \sum_i \alpha_i^n, \tag{4.3.3}$$

and using the relation

$$\exp \sum \frac{1}{n} x^n S^n(\alpha) = 1 + h_1 x + h_2 x^2 + h_3 x^3 + \dots \tag{4.3.4}$$

we can express the S_n as polynomials in the h_n (with integer coefficients), and the h_n as polynomials (with rational coefficients) in the S_n, so sums of products of the S_n also generate $\mathcal{S}(\alpha)$.

While it is not obvious, the Schur functions can be written in terms of the h_n as

$$\Phi_{\{\lambda\}}(\alpha) = \det |h_{\lambda_s - s + t}|. \tag{4.3.5}$$

For example

$$\Phi_{\{p\}} = h_p, \tag{4.3.6}$$

and

$$\Phi_{\{p,q,r\}} = \begin{vmatrix} h_p & h_{p+1} & h_{p+2} \\ h_{q-1} & h_q & h_{q+1} \\ h_{r-2} & h_{r-1} & h_r \end{vmatrix}. \tag{4.3.7}$$

Notice the pattern: the sequence λ_i provides the suffices on the entries on the diagonal. The suffices increase to the right and decrease to the left. We set $h_0 = 1$ and $h_n = 0$ if $n < 0$. We will establish these identities in a later section of this chapter.

We can also easily evaluate

$$\Phi_{\{1^N\}}(z) = z_1 z_2 \dots z_N = a_N, \tag{4.3.8}$$

and this is an example of the expression for $\Phi_{\{\lambda\}}$ in terms of the a_n. In these formulae we use the *conjugate partition*, the Young tableau with the rows and columns interchanged, and construct the determinant analogously to the expression in terms of the h_n's, e.g.

$$\Phi_{\{322\}} = \begin{vmatrix} h_3 & h_4 & h_5 \\ h_1 & h_2 & h_3 \\ 1 & h_1 & h_2 \end{vmatrix} = \begin{vmatrix} a_3 & a_4 & a_5 \\ a_2 & a_3 & a_4 \\ 0 & 1 & a_1 \end{vmatrix}. \tag{4.3.9}$$

In the first expression the diagonal suffixes $\{3, 2, 2\}$ come from the diagram

□□□
□□
□□

and in the second from the conjugate partition $\{3, 3, 1\}$

□□□
□□□
□

None of the relations between the h_n, a_n, S_n, and $\Phi_{\{\lambda\}}$ depend explicitly on N. This is a great advantage because they provide formulae that do not grow in complexity as N becomes large. There *is* a problem when N is finite however: the a_i for $i > N$ are zero, as are the Schur functions with more than N rows in the partition $\{\lambda\}$. Similarly only the first N of the h_n and S_n's are functionally independent. To overcome these constraints we can take N to be larger than any n we might be interested in. In this case the polynomials become functionally independent, and we speak of the ring of *universal symmetric polynomials*.

The Schur functions form a *linear basis* for the ring $\mathcal{S}(\alpha)$. Any element of $\mathcal{S}(\alpha)$ is a linear combination of $\Phi_{\{\lambda\}}$'s. To prove this observe that any symmetric polynomial in the α's can be converted to an antisymmetric polynomial by multiplying by the Vandermonde determinant $D(\alpha)$. Then antisymmetrising each monomial in the the resulting expression converts it to a $\Psi_{\{\lambda\}}(\alpha)$. Finally dividing out the catalytic $D(\alpha)$ returns the symmetric polynomial as a sum of $\Phi_{\{\lambda\}}$.

There is a very useful identity connecting the Φ's and the S_k's: take two sets of indeterminates, α_i and β_i, then

$$\exp \sum_k \frac{1}{k} S_k(\alpha) S_k(\beta) = \sum_{\{\lambda\}} \Phi_{\{\lambda\}}(\alpha)\Phi_{\{\lambda\}}(\beta). \tag{4.3.10}$$

This result will be proved in the appendix. The proof depends on a trivial modification of (1.3.12)

$$\det \left| \frac{1}{1-\alpha_s\beta_t} \right| = \frac{D(\alpha)D(\beta)}{\prod_{ij}(1-\alpha_i\beta_j)}. \tag{4.3.11}$$

Once we have the relation (4.3.10) we can take all but one, say $\beta_1 = x$, of the β_i to be zero. We see that (4.3.2) is a special case of (4.3.10).

4.4. $\mathcal{S}(z)$ as a Hilbert Space

The algebra $\mathcal{S}(z)$ is both a vector space spanned by the $\Phi_{\{\lambda\}}$ and a ring generated by the $S_n(z)$. We now define an inner product on $\mathcal{S}(z)$ which serves to make it into a Hilbert Space. For this purpose we wish the functions $\Phi_{\{\lambda\}}$ to form a linearly independent basis. We could achieve this by taking the number N of the z_i to be infinite. It is technically simpler to proceed as described in the last section and merely take N sufficiently large that all the states we actually use are independent. This avoids all problems of convergence of infinite determinants. It is also a physically reasonable approach since, at least in condensed matter applications, the Fermi/Dirac sea always has finite depth. What we need in practice is that the energy scale of relevant physical processes be much smaller than the bandwidth.

The natural inner product on $\mathcal{S}(z)$ is the one it inherits from the N-body Hilbert space spanned by the fermion occupation number eigenstates $\Psi_{\{\lambda\}}(z)$. We define

$$\langle \Phi_{\{\lambda\}} | \Phi_{\{\lambda'\}} \rangle_F = \frac{1}{N!} \int d\theta_1 \ldots d\theta_N |D(z)|^2 \overline{\Phi_{\{\lambda\}}}(e^{i\theta}) \Phi_{\{\lambda'\}}(e^{i\theta}). \tag{4.4.1}$$

The measure factor $|D(z)|^2$ cancels the denominator of the Schur functions so

$$\langle \Phi_{\{\lambda\}} | \Phi_{\{\lambda'\}} \rangle_F = \langle \Psi_{\{\lambda\}} | \Psi_{\{\lambda'\}} \rangle = \delta_{\{\lambda\}\{\lambda'\}}. \tag{4.4.2}$$

Thus the Schur functions form an orthonormal basis with regard to the inner product $\langle , \rangle_F$.

The integration in (4.4.1) is that induced on the maximal torus of $U(N)$ by the Haar measure. Consequently (4.4.2) coincides with the character orthogonality relation from the representation theory of $U(N)$ [6].

We can define a superficially different inner product on $\mathcal{S}(z)$ by defining it on polynomials in the $S_n(z)$. Regard the S_n as being independent complex variables and set

$$\langle f(S)|g(S)\rangle_B = \int \prod_k \left[\frac{d^2 S_k}{\pi k}\right] \overline{f(S)} g(S) e^{-\sum_k \frac{1}{k}|S_k|^2}. \tag{4.4.3}$$

In particular the product of the S_n's themselves is

$$\langle S_n|S_{n'}\rangle_B = n\delta_{n,n'}. \tag{4.4.4}$$

A general element in the ring is a sum of products of the S_n such as

$$S_{(l)}(z) \equiv S_1^{l_1} S_2^{l_2} \dots S_n^{l_n} \tag{4.4.5}$$

and the inner product of two such monomials is

$$\langle S_{(l)}|S_{(l')}\rangle_B = (1^{l_1} l_1! 2^{l_2} l_2! \dots n^{l_n} l_n!)\delta_{(l)(l')}. \tag{4.4.6}$$

This second inner product is essentially that of a bosonic Bargman-Fock space where a commuting family of creation operators $\hat{a}_n^\dagger$ is represented by multiplication by S_n. The corresponding annihilation operators, $\hat{a}_n$, (not to be confused with the elementary symmetric functions a_n) are their Hermitian adjoints with respect to this product. These are easily seen to be the derivatives $n\partial_{S_n}$. Our Hilbert space is therefore isomorphic to the space created from a cyclic vector $|0\rangle$ by application of bosonic $\hat{a}_n^\dagger$'s whose commutation relations are

$$[a_n, a_{n'}^\dagger] = n\delta_{n,n'} \qquad n > 0. \tag{4.4.7}$$

If we define $a_{-n} = a_n^\dagger$ we can write

$$[a_n, a_{n'}] = n\delta_{n+n',0}, \tag{4.4.8}$$

These are the commutation relations of a level-one abelian Kac-Moody algebra (For a review see [7]; also see [8]). We will soon be able to identify the $\hat{a}_n$ as the Fourier components of the charge $\psi_R^\dagger \psi_R$.

The remarkable fact which lies behind bosonization is that these two inner products, $\langle , \rangle_F$, and $\langle , \rangle_B$, are identical. To prove this, use the reproducing kernel identity

$$\int \left[\prod \frac{d^2 S_k(z)}{\pi k}\right] F(S_k(z)) e^{-\sum_k \frac{1}{k}|S_k(z)|^2} e^{+\sum_k \frac{1}{k}\overline{S_k(z)} S_k(z')} = F(S_k(z')) \tag{4.4.9}$$

and the identity from the last section

$$\exp \sum_k \frac{1}{k} \overline{S_k(z)} S_k(z') = \sum_{\{\lambda\}} \overline{\Phi_{\{\lambda\}}(z)} \Phi_{\{\lambda\}}(z') \tag{4.4.10}$$

with the choice $F(S_k(z)) = \Phi_{\{\mu\}}(z)$.

Since the $\Phi_{\{\lambda\}}$ form an independent set, comparing coefficients gives

$$\langle \Phi_{\{\lambda\}} | \Phi_{\{\lambda'\}} \rangle_B = \int \left[\prod \frac{d^2 S_k}{\pi k} \right] \overline{\Phi_{\{\lambda\}}} \Phi_{\{\lambda'\}} e^{-\sum_k \frac{1}{k} |S_k|^2} = \delta_{\{\lambda\}\{\lambda'\}}. \tag{4.4.11}$$

So the $\Phi_{\{\lambda\}}$ which form an orthonormal set with respect to $\langle , \rangle_F$ also do so with respect to $\langle , \rangle_B$. Thus $\langle a, b \rangle_F = \langle a, b \rangle_B$ for all $a, b \in \mathcal{S}$.

An alternative demonstration of the equivalence of the two inner products is via Frobenius' famous reciprocity formula connecting the characters of the permutation group $\mathcal{S}_N$ with the characters of $GL(N)$ [9, 10]. Frobenius' formula asserts that

$$S_{(l)} = \sum_{\{\lambda\}} \chi_{(l)}^{\{\lambda\}} \Phi_{\{\lambda\}}(z) \tag{4.4.12}$$

where the $\chi_{(l)}^{\{\lambda\}}$ are the characters of the representation $\{\lambda\}$ of the permutation group on $l_1 + 2l_2 + 3l_3 + \ldots = |\{\lambda\}|$ symbols. The conjugacy classes of the group are labeled by (l). As group characters the $\chi_{(l)}^{\{\lambda\}}$ obey the orthogonality conditions

$$\frac{1}{g} \sum_{(l)} g_{(l)} \chi_{(l)}^{\{\lambda\}} \chi_{(l)}^{\{\lambda'\}} = \delta_{\{\lambda\}\{\lambda'\}} \tag{4.4.13}$$

where $g = |\{\lambda\}|!$ is the order of the permutation group and

$$g_{(l)} = \frac{g}{1^{l_1} 2^{l_2} \ldots l_1! l_2! \ldots l_n!} \tag{4.4.14}$$

is the number of permutation group elements in the conjugacy class. Eq. (4.4.15) can be now be inverted to give $\Phi_{\{\lambda\}}$ in terms of the S_k:

$$\Phi_{\{\lambda\}}(z) = \frac{1}{g} \sum_{(l)} g_{(l)} \chi_{(l)}^{\{\lambda\}} S_{(l)} \tag{4.4.15}$$

and then (4.4.6) and (4.4.13) yield (4.4.11). This demonstration is not as independent as it seems — the conventional proof of (4.4.12) depends on (4.4.10).

The formulae (4.4.12) and (4.4.15) relating the bosonic and fermionic bases of the Hilbert space are interesting in their own right. It must be well-known that the unitary transformation linking the bosonic and fermion bases of the Hilbert space is given by Frobenius' formula, but I have not found it stated explicitly anywhere.

The fruit of our labours is two equivalent bases for the many-body Hilbert space: one, labeled by the fermion occupation numbers, is associated with the Schur functions, and one, a bosonic Fock space, is built by the operators $\hat{a}_n^\dagger$ which correspond to the S_n's. The next step is the identification of the $\hat{a}_n$ operators with the Fourier components of the currents from chapter 1.

4.5. The $LU(1)$ Group Action

The chiral charges $\psi_R^\dagger\psi_R = j_R$ can be exponentiated to generate a group of position dependent "gauge transformations", which act on the multi-particle Hilbert space. These group elements can be applied to the vacuum to form a family of coherent states from which we can construct a bosonic coherent state path integral. This is done in [rep. 29] using the ideas developed in [rep. 26,28]. A general reference for this section is [11].

Roughly speaking we form the formally unitary operator $U = \exp(i\int j(\theta)\phi(\theta)d\theta)$ parameterized by the function $\phi(\theta)$, but to make this defined we need to normal-order the exponential. We can gain insight into this construction, and as a byproduct identify the Fourier components of $j(\theta)$ with the $\hat{a}_n$ from the last section, by initially defining the action of this *loop group* $LU(1)$ on the single-particle states.

An element of the loop group is a mapping from the circle on which the particles live to the group $U(1)$. We can regard the element as a position-dependent phase $e^{i\phi}$. It acts on the single-particle states by multiplying each $\psi_n(\theta)$ by the common phase $e^{i\phi(\theta)}$. If we then substitute the new wavefunctions $\psi_n'(\theta) = e^{i\phi(\theta)}\psi_n(\theta)$ into the Slater determinant, and expand out the resulting states in terms of the $\Psi_{\{\lambda\}}$, we induce a unitary action, $U(e^{i\phi(\theta)})$, of $LU(1)$ on the space of many-body states.

Perhaps surprisingly, in the limit of a deep Dirac sea, this induced many-body group action does not have the same commutation relations as the single-particle action but forms instead a *projective* unitary representation of $LU(1)$. We find that

$$U(e^{i\phi_1})U(e^{i\phi_2}) = e^{iS(\phi_1,\phi_2)}U(e^{i\phi_1+i\phi_2}) \tag{4.5.1}$$

where $S(\phi_1, \phi_2)$ is a *cocycle* — i.e., it satisfies the consistency condition

$$S(\phi_1,\phi_2) + S(\phi_1+\phi_2,\phi_3) - S(\phi_1,\phi_2+\phi_3) - S(\phi_2,\phi_3) = 0. \tag{4.5.2}$$

The cocycle condition is needed (as one sees by evaluating $U(e^{i\phi_1})U(e^{i\phi_2})U(e^{i\phi_3})$ in two distinct ways) for the group action to be associative.

Th factor $S(\phi_1, \phi_2)$ originates in the way we will treat the infinitely deep Dirac sea. For the infinitesimal elements which can be identified with the Lie algebra of $LU(1)$ it reduces to the Schwinger term in the commutator of the charges

$$[j_R(\theta), j_R(\theta')] = -\frac{1}{2\pi}\partial_\theta\delta(\theta-\theta'). \tag{4.5.3}$$

In the mathematics literature S is usually called the Kac-Peterson cocycle although, in the current algebra context, it was originally discovered by Jordan.

To construct the action of $LU(1)$ on the Fock space it is convenient to break the function ϕ into four parts

$$e^{i\phi(z)} = e^{i\alpha}\cdot z^m \cdot e^{i\phi^+(z)}\cdot e^{i\phi^-(z)}. \tag{4.5.4}$$

The factor $e^{i\alpha}$ is a z-independent phase z^m (with m an integer), representing the winding number of the map $z \to e^{i\phi}$ as z circles the origin, and

$$i\phi^+ = \sum_{n>0}\frac{1}{n}s_n z^n \tag{4.5.5}$$

$$i\phi^- = \sum_{n<0} \frac{1}{n} s_n z^n \tag{4.5.6}$$

have Fourier/Laurent expansions with only positive or negative frequency terms respectively — so their action on the space of one-particle states z_n is represented by an upper, or lower, triangular matrix with 1's on the diagonal.

We will deal with the action of each of the factors separately.

i) Action of $U(e^{i\phi^+})$

Firstly notice that when we replace each z_i^n in the determinant part of Ψ_0 by $z_i^n e^{i\phi^+(z_i)}$ the same factor occurs for every element of a given row. The effect on the whole Slater determinant is then to multiply it by

$$\exp \sum_{n>0} \frac{1}{n} s_n S_n(z) \tag{4.5.7}$$

where $S_n(z) = \sum_i z_i^n$ are the power sums. In this way we identify multiplication of the many-body wavefunction by the symmetric function S_n with the action of j_n.

We could also expand the product out as a sum of the $\Psi_{\{\lambda\}}$.

$$\exp\left(\sum_{n>0} \frac{1}{n} s_n S_n(z)\right) \Psi_0(z) = \sum_{\{\lambda\}} A_{\{\lambda\}} \Psi_{\{\lambda\}} \tag{4.5.8}$$

To find the $A_{\{\lambda\}}$ we introduce the coefficients h_n by

$$e^{i\phi^+(z)} = 1 + h_1 z + h_2 z^2 + \dots . \tag{4.5.9}$$

We also set $h_0 = 1$ and $h_n = 0$, $n < 0$. If the s_n were the power sums of some set of variables ζ, these h_n would be the corresponding homogeneous product sums.

Using the column linearity of the determinants, e.g.,

$$\begin{vmatrix} z_1^{N-1}(1+h_1 z_1) & z_1^{N-2} & \dots & 1 \\ z_2^{N-1}(1+h_1 z_2) & z_2^{N-2} & \dots & 1 \\ \vdots & \vdots & \ddots & \vdots \\ z_N^{N-1}(1+h_1 z_N) & z_N^{N-2} & \dots & 1 \end{vmatrix} = \begin{vmatrix} z_1^{N-1} & z_1^{N-2} & \dots & 1 \\ z_2^{N-1} & z_2^{N-2} & \dots & 1 \\ \vdots & \vdots & \ddots & \vdots \\ z_N^{N-1} & z_N^{N-2} & \dots & 1 \end{vmatrix} + h_1 \begin{vmatrix} z_1^{N} & z_1^{N-2} & \dots & 1 \\ z_2^{N} & z_2^{N-2} & \dots & 1 \\ \vdots & \vdots & \ddots & \vdots \\ z_N^{N} & z_N^{N-2} & \dots & 1 \end{vmatrix} \tag{4.5.10}$$

it is straightforward to identify the coefficient of $\Psi_{\{\lambda\}}$. For example with $\lambda_1 = 3, \lambda_2 = 2, \lambda_3 = 2$ we have

$$A_{\{322\}} = \begin{vmatrix} h_3 & h_4 & h_5 \\ h_1 & h_2 & h_3 \\ 1 & h_1 & h_2 \end{vmatrix} \tag{4.5.11}$$

since we can add three powers to the first column by using h_3, and two powers to each of the next two columns by using h_2; alternatively we can use one h_4 in the second column, one h_1 in the first, and one h_2 in the third to get the same result — and so on. In general, using the notation $|a_{ij}|$ for the determinant of the matrix a_{ij},

$$A_{\{\lambda\}} = \det |h_{\lambda_i - i + j}| \, . \qquad (4.5.12)$$

If as suggested above, the coefficients s_n in (4.5.4) are identified with some power sums $S_n(\zeta)$ and the h_n with the homogeneous product sums of these ζ's, then (4.4.10) shows that the coefficients $A_{\{\lambda\}}$ are the Schur functions, $\Phi_{\{\lambda\}}(\zeta)$. We have thus established the relation (4.3.5).

$$\Phi_{\{\lambda\}}(\zeta) = \det |h_{\lambda_i - i + j}| \, , \qquad (4.5.13)$$

If we act with $U(e^{i\phi^+}(z))$ on the excited state $\Psi_{\{\mu\}}$, then we define the coefficient functions $\Phi_{\{\lambda/\mu\}}$ by

$$\left(\exp \sum_{n>0} \frac{1}{n} S_n(\zeta) S_n(z) \right) \Psi_{\{\mu\}}(z) = \sum_{\{\lambda\}} \Phi_{\{\lambda/\mu\}}(\zeta) \Psi_{\{\lambda\}}(z) \qquad (4.5.14)$$

These are the "relative" Schur functions. They are zero unless $|\lambda| \geq |\mu|$. Again by expanding the determinants in columns we find

$$\Phi_{\{\lambda/\mu\}}(\zeta) = \left| h_{\lambda_i - \mu_j - i + j} \right| \qquad (4.5.15)$$

ii) Action of $U(e^{i\phi^-})$

Define the coefficients g_n by

$$e^{i\phi^-} = \sum g_n z^{-n} \qquad (4.5.16)$$

and consider the effect on the state $|\{\lambda\}\rangle$ of replacing z^n by $z^n(\sum g_n z^{-n})$. If we do this in the ground state $|0\rangle$, we will have no effect on the powers near the left hand columns of the determinant, since the reduction of any powers in these columns will always yield two that are identical. There will be effects at the far right of the determinant, but they are infinitely deep down in the Dirac sea. We now make a crucial move. Being interested only in effects near the Fermi surface, and the right hand edge of the determinant being only a cut-off point, we deliberately keep only the left hand excitations in our expansion of the determinant, and in this way define the action of $U(e^{i\phi^-})$ on the many body state. This omission of the terms from the right side of the determinant is the origin of non-commutativity of the many-body action and the ultimate source of the cocycle.

On excited states $U(e^{i\phi^-})$ does have an effect. Again expanding in columns and keeping only left hand edge effects, we find

$$\begin{aligned} U(e^{i\phi^-})\Phi_{\{\lambda\}}(z) &= \sum_{\{\mu\}} |g_{\lambda_s - \mu_t - s + t}| \Phi_{\{\mu\}} \\ &= \sum_{\{\mu\}} \Phi_{\{\lambda/\mu\}}(\zeta) \Phi_{\{\mu\}}(z) \end{aligned} \qquad (4.5.17)$$

(Note the interchange $\{\mu/\lambda\} \to \{\lambda/\mu\}$ compared to the previous use of the relative Schur functions.)

We can rewrite this by first observing that the product of two Schur functions is a sum of Schur functions

$$\Phi_{\{\lambda\}}\Phi_{\{\mu\}} = \sum_{\{\lambda\}} \Gamma_{\lambda\mu\nu}\Phi_{\{\nu\}} \tag{4.5.18}$$

where the coefficients $\Gamma_{\lambda\mu\nu} = \langle\{\nu\}|\Phi_{\{\lambda\}}|\{\mu\}\rangle$ are integers given by the Littlewood-Richardson rule. Using these coefficients and the original definition:

$$\left(\sum_{\{\nu\}} \Phi_{\{\nu\}}(x)\Phi_{\{\nu\}}(y)\right) \Phi_{\{\lambda\}}(y) = \sum_{\{\mu\}} \Phi_{\{\mu/\lambda\}}(x)\Phi_{\{\mu\}}(y) \tag{4.5.19}$$

we can express $\Phi_{\{\lambda/\mu\}}$ as

$$\Phi_{\{\lambda/\mu\}} = \sum_{\{\nu\}} \Gamma_{\nu\lambda\mu}\Phi_{\{\nu\}} \tag{4.5.20}$$

We now introduce the derivatives

$$n\frac{\partial}{\partial S_n} = S_n^\dagger \tag{4.5.21}$$

which are the Hermitian adjoints, with respect to the Bargman-Fock inner product, of the operation of multiplication by S_n. Consider the operator $\Phi^\dagger_{\{\lambda\}}(x) = \Phi_{\{\lambda\}}(\partial_x)$ which is obtained from $\Phi_{\{\lambda\}}(x)$, regarded as a function of $S_n(x)$, by replacing the S_n's by their adjoints. Then

$$\begin{aligned}\Phi_{\{\lambda\}}(\partial_x)\sum_{\{\mu\}} \Phi_{\{\mu\}}(x)\Phi_{\{\mu\}}(y) &= \Phi_{\{\lambda\}}(\partial_x)\exp\left(\sum_n \frac{1}{n} S_n(x)S_n(y)\right)\\ &= \Phi_{\{\lambda\}}(y)\exp\left(\sum_n \frac{1}{n} S_n(x)S_n(y)\right)\\ &= \sum_{\{\mu\}} \Phi_{\{\mu/\lambda\}}(x)\Phi_{\{\mu\}}(y)\end{aligned} \tag{4.5.22}$$

so, from the linear independence of the $\Phi_{\{\lambda\}}$, we deduce that

$$\Phi_{\{\lambda\}}(\partial_x)\Phi_{\{\mu\}}(x) = \Phi^\dagger_{\{\lambda\}}(x)\Phi_{\{\mu\}}(x) = \Phi_{\{\mu/\lambda\}}(x). \tag{4.5.23}$$

As a result of these definitions we find that, acting on the Schur functions which create the excited states, we have

$$U\left(\exp\sum \frac{1}{n}S_n(\zeta)z^{-n}\right) = \sum_{\{\nu\}} \Phi_{\{\nu\}}(\zeta)\Phi_{\{\nu\}}(\partial_z) = \exp\left(\sum \frac{1}{n}S_n(\zeta)\left(n\frac{\partial}{\partial S_n}\right)\right) \tag{4.5.24}$$

Because of the derivative operators $U(e^{i\phi^+})$ and $U(e^{i\phi^-})$ no longer commute, but almost do so, up to a factor which is the Kac-Peterson cocycle.

iii) Action of $U(e^{i\alpha})$ and $U(z^n)$

The last parts of the loop group to consider are the action of the n-fold winding produced by taking $e^{i\phi} = z^n$, and the effect induced by the multiplication of the single particle states by a constant phase. It should be obvious that on multiplying every state in the Slater determinant by z^n, and following the previous philosophy of forgetting about the effect on the right hand edge of the determinant, we are just increasing the number of electrons in the state by n. As for the constant phase, we adopt a "normal ordering" convention and declare the N fermion state to be neutral, but the $N+n$ fermion state to have charge n. This means that when the single particle states are multiplied by $e^{i\alpha}$, the many body state is defined to pick up a phase $e^{in\alpha}$. We require therefore that

$$U(e^{i\alpha})|0\rangle = |0\rangle \tag{4.5.25}$$

and

$$U(e^{i\alpha})U(z^n) = e^{in\alpha}U(z^n)U(e^{i\alpha}). \tag{4.5.26}$$

This may be achieved by introducing an additional variable S_0 and setting

$$U(e^{i\alpha}) = \exp\left(i\alpha\frac{\partial}{\partial S_0}\right) \tag{4.5.27}$$

and

$$U(z^n) = \exp nS_0, \tag{4.5.28}$$

but in the notation inherited from string theory one uses

$$U(e^{i\alpha}) = e^{i\alpha p_0} \quad \text{and} \quad U(z^n) = e^{inq_0} \tag{4.5.29}$$

with

$$[q_0, p_0] = i, \tag{4.5.30}$$

and this is the notation I prefer. Both p_0 and q_0 are Hermitian.

Appendix 4A. The Very Useful Identity

In this appendix I will sketch the proof of the very useful identity

$$\exp\left(\sum_{n=1}^{\infty}\frac{1}{n}S_n(z)S_n(\zeta)\right) = \sum_{\{\lambda\}}\Phi_{\{\lambda\}}(z)\Phi_{\{\lambda\}}(\zeta). \tag{4A.1}$$

Now

$$\begin{aligned}\sum_{1}^{\infty}\frac{1}{n}S_n(z)S_n(\zeta) &= \sum_{n=1}^{\infty}\frac{1}{n}\sum_i z_i^n\sum_j \zeta_j^n \\ &= -\sum_{i,j}\ln(1-z_i\zeta_j) \\ &= -\ln\prod_{i,j}(1-z_i\zeta_j).\end{aligned} \tag{4A.2}$$

Thus the left hand side of (4A.1) is

$$\frac{1}{\prod_{i,j}(1 - z_i \zeta_j)}. \tag{4A.3}$$

Now we use Cauchy's identity in the form

$$\det \left| \frac{1}{1 - z_i \zeta_j} \right| = \frac{D(z)D(\zeta)}{\prod_{ij}(1 - z_i \zeta_j)}, \tag{4A.4}$$

to show that

$$\exp\left(\sum_{n=1}^{\infty} \frac{1}{n} S_n(z) S_n(\zeta) \right) = \frac{1}{D(z)D(\zeta)} \det \left| \frac{1}{1 - z_i \zeta_j} \right|. \tag{4A.5}$$

We expand out the entries in the determinant on the right hand side of (4A.5) in a geometric series and rearrange the monomial terms into $\Psi_{\{\lambda\}}(z)$ and $\Psi_{\{\lambda\}}(\zeta)$'s. Dividing by the D's completes the derivation.

Appendix 4B. Jacobi's Triple Product Formula

The combinatorics of bosonization provides a simple proof of the Jacobi triple product formula which is important in the theory of theta functions.

Consider a Fermi system with energy levels $\epsilon_k = 2k + 1$, $k \in Z$. In the ground-state all negative energy states are occupied. We measure the energy E and the particle number n relative to this state.

Now we calculate the partition function

$$\mathcal{Z}(\beta, x) = \mathrm{Tr}(e^{-\beta H} x^Q), \tag{4B.1}$$

where H and Q are the operators whose eigenvalues are E and n respectively.

There are two ways to evaluate $\mathcal{Z}(\beta, x)$. The first is the "bosonic" route, and begins by noticing that all the neutral, $n = 0$, excitations are accounted for by rearranging the particles occupying the sea, and that this is given by the sum over partitions as described in (4.2.5). States with charge n are then obtained by adding n extra particles in the levels $\epsilon_k = 1, 3, \ldots, 2n-1$. Since the sum of the first n odd numbers is n^2, this state has energy $E = n^2$. We again sum over rearrangements and find

$$\mathcal{Z}(\beta, x) = \sum_{n=-\infty}^{\infty} e^{-\beta n^2} x^n \prod_{n=1}^{\infty} \frac{1}{(1 - e^{-2\beta n})}. \tag{4B.2}$$

The second, "fermionic", route is to recognize that every level ϵ_k can be either occupied or empty. So

$$\mathcal{Z}(\beta, x) = \prod_{n=1}^{\infty} (1 + x e^{-\beta(2n-1)})(1 + x^{-1} e^{-\beta(2n-1)}). \tag{4B.3}$$

The first set of factors counts the energy cost of occupying an empty positive energy state and the second the (positive) energy cost of emptying an occupied negative energy state.

Equating the two expressions gives

$$\sum_{n=-\infty}^{\infty} e^{-\beta n^2} x^n = \prod_{n=1}^{\infty} (1 - e^{-2\beta n})(1 + x e^{-\beta(2n-1)})(1 + x^{-1} e^{-\beta(2n-1)}). \tag{4B.4}$$

This is the Jacobi triple product formula.

References

[1] L. A. Dickey, *Soliton Equations and Hamiltonian Systems* (World Scientific, Singapore, 1991).

[2] D. E. Littlewood, *University Algebra*, 2nd ed. (William Heinemann, London 1961, also Dover Editions, 1970).

[3] I. G. Macdonald, *Symmetric Functions and Hall Polynomials* (Oxford University Press, 1979).

[4] M. Hamermesh, *Group Theory and its Application to Physical Problems* (Addison Wesley, 1962).

[5] D. E. Littlewood, *The Theory of Group Characters and Matrix Representations and Groups*, 2nd ed. (Oxford, 1950).

[6] H. Weyl, *The Classical Groups, their Invariants and Representations* (Princeton, 1946).

[7] P. Goddard and D. Olive, *Int. J. Mod. Phys.* **A1** (1986) 303.

[8] V. G. Kac *Infinite Dimensional Lie Algebras* (Cambridge University Press, 1985).

[9] M. Hamermesh *op. cit.*, p. 197.

[10] H. Weyl, *The Theory of Groups and Quantum Mechanics* (Dover Editions, 1950), p. 331.

[11] A. Pressley and G. Segal, *Loop Groups* (Clarendon Press, Oxford, 1986).

Reprinted Papers

Progress of Theoretical Physics Vol. 5, No. 4, July~August, 1950

Remarks on Bloch's Method of Sound Waves applied to Many-Fermion Problems

Sin-itiro TOMONAGA

Institute for Advanced Study
Princeton, N. J.
U.S.A.*

(Received June 29, 1950)

Abstract

The fact implied by Bloch several years ago that in some approximate sense the behavior of an assembly of Fermi particles can be described by a quantized field of sound waves in the Fermi gas, where the sound field obeys Bose statistics, is proved in the one-dimensional case. This fact provides us with a new possibility of treating an assembly of Fermi particles in terms of the equivalent assembly of Bose particles, namely, the assembly of sound quanta. The field equation for the sound wave is found to be linear irrespective of the absence or presence of mutual interaction between particles, so that this method is a very useful means of dealing with many-Fermion problems. It is also applicable to the case where the interparticle force is not weak. In the case of force of too short a range this method fails.

§ 1. Introduction and summary.

The well-known method of Thomas and Fermi provides us with a very practical approximate treatment of many-Fermion problems. Because in this approximation, however, each particle is supposed to move independently, the effect of interaction between particles being simply replaced by an average field of force, one cannot speak of correlations between particles in this rough approximation. This simple method does not apply to problems in which inter-particle correlation plays an important role. A step toward the improvement of the method so as to include the correlation was taken by Euler[1], who calculated the effect of inter-particle interaction, which causes the correlation, by perturbation theory assuming the interaction to be small.

The calculation of such a type is carried out in the following manner: In the zeroth approximation each particle is in some one-particle quantum state, which we shall call a " level ". Let us consider the lowest state for the sake of definitness. In this state all levels up to some highest one, which shall be called the " Fermi maximum ", are each filled by one particle. The perturbation energy, the energy due to the inter-particle force, has non-vanishing matrix

* Now returned to the Tokyo University of Education (Tokyo Bunrika Daigaku), Tokyo.

elements which cause virtual transitions to states in which two particles are excited simultaneously to levels higher than the Fermi maximum.

The state which results from the inter-particle force is thus such one which is a superposition of the zero-order state and various excited states in which holes and particles are present in some levels below and above the Fermi maximum respectively. In the lowest approximation of the perturbation calculation, the numbers of holes and excited particles are two; but, if the calculation is carried out to higher order, which is necessary when the inter-particle force is not small, there appear states in which a considerable number of holes and excited particles are present. The amplitudes of such highly excited states will be appreciable if the inter-particle force is strong.

Such a mixture of excited states gives rise to correlation between particles. We can in this way include the effects of correlations in treating certain problems. In the case of strong inter-particle forces, however, it is necessary to carry through the perturbation calculation up to a very high order, and this is too involved to be practicable. Recently, Nogami[2] proposed a method which applies also to cases of rather strong inter-particle forces. Still his method requires some kind of weakness of interaction because he had to neglect the interactions of holes and of excited particles with each other as well as the interactions between holes and exited particles. This neglect cannot be justified when too many holes and excited particles are present.

In such a situation it is desirable to find some approximate method of dealing with many-Fermion problems different from the perturbation method.

In his famous work on the stopping power of charged particles, Bloch[3] has treated the excited states of the Fermi gas not as states with holes and excited particles, but as states in which the gas oscillates. In this work it was not necessary to treat the oscillation quantum-theoretically. In a later paper[4], he also dealt with a problem in which the quantum aspect of the oscillation was essential. This was a problem in which the density fluctuation played a role. He showed in this work that the density fluctuation of a degenerate Fermi gas (in his theory it was sufficient to treat the gas without inter-particle force) can be calculated in two different ways giving the same result. The one was the orthodox method whereby one calculates directly the expectation value of the operator having expressed in terms of the quantized field variables ψ and ψ^* describing the assembly of the Fermi particles. The other method was to calculate the zero-point amplitudes of sound waves in the gas, the equations of motion for the sound waves being properly chosen. He showed that the correct value of the fluctuation was obtained in this way if the choice of the equations of motion for the sound waves was properly made and if the zero-point amplitudes of the waves had such values as would be expected for a sound field obeying Bose statistics.

If it is proved that the excited states of an assembly of Fermi particles can be in fact described as an excitation of sound waves, where the sound can be described by a Bose field, it will provide us with a new method of treating many-Fermion problems, not dealing directly with the assembly of the Fermi particles, but dealing with the equivalent assembly of Bose particles, i. e. the assembly of sound quanta.

There is a prospect that in this latter method the assumption of the weakness of inter-pariicle forces will not be essential. This prospect lies in the following situation: It is expected that the equations of motion for the sound waves will be linear in the field variables describing the sound field; otherwise we could not speak of waves at all. The field variables for the sound field will be the density ρ of the gas and its properly defined canonical conjugate. Now the essential point is that the linearity of the problem will not be destroyed even when an inter-paiticle force is present, because the interaction energy between particles is bilinear in ρ. This fact is in marked contrast to the circumstance that in terms of ψ and ψ^* the interaction energy contains four ψ's, or, more precisely, two ψ's and twe ψ^{*}'s; thus the field equations for ψ and ψ^* are no longer linear when an inter-particle force is present. This circumstance made it very difficult to treat many-Fermion problems with inter-particle forces. This difficulty will disappear when we deal with the problem in terms of the ρ field but not of the ψ field.

The purpose of this paper is to show that these expectations are really fulfilled. It will be shown that in some approximation, which does not necessarily require the weakness of the interaction, excitated states of Fermi gas are in fact equivalent to corresponding excitations of sound waves, and that the sound is describable by a Bose field whose field equations are linear in the field variables irrespective of the absence or presence of interparticle forces. Thus, Bloch's method of sound waves will be a very useful method for many-Fermion pioblems.

The possibility of this new method was found independently by Bohm[6], who discusses a very interesting phenomenon of plasma-like oscillations in a degenerate electron gas from a very similar point of view. He, too, gives a proof of the fact that these oscillations are describable by a Bose field in some approximation. He utilizes further the linearity of the field equations to study such a pure correlation phenomenon as plasma oscillations of an electron gas.

The present paper is of a rather more mathematical nature than physical in that it aims mainly at the analysis of the mathematical structure of the method, clarifying the underlying assumptions and the limit of applicability. A mathematically closed and clear-cut presentation of the theory is achieved, however, at the expense of physical usefulness, because, thus far, the author has succeded only in giving a complete formulation for a one-dimensional assembly of particles.

It is rather certain that a similar method applies to the three dimensional case too—this, indeed, has been done by Bohm—but the situation is more complicated in this case.

The mathematicl relation between the field of sound quanta on the one hand and the original field of Fermi particles on the other is very similar to the relation between the field of light quanta and the field of neutrinos in the neutrino theory of light[6]. One will find everywhere a marked parallelism between our theory and the neutrino theory of light.

The discussions will be performed in several steps: In § 2 we shall prove that the sound can be described by a Bose field under some assumptions imposed on the states under consideration. Then we shall set up in § 3 the Hamiltonian for the sound field. We shall see that the Hamiltonian is in fact bilinear in the field variables, so that the whole problem is liner. In § 4 we shall show that this Hamiltonian is really equal to the original Hamiltonian for the assembly of Fermi particles, and, therefore, the assembly of sound quanta is equivalent to the original assembly of Fermi particles. In § 5 we will go over to the solution of the eigenvalue problem. This can be done very simply by finding the normal coordinates for the sound field, because the field equations are linear. In § 6 we shall give several general formulae of physical interest which are derived directly by our method. A criterion for the applicability of the method will also be given in this section. In the last section we shall briefly mention the bearing of our results on the plasma oscillations treated by Bohm. Also, the relation between the two kinds of descriptions of the system, one as an assembly of Fermi particles and the other as an assembly of sound quanta, is discussed briefly.

In § 6 we shall see that our method does not work if the inter-particle force is of short range. The range of force must be larger than four times the mean distance between particles. Since sound waves with wave length shorter than the mean distance between particles have no meaning, it is quite conceivable that the method of sound waves fails in describing the event occuring in a small space region comparable with the mean distance between particles. This is the reason why the method fails in the case of short range force. The method is, on the other hand, very suitable for dealing with the case of long-range force in which a considerable number of holes and excited particles are present in the neighborhood of the Fermi maximum. Thus the present method covers a field where the known methods have failed.

§ 2. Approximate commutation relations for the density field.

Let us consider an assembly of Fermi particles in a one-dimensional "box" of length L. Let $\psi(x)$ and $\psi^*(x)$ be the quantized wave functions describing the assembly. The density $\rho(x)$ of the particles is then given by

$$\rho(x)=\psi^*(x)\psi(x). \tag{2.1}$$

We introduce the Fourier transforms ψ_n and ψ_n^* of $\psi(x)$ and $\psi^*(x)$ respectively. They are defined by

$$\begin{cases} \psi(x)=\dfrac{1}{\sqrt{L}}\sum_n \psi_n \exp\left(\dfrac{2\pi i}{L}nx\right), \\ \psi^*(x)=\dfrac{1}{\sqrt{L}}\sum_n \psi_n^* \exp\left(-\dfrac{2\pi i}{L}nx\right), \end{cases} \qquad n=0,\ \pm1,\ \pm2,\cdots. \tag{2.2}$$

The Fourier transform of $\rho(x)$, which is defined by

$$\rho(x)=\frac{1}{L}\sum_n \rho_n \exp\left(\frac{2\pi i}{L}nx\right), \tag{2.3}$$

is evidently given by

$$\rho_n=\sum_{n'}\psi^*_{n'}\psi_{n'+n}=\sum_{\bar{n}}\psi^*_{\bar{n}-\frac{n}{2}}\psi_{\bar{n}+\frac{n}{2}}, \tag{2.4}$$

where $\bar{n}$ takes integral values when n is even and half-odd integral values when n is odd.

Our purpose is to describe the system in terms of the density field $\rho(x)$ instead of describing it by the wave field $\psi(x)$. The first task is to find the commutation relations between the field quantity $\rho(x)$ and its properly defined canonical conjugate. For this purpose we separate each ρ into two parts ρ_n^+ and ρ_n^- by means of

$$\begin{cases} \rho_n^+=\sum_{\bar{n}>0}\psi^*_{\bar{n}-\frac{n}{2}}\psi_{\bar{n}+\frac{n}{2}}, \\ \rho_n^-=\sum_{\bar{n}<0}\psi^*_{\bar{n}-\frac{n}{2}}\psi_{\bar{n}+\frac{n}{2}}. \end{cases} \tag{2.5}$$

We have evidently

$$\rho_n=\rho_n^+ +\rho_n^- \tag{2.6}$$

(in case of even n the term with $\bar{n}=0$ is absent in (2.6), but this does not cause any serious error.) We have further

$$\rho_n^+=\rho_{-n}^{+*}, \qquad \rho_n^-=\rho_{-n}^{-*}. \tag{2.7}$$

It will be seen later that the separation of ρ_n by (2.6) corresponds to the separation of the field into parts with positive and negative frequencies, which is the usual procedure in field theory. (See (3.1) and (3.2)).

We now examine the commutation relations of the ρ's. The commutation relations between the ψ's and ψ^*'s are

$$\begin{cases} [\psi_n^*,\ \psi_{n'}]_+=\delta_{n,n'}, \\ [\psi_n^*,\ \psi_{n'}^*]_+=[\psi_n,\ \psi_{n'}]_+=0. \end{cases} \tag{2.8}$$

From these relations we get the commutation relations of the ρ's. For the ρ^+'s they are

$$[\rho_n^+,\ \rho_{n'}^+]=\begin{cases} \sum\limits_{-\frac{n}{2}<\bar{n}\leq-\frac{|n'|}{2}}\psi^*_{\bar{n}-\frac{n}{2}-\frac{n'}{2}}\psi_{\bar{n}+\frac{n}{2}+\frac{n'}{2}} & \text{for } n>0,\ n'\leq -n, \\ \sum\limits_{\frac{n'}{2}<\bar{n}\leq\frac{n}{2}}\psi^*_{\bar{n}-\frac{n}{2}-\frac{n'}{2}}\psi_{\bar{n}+\frac{n}{2}+\frac{n'}{2}} & \text{for } n>0,\ -n\leq n'\leq n, \\ -\sum\limits_{\frac{n}{2}<\bar{n}\leq\frac{n'}{2}}\psi^*_{\bar{n}-\frac{n}{2}-\frac{n'}{2}}\psi_{\bar{n}+\frac{n}{2}+\frac{n'}{2}} & \text{for } n>0,\ n'\geq n, \end{cases} \tag{2.9}$$

which, for the special cases of $n'=-n$, reduce to

$$[\rho_n^+,\ \rho_{-n}^+]=\sum_{-\frac{n}{2}<\bar{n}\leqq\frac{n}{2}} \psi_{\bar{n}}^*\psi_{\bar{n}} \qquad \text{for } n>0. \tag{2.9'}$$

In the same way, we get for the ρ^-'s

$$[\rho_n^-,\ \rho_{n'}^-]=\begin{cases} -\sum\limits_{\frac{n'}{2}\leqq\bar{n}<\frac{n}{2}} \psi^*_{\bar{n}-\frac{n}{2}-\frac{n'}{2}}\psi_{\bar{n}+\frac{n}{2}+\frac{n'}{2}} & \text{for } n>0,\ n'\geqq -n, \\ -\sum\limits_{-\frac{n}{2}\leqq\bar{n}<-\frac{n'}{2}} \psi^*_{\bar{n}-\frac{n}{2}-\frac{n'}{2}}\psi_{\bar{n}+\frac{n}{2}+\frac{n'}{2}} & \text{for } n>0,\ -n\leqq n'\leqq n, \\ \sum\limits_{-\frac{n'}{2}\leqq\bar{n}<-\frac{n}{2}} \psi^*_{\bar{n}-\frac{n}{2}-\frac{n'}{2}}\psi_{\bar{n}+\frac{n}{2}+\frac{n'}{2}} & \text{for } n>0,\ n'\geqq n, \end{cases} \tag{2.10}$$

and

$$[\rho_n^-,\ \rho_{-n}^-]=-\sum_{-\frac{n}{2}\leqq\bar{n}<\frac{n}{2}} \psi_{\bar{n}}^*\ \psi_{\bar{n}} \qquad n>0. \tag{2.10'}$$

The commutation relations between the ρ^+'s and ρ^-'s are found to be

$$[\rho_n^-,\ \rho_{n'}^+]=\begin{cases} \sum\limits_{-\frac{n}{2}<\bar{n}<\frac{n'}{2}} \psi^*_{\bar{n}-\frac{n}{2}-\frac{n'}{2}}\psi_{\bar{n}+\frac{n}{2}+\frac{n'}{2}} & \text{for } -n<n', \\ -\sum\limits_{\frac{n}{2}<\bar{n}<-\frac{n'}{2}} \psi^*_{\bar{n}-\frac{n}{2}-\frac{n'}{2}}\psi_{\bar{n}+\frac{n}{2}+\frac{n'}{2}} & \text{for } n<-n', \end{cases} \tag{2.11'}$$

and

$$[\rho_n^-,\ \rho_{-n}^+]=0. \tag{2.11'}$$

We now show that the commutation relations (2.9)—(2.11′) can be replaced by simpler ones if the states under consideration satisfy, at least approximately, some conditions which will be specified below. These simplified commutation relations tell us that the density field can be regarded as a Bose-field under these restricting conditions.

We first consider the case of the ideal Fermi gas in which there are no interactions between particles. If the gas is not excited too highly, only particles in the neighborhood of the Fermi maximum are raised to higher levels. There exist holes and excited particles only in the neighborhood of the surface of the Fermi sea. Now, in the case of a non-ideal Fermi gas, the inter-particle forces cause virtual transitions of particles. Thus extra holes and exited particles appear. But, if the range of the inter-particle force is not too short and the force itself is not too strong, these virtual holes and excited particles are still present only in the neighborhood of the Fermi maximum. Such are the states to which we will confine ourselves.

If we confine ourselves to states of such type, we can simplify the commutation relations (2.9)—(2.11′) in the following manner.

Let us consider, for instance, the first commutation relations $[\rho_n^+,\ \rho_{n'}^+]$ for which $n>0$ and $-n<n'<n$. We notice that the expression on the right-

hand side is a sum of operators each bringing one particle from the level $\bar{n}+\frac{n}{2}+\frac{n'}{2}$ to the level $\bar{n}-\frac{n}{2}-\frac{n'}{2}$. Because the summation over $\bar{n}$ is extended only between $\frac{n'}{2}$ and $\frac{n}{2}$, the final levels $\bar{n}-\frac{n}{2}-\frac{n'}{2}$ lie in a limited interval between $-\frac{n}{2}$ and $-\frac{n'}{2}$. Now, let $n_{\max}$ denote the value of $|n|$ at the Fermi maximum. Then, if n and $|n'|$ are both sufficiently small compared with $n_{\max}$ (the discussion about how small they should be will be given below), the levels $-\frac{n}{2}$ and $-\frac{n'}{2}$ both lie deep in the bottom of the Fermi sea where there are holes. In such a case the operator $\psi^+_{\bar{n}-\frac{n}{2}-\frac{n'}{2}}\psi_{\bar{n}+\frac{n}{2}+\frac{n'}{2}}$ will give a vanishing result because the final level is occupied. Thus, for the states under considration our commutators $[\rho_n^+, \rho_{n'}^+]$ are equivalent to zero.

We next consider $[\rho_n^+, \rho_{-n}^+]$. In this case, the right-hand side is

$$\sum_{-\frac{n}{2}<\bar{n}\leq\frac{n}{2}} \psi_{\bar{n}}{}^*\psi_{\bar{n}} = \sum_{-\frac{n}{2}<\bar{n}\leq\frac{n}{2}} N_{\bar{n}}\ , \tag{2.9''}$$

where $N_{\bar{n}}$ is the occupation number of the level $\bar{n}$. Since the level $\bar{n}$, which lies between $-\frac{n}{2}$ and $\frac{n}{2}$, lies deep in the Fermi sea if n is small compared with $n_{\max}$, it is occupied by one particle. Then the sum $\sum N_{\bar{n}}$ is simply equal to the number of levels between $-\frac{n}{2}$ and $\frac{n}{2}$, which is just n. So we find that $[\rho_n^+, \rho_{-n}^+]$ is equivalent to n.

A similar consideration applies to the remaining commutators. We can see that the following commutation relations hold in the sense of equivalence:

$$\begin{aligned}
&[\rho_n^+, \rho_{n'}^+]=n\delta_{n,-n'}\\
&[\rho_n^-, \rho_{n'}^-]=-n\delta_{n,-n'}\\
&[\rho_n^+, \rho_{n'}^-]=0.
\end{aligned} \tag{2.12}$$

In order that these simpler commutation relations can be used instead of the original ones, it was necessary that $|n|$ and $|n'|$ be sufficiently small compared with $n_{\max}$. We now discass this point more quantitatively.

First we notice that the total number N of the particles is related to $n_{\max}$ by

$$N=2n_{\max}+1, \tag{2.13}$$

which we shall make use of in later considerations.

We now assume that in the states under consideration there are no holes under the level specified by $|n|=\alpha n_{\max}$, α being a positive number less than unity. Then, we can easily see that

$$\frac{3}{2}|n|,\ \frac{3}{2}|n'| < a\, n_{max} \tag{2.14}$$

is the required condition for the validity of the simplar commutation relations. (The factor 3/2 is required in order that (2.11) be equivalent to the third relation of (2.12)).

So far as the present consideration is concerned, there is no restriction on a. But we shall see later that a must be 3/4 in order that the whole treatment work consistently. The required conditions will be found to be the following:

(I) In the region $|n| < \frac{3}{4} n_{max}$, there should be no holes.

(II) In the region $|n| > \frac{5}{4} n_{max}$, there should be no excited particles.

(III) The absolute values of n in ρ_n^+ and ρ_n^- should not exceed $\frac{1}{2} n_{max}$:

$$|n| < \frac{1}{2} n_{max}. \tag{2.15}$$

Our method works when and only when these conditions are satisfied. The conditions (I) and (II) restrict the states; therefore, it is always necessary in applying our method to verify whether the states obtained as an answer do really satisfy these conditions. The condition (III) requires that no sound waves having shorter wave length than $2L/n_{max}$ play a role in the problem. This condition will be satisfied if the range of inter-particle force is sufficiently long. The reason for the the necessity of the special choice of $a=3/4$ will be given later.

§ 3. Equation of motion and Hamiltonian for the density field.

We first consider the case of non-interacting particles. In this case the change with time of ρ_n^+ and ρ_n^- can be obtained easily because we know the change with time of ψ_n and $\psi_n{}^*$. In the general case we them find a very complicated time dependence of the ρ's. The motions of ρ_n^+ and ρ_n^- are by no means simple harmonic because each term of $\sum \psi^*_{\bar{n}-\frac{n}{2}}\, \psi_{\bar{n}+\frac{n}{2}}$ has a frequency which depends not only on n but also on $\bar{n}$. This fact means that the density does by no means behave like waves.

However, we may here make use of our conditions imposed on the states. According to them holes and excited particles are present in a narrow interval only, from $(3/4)n_{max}$ to $(5/4)n_{max}$, in the neighborhood of n_{max}. This results in non-vanishing matrix elements in $\sum \psi^*_{\bar{n}-\frac{n}{2}}\, \psi_{\bar{n}+\frac{n}{2}}$ being contributed solely by terms with $\bar{n} \pm \frac{n}{2}$ nearly equal to n_{max} or $-n_{max}$ according as we are dealing with ρ_n^+ or ρ_n^-. We may then approximate the frequencies of ψ^* and ψ by their expansion in the neighborhood of $\bar{n}\pm\frac{n}{2}=n_{\max}$ or $\bar{n}\pm\frac{n}{2}=-n_{\max}$ respectively. The frequency of $\psi_{\bar{n}+\frac{n}{2}}$ is, for instance,

$$\frac{1}{2\pi\hbar}\frac{1}{2m}\left(\frac{2\pi\hbar}{L}\right)^2\left(\bar{n}+\frac{n}{2}\right)^2,$$

m being the mass of the particles. If we here expand $\left(\bar{n}+\frac{n}{2}\right)^2$ in the neighborhood of $\bar{n}+\frac{n}{2}=n_{\max}$, we get

$$\begin{aligned}\left(\bar{n}+\frac{n}{2}\right)^2&=\left\{n_{\max}+\left(\bar{n}+\frac{n}{2}-n_{\max}\right)\right\}^2\\&=n_{\max}^2+2n_{\max}\left(\bar{n}+\frac{n}{2}-n_{\max}\right)+\cdots.\end{aligned}$$

Neglecting the small term $\left(\bar{n}+\frac{n}{2}-n_{\max}\right)^2$, the time dependence of $\psi_{\bar{n}+\frac{n}{2}}$ becomes

$$\psi_{\bar{n}+\frac{n}{2}}\approx\exp i\left[+\frac{1}{\hbar}\left(\frac{2\pi\hbar}{L}\right)^2\frac{n_{\max}^2}{2m}-\frac{1}{\hbar}\left(\frac{2\pi\hbar}{L}\right)^2\frac{n_{\max}}{m}\left(\bar{n}+\frac{n}{2}\right)\right]t.$$

In the same way we have

$$\psi_{\bar{n}-\frac{n}{2}}\approx\exp i\left[-\frac{1}{\hbar}\left(\frac{2\pi\hbar}{L}\right)^2\frac{n_{\max}^2}{2m}+\frac{1}{\hbar}\left(\frac{2\pi\hbar}{L}\right)^2\frac{n_{\max}}{m}\left(\bar{n}-\frac{n}{2}\right)\right]t.$$

Then, combining these two, we get a time dependence for ρ_n^+ of the form

$$\rho_n^+\approx\exp\left[-\frac{i}{\hbar}\left(\frac{2\pi\hbar}{L}\right)^2\frac{n_{\max}}{m}n\right]t. \tag{3.1}$$

By the same consideration using the expansion in the neighborhood of $\bar{n}\pm\frac{n}{2}=-n_{\max}$, we get

$$\rho_n^-\approx\exp\left[+\frac{i}{\hbar}\left(\frac{2\pi\hbar}{L}\right)^2\frac{n_{\max}}{m}n\right]t. \tag{3.2}$$

From (3.1) and (3.2) we see that ρ^+ and ρ^- satisfy the equations of motion

$$\begin{cases}\dot{\rho}_n^+=-\dfrac{i}{\hbar}\left(\dfrac{2\pi\hbar}{L}\right)^2\dfrac{n_{\max}}{m}n\rho_n^+\\[2ex]\dot{\rho}_n^-=+\dfrac{i}{\hbar}\left(\dfrac{2\pi\hbar}{L}\right)^2\dfrac{n_{\max}}{m}n\rho_n^-.\end{cases} \tag{3.3}$$

We notice here that the commutation relations (2.12) lead to the fact that the canonically conjugate momentum for

$$\rho_n=\rho_n^++\rho_n^- \tag{3.4}$$

is given by

$$\pi_n = \frac{i}{2}\frac{1}{n}\left(\rho_{-n}^{+} - \rho_{-n}^{-}\right). \tag{3.5}$$

As can be easily verified, we see that

$$[\rho_n, \pi_{n'}] = i\delta_{n,n'}. \tag{3.6}$$

By the equation of motion we then find

$$\pi_n = \frac{\hbar}{2}\left(\frac{L}{2\pi\hbar}\right)^2 \frac{m}{n_{\max}} \frac{1}{n^2}\dot{\rho}_{-n}. \tag{3.5'}$$

The last relation gives a physical meaning to our canonical momenta: As usual, momenta are proportional to the time derivatives of the coordinates. Though the equations of motion of the form of (3.3) hold only for the case of non-interacting particles, we shall see later that (3.5′) holds more generally.

We shall now set up the Hamiltonian for the ρ field. The Hamiltonian is determined by the requirement that

$$\frac{\hbar}{i}\dot{\rho}_n^{\pm} = [\mathfrak{H}, \rho_n^{\pm}] \tag{3.7}$$

yield the equations of motion (3.3) wtih the help of the commutation relations (2.12). We then find that the Hamiltonian has a very simple form:

$$\mathfrak{H} = \left(\frac{2\pi\hbar}{L}\right)^2 \frac{n_{\max}}{m} \sum_{n>0} (\rho_{-n}^{+}\rho_n^{+} + \rho_n^{-}\rho_{-n}^{-}). \tag{3.8}$$

The order of non-commuting factors is here specified by the condition that $\mathfrak{H}$ has no zero point value.

Though we have set up the Hamiltonian in this way, it is by no means self-evident that this Hamiltonian gives in fact the energy of the system (or eventually energy plus some additive constant). That this is actually the case will be proved in the next section.

So far we have neglected the mutual interaction between particles. It is quite simple to introduce the interaction into the theory, when the interaction force is of the ordinary type, neither of exchange type nor velocity dependent. In this case the interaction energy has the form

$$H_{\text{int}} = \frac{1}{2}\iint \rho(x)\rho(x')J(|x-x'|)dxdx' - \frac{1}{2}\int \rho(x)J(0)dx, \tag{3.9}$$

$J(|x-x'|)$ being the potential of the inter-particle force. The term $\frac{1}{2}\int \rho(x)J(0)dx$ is subtracted in order to remove the interaction of a particle with itself. In terms of the Fourier transform we find

$$H_{\text{int}} = \frac{1}{2}\sum_n{}' \rho_n\rho_{-n}J_n + \frac{1}{2}\rho_0^2 J_0 - \frac{1}{2}\rho_0 J(0), \tag{3.10}$$

where J_n is the matrix element of the potential defined by

$$J_n=\frac{1}{L}\int J(x)\exp\left(\frac{2\pi i}{L}nx\right)dx=J_{-n}. \tag{3.11}$$

The prime on the $\sum'$ symbol means that the term with $n=0$ should be omitted. ρ_0 is the quantity defined by

$$\rho_0=\sum_{\bar{n}}\psi_{\bar{n}}^*\psi_{\bar{n}}=N, \tag{3.12}$$

i.e. the total number of particles.

Since H_{int} commutes with each ρ_n, which is the sum $\rho_n^+ + \rho_n^-$, the equation of motion for ρ_n is not affected by the interaction, while the equations of motion for ρ_n^+ and ρ_n^- separately are affected by the interaction. This fact means that the canonically conjugate momenta π_n are still given by (3.5′) irrespective of the presence or absence of the interaction.

In terms of ρ^+ and ρ^- the interaction Hamiltonian can be expressed as

$$\begin{aligned}H_{\text{int}}=&\sum_{n>0}J_n(\rho_{-n}^+\rho_n^+ + \rho_n^-\rho_{-n}^- + \rho_n^+\rho_{-n}^- + \rho_n^-\rho_{-n}^+)\\&+\sum_{n>0}nJ_n+\frac{1}{2}N^2J_0-\frac{1}{2}NJ(0).\end{aligned} \tag{3.13}$$

The term $\sum_{n>0}nJ_n$ appeared when we performed the rearrangement of factors in $\rho_n^+\rho_{-n}^+$ and $\rho_{-n}^-\rho_n^-$ into the correct order.

It is the essential point of our method that the energy, the "kinetic part" $\mathfrak{H}$ as well as the "potential part" H_{int}, is bilinear in the $\rho^{\pm}$'s. This fact is in marked contrast to the fact that, in terms of ψ and ψ^*, only the kinetic part is bilinear; the potential part contains four ψ's. This latter fact makes it necessary to solve a complicated non-linear problem if one wishes to deal with a many-particle problem with inter-particle force. Now that we have found the Hamiltonian to be bilinear in the ρ's, irrespective of the presence or absence of the inter-particle force, we have no such difficulty if we deal with the problem in terms of the ρ field not in terms of the ψ field. The problem is simply to perform the principal-axes transformation of the bilinear form, i. e. to find the normal coordinates for the ρ field.

Before we enter into this problem, we shall show that the Hamiltonian $\mathfrak{H}$ is in fact equal to the kinetic energy of the system minus some constant which depends only on the total number of particles.

§ 4. Energy and momentum of the system.

In this section we give the proof that the Hamiltonian $\mathfrak{H}$ really gives the kinetic energy of the system. This can be done by a straightforward calculation in the following manner.

For this purpose we first examine $\sum\rho_{-n}^+\rho_n^+$. We examine the diagonal and the non-diagonal parts of $\sum \rho^+_{-n}\rho^+_n$ separately, where the matrix is supposed to be referred to in the representation in which the occupation numbers of the particles are diagonal.

We have

$$\begin{aligned}\sum_{\frac{n_{\max}}{2}>n>0} \rho^+_{-n}\rho^+_n &= \sum_{\frac{n_{\max}}{2}>n>0}\ \sum_{\bar n>0}\sum_{\bar n'>0} \psi^*_{\bar n+\frac n2}\psi_{\bar n-\frac n2}\psi^*_{\bar n'-\frac n2}\psi_{\bar n'+\frac n2}\\ &= \sum_{\frac{n_{\max}}{2}>n>0} \Big(\sum_{\bar n>0}\sum_{\bar n'>0}\Big)' \psi^*_{\bar n+\frac n2}\psi_{\bar n-\frac n2}\psi^*_{\bar n'-\frac n2}\psi_{\bar n'+\frac n2}\\ &\quad + \sum_{\frac{n_{\max}}{2}>n>0}\ \sum_{\bar n>0} (1-N_{\bar n-\frac n2})N_{\bar n+\frac n2}. \end{aligned} \tag{4.1}$$

The prime on $(\sum\sum)'$ means that terms with $\bar n=\bar n'$ are omitted. By this omission the first sum on the right-hand side of (4.1) gives the non-diagonal part and the second sum the diagonal part of $\sum \rho^+_{-n}\rho^+_n$. The summation over n is to be extended only up to $n_{\max}/2$ because of our condition (2.15) (III).

We first observe the diagonal part

$$D^+ = \sum_{\frac{n_{\max}}{2}>n>0}\ \sum_{\bar n>0} (1-N_{\bar n-\frac n2})N_{\bar n+\frac n2}. \tag{4.2}$$

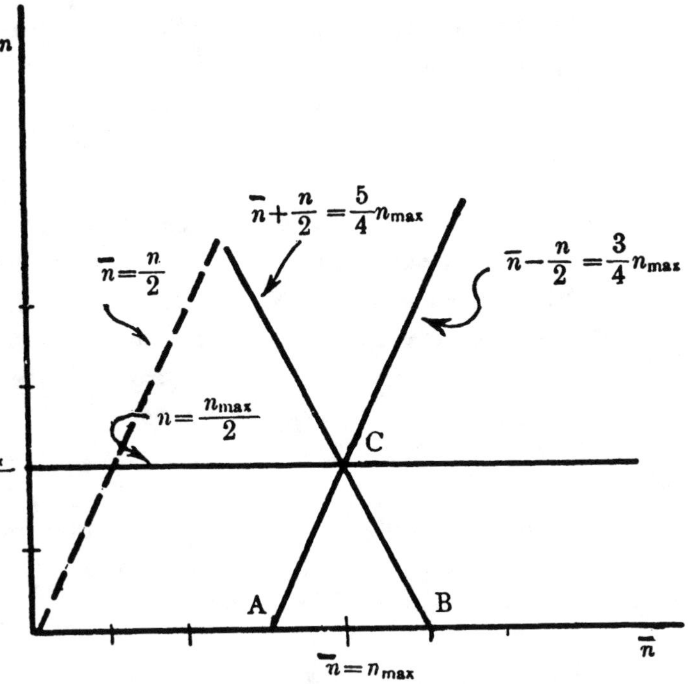

Fig. 1.

We make use of the fact that, because of the first two conditions of (2.15), the summand $(1-N_{\bar n-\frac n2})N_{\bar n+\frac n2}$ has a non-vanishing value only at points $(n, \bar n)$ lying inside the triangle ABC (see Fig. 1) enclosed by the three straight lines

$$\begin{cases} \bar{n}+\dfrac{n}{2}=\dfrac{5}{4}n_{\max} \\ \bar{n}-\dfrac{n}{2}=\dfrac{3}{4}n_{\max} \\ n=0\ . \end{cases}$$

Since the summand vanishes outside this triangle, two of the boundaries of the domain of summation

$$n=\frac{n_{\max}}{2} \qquad \text{and} \qquad \bar{n}=0 \tag{4.3}$$

can be changed at our will so long as the changed boundaries lie outside the triangle. So we may use the boundaries

$$n=\infty \qquad \text{and} \qquad \bar{n}=\frac{n}{2} \tag{4.3'}$$

instead of the original ones. Then we have

$$D^{+}=\sum_{\infty>n>0}\ \sum_{\bar{n}\geq\frac{n}{2}}(1-N_{\bar{n}-\frac{n}{2}})N_{\bar{n}+\frac{n}{2}}$$

$$=\sum_{n>0}\ \sum_{\bar{n}'\geq n}(1-N_{\bar{n}'-n})N_{\bar{n}'}. \tag{4.4}$$

We now notice that

$$\sum_{n>0}\ \sum_{\bar{n}'\geq n}N_{\bar{n}'}=\sum_{\bar{n}'\geq 0}\bar{n}'N_{\bar{n}'} \tag{4.5}$$

and

$$\sum_{n>0}\ \sum_{\bar{n}'\geq n}N_{\bar{n}'-n}N_{\bar{n}'}=\frac{1}{2}(\sum_{\bar{n}\geq 0}N_{\bar{n}})^{2}-\frac{1}{2}\sum_{\bar{n}\geq 0}N_{\bar{n}}$$

$$=\frac{1}{2}n_{\max}(n_{\max}+1). \tag{4.5'}$$

We then get from (4.4)

$$D^{+}=\sum_{n\geq 0}nN_{n}-\frac{1}{2}n_{\max}(n_{\max}+1). \tag{4.6}$$

In the same way we get the corresponding result for $\sum\rho_{\bar{n}}\rho_{-\bar{n}}$

$$D^{-}=\sum_{n\leq 0}|n|N_{n}-\frac{1}{2}n_{\max}(n_{\max}+1). \tag{4.7}$$

We now go over to the non-diagonal part:

$$C^{+}=\sum_{\frac{n_{\max}}{2}>n>0}(\sum_{\bar{n}>0}\ \sum_{\bar{n}'>0})'\ \psi^{*}_{\bar{n}+\frac{n}{2}}\psi_{\bar{n}-\frac{n}{2}}\psi^{*}_{\bar{n}'-\frac{n}{2}}\psi_{\bar{n}'+\frac{n}{2}}, \tag{4.8}$$

and show that C^{+} vanishes.

By writing $-n$ instead of n, $\bar{n}'$ instead of $\bar{n}$ and $\bar{n}$ instead of $\bar{n}'$, we see that C^+ can be written also in the following form:

$$C^+ = \sum_{-\frac{n_{max}}{2}<n<0} \Big(\sum_{\bar{n}>0} \sum_{\bar{n}'>0}\Big)' \psi^*_{\bar{n}+\frac{n}{2}} \psi_{\bar{n}-\frac{n}{2}} \psi^*_{\bar{n}'-\frac{n}{2}} \psi_{\bar{n}'+\frac{n}{2}}, \tag{4.8$'$}$$

where we have used the fact that $\psi^*_{\bar{n}+\frac{n}{2}} \psi_{\bar{n}-\frac{n}{2}}$ commutes with $\psi^*_{\bar{n}'-\frac{n}{2}} \psi_{\bar{n}'+\frac{n}{2}}$ since $\bar{n} \neq \bar{n}'$. From (4.8) and (4.8′) we get

$$C^+ = \frac{1}{2} \sum'_{-\frac{n_{max}}{2}<n<\frac{n_{max}}{2}} \Big(\sum_{\bar{n}>0} \sum_{\bar{n}'>0}\Big)' \psi^*_{\bar{n}+\frac{n}{2}} \psi_{\bar{n}-\frac{n}{2}} \psi^*_{\bar{n}'-\frac{n}{2}} \psi_{\bar{n}'+\frac{n}{2}}. \tag{4.8$''$}$$

Now a considerasion similar to the above one shows that, because of the condition (2.15), we can change the boundaries of summation

$$n = \pm\frac{n_{max}}{2}, \quad \bar{n}=0 \quad \text{and} \quad \bar{n}'=0 \tag{4.9}$$

into new ones

$$n = \pm\infty, \qquad \bar{n} = -\frac{n}{2}, \qquad \bar{n}' = \frac{n}{2}. \tag{4.9$'$}$$

We then get

$$C^+ = \frac{1}{2} \sum'_{-\infty<n<\infty} \Big(\sum_{\bar{n}\geq-\frac{n}{2}} \sum_{\bar{n}'\geq\frac{n}{2}}\Big)' \psi^*_{\bar{n}+\frac{n}{2}} \psi_{\bar{n}-\frac{n}{2}} \psi^*_{\bar{n}'-\frac{n}{2}} \psi_{\bar{n}'+\frac{n}{2}}; \tag{4.10}$$

but this C^+ vanishes as is shown in the following manner.

By rearranging factors in (4.10) we get

$$C^+ = -\frac{1}{2} \sum'_{-\infty<n<\infty} \Big(\sum_{\bar{n}\geq-\frac{n}{2}} \sum_{\bar{n}'\geq\frac{n}{2}}\Big)' \psi^*_{\bar{n}'-\frac{n}{2}} \psi_{\bar{n}-\frac{n}{2}} \psi^*_{\bar{n}+\frac{n}{2}} \psi_{\bar{n}'+\frac{n}{2}}. \tag{4.10$'$}$$

If we put here

$$\begin{cases} l = \bar{n}' - \bar{n} \\ \bar{l}' = \frac{1}{2}(\bar{n}' + \bar{n} + n) \\ \bar{l} = \frac{1}{2}(\bar{n}' + \bar{n} - n), \end{cases}$$

we can write (4.10′) in the form:

$$C^+ = -\frac{1}{2} \sum'_{-\infty<l<\infty} \Big(\sum_{\bar{l}\geq-\frac{l}{2}} \sum_{\bar{l}'\geq\frac{l}{2}}\Big)' \psi^*_{\bar{l}+\frac{l}{2}} \psi_{\bar{l}-\frac{l}{2}} \psi^*_{\bar{l}'-\frac{l}{2}} \psi_{\bar{l}'+\frac{l}{2}}. \tag{4.11}$$

That the domains of summation over l, $\bar{l}$ and $\bar{l}'$ are really as specified by the inequalities under the summation symbols can be easily shown. One has only to notice the identities

$$\begin{cases} \bar{n}+\dfrac{n}{2}=\bar{l}'-\dfrac{l}{2} \\ \bar{n}'-\dfrac{n}{2}=\bar{l}+\dfrac{l}{2}. \end{cases}$$

Comparing (4.11) with (4.10) we find $C^+=-C^+$, which means that

$$C^+=0. \tag{4.12}$$

A similar consideration applies to $\sum \rho_n^- \rho_{-n}^-$. We find that the non-diagonal part of $\sum \rho_n^- \rho_{-n}^-$ vanishes:

$$C^-=0. \tag{4.13}$$

This rather lengthy proof that $C^+=C^-=0$ can be summarized in the following way. Each term in C^+ or C^- corresponds to a simultaneous transition of two particles. If there is a term corresponding to a transition $a \to b,\ c \to d$, there is always a term corresponding to the transition $a \to d,\ c \to b$. Because of the Fermi statistics, the latter term has a sign which is opposite to the former so that they cancel each other when added together.

Substituting (4.6), (4.7), (4.12) and (4.13) into $\mathfrak{H}$, we obtain

$$\mathfrak{H}=\left(\frac{2\pi\hbar}{L}\right)^2 \frac{n_{\max}}{m}\{\sum_n |n|\, N_n - n_{\max}(n_{\max}+1)\}$$

$$=\left(\frac{2\pi\hbar}{L}\right)^2 \frac{n_{\max}}{m}\{\sum_{|n|\leqq n_{\max}} |n|\,(N_n-1)+\sum_{|n|>n_{\max}} |n|\, N_n\}. \tag{4.14}$$

On the other hand the kinetic energy of the system is evidently

$$H_{\text{kin}}=\left(\frac{2\pi\hbar}{L}\right)^2 \frac{1}{2m} \sum_n n^2 N_n. \tag{4.15}$$

In the state of perfect degeneracy, i. e., in the state where all levels between $-n_{\max}$ and $n_{\max}$ are occupied and all other levels are empty, this energy becomes

$$H_0=\left(\frac{2\pi\hbar}{L}\right)^2 \frac{1}{2m} \sum_{|n|\leqq n_{\max}} n^2. \tag{4.16}$$

We shall now show that, in the same approximation, which is allowed by our condition (2.15), our Hamiltonian $\mathfrak{H}$ is equal to the energy (4.15) minus the constant energy (4.16) : $\mathfrak{H}=H_{\text{kin}}-H_0$.

We calculate $H_{\text{kin}}-H_0$ in the following way: From (4.15) and (4.16) we get

$$H_{\text{kin}}-H_0=\left(\frac{2\pi\hbar}{L}\right)^2 \frac{1}{2m}\{\sum_{|n|\leqq n_{\max}} n^2(N_n-1)+\sum_{|n|>n_{\max}} n^2 N_n\}. \tag{4.17}$$

Making use of the assumption that holes and excited particles are present only in the neighborhood of $\pm n_{\max}$, we see that N_n-1 and N_n differ from zero only when n lies in the neighborhood of $\pm n_{\max}$. This allows us to use instead of n^2 its expansion:

$$\begin{aligned} n^2 &= \{\pm n_{\max} + (n \mp n_{\max})\}^2 \\ &\approx n_{\max}^2 \pm 2n_{\max}(n \mp n_{\max}) \\ &= -n_{\max}^2 + 2n_{\max}|n|. \end{aligned}$$

Substituting this in (4.17), we find

$$H_{\text{kin}} - H_0 = -\left(\frac{2\pi\hbar}{L}\right)^2 \frac{n_{\max}^2}{m}\{N - (2n_{\max}+1)\} + \mathfrak{H}. \tag{4.17'}$$

Then the use of (2.13) gives immediately the required result:

$$H_{\text{kin}} - H_0 = \mathfrak{H}. \tag{4.18}$$

This result means that our Hamiltonian $\mathfrak{H}$ is really the kinetic energy of the assembly minus a constant which depends only on the total number of particles. This constant is the energy of the assembly in the case of perfect degeneracy and thus $\mathfrak{H}$ can be interpreted as the deviation of the kinetic energy from this standard value. We express this fact by simply calling $\mathfrak{H}$ " the excitation kinetic energy ".

The total energy of the system is evidently

$$H = \mathfrak{H} + H_{\text{int}} + H_0 \tag{4.19}$$

when there are inter-particle forces.

The relations (4.6) and (4.7) can be used to express the momentum of the assembly in terms of ρ. Let the momentum be denoted by G. Then

$$G = \left(\frac{2\pi\hbar}{L}\right) \sum_{n>0} (\rho_{-n}^{+}\rho_{n}^{+} - \rho_{n}^{-}\rho_{-n}^{-}). \tag{4.20}$$

The expression for G applies irrespective of the presence or absence of the inter-particle force.

One sees that for the proof of the relation (4.18) the possibility of change of domains of summation, the replacement of (4.3) by (4.3′) and the replacement of (4.9) by (4.9′), were essential. These are possible only when the three conditions in (2.15) are simultaneously satisfied. This is the reason why we had to choose the special value of $a=3/4$. It is easy to see that another choice of a would make it impossible to change the summation domains so that we would not get the relation (4.18).

§ 5. Solution of the eigen-value problem.

We now go over to the solution of the eigen-value problem of our Hamiltonian. As mentioned before, this can be done by performing the principal-axes transformation of the bilinear form representing the Hamiltonian.

Adding (3.8), (3.13) and H_0, the total Hamiltonian is obtained:

$$H=\mathfrak{H}+H_{\rm int}+H_0$$
$$=\sum_{n>0}\left\{\left(\frac{2\pi\hbar}{L}\right)^2\frac{n_{\rm max}}{m}+J_n\right\}(\rho_{-n}^+\rho_n^+ +\rho_n^-\rho_{-n}^-)$$
$$+\sum_{n>0}J_n(\rho_n^+\rho_{-n}^- +\rho_n^-\rho_{-n}^+)$$
$$+\sum_{n>0}nJ_n+\frac{1}{2}N^2J_0-\frac{1}{2}NJ(0)+H_0. \tag{5.1}$$

We introduce real coordinates Q_n and real momenta P_n by means of

$$\begin{cases}\rho_n^+ =\sqrt{\frac{n}{2}}(Q_n+iP_n)\\ \rho_{-n}^+=\sqrt{\frac{n}{2}}(Q_n-iP_n)\\ \rho_n^- =\sqrt{\frac{n}{2}}(Q_{-n}-iP_{-n})\\ \rho_{-n}^-=\sqrt{\frac{n}{2}}(Q_{-n}+iP_{-n}),\end{cases}$$

where the suffix n is considered positive. Then in terms of P's and Q's the Hamiltonian is expressed as

$$H=\sum_n|n|\left\{\left(\frac{2\pi\hbar}{L}\right)^2\frac{n_{\rm max}}{m}+J_n\right\}\left(\frac{1}{2}P_n^2+\frac{1}{2}Q_n^2-\frac{1}{2}\right)$$
$$+\frac{1}{2}\sum_n|n|J_n(Q_nQ_{-n}-P_nP_{-n})$$
$$+\frac{1}{2}\sum_n|n|J_n+\frac{1}{2}N^2J_0-\frac{1}{2}NJ(0)+H_0. \tag{5.3}$$

In this expression the suffix n takes both positive and negative values and the summation $\sum_n$ is extended over positive as well as negative valus of n.

The commutation relations for the P's and Q's are

$$\begin{cases}[Q_n, P_{n'}]=i\,\delta_{n,n'}\\ [Q_n, Q_{n'}]=[P_n, P_{n'}]=0.\end{cases} \tag{5.4}$$

The transformation of H into principal form can be performed by defining new variables q_n and p_n by means of the canonical transformation

$$P_n=\frac{1}{2}\left\{\left(\frac{T_n+2U_n}{T_n}\right)^{1/4}+\left(\frac{T_n}{T_n+2U_n}\right)^{1/4}\right\}p_n+\frac{1}{2}\left\{\left(\frac{T_n+2U_n}{T_n}\right)^{1/4}-\left(\frac{T_n}{T_n+2U_n}\right)^{1/4}\right\}p_{-n}$$
$$Q_n=\frac{1}{2}\left\{\left(\frac{T_n+2U_n}{T_n}\right)^{1/4}+\left(\frac{T_n}{T_n+2U_n}\right)^{1/4}\right\}q_n-\frac{1}{2}\left\{\left(\frac{T_n+2U_n}{T_n}\right)^{1/4}-\left(\frac{T_n}{T_n+2U_n}\right)^{1/4}\right\}q_{-n} \tag{5.5}$$

with the abbreviations

$$\begin{cases} T_n=\left(\dfrac{2\pi\hbar}{L}\right)^2\dfrac{n_{\max}}{m}|n|, \\ U_n=|n|J_n\,. \end{cases} \tag{5.6}$$

It can be easily seen that the p_n and q_n satisfy

$$\begin{cases} [q_n, p_{n'}]=i\delta_{n,n'}\,, \\ [q_n, q_{n'}]=[p_n, p_{n'}]=0, \end{cases} \tag{5.7}$$

and that H is transformed into

$$H=\sum_n \sqrt{T_n}\left\{\sqrt{T_n+2U_n}\left(\frac{1}{2}p_n^2+\frac{1}{2}q_n^2-\frac{1}{2}\right)+\frac{1}{2}\left(\sqrt{T_n+2U_n}-\sqrt{T_n}\right)\right\}$$

$$+\frac{1}{2}N^2J_0-\frac{1}{2}NJ(0)+H_0\,. \tag{5.8}$$

The momentum G, expressed in terms of q's and p's, is

$$G=\left(\frac{2\pi\hbar}{L}\right)\sum_n n\left(\frac{1}{2}p_n^2+\frac{1}{2}q_n^2-\frac{1}{2}\right). \tag{5.9}$$

The expressions (5.8) and (5.9) show that our system is in fact equivalent to an assembly of uncoupled harmonic oscillators, or, better, to an assembly of uncoupled sound quanta. Each sound quantum has momentum and energy given by

$$\text{momentum}=\left(\frac{2\pi\hbar}{L}\right)n \qquad n=0,\ \pm 1,\ \pm 2,\cdots \tag{5.10}$$

and

$$\text{energy}=\sqrt{T_n(T_n+2U_n)}$$

$$=\left[\left(\frac{2\pi\hbar}{L}\right)^2\frac{n_{\max}}{m}\left\{\left(\frac{2\pi\hbar}{L}\right)^2\frac{n_{\max}}{m}+2J_n\right\}\right]^{1/2}|n|, \tag{5.10'}$$

and the number of sound quanta with the momentum $\left(\frac{2\pi\hbar}{L}\right)n$ is represented by the operator

$$M_n=\frac{1}{2}p_n^2+\frac{1}{2}q_n^2-\frac{1}{2}. \tag{5.11}$$

In terms of M_n the energy and the momentum are

$$H=\sum_n \sqrt{T_n}\left\{\sqrt{T_n+2U_n}\,M_n+\frac{1}{2}\left(\sqrt{T_n+2U_n}-\sqrt{T_n}\right)\right\} \tag{5.8'}$$

$$+\frac{1}{2}N^2J_0-\frac{1}{2}NJ(0)+H_0\,,$$

and

$$G=\left(\frac{2\pi\hbar}{L}\right)\sum_n n\, M_n \tag{5.9'}$$

respectively.

In the wave language the relations (5.10) and (5.11') state that the sound wave with the wave length

$$\lambda_n=\frac{L}{|n|} \tag{5.12}$$

has the frequency

$$\nu_n=\frac{1}{2\pi}\left[\left(\frac{2\pi}{L}\right)^2\frac{n_{\max}}{m}\left\{\left(\frac{2\pi\hbar}{L}\right)^2\frac{n_{\max}}{m}+2J_n\right\}\right]^{1/2}|n|\,. \tag{5.12'}$$

The phase velocity of the sound wave is

$$v_n=\left[\left(\frac{2\pi\hbar}{L}\right)^2\frac{n_{\max}}{m}\left\{\left(\frac{2\pi\hbar}{L}\right)^2\frac{n_{\max}}{m}+2J_n\right\}\right]^{1/2}. \tag{5.13}$$

The phase velocity given by (5.13) is dependent on n and hence on λ_n when J_n does not vanish. This means that, when the inter-particle force is present, our Fermi gas is dispersive. (5.13) shows further that the phase velocity of the sound waves increases or decreases from its value in the ideal gas according as the inter-particle force is repulsive or attractive.

The formula (5.8') enables us immediately to calculate the energy in various stationaly states. In the lowest state, for instance, where there are no sound quanta, we get

$$E=\frac{1}{2}\sum_n\sqrt{T_n}\left(\sqrt{T_n+2U_n}-\sqrt{T_n}\right)$$
$$+\frac{1}{2}N^2J_0-\frac{1}{2}NJ(0)+H_0\,. \tag{5.14}$$

In the case of an attractive inter-particle force it may happen that E, considered as a function of L, has a minimum for some value of L. Then this will mean that the assembly is capable of forming a stable aggregate.

In the case where J_n is very small, we can expand $\sqrt{T_n+2U_n}$ and thus express E as a power series in J_n. We shall then find that the term linear in J_n together with the term H_0 gives the energy in the usual Tomas-Fermi ap-approximation including the effect of exchange. The energy quadratic in J_n corresponds to the energy obtained by Euler as the second order term of a perturbation calculation.

§ 6. Correlation between particles. Transitions caused by an external perturbing force. Criterion for applicability of the method.

In this section we shall mention some general results of physical interest which can be obtained directly by our theory. We shall give also a criterion for the applicability of our method.

(I) Correlation of position of particles.

As is known there is no correlation between particles in an ideal gas besides that due to the exclusion principle. The interaction between particles causes an extra correlation, which can be calculated very simply.

Mathematically the correlation of position can be exppessed by the quantity

$$C(\xi)=\frac{1}{N^2}\left\langle\int\rho(x)\rho(x+\xi)dx\right\rangle_{Av}, \tag{6.1}$$

$C(\xi)d\xi$ giving the probability of finding another particle at a distance between ξ and $\xi+d\xi$ from a particle. This correlation function can be calculated in the following manner.

Expressing (6.1) in terms of the Fourier transform of $\rho(x)$, we have

$$C(\xi)=\frac{1}{L}+\frac{1}{2N^2L}\sum_{n>0}\langle\rho_n\rho_{-n}\rangle_{Av}\cos\left(\frac{2\pi n}{L}\xi\right), \tag{6.2}$$

so that our task is to find $\langle\rho_n\rho_{-n}\rangle_{Av}$. In terms of the q's and p's defined by (5.2) and (5.5) the operator $\rho_n\rho_{-n}$ is

$$\rho_n\rho_{-n}=2n\sqrt{\frac{T_n}{T_n+2U_n}}\left\{\frac{1}{2}(p_n^2+q_n^2)+\frac{1}{2}(p_{-n}^2+q_{-n}^2)+q_nq_{-n}-p_np_{-n}\right\}. \tag{6.3}$$

It then follows that

$$\langle\rho_n\rho_{-n}\rangle_{Av}=2n\sqrt{\frac{T_n}{T_n+2U_n}}(M_n+M_{-n}+1), \tag{6.4}$$

which gives

$$C(\xi)=\frac{1}{L}+\frac{1}{N^2L}\sum_n\left(M_n+\frac{1}{2}\right)\sqrt{\frac{T_n}{T_n+2U_n}}|n|\cos\left(\frac{2\pi n}{L}\xi\right). \tag{6.5}$$

The correlation function $C(\xi)$ consists of two parts. The one is the part expressing the correlation which already exists in the absence of the inter-particle force. This part is due solely to the exclusion principle. This part of the correlation can be obtained by putting $U_n=0$ in (6.5). Denoting this part of $C(\xi)$ by $C_0(\xi)$, we have

$$C_0(\xi)=\frac{1}{L}+\frac{1}{N^2L}\sum_n\left(M_n+\frac{1}{2}\right)|n|\cos\left(\frac{2\pi n}{L}\xi\right). \tag{6.6}$$

The second part of $C(\xi)$ is that part of the correlation which is purely due to inter-particle forces. Denoting this second part by $C_1(\xi)$, we get

$$C_1(\xi)=\frac{1}{N^2L}\sum_n\left(M_n+\frac{1}{2}\right)\left(\sqrt{\frac{T_n}{T_n+2U_n}}-1\right)|n|\cos\left(\frac{2\pi n}{L}\xi\right). \tag{6.7}$$

(6.7) can also be written as

$$C_1(\xi)=\frac{\partial}{\partial\xi}\left\{\frac{1}{2\pi N^2}\sum_n\left(M_n+\frac{1}{2}\right)\left(\sqrt{\frac{T_n}{T_n+2U_n}}-1\right)\sin\left(\frac{2\pi|n|}{L}\xi\right)\right\}. \tag{6.7'}$$

Then the quantity

$$D_1(\xi)=\frac{1}{2\pi N^2}\sum_n\left(M_n+\frac{1}{2}\right)\left(\sqrt{\frac{T_n}{T_n+2U_n}}-1\right)\sin\left(\frac{2\pi|n|}{L}\xi\right) \tag{6.7''}$$

can be interpreted as the change of probability due to the inter-particle force of finding another particle within the distance ξ from a particle.

(II) Matrix element of an external perturbing force.

Suppose an external perturbing force is impressed upon the system. Then the perturbation will cause transitions of the system. As is well known, the transition probabilities depend essentially on the matrix elements of the perturbing energy. Now, it is very simple to calculate these matrix elements.

Let the potential of the perturbing force be $V(x)$. Then the perturbation energy is

$$\begin{aligned}H'&=\int\rho(x)V(x)\,dx\\&=\sum_{n>0}\{V_n(\rho_n^++\rho_n^-)+V_{-n}(\rho_{-n}^++\rho_{-n}^-)\},\end{aligned} \tag{6.8}$$

where V_n and V_{-n} are the matrix elements of $V(x)$ defined by

$$\left\{\begin{aligned}&V_n=\frac{1}{L}\int V(x)\,\exp\left(\frac{2\pi i}{L}nx\right)dx\\&V_{-n}=V_n^*.\end{aligned}\right. \tag{6.9}$$

In terms of the q's and p's it is found that

$$\begin{aligned}H'=\sum_{n>0}\left(\frac{T_n}{T_n+2U_n}\right)^{1/4}&\left[\sqrt{\frac{n}{2}}V_n\{(q_n+ip_n)+(q_{-n}-ip_{-n})\}\right.\\&\left.+\sqrt{\frac{n}{2}}V_{-n}\{(q_{-n}+ip_{-n})+(q_n-ip_n)\}\right].\end{aligned} \tag{6.10}$$

Now, the matrix elements of $q_{\pm n}+ip_{\pm n}$ as well as $q_{\pm n}-ip_{\pm n}$ are all known. They are the matrix elements for harmonic oscillators. Our result (6.10) shows that the matrix elements of H' are given simply by multiplying these universal matrix elements by the factor $(T_n/T_n+2U_n)^{1/4}\sqrt{n/2}\,V_{\pm n}$. When the inter-particle force is absent, the factors are simply $\sqrt{n/2}\,V_{\pm n}$. Our result shows that the effect of the inter-particle force is simply to replace $\sqrt{n/2}\,V_{\pm n}$ by $(T_n+2U_n)^{1/4}\times\sqrt{n/2}\,V_{\pm n}$. The effect of the inter-particle force can thus be dealt with as an apparent change of the perturbing potential by the factor $(T_n/T_n+2U_n)^{1/4}$, the effective potential being expressed as

$$V_{\pm n}^{\text{eff}}=\left(\frac{T_n}{T_n+2U_n}\right)^{1/4}V_{\pm n}. \tag{6.11}$$

The factor $\left(\frac{T_n}{T_n+2U_n}\right)^{1/4}$ is larger or smaller than unity according as the inter-particle force is attractive or repulsive. This means that in the case of an attractive inter-particle force the effect of an external perturbing force becomes larger, and in the case of a repulsive one the perturbing force becomes less effective as compared with the case of an ideal gas.

(III) Criterion for the applicability of the method.

Because our theory is based on the fundamental conditions of (2.15), it is necessary to see under what circumstances we can be sure that these conditions are satisfied, at least approximately.

It is first clear that the range of the inter-particle force should not be too short in order that the third condition of (2.15) be satisfied. This condition requires that sound waves with wave lengths shorter than $2L/n_{\max}$ play no role. Let us fist consider the lowest state. Then this condition requires that in the sum $\sum\limits_n \sqrt{T_n}\,(\sqrt{T_n+2U_n}-\sqrt{T_n})$ of (5.14) no contribution should arise from terms with $n>n_{\max}/2$. This requires that J_n is negligibly small for $n>n_{\max}/2$. Thus, roughly speaking, the range of the force should be longer than $2L/n_{\max}$ or larger than four times the mean distance of particles.

In the excited states we have to impose a further condition: There should not be any sound quanta with wave length shorter than $2L/n_{\max}$.

In order to see whether or not the first two conditions of (2.15) are satisfied, it is necessary to examine the eigen function which is obtained and see how the holes and excited particles are distributed among the levels. If this eigen function is such that the probability distribution of holes and excited particles is essentially limited between the levels $(3/4)n_{\max}$ and $(5/4)n_{\max}$, the solution obtained will be a good approximation. There is a rough but simple criterion to see under what circumstances such will be the case.

It is to calculate the mean excitation kinetic energy per particle. If this energy is too large, we must expect that there will be a large number of highly excited particles and many holes lying deep in the Fermi sea. If this energy is small, we are sure that the excited particles and holes are present only in the neighborhood of the Fermi maximum.

The excitation kinetic energy per particle is given by

$$\frac{1}{N}\langle\mathfrak{H}\rangle_{Av}=\frac{1}{2n_{\max}+1}\left(\frac{2\pi\hbar}{L}\right)^2\frac{n_{\max}}{m}\sum_{n>0}\langle\rho_{-n}^{+}\rho_n^{+}+\rho_n^{-}\rho_{-n}^{-}\rangle_{Av}, \tag{6.12}$$

where we have

$$\rho_{-n}^{+}\rho_{n}^{+}+\rho_{n}^{-}\rho_{-n}^{-}=\frac{n}{2}\Big[\Big(\sqrt{\frac{T_n+2U_n}{T_n}}+\sqrt{\frac{T_n}{T_n+2U_n}}\Big)\Big\{\frac{1}{2}(p_n^2+q_n^2)+\frac{1}{2}(p_{-n}^2+q_{-u}^2)\Big\}$$
$$+\Big(\sqrt{\frac{T_n+2U_n}{T_n}}-\sqrt{\frac{T_n}{T_n+2U_n}}\Big)(p_n p_{-n}-q_n q_{-n})-2\Big].$$

Taking the expectation value, we then get

$$\frac{1}{N}\langle\mathfrak{H}\rangle_{Av}=\Big(\frac{2\pi\hbar}{L}\Big)^2\frac{1}{2m}\sum_n\Big\{\Big(\sqrt{\frac{T_n+2U_n}{T_n}}+\sqrt{\frac{T_n}{T_n+2U_n}}\Big)M_n$$
$$+\frac{1}{2}\Big(\sqrt{\frac{T_n+2U_n}{T_n}}+\sqrt{\frac{T_n}{T_n+2U_n}}-2\Big)\Big\}\frac{|n|}{2}; \tag{6.13}$$

in particular, for the lowest state

$$\frac{1}{N}\langle\mathfrak{H}\rangle_{Av}=\Big(\frac{2\pi\hbar}{L}\Big)^2\frac{1}{8m}\sum_n\Big(\sqrt{\frac{T_n+2U_n}{T_n}}+\sqrt{\frac{T_n}{T_n+2U_n}}-2\Big)|n|. \tag{6.13'}$$

We must now determine the value of this mean excitation energy below which our method works. In order to find this critical value, we consider a special state in which all particles in the levels between $(3/4)n_{\max}$ and $n_{\max}$ are raised to the levels between $n_{\max}$ and $(5/4)n_{\max}$. In this state the mean excitation energy per particle is $\Big(\frac{2\pi\hbar}{L}\Big)^2\frac{1}{m}\frac{n_{\max}^2}{4^2}$. If the mean excitation energy is larger than this amount in some state, it is certain that in this state some particles are excited to levels higher than $(5/4)n_{\max}$ or some holes are present in levels below $(3/4)n_{\max}$. So our method cannot work consistently in such a case. We find in this way that the necessary condition for the applicability of our method is

$$\frac{1}{N}\langle\mathfrak{H}\rangle_{Av}<\Big(\frac{2\pi\hbar}{L}\Big)^2\frac{1}{m}\frac{n_{\max}^2}{16}. \tag{6.14}$$

In the lowest state (6.14) is

$$\sum_n\Big(\sqrt{\frac{T_n+2U_n}{T_n}}+\sqrt{\frac{T_n}{T_n+2U_n}}-2\Big)|n|<\frac{n_{\max}^2}{2}. \tag{6.15}$$

Of course the condition (6.14) or (6.15) is only necessary but not sufficient. There may be cases in which this condition is satisfied and still our method does not work. But such circumstances will occur rather exceptionally.

The condition (6.15) is certainly satisfied if the inter-particle force is so small that we can expand the square root of the form $\sqrt{T_n+2U_n}$ as a power series in J_n and use only a few terms. This condition is just the one required by the validity of the perturbation theory. But the condition (6.15) is weaker than the condition arising in the perturbation theory, because (6.15) does not require that J_n be uniformly small for all values of n. It requires only the smallness of the sum as a whole, and therefore our method has a wider domain of applicability than the perturbation theory. Because of the factor n outside the bracket, terms with n small do not contribute much to the sum on the left-hand side of (6.15), so that J_n with small n may even be very large, provided that T_n+2U_n does not become negative. This fact means that our method is especially suitable for long-range repulsive interactions. As has been first noticed and worked out by Bohm, the method of sound waves provides a very promising means for the study of an assembly of electrons interacting with each other through a repulsive Coulomb force.

§ 7. Concluding remarks.

In concluding the paper, we shall give a few disconnected remarks.

(I) Plasma-like oscillations.

In order to relate the results of our considerations to the work of Bohm, we shall mention briefly the bearing of our results on plasma-like oscillations of a degenerate electron gas. In this case the inter-particle force is the Coulomb repulsion. By Coulomb force is meant here the fact that J_n is proportional to to $1/n^2$. (Notice that the Poisson equation for a point source is $n^2 J_n=$const. when expressed in terms of the Fourier transform of the potential). According to (5.12′) we find that in this case V_n is independent of n for small n, i.e. for n so small that we can neglect $\left(\frac{2\hbar\pi}{L}\right)^2\frac{n_{\max}}{m}$ as compared with $2J_n=\text{const}/n^2$. This means that the frequency of the wave is independent of the wave length, which is characteristic of plasma oscillations.

Another remarkable fact which occurs in the case of Coulomb repulsion is the following. According to (6.11) we find that the matrix elements of the external perturbing force become very small for small n. This means that the plasma oscillations are disturbed very little by external perturbing forces. This fact has been anticipated by Bohm in his paper about superconductivity.[7]

(II) Relation between the two kinds of descriptions.

As we have seen, the system can be described either as an assembly of Fermi particles, or as an assembly of sound quanta. Then the question arises: What relation will exist between the states, in one of which the occupation numbers of the Fermi particles have some specified values, and in the other of which the occupation numbers of the sound quanta have some specified values.[8] As the operator $\frac{1}{2}(p_n^2+q_n^2)-\frac{1}{2}$, which represents the number of sound quanta, and the operator $\psi_n^*\psi_n$, which represents the number of Fermi particles, do not commute with each other, it is clear that one cannot assign numerical values simultaneously to both kinds of occupation numbers. In other words, a state in which the occupation numbers of sound quanta have some definite values is a complicated superposition of various states, each of which is specified by a different set of which is specified by a different set of values of occupation numbers of Fermi particles, and *vice versa.* The statistical relation between the two kinds of occupation numbers will be obtained if one can determine the amplitude of each state in this superposition. We shall describe here briefly the general prescription to get this statistical relation.

First consider the case of a system of non-interacting particles. In this case, we can use $\frac{1}{2}(P_n^2+Q_n^2)-\frac{1}{2}$ instead of $\frac{1}{2}(p_n^2+q_n^2)-\frac{1}{2}$. The operator $\mathfrak{H}$, representing the energy of the system, is a function of the number of sound quanta. At the same time, it is a function of the number of the Fermi particles too. This means that a state in which the occupation numbers of sound quanta have some specified values, and a states in which the occupation numbers of Fermi particles have some specified values, are both some eigen-states of the energy operator. This fact results in all states in the superposition mentioned above belonging to the same eigenvalue of the energy. Now, because the number of linearly independent states belonging to one eigenvalue of $\mathfrak{H}$ is finite, one has only to solve a secular equation of finite degree in order to determine the coefficients in the superposition under consideration. So it is always possible, at least in principle, to answer the question about the statistical relation between the two kinds of occupation numbers.

In the case of interacting particles, we must first express the state in which the occupation numbers of the sound quanta have definite values as a superposition of various states, each having some definite set of values of $\frac{1}{2}(P_n^2+Q_n^2)-\frac{1}{2}$. The problem is none other than to expand Hermite functions of q's in terms of other Hermite functions of Q's. We then apply the method mentioned above to each term of this superposition. In this way we find the required statistical relation for the general case.

In a stationary state the occupation numbers of the sound quanta have some definite values, but the occupation numbers of the Fermi particles do not. The stationary state is a very complicated superposition of states with various numbers of holes and excited particles at various levels. This means that the stationary state is very far from a state where only one particle is excited to some higher level. This fact corresponds to Bohr's statement[9] that the one-particle model is a very bad approximation for the stationary states of such an assembly.

(III) A possible application to the case of exchange forces.

Though our method is applicable only to the case of ordinary forces, there seems to be a hope of applying it also to the case of exchange forces. It is to combine it with Ritz's procedure using a trial function whose form is suggested by our method when applied to ordinary forces. This procedure means physically to replace the exchange force by an equivalent ordinary force in such a way as to gives the best approximation.

Unfortunately, the mathematical structure of the method when generalized to the three-dimensional case is not simple and the author has not yet accomplished it. So in this paper we must be content only to present considerations of a rather mathematical nature without entering into real physical problems.

Acknowledgement.

I would like to express my sincere thanks to Professor J. R. Oppenheimer for giving me the opportunity of working at the Institure for Advanced Study and for his kind interest in this work. I am also indebted to Professor D. Bohm for telling me his interesting results before publication and for fruitful conversation.

References

1) H. Euler, ZS. Phys. **105** (1937), 553.
2) M. Nogami, Prog. Theor. Phys. **5** (1950), 65.
3) F. Bloch, ZS. Phys. **81** (1933), 363.
4) F. Bloch, Helv. Phys. Acta **7** (1934), 385.
5) D. Bohm, yet unpublished.
6) P. Jordan, ZS. Phys. **93** (1935), 464; **98** (1936), 759; **99** (1936), 109;
R. de L. Kronig, Physica **2** (1935), 491, 854, 968.
7) D. Bohm, Phys. Rev. **75** (1949), 502.
8) R. de L. Kronig, loc. cit.
9) N. Bohr, Nature **137** (1931), 344.

Particle states of a quantized meson field

By T. H. R. Skyrme
Atomic Energy Research Establishment, Harwell

(*Communicated by Sir William Penney, F.R.S.—Received* 27 *January* 1961)

A simple non-linear field theory is considered as the model for a recently proposed classical field theory of mesons and their particle sources. Quantization may be made according to canonical procedures; the problem is to show the existence of quantum states corresponding with the particle-like solutions of the classical field equations. A plausible way to do this is suggested.

1. Introduction

In previous communications (Skyrme 1958, 1959, and 1961, referred to here as I, II and III, respectively) the author has considered a particular class of non-linear field theories, that may be relevant to the description of the strongly interacting elementary particles, mesons and baryons. As a contribution towards understanding the quantized form of the field theory proposed in III, a much simpler model is considered here.

This model is essentially the same as that considered in the latter part of I, with some alteration of detail. The aim here is to formulate the problems as precisely as possible and to suggest the direction in which solutions may be found. Difficulties are associated with various limiting operations and these have not been resolved.

The classical model field theory is simple and well defined with solutions of mesonic and particle types. There is a conserved particle current associated with the multiple-valuedness of an angular representation of the field. A Hamiltonian can be found and the theory quantized formally in the usual way. Meson-like quantum states can be constructed in a perturbation series but there is no simple correspondence with the classical theory for particle-like states. If such states exist, and *a priori* this seems reasonable, they will be anomalous solutions with some resemblance to 'superconductor' solutions that have been considered by Nambu (1960) and Goldstone (1961).

The transformation that would be expected to give such solutions leads naturally to particle operators almost of neutrino character, interacting indefinitely weakly with the rest of the field. These operators can be introduced as auxiliary variables describing the multiple valuedness of the angular field representation. A further transformation introduces coupling to the meson field, leading to a Hamiltonian very similar to that for particles in the classical theory. From this Hamiltonian particle-like solutions, with non-zero mass, can apparently be constructed in a self-consistent perturbation series; these provide the closest correspondence with the classical solution.

The field is defined in § 2, and the classical aspect reviewed in § 3. Some analogies are discussed in § 4 and these lead to a tentative analysis of the quantized field theory outlined in §§ 5 and 6.

2. Field definitions

The model involves one angular variable α in one dimension x of space (apart from time), analogous to the three angular variables and three space-dimensions of the theory formulated in III. In terms of two real field variables

$$\phi_3 = \sin\alpha \quad \text{and} \quad \phi_4 = \cos\alpha, \tag{1}$$

the postulated Lagrangian density is

$$\mathscr{L} = -\frac{\epsilon}{2\pi}\left[\frac{1}{2}\left(\frac{\partial\phi_\rho}{\partial x_\mu}\right)^2 + \kappa^2(1-\phi_4)\right] \tag{2}$$

$$= -\frac{\epsilon}{4\pi}\left(\frac{\partial\alpha}{\partial x_\mu}\right)^2 - \frac{\epsilon\kappa^2}{2\pi}(1-\cos\alpha). \tag{3}$$

Here the indices ρ and μ are summed over the values 3 and 4, for the sake of analogy, with $x_3 = x$ and $x_4 = ict$; subsequently c is put equal to unity. κ is a constant reciprocal length, and ϵ an energy scaling constant of dimensions $\hbar c$ if $\int \mathscr{L}\,\mathrm{d}x$ is an energy. The two terms correspond to the two terms in equation (23) of III, but the difference in numbers of dimensions and of degree in derivatives means that the second term here destroys the rotational symmetry of the former. This is an imperfection of the model which seems inevitable: it is not now suggested, as in I, that the rotational symmetry of the full theory should be limited, at least for the strongest interactions.

The boundary condition will be, as in III, that

$$\phi_4 = 1 \quad \text{at infinity} \quad (x = \pm\infty). \tag{4}$$

The field equation derived from (3) is

$$\Box\alpha - \kappa^2\sin\alpha = 0, \tag{5}$$

where here

$$\Box = \frac{\partial^2}{\partial x^2} - \frac{\partial^2}{\partial t^2}.$$

The energy-momentum tensor of the field is defined in the usual way, with the usual conservation laws. In addition there is a conserved particle current j_μ, with

$$j_3 = -\frac{\mathrm{i}}{2\pi}\frac{\partial\alpha}{\partial x_4}, \quad j_4 = \frac{\mathrm{i}}{2\pi}\frac{\partial\alpha}{\partial x_3} = \mathrm{i}j_0, \tag{6}$$

obviously satisfying

$$\partial j_\mu/\partial x_\mu = 0. \tag{7}$$

The particle number

$$N = \int j_0\,\mathrm{d}x = (1/2\pi)\,[\alpha(\infty) - \alpha(-\infty)]. \tag{8}$$

The canonically conjugate field variable is

$$\beta = \frac{\partial\mathscr{L}}{\partial\dot\alpha} = \frac{\epsilon}{2\pi}\int\frac{\partial\alpha}{\partial t}, \tag{9}$$

and the Hamiltonian has the density

$$\mathscr{H} = \frac{\pi}{\epsilon}\beta^2 + \frac{\epsilon}{4\pi}\left(\frac{\partial\alpha}{\partial x}\right)^2 + \frac{\epsilon\kappa^2}{2\pi}(1-\cos\alpha). \tag{10}$$

3. Classical solutions

The field equation (5) has wave-like solutions in which α is a function of $(kx-\omega t)$ with $\omega^2-k^2>0$; these correspond to the elementary solutions of the linearized equation

$$\Box\alpha-\kappa^2\alpha=0. \tag{11}$$

As was noted in I, equation (5) also has particle-like solutions in which α is a function of $(x-vt)/(1-v^2)^{\frac{1}{2}}$. In particular the static solution

$$\tan\tfrac{1}{4}\alpha=\exp\left[\pm\kappa(x-x_0)\right] \tag{12}$$

describe a particle $(+)$ or anti-particle $(-)$ at the point $x=x_0$. For these the particle number N, equation (8), has the values ± 1.

The field equation is sufficiently simple that it is practicable to examine numerically the interaction of wave-packets of these types, and computations will be reported elsewhere. The aim of this paper, however, is the formal elucidation of a similar solution in the quantized theory.

The characteristic of a particle in this theory is that α changes by 2π in a small distance, and it is natural to introduce such steps explicitly into the field equations. Suppose that $\omega(x)$ is some function varying smoothly from 0 to 1 around the point $x=0$, tending in a limit to the step-function $\theta(x)$ equal to 1 or 0 according as $x>$ or <0.

To consider the motion of a particle at the point $x=X(t)$ the transformation

$$\alpha\rightarrow\alpha+2\pi\omega(x-X) \tag{13}$$

is made, with particle velocity

$$v=\mathrm{d}X/\mathrm{d}t. \tag{14}$$

To the derivative terms in the Lagrangian $L_0=\int\mathscr{L}\,\mathrm{d}x$ must then be added

$$L'=-\epsilon\int\left(\frac{\partial\alpha}{\partial x}+v\frac{\partial\alpha}{\partial t}\right)\omega'(x-X)\,\mathrm{d}x-\epsilon\pi(1-v^2)\int\omega'^2(x-X)\,\mathrm{d}x. \tag{15}$$

The $\cos\alpha$ term will also be modified but the change will become infinitesimal as $\omega(x)$ tends to the step-function $\theta(x)$, and will not be written explicitly. L' gives a source term for the α-field arising from the derivative interaction. The variational equation for X is, however, satisfied identically when substitution is made from the field equation, as is to be expected because additional equations cannot be found without imposing some constraint on the variation of X.

A natural condition to impose in the classical theory is

$$\alpha(X)=\pi. \tag{16}$$

Time derivatives of this equation, $\partial x/\partial t+v(\partial\alpha/\partial x)=0$ at $x=X$, etc., then give reasonable equations of motion for the particle co-ordinate X. The inertia of the source is proportional to $(\partial\alpha/\partial x)_X$; when this vanishes the particle will be on the verge of annihilation or creation with a partner antiparticle at the same point.

The terms (15) are not Lorentz covariant because of the non-relativistic form-factor $\omega(x)$, but the result of any calculation will be invariant as is the basic theory;

the form factor is merely introduced for calculational convenience. In the limit $\omega \to \theta$ and $\omega' \to \delta$, the first term is covariant and the second may formally so be made by division with $(1-v^2)^{\frac{1}{2}}$ to allow for the contraction of the moving source. The terms may then be written

$$L' = -(1-v^2)^{\frac{1}{2}}\left\{M_0 + \frac{\epsilon}{(1-v^2)^{\frac{1}{2}}}\left(\frac{\partial \alpha}{\partial x} + v\frac{\partial \alpha}{\partial t}\right)_X\right\}, \tag{17}$$

where M_0 is an invariant bare mass constant, formally infinite; the divergence is cancelled by that of the self-mass which arises from the coupling to the meson field as was seen in I, § 7.

With this addition to the Lagrangian the conjugate variable becomes

$$\beta = \frac{\epsilon}{2\pi}\frac{\partial \alpha}{\partial t} - \epsilon v \delta(x-X) \tag{18}$$

and formally the Hamiltonian is

$$\begin{aligned} H &= vP + \int \beta \frac{\partial \alpha}{\partial t}\,\mathrm{d}x - L_0 - L' \\ &= H_0 + vP + M_0(1-v^2)^{\frac{1}{2}} + \left[\epsilon \frac{\partial \alpha}{\partial X} + 2\pi v \beta\right]_{x=X} + \frac{M_0 v^2}{(1-v^2)^{\frac{1}{2}}}, \end{aligned} \tag{19}$$

the last term arising from the square of the additional term in (18) associated with the derivative interaction. This indeed is the correct expression for the total field energy if the momentum

$$P = M_0 v/(1-v^2)^{\frac{1}{2}} \tag{20}$$

but this relation could not be obtained in the usual way because X is a redundant variable of the system.

4. The quantized field

The Hamiltonian (10) with the canonical commutation relations

$$[\alpha(x), \beta(x')] = \mathrm{i}\hbar\,\delta(x-x') \tag{21}$$

may be solved by a formal perturbation series starting with the linearized form describing a meson field of mass $\hbar\kappa/c$. The only (logarithmic) divergences arise from the meson-meson interactions in the last term which could be handled by the usual renormalization methods; these are not particularly significant, however, as the structure of this term in the model is arbitrary. The important interesting question is whether states exist analogous to the classical particle solutions with $N = \pm 1$. By symmetry the vacuum should have $N = 0$, above which there should exist some lowest state with $N = \pm 1$, interpretable as a particle. The plan is to look for a transformation of the Hamiltonian which will exhibit some particle state as a start for a perturbation analysis; the following analogies suggest a possible approach.

In two dimensions the analogue of Dirac's equation can be written in the usual form

$$\left(\gamma_\mu \frac{\partial}{\partial x_\mu} + \frac{Mc}{\hbar}\right)\psi = 0, \tag{22}$$

where ψ is a two-component field, the index μ is summed only over 3 and 4, and $\gamma_3, \gamma_4, \gamma_5$ form a triad of anticommuting Pauli matrices. The field Hamiltonian is

$$\int \psi^\dagger \left\{ \gamma_5 \left(-i\hbar c \frac{\partial}{\partial x} \right) + \gamma_4 Mc^2 \right\} \psi \, dx. \tag{23}$$

γ_5 is both the velocity and the analogue of its namesake in 4 dimensions (where $\boldsymbol{\alpha} = -\gamma_5 \boldsymbol{\sigma}$).

The exponential type of interaction discussed in I and II arises when M is replaced by $M \exp(-i\gamma_5 \alpha)$; after the transformation $\psi \to \exp(\frac{1}{2}i\gamma_5\alpha)\psi$ the particle Lagrangian density is

$$-\bar{\psi} \left(\gamma_\mu \hbar c \frac{\partial}{\partial x_\mu} + Mc^2 + \tfrac{1}{2} i\hbar c \gamma_\mu \gamma_5 \frac{\partial \alpha}{\partial x_\mu} \right) \psi \tag{24}$$

with an interaction term similar to (15).

The transformation (13) introduced to describe a classical particle suggests the general separation of the field

$$\alpha(x) = \hat{\alpha}(x) + 2\pi N(x), \tag{25}$$

where $\hat{\alpha}(x)$ lies in the standard zone, $-\pi < \hat{\alpha} < \pi$, and $N(x)$ is an integer at any point x. Then the meson Lagrangian

$$L_0(\alpha) = L_0(\hat{\alpha}) - \epsilon \frac{\partial \hat{\alpha}}{\partial x_\mu} \frac{\partial N}{\partial x_\mu} - \epsilon\pi \left(\frac{\partial N}{\partial x_\mu} \right)^2. \tag{26}$$

Comparison of (24) and (26) indicates a correspondence, that is

$$\left.\begin{aligned} \frac{\partial N}{\partial x_\mu} &\leftrightarrow \frac{\hbar c}{2\epsilon} \bar{\psi} i \gamma_\mu \gamma_5 \psi, \\ \frac{\partial N}{\partial x} &\leftrightarrow \frac{\hbar c}{2\epsilon} \psi^\dagger \psi, \\ \frac{\partial N}{\partial t} &\leftrightarrow \frac{\hbar c}{2\epsilon} \psi^\dagger \gamma_5 \psi. \end{aligned}\right\} \tag{27}$$

Now $\partial N/\partial x$ is the particle number density of the meson field; the correspondence (27) would be natural if the constant $\epsilon = \frac{1}{2}\hbar c$. This leads to the conjecture that the quantized field has particle solutions if and only if this condition is satisfied.

The interaction terms arise from the changes

$$\left.\begin{aligned} \alpha &\to \alpha + 2\pi \int^x \psi^\dagger \psi \, dx', \\ \beta &\to \beta - \tfrac{1}{2} \psi^\dagger \gamma_5 \psi, \end{aligned}\right\} \tag{28}$$

which would follow from unitary transformation by

$$S = \exp \left\{ \tfrac{1}{2} i \int \alpha \psi^\dagger \gamma_5 \psi \, dx + 2\pi i \int \beta \left(\int^x \psi^\dagger \psi \, dx' \right) dx \right\}. \tag{29}$$

On the ψ-field this transformation would introduce the factor

$$\exp \left\{ -\tfrac{1}{2} i \gamma_5 \alpha(x) - 2\pi i \int_x^\infty \beta(x') \, dx' \right\}. \tag{30}$$

5. Particle operators

Consider therefore in connexion with the quantum Hamiltonian (10) the operators

$$U_\gamma = \exp\left\{\mathrm{i}\epsilon\gamma\int\alpha(x')\,\omega'(x'-x)\,\mathrm{d}x' + 2\pi\mathrm{i}\int\beta(x')\,\omega(x'-x)\,\mathrm{d}x'\right\}, \tag{31}$$

where γ has the eigenvalues ± 1 of γ_5 and $\omega(x)$ is an approximation to the step-function as in § 3; both $\hbar$ and c will now be put equal to unity.

They satisfy the equation

$$\frac{\mathrm{i}\partial U}{\partial t} = [U, H] = -\mathrm{i}\gamma\frac{\partial U}{\partial x} + \frac{\epsilon\kappa^2}{2\pi}\int[\cos\alpha(x') - \cos\{\alpha(x') + 2\pi\omega(x'-x)\}]\,\mathrm{d}x'. \tag{32}$$

As ω tends to the step-function the last term becomes proportional to

$$\int[1-\cos 2\pi\omega(t)]\,\mathrm{d}t,$$

tending to zero. In this limit then the operators satisfy a neutrino-like equation.

The commutation relations of the U are easily evaluated

$$U_\gamma(x)\,U_{\gamma'}(x') = U_{\gamma'}(x')\,U_\gamma(x)\exp(2\pi\mathrm{i}\epsilon G), \tag{33}$$

with

$$G = \int[\gamma'\omega(t+x'-x) - \gamma\omega(t+x-x')]\,(\mathrm{d}\omega/\mathrm{d}t)\,\mathrm{d}t. \tag{34}$$

For $\gamma = -\gamma'$, $G = \pm 1$, independent of the shape of ω; for $\gamma = \gamma'$, $G \to \pm 1$ as ω tends to the step-function, except for $x = x'$ when $G = 0$. Therefore in the limit the U 'almost always' anticommute, if $\epsilon = \frac{1}{2}\hbar c$, providing another motive for this choice.

Again with $\epsilon = \frac{1}{2}$, U is a double-valued function only of the basic fields ϕ_ρ, but for a general choice it would be indefinitely many valued. From such a point of view the particle states of the system arise when the state-function is allowed to be a double-valued function of the field, at all points of space independently. Whatever may be the basic reason for this determination of the energy scale, it will now be assumed fixed thus.

In the limit, if the last term in (31) is negligible, the operators U may be used to construct states with additional 'neutrinos'; for if ϕ is any eigenstate of H with energy E and particle number N, the state

$$\left[\int U_\gamma^\dagger(x)\,\mathrm{e}^{\mathrm{i}kx}\,\mathrm{d}x\right]\phi$$

has energy $E+\gamma k$ and particle number $N+1$. Whether these limiting states really 'exist' is in fact rather academic because in this context the 'neutrinos' are completely non-interacting and so effectively unobservable.

These operations seem to be the obvious ones needed to describe the possible multiple-valuedness of the angular variables; the next problem is to find how to use them to build up massive particle states. The operators can be introduced as additional co-ordinates for the system; if $\psi_\pm$, labelled by the eigenvalues of γ_5, are the anticommuting field operators of an assumed neutrino-like field with Hamiltonian

$$H_\nu = \int\psi^\dagger\left(-\mathrm{i}\gamma_5\frac{\partial}{\partial x}\right)\psi\,\mathrm{d}x, \tag{35}$$

then the commuting products $(U^{\dagger}\psi)$ of corresponding operators satisfy

$$[H+H_{\nu}, U^{\dagger}\psi] = \pm\,\mathrm{i}(\partial/\partial x)\,(U^{\dagger}\psi) \tag{36}$$

so that $H^* = H + H_{\nu}$ commutes with the operators $\int U^{\dagger}\psi\,\mathrm{d}x$. Similarly

$$N^* = N + \int \psi^{\dagger}\psi$$

commutes with $U^{\dagger}\psi$.

If ψ is any eigenstate of H^* then by operation with a function F of the $\int U^{\dagger}\psi\,\mathrm{d}x$ other states can be generated with different numbers of neutrinos, replacing steps in the meson field. By projection on to the neutrino vacuum a state is formed which will then be an eigenstate of H. For if χ is the neutrino vacuum and

$$\phi = \chi^{\dagger}F\psi, \tag{37}$$

then

$$H\phi = \chi^{\dagger}(H+H_{\nu})\,F\psi = \chi^{\dagger}FH^*\psi \tag{38}$$

and the particle number is likewise the eigenvalue of N^*.

Since H^* is separated the transformation is in some way trivial but it provides a convenient framework for the representation of anomalous solutions. The form of F does not appear to be important, so long as the projected state ϕ is not null.

6. Quantum particles

The aim is to find an eigenstate of H that describes a particle. Such a state could be generated from various states ψ with Hamiltonian H^* by the projection (37). It should be possible to find one such state that is a functional only of the reduced variables $\hat{a}(x)$, of equation (25), with $N = 0$, and of the 'neutrino' field ψ.

This will be sought by transformation of H^* into a form similar to that which arises naturally in the classical description of a particle, as equation (19). The formal possibility of a (divergent) bare mass for the sources may be introduced by the canonical transformation of the particle field:

$$\left.\begin{aligned} \psi_+ &\rightarrow \exp\left[\mathrm{i}\pi\int_{-\infty}^{x} \psi_-^{\dagger}\,\psi_-\,\mathrm{d}x\right]\psi_+,\\ \psi_- &\rightarrow \exp\left[\mathrm{i}\pi\int_{x}^{\infty} \psi_+^{\dagger}\,\psi_+\,\mathrm{d}x\right]\psi_-, \end{aligned}\right\} \tag{39}$$

where $\psi_{\pm}$ are the components of ψ with $\psi_5 = \pm 1$. This sends

$$H_{\nu} \rightarrow H_{\nu} + \tfrac{1}{2}\pi\int \psi^{\dagger}(1+\gamma_5)\,\psi\psi^{\dagger}(1-\gamma_5)\,\psi\,\mathrm{d}x. \tag{40}$$

The additional contact interaction term equivalent with

$$\tfrac{1}{4}\pi\int \psi\gamma_{\mu}(1+\gamma_5)\,\psi\overline{\psi}\gamma_{\mu}(1-\gamma_5)\,\psi\,\mathrm{d}x \tag{41}$$

is still consistent with zero mass, but embraces also the possibility of a divergence.

Next the coupling to the meson field is introduced by the second canonical transformation

$$\left.\begin{aligned} \psi &\rightarrow \exp\left(\tfrac{1}{2}\mathrm{i}\gamma_5\hat{\alpha}\right)\psi,\\ \beta &\rightarrow \beta + \tfrac{1}{2}\psi^{\dagger}\gamma_5\psi. \end{aligned}\right\} \tag{42}$$

16-2

This gives the new Hamiltonian, a transform of H^*,

$$H_1 = H_0(\alpha)+\int\psi^\dagger\gamma_5\left(-\mathrm{i}\frac{\partial}{\partial x}\right)\psi\,\mathrm{d}x+\tfrac{1}{2}\pi\int\psi^\dagger(1+\gamma_5)\,\psi\psi^\dagger(1-\gamma_5)\,\psi\,\mathrm{d}x$$
$$+\int\psi^\dagger\left(\frac{1}{2}\frac{\partial\alpha}{\partial x}+2\pi\beta\gamma_5\right)\psi\,\mathrm{d}x+\tfrac{1}{2}\pi\int(\psi^\dagger\gamma_5\psi)^2\,\mathrm{d}x. \tag{43}$$

There is a term-by-term correspondence with the classical form (19); γ_5 replaces v and the third and fifth terms can be combined to give

$$\tfrac{1}{2}\pi\int(\psi^\dagger\psi)^2\,\mathrm{d}x, \tag{44}$$

corresponding to the bare mass term $M_0(1-v^2)^{-\frac{1}{2}}$.

The Hamiltonian H_1 might be derived in the conventional way from an extended Lagrangian density

$$\mathscr{L}_1 = -\frac{1}{8\pi}\left(\frac{\partial\hat{\alpha}}{\partial x_\mu}+2\pi\bar{\psi}\mathrm{i}\gamma_\mu\gamma_5\psi\right)^2-\frac{\kappa^2}{4\pi}(1-\cos\alpha)-\bar{\psi}\gamma_\mu\frac{\partial}{\partial x_\mu}\psi. \tag{45}$$

This Lagrangian defines a theory that is at least renormalizable in the conventional manner since the worst possible types of divergency are logarithmic (the derivative interaction is compensated by the fewer number of space dimensions). It has 'γ_5-invariance' because there is no explicit mass-like term, and indeed it still describes massless non-interacting neutrino-like particles as the normal type of particle solution. But it is now possible in this form to consider also whether self-consistent solutions are possible with non-zero mass.

Suppose that the field ψ can describe particles with mass M, and that μ is the renormalized meson mass. Then, in second order, the mass correction terms will equate $\int\bar{\psi}\Delta M\psi$ with the one-particle part of

$$\tfrac{1}{2}\pi\int(\bar{\psi}\mathrm{i}\gamma_\mu\gamma_5\psi)^2\,\mathrm{d}x-\tfrac{1}{8}\mathrm{i}\left[\int\left(\bar{\psi}\mathrm{i}\,\psi_\mu\gamma_5\psi\frac{\partial\alpha}{\partial x_\mu}\right)\mathrm{d}x\right]^2. \tag{46}$$

If p is the momentum of the external particle line, q that of the internal particle and $k = p-q$ that of the meson, the usual methods give

$$\Delta M = \pi\int\frac{\mathrm{d}^2q}{(2\pi)^2}\mathrm{i}\gamma_\mu\gamma_5[\epsilon+\mathrm{i}(M+\mathrm{i}\gamma q)]^{-1}\mathrm{i}\gamma_\mu\gamma_5$$
$$-\tfrac{1}{4}\mathrm{i}\int\frac{\mathrm{d}^2k}{(2\pi)^2}[4\pi/\epsilon+\mathrm{i}(k^2+\mu^2)]\,\mathrm{i}\gamma k\gamma_5[\epsilon+\mathrm{i}(M+\mathrm{i}\gamma q)]^{-1}\mathrm{i}\gamma k\gamma_5. \tag{47}$$

Since
$$\mathrm{i}\gamma k(M+\mathrm{i}\gamma q)\,\mathrm{i}\gamma k = -k^2(M-\mathrm{i}\gamma p)-\mathrm{i}\gamma k(q^2-p^2), \tag{48}$$

the second term is equal, for a real state with $M+\mathrm{i}\gamma p = 0$, to

$$2\pi\mathrm{i}M\int\frac{\mathrm{d}^2k}{(2\pi)^2}k^2[k^2+\mu^2-\mathrm{i}\epsilon]^{-2}[q^2+M^2-\mathrm{i}\epsilon]^{-1}. \tag{49}$$

The divergent part of this integral is cancelled exactly by the first term of (46), just as in the classical theory the interaction cancels the divergence of the bare mass (§ 3). The remainder gives

$$\Delta M/M = -2\pi\mathrm{i}\mu^2\int\frac{\mathrm{d}^2k}{(2\pi)^2}[k^2+\mu^2-\mathrm{i}\epsilon]^{-1}[q^2+M^2-\mathrm{i}\epsilon]^{-1}$$
$$= \left(\frac{4M^2}{\mu^2-1}\right)^{-\frac{1}{2}}\tan^{-1}\left(\frac{4M^2}{\mu^2-1}\right)^{\frac{1}{2}}. \tag{50}$$

Thus to 'second order' the consistency condition $\Delta M = M$ is satisfied either by $M = 0$, the neutrino-like solutions, or by $M = \frac{1}{2}\mu$; for comparison the classical theory, with $\epsilon = \frac{1}{2}\hbar c$, has $M = (2/\pi)(\hbar\kappa/c)$. In the same order the meson mass correction is generally finite though it diverges for the particular value $M = \frac{1}{2}\mu$ because then the meson is almost unstable against disintegration into a particle pair. There are divergent corrections that arise from the $\cos\alpha$ term but these are peculiar to the model, and probably not of fundamental significance. Apart from the latter it is plausible that higher-order renormalizations may be finite and lead to a sensible description of a massive particle state.

7. Comments

I have considered the system described by the Hamiltonian (10). Classically there are meson-like and particle-like solutions; the fundamental problem is to study the nature of the quantized solutions, and in particular to show the existence of particle-like states. I have outlined a method for doing this which needs critical investigation on a number of points:

(1) Whether the condition $\epsilon = \frac{1}{2}\hbar c$ is essential for the existence of particle-like states; and if it is, the significance. It may be related either to the nature, geometrical interpretation, of the fields ϕ, or to some deeper interpretation of quantum theory.

(2) The nature of the limiting operations in which the form-factor $\omega(x)$ tends to a step-function, and the existence of the 'neutrino' operators U in this limit.

(3) The introduction of additional neutrino co-ordinates ψ with the projection operation F, and its compatibility with the above limiting operation.

(4) The significance of the transformation (40) which introduces the divergent self-mass associated with a point-singularity.

The final stage, the effective Lagrangian (46) and Hamiltonian (44), has an attractive resemblance to the classical theory; I have not, however, found a satisfactory proof that its solutions will project into ones of the original mesonic Lagrangian (3) and Hamiltonian (10).

An alternative discussion of the quantization that seems particularly appropriate to the problem of relating quantum and classical solutions might involve the use of Feymann path integrals for the field variables separated as in equation (25); the condition $\epsilon = \frac{1}{2}\hbar c$ would then enter in a simple and central way. I have, however, found considerable difficulty in formulating the limiting operations that are needed.

References

Goldstone, J. 1961 *Nuovo Cimento*, **19**, 154.
Nambu, Y. 1960 *Chicago Rep.* no. 60/21.
Skyrme, T. H. R. 1958 *Proc. Roy. Soc.* A, **247**, 260.
Skyrme, T. H. R. 1959 *Proc. Roy. Soc.* A, **252**, 236.
Skyrme, T. H. R. 1961 *Proc. Roy. Soc.* A, **260**, 127.

JOURNAL OF MATHEMATICAL PHYSICS VOLUME 6, NUMBER 2 FEBRUARY 1965

Exact Solution of a Many-Fermion System and Its Associated Boson Field

DANIEL C. MATTIS
International Business Machines Corp., Thomas J. Watson Research Center, Yorktown Heights, New York

AND

ELLIOTT H. LIEB*
Belfer Graduate School of Science, Yeshiva University, New York, New York
(Received 22 September 1964)

Luttinger's exactly soluble model of a one-dimensional many-fermion system is discussed. We show that he did not solve his model properly because of the paradoxical fact that the density operator commutators $[\rho(p), \rho(-p')]$, which always vanish for any finite number of particles, no longer vanish in the field-theoretic limit of a filled Dirac sea. In fact the operators $\rho(p)$ *define* a boson field which is *ipso facto* associated with the Fermi–Dirac field. We then use this observation to solve the model, and obtain the exact (and now nontrivial) spectrum, free energy, and dielectric constant. This we also extend to more realistic interactions in an Appendix. We calculate the Fermi surface parameter $\bar{n}_k$, and find: $\partial\bar{n}_k/\partial k|_{k_F} = \infty$ (i.e., there exists a sharp Fermi surface) only in the case of a sufficiently weak interaction.

I. INTRODUCTION

THE search for a soluble but realistic model in the many-electron problem has been just about as unfruitful as the historic quest for the philosopher's stone, but has equally resulted in valuable byproducts. For example, 15 years ago Tomonaga[1] published a theory of interacting fermions which was soluble only in one dimension with the provision that certain truncations and approximations were introduced into his operators. Nevertheless he had success in showing approximate boson-like behavior of certain collective excitations, which he identified as "phonons." (Today we would denote these as "plasmons," following the work of Bohm and Pines.[2]) Lately, Luttinger[3] has revived interest in the subject by publishing a variant model of spinless and massless one-dimensional interacting fermions, which demonstrated a singularity at the Fermi surface, compatible with the results of the modern many-body perturbation theory.[4]

Unfortunately, in calculating the energies and wavefunctions of his model Hamiltonian, Luttinger fell prey to a subtle paradox inherent in quantum field theory[5] and *therefore did not achieve a correct solution of the problem he himself had posed.* In the present paper we shall give the solution to his interesting problem and calculate the free energy. We shall show the existence of collective plasmon modes, and shall calculate the singularity at the Fermi surface (which may in fact disappear if the interaction is strong enough), the energy of the plasmons, and the (nontrivial) dielectric constant of the system. In an Appendix we shall show how the model may be generalized in such a manner as to remove certain restrictions on the interactions which Luttinger had found necessary to impose.

It is fortunate that solid-state and many-body theorists have so far been spared the plagues of quantum field theory. Second quantization has been often just a convenient bookkeeping arrangement to save us from writing out large determinantal wavefunctions. However there is a difference between very large determinants and *infinitely* large ones; we shall show that one of the important differences *is the failure of certain commutators to vanish* in the field-theoretic limit when common sense and experience based on finite N tells us they *should* vanish! (Here N refers to the number of particles in the field.)

* Research supported by the U. S. Air Force Office of Scientific Research.

[1] S. Tomonaga, Progr. Theoret. Phys. (Kyoto) 5, 544 (1950).

[2] D. Bohm and D. Pines, Phys. Rev. 92, 609 (1953).

[3] J. M. Luttinger, J. Math. Phys. 4, 1154 (1963). Note that we set his $v_0 = 1$, thereby fixing the unit of energy. References to this paper will be frequent, and will be denoted by L (72), for example, signifying his Eq. (72).

[4] J. M. Luttinger and J. C. Ward, Phys. Rev. 118, 1417 (1960).

[5] Luttinger made a transformation, L (8), which was canonical in appearance only. But in the language of G. Barton [*Introduction to Advanced Field Theory,* (Interscience Publishers, Inc., New York, 1963), pp. 126 *et seq.*] this transformation connected two "unitarily inequivalent" Hilbert spaces, which has as a consequence that commutators, among other operators, must be reworked so as to be well-ordered in fermion field operators. It was first observed by Julian Schwinger [Phys. Rev. Letters 3, 296 (1959)] that the very fact that one postulates the existence of a ground state (i.e., the filled Fermi sea) *forces* certain commutators to be nonvanishing even though in first quantization they automatically vanish. The "paradoxical contradictions" of which Schwinger speaks seem to anticipate the difficulties in the Luttinger model.

We shall show that these nonvanishing commutators *define* boson fields which must *ipso facto* always be associated with a Fermi–Dirac field, and we shall use the ensuing commutation relations to solve Luttinger's model exactly. Because this model is soluble both in the Hilbert space of finite N and also in the Hilbert space $N = \infty$, with different physical behavior in each, we believe it has applications to the *theory of fields* which go beyond the study of the many-electron problem. The model can be extended to the case of electrons with spin. This has interesting consequences in the band *theory of ferromagnetism*, as will be discussed in some detail in an article under preparation.[5a]

II. MODEL HAMILTONIAN

We recall Luttinger's Hamiltonian[3] and recapitulate some of his results:

$$H = H_0 + H', \tag{2.1}$$

where the "unperturbed" part is

$$H_0 = \int_0^L dx\ \psi^+(x)\sigma_3 p\psi(x) \tag{2.2a}$$

$$= \sum_k (a_{1k}^* a_{1k} - a_{2k}^* a_{2k})k, \tag{2.2b}$$

and the interaction is

$$H' = 2\lambda \iint_0^L dx\, dy\ \psi_1^+(x)\psi_1(x) \times V(x-y)\psi_2^+(y)\psi_2(y) \tag{2.3a}$$

$$= \frac{2\lambda}{L} \sum \delta_{k_1+k_3,k_2+k_4} v(k_3 - k_4) \times a_{1k_1}^* a_{1k_2} a_{2k_3}^* a_{2k_4}. \tag{2.3b}$$

Here ψ is a two-component field and the form (b) of the operator is obtained from (a) by setting

$$\psi = \frac{1}{\sqrt{L}} \sum_k e^{ikx} \begin{pmatrix} a_{1k} \\ a_{2k} \end{pmatrix}$$

and

$$\psi^+ = \frac{1}{\sqrt{L}} \sum_k e^{-ikx} (a_{1k}^*, a_{2k}^*), \tag{2.4}$$

with a_{jk}'s defined to be anticommuting fermion operators which obey the usual relations

$$\begin{gathered} a_{jk}a_{j'k'} + a_{j'k'}a_{jk} \equiv \{a_{jk}, a_{j'k'}\} = 0 \\ \{a_{jk}^*, a_{j'k'}^*\} = 0, \text{ and } \{a_{jk}, a_{j'k'}^*\} = \delta_{jj'}\delta_{kk'}. \end{gathered} \tag{2.5}$$

Luttinger noted that for an appropriate operator S_0, the canonical transformation

$$\tilde{H} = e^{i\lambda S_0} H e^{-i\lambda S_0} \tag{2.6}$$

gave the result that

$$\tilde{H} = H_0, \tag{2.7}$$

and consequently that *the spectrum of* $H = H_0 + H'$ *was the same as that of* H_0, *independent of the interaction* $V(x - y)$. This can be explicitly verified for his choice of

$$S_0 = \iint_0^L dx\, dy\ \psi_1^+(x)\psi_1(x)E(x-y)\psi_2^+(y)\psi_2(y), \tag{2.8}$$

where $E(x)$, not to be confused with the energy E, is defined by:

$$\partial E(x-y)/\partial x \equiv V(x-y), \tag{2.9}$$

assuming that

$$\bar{V} \equiv \frac{1}{L}\int_0^L V(x)\, dx = 0. \tag{2.10}$$

In the Appendix we shall show among other things how to generalize to $\bar{V} \neq 0$. It is also simple and instructive to verify Eqs. (2.6) and (2.7) somewhat differently by using the *first* quantization,

$$H_0 = -i\sum_{n=1}^N \frac{\partial}{\partial x_n} + i\sum_{m=1}^M \frac{\partial}{\partial y_m} \tag{2.11}$$

and

$$H' = 2\lambda \sum_{n=1}^N \sum_{m=1}^M V(x_n - y_m), \tag{2.12}$$

where N and M are, respectively, the total number of "1" particles and "2" particles, with coordinates x_n and y_m, respectively. The properly antisymmetrized wavefunctions are given by

$$\Psi = \det |e^{ik_ix_j}| \det |e^{iq_iy_j}| \times \exp\left\{\sum_{n=1}^N \sum_{m=1}^M i\, E(x_n - y_m)\right\}. \tag{2.13}$$

Using Eqs. (2.9) and (2.10), Ψ is readily seen to obey Schrödinger's equation

$$H\Psi = E\Psi \tag{2.14}$$

with just the unperturbed eigenvalue

$$E = \sum_{n=1}^N k_n - \sum_{m=1}^M q_m. \tag{2.15}$$

The wavenumbers are of the form

$$k_i \text{ or } q_i = 2\pi \text{ integer}/L, \tag{2.16}$$

as required for periodic boundary conditions. This is in exact agreement with the results of Ref. 3, and can also be checked in perturbation theory; first-

[5a] D. Mattis, Physics 1, 184 (1964).

order perturbation theory also gives vanishing results, and indeed, it is easy to verify that to every order in λ the cancellation is complete, in accordance with the exact result given above.

Up to this point, Luttinger's analysis (which we have briefly summarized) is perfectly correct. It is the next step that leads to difficulty. The Hamiltonian discussed so far has no ground-state energy; in order to remove this obstacle, and thereby establish contact with a real electron gas, Luttinger proposed modifying the model by "filling the infinite sea" of negative energy levels (i.e., all states with $k_1 <$ and $q_2 > 0$). Following L(8) we define b's and c's obeying the usual anticommutators, such that

$$a_{1k} = \begin{cases} b_k & k \geq 0 \\ c^*{}_k & k < 0, \end{cases}$$

and

$$a_{2k} = \begin{cases} b_k & k < 0 \\ c_k^* & k \geq 0. \end{cases} \tag{2.17}$$

Using this notation the total particle-number operator becomes

$$\mathfrak{N} = \sum_{\text{all } k} b_k^+ b_k - c_k^+ c_k \tag{2.17a}$$

(i.e., the number of particles minus the number of holes).

Since the Hamiltonian commutes with $\mathfrak{N}$ we can demand that $\mathfrak{N}$ have eigenvalue N_0. In the noninteracting ground state there are no holes and the b particles are filled from $-k_F$ to k_F where $k_F = \pi(N_0/L) = \pi\rho$. The noninteracting ground-state energy is $N_0\pi\rho +$ energy of the filled sea (W).

The kinetic energy assumes the form

$$H_0 = \sum_{\text{all } k} (b_k^* b_k + c_k^* c_k) \, |k| + W, \tag{2.18}$$

where

$$W = (\sum_{k<0} k - \sum_{k>0} k) \tag{2.18a}$$

is the infinite energy of the filled sea, an uninteresting c number which we drop henceforth in accordance with Luttinger's prescription. The interaction [H', Eq. (2.3) and the operator S_0, Eq. (2.8)] can also be expressed in the new language by means of the substitution (2.17). The reader will no doubt be surprised, as indeed we were, to find that now with the new operators, *Eq.* (2.7), *with* $\tilde{H}$ *defined in* (2.6), *is no longer obeyed.*

Upon further reflection one sees that this must be so, on the basis of very general arguments. In the new Hilbert space defined by the transformation to the particle–hole language (2.17), H is no longer unbounded from below and now has a ground state. A general and inescapable *concavity theorem* states that if $E_0(\lambda)$ is the ground-state energy in the presence of interactions, (2.3), then

$$\partial^2 E_0(\lambda)/\partial\lambda^2 < 0. \tag{2.19}$$

This inequality is incompatible with the previous result, viz. all E = independent of λ, which was possible only in the strange case of a system without a ground state.

The same thing can be seen more trivially using second-order perturbation theory (first-order perturbation theory vanishes). It is easily seen that

$$E_0^{(2)} = -\left(\frac{2\lambda}{L}\right)^2 \sum_k \frac{|v(k)|^2}{2k} n_1(k) n_2(-k), \tag{2.20}$$

where $n_1(k)$ and $n_2(k)$ are the number of ways of shifting a particle of type "1" and type "2" respectively by an amount k to an unoccupied state. A simple geometric exercise will convince the reader of the following facts: (1) if we start with a state having a finite number of particles, then n_1 and n_2 are *always* even functions of k (i.e., there are just as many ways to increase the momentum by k as to decrease it by the same amount.) (2) If we start with a filled infinite sea then there is no way to decrease the momentum of the "1" particles nor to increase the momentum of "2" particles. Hence for this second case $n_1(k)n_2(-k)$ is nonzero only for $k > 0$. Thus $E_0^{(2)}$ vanishes for a state with a finite number of particles, but it is negative for a filled sea.

If the reader is unconvinced by perturbation theory, then he can easily prove that E_0 is lowered by doing a variational calculation.

What has gone wrong? We turn to some algebra to resolve this paradox, and following this, present a solution of the field-theoretic problem defined by $H_0 + H'$ in the representation of b's and c's.

III. CASE OF THE FILLED DIRAC SEA

The various relevant operators are given below; the form (a) of each equation will *not* be used in the bulk of the paper, and is just given here for completeness. In the following equations, $p > 0$.

$$\rho_1(+p) \equiv \sum_k a_{1\,k+p}^* a_{1\,k} \tag{3.1a}$$

$$= \sum_{k<-p} c_{k+p} c_k^* + \sum_{-p\leq k<0} b_{k+p}^* c_k^* + \sum_{k\geq 0} b_{k+p}^* b_k, \tag{3.1b}$$

$$\rho_1(-p) \equiv \sum_k a_{1\,k}^* a_{1\,k+p} \tag{3.2a}$$

$$= \sum_{k<-p} c_k c_{k+p}^* + \sum_{-p\leq k<0} c_k b_{k+p} + \sum_{k\geq 0} b_k^* b_{k+p}, \tag{3.2b}$$

$$\rho_2(+p) \equiv \sum_k a^*_{2\,k+p} a_{2\,k} \tag{3.3a}$$

$$= \sum_{k<-p} b^*_{k+p} b_k + \sum_{-p\le k<0} c_{k+p} b_k + \sum_{k\ge 0} c_{k+p} c^*_k, \tag{3.3b}$$

$$\rho_2(-p) \equiv \sum_k a^*_{2\,k} a_{2\,k+p} \tag{3.4a}$$

$$= \sum_{k<-p} b^*_k b_{k+p} + \sum_{-p\le k<0} b^*_k c^*_{k+p} + \sum_{k>0} c_k c^*_{k+p}. \tag{3.4b}$$

Equations (3.1a)–(3.4a) give the density operators in the original representation, so let us calculate in this language a commutator such as (assume $p \ge p' \ge 0$ for definiteness)

$$[\rho_1(-p),\ \rho_1(p')] = \sum_{k,k'} [a^*_{1\,k} a_{1\,k+p},\ a^*_{1\,k'+p'} a_{1\,k'}]$$

$$= \sum_{k=-\infty}^{+\infty} a^*_{1\,k} a_{1\,k+p-p'} - \sum_{k=-\infty}^{+\infty} a^*_{1\,k+p'} a_{1\,k+p} = 0. \tag{3.5}$$

The zero result could have been expected by writing the operators in first quantization:

$$\rho_1(-p) = \sum_n e^{-ipx_n} \quad \text{and} \quad \rho_2(p) = \sum_m e^{ipy_m}, \tag{3.6}$$

whence they evidently commute. Nevertheless, the zero result is achieved in (3.5) only through the almost "accidental" cancellation of two operators, each of which may diverge in the field-theory limit when $N = \infty$. We now show that in that limit the operators in fact no longer cancel, by evaluating the commutator using form (b) for the density operators. It is a matter of only some minor manipulation to obtain the important new result:

$$[\rho_1(-p),\ \rho_1(p')] = [\rho_2(p),\ \rho_2(-p')]$$

$$= \delta_{p,p'} \sum_{-p<k<0} 1 = \frac{pL}{2\pi}\, \delta_{p,p'}, \quad (p' > 0). \tag{3.7a}$$

In addition,

$$[\rho_1(p),\ \rho_2(p')] = 0. \tag{3.7b}$$

A quick check is provided by evaluating the vacuum expectation value

$$\langle 0|\ [\rho_1(-p),\ \rho_1(p)]\ |0\rangle$$

$$= \sum_{-p<k,k'<0} \langle 0|\ c_k b_{k+p} b^*_{k'+p} c^*_{k'}\ |0\rangle = pL/2\pi, \tag{3.8}$$

which is exactly what is expected on the basis of the previous equation. Evidently the form (b) of the operators $(2\pi/pL)^{+\frac{1}{2}}\rho_1(+p)$ and $(2\pi/pL)^{+\frac{1}{2}}\rho_2(-p)$ have properties of boson raising operators [call them $A^*(p)$ and $B^*(-p)$] and $(2\pi/pL)^{+\frac{1}{2}}\rho_1(-p)$ and $(2\pi/pL)^{+\frac{1}{2}}\rho_2(+p)$ have properties of boson lowering operators [$A(p)$ and $B(-p)$], i.e.,

$$[A, B] = [A^*, B] = 0,$$
$$[A(p),\ A^*(p')] = [B(-p), B^*(-p)] = \delta_{p,p'}. \tag{3.9}$$

The B field is the continuation of the A field to negative p; therefore together they form a *single* boson field defined for all p.

The relationship of the $\rho(p)$'s to Luttinger's $N(x)$'s, L(25), is obtained by using (2.4):

$$N_1(x) = \psi^*_1(x)\psi_1(x) = \frac{1}{L}\sum_p \rho_1(p) e^{-ipx},$$
$$N_2(x) = \psi^*_2(x)\psi_2(x) = \frac{1}{L}\sum_p \rho_2(p) e^{-ipx}. \tag{3.10}$$

IV. SOLUTIONS OF THE MODEL HAMILTONIAN

Before making use of the results of the previous section, we remark that $\rho_1(+p)$ and $\rho_2(-p)$ are exact raising operators of H_0, and $\rho_1(-p)$ and $\rho_2(p)$ are exact lowering operators of H_0 corresponding to excitation energies p. That is,

$$[H_0,\ \rho_1(\pm p)] = \pm p\rho_1(\pm p),$$
$$[H_0,\ \rho_2(\pm p)] = \mp p\rho_2(\pm p). \tag{4.1}$$

The identification of the ρ's with boson operators made in the previous section suggested to us the possibility of constructing a new operator T which obeys the same equations (4.1), as H_0. This is indeed possible, if we define T as follows:

$$T \equiv \frac{2\pi}{L} \sum_{p>0} \{\rho_1(p)\rho_1(-p) + \rho_2(-p)\rho_2(p)\} \tag{4.2}$$

[the ρ's being defined here and in the remainder of the paper by Eqs. (3.1b)—(3.4b), i.e., always in the hole-particle representation]. It follows that

$$[T,\ \rho_1(\pm p)] = \pm p\rho_1(\pm p) \tag{4.3}$$

as required, and similarly for $\rho_2(\mp p)$. Therefore, let us decompose H into two parts

$$H = H_1 + H_2 \tag{4.4}$$

with

$$H_1 = H_0 - T = \Big\{ \sum_k |k|\ (b^*_k b_k + c^*_k c_k)$$
$$- \frac{2\pi}{L} \sum_{p>0} \{\rho_1(p)\rho_1(-p) + \rho_2(-p)\rho_2(p)\} \Big\}, \tag{4.5}$$

and

$$H_2 = H' + T$$
$$= \frac{1}{L}\Big[2\lambda \sum_{p>0} \{v(p)\rho_1(-p)\rho_2(p) + v(-p)\rho_1(p)\rho_2(-p)\}$$
$$+ 2\pi \sum_{p>0} \{\rho_1(p)\rho_1(-p) + \rho_2(-p)\rho_2(p)\} \Big] \tag{4.6}$$

with $v(p)$ = real, even function of p. By actual construction, all the ρ operators which appear in H_2

commute with H_1. This will be an important feature in constructing an exact solution of the model. We define an Hermitian operator S,

$$S = \frac{2\pi i}{L} \sum_{\text{all } p} \frac{\varphi(p)}{p} \rho_1(p)\rho_2(-p), \tag{4.7}$$

where $\varphi(p)$ is also a real, even, function of p to be determined subsequently by imposing a condition that the unitary transformation e^{iS} diagonalize H_2. First we evaluate the effect of such a transformation on various operators. It commutes with H_1,

$$e^{iS}H_1e^{-iS} = H_1 = H_0 - T, \tag{4.8}$$

because both ρ_1 and ρ_2 appearing in S commute with H_1, as noted above. In the following, p can have either sign:

$$e^{iS}\rho_1(p)e^{-iS} = \rho_1(p)\cosh\varphi(p) + \rho_2(p)\sinh\varphi(p), \tag{4.9}$$

$$e^{iS}\rho_2(p)e^{-iS} = \rho_2(p)\cosh\varphi(p) + \rho_1(p)\sinh\varphi(p). \tag{4.10}$$

We have verified that this transformation is a proper unitary transformation and preserves commutation relations (3.7) as well as anticommutation relations (2.5), and the reader may easily check this point. H_2 is brought into canonical form by requiring that in (exp iS) H_2 (exp $-iS$) there be no cross terms such as $\rho_1(p)\rho_2(-p)$. This leads to the equation

$$\tanh 2\varphi = -\lambda v(p)/\pi, \tag{4.11}$$

which cannot be obeyed unless

$$|\lambda v(p)| < \pi \quad \text{for all} \quad p. \tag{4.12}$$

Equation (4.12) serves to limit the magnitude of potentials capable of having well-behaved solutions (e.g., a real ground-state energy). For the more realistic potentials discussed in the Appendix, there is also a more realistic bound on $v(p)$: there, $v(p)$ may not be *too* attractive, but it can have any magnitude when it is repulsive, i.e., positive.

With the choice of φ in (4.11), the evaluation of H_2 becomes

$$e^{iS}H_2e^{-iS} = \frac{2\pi}{L}\sum_{p>0}\operatorname{sech} 2\varphi(p)\{\rho_1(p)\rho_1(-p) + \rho_2(-p)\rho_2(p)\} - \sum_{p>0} p(1 - \operatorname{sech} 2\varphi). \tag{4.13a}$$

The second term is the vacuum renormalization energy

$$W_1 = -\sum_{p>0} p(1 - \operatorname{sech} 2\varphi) = \frac{L}{2\pi}\int_0^\infty dp\, p\left\{\left(1 - \frac{\lambda^2 v^2(p)}{\pi^2}\right)^{\frac{1}{2}} - 1\right\}. \tag{4.13b}$$

It may be expanded in powers of λ to effect a comparison with Goldstone's many-body perturbation theory[4]; we have checked that they agree to third order.

The problem is now formally solved, for we can find all the eigenfunctions and eigenvalues by studying Eqs. (4.4), (4.8), and (4.13). First notice that the operator T does not depend upon the interaction and that if there is *no interaction* we could write the Hamiltonian either as

$$H = H_0, \tag{4.14a}$$

or as

$$H = (H_0 - T) + T = H_1 + H_2. \tag{4.14b}$$

Since H_1 and H_2 commute, every eigenstate, Ψ, of H may be assumed to be an eigenfunction of H_1 and H_2 separately. Moreover, Ψ may also be assumed to be an eigenfunction of each $\alpha_p = A_p^+A_p$ and $\beta_p = B_{-p}^+B_{-p}$ for all $p > 0$, since these operators commute with H and $\mathfrak{N}$.

Evidently (4.14a) and (4.14b) provide two different ways of viewing the noninteracting spectrum. H_0 is quite degenerate: the raising operators of H_0 are the b^+'s and c^+'s. By requiring that Ψ also be an eigenstate of α_p, β_p and H, we are merely attaching quantum numbers to the degenerate levels of H_0. If $\alpha_p\Psi = n_p\Psi$ and $\beta_p\Psi = m_p\Psi$ (where n_p and m_p are of course integers), we say that we have n_p plasmons of momentum p and m_p plasmons of momentum $-p$. With no interaction the energy of a plasmon is

$$\epsilon(p) = |p|. \tag{4.15}$$

We may speak of H_1 as the quasiparticle part of the Hamiltonian; in H_1 the operator T plays the role of subtracting the plasmon part of the energy from H_0.

When we turn on the interaction, the above description of the energy levels is still valid, except that now we are *forced* to use the form (4.14b) because H_2 is no longer T. The degeneracy of H is partially removed by the interaction, because now the energy of a plasmon is

$$\epsilon'(p) = |p| \operatorname{sech} 2\varphi(p). \tag{4.16}$$

Notice that the plasmon energy is always *lowered* [and therefore the plasmons cannot propagate faster than the speed of light $c = 1$, i.e., $d\epsilon'/dp \leq 1$. In the more realistic case discussed in the Appendix, the plasmon energy *can* be increased by the interaction although $d\epsilon'/dp \leq 1$ is always obeyed.] by the interaction; if (4.12) is violated the plasmon energy is no longer real and the system becomes unstable. Note, there are no plasmons in the ground state, so that W_1 (4.13), is the shift in the ground-state energy of the system.

There is one important point, however, that requires some elucidation. We would like to be able to say that in view of the fact that H_1, $\alpha(p)$, and $\beta(p)$

conserve particle number, the most general energy level of H (fixed N_0) is the sum of *any* energy of H_1 (same N_0, and no plasmons) plus *any* (plasmon) energy of H_2 (note: the plasmon spectrum is independent of N_0). Were we dealing with a finite-dimensional vector space, such a statement would not be true, for even though H_1 and H_2 commute they could not possibly be independent. Thus, if H_2 had n eigenvalues $e_1, \cdots, e_n$, and if H_1 had an equal number $E_1, \cdots, E_n$ the general total eigenvalue would not be *any* combination of $e_i + E_i$ for this would give too many values (viz. n^2 instead of n.) But we are dealing with an infinite-dimensional Hilbert space and the additivity hypothesis is in fact true for the present model.

To prove this assertion we consider any eigenstate Ψ which is necessarily parameterized by the integers n_p and m_p. Consider the state $\Phi = \{\prod_p (A_p)^{n_p}(B_p)^{m_p}\}\Psi$. The state Φ is nonvanishing and has quantum numbers $n_p = 0 = m_p$. It is also an eigenstate of H_1 with energy $E_1(\Psi)$. In addition (and this is the important point) the state Ψ may be recovered from Φ by the equation

$$\Psi = \text{const} \times \{\prod_p (A_p^+)^{n_p}(B_p^+)^{m_p}\}\Phi.$$

To every state Ψ, therefore, there corresponds a *unique* state Φ from which it may be obtained using raising operators. Conversely, to any eigenstate of H_1 (for fixed N_0) we may apply raising operators as often as we please and obtain a new (nonvanishing) eigenstate. Thus the general energy is an arbitrary sum of quasiparticle and plasmon energies.

It may be wondered where we used the fact that the Hilbert space is infinite-dimensional in the above proof. The answer lies in the boson commutation relations of the A's and B's. It is impossible to have such relations in a finite-dimensional vector space.

The eigenvalues corresponding to these states Φ will be labeled in some order, E_i $(i = 1, 2, \cdots)$, so that the total canonical partition function $Z(\lambda)$ and the free energy $F(\lambda)$ are given by

$$Z(\lambda) = e^{-F(\lambda)/kT} = \left(\sum_i e^{-E_i/kT}\right)\left(e^{-W_1/kT}\right) \prod_{\substack{\text{all } p \\ \neq 0}} \left(\sum_{n=0}^{\infty} e^{-n\epsilon'(p)/kT}\right). \tag{4.17}$$

The first factor is difficult to evaluate directly. However it can be obtained circuitously by noting that the energies E_i are independent of λ and therefore

$$Z(0) = e^{-F(0)/kT} = \left(\sum_i e^{-E_i/kT}\right) \prod_{\substack{\text{all } p \\ \neq 0}} \left(\sum_{n=0}^{\infty} e^{-n\epsilon(p)/kT}\right). \tag{4.18}$$

But the second factor can be trivially evaluated, as can $F(0)$ = free energy of noninteracting fermions. Therefore we use (4.18) to eliminate the trace involving the E_i's in (4.17), with the final result:

$$F(\lambda) = F(0) + W_1 + 2kT \sum_{p>0} \ln \{(1 - e^{-\epsilon'(p)/kT})/(1 - e^{-\epsilon(p)/kT})\}, \tag{4.19}$$

where ϵ and ϵ' are given in (4.15) and (4.16). It is noteworthy that the ground state and free energy both diverge in the case of a δ-function potential.

V. EVALUATION OF THE MOMENTUM DISTRIBUTION

In this section we calculate the mean number of particles with momentum k. This quantity is $\bar{n}_k$ and is the expectation value of

$$n_k = b_k^+ b_k \tag{5.1}$$

in the ground state. Since $\bar{n}_k$ is an even function of k we need only consider $k > 0$, and it is further convenient to introduce a Fourier transform so that [using (2.4)]

$$\bar{n}_k = \frac{1}{L} \iint_0^L ds\, dt\, e^{ik(s-t)} I(s, t). \tag{5.2}$$

Here

$$I(s, t) = \langle\Psi|\, \psi_1^+(s)\psi_1(t)\, |\Psi\rangle = \langle\Psi_0|\, e^{iS}\psi_1^+(s)e^{-iS}e^{iS}\psi_1(t)e^{-iS}\, |\Psi_0\rangle, \tag{5.3}$$

where S is given by (4.7), Ψ is the new ground state, and Ψ_0 is the noninteracting ground state which is filled with b particles between $-k_F$ and k_F and has no holes (or c particles). This assignment depends on there having been no level crossing, which can be readily verified using (4.7)–(4.13).

In order to calculate the quantity $e^{iS}\psi_1(t)e^{-iS}$ we introduce the auxiliary operator

$$f_\sigma(t) = e^{i\sigma S}\psi_1(t)e^{-i\sigma S}, \tag{5.4}$$

where σ is a c number. We observe that $f_1(t)$ is the desired quantity while

$$f_0(t) = \psi_1(t). \tag{5.5}$$

In addition,

$$\partial f/\partial\sigma = e^{i\sigma S} i[S, \psi_1(t)]e^{-i\sigma S} = e^{i\sigma S}[2\pi/L \sum_p \rho_2(-p)\varphi(p)p^{-1}e^{ipt}]e^{-i\sigma S} f_\sigma(t), \tag{5.6}$$

where we have used the commutation relations (3.7) as well as the fact that ψ_1 commutes with ρ_2. Equation (5.6) is a differential equation for $f_\sigma(t)$ and (5.5) is the boundary condition. The solution is

$$f_\sigma(t) = W_\sigma(t)R_\sigma(t)\psi_1(t), \tag{5.7}$$

where

$$W_\sigma(t) = \exp\{2\pi/L \sum_{p>0} [\rho_1(-p)e^{ipt} - \rho_1(p)e^{-ipt}]p^{-1}[\cosh \sigma\varphi(p) - 1]\} \tag{5.8}$$

and

$$R_\sigma(t) = \exp\{2\pi/L \sum_{p>0} [\rho_2(-p)e^{ipt} - \rho_2(p)e^{-ipt}]p^{-1} \sinh \sigma\varphi(p)\} \tag{5.9}$$

The reader may verify that (5.7) satisfies (5.5) and (5.6) by using the commutation relations (3.7). We recall the well-known rule that

$$\exp(A + B) = \exp(A)\exp(B)\exp(-1/2[A, B]) \tag{5.10}$$

when $[A, B]$ commutes with A and B. From here on we shall set $\sigma = 1$ and drop it as a subscript. We note that since $\rho_1(p)^+ = \rho_1(-p)$ and $\rho_2(p)^+ = \rho_2(-p)$,

$$R^+(t) = R^{-1}(t) \quad \text{and} \quad W^+(t) = W^{-1}(t). \tag{5.11}$$

We also note that R and W commute with each other. Thus, (5.3) becomes

$$I(s, t) = \langle\Psi_0|\, \psi_1^+(s)R^{-1}(s)W^{-1}(s)W(t)R(t)\psi_1(t)\, |\Psi_0\rangle = I_1(s, t)I_2(s, t), \tag{5.12}$$

where

$$I_1(s, t) = \langle\Psi_1|\, \psi_1^+(s)W^{-1}(s)W(t)\psi_1(t)\, |\Psi_1\rangle, \quad I_2(s, t) = \langle\Psi_2|\, R^{-1}(s)R(t)\, |\Psi_2\rangle. \tag{5.13}$$

We have used the fact that the ground state is a product state: $\Psi_0 = \Psi_1 * \Psi_2$ where Ψ_1 is a state of the "1" field and Ψ_2 is a state of the "2" field. Ψ_1 is filled with b particles up to $+k_F$ and has no c particles; Ψ_2 is filled with b particles down to $-k_F$ and has no c particles.

Now, using the definition (5.8) and the rule (5.10) we easily find that

$$W^{-1}(s)W(t) = W_-(s, t)W_+(s, t)Z_1(s, t), \tag{5.14}$$

with

$$W_+(s, t) = \exp\{2\pi/L \sum_{p>0} \rho_1(-p)[\cosh\varphi(p) - 1] \times p^{-1}(e^{ipt} - e^{ips})\},$$

$$W_-(s, t) = \exp\{2\pi/L \sum_{p>0} \rho_1(p)[\cosh\varphi(p) - 1] \times p^{-1}(e^{-ips} - e^{-ipt})\},$$

$$Z_1(s, t) = \exp\{2\pi/L \sum_{p>0} [\cosh\varphi(p) - 1]^2 \times p^{-1}(e^{ip(s-t)} - 1)\}. \tag{5.15}$$

Likewise,

$$R^{-1}(s)R(t) = R_-(s, t)R_+(s, t)Z_2(s, t), \tag{5.16}$$

with

$$R_+(s, t) = \exp\{2\pi/L \sum_{p>0} \rho_2(p)[\sinh\varphi(p)] \times p^{-1}(e^{-ips} - e^{-ipt})\},$$

$$R_-(s, t) = \exp\{2\pi/L \sum_{p>0} \rho_2(-p)[\sinh\varphi(p)] \times p^{-1}(e^{ipt} - e^{ips})\},$$

$$Z_2(s, t) = \exp\{2\pi/L \sum_{p>0} [\sinh\varphi(p)]^2 \times p^{-1}(e^{ip(t-s)} - 1)\}. \tag{5.17}$$

We see at once from the definition (3.1b), (3.2b), of $\rho_1(p)$ that, for $p > 0$, $\rho_1(-p)\,|\Psi_1\rangle = 0$. Similarly $\langle\Psi_1|\,\rho(p) = 0$, $\rho_2(p)\,|\Psi_2\rangle = 0$, and $\langle\Psi_2|\,\rho_2(-p) = 0$. Hence,

$$I_2(s, t) = Z_2(s, t)$$

and

$$I_1(s, t) = Z_1(s, t)\langle\Psi_1|\, W_-^{-1}\psi_1^+(s)W_-W_+\psi_1(t)W_+^{-1}\, |\Psi_1\rangle. \tag{5.18}$$

If we now define

$$h_+(y) = 2\pi/L \sum_{p>0} [\cosh\varphi(p) - 1] \times p^{-1}(e^{ipt} - e^{ips})e^{-ipy},$$

$$h_-(y) = 2\pi/L \sum_{p>0} [\cosh\varphi(p) - 1] \times p^{-1}(e^{-ipt} - e^{-ips})e^{ipy}, \tag{5.19}$$

combining (3.10) and (5.15) we have that

$$W_+(s, t) = \exp \int_0^L N_1(y)h_+(y)\,dy, \quad W_-(s, t) = \exp -\int_0^L N_1(y)h_-(y)\,dy. \tag{5.20}$$

Since

$$[\psi_1(x), N_1(y)] = \delta(x - y)\psi_1(x), \quad [\psi_1^+(x), N_1(y)] = -\delta(x - y)\psi_1^+(x), \tag{5.21}$$

it follows that

$$W_+(s, t)\psi_1(t)W_+^{-1}(s, t) = \psi_1(t)\exp[-h_+(t)], \quad W_-^{-1}(s, t)\psi_1^+(s)W_-(s, t) = \psi_1^+(s)\exp[+h_-(s)]. \tag{5.22}$$

Finally,

$$\langle\Psi_1|\ \psi_1^+(s)\psi_1(t)\ |\Psi_1\rangle = 1/L \sum_{p\leq k_F} e^{ip(t-s)} \equiv Z_3(s,\ t). \qquad (5.23)$$

Combining all these results, we conclude that

$$I(s,\ t) = Z_0(s,\ t)Z_1(s,\ t)Z_2(s,\ t)Z_3(s,\ t), \qquad (5.24)$$

where

$$Z_0(s,\ t) = \exp\,(h_-(s) - h_+(t)) = \exp\,\{-4\pi/L \sum_{p>0} [\cosh \varphi(p) - 1] \times (1 - e^{ip(s-t)})\}. \qquad (5.25)$$

In order to make a comparison with Luttinger's calculation of $\bar{n}_k$, we first observe that the functions $Z_i(s,\ t)$ are really functions of $r = s - t$ and that they are periodic in s and t in $(0,\ L)$. We then define the functions $G(r)$ and $Q(r)$ as follows:

$$\exp\,[-Q(r)] \equiv G(r) \equiv Z_0(r)Z_1(r)Z_2(r). \qquad (5.26)$$

Substituting (5.26), (5.24), and (5.23) into (5.2) we obtain

$$\bar{n}_k = 2\pi/L \sum_{p\leq k_F} F(k-p), \qquad (5.27)$$

where

$$F(k) = 1/2\pi \int_{-\frac{1}{2}L}^{\frac{1}{2}L} dr\, e^{ikr}e^{-Q(r)} \qquad (5.28)$$

$$\cong 1/2\pi \int_{-\infty}^{\infty} dr\, e^{ikr}e^{-Q(r)}. \qquad (5.29)$$

In (5.29) we have passed to the bulk limit N, $L\rightarrow\infty$, not an approximation.

At this point our expression for $\bar{n}_k$ is formally the same as Luttinger's [cf. L (52), L (69)]. The difference is that our Q is different from his. He obtains Q by evaluating an infinite Toeplitz determinant with the result that [L (70)]

$$Q(r) = \lambda^2/2\pi^2 \int_0^\infty dp\, \frac{1-\cos pr}{p}\, |v(p)|^2. \quad \text{(Luttinger)} \qquad (5.30)$$

Our Q, which is the correct one to use, is obtained by combining (5.15), (5.17), and (5.25), replacing sums by integrals in the usual way, and using the definition (4.11) of $\varphi(p)$. The result is

$$Q(r) = \lambda^2/2\pi^2 \int_0^\infty dp\, \frac{1-\cos pr}{p}\, |u(p)|^2, \qquad (5.31)$$

where

$$|u(p)|^2 = (2\pi^2/\lambda^2)\{(1-(\lambda v(p)/\pi)^2)^{-\frac{1}{2}} - 1\}. \qquad (5.32)$$

It is worth noting that (5.30) agrees with (5.31) to leading order in λ^2.

Since we have not yet specified $v(p)$, we may now follow Luttinger's discussion from this point on with the proviso that we use the correct (λ dependent) $u(p)$ instead of $v(p)$. The reader is referred to pages 1159 and 1160 of Luttinger's paper.

There are two main conclusions one can draw. The first is that if we start with a δ-function interaction [so that $v(p)$ and hence $u(p)$] are constants, it can be shown that $\bar{n}_k = \frac{1}{2}$ for all k. Such a result is quite unphysical, but it is not unreasonable because the ground-state energy W (4.13a) diverges when $v(p) =$ constant at large p. Also, the result would be the same if we started with the more physical interaction

$$H' = 1/L \sum_p \{\rho_1(p) + \rho_2(p)\}\{\rho_1(-p) + \rho_2(-p)\}v(p)$$

discussed in the Appendix. This is indeed unfortunate, because relativistic field theories usually begin with local (δ-function) interactions.

The second conclusion is that if one makes a reasonable assumption about $v(p)$, and hence about $u(p)$ and $Q(r)$, one finds that for k in the vicinity of k_F, $\bar{n}_k$ behaves like

$$\bar{n}_k \sim d - e\,|k-k_F|^{2\alpha}\,\sigma(k-k_F), \qquad (5.33)$$

where

$$\sigma(k) = 1, \quad k>0; \qquad = -1, \quad k<0 \qquad (5.34)$$

and d, e, and α are certain positive constants. Now in Luttinger's calculation

$$\alpha = \lambda^2/4\pi^2 v(0)^2, \quad \text{(Luttinger)} \qquad (5.35)$$

[cf. L(75)], where $v(0) \equiv \lim_{p\rightarrow 0} v(p)$.

If $2\alpha < 1$, then the conclusion to be drawn is that although the interaction removes the discontinuity in $\bar{n}_k$ at the Fermi surface, we are left with a function that has an infinite slope there. There is, so to speak, a residual Fermi surface. In Sec. IV of his paper, Luttinger shows that at least for one example of $v(p)$ perturbation theory gives the same qualitative result as (5.33) with the same value of α, (5.35).

If, on the other hand, $2\alpha > 1$ then there is no infinite derivative at the Fermi surface. $\bar{n}_k$ is perfectly smooth there (although, technically speaking, it is nonanalytic unless $2\alpha =$ odd integer.) In this case virtually all trace of the Fermi surface has been eliminated. But notice that the correct α to use is obtained by replacing $v(0)$ by $u(0) \equiv \lim_{p\rightarrow 0} u(p)$ in (5.35), i.e.,

$$2\alpha = \{1 - [\lambda v(0)/\pi]^2\}^{-\frac{1}{2}} - 1. \qquad (5.36)$$

Thus, even subject to the requirement that $|\lambda v(0)|$ be less than π, 2α can become as large as one pleases. Yet perturbation theory predicts (5.35) which yields 2α always less than $\frac{1}{2}$.

We may conclude that a strong enough interaction can eliminate the Fermi surface, while perturbation theory predicts that is always there.

VI. DIELECTRIC CONSTANT

Because the response to external fields of wave vector q only depends on an interaction expression linear in the density operators, we can immediately obtain for the generalized static susceptibility function or *dielectric constant* (response ÷ driving force), for any temperature, T

$$\chi_\lambda(q, T) = \chi_0(q, T)\{\sinh\varphi(q) + \cosh\varphi(q)\}^2 \cosh 2\varphi_q$$
$$= \chi_0(q, T)\frac{1}{1 + \lambda v(q)/\pi} \qquad (6.1)$$

in terms of the "unperturbed" susceptibility $\chi_0(q, T)$. It is also a simple exercise to calculate exactly the time dependent susceptibility in terms of the "unperturbed" quantity.

It is interesting to note that the susceptibility can diverge (which is symptomatic of a phase transformation) only for

$$\lambda v(q) \to -\pi, \qquad (6.2)$$

i.e. only for sufficiently *attractive* interactions and not for repulsive $[v(q) > 0]$ interactions.

Recently Ferrell[6] advanced plausible arguments why a one-dimensional metal cannot become superconducting. We can prove this rigorously in the present model. The electron–phonon interaction is

$$H_{\mathrm{el-ph}} = \sum_p g(p)[\rho_1(p) + \rho_2(p)]\cdot[\xi_p + \xi_{-p}^+], \qquad (6.3)$$

where ξ and ξ^+ are the phonon field operators. In the "filled-sea" limit this coupling is bilinear in harmonic-oscillator operators, and therefore the Hamiltonian continues to be exactly diagonalizable. The new normal modes can be calculated and there is found to be no phase transition at any finite temperature.

APPENDIX

We shall be interested in extending Luttinger's model in two ways. Firstly, we note that the restriction $\bar{V} = 0$ is really not necessary. Turning back to Eqs. (2.13) *et seq.* we impose periodic boundary conditions $\Psi(\cdots, x_i + L, \cdots) = \Psi(\cdots, x_i, \cdots)$, and find that

$$(q + N\lambda\bar{V}) \text{ and } (k + M\lambda\bar{V}) = 2\pi/L \times \text{integer} \quad (A1)$$

replace the usual condition (2.16), where N = number of "1" particles and M = number of "2" particles. However, when $N, M \to \infty$ in the field-theoretic limit the problem evidently becomes ill-defined unless $\bar{V} \equiv 0$.

A less trivial observation concerns the form of the interaction potential. There is no reason to restrict it to the form $\propto \rho_1\rho_2$, and in fact the more realistic two-body interaction

$$H' = \frac{\lambda}{L}\sum_p v(p)\{\rho_1(-p) + \rho_2(-p)\}\{\rho_1(p) + \rho_2(p)\} \qquad (A2)$$

is fully as soluble as the one assumed in the text, for any strength positive $v(p)$, and provided only

$$\lambda v(p) > -\tfrac{1}{2}\pi, \qquad (A3)$$

i.e. provided no Fourier component is *too* attractive. The shift in the ground-state energy is now given by

$$W_2 = \sum_{p>0} p\left\{\left(1 + \frac{2\lambda v(p)}{\pi}\right)^{\frac{1}{2}} - 1\right\}. \qquad (A4)$$

The plasmon energy is now

$$\epsilon''(p) \equiv |p|\,(1 + 2\lambda v(p)/\pi)^{\frac{1}{2}} \qquad (A5)$$

and *for the important case of the Coulomb repulsion*, $v(p) = p^{-2}$, the plasmons describe a relativistic boson field with mass

$$m^* \equiv (2\lambda/\pi)^{\frac{1}{2}} \qquad (A6)$$

and dispersion

$$\epsilon''(p) = (p^2 + m^{*2})^{\frac{1}{2}}. \qquad (A7)$$

Here, too, $d\epsilon''/dp < 1$.

ACKNOWLEDGMENTS

We thank D. Jepsen, L. Landovitz, and T. Schultz for interesting discussions during various stages of this work. We especially thank J. Luttinger for performing a detailed review of his own work and ours, and thus corroborating the results of the present article. While reading our manuscript, Professor Luttinger called our attention to Schwinger's relevant, indeed prophetic remarks (cf. Ref. 5), and we thank him for this reference. We wish to thank Dr. A. S. Wightman, for pointing out to us that P. Jordan first discovered the bosons associated with fermion fields (in his attempts to construct a neutrino theory of light), and for a list of references.[7]

[6] R. A. Ferrell, Phys. Rev. Letters 13, 330 (1964).

[7] P. Jordan, Z. Physik 93, 464 (1935); 98, 759 (1936); 99, 109 (1936); 102, 243 (1936); 105, 114 (1937); 105, 229 (1937). M. Born and N. Nagendra-Nath, Proc. Ind. Acad. Sci. 3, 318 (1936). A. Sokolow, Phys. Z. der Sowj. 12, 148 (1937).

PHYSICAL REVIEW VOLUME 182, NUMBER 2 10 JUNE 1969

Tomonaga's Model and the Threshold Singularity of X-Ray Spectra of Metals

K. D. SCHOTTE* AND U. SCHOTTE†
University of California, San Diego, La Jolla, California 92037
(Received 3 February 1969)

Singularities near the threshold of the soft x-ray spectra of metals have been predicted by Mahan and have recently been calculated by Nozières *et al.* using the model of a localized core hole. We show that the singular behavior can be understood in terms of density waves of the conduction electrons which are excited when, in the absorption process, the core hole is created, providing an attractive potential for the conduction electrons. For the description of the conduction electrons in terms of density waves, Tomonaga's model is adopted.

1. INTRODUCTION

In the x-ray absorption process, a deep-lying core electron is excited into the conduction band by an incoming x-ray photon. The core hole left behind acts as a one-body potential on the conduction electrons. A simplified model has been used to describe the situation[1]:

$$H=\sum_k \epsilon_k a_k^\dagger a_k+E_0 b^\dagger b+\frac{1}{N}\sum_{k,k'} V_{kk'}a_k^\dagger a_{k'}bb^\dagger. \quad (1)$$

Here the hole (described by $b^\dagger, b$) is represented by a single nondegenerate level E_0 with infinite lifetime. It is assumed to interact with the free conduction electrons (described by $a_k^\dagger$, a_k) via a contact potential $V_{kk'}=V$; the interaction between conduction electrons has been neglected.

We confine ourselves to the discussion of the adsorption process. In this case the transition role according to the Golden Rule is

$$W(\omega)=2\pi\sum_f|\sum_k w_k(\omega)\langle f|a_k^\dagger b|i\rangle|^2\delta(\bar{E}_i+\omega-\bar{E}_f). \quad (2)$$

For simplicity the matrix elements of the dipole operator $w_k(\omega)$ will be regarded as constant. In an equivalent one-body description, the transition rate is given by

$$W(\omega)=2\pi w^2\sum_f\left|\left\langle f_{n+1}\left|\frac{1}{\sqrt{N}}\sum_k a_k^\dagger\right|i_n\right\rangle\right|^2 \times\delta(E_i-E_f+E_0+\omega). \quad (3)$$

The initial state

$$|i_n\rangle=\prod_{k=0}^{k_F} a_k^\dagger|0\rangle$$

is the n-particle ground state of the Hamiltonian

$$H_i=\sum_k \epsilon_k a_k^\dagger a_k, \quad (4)$$

with the energy $E_i=\bar{E}_i-E_0$. The final states $|f_{n+1}\rangle$ are the $(n+1)$-particles eigenstates of the Hamiltonian

$$H_f=\sum_k \epsilon_k a_k^\dagger a_k+\frac{V}{N}\sum_{k,k'} a_k^\dagger a_{k'}. \quad (5)$$

The final states $|f_{n+1}\rangle$ hold the clue to the problem. According to Anderson,[2] the overlap between the initial and any final state containing a finite number of electron-hole pairs is zero in the limit of infinite volume.

So one has the choice of doing the calculations with finite volume, performing the limiting process to infinite volume in the end, or of circumventing the problem to determine the final states by using Green's-function techniques.[1] However, the physical origin of the singular behavior of the transition rate near the threshold[1,3] does not seem to be clearly understood.

In this work, we offer another way of calculating the response of the free electrons to the sudden switching on of the core hole potential. We will use the Tomonaga model[4] by which we can describe the excitations of the Fermi sea in terms of density waves. On the one hand, we circumvent the difficulty posed by the Anderson theorem, and, on the other hand, we have a plausible physical interpretation of what is going on.

2. FORMULATION OF THE PROBLEM IN TERMS OF TOMONAGA'S BOSONS

We introduce the density operator

$$\rho_k=\sum_{k_1=0}^{k_D-k}\frac{1}{\sqrt{N}}a_{k_1}^\dagger a_{k_1+k},$$
$$\rho_{-k}=\sum_{k_1=k}^{k_D}\frac{1}{\sqrt{N}}a_{k_1}^\dagger a_{k_1-k},\quad k\geqslant 0 \quad (6)$$

and consider a band of width D, assuming a constant density of states within the band. $k=0$ and $k_D=k$ are the momenta at the bottom and at the top of band, respectively. The electron energies are given by

$$\epsilon_k=(k-k_F)/\rho, \quad (7)$$

where ρ is the density of states.

---

* Supported by the Deutsche Forschungs-Gemeinschaft. Supported by the U. S. National Science Foundation and the Office of Naval Research.

† U. S. gratefully acknowledges a Travel Grant from the Deutsche Forschungs-Gemeinschaft. On leave of absence of the Institut für Theoretische Physik der Universität zu Köln.

[1] P. Nozières and C. T. de Dominicis, Phys. Rev. 178, 1084 (1969); B. Roulet, I. Gavoret, and P. Nozières, *ibid.* 178, 1252 (1969).

[2] P. W. Anderson, Phys. Rev. Letters 18, 1049 (1967).

[3] G. D. Mahan, Phys. Rev. 163, 612 (1967).

[4] S. Tomonaga, Progr. Theoret. Phys. (Kyoto) 5, 544 (1950).

Actually, in the Tomonaga model the free-particle energy has to be proportional to the momentum. Another important feature of the Tomonaga model is its one-dimensionality. But the problem considered here can be looked upon as a one-dimensional one because it involves only s-wave scattering. So the summation in (6) over k_1 is to be understood as a summation over energy shells. We examine now the commutation relations of the ρ_k's:

$$[\rho_k,\rho_{k'}]=0 \quad \text{for} \quad k, k'>0 \quad \text{and} \quad k, k'<0, \tag{8}$$

$$[\rho_k,\rho_{-k'}]=\frac{1}{N}\sum_{k_1=0}^{k'} a_{k_1}^\dagger a_{k_1+k-k'} - \frac{1}{N}\sum_{k_1=k_D-k}^{k_D-k+k'} a_{k_1}^\dagger a_{k_1+k-k'} \quad \text{for} \quad k>k'>0, \tag{9}$$

$$[\rho_k,\rho_{-k}]=\frac{1}{N}\sum_{k_1=0}^{k} a_{k_1}^\dagger a_{k_1} - \frac{1}{N}\sum_{k_1=k_D-k}^{k_D} a_{k_1}^\dagger a_{k_1}. \tag{10}$$

We adopt Tomonaga's approximation and substitute these complicated commutation relations by simpler ones:

$$[\rho_k,\rho_{-k'}]=k\delta_{kk'}. \tag{11}$$

For a detailed discussion of this important step we refer to Tomonaga's paper. In short, the simpler commutation relations may be used if only a certain subspace of all possible states has to be considered. These are the states which do not have unoccupied levels deep in the bottom of the band or occupied levels high above the Fermi energy at the top of the band, such that the parts left out of the commutator do not contribute very much. In addition, we have to assume that V be very small compared to the bandwidth. On the one hand we had to consider a contact potential to keep mathematics simple, but, on the other hand, a contact potential causes excitations deep below and high above the Fermi level (or short density waves), which is explicitly excluded in the model. So we must expect all our results to be valid only for small V.

With (11) we calculate the commutator

$$[H_i,\rho_k]=-(k/\rho)\rho_k, \tag{12}$$

where use has been made of the linear energy-momentum dependence, and we conclude that H_i has the form[5]

$$\tilde{H}_i=\sum_{k>0}\frac{1}{\rho}\rho_{-k}\rho_k. \tag{13}$$

The transcription of H_f [Eq. (5)] in terms of bosons also leads to a very simple expression,

$$\tilde{H}_f=\sum_{k>0}\frac{1}{\rho}\rho_{-k}\rho_k+\frac{V}{\sqrt{N}}\sum_{k>0}(\rho_k+\rho_{-k}). \tag{14}$$

[5] The tacit assumption is the completeness of the k, which is only fulfilled in the subspace of s-wave states.

It is convenient to introduce the normalized boson operators ($k>0$)

$$b_k=(1/\sqrt{k})\rho_k, \quad b_k^\dagger=(1/\sqrt{k})\rho_{-k}, \tag{15}$$

which obey the usual boson commutation relations

$$[b_k,b_{k'}^\dagger]=\delta_{kk'}. \tag{16}$$

In terms of bosons

$$\tilde{H}_i=\sum_{k>0}\frac{k}{\rho}b_k^\dagger b_k, \tag{17}$$

and

$$H_f=\sum_{k>0}\frac{k}{\rho}\left(b_k^\dagger+\frac{V}{\sqrt{(kN)}}\right)\left(b_k+\frac{V}{\sqrt{(kN)}}\right)-\frac{V^2\rho}{N}\sum_k 1. \tag{18}$$

$\tilde{H}_i$ and $\tilde{H}_f$ describe a set of harmonic oscillators. The effect of the potential is a shift of the zero points of the harmonic oscillators. As the transformation procedure is only correct up to a constant we will drop the constant term on the right-hand side of Eq. (18) in the following: In order to calculate the transition rate (3) we have to express the operator

$$a^\dagger=\frac{1}{\sqrt{N}}\sum_{k_1} a_{k_1}^\dagger$$

in terms of harmonic-oscillator coordinates. This is a rather delicate problem. First we examine the commutation relations between $a^\dagger$ and the density operators ρ_k. We find

$$[\rho_k,a^\dagger]=\frac{1}{\sqrt{N}}a^\dagger-\frac{1}{N}\sum_{k_1=k_D-k}^{k_D} a_{k_1}^\dagger \quad \text{for} \quad k>0, \tag{19}$$

$$[\rho_{-k},a^\dagger]=\frac{1}{\sqrt{N}}a^\dagger-\frac{1}{N}\sum_{k_1=0}^{k} a_{k_1}^\dagger \quad \text{for} \quad k>0. \tag{20}$$

We simplify these commutation relations in the spirit of Tomonaga's approximation, i.e., the matrix elements

$$\left\langle f_{n+1}\left|\frac{1}{N}\sum_{k_1=0}^{k} a_{k_1}^\dagger\right|i_n\right\rangle$$

are considered to be small. This is correct within the Tomonaga model and means that the energy of the final states has to be small compared to the bandwidth. So, we shall use

$$[\rho_k,a^\dagger]=(1/\sqrt{N})a^\dagger \quad \text{for all } k>0 \tag{21}$$

instead of the exact commutation relations [Eqs. (19) and (20)]. Defining

$$U=\exp[\sum_k \alpha_k(b_k^\dagger-b_k)], \quad \alpha_k \text{ real}, \tag{22}$$

we have a canonical transformation which acts as a translation operator on any function of $b_k^\dagger$ and b_k, namely,

$$U^\dagger b_k U = b_k + \alpha_k \quad \text{or} \quad [b_k, U] = \alpha_k U. \qquad (23)$$

This relation is identical to (21), if we put $\alpha_k = 1/\sqrt{(kN)}$ and

$$a^\dagger \propto \exp\left[\sum_k \frac{1}{\sqrt{(kN)}}(b_k{}^\dagger - b_k)\right]. \qquad (24)$$

3. TRANSITION PROBABILITY IN THE DENSITY WAVE MODEL

Before calculating the transition rate, let us ask how to interpret the Golden Rule [Eq. (3)] in the new description. The initial state is the lowest eigenstate of a set of harmonic oscillators [Eq. (17)]—the analog to the "quiescent" Fermi sea. The final states are all possible excitations of another set of oscillators with shifted zero points [see Eq. (18)]. The ground state of this second set is also the analog to a quiescent Fermi sea, which is (in the language of the fermion description) a Slater determinant of scattering waves. Excitations in both oscillator systems can be pictured as density waves.

According to (24), $a^\dagger$ acts as a shift operator distorting, say, the initial quiescent Fermi sea. That is, by injecting the core-state electron into the band, a local inhomogeneity of the electron density is created, by which each oscillator is elongated by a small amount. At the same time, the oscillators have to react to the core-hole potential V produced in the absorption process. This is described by the zero-point shift of the second set of oscillators. As will be seen later these two effects are additive. This picture of the problem has some similarity to the "small polaron," where now the real lattice is replaced by the band electrons, and the real phonons are replaced by Tomonaga's "phonons."

Instead of the transition rate [Eq. (3)], we calculate the correlation function

$$\mathfrak{F}(t) = \langle i|\exp(+i\tilde{H}_i t)\ a \exp(-i\tilde{H}_f t)\ a^\dagger|i\rangle, \qquad (25)$$

which is related to $W(\omega)$ by

$$W(\omega) \propto \operatorname{Im}\frac{i}{\pi}\int_0^\infty e^{i(\omega+E_0)t}\mathfrak{F}(t)dt. \qquad (26)$$

The canonical transformation

$$U_V = \exp\left[V\rho \sum_k \frac{1}{\sqrt{(kN)}}(b_k{}^\dagger - b_k)\right] \qquad (27)$$

transforms $\tilde{H}_i$ into $\tilde{H}_f$:

$$\tilde{H}_f = U_V{}^\dagger \tilde{H}_i U_V, \qquad (28)$$

so that

$$\mathfrak{F}(t) = \langle i|\exp(+i\tilde{H}_i t) a U_V{}^\dagger \exp(-i\tilde{H}_i t) U_V a^\dagger|i\rangle, \qquad (29)$$

with

$$U a^\dagger \equiv B = \exp\left[(1+V\rho)\sum_k \frac{1}{\sqrt{(kN)}}(b_k{}^\dagger - b_k)\right] \qquad (30)$$

and

$$B(t) = \exp\left[(1+V\rho)\sum_k \frac{1}{\sqrt{(kN)}} \times (b_k{}^\dagger e^{+i(k/\rho)t} - b_k e^{-i(k/\rho)t})\right]. \qquad (31)$$

We have

$$\mathfrak{F}(t) = \langle i|B^\dagger(t)B(0)|i\rangle, \qquad (32)$$

which is equal to[6]

$$\mathfrak{F}(t) = \exp\left((1+V\rho)^2\frac{1}{N}\sum_{k>0}\frac{1}{k}(e^{-i(k/\rho)t}-1)\right); \qquad (33)$$

converting the sum in an integral, introducing a cutoff[7]

$$\frac{1}{N}\sum_{k>0}^{k_{\max}}\frac{1}{k}(e^{-i(k/\rho)t}-1) = \int_0^{itk_{\max}(1/\rho)}\frac{e^{-x}-1}{x}dx, \qquad (34)$$

we find as the leading term for large t

$$\mathfrak{F}(t) \sim [t^{(1+V\rho)^2}]^{-1}. \qquad (35)$$

Inserting $-V\rho = \delta_B/\pi$, where δ_B is the phase shift in Born approximation, we finally get for the energy dependence near the threshold

$$W(\omega) \sim (\omega + E_0)^{-[2\delta_B/\pi - (\delta_B/\pi)^2]}. \qquad (36)$$

Thus, we have recovered the main feature of Nozière's result, namely, the singular threshold behavior of the response function. But in the correct answer δ_B is replaced by δ, the exact phase shift. In this context it is amusing to note that in the Luttinger model,[8] where an infinite energy spectrum with a linear dispersion is assumed from the start, δ_B is the correct phase shift. The Schrödinger equation (for s waves) for a linear dispersion reads

$$[-(i/\rho)(d/dx) + V(x)]\Psi(x) = (E - E_F)\Psi(x), \qquad (37)$$

and the solution is given by

$$\Psi(x) = \exp\left[ikx - i\rho\int_0^x V(x)dx\right], \qquad (38)$$

so that

$$\Psi(x) \sim e^{ikx+i\delta} \quad \text{as} \quad x \to \infty, \qquad (39)$$

with

$$\delta = -\rho\int_0^\infty V(x)dx = -\pi\rho V \qquad (40)$$

---

[6] To evaluate an expression of the form $\langle e^A e^B\rangle$ [compare (32)] one makes use of the well-known relations $e^A e^B = e^{A+B+(1/2)[A,B]}$, which holds if $[A,B]$ is a c number, and $\langle e^{L(b^\dagger,b)}\rangle = \exp\frac{1}{2}\langle L^2(b^\dagger,b)\rangle$, where L is any linear combination of Bose operators.

[7] According to the Tomonaga model, $k_{\max} \approx \frac{1}{2}k_f$.

[8] J. M. Luttinger, J. Math. Phys. 4, 1154 (1963).

for $V(x)=2\pi V\delta(x)$, which has been used in the calculations above.

If one wants to stick to a more realistic model with a finite energy band, there are different ways to improve our result. The Luttinger model indicates that the incomplete result is in part due to the linear energy-momentum dispersion. A more realistic, namely, the quadratic dispersion leads to anharmonic terms, as can be seen from Schick's[9] paper. Another way would be to use the energy as variable instead of the momentum. But while this leaves the kinetic-energy term simple, the potential energy V would be very difficult to handle, again leading to anharmonic terms. An important point any way is to introduce a more realistic potential of arbitrary strength. But whatever one tries for larger V, one is soon struck with anharmonic effects. This should be sufficient to elucidate the situation. As there are other ways to calculate the exact transition rate,[1,10] we did not try to solve the anharmonic-oscillator problem.

ACKNOWLEDGMENTS

The authors are indebted to Professor W. Kohn for bringing the Tomonaga model to their attention, and to Professor L. Sham for critically reading the manuscript.

[9] M. Schick, Phys. Rev. 166, 404 (1968).

[10] K. D. Schotte (to be published).

New wave-operator identity applied to the study of persistent currents in 1D

Daniel C. Mattis*

Belfer Graduate School of Science, Yeshiva University, New York, New York 10033
(Received 31 October 1973)

We show that a large class of backward-scattering matrix elements involving $\Delta k \sim \pm 2k_F$ vanish for fermions interacting with two-body attractive forces in one dimension. (These same matrix elements are finite for noninteracting particles and infinite for particles interacting with two-body repulsive forces.) Our results demonstrate the possibility of persistent currents in one dimension at $T = 0$, and are a strong indication of a metal-to-insulator transition at $T = 0$ for repulsive forces. They are obtained by use of a convenient representation of the wave operator in terms of density-fluctuation operators.

INTRODUCTION

It is usual to express the density-fluctuation operators[1] $\rho(p)$ as bilinear forms of the wave operator $\Psi(x)$. We have recently succeeded in inverting the process, expressing the wave operator as an exponential form of the density fluctuation operators, in the special case of Luttinger's soluble model of interacting fermions in one dimension[2]. While certain aspects of our procedure could obviously be used in other applications[3] or even adapted to the case of electrons in three dimensions, we limit the present application to the challenging question, of whether persistent currents (i.e., *supercurrents*) can exist in one dimension despite arbitrary random scattering potentials. The surprising result is that, for sufficiently attractive two-body forces, a current-carrying state at $T = 0$ can have infinite lifetime regardless of the strength of the scattering mechanisms. Therefore, it is proved rigorously that superconductivity can exist, at $T = 0$, in one dimension, despite the well-known lack of long range order. We also find the converse, that for sufficiently repulsive two-body forces, the lifetime of a current-carrying state at $T = 0$ tends to zero, and the system acquires the attributes of an insulator. The nontrivial generalization of these results to finite temperature is the subject of an ongoing, separate, study.

DETAILS OF THE MODEL

We first recall certain aspects of the soluble many-fermion model[2] under scrutiny. It consists of right-going particles (labeled 1) having constant velocity v_0, and left-going particles (labeled 2) with velocity $-v_0$, with interactions characterized by a two-body potential $V(x - x')$ and coupling constant λ, obeying a Hamiltonian:

$$\mathcal{H} = v_0 \sum_k k(n_{1k} - n_{2k}) + (\lambda/L) \sum_p U(p)[\rho_1(p) + \rho_2(p)][\rho_1(-p) + \rho_2(-p)], \quad (1)$$

where p, k refer to wave numbers, $U(p)$ is the Fourier transform of $V(x - x')$ and the various operators are

$$n_{ik} = a^{+}_{ik} a_{ik}, \quad \rho_i(p) = \sum_k a^{+}_{i\, k+p} a_{ik}$$
$$\Psi_i(x) = L^{-1/2} \sum_k a_{ik} e^{ikx} \quad (2)$$

with L = dimension of the space, for purposes of box normalization. The particle-current operator j_{op} takes the form

$$j_{\mathrm{op}} = V_0 \sum_k (n_{1k} - n_{2k}). \quad (3)$$

If, for example, at $T = 0$ we set the coupling constant $\lambda = 0$, we find for the eigenvalue of (3) the value

$$j = v_0 (L/2\pi)(k_{1F} + k_{2F}), \quad (4)$$

for, at $T = 0$, the ground state of the non-interacting particles is described by occupation numbers $n_{1k} = 1$ for $-\infty < k < k_{1F}$ and $n_{1k} = 0$ for $k > k_{1F}$, together with $n_{2k} = 1$ for $k_{2F} < k < +\infty$, and $n_{2k} = 0$ for $k < k_{2F}$. In the ground state, $k_{2F} = -k_{1F}$ and no current flows. In general, however, we can have $k_{1F} \neq -k_{2F}$ and the current eigenvalue j will be nonzero. This conclusion is unaffected by the interactions when $\lambda \neq 0$, for j_{op} commutes with both parts of the Hamiltonian $\mathcal{H}$ separately, and j is therefore a good quantum number until a mechanism for decay of the current is introduced into the Hamiltonian.

Accordingly, we introduce a mechanism allowing electrons to be backward scattered from one branch to the other, in order to test the hypothesis of persistent currents. For definiteness, consider a one-body scattering Hamiltonian $\mathcal{H}'$:

$$\mathcal{H}' = \int dx [W(x) \Psi_2^{+}(x) \Psi_1(x) + \text{h.c.}], \quad (5)$$

where $W(x)$ is a random potential. Because j_{op} does not commute with $\mathcal{H}'$, j is no longer a constant of the motion, and generally decays exponentially:

$$j(t) = j(0) \exp(-t/\tau), \quad (6)$$

where τ = lifetime of the current, is a measure of the strength of the scattering potential $W(x)$ and of the effective density of one-particle states. To probe the latter, we compute the matrix element:

$$M(i \to f) \equiv \langle f | \Psi_2^{+}(x) \Psi_1(x) | i \rangle, \quad (7)$$

in which $|i\rangle$ is the initial, exact, eigenstate of both $\mathcal{H}$ and j_{op}, and $|f\rangle$ is the final eigenstate of these operators. It will be appreciated that if the initial eigenvalue of j_{op} is j, the final eigenvalue is $j - 2v_0$. The matrix elements M enable us to compute the structure of $\mathcal{H}'$ in the Hilbert space of the eigenstates of $\mathcal{H}$. Some of them could be finite, as for noninteracting particles, and then we would have normal decay. But if we find that the matrix elements connecting low-lying states are *all* zero, then there can be no scattering, and the existence of persistent currents is demonstrated. If, on the other hand, some matrix elements are infinite, then we may conclude either that the lifetime τ of a current is zero, or, more accurately, that the effects of $\mathcal{H}'$ are too profound to be taken into account by perturbation theory (for it causes an insulator phase to replace the metallic phase, and this should presumably be taken into account before the effects of the two-body forces.) In the following, we shall find all these possibilities to be realizable, depending on the sign and magnitude of $U(0) = \int dx\, V(x)$.

PRELIMINARY COMPUTATIONS

We start by recalling the unitary transformation S which renders $\exp(iS)\mathcal{H}\exp(-iS)$ diagonal. It has the form[2]

$$S = (2\pi i/L)\sum_{\text{all}\, p} p^{-1}\varphi(p)\rho_1(p)\rho_2(-p). \tag{8}$$

We recall that, owing to the peculiarities of a filled Fermi sea, the ρ's do not all commute, but obey the commutation relations

$$[\rho_i(p), \rho_j(-p)] = \epsilon_i \delta_{ij}\, pL/2\pi \tag{9}$$

in which $\epsilon_1 = -1$ and $\epsilon_2 = +1$. The correct value of φ to diagonalize $\mathcal{H}$ is found to be

$$\varphi(p) = -\tfrac{1}{4}\ln[1 + 2\lambda u(p)], \tag{10}$$

where $u(p) \equiv U(p)/\pi v_0$, so that

$$e^{iS}\mathcal{H}e^{-iS} = (2\pi v_0/L)\sum_{p>0}[1 + 2\lambda u(p)]^{1/2} \times [\rho_1(p)\rho_1(-p) + \rho_2(-p)\rho_2(p)] + W_1. \tag{11}$$

Making use of the commutation relations (9), one sees that $\mathcal{H}$ is reduced to a set of noninteracting harmonic oscillators having characteristic energy $E(p) = v_0 p\,[1 + 2\lambda u(p)]^{1/2}$. W_1 is the vacuum renormalization energy,

$$W_1 = \frac{Lv_0}{2\pi}\int_0^\infty dp\, p\{[1 + 2\lambda u(p)]^{1/2} - 1 - \lambda u(p)\}. \tag{12}$$

To obtain the effect of S on $\mathcal{H}'$, we have found the following new operator identities to be extremely convenient:

$$\Psi_1(x) \Leftrightarrow \frac{e^{ik_{1F}x}}{L^{1/2}}\exp\left(\frac{-2\pi}{L}\sum_{p>0}p^{-1}\rho_1(p)e^{-ipx}\right) \times \exp\left(\frac{2\pi}{L}\sum_{p>0}p^{-1}\rho_1(-p)e^{ipx}\right) \tag{13a}$$

and

$$\Psi_2(x) \Leftrightarrow \frac{e^{ik_{2F}x}}{L^{1/2}}\exp\left[i\pi\int dx'\,\Psi_1^*(x')\Psi_1(x')\right] \times \exp\left(\frac{-2\pi}{L}\sum_{p>0}p^{-1}\rho_2(-p)e^{ipx}\right) \times \exp\left(\frac{2\pi}{L}\sum_{p>0}p^{-1}\rho_2(p)e^{-ipx}\right) \tag{13b}$$

with $p = \text{integer} \times 2\pi/L$. The double arrows indicate that the identities hold in a special sense only; supposing $|F,N\rangle$ to be the ground state Fermi sea corresponding to N particles of type 1 and Ω_1 to be an arbitrary function of the $\rho_1(\pm p)$ operators, we have

$$\Psi_1(x)\Omega_1|F, N+1\rangle = \frac{e^{ik_{1F}x}}{L^{1/2}}\exp\left(\frac{-2\pi}{L}\sum_{p>0}p^{-1}\rho_1(p)e^{-ipx}\right) \times \exp\left(\frac{2\pi}{L}\sum_{p>0}p^{-1}\rho_1(-p)e^{ipx}\right)\Omega_1|F,N\rangle \tag{14}$$

and similarly for $\Psi_2(x)$ and for Hermitean conjugate operators $\Psi_i^*(x)$. Since any state in our Hilbert space can be written in the form $\Omega_1|F,N\rangle$, Eqs. (13), and their Hermitean conjugate relations, are operator identities in every practical sense. We note also that they are kinematical identities, independent of the nature of the dynamical interactions or of the magnitude of the coupling constant.

Applying the unitary transformation S to the wave operators,

$$\Psi_i(x) \to \exp(iS)\Psi_i(x)\exp(-iS), \tag{15}$$

we obtain expressions that are readily evaluated using the bosonlike commutation relations, Eq.(9). We cast such expressions into normal ordering [$\rho_1(-p)$ to the right of $\rho_1(+p)$, $\rho_2(p)$ to the right of $\rho_2(-p)$, with $p>0$]. As an example, consider the bilinear form:

$$\begin{aligned}\Psi_1^\dagger(x')\Psi_1(x) \to \frac{1}{L}&\exp[ik_{1F}(x-x')]\exp\left(\frac{2\pi}{L}\sum_{p>0}p^{-1}e^{ip(x'-x)}\right)\\ &\times\exp\left(-\frac{4\pi}{L}\sum_{p>0}p^{-1}[1-\cos p(x-x')]\sinh^2\varphi_p\right)\\ &\times\exp\left(\frac{2\pi}{L}\sum_{p>0}p^{-1}\rho_2(-p)(e^{ipx}-e^{ipx'})\sinh\varphi_p\right)\\ &\times\exp\left(-\frac{2\pi}{L}\sum_{p>0}p^{-1}\rho_2(p)(e^{-ipx}-e^{-ipx'})\cosh\varphi_p\right)\\ &\times\exp\left(-\frac{2\pi}{L}\sum_{p>0}p^{-1}\rho_1(p)(e^{-ipx}-e^{-ipx'})\cosh\varphi_p\right)\\ &\times\exp\left(\frac{2\pi}{L}\sum_{p>0}p^{-1}\rho_1(-p)(e^{ipx}-e^{ipx'})\cosh\varphi_p\right)\end{aligned} \tag{16}$$

This generalizes an earlier result,[4] the calculation of the ground-state expectation value by an entirely different and more laborious technique:

$$\langle F|\Psi_1^*(x')\Psi_1(x)|F\rangle = \frac{1}{L}e^{ik_{1F}(x-x')}\Sigma(x'-x) \times \exp\left(\frac{-4\pi}{L}\sum_{p>0}p^{-1}[1-\cos p(x-x')]\sinh^2\varphi_p\right). \tag{17}$$

Here, and elsewhere, the following identity proves helpful:

$$\Sigma(R) \equiv \exp\left[\frac{2\pi}{L}\sum_{p>0}p^{-1}e^{ipR}\right] = \sum_{p>0}^{\infty}e^{ipR} = [1 - e^{i2\pi R/L}]^{-1} \tag{18}$$

We have denoted this quantity $\Sigma(R)$ for typographical convenience.

SCATTERING MATRIX ELEMENT

The state of lowest energy carrying a current j is denoted the ground state for current j, and symbolized $|F;j\rangle$. At $T = 0$ one may always assume the initial state to be a state of this type.

We therefore calculate the transition matrix element from an initial state, the ground state of current $j > 0$, to a final state, which can be either the ground state of current $j - 2v_0$, or any excited state of the same current. The total rate of decay out of the initial state into the final states, subject to the requirement of conservation of energy, determines the lifetime τ of the current j. It shall, however, not be necessary for us to calculate τ in any detail in cases when $U(0) \neq 0$, for we shall find $\tau = 0$ when $U(0) > 0$ and $\tau = \infty$ when $U(0) < 0$.

We apply (15) to the right-hand sides of Eqs.(13) and their Hermitean conjugates, to obtain

$$\Psi_2^\dagger(x)\Psi_1(x) \to (e^{i(k_{1F}-k_{2F})x}/L)e^{-\alpha}e^{B_2^\dagger}e^{-B_2}e^{-A_1^\dagger}e^{A_1}, \tag{19}$$

in which we note that the phase factor $k_{1F}-k_{2F}\sim 2k_F$ corresponds to backward scattering across the Fermi surface, and

$$\begin{aligned}\alpha &= \frac{2\pi}{L}\sum_{p>0} p^{-1}(e^{2\varphi(p)}-1) = \int_0^\infty dp\, p^{-1}(e^{2\varphi(p)}-1)\\ &= \int_0^\infty dp\, p^{-1}\{[1+2\lambda u(p)]^{-1/2}-1\},\\ A_1 &= \frac{2\pi}{L}\sum_{p>0} p^{-1}\rho_1(-p)e^{ipx}e^{\varphi(p)},\\ B_2 &= \frac{2\pi}{L}\sum p^{-1}\rho_2(p)e^{-ipx}e^{\varphi(p)}.\end{aligned} \tag{20}$$

Therefore the ground state-ground state matrix element, which we write $M(F\to F)$ in an obvious notation, takes on the value:

$$M(F\to F) = (1/L)e^{i(k_{1F}-k_{2F})x}e^{-\alpha}. \tag{21}$$

The magnitude of $M(F\to F)$ depends on α, and this in turn depends sensitively on $\varphi(0)\equiv \lim_{p\to 0}\varphi(p)$. A two-body interaction which is attractive on the whole has $U(0) < 0$, hence, by Eq. (10), $\varphi(0) > 0$. Such an interaction implies a positive α which is logarithmically divergent $(+\infty)$, and thus a vanishing matrix element. Similarly, a two-body interaction which is repulsive on the whole implies a negatively divergent value of α, hence an infinite matrix element. When both $\varphi(0)$ and $\varphi(\infty)$ are zero, the integral defining α is well-behaved and the matrix element is finite. It should be noted that any two-body interaction $V(x-x')$, the spatial integral of which is nonzero, corresponds to a Fourier transform $U(p\to 0)\neq 0$, hence to a divergent α (negatively or positively divergent according as to whether the interaction is repulsive or attractive). In all such cases the matrix element $M(F\to F)$ is nonanalytic in the coupling constant λ at $\lambda = 0$, despite the persistence of a "sharp Fermi surface" to finite values of λ (cf. discussion in Ref. 4). In the case of potentials which are neither repulsive nor attractive on the whole, $U(p\to 0) = 0$, α is finite and is a continuous function of λ. In such cases only is the decay of an induced current qualitatively the same as for noninteracting particles.

STRUCTURED FINAL STATES

Concerning the divergence in α arising primarily from long wavelengths $(p\to 0)$, it is legitimate to wonder whether it is not possible to cancel this divergence through an appropriate linear combination of low-lying excited states. We shall examine two typical compound final states in some detail:

$$\langle Q^{(1)}| \equiv \langle F; j-2v_0|a_{1k_{1F}}a^\dagger_{1(k_{1F}-Q)}, \tag{22a}$$

$$\langle Q^{(2)}| \equiv \langle F; j-2v_0|a^\dagger_{2k_{2F}}a_{2(k_{2F}-Q)}. \tag{22b}$$

These have the advantage of being eigenstates of the free-fermion Hamiltonian $(\lambda = 0)$. It is of interest to see whether the matrix elements $M(F\to Q^{(i)})$ vanish or diverge under the same conditions as $M(F\to F)$. We start the analysis under the supposition that the forces are essentially attractive $[\varphi(p) > 0]$; a separate analysis will follow in the case of essentially repulsive forces.

After some elementary manipulations based on Eqs. (2) and (13), we obtain

$$\begin{aligned}M(F\to Q^{(1)}) &= (1/L)e^{-\alpha}e^{i(k_{1F}-k_{2F}-Q)x}\\ &\times (1/L)\int dR\, e^{-iQR}\Sigma(R)\\ &\times \exp\Big((2\pi/L)\sum_{p>0}p^{-1}e^{ipR}(e^{\varphi(p)}-1)\Big)\end{aligned} \tag{23a}$$

and

$$M(F\to Q^{(2)}) = e^{2iQx}M(F\to Q^{(1)}). \tag{23b}$$

It is therefore sufficient to study the behavior of $M(F\to Q^{(1)})$. If $\varphi(0)\neq 0$ the sum in the exponential is logarithmically divergent, and we manipulate it so as to combine it with the divergent expression in α.

Thus

$$\begin{aligned}&\frac{2\pi}{L}\sum_{p>0}p^{-1}e^{ipR}(e^{\varphi(p)}-1)\\ &= \frac{2\pi}{L}\sum_{p>0}p^{-1}(e^{\varphi(p)}-1) - \frac{2\pi}{L}\sum_{p>0}p^{-1}(1-e^{ipR})(e^{\varphi(p)}-1)\end{aligned} \tag{24}$$

Finally,

$$\begin{aligned}M(F\to Q^{(1)}) &= \frac{1}{L}e^{-\alpha'}e^{i(k_{1F}-k_{2F}-Q)x}\,\frac{1}{L}\int dR\, e^{-iQR}\Sigma(R)\\ &\times \exp\Big(-\int_0^\infty dp\, p^{-1}(1-e^{ipR})(e^{\varphi(p)}-1)\Big)\end{aligned} \tag{25}$$

$$\equiv \frac{1}{L}e^{-\alpha'}e^{i(k_{1F}-k_{2F}-Q)x}I(Q), \tag{25a}$$

where

$$\alpha' \equiv \int_0^\infty dp\, p^{-1}e^{\varphi(p)}(e^{\varphi(p)}-1). \tag{26}$$

We note that although $\alpha' < \alpha$ [for the case under consideration, viz., $\varphi(p) > 0$], it is nonetheless infinite when $\varphi(0)\neq 0$. It remains only to study the behavior of $I(Q)$, and to verify that all quantities reduce to the appropriate value when the interaction is turned off. For this purpose, it is most convenient to expand the exponential in a power series about $R = 0$, retaining up to quadratic terms. Thus,

$$\int_0^\infty dp\, p^{-1}(1-e^{ipR})(e^{\varphi(p)}-1) = -i\gamma R + \tfrac{1}{2}\delta R^2 + O(R^3), \tag{27}$$

Where

$$\gamma \equiv \int_0^\infty dp(e^{\varphi(p)}-1), \quad \delta \equiv \int_0^\infty dp\, p(e^{\varphi(p)}-1), \tag{28}$$

both positive quantities in the case under consideration. Then,

$$\begin{aligned}I(Q) &= \frac{1}{L}\int dR\, e^{-iQR}\Sigma(R)e^{i\gamma R-\delta R^2/2}\\ &= \frac{1}{L}\sum_{p>0}\int_{-\infty}^{\infty} dR\, e^{i(p-Q+\gamma)R}e^{-\delta R^2/2}\\ &= \frac{1}{L}\sum_{p>0}\Big(\frac{2\pi}{\delta}\Big)^{1/2}e^{-(p+\gamma-Q)^2/2\delta}\\ &= \frac{1}{(2\pi\delta)^{1/2}}\int_0^\infty dp\, e^{-(p+\gamma-Q)^2/2\delta}.\end{aligned} \tag{29}$$

[In the limit $\lambda\to 0$, both γ and δ vanish and, for any finite positive Q, $I = 1$, and $M(F\to Q^{(1)})$ tends to what is obviously the correct value for free particles.] For any $\lambda > 0, I$ is finite and in the case of $\varphi(0) > 0$ *the matrix element* $M(F \to Q^{(i)})$ *vanishes just as did* $M(F \to F)$.

We now verify that, for repulsive forces, the matrix element diverges. It must first be understood that when $\varphi(p) < 0$, the main contribution to the spatial integral in (25) is from a region near $R = \pm\frac{1}{2}L$. Because of periodic boundary conditions, we have

$$e^{ip(R\pm L/2)} = -e^{ipR}, \quad \sum(R\pm L/2) = (1 + e^{i2\pi R/L})^{-1}. \tag{30}$$

Therefore we cast (25) in the form

$$M(F \to Q^{(1)}) = (1/L)\, e^{-\alpha''}\, e^{i(k_{1F}-k_{2F}-Q)x} J(Q), \tag{25b}$$

where

$$\alpha'' \equiv \int_0^\infty dp\, p^{-1}(e^{\varphi(p)} - 1)(e^{\varphi(p)} + 2)$$

and

$$J(Q) \equiv -\frac{1}{L}\int dR\, e^{-iQR}(1 + e^{i2\pi R/L})^{-1} \times \exp\left(\int_0^\infty dp\, p^{-1}(1 - e^{ipR})(e^{\varphi(p)-1})\right) \tag{31}$$

We can evaluate $J(Q)$ by the same methods in (27-29), and show it is finite. *Thus, the divergence* $[\alpha'' \to -\infty$ *whenever* $\varphi(0) < 0]$ *is again confirmed.*

RECAPITULATION AND FUTURE APPLICATIONS

We have found a representation for fermion wave operators in a specific one-dimensional model, in terms of density fluctuation operators, which enables the exact evaluation of rather complicated matrix elements. In applying this to the problem of persistent current we observed that in the case of repulsive two-body forces, $U(0) > 0$, the scattering matrix elements due to impurities become infinite, and in the case of attractive two-body forces, $U(0) < 0$ they vanish. It should be noted that neither the ground state energy W_1, Eq. (12), nor the sharpness of the Fermi surface[4], are so singularly dependent on $U(0)$.

Our finding of what is tantamount to superconductivity, for electrons interacting with attractive forces, is in harmony with the well-known results of the BCS theory of superconductivity for three-dimensional systems. Recently, Heeger and his collaborators[5] have found anomalously large conductivity in certain linear chain molecules (TTF-TCNQ) near a finite temperature ~ 58°K, followed by a rapid decrease in conductivity as the temperature is further decreased. For these experimental facts to be explained on the basis of any one-dimensional model requires a calculation at finite temperature, and, possibly also, considerations of the electron spin and the electron-phonon interactions.

Note added in proof: We have now succeeded in evaluating τ at finite temperature and in taking the electron-phonon forces explicitly into account. (Full details have been submitted for publication elsewhere).

ACKNOWLEDGMENT

It is a pleasure to thank Mr. R. Peña for his helpful collaboration in the early stages of this work.

*Supported in part by Air Force Contract AFOSR-72-2153.

[1]Also often denoted "current operators" as in "current algebras." We reserve that nomenclature for the quantity defined in Eq. (3).

[2]A model introduced by J. M. Luttinger, J. Math. Phys. **4**, 1154 (1963), given an exact solution by D. Mattis and E. Lieb, J. Math. Phys. **6**, 304 (1965), and which is analogous to the Thirring model; see discussion in E. Lieb and D. Mattis, *Mathematical Physics in One Dimension* (Academic, New York, 1966), Chap. 4.
[3]Such as the Tomonoga model; see discussion, references and reprints in Lieb and Mattis, Ref. 2 (1966) and subsequent work by H. Gutfreund and M. Sehick, Phys. Rev. **168**, 418 (1968).
[4]D. Mattis and E. Lieb, J. Math. Phys. **6**, 304 (1965), Sec. V. Eqs. (5.3) *et seq.*
[5]L. B. Coleman, J. J. Cohen, D. J. Sandman, F. G. Yamagashi, A. F. Garito and A. J. Heeger, Solid State Comm. **12**, 1125 (1973).

PHYSICAL REVIEW B VOLUME 9, NUMBER 7 1 APRIL 1974

Single-particle states, Kohn anomaly, and pairing fluctuations in one dimension

A. Luther*
Lyman Laboratory, Harvard University, Cambridge, Massachusetts 02138

I. Peschel
Institut Max von Laue-Paul Langevin, D-8046 Garching, Germany

(Received 15 October 1973)

We compute the single-particle spectral density, susceptibility near the Kohn anomaly, and pair propagator for a one-dimensional interacting-electron gas. With an attractive interaction, the pair propagator is divergent in the zero-temperature limit and the Kohn singularity is removed. For repulsive interactions, the Kohn singularity is stronger than the free-particle case and the pair propagator is finite. The low-temperature behavior of the interacting system is not consistent with the usual Ginzburg-Landau functional because the frequency, temperature, and momentum dependences are characterized by power-law behavior with the exponent dependent on the interaction strength. Similarly, the energy dependence of the single-particle spectral density obeys a power law whose exponent depends on the interaction and exhibits no quasiparticle character. Our calculations are exact for the Luttinger or Tomonaga model of the one-dimensional interacting system.

I. INTRODUCTION

Interest in one-dimensional systems has recently been regenerated, largely in response to experiments on organic complexes of predominantly one-dimensional electronic character.[1,2] Some of these experiments are interpreted as providing evidence for the Peierls instability,[3] and a theory for understanding the one-dimensional fluctuations near such an instability has been proposed.[4] It has also been suggested that the dramatic increase in conductivity reported for the tetrathiofulvalinium-tetracyanoquinodimethan (TTF-TCNQ) system[1] could be caused by one-dimensional fluctuations of a superconducting order parameter.[5]

These interpretations are based, at least implicitly, on a picture of one-dimensional electronic states for which the electron-electron interaction causes no qualitative changes to the single-particle states beyond simple renormalizations. Since the work of Mattis and Lieb,[6] however, it has been known that the usual quasiparticle picture does not apply to one-dimensional interacting systems. Conventional perturbation or fluctuation theories are therefore not obviously meaningful. We attempt here an understanding of this difficulty by calculating the single-particle Green's function $G(q,\omega)$, the particle-hole susceptibility $\chi(q,\omega)$ near the Kohn anomaly, and the pair propagator $P(q,\omega)$.

The character of the superconducting fluctuations are described by P, and it is an obvious response function to study. From G, the spectral weight of the single-particle excitations is determined, and the importance of interactions in causing departures from the quasiparticle picture directly evaluated. From $\chi(q,\omega)$ at momenta near twice the Fermi momentum $q\simeq 2k_F$, the low-lying excitations of the interacting-electron gas are computed and the modification of the Peierls instability due to interactions is discussed. Our calculations for G, χ, and P are exact within the Luttinger or Tomonaga model of the interacting-electron gas.

In order to compute these functions, it is necessary to go beyond previous considerations of such one-dimensional systems, because the time evolution of the single-fermion operators is required. It is, in general, extremely complicated to work directly with these operators. We circumvent this difficulty by finding an operator in the boson space which satisfies the proper commutation and anticommutation relations and has the same expectation values as the original fermion operators. The computation of G, χ, and P is thereby reduced to a solvable boson problem.

Some of our new results could have been anticipated from the solution of Mattis and Lieb.[6] The spectral function, given by $\mathrm{Im}G(q,\omega)$, has a δ-function peak at zero temperature for free particles. With interactions present, this peak vanishes, and $\mathrm{Im}G(k_F,\omega)$ exhibits a power-law behavior extending to energies of the order of the bandwidth. The exponent of this power law depends on the interaction strength but not its sign. One expects to find no quasiparticle peak for this system since the discontinuity in the $T=0$ occupation number has vanished. The entire spectral weight must therefore be contained in the "incoherent background." In spite of this power law, the specific heat at low temperatures remains linear[6] in T.

The result for $\chi(q,\omega=0)$ is somewhat surprising. Smearing out the $T=0$ occupation number might be expected to weaken the logarithmic Kohn singularity at $2k_F$, just as thermal smearing of free particles converts this singularity to a $\ln T$ behavior. In fact, for *repulsive* interactions, a *stronger* power-law singularity is found, with the exponent of that power law dependent on the interaction strength. Attractive interactions do indeed give a finite $\chi(2k_F, 0)$ at $T=0$, but it is substantially smaller than one would guess from the smearing of the occupation number. Even for very weak interactions, the departures from the free-particle susceptibility are quite pronounced.

The $T=0$ dynamic susceptibility near $2k_F$ exhibits a threshold singularity for frequences at the edge of the continuum, reminiscent of the x-ray-threshold edge.[7] The edge singularity satisfies a power law, and is enhanced (divergent) for repulsive interactions and suppressed for attractive interactions. In both cases the exponent of the power law depends on the interaction strength. As in the x-ray problem, these singularities arise because the interactions cause a single-particle excitation to be "dressed" with an infinite number of particle-hole pairs. These pairs are just the large-wavelength boson excitations of the Luttinger or Tomonaga models[8] and conspire to drastically modify the low-lying excitations around $2k_F$.

The pair propagator $P(q,\omega)$ is found to be the same as $\chi(q-2k_F,\omega)$, provided the sign of the interaction strength is reversed. An attractive interaction thus produces a power-law divergence in $P(q=0,\omega=0)$ as the temperature is lowered, indicative of large pairing fluctuations. However, the functional dependence on ω, q, and T is not of the Ginzburg-Landau type, due to the power-law behavior mentioned above.

We find a new characteristic temperature, T_0, below which these interaction effects are important. This is estimated to be $T_0 \sim We^{-1/\gamma}$, where γ is an interaction parameter and W is the order of the bandwidth. Although it is possible that the models we discuss may be an inadequate description of some systems, we argue that the fluctuation effects calculated here will still be important for temperatures less than T_0. For this situation, our solutions provide a better starting point than those which involve the quasiparticle assumption.

II. SINGLE-PARTICLE GREEN'S FUNCTION

A. Field operators and Luttinger model

In this section, the formal computation of $G(q,\omega)$ is given. It proves to be simpler to work with the Fourier transform function $G(x,t)$, which is defined by

$$G(x,t) = -i\Theta(t)\langle[\psi_1(x,t),\psi_1^\dagger]_+\rangle\ , \tag{1}$$

where $\psi_1(x,t)$ is the field operator for the fermions (time evolved with the full Hamiltonian H), the subscript + indicates the anticommutator, and the angular brackets denote an average in the density matrix $e^{-\beta H}$. The Hamiltonian for the Luttinger model is

$$H = v_F\sum_k k(a_{1,k}^\dagger a_{1,k} - a_{2,k}^\dagger a_{2,k}) + \sum_p V_p\rho_1(p)\rho_2(-p)\ , \tag{2}$$

where sums are over the discrete indices of one-dimensional plane waves on a line of length L, $\psi_1(x)=L^{-1/2}\sum_k e^{ikx}a_{1,k}$, $\psi_2(x)=L^{-1/2}\sum_k e^{ikx}a_{2,k}$, V_p is the Fourier transform of the two-body interaction, it is assumed all states below the energy k_Fv_F are filled, and $k_F=mv_F$ is the Fermi momentum ($\hbar=1$). The objects $\rho_1(p)$ and $\rho_2(p)$ are density operators, defined by

$$\begin{aligned}\rho_1(p)&=\sum_k a_{1,k+p}^\dagger a_{1,k}\ ,\quad \rho_1(-p)=\sum_k a_{1,k}^\dagger a_{1,k+p}\ ,\\ \rho_2(p)&=\sum_k a_{2,k+p}^\dagger a_{2,k}\ ,\quad \rho_2(-p)=\sum_k a_{2,k}^\dagger a_{2,k+p}\ ,\end{aligned} \tag{3a}$$

where $p>0$. They satisfy a boson algebra,

$$\begin{aligned}&[\rho_1(-p),\rho_1(p')]=[\rho_2(p),\rho_2(-p')]=(pL/2\pi)\delta_{p,p'}\ ,\\ &[\rho_1(p),\rho_2(p')]=0\ .\end{aligned} \tag{3b}$$

These relations, together with

$$\Big[\rho_1(p),v_F\sum_k k a_{1,k}^\dagger a_{1,k}\Big]=-v_Fp\rho_1(p) \tag{4}$$

and the corresponding result for $\rho_2(p)$, permit the exact diagonalization of H. The transformation which effects this diagonalization is given by $e^{iS}He^{-iS}=H_D$, where $S=2\pi iL^{-1}\sum_p p^{-1}\varphi(p)\rho_1(p)\rho_2(-p)$, and $\tanh 2\varphi(p)=-V_p(\pi v_F)^{-1}$. The reader is referred to the paper by Mattis and Lieb[6] for the details of this solution and the proof of the above statements. We consider now the extension of this solution to find $G(x,t)$.

A direct computation of the equation of motion for the operator $a_{1,k}$ shows that it couples to the operator $a_{1,k}\rho_2(p)$. The equation of motion for this new operator couples to yet more complicated operators, and one quickly obtains an infinite set of coupled operator equations. Clearly this brute-force approach is extremely awkward, if not impossible.

Such difficulties can be circumvented by using a new operator $O_1(x)$, which has the same equation of motion as the field operator, $\psi_1(x)$, and the same commutation relation with the density operators. This operator is defined by

$$O_1(x)=(2\pi\alpha)^{-1/2}e^{ik_Fx+\phi_1(x)}\ , \tag{5}$$

where

$$\phi_1(x)=2\pi L^{-1}\sum_{k>0}k^{-1}e^{-\alpha k/2}[\rho_1(-k)e^{ikx}-\rho_1(k)e^{-ikx}]$$

and α is a cutoff parameter which is determined below. Since $O_1(x)$ is expressed entirely in terms of density operators, it is obvious that correlation functions containing only products of $O_1(x, t)$ and its Hermitean conjugate can be evaluated, using the same transformations which diagonalize H.

It remains to be shown that these new correlation functions are equal to those defined with the original $\psi_1(x)$ fields. The commutation relations with density operators are[9]

$$[\psi_1(x), \rho_1(k)] = e^{ikx}\psi_1(x) , \tag{6a}$$

$$[O_1(x), \rho_1(k)] = e^{ikx}O_1(x) , \tag{6b}$$

and the equations of motion are

$$\frac{d}{dt}\psi_1(x, t) = -v_F\frac{d}{dx}\psi_1(x, t) - i\sum_p V_p e^{ipx}\psi_1(x, t)\rho_2(-p, t) , \tag{7a}$$

$$\frac{d}{dt}O_1(x, t) = -v_F\frac{d}{dx}O_1(x, t) - i\sum_p V_p e^{ipx}O_1(x, t)\rho_2(-p, t) - 2\pi i L^{-1}\sum_k e^{-\alpha k}O_1(x, t) . \tag{7b}$$

The last term in Eq. (7b) represents a constant energy shift, which can be removed by choosing a new zero of energy. Since this reference energy is irrelevant for the computation of $G(q, \omega)$, we simply discard it.

The transformation e^{iS}, which diagonalized H, involves only density operators. According to Eq. (6) $e^{iS}\psi_1(x)e^{-iS}$ will therefore transform exactly as $e^{iS}O_1(x)e^{-iS}$. The former has been evaluated previously,[6] with the result

$$e^{iS}\psi_1(x)e^{-iS} = e^{w_1(x)}e^{r_2(x)}\psi_1(x) , \tag{8a}$$

where

$$w_1(x) = 2\pi L^{-1}\sum_{k>0} k^{-1}(\cosh\varphi - 1) \times [\rho_1(-k)e^{ikx} - \rho_1(k)^{-ikx}]$$

and

$$r_2(x) = 2\pi L^{-1}\sum_{k>0} k^{-1}\sinh\varphi \times [\rho_2(-k)e^{ikx} - \rho_2(k)e^{-ikx}] .$$

The same procedure applied to the $O_1(x)$ operator gives precisely the same result

$$e^{iS}O_1(x)e^{-iS} = e^{w_1(x)}e^{r_2(x)}O_1(x) . \tag{8b}$$

Finally, consider the computation of the correlation functions $\langle G|\psi_1(x, t)\psi_1^\dagger|G\rangle$ and $\langle G|O_1(x, t)O_1^\dagger|G\rangle$ where $|G\rangle$ is the ground state of H. Let $|0\rangle$ be the ground state of H_D, which is related to $|G\rangle$ by $|G\rangle = e^{-iS}|0\rangle$. Using $e^{iS}He^{-iS} = H_D$, the correlation functions can be written

$$\langle G|\psi_1(x, t)\psi_1^\dagger|G\rangle = \langle 0|e^{iH_Dt}e^{iS}\psi_1(x)e^{-iS}e^{-iH_Dt}e^{iS}\psi_1^\dagger e^{-iS}|0\rangle , \tag{9}$$

with an identical equation for $\langle G|O_1(x, t)O_1^\dagger|G\rangle$. The relations established by Eqs. (6)–(8) ensure the equality $\langle G|\psi_1(x, t)\psi_1^\dagger|G\rangle = \langle G|O_1(x, t)O_1^\dagger|G\rangle$, provided that $\langle 0|\psi_1(x)\psi_1^\dagger|0\rangle = \langle 0|O_1(x)O_1^\dagger|0\rangle$. As shown in Ref. 6, $|0\rangle$ is just the filled Fermi sea with no boson excitations present, so that the evaluation of these latter expectation values is trivial,[10] giving (at $T = 0$)

$$\langle 0|\psi_1(x)\psi_1^\dagger|0\rangle = L^{-1}\sum_{k>k_F} e^{ikx} , \tag{10a}$$

$$\langle 0|O_1(x)O_1^\dagger|0\rangle = (2\pi\alpha)^{-1}e^{ik_Fx}\exp 2\pi L^{-1}\sum_{k>0} k^{-1}e^{-\alpha k}(e^{ikx} - 1) = (2\pi)^{-1}(\alpha - ix)^{-1}e^{ik_Fx} . \tag{10b}$$

In Eq. (10b), we have taken the limit $\sum_k \to L(2\pi)^{-1}\int dk$. If an infinitesimal imaginary part is added to k in Eq. (10a), to ensure convergence for large k, we have agreement provided the limit $\alpha \to 0$ is taken. The correlation functions $\langle G|\psi_1(x, t)\psi_1^\dagger|G\rangle$ and $\langle G|O_1(x, t)O_1^\dagger|G\rangle$ are therefore identical, in this limit.

There is another sense in which these correlation functions are equal which helps in the computation of χ and P. If the density of states is not a constant, as in the Luttinger model, but is chosen such that $\sum_k \to L(2\pi)^{-1}\int dk e^{-\alpha|k-k_F|}$ in Eq. (10a) there is an equality between Eqs. (10a) and (10b) even for finite α. We therefore interpret α^{-1} as the bandwidth (in momentum units) and expect this quantity to appear in our calculations whenever the bandwidth or excitations far away from the Fermi surface are important.

In a formalistic sense, our solution for dynamical quantities are exact solutions of the Luttinger model only in the limit $\alpha \to 0$. Finite α corresponds to an alteration of the energy spectrum far from the Fermi surface, as is done in the Tomonaga model. However, certain properties of the Luttinger model, such as the Kohn singularity at $2k_F$, depend only on the low-lying states. Our results for these properties are thus exact, in a very real sense, despite the necessity of a cutoff. The situation is similar to the Kondo problem,[10] where infrared singularities are determined only by the low-lying states, but a cutoff at the bandwidth is needed for convergence.

These considerations are readily extended to other correlation functions and finite temperatures. In all cases relevant for this paper, replacement of $\psi_1(x)$ by $O_1(x)$ does not change the correlation functions. As is known in the Tomonaga model for the Kondo problem, this is not always the case, and proper caution must be exercised for other correlation functions.

B. Evaluation of Green's function $G(x,t)$

Applying this method to the computation of $G(x, t)$, we find, from Eqs. (1) and (5),

$$G(x,t) = -i\Theta(t)(2\pi\alpha)^{-1}e^{ik_Fx}\langle[e^{\Phi_1(x,t)}, e^{-\Phi_1}]_+\rangle$$
$$= -i\Theta(t)(2\pi\alpha)^{-1}e^{ik_Fx}\langle e^{iHt}e^{\Phi_1(x)}e^{-iHt}e^{-\Phi_1} + (t\to -t, x\to -x)\rangle . \quad (11)$$

Inserting the diagonalizing transformation, e^{iS} and using Eq. (8), this expression can be written

$$G(x,t) = -i\Theta(t)(2\pi\alpha)^{-1}e^{ik_Fx} \times[\langle e^{w_1(x,t)}e^{r_2(x,t)}e^{\Phi_1(x,t)}e^{-\Phi_1}e^{-r_2}e^{-w_1}\rangle_D + (t\to -t, x\to -x)] , \quad (12)$$

where the time evolution and averaging is now performed using H_D, which is *diagonal* in the density operators. Equation (12) is thus a standard form, which is readily evaluated[11] to give the result ($T=0$)

$$G(x,t) = -i\Theta(t)(2\pi\alpha)^{-1}\exp(ik_Fx) \times\{\exp[Q_1(x,t)+Q_2(x,t)] + (t\to -t, x\to -x)\} , \quad (13)$$

where

$$Q_1(x,t) = 2\pi L^{-1}\sum_{k>0} k^{-1}(\cosh\varphi - 1 + e^{-\alpha k})^2 \times(e^{ikx - i\epsilon_k t} - 1) ,$$
$$Q_2(x,t) = 2\pi L^{-1}\sum_{k>0} k^{-1}(\sinh\varphi)^2(e^{-ikx - i\epsilon_k t} - 1) ,$$

and $\epsilon_k = v_F k\,\mathrm{sech}\varphi(k)$. In order to complete the solution, we use the identity

$$\alpha^{-1} = (\alpha - is)^{-1}e^{-2\pi L^{-1}\sum_{k>0} k^{-1}e^{-\alpha k}(e^{ikS}-1)}$$

and take the limit $\alpha\to 0$. For the case $t=0^+$, we find

$$G(x,0) = (2\pi)^{-1}e^{ik_Fx}[e^{Q(x)}(x+i0)^{-1} + (x\to -x)] , \quad (14)$$

where

$$Q(x) = 4\pi L^{-1}\sum_{k>0} k^{-1}\sinh^2\varphi(\cos kx - 1) .$$

The correlation function $\langle O_1(x)O_1^\dagger\rangle$ is thus $(2\pi x)^{-1} \times e^{ik_Fx + Q(x)}$, identical to the result of Mattis and Lieb[6] for $\langle\psi_1(x)\psi_1^\dagger\rangle$.

Equation (13) is the formal solution for $G(x, t)$ for an arbitrary interaction consistent with the limitation that φ is real. The sum rule on the spectral function $1 = -\pi^{-1}\int_{-\infty}^{\infty}d\omega\,\mathrm{Im}G(q,\omega)$ for all q is a check of the equal-time anticommutator. Since our solution reproduces the exact result in that limit, it follows without computation that this sum rule is satisfied. In the following, the evaluation $G(q,\omega)$ for a special form of V_p is given. It is an approximation to Eq. (13), and therefore does not satisfy the sum rule exactly, although it does lend considerable insight into the exact structure of $G(q,\omega)$.

In order to proceed with the Fourier transformation leading to $G(q,\omega)$, it is necessary to assume a particular form for the interaction V_p. This is taken to be $\sinh^2\varphi(p) = \gamma e^{-pr}$, where r is the range of the interaction and γ can be related to $V_{p=0}$ through $\tanh 2\varphi(0) = -V_0(\pi V_F)^{-1}$. The excitation spectrum $\epsilon_p = v_F p\,\mathrm{sech}\varphi(p)$ is approximated by $\epsilon_p = cp$, where c is the renormalized Fermi velocity, $c = v_F\mathrm{sech}\varphi(0)$. We are primarily interested in excitations near the Fermi level and their consequences; the precise details of the large momentum states are not of interest here. Results which depend on r cannot be regarded as "universal," since another form for V_p could give a different answer. However, certain features will be independent of r, and these we argue to be a general property of the system.

With this assumed V_p and ϵ_p, the integrals for $Q_1(x,t)$ and $Q_2(x,t)$, are easily evaluated, leading to the result at zero temperature:

$$G(x,t) = i\Theta(t)\pi^{-1}e^{ik_Fx} \times\mathrm{Im}\left\{(x-ct+i0)^{-1}\left[\left(\frac{x}{r}\right)^2 + \left(1+\frac{ict}{r}\right)^2\right]^{-\gamma}\right\} . \quad (14')$$

The power-law behavior, induced by the interactions, is a "universal" feature of this model.

Fourier transforming Eq. (14') is tedious but straightforward. Taking the imaginary part of the two terms in the curly brackets, we find $G(q,\omega) = G_1(q,\omega) + G_2(q,\omega)$, where

$$G_1(q',\omega) = -ic^{-1}\int_0^\infty dt\int_{-\infty}^{\infty}dx\,e^{i\omega t - iq'x}\delta(s) \times\mathrm{Re}[F(s)F(s')] ,$$
$$G_2(q',\omega) = -i(\pi c)^{-1}\int_0^\infty dt\int_{-\infty}^{\infty}dx\,e^{i\omega t - iq'x}s^{-1} \times\mathrm{Im}[F(s)F(s')] . \quad (15)$$

Here, we have defined $q' = q - k_F$, $s = t - x/c$, $s' = x/c + t$, and $F(s) = (1 + isc/r)^{-\gamma}$. The first of these can be written

$$G_1(q',\omega) = -i\left(\frac{r}{c}\right)^\gamma\int_0^\infty dt\,e^{i(\omega - cq')t}\,\mathrm{Re}\left(\frac{r}{c}+2it\right)^{-\gamma} . \quad (16)$$

For $\omega - cq'$ sufficiently small, only the long-time behavior is important in this integral, and r/c can be neglected in comparison with t, provided $\gamma < 1$.

The integral can then be reduced to a Γ function, by changing variables, $i(\omega - cq')t \to -t'$, and rotating the contour of integration in the t' plane, so that the integral on t' runs from 0 to $+\infty$. The result is

$$G_1(q',\omega) \cong \frac{\Gamma(1-\gamma)}{\omega - cq'}\left(\frac{r(\omega - cq')}{2ic}\right)^{\gamma} \cos\tfrac{1}{2}\pi\gamma . \quad (17)$$

For the spectral function, we require only $\mathrm{Im}G(q',\omega)$. Using the property $G_2(q',\omega) = G_2^*(-q',-\omega)$, the limits of integration over t in Eq. (15) may be extended to $-\infty$. Changing variables to s and s' the integral for $\mathrm{Im}G_2(q',\omega)$ becomes

$$\mathrm{Im}G_2(q',\omega) = -(4\pi)^{-1}\mathrm{Im}\Big[\int_{-\infty}^{\infty} ds s^{-1} e^{is(\omega - cq')/2} F(s) \times \int_{-\infty}^{\infty} ds' e^{is'(\omega + cq')/2} F(s') + (\omega \to -\omega, q' \to -q')\Big] . \quad (18)$$

These integrals may be evaluated in the low-energy region, using the approximation discussed above Eq. (17). The result is

$$\mathrm{Im}G_2(q',\omega) = -2(\pi\gamma)^{-1}\left(\frac{r}{c}\right)^{2\gamma}\Gamma^2(1-\gamma) \times [\mathrm{Im}(\omega - cq')^{\gamma}\mathrm{Im}(\omega - cq')^{\gamma-1} + (\omega \to -\omega, q' \to -q')] . \quad (19)$$

For the momentum at the Fermi surface, $q' = 0$, the spectral function $-\pi^{-1}\mathrm{Im}G(0,\omega + i0)$ is given by adding Eqs. (17) and (19) with the result

$$-\pi^{-1}\mathrm{Im}G(0,\omega + i0) \cong \frac{r}{c}\left(\frac{\omega r}{2c}\right)^{\gamma-1}\Gamma(1-\gamma)\sin\tfrac{1}{2}\pi\gamma + \left(\frac{2r}{\gamma c}\right)\left(\frac{\omega r}{2c}\right)^{2\gamma-1} \times [\Gamma(1-\gamma)\sin\pi\gamma]^2 . \quad (20)$$

This holds for $\omega \ll c/r$, i.e., for energies much less than that corresponding to the range of the two-body force. In the opposite limit, the spectral function behaves as ω^{-2}, as can be seen directly from Eqs. (16) and (18).

This result has an interesting interpretation. The quasiparticle picture would lead to a spectral function of the form $-\pi^{-1}\mathrm{Im}G(0,\omega + i0) = Z\delta(\omega) + b(\omega)$, where Z is a single-particle residue and $b(\omega)$ is the incoherent background, whose integrated strength is $(1 - Z)$. Equation (20) does not contain the δ-function and evidently $Z = 0$.

It is interesting to consider the difficulties which occur if $G(q,\omega)$ is studied using perturbation theory. Leading divergences of the form $\gamma(\gamma\ln\omega)^n$ are encountered, along with other combinations of γ and $\ln\omega$. Partial summations, of the type often applied to such divergent series, might be expected to remove these singularities, replacing divergent quantities such as $\gamma^2\ln\omega$ by $\gamma^2\ln(\omega^2 + c')$, where c' is some constant. Such partial summations would be highly misleading here, and we believe our result should serve as a warning for the qualitatively different effects of fluctuations in one-dimensional interacting systems.

III. SUSCEPTIBILITY OF LUTTINGER AND TOMONAGA MODELS

As discussed by many authors, the susceptibility of the one-dimensional free-electron gas diverges at twice the Fermi momentum, as the temperature is lowered to zero. The divergence is caused by transitions between states at opposite ends of the Fermi line and is properly classified as an infrared singularity. In mean-field theories, this behavior is responsible for the Peierls lattice instability, in which a phonon frequency vanishes according to the perturbation equation

$$\omega_q^2 = \omega_{0,q}^2 - \omega_{0,q} g_L^2 \chi(q,\omega) , \quad (21)$$

where $\omega_{0,q}$ is the "unperturbed" phonon frequency and g_L the coupling constant. This soft-mode instability is therefore intimately connected with a large susceptibility.

Section II implies that departures from the free-particle result for χ might be significant, even in the absence of electron-lattice coupling. In order to understand these modifications, we study χ for the interacting gas, without coupling to the phonons. In the concluding section (Sec. IV) some possible modifications which occur when the electron-phonon coupling is introduced are briefly mentioned.

The first problem to consider is the proper definition of χ for these one-dimensional models. In the Luttinger model, the density operator which causes transitions across the Fermi line is defined by $\sigma(q) = \sum_k a_{1,k+q}^\dagger a_{2,k}$. There are additional excitations in this model, arising from transitions within the 1 or 2 branches, which are given by the usual density operators $\rho_1(q)$ and $\rho_2(q)$. The susceptibilities corresponding to these excitations have already been calculated,[6] and are approximately constant near $q = 2k_F$, with no dramatic frequency, momentum, or temperature dependence. We will therefore neglect these processes here, and concentrate entirely on that susceptibility exhibiting the infrared singularity, defined by

$$\chi(x,t) = -i\Theta(t)\langle[\sigma(x,t),\sigma^\dagger]\rangle , \quad (22)$$

where $\sigma(x) = L^{-1}\sum_q \sigma(q)e^{iqx} = \psi_1^\dagger(x)\psi_2(x)$ and time evolution, as well as thermal averaging, is with the full H of Eq. (2).

The computation of χ can be readily performed, using the operator $O_1(x)$ [Eq. (5)] for $\psi_1(x)$, and the corresponding operator $O_2(x) = (2\pi\alpha)^{-1/2}e^{-ik_Fx + \phi_2(x)}$, where

$$\phi_2(x) = 2\pi L^{-1}\sum_{k>0} k^{-1}e^{-\alpha k/2}[\rho_2(-k)e^{ikx} - \rho_2(k)e^{-ikx}]$$

for $\psi_2(x)$. After some lengthy, but straightforward manipulations,[10] the following result is found:

$$\chi(x,t) = -i\Theta(t)(2\pi\alpha)^{-2}\exp(2ik_F x) \times[\exp[U(x,t)+U(x,-t)] - (x\to -x, t\to -t)], \quad (23)$$

where

$$U(x,t) = 2\pi L^{-1}\sum_k k^{-1}U_k^2[n_k(e^{-ikx+i\epsilon_k t}-1) + (1+n_k)(e^{-ikx-i\epsilon_k t}-1)],$$

$n_k=(e^{\epsilon_k/T}-1)^{-1}$, and $U_k = e^{-\alpha k}+\cosh\varphi+\sinh\varphi-1$. In order to proceed further, information about the momentum dependence of U_k^2 or V_k is necessary. A convenient form for our purposes is $U_k^2=e^{-2\alpha k}$ $-ge^{-kr}$, where r is the range and g is a measure and has the sign of the $k=0$ potential strength. The spectrum is taken to be linear, $\epsilon_k = ck$. It should be noted that this form for U_k^2 implies a slightly different V_k than was used in Sec. II.

It is also necessary to specify a density of states, because in the limit $\alpha\to 0$, $e^{U(x,t)}$ diverges at short distance and time. Those short distance singularities reflect very large momentum excitations in the Luttinger model, which do not occur in a real system with finite bandwidth. We choose a density of states corresponding to finite α, as discussed in Sec. II. This choice does not affect the "universal" nature of the infrared singularity which arises from *large* x and t. For simplicity, we take $2\alpha = r$. All of these specilizations are consistent with our focus on the contributions of the low-lying states to $\chi(q,\omega)$.

The integral for $U(x,t)$ is a standard form, which can be evaluated, giving

$$\chi(x,t) = -i\Theta(t)(2\pi\alpha)^{-2}e^{2ik_F x}\left[\left(\frac{\Gamma(x_0-isT)\Gamma(1+x_0+isT)\Gamma(x_0-is'T)\Gamma(1+x_0+is'T)}{\Gamma^2(x_0)\Gamma^2(1+x_0)}\right)^{1-\epsilon} - (s\to -s, s'\to -s')\right], \quad (24)$$

where $x_0 \equiv rT/c$, $s=t-x/c$, and $s'=t+x/c$. Using the properties of the Γ function, Eq. (24) can be simplified to

$$\chi(x,t) \cong 2(2\pi\alpha)^{-2}x_0^{2-2\epsilon}e^{2ik_F x}\Theta(t) \times\left(\frac{\pi sT}{\sinh\pi sT}\right)^{1-\epsilon}\left(\frac{\pi s'T}{\sinh\pi s'T}\right)^{1-\epsilon} \times \mathrm{Im}(x_0-isT)^{\epsilon-1}(x_0-is'T)^{\epsilon-1}, \quad (25)$$

where the approximation indicates that x_0 has been dropped in comparison with 1. It is this form which is most convenient for the following discussion.

The Fourier transform of Eq. (25) involves a time integral from 0 to $+\infty$. Using the symmetry properties of the integrand, it is easy to show that the limits may be extended to $-\infty$ in the integral for $\mathrm{Im}\chi(q,\omega)$. The integration variables may then be changed from x and t to s and s , with the result

$$\mathrm{Im}\chi(Q,\omega) = -(4\pi^2 c)^{-1}(r/c)^{-2\epsilon}[F(\omega-cQ)F(\omega+cQ) - F(-\omega+cQ)F(-\omega-cQ)], \quad (26)$$

where

$$F(z) = \int_{-\infty}^{\infty} ds\left(\frac{\pi Ts}{\sinh\pi Ts}\right)^{1-\epsilon} e^{isz/2}\left(\frac{r}{c}-is\right)^{\epsilon-1}$$

and $Q=q-2k_F$. $F(z)$ may be evaluated by making use of the approximation $\pi Ts/\sinh\pi Ts \cong e^{-\theta|s|}$, where $\theta \cong \frac{1}{2}\pi T$ is chosen to simulate the exact temperature dependence for the free gas, $g=0$. We then neglect r/c in comparison to s, which is justified provided $zr \ll c$ and $Tr \ll c$. For the case $g<0$, it is necessary to first integrate by parts before neglecting r/c, to circumvent a spurious divergence at $s=0$. The result is

$$F(z) = -2\Gamma(g)\mathrm{Im}(z/2-i\theta')^{-\epsilon}, \quad (27)$$

where $\theta'=\theta(1-g)$. This approximation is good for both positive and negative z, and $g<1$. In the zero-temperature limit, $\mathrm{Im}\chi(Q,\omega)$ is given simply by the equations

$$\mathrm{Im}\chi(Q,\omega)=0, \quad |\omega|<c|Q|$$
$$\mathrm{Im}\chi(Q,\omega) = (\pi^2 c)^{-1}(r/c)^{-2\epsilon}\Gamma^2(g)\sin^2\pi g \times[\tfrac{1}{4}(\omega^2-c^2Q^2)]^{-\epsilon}, \quad |\omega|>c|Q|. \quad (28)$$

As could have been anticipated, the power-law behavior in the Green's function [Eq. (20)] has a counterpart in the susceptibility. Here, however, the sign of the potential is important—repulsive interactions ($g>0$) lead to a divergence at the edge of the continuum $\omega=\pm cQ$, while attractions cause $\mathrm{Im}\chi$ to vanish there.

The properties of $\mathrm{Re}\chi(Q,\omega)$ at $T=0$, can be understood by considering the Kramers-Kronig dispersion relation using Eq. (28). For $g>0$, the divergence at $\omega=\pm cQ$ also implies a divergence in $\mathrm{Re}\chi$ along that edge. In particular, $\mathrm{Re}\chi(Q,0)$ diverges as $Q^{-2\epsilon}$. A finite temperature will cause rounding of these divergences, but a peak at the edge will remain.

This dispersion integral for $\mathrm{Re}\chi(Q,\omega)$ can be evaluated, giving the result ($g>0$)

$$\mathrm{Re}\chi(Q,\omega)=(2\pi^2c)^{-1}\Gamma^2(q)\left(\frac{r}{c}\right)^{-2g}\times\mathrm{Im}\left[\left(\frac{\omega+cQ}{2}-i\theta'\right)^{-g}\times\left(\frac{\omega-cQ}{2}-i\theta'\right)^{-g}+(\omega\to-\omega)\right]. \quad (29)$$

Note the divergence of the static susceptibility at $Q=0$ is not logarithmic, but T^{-2g}, and the curvature $(\partial^2/\partial Q^2)\mathrm{Re}\chi(Q,0)|_{Q=0}$ diverges as T^{-2-2g}.

For very small g, it is necessary to modify Eq. (29) to account for the low-frequency restriction used to derive Eq. (28). This can be approximately treated with a cutoff in the dispersion integral, at $\omega\cong c/\alpha$. A correction term in Eq. (29) thereby results which is equal to $-\mathrm{Re}\chi(Q,c/\alpha)$. For example, the $T=0$ and $Q=0$ susceptibility becomes

$$\mathrm{Re}\chi(0,\omega)=(2\pi^2c)^{-1}\Gamma^2(g)\sin 2\pi g\left(\frac{r}{2c}\right)^{-2g}\times\left[\omega^{-2g}-\left(\frac{c}{r}\right)^{-2g}\right], \quad (30)$$

which reduces to $2(\pi c)^{-1}\ln(c/\omega r)$ in the limit $g\to 0$, the correct result for the free gas.

For attractive interactions ($g<0$), $\mathrm{Im}\chi(Q,\omega)$ vanishes as $\omega\to cQ$, therefore $\mathrm{Re}\chi(Q,\omega)$ will be finite. For large ω, Eq. (28) is proportional to $\omega^{2|g|}$, and it is thus necessary to use the cutoff procedure discussed above. The dispersion integral for the static susceptibility then gives the result

$$\mathrm{Re}\chi(Q,0)=[\pi c\Gamma^2(1+|g|)]^{-1}I(\tfrac{1}{2}Q\alpha), \quad (31)$$

where

$$I(x)=2^{-2|g|}\Gamma(2|g|,2x)+x^{2|g|}\int_1^\infty(du/u)e^{-2xu}\times[(u^2-1)^{|g|}-u^{2|g|}],$$

and $\Gamma(a,b)$ is the incomplete Γ function. For Q near zero, the susceptibility has a power-law Q dependence, but is finite and has a cusp maximum at $Q=0$. Similar results for the ω and T dependence can be readily derived.

IV. PAIRING FLUCTUATIONS

The conventional theory of pairing fluctuations involves the two-particle propagator or pair susceptibility. Successful theories of one-dimensional fluctuations for thin whiskers have been given, using a Ginzburg-Landau functional to describe these order-parameter fluctuations. An extension of these ideas to one-dimensional systems has also been proposed.[5] However, there is a fundamental distinction between a three-dimensional system, with one-dimensional order-parameter fluctuations, and a genuine one-dimensional system. The former is three dimensional from the point of view of interparticle interactions and can support conventional one-dimensional fluctuations because of the large coherence length. We shall see that the latter can be quite different.

The difference is illustrated by the pair propagator $P(q,\omega)$, defined by

$$P(q,\omega)=\int_0^\infty(dt/i)e^{i\omega t}\langle[P_q(t),P_{-q}]\rangle, \quad (32)$$

where

$$P_q=\int dx e^{iqx}L^{-1}\sum_k a^\dagger_{1,k+q}a^\dagger_{2,-k}$$

is the pair operator, spin labels have been suppressed, and time evolution as well as thermal averaging is with the full H of Eq. (2). $P(q,t)$ is the Fourier transform of $P(x,t)$, defined by

$$P(x,t)=-i\Theta(t)\langle[\psi_1(x,t)\psi_2(x,t),\psi_2^\dagger\psi_1^\dagger]\rangle.$$

Using the operators $O_1(x,t)$ and $O_2(x,t)$ as in Eq. (23), the computation of $P(x,t)$ is reduced to familiar algebra.[10] The result is

$$P(x,t)=-i\Theta(t)(2\pi\alpha)^{-2}\{\exp[U(x,t)+U(x,-t)]-(x\to-x,t\to-t)\}, \quad (33)$$

provided $\sinh\varphi$ in the definition of $U(x,t)$ following Eq. (23) is replaced by $-\sinh\varphi$. Except for the phase factor, these two equations are equal, thus $P(q,\omega)=\chi(q-2k_F,\omega)$ with a change in the sign of the interaction potential. The results of Sec. IV can therefore be used to determine $P(q,\omega)$.

An attractive interaction produces a divergent $P(0,0)$ as $T\to 0$, indicative of strong pairing fluctuations at low temperatures. Indeed these can be expected to produce enhancement of transport coefficients. It is interesting to note that either P or χ is divergent for the interacting system, as $T\to 0$, indicative of a strong tendency towards either a particle-particle or particle-hole instability.

V. DISCUSSION

Before exploring the consequences of our results, it is necessary to recall the relation of the Luttinger and Tomonaga models to real one-dimensional systems. The solution discussed here would only be of academic interest if these models proved to be deficient in some important respect. It is generally believed that these models are good descriptions of reality provided the interaction is sufficiently long ranged. That can be understood from the form of the interaction Hamiltonian, Eq. (2). Consider a pair state in the real system, consisting of particles of momenta near $+k_F$ and $-k_F$. This pair state can scatter to another pair state with momenta near the original, which is a small momentum process. A large momentum scattering may occur, with $k_F \to -k_F$ and $-k_F \to k_F$. With Eq. (2), the corresponding pair state is constructed by a particle in branch 1 near $+k_F$, and a particle in branch 2 near $-k_F$. This pair state scatters according to $1 \to 1$ and $2 \to 2$, which is the small-momentum process, but the $1 \to 2$ and $2 \to 1$ large-momentum process does not occur.

In order for the large-momentum processes to be negligible, it is necessary for the interaction strength to be very small at $\sim 2k_F$, which is the typical momentum transfer involved for the $1 \to 2$, $2 \to 1$ pair transition at the Fermi energy. However, our results have important application even if this is not the case. Any treatment which omits the small-momentum processes will be plagued with $\ln\omega$ singularities, as discussed in Sec. II. A perturbation expansion about our solution might be expected to be a reasonable approximation for these systems with appreciable interaction strength at large momenta.

It is important to recognize the limitations of the quasiparticle approximations used in a recent fluctuation theory of the Peierls instability. Examination of Eq. (29) indicates that departures from free-particle behavior in the small-Q and small-ω region occur when $(Tc^{-1}\alpha)^{2\epsilon}$ differs appreciably from unity. Because $c\alpha^{-1}$ is of the order of a bandwidth, $T\alpha c^{-1}$ is typically 10^{-3} or smaller, leading to significant deviations from unity for quite modest interactions, $g\sim 0.2$. Requiring $(T\alpha c^{-1})^{2\epsilon}\sim e^{-1}$, determines a characteristic $T_0 \sim c\alpha^{-1}e^{-1/2\epsilon}$.

Also, the effects of weak interactions between one-dimensional strands or finite length might be expected to be important in real systems. Although we have not analyzed these in detail, it is plausible that they could introduce an "effective temperature", T^*, into the problem, that is, a low-energy cutoff which removes the low-temperature singularities. For the finite chain, $T^* \sim 2\pi L^{-1}c$, while T^* for interchain coupling is presumably of the order of the coupling strength itself. It is still quite easy for $(T^*c^{-1}\alpha)^{2\epsilon}$ to depart significantly from unity, indicating the interactions studied in this paper to be important. These questions, however, deserve a more careful analysis than presented here.

It is interesting to consider the complete Hamiltonian for the interacting system coupled to the phonons.[12] Using the field operators $O_1(x)$ and $O_2(x)$ defined by Eq. (5) and in the text preceeding Eq. (22), we find

$$H_L = H + \sum_{q,\lambda} \omega_{q\lambda} b^\dagger_{q\lambda} b_{q\lambda} + \sum_{q,\lambda} \int dx e^{i\vec{q}\cdot\vec{x}} g_L(q,\lambda) \times [O_1^\dagger(x) O_2(x)(b_{q\lambda} + b^\dagger_{-q\lambda}) + \text{H.c.}] , \qquad (34)$$

where H is the Hamiltonian of Eq. (2), $\omega_{q\lambda}$ and $b_{q\lambda}$ refer to the lattice phonons, the $g_L(q,\lambda)$ is the coupling constant. Following the discussion of Sec. II, and Ref. (6), H can be represented entirely in terms of the boson density operators $\rho_1(p)$ and $\rho_2(p)$. Equation (34) is thus a Hamiltonian defined entirely in the space of these bosons. This new model for the Peierls instability incorporates all the physics of the original system, but has eliminated all reference to the fermion operators. It can be shown, to lowest order in $g_L(g,\lambda)$, that the phonon mass operator is proportional to $\chi(q,\omega)$, indicated by Eq. (21). Equation (34) appears to be particularly suitable for studying corrections to this lowest-order approximation and exploring the real nature of the Peierls instability.

Note added in manuscript. After submitting this paper for publication, we were informed that D. C. Mattis has independently introduced an operator equivalent to $O_1(x)$ [Eq. (5)] in the limit $\alpha \to 0$. He used this operator to study backward scattering from impurities in the Luttinger model.[13]

Note added in proof. An earlier calculation of $G(x,t)$ has been given by Carl Dover.[14] In comparison, our method has the virtue of simplicity and easy extension to finite temperature. The first published prediction that χ and P exhibit power laws, is due to Solyom,[15] using arguments based on perturbation theory, in agreement with our exact solution in the weak coupling limit.

ACKNOWLEDGMENTS

We wish to acknowledge many stimulating discussions with P. Fulde and W. Dieterich. One of us (A. L.) wishes to acknowledge the hospitality of the Institut Max von Laue-Paul Langevin where this work originated.

*Supported in part by the National Science Foundation.

[1]L. B. Coleman, M. J. Cohen, D. J. Sandman, F. G. Yamagishi, A. F. Garito, and A. J. Heeger, Solid State Commun. 12, 1125 (1973).
[2]H. R. Zeller, Adv. Solid State Phys. 13, 31 (1973).
[3]M. J. Rice and S. Strässler, Solid State Commun. 13, 125 (1973).
[4]P. A. Lee, T. M. Rice, and P. W. Anderson, Phys. Rev. Lett. 31, 462 (1973).
[5]P. W. Anderson, P. A. Lee, and M. Saitoh, Solid State Commun. 13, 697 (1973).
[6]Daniel C. Mattis and Elliot H. Lieb, J. Math. Phys. 6, 304 (1965). These methods have been extended to the Tomonaga model by H. Gutfreund and M. Schick [Phys. Rev. 168, 418 (1968)]..
[7]P. Nozières and C. T. de Dominicis, Phys. Rev. 178, 1097 (1969).
[8]K. D. Schotte and U. Schotte, Phys. Rev. 182, 479 (1969).
[9]To be precise the operator identities involving $O_1(x)$ are

only exact in the limit $\alpha \to 0$, which is in fact taken in subsequent calculations.

[10]K. D. Schotte, Z. Phys. 230, 99 (1970). M. Blume, V. J. Emery, and A. Luther, Phys. Rev. Lett. 25, 450 (1970).

[11]We make use of the well-known operator identities $e^A e^B = \exp(A + B + [A, B]/2)$ and $\langle e^A \rangle = e^{\langle A^2 \rangle /2}$, where A and B are linear combinations of boson creation and destruction operators and the average is in a harmonic oscillator density matrix.

[12]The effects of phonons on BCS pairing in a one-dimensional system have been considered by B. R. Patton and L. J. Sham [Phys. Rev. Lett. 31, 631 (1973)].

[13]Daniel C. Mattis, report of work prior to publication.

[14]Carl B. Dover, Ann. Phys. (N. Y.) 50, 500 (1968).

[15]J. Solyom, J. Low-Temp. Phys. 12, 529 (1973).

Backward Scattering in the One-Dimensional Electron Gas*

A. Luther†
Lyman Laboratory, Harvard University, Cambridge, Massachusetts 02138

and

V. J. Emery
Brookhaven National Laboratory, Upton, New York 11973
(Received 16 July 1974)

An exact solution to the one-dimensional electron gas with a particular attractive-interaction strength for scattering across the Fermi "surface" is given. It is shown that conductivity enhancement occurs for physically interesting values of the coupling constants. Scaling arguments are advanced to demonstrate that this solution applies generally for attractive backward scattering. In addition, the spinless problem is solved exactly for arbitrary couplings.

Progress towards an understanding of the equilibrium and transport properties of quasi-one-dimensional conductors has been hampered by a lack of knowledge about the underlying interacting electron system. Although the Tomonaga and Luttinger models[1] have provided some insight, their generality can be questioned as a result of their neglect of interactions near twice the Fermi momentum, $2k_F$, which are responsible for backward scattering. This note reports an exact solution to the more general problem with an attractive interaction at $2k_F$, and uses it to construct a qualitative picture for the general interacting one-dimensional system.

Several properties of the exact solution are particularly significant. It requires an attractive interaction at $2k_F$ of a specific strength but the small-momentum interaction can be arbitrary, a situation of sufficient generality to be of interest for experiments on the quasi-one-dimensional systems. Depending upon the sign and magnitude of the small-momentum part, a large conductivity enhancement can occur as the temperature tends to zero, in contrast to a recent approximate treatment.[2] We compute the temperature dependence of the conductivity as well as other physically important response functions which describe the low-temperature pairing, charge and spin fluctuations, and magnetic susceptibility.

Together with the renormalization-group calculations of Menyhárd and Sólyom,[3] our solution can be used to provide a complete, if qualitative, picture of the one-dimensional interacting gas. This approach is analogous to the scaling argument of Anderson, Yuval, and Hamann for the Kondo problem.[4]

The model we solve is a logical extension of the Luttinger or Tomonaga model to include spin as well as interactions which scatter particles from $+k_F$ to $-k_F$. The Hamiltonian is the sum of free-particle kinetic energy, linearized near k_F, plus the usual small-momentum interaction,[5]

$$\mathcal{H}_s = v_F \sum_{k,s} k(a_{k,s}^\dagger a_{k,s} - b_{k,s}^\dagger b_{k,s}) + 2L^{-1}\sum_k V\rho_1(k)\rho_2(-k), \tag{1}$$

where the operators $a_{k,s}$ ($b_{k,s}$) describe spin-$\frac{1}{2}$ fermions with momentum k ($-k$); $\rho_1(k)$ and $\rho_2(k)$ are density operators, $\rho_1(k) = 2^{-1/2}\sum_{ps} a_{p+k,s}^\dagger a_{p,s}$ and $\rho_2(k) = 2^{-1/2}\sum_{ps} b_{p+k,s}^\dagger b_{p,s}$. It is helpful in our subsequent discussion to introduce the spin-density operators $\sigma_1(k) = 2^{-1/2}\sum_{ps} s a_{p+k,s}^\dagger a_{ps}$ and $\sigma_2(k) = 2^{-1/2}\sum_{ps} s b_{p+k,s}^\dagger b_{p,s}$ with $s = \pm 1$. The operators ρ and σ satisfy the algebra of the Luttinger model, $[\rho_1(-k), \rho_1(+k')] = [\rho_2(+k), \rho_2(-k')] = \delta_{kk'} kL/2\pi$, together with identical equations for the σ commutators, $[\sigma_1(-k), \sigma_1(+k')] = [\sigma_2(+k), \sigma_2(-k')] = \delta_{kk'} kL/2\pi$, while all others vanish. The length of the sample is L and all states below $v_F k_F$ are filled. The kinetic energy term in Eq. (1) can be written in terms of these operators, by making use of well-known identities extended to the spin-$\frac{1}{2}$ case:

$$v_F \sum_{k,s} k(a_{k,s}^\dagger a_{k,s} - b_{k,s}^\dagger b_{k,s}) - 2\pi v_F L^{-1}\sum_k [\rho_1(k)\rho_1(-k) + \sigma_1(k)\sigma_1(-k) + \rho_2(-k)\rho_2(k) + \sigma_2(-k)\sigma_2(k)],$$

indicating a separation into density and spin-density operators.

The large-momentum-transfer terms are described by an additional $\mathcal{H}_L$ in the form[6]

$$\mathcal{H}_L = \sum_{s,s'} \int dx\, \Psi_{1,s}^\dagger(x)\Psi_{2,s'}^\dagger(x)\Psi_{1,s'}(x)\Psi_{2,s}(x)[U_\parallel \delta_{s,s'} + U_\perp \delta_{s,-s'}], \tag{2}$$

where $\Psi_{1,s}(x) = L^{-1/2}\sum_k \exp(ikx) a_{k,s}$ and $\Psi_{2,s}(x) = L^{-1/2}\sum_k \exp(ikx) b_{k,s}$. The $U_\parallel$ term can be written in terms of the density and spin-density operators by a permutation of the inner two operators, and becomes equal to

$$-U_\parallel L^{-1}\sum_k \rho_1(k)\rho_2(-k) - U_\parallel L^{-1}\sum_k \sigma_1(k)\sigma_2(-k).$$

The $U_\perp$ term contains field operators of opposite spin, for which we use the boson representation[7,8]:

$$\Psi_{j,s}(x) = (2\pi\alpha)^{-1/2}\exp\{\pm[ik_F x + 2\pi L^{-1}\sum_k k^{-1}\exp(-\tfrac{1}{2}\alpha|k| - ikx)\rho_{j,s}(k)]\}, \tag{3}$$

where $v_F\alpha^{-1}$ is the bandwidth, $\rho_{1,s}(k) = \sum_p a_{p+k,s}^\dagger a_{p,s}$, $\rho_{2,s}(k) = \sum_p b_{p+k,s}^\dagger b_{p,s}$, and the plus (minus) sign goes with $j = 1$ ($j = 2$). This relationship has been discussed in detail for the spinless case[7] and the generalization here is trivial, although the notation is somewhat cumbersome. This permits us to write the products of field operators in the $U_\perp$ term as $(2\pi\alpha)^{-2}\exp\{2^{-1/2}[\varphi_1(x) + \varphi_2(x)]\}$, with

$$\varphi_j(x) = 2\pi L^{-1}\sum_k k^{-1}\exp(\tfrac{1}{2}\alpha|k| - ikx)\sigma_j(k). \tag{4}$$

The total Hamiltonian $\mathcal{H}_s + \mathcal{H}_L$ can now be written as the sum $\mathcal{H}_0 + \mathcal{H}_1$, where

$$\mathcal{H}_0 = 2\pi v_F L^{-1}\sum_k [\rho_1(k)\rho_1(-k) + \rho_2(-k)\rho_2(k)] + L^{-1}\sum_k (2V - U_\parallel)\rho_1(k)\rho_2(-k), \tag{5}$$

and

$$\mathcal{H}_1 = 2\pi v_F L^{-1}\sum_k [\sigma_1(k)\sigma_1(-k) + \sigma_2(-k)\sigma_2(k)] - L^{-1}\sum_k U_\parallel \sigma_1(k)\sigma_2(-k) + U_\perp(2\pi\alpha)^{-2}\int dx\,\{\exp(2^{-1/2}[\varphi_1(x) + \varphi_2(x)]) + \text{H.c.}\}, \tag{6}$$

and a separation into density and spin-density operators has been achieved. Obviously $[\mathcal{H}_1, \mathcal{H}_0] = 0$. It is interesting to note that the most general Hamiltonian for spinless fermions, $\sigma(k) = 0$, including large-momentum interactions is still of the Tomonaga-Luttinger form, for which the thermodynamics[5] and correlation functions[7] are known.

This separation into $\mathcal{H}_1$ and $\mathcal{H}_0$ is essential because $\mathcal{H}_0$ is a quadratic form in boson operators and may be diagonalized by means of a canonical transformation. We now show that for a particular value of $U_\parallel$, $\mathcal{H}_1$ may also be diagonalized. Since the part of $\mathcal{H}_1$ which does not involve $U_\perp$ is quadratic it may be diagonalized by a transformation $e^{iS}\mathcal{H}_1 e^{-iS} = \mathcal{H}_1'$, with $S = 2\pi i L^{-1}\varphi\sigma_1(k)\sigma_2(-k)$, and $\tanh 2\varphi = -U_\parallel \times (2\pi v_F)^{-1}$. This transformation changes the velocity to $v_F' = v_F \operatorname{sech} 2\varphi$ and the exponent of the $U_\perp$ term to $2^{-1/2}e^{\varphi}[\varphi_1(x)+\varphi_2(x)]$. The point now is that if $2^{-1/2}e^{\varphi} = 1$, comparison of Eqs. (3) and (6) shows that the $U_\perp$ term is just the boson representation of a product of spinless Fermi fields $\Psi_2^\dagger(x)\Psi_1(x)\exp(-2i \times k_F x)$. Writing the kinetic energy also in fermion representation, we find that $\mathcal{H}_1'$ is quadratic and solvable:

$$\mathcal{H}_1' = v_F' \sum_k k(a_k^\dagger a_k - b_k^\dagger b_k) + U_\perp(2\pi\alpha)^{-1}\sum_k (a_k^\dagger b_{k-2k_F} + \text{H.c.}). \tag{7}$$

It is possible to make $2^{-1/2}e^{\varphi} = 1$ by choosing $U_\parallel$ so that $\tanh 2\varphi = -U_\parallel(2\pi v_F)^{-1} = \frac{3}{5}$. Equation (7) has the eigenvalue spectrum $v_F k_F \pm [(k-k_F)^2 + \Delta^2]^{1/2}$ and $v_F k_F \pm [(k+k_F)^2+\Delta^2]^{1/2}$, where $\Delta = U_\perp \times (2\pi\alpha)^{-1}$, indicating the appearance of gaps at $+k_F$ and $-k_F$. These gaps have important consequences for the properties of our solution.

The total Hamiltonian $\mathcal{H}' = e^{iS}\mathcal{H}e^{-iS} = \mathcal{H}_0 + \mathcal{H}_1'$ is now diagonalized, and it is straightforward to compute the free energy and correlation functions in the usual fashion. We may distinguish two regions: low temperatures, $T \ll \Delta$, for which the gap at the Fermi energy means that only excitations in $\mathcal{H}_0$ (and consequently density operators) are important; and high temperatures, $T \gg \Delta$, when the gap, and consequently $U_\perp$, are negligible. (A similar division applies to the frequency and momentum dependence.) In either case, the Hamiltonian is of the Tomonaga form, and the analytic behavior of correlation functions is therefore known from previous work.[7]

The properties of our solution are best expressed by the exponents characterizing the different spectral functions. These exponents are defined by $\operatorname{Im}C(\omega) \propto \omega^{\mu}$, where C is the correlation function in question. Because the analytic form of these functions is of Tomonaga-Luttinger form, the temperature and momentum dependence is also determined by this exponent. The total scale dimension of these correlation functions is given by $\mu + 2$, the extra 2 resulting from two Fourier transforms in the definition of C. The exponents are listed in Table I, along with the temperature dependence of the magnetic susceptibility, for both low and high temperatures.

The electrical-conductivity relaxation rate in the impurity-dominated region is determined by the spectral function for the parallel-spin $2k_F$ susceptibility,[9] the characteristic energy W responsible for the interaction, and the free-particle conductivity σ_0. The result in Table I then means that $\sigma = ne^2\tau/m = \sigma_0(T/W)^{-\mu}$, which will be enchanced ($\mu > 0$) whenever $v' < -\frac{3}{5}$. Under these conditions, the spectral density at $2k_F$ vanishes, causing the reduction in scattering from the impurity.

It is interesting to compare with the renormalization-group approach of Menyhárd and Sólyom.[3] They considered the special case $U_\parallel = U_\perp = g_1$ and, for $g_1 < 0$, their exponents are qualitatively different from ours because they find that g_1 scales onto a fixed point ($g_1 = -2\pi v_F$) which is outside the weak-coupling region for which their perturbation expansions are adequate. A more credible use of the renormalization-group equations, analogous to scaling theories of the Kondo problem,[4] uses the fact that, provided no fixed point intervenes, the coupling constant would scale through the value $g_1 = -\frac{6}{5}\pi v_F$ at which our exact solution

TABLE I. Exponents of the spectral functions, defined in text, and the low- and high-temperature behavior of the magnetic susceptibility. The parameters $v' = V(\pi v_F)^{-1} + \frac{3}{5}$ and $\Delta = U_\perp(2\pi\alpha)^{-1}$.

Spectral function and operator	Low-temperature ($T \gg \Delta$) exponent	High-temperature ($T \gg \Delta$) exponent
$2k_F$ susceptibility, $\Psi_{1,+}^\dagger\Psi_{2,+}$	$-2 + [(1-v')/(1+v')]^{1/2}$	$-\frac{3}{2} + [(1-v')/(1+v')]^{1/2}$
$2k_F$ susceptibility, $\Psi_{1,+}^\dagger\Psi_{2,-}$	$-2 + [(1-v')/(1+v')]^{1/2}$	$[(1-v')/(1+v')]^{1/2}$
Singlet pairing, $\Psi_{1,+}^\dagger\Psi_{2,-}^\dagger$	$-2 + [(1+v')/(1-v')]^{1/2}$	$-\frac{3}{2} + [(1+v')/(1-v')]^{1/2}$
Triplet pairing, $\Psi_{1,+}^\dagger\Psi_{2,+}^\dagger$	$-2 + [(1+v')/(1-v')]^{1/2}$	$[(1+v')/(1-v')]^{1/2}$
Susceptibility, $\sigma_1+\sigma_2$	$(\pi v_F')^{-1}(2\pi\beta\Delta)^{1/2}e^{-\beta\Delta}$	$2(\pi v_F)^{-1}(1-U_\parallel/2\pi v_F)^{-1}$

may be used. This suggests that the gap Δ exists for the weaker-coupling problems, $0 > g_1 > -\frac{6}{5}\pi v_F$, as well. If this gap exists, the low-temperature exponents given in Table I are still correct, with a new $v'2\pi v_F = 2V - U_\parallel$. The high-temperature exponents are more complicated but can be computed by setting $U_\perp = 0$, and following the usual procedure for the Tomonaga-model correlation functions.[7]

These exponents give substantial physical insight into the behavior of the system. For example, if the conductivity exponent is negative at high temperatures and positive at low temperatures a conductivity peak occurs at a temperature of the order of the gap, where the two power laws "cross over". The magnitude of this peak depends on W and $U_\perp$ as well as the difference $\delta\mu$ between the two exponents. For $U_\parallel(\pi v_F)^{-1} = -\frac{6}{5}$, there can be a rather small peak because $\delta\mu$ is only $\frac{1}{2}$. We have calculated $\delta\mu$ for smaller $|U_\parallel|$, assuming that the scaling arguments are correct, and find that $\delta\mu$ approaches $\frac{6}{5}$ as $|U_\parallel| \to 0$. This is a new mechanism for a conductivity peak which merits a more detailed discussion than can be presented here.

Spin-flip impurity scattering will have a temperature dependence given by the opposite-spin $2k_F$ susceptibility. Here the difference in exponents is large, indicating a stronger scattering at low temperatures which would dominate other mechanisms.

At the risk of making this problem appear more complicated (but even more intriguing), we now point out the reason behind our notation $U_\parallel$ and $U_\perp$. These have been chosen to emphasize the connection to the Kondo parameters $-J_\parallel$ and $J_\perp$. The Hamiltonian in Eq. (6) has many similarities to the Tomonaga version of the Kondo Hamiltonian, and we expect many of the renormalization-group arguments to be applicable here as well. In particular, the weak-coupling scale energy for $\mathcal{H}_1$ is $|U|^{1/2}\exp(-|U|^{-1})$, for $U_\parallel = U_\perp = U < 0$, as in the Kondo problem.

The inclusion of electron-phonon coupling raises many questions for which there are only qualitative answers at present. In the adiabatic model, which views these interactions as giving rise to additional electron-electron coupling constants, the question of resistivity or conductivity enhancement depends sensitively on the relative signs and magnitudes of all coupling constants. Furthermore, if the phonon frequencies shift with temperature, these coupling constants also vary, and the calculation of the appropriate exponents becomes complicated. This problem is currently under investigation.

Much of this work was carried out while A. L. was a summer visitor at Brookhaven National Laboratory. The model solved in this paper appears to be related to generalizations of the Thirring model,[10] as pointed out to us by S. Coleman.

---

*Work supported in part by the U. S. Atomic Energy Commission.

†Research supported in part by the National Science Foundation under Grant No. GH 32774.

[1]E. H. Lieb and D. C. Mattis, *Mathematical Physics in One Dimension* (Academic, New York, 1966).

[2]H. Fukuyama, M. Rice, and C. Varma, to be published.

[3]N. Menyhárd and J. Sólyom, J. Low Temp. Phys. 12, 529 (1973); J. Sólyom, J. Low Temp. Phys. 12, 547 (1973).

[4]P. W. Anderson, G. Yuval, and D. R. Hamann, Phys. Rev. B 1, 4464 (1970).

[5]D. C. Mattis and E. H. Lieb, J. Math. Phys. 6, 304 (1965).

[6]There are many equivalent representations of these terms, and our notation is convenient for later purposes. The Hubbard model requires selecting $V = U_\parallel = U_\perp$, for which case the parallel-spin interactions cancel, while the notation of Ref. 3 corresponds to the choice $U_\parallel = U_\perp = g_1$ and $V = g_2$.

[7]A. Luther and I. Peschel, Phys. Rev. B 9, 2911 (1974).

[8]D. Mattis, J. Math. Phys. 15, 609 (1974).

[9]A. Luther and I. Peschel, Phys. Rev. Lett. 32, 992 (1974).

[10]R. Dashen and Y. Frishman, Phys. Lett. 46B, 439 (1973).

PHYSICAL REVIEW D VOLUME 11, NUMBER 8 15 APRIL 1975

Quantum sine-Gordon equation as the massive Thirring model*

Sidney Coleman
Lyman Laboratory of Physics, Harvard University, Cambridge, Massachusetts 02138
(Received 6 January 1975)

The sine-Gordon equation is the theory of a massless scalar field in one space and one time dimension with interaction density proportional to $\cos\beta\varphi$, where β is a real parameter. I show that if β^2 exceeds 8π, the energy density of the theory is unbounded below; if β^2 equals 4π, the theory is equivalent to the zero-charge sector of the theory of a free massive Fermi field; for other values of β, the theory is equivalent to the zero-charge sector of the massive Thirring model. The sine-Gordon soliton is identified with the fundamental fermion of the Thirring model.

I. INTRODUCTION

The sine-Gordon equation is the sophomoric but unfortunately standard name for the theory of a single scalar field in one space and one time dimension, with dynamics determined by the Lagrangian density[1]

$$\mathcal{L}=\tfrac{1}{2}\partial_\mu\varphi\partial^\mu\varphi+\frac{\alpha_0}{\beta^2}\cos\beta\varphi+\gamma_0\ . \tag{1.1}$$

Here the sum over repeated indices is implied, and α_0, β, and γ_0 are real parameters. Since we are free to redefine the sign of φ, we can always take β to be positive. Likewise, since we are free to shift φ by π/β, we can always take α_0 to be positive. We can gain some rough insight into the physical meaning of these parameters if, in the classical theory, we expand about the configuration of minimum energy ($\varphi=0$):

$$\frac{\alpha_0}{\beta^2}\cos\beta\varphi=\frac{\alpha_0}{\beta^2}-\frac{\alpha_0}{2}\varphi^2+\frac{\alpha_0\beta^2}{4!}\varphi^4+\cdots. \tag{1.2}$$

From this we see that if we wish to adjust our zero of energy density so that the minimum energy is zero we must choose

$$\gamma_0=-\frac{\alpha_0}{\beta^2}\ . \tag{1.3}$$

Also, α_0 is the "squared mass" (actually, since we are discussing a classical theory, squared inverse wavelength) associated with the spectrum of small oscillations about the minimum, and β is a parameter that measures the strength of the interactions between these small oscillations.

Of course, this is just a linearized analysis. The exact classical sine-Gordon equation has been extensively studied, and there exists an enormous literature concerning it.[2] Recently, some attempts have been made to use the known properties of the classical theory as a starting point for the investigation of the quantum theory.[3] This paper is a report on an attempt to investigate the properties of the quantum theory directly, without reference to the classical theory. The main results of the investigation are the following:

(1) As for any theory of a scalar field in two dimensions with nonderivative interactions, all divergences that occur in any order of perturbation theory can be removed by normal-ordering the Hamiltonian. For the sine-Gordon equation, this normal-ordering is equivalent to a multiplicative renormalization of α_0 and an additive renormalization of γ_0; β is not renormalized.[4]

(2) If β^2 exceeds 8π, the energy per unit volume is unbounded below, and the theory has no ground state.

(3) If β^2 is less than 8π, the theory is equivalent to the charge-zero sector of the massive Thirring model. This is a surprise, since the Thirring model is a canonical field theory whose Hamiltonian is expressed in terms of fundamental Fermi fields only.[5]

(4) In the special case $\beta^2=4\pi$, the sine-Gordon equation describes the charge-zero sector of a *free* massive Dirac field theory.

Further explanation of the last two points may be helpful to the reader.

The (massless) Thirring model[6] is a theory of a single Dirac field in one space and one time dimension, with dynamics determined by the Lagrangian density

$$\mathcal{L}=\bar\psi i\gamma_\mu\partial^\mu\psi-\tfrac{1}{2}gj^\mu j_\mu\ , \tag{1.4}$$

where

$$j^\mu=\bar\psi\gamma^\mu\psi\ , \tag{1.5}$$

and g is a free parameter, the coupling constant. The formal definition of j_μ as a product of Dirac fields is plagued with ambiguities, just as in four-dimensional theories, but, again just as in four-dimensional theories, these ambiguities can be resolved by demanding that j_μ obey the proper Ward identities.[7] Once this has been done, the Hamiltonian derived from Eq. (1.4) requires no further renormalizations. The model is exactly soluble and is sensible for g greater than minus π.

Within the massless Thirring model, it is possible to define a renormalized scalar density

$$\sigma = Z\bar{\psi}\psi , \tag{1.6}$$

where Z is a cutoff-dependent constant. The massive Thirring model is formally defined by adding a term proportional to σ to the Lagrangian density (1.4):

$$\mathcal{L} \to \mathcal{L} - m'\sigma . \tag{1.7}$$

Here m' is simply a real parameter; it is not to be identified with the mass of any presumed one-particle state. The massive Thirring model is not exactly soluble, and, to my knowledge, it is unknown whether (1.7) defines a physically sensible theory for any value of g. However, it is certainly true that every term in the perturbation series for the Green's functions of the theory in powers of m' is well defined, except for the trivial infrared problems associated with performing a mass perturbation expansion about a massless theory. These can be circumvented by a standard trick. Instead of (1.7), we consider the theory defined by

$$\mathcal{L} = \bar{\psi} i\gamma^{\mu}\partial_{\mu}\psi - \frac{g}{2} j^{\mu} j_{\mu} - m'\sigma f(x) , \tag{1.8}$$

where f is some function of space-time with compact support. The infrared divergences now disappear. A (probably unworkable) prescription for solving the theory would be to first sum up the perturbation series in m' with fixed f, and only then go to the limit $f=1$.

I can now explain more clearly results (3) and (4) above. The perturbation series described above for the massive Thirring model is term-by-term identical with a perturbation series (in α_0) for the sine-Gordon equation, if the following identifications are made between the two theories:

$$4\pi/\beta^2 = 1 + g/\pi , \tag{1.9}$$

$$-\frac{\beta}{2\pi}\epsilon^{\mu\nu}\partial_{\nu}\varphi = j^{\mu} , \tag{1.10}$$

$$\frac{\alpha_0}{\beta^2}\cos\beta\varphi = -m'\sigma . \tag{1.11}$$

Of course, this analysis does not show that either the massive Thirring model or the sine-Gordon equation exists, in the strict sense of constructive field theory. I believe it is fair to say, though, that it does show that if either theory exists, it is equivalent to the other. Please note that Eq. (1.9) implies that if $\beta^2 = 4\pi$, then $g=0$. This is result (4).

The analysis that leads to these results is explained in Secs. II, III, and IV below. Section V contains some remarks about the meaning of the results, and some conjectures.

II. THE HAMILTONIAN

For any scalar field theory in two dimensions with nonderivative interactions, the only ultraviolet divergences that occur in any order of perturbation theory come from graphs that contain a closed loop consisting of a single internal line, that is to say, graphs in which two fields at the same vertex are contracted with each other. Thus, for example, the graph of Fig. 1(a) is ultraviolet-divergent, while that of Fig. 1(b) is not. Thus, all ultraviolet divergences can be removed by normal-ordering the interaction Hamiltonian in the interaction picture.

However (at least for the moment), we are not interested in the interaction picture but in the Schrödinger picture. It is easy to see what the normal-ordering prescription corresponds to in this picture. Schrödinger-picture operators are given as functions of the field $\varphi(x)$ and the canonical momentum density $\pi(x)$, where x is the spatial coordinate. If we define operators $a(k,m)$ by

$$\varphi(x) = \int \frac{dk}{2\pi}\left[\frac{1}{2\omega(k,m)}\right]^{1/2}[a(k,m)e^{-ikx} + a^{\dagger}(k,m)e^{ikx}] , \tag{2.1}$$

and

$$\pi(x) = i\int \frac{dk}{2\pi}\left[\frac{\omega(k,m)}{2}\right]^{1/2}[a(k,m)e^{-ikx} - a^{\dagger}(k,m)e^{ikx}] , \tag{2.2}$$

where

$$\omega(k,m) = (k^2 + m^2)^{1/2} , \tag{2.3}$$

then the normal-ordered Schrödinger operator corresponds to the operator rearranged with all the a's on the right and all the $a^{\dagger}$'s on the left.

However, this prescription is ambiguous, because it does not tell us what m is. Of course, if we are really doing interaction-picture perturbation theory, it would be senseless to choose m to be other than that mass which occurs in the

(a)

(b)

FIG. 1. Two typical graphs in a scalar field theory in two dimensions with nonderivative interactions; (a) is logarithmically divergent; (b) is convergent.

free Hamiltonian. However, for the sine-Gordon equation, it is not clear what is the most profitable way to divide the Hamiltonian into a free and an interaction part, or, indeed, whether any such division is profitable. Therefore, for the moment at least, we will not specify m, and we will denote by N_m the normal-ordering operation defined by the mass m.

[This is off the main line of the argument, but it is amusing to consider what would happen, in perturbation theory, if we chose m to be different from μ, the mass in the free Hamiltonian. In this case, it is easy to see that the divergent loop integral of Fig. 1(a) is replaced by a convergent integral according to the following prescription:

$$\int \frac{d^2k}{k^2+\mu^2} \to \int d^2k\left(\frac{1}{k^2+\mu^2}-\frac{1}{k^2+m^2}\right). \tag{2.4}$$

Note that if $m=\mu$ (conventional normal-ordering), the graph is canceled completely (the usual result).]

So much for generalities. We now turn to the sine-Gordon Hamiltonian density,

$$\mathcal{H}=-\tfrac{1}{2}\pi^2+\tfrac{1}{2}\left(\frac{\partial\varphi}{\partial x}\right)^2-\frac{\alpha_0}{\beta^2}\cos\beta\varphi-\gamma_0\,. \tag{2.5}$$

Let us begin with the cosine term. Wick's theorem tells us that, for a free field of mass m, and for any space-time function $J(x)$,

$$\exp\left[i\int J(x)\varphi(x)d^2x\right]=N_m\exp\left[i\int J(x)\varphi(x)d^2x\right]\exp\left[-\tfrac{1}{2}\int J(x)\Delta(x-y;m)J(y)d^2x\,d^2y\right], \tag{2.6}$$

where Δ is the free-field two-point Wightman function. [If we replace the unordered exponential in Eq. (2.5) by a time-ordered exponential, the Wightman function is replaced by the Feynman propagator.] For small spacelike separation

$$\Delta(x;m)=-\frac{1}{4\pi}\ln cm^2x^2+O(x^2)\,, \tag{2.7}$$

where c is a numerical constant (related to Euler's constant) and

$$x^2=-x_\mu x^\mu\,, \tag{2.8}$$

and is positive for spacelike separation. We cut off the theory by replacing $\Delta(x;m)$ by

$$\Delta(x;m)-\Delta(x;\Lambda)\equiv\Delta(x;m;\Lambda)\,, \tag{2.9}$$

where Λ is a large mass, the cutoff. This is nonsingular at the origin,

$$\Delta(0;m;\Lambda)=-\frac{1}{4\pi}\ln\frac{m^2}{\Lambda^2}\,. \tag{2.10}$$

We can now use Eq. (2.6) for J a δ function. We find

$$e^{i\beta\varphi}=N_m\left(\frac{m^2}{\Lambda^2}\right)^{\beta^2/8\pi}e^{i\beta\varphi}\,. \tag{2.11}$$

This has been derived in the interaction picture, but must also be true in the Schrödinger picture, since it only involves fields at the same time. Thus, if we define, for some arbitrarily chosen mass m,

$$\alpha=\alpha_0\left(\frac{m^2}{\Lambda^2}\right)^{\beta^2/8\pi}, \tag{2.12}$$

then

$$\alpha_0\cos\beta\varphi=\alpha N_m\cos\beta\varphi\,. \tag{2.13}$$

Of course, if we had chosen a different mass (call it μ) to define the theory, this would just have led to a finite multiplicative redefinition of α. Indeed, from Eq. (2.11),

$$N_m\cos\beta\varphi=\left(\frac{\mu^2}{m^2}\right)^{\beta^2/8\pi}N_\mu\cos\beta\varphi\,. \tag{2.14}$$

This identity will be very important to us shortly.

Much less care needs to be taken in normal-ordering the remainder of the Hamiltonian density, since this is just a quadratic form in the fundamental fields, and therefore the only effect of normal-ordering is to add a constant. For brevity of subsequent notation, we define

$$\mathcal{H}_0=\tfrac{1}{2}\pi^2+\tfrac{1}{2}\left(\frac{d\varphi}{dx}\right)^2. \tag{2.15}$$

Then, from Eqs. (2.1) and (2.2),

$$\mathcal{H}_0=N_m\mathcal{H}_0+E_0(m)\,, \tag{2.16}$$

where

$$E_0(m)=\int\frac{dk}{8\pi}\,\frac{2k^2+m^2}{\omega(k,m)}\,. \tag{2.17}$$

Thus, if we define

$$\gamma=\gamma_0+E_0(m)\,, \tag{2.18}$$

then

$$\mathcal{H}_0-\gamma_0=N_m\mathcal{H}_0-\gamma\,. \tag{2.19}$$

Parallel to Eq. (2.14) we have

$$\begin{aligned}N_m\mathcal{H}_0&=N_\mu\mathcal{H}_0+E_0(\mu)-E_0(m)\\&=N_\mu\mathcal{H}_0+\frac{1}{8\pi}(\mu^2-m^2)\,.\end{aligned} \tag{2.20}$$

Assembling all this, we find the cutoff-independent form of the Hamiltonian density:

$$\mathcal{H} = N_m\left(\mathcal{H}_0 - \frac{\alpha}{\beta^2}\cos\beta\varphi - \gamma\right) . \tag{2.21}$$

where α, β, and γ are finite parameters. This is result (1) of the Introduction.

III. A VARIATIONAL COMPUTATION

Considerable insight into the physical import of the reordering equations, (2.14) and (2.20), can be obtained by considering the following trivial problem. Suppose we are given the Hamiltonian density

$$\mathcal{H} = N_m(\mathcal{H}_0 + \tfrac{1}{2}\beta^2\varphi^2) . \tag{3.1}$$

This is a free-field Hamiltonian density, normal-ordered with a perverse mass; that is to say, m is not necessarily equal to β.

Let us attempt to find the ground state of this theory by the Rayleigh-Ritz variational method. For our trial states we will use the vacuum states appropriate to a free field of mass μ. These states are defined by

$$a(k, \mu)|0, \mu\rangle = 0 . \tag{3.2}$$

The computation is made trivial by reordering. Expanding Eq. (2.14) in powers of β^2, we find

$$N_m(\tfrac{1}{2}\beta^2\varphi^2) = N_\mu(\tfrac{1}{2}\beta^2\varphi^2) - \frac{\beta^2}{8\pi}\ln\frac{\mu^2}{m^2} . \tag{3.3}$$

Thus, from this and Eq. (2.20),

$$\mathcal{H} = N_\mu(\mathcal{H}_0 + \tfrac{1}{2}\beta^2\varphi^2) + \frac{1}{8\pi}\left(\mu^2 - m^2 - \beta^2\ln\frac{\mu^2}{m^2}\right) , \tag{3.4}$$

whence

$$\langle 0, \mu|\mathcal{H}|0, \mu\rangle = \frac{1}{8\pi}\left(\mu^2 - m^2 - \beta^2\ln\frac{\mu^2}{m^2}\right) . \tag{3.5}$$

As a function of μ, the right-hand side of this equation assumes its minimum value when μ is β. This is, of course, the correct result.

Now let us perform exactly the same computation for the sine-Gordon equation:

$$\mathcal{H} = N_m\left(\mathcal{H}_0 - \frac{\alpha}{\beta^2}\cos\beta\varphi - \gamma\right) . \tag{2.21}$$

From Eqs. (2.14) and (2.20) we find

$$\mathcal{H} = N_\mu\left[\mathcal{H}_0 - \frac{\alpha}{\beta^2}\left(\frac{\mu^2}{m^2}\right)^{\beta^2/8\pi}\cos\beta\varphi + \frac{1}{8\pi}(\mu^2 - m^2) - \gamma\right] , \tag{3.6}$$

whence

$$\langle 0, \mu|\mathcal{H}|0, \mu\rangle = -\frac{\alpha}{\beta^2}\left(\frac{\mu^2}{m^2}\right)^{\beta^2/8\pi} + \frac{1}{8\pi}(\mu^2 - m^2) - \gamma . \tag{3.7}$$

In contrast to Eq. (3.5), the right-hand side of this equation is unbounded below, as μ goes to infinity, if β^2 exceeds 8π. Thus, in this case, the energy of the theory (even if we restrict it to a box of finite volume) is unbounded below. The theory has no ground state, and is physically nonsensical. This is result (2) of the Introduction.

IV. PERTURBATION THEORY

Following the plan set forth in Sec. I, we will now construct a perturbation series (in α) for the Green's functions of the quantum sine-Gordon equation. The Green's functions we will choose to compute will be those corresponding to the vacuum expectation value of the time-ordered product of a string of interaction Hamiltonian densities. The interaction density is a somewhat unconventional choice for the local field that characterizes the theory, but, in principle, there is no reason to prefer any local field to any other for this purpose, and by choosing the interaction density we will be able to shorten our labors significantly.

Before beginning the computation, we will have to straighten out the problems associated with the infrared divergences which plague all perturbative calculations about a massless field theory. These problems are notoriously acute for a massless scalar field theory in two dimensions. For example, the very expression for the field in terms of annihilation and creation operators, Eq. (2.2), is infrared-divergent; as a consequence of this, it is not possible to define $\varphi(x)$ as a Wightman field in the conventional sense.[8] (This does not mean that the theory of a free massless scalar meson in two dimensions is nonsense. For example, $\partial_\mu\varphi$ is perfectly well defined—the extra power of momentum brought in by differentiation eliminates the infrared divergence—and can be used as the Wightman field.)

The simplest way to circumvent these problems is to give the scalar meson a small mass μ. This will enable us to compute the terms in the perturbation expansion by conventional techniques (e.g., Wick's theorem). We will then send μ to zero. Of course, this will still leave us with the usual infrared problems, just like those which arise for the Thirring model, discussed in Sec. I. We will handle these by the method described previously, i.e., by multiplying the interaction by a space-time function of compact support, $f(x)$.

Thus, we are led to the Hamiltonian density

$$\mathcal{H} = N_m\left[\mathcal{H}_0 + \tfrac{1}{2}\mu^2\varphi^2 + f(x)\left(-\frac{\alpha}{\beta^2}\cos\beta\varphi - \gamma\right)\right] . \tag{4.1}$$

To find the desired Green's functions we must (1) compute the vacuum persistence amplitude to all orders in perturbation theory; (2) send μ to zero, keeping f fixed; (3) sum up the series; (4) send f to one; and (5) compute appropriate variational derivatives with respect to f about $f=1$. The last three steps are beyond my analytic ability; fortunately, they are not necessary for our purposes, for we will establish the identity with the perturbation series (in m') for the massive Thirring model at the end of step (2).

It is instructive to compute a slightly more general object than is needed for our immediate purposes. Consider

$$T\langle 0,\mu|\prod_i N_m e^{i\beta_i\varphi(x_i)}|0,\mu\rangle \ , \tag{4.2}$$

where φ is a free field of mass μ, the β's are real parameters, and the x's are space-time points (within the support of f). Because the x's are restricted to a finite region, we can uniformly apply the short-distance approximation for the Feynman propagator. This is identical to Eq. (2.7), except for factors of $i\epsilon$, which I will suppress anyway, for notational simplicity:

$$\Delta_F(x) = -\frac{1}{4\pi}\ln c\mu^2x^2 \ . \tag{4.3}$$

If we use the identity

$$N_m e^{i\beta\varphi} = \left(\frac{\mu^2}{m^2}\right)^{\beta^2/8\pi} N_\mu e^{i\beta\varphi} \ , \tag{4.4}$$

we can compute (4.2) directly, using Wick's theorem, Eq. (2.6), and ignoring all terms involving contractions of fields at the same point. The result is easily seen to be

$$\left(\frac{\mu^2}{m^2}\right)^{(\Sigma\beta_i^2/8\pi)} \prod_{i>j} [c\mu^2(x_i-x_j)^2]^{\beta_i\beta_j/4\pi} \ . \tag{4.5}$$

The critical point is that this expression is proportional to

$$\mu^{(\Sigma\beta_i)^2/4\pi} \ . \tag{4.6}$$

Thus, if the β's sum to zero, (4.5) is independent of μ; if they do not sum to zero, (4.5) vanishes as μ goes to zero. This result is an important consistency check on our computation, for it is just what we should expect in a massless free scalar field theory in two dimensions. The Lagrangian for such a theory is invariant under the transformation

$$\varphi \to \varphi + \lambda \ , \tag{4.7}$$

where λ is a real parameter. Formally, this implies

$$e^{i\beta\varphi} \to e^{i\beta\lambda}e^{i\beta\varphi} \ , \tag{4.8}$$

which in turn implies that the vacuum expectation values vanish unless the sum of the β's is zero—exactly what we have found. This argument breaks down in greater than two dimensions because the symmetry (4.7) is spontaneously broken. However, in two dimensions there is no spontaneous breakdown of continuous symmetries,[9] so the argument should be valid—and it is.

Now let us specialize these results to the sine-Gordon equation. The object of interest is

$$\frac{\alpha}{\beta^2}N_m\cos\beta\varphi = \frac{\alpha}{2\beta^2}(A_+ + A_-) \ , \tag{4.9}$$

where

$$A_\pm = N_m e^{\pm i\beta\varphi} \ . \tag{4.10}$$

In the limit of vanishing μ, the only nonzero terms in the perturbation expansion are those with equal numbers of A_+'s and A_-'s. These terms are given by

$$T\langle 0|\prod_{i=1}^{n} A_+(x_i)A_-(y_i)|0\rangle = \frac{\prod_{i>j}[(x_i-x_j)^2(y_i-y_j)^2c^2m^4]^{\beta^2/4\pi}}{\prod_{i,j}[cm^2(x_i-y_j)^2]^{\beta^2/4\pi}} \ . \tag{4.11}$$

Now let us construct the corresponding perturbation series (in m) for the massive Thirring model. Here the object of interest is

$$m'\sigma = m'(\sigma_+ + \sigma_-) \ , \tag{4.12}$$

where

$$\sigma_\pm = \tfrac{1}{2}Z\bar\psi(1\pm\gamma_5)\psi \ , \tag{4.13}$$

and Z is the (cutoff-dependent) multiplicative renormalization constant referred to in Sec. I. Because the massless Thirring model is chirally invariant, the only nonzero terms in the perturbation expansion are those with equal number of σ_+'s and σ_-'s. These are best calculated in a basis in which

$$\gamma_0 = \sigma_x, \quad \gamma_1 = i\sigma_y, \quad \gamma_5 = -\sigma_z \ . \tag{4.14}$$

In this basis

$$\sigma_+ = Z\psi_1^\dagger\psi_2, \quad \sigma_- = Z\psi_2^\dagger\psi_1 \ , \tag{4.15}$$

where the subscripts indicate the components of the Dirac spinor.

The expressions we need for the computation can be found in the beautiful paper of Klaiber,[6] which contains closed forms for the vacuum expectation values of arbitrary strings of ψ_i's and $\psi_i^\dagger$'s. Klaiber's formulas are lengthy, and I will not reproduce them here, but merely state the consequences of them that are relevant to our purposes. All of these can be obtained from Klaiber's work by trivial manipulations; for the skeptical reader who wishes to verify that I have made no algebraic errors, I will adhere as closely as possible to Klaiber's notation.

In an arbitrary Green's function, if we bring a $\psi_1^\dagger$ and a ψ_2 to the same space-time point, the Green's function becomes singular. However, if we define,

$$\sigma_+(x) \propto \lim_{y\to x} (x-y)^{2\delta}\psi_1^\dagger(x)\psi_2(y)\ , \tag{4.16}$$

then σ_+ has finite matrix elements. Here Klaiber's parameter δ is given by

$$\delta = \frac{g}{4\pi}\,\frac{2\pi+g}{\pi+g}\ , \tag{4.17}$$

where g is the coupling constant as defined in Sec. I, and the proportionality sign is to indicate that σ_+ is defined only up to a finite multiplicative renormalization. The local field σ_- is defined by the adjoint equation.

With this definition, it follows from Klaiber's formulas that

$$\langle 0|T\prod_{i=1}^{n}\sigma_+(x_i)\sigma_-(y_i)|0\rangle$$
$$= (\tfrac{1}{2})^{2n}\frac{\prod_{i>j}[(x_i-x_j)^2(y_i-y_j)^2M^4]^{(1+b/\pi)}}{\prod_{i,j}[M^2(x_i-y_j)^2]^{(1+b/\pi)}}\ , \tag{4.18}$$

where M is an arbitrary mass, the reflection of the arbitrary finite renormalization in the definition of the σ's, and Klaiber's parameter b is given by

$$b = -\frac{g}{1+g/\pi}\ . \tag{4.19}$$

This equation is identical with Eq. (4.7) if we make the identifications

$$-\sigma_\pm = \tfrac{1}{2}A_\pm\ , \tag{4.20}$$

$$M^2 = cm^2\ , \tag{4.21}$$

and

$$\frac{1}{1+g/\pi} = \frac{\beta^2}{4\pi}\ . \tag{4.22}$$

Thus, the two perturbation theories are identical if we choose

$$m' = \alpha/\beta^2\ , \tag{4.23}$$

whence

$$-m'\sigma = \frac{\alpha}{\beta^2}N_m\cos\beta\varphi\ . \tag{4.24}$$

Equations (4.22) and (4.24) are Eqs. (1.9) and (1.11) of the introduction. These equations do not depend on properly matching the renormalization conventions in the two theories, Eq. (4.21). In contrast, Eqs. (4.20) and (4.23) do depend on proper matching, and therefore have no convention-independent meaning.

Now let us turn to the current. In the Thirring model, Klaiber has shown that

$$[j^\mu(x), \psi(y)] = -[g^{\mu\nu} + (1+g/\pi)^{-1}\epsilon^{\mu\nu}\gamma_5]\times\psi(y)D_\nu(x-y)\ , \tag{4.25}$$

where D_ν is the gradient of the massless free scalar commutator. At equal times,

$$D_\nu(0, x^1) = g_{\nu 0}\delta(x^1)\ . \tag{4.26}$$

Thus, the current is properly normalized; that is to say, if we identify the conserved charge associated with the current as fermion number, the field carries fermion number minus one. From Eq. (4.21) and the definition of the σ's, Eq. (4.12), it follows that

$$[j^\mu(x), \sigma_\pm(y)] = \mp 2\left(1+\frac{g}{\pi}\right)^{-1}\epsilon^{\mu\nu}D_\nu(x-y)\sigma_\pm(y)\ . \tag{4.27}$$

This equation completely defines j^μ in the sector in which we are interested, the set of all states that can be created from the vacuum by application of the σ's in the massless Thirring model.

In the massless scalar theory defined in the first part of this section, $\partial_\nu\varphi$ is a free field. Thus, it is trivial to compute its commutators with the A's; we find

$$[\partial_\nu\varphi(x), A_\pm(y)] = \pm\beta D_\nu(x-y)A_\pm(y)\ . \tag{4.28}$$

Since the A's are identified with the σ's, it follows that

$$j^\mu = -\frac{2}{\beta(1+g/\pi)}\epsilon^{\mu\nu}\partial_\nu\varphi$$
$$= -\frac{\beta}{2\pi}\epsilon^{\mu\nu}\partial_\nu\varphi\ . \tag{4.29}$$

This is the last of the results announced in the Introduction and completes the argument.

V. PUZZLING QUESTIONS, BUT NOT BEYOND ALL CONJECTURE

A. Bounds on coupling constants

We have seen that the sine-Gordon equation is physically nonsensical if β^2 exceeds 8π. I have not been able to show that it is physically sensible for any β^2 less than 8π, except for two points: zero, where the theory is a free massive Bose field theory, and 4π, where the theory is a free massive Fermi field theory. The natural conjecture is that the theory is sensible for β^2 less than some upper bound lying somewhere between 4π and 8π, but I know of no evidence to support this conjecture.

In the language of the Thirring model, $\beta^2 < 8\pi$ corresponds to

$$g > -\pi/2 \ . \tag{5.1}$$

From the viewpoint of the Thirring model, this is a surprising result, for this bound is stronger than the corresponding one for the massless model, $g > -\pi$. Thus, there is a whole range of coupling constants for which the massless Thirring model is perfectly sensible, but for which the addition of a mass term, no matter how small its coefficient, causes the bottom to drop out of the energy spectrum. Is it possible to understand this phenomenon purely in Thirring-model language? Yes, it is; here is the argument:

The massless Thirring model is exactly scale-invariant; there is a conserved scale current and a unitary dilatation operator. This implies that the Hamiltonian density of the massless model must transform under dilatations according to

$$\lambda: \ \mathcal{H}^{(0)}(x) \to \lambda^2 \mathcal{H}^{(0)}(\lambda x) \ , \tag{5.2}$$

where λ is the dilatation parameter. As can be seen from Eq. (4.16), the field σ has anomalous dimension b/π; thus,

$$\lambda: \ \sigma(x) \to \lambda^{1/(1+g/\pi)} \sigma(\lambda x) \ . \tag{5.3}$$

Now let us imagine performing a variational estimate of the ground-state energy of the massive model, using as our trial states a family of translation-invariant states which are dilatation transforms of each other. If we label these states by the dilatation parameter λ, then

$$\langle \lambda | \mathcal{H} | \lambda \rangle = \lambda^2 \langle 1 | \mathcal{H}^{(0)} | 1 \rangle + m' \lambda^{1/(1+g/\pi)} \langle 1 | \sigma | 1 \rangle \ . \tag{5.4}$$

If g is less than $-\pi/2$, the second term dominates as λ goes to infinity. Thus, we are in trouble if the second term is negative. But since σ can be turned into $-\sigma$ by a chiral rotation, we can always find a state such that the expectation value of σ is negative. Q.E.D.

B. Metamorphosis of fermions into bosons

One of the most striking features of the results established here is that a theory which is "obviously" a theory of fermions is equivalent to a theory which is "obviously" a theory of bosons.[10] This peculiar result should seem even more peculiar to you if you have followed the proof of it in Sec. IV, for the proof rested on identifying a certain sector of the massless Thirring model with a certain sector of massless free scalar theory. That *these* theories are equivalent in any sense seems too preposterous to believe. Is there no way of telling fermions from bosons? The answer is no, there is not—if one can measure only a restricted set of local fields, *and* if the particles are massless, *and* if the world is two-dimensional.

This can most easily be seen by a simpler and less singular theory than either of the two at hand: the theory of a free massless Dirac field. Let us suppose that, in such a theory, we can only measure the Green's functions associated with local operators of charge zero. That is to say, we can only make particle-antiparticle pairs out of the vacuum, not single particles. This is not an unrealistic restriction; it is precisely that restriction which we face in the real four-dimensional world, where we can only measure charge-zero observables. Nevertheless, in the real world, we have no difficulty in determining that electrons are fermions and charged pions are bosons. The reason is that after we create a pair in the real world, the components of the pair separate; thus, if we wait long enough, we can arrange matters so one of the components of the pair is in our laboratory and the other is at the orbit of Pluto and traveling outward. However, for massless particles in two dimensions, it is quite possible to make a pair that never separates. Such a pair consists of two particles moving in the same direction. The wave functions do not spread; they just move on steadily at the speed of light, and the particles never get away from each other. If the particles had a mass, or if the world were of greater than two dimensions, this would not be possible.

Another way of seeing the same thing is to study the charge-zero sector of the Fock space of the massless Dirac theory. This contains states corresponding to a fermion and antifermion, each in a normalizable state with spatial momentum support restricted to the positive axis. Even though this is a normalizable state, it is still an eigenstate of $P_\mu P^\mu$, with eigenvalue zero; all of the two-momenta in the supports of the individual particle states are aligned null vectors. Thus, if we adopt the usual definition of a particle, a normalizable eigenstate of P^2, we must say that the Fock space of a massless Dirac field contains massless Bose particles. Again, this is only possible for mass zero and dimension two.

However, all this pathology disappears, even in two dimensions, if the particles are massive. Then pairs will separate, and there are no two-particle normalizable mass eigenstates. Thus, the problems of interpretation discussed above are not really relevant to the theories considered here, theories of massive particles. Here the same kind of experiments that tell us unambiguously in four dimensions that electrons are fermions should tell us unambiguously that, for example, the only particles in the sine-Gordon equation with $\beta^2=4\pi$ are free massive fermions.[11]

C. Classical theory, quantum solitons, and a conjecture

To obtain insight into the connection between the quantum sine-Gordon equation and its classical limit, let us introduce the rescaled variables

$$\varphi' = \beta\varphi, \quad \gamma_0' = \gamma_0/\beta^2 \ . \tag{5.5}$$

In terms of these variables, the Lagrangian density is

$$\mathcal{L} = \frac{1}{\beta^2}[\tfrac{1}{2}(\partial_\mu\varphi')^2 + \alpha_0\cos\varphi' + \gamma_0'] \ . \tag{5.6}$$

Since, in classical physics, multiplying the Lagrangian density by a constant has no effect on the physics (other than trivially redefining the scale of energy), β is an irrelevant parameter in the classical sine-Gordon equation. In the quantum theory, as we have seen, this is not so; this is because the relevant object for quantum physics is $\mathcal{L}/\hbar$, and rescaling the Lagrangian is equivalent to rescaling $\hbar$. (We have obscured this until now by choosing our units so $\hbar=1$.) Thus, in the quantum theory, the relevant parameter is $\beta^2\hbar$. Therefore, semiclassical approximation schemes, such as the expansion of Goldstone and Jackiw,[3] or the modified WKB approximation of Dashen, Hasslacher, and Neveu,[3] being small-$\hbar$ approximations, are necessarily also small-β approximations.[12] Since small β is large g, these methods give us information about the Thirring model for *large* coupling constants. What is this information?

The classical equation possesses a time-independent solution of finite energy, the famous soliton

$$\varphi'(x) = 4\tan^{-1}\exp(x\sqrt{\alpha_0}) \equiv f(x) \ . \tag{5.7}$$

Of course, spatial translations and Lorentz transformations of this solution are also solutions. The semiclassical analyses show that these solutions correspond to a particle in the quantum theory, the quantum soliton. In the leading approximation, the mass of the quantum soliton is proportional to $1/\beta^2$, and the states of the quantum soliton correspond to certain superpositions of coherent states of the φ field, such that for a one-soliton state approximately localized at the point a,

$$\langle\varphi'(x)\rangle = f(x-a) \ . \tag{5.8}$$

All one-soliton states are orthogonal to states made by applying local polynomials in φ to the vacuum state. [This is because all such states have the property that the expectation value of φ' goes to zero as x goes to plus or minus infinity, while (5.8) goes to zero at minus infinity but goes to 2π for plus infinity.] However, states containing equal numbers of solitons and antisolitons do not have this property. For a state consisting of a widely separated soliton-antisoliton pair

$$\langle\varphi'(x)\rangle \approx f(x-a) - f(x+a) \ , \tag{5.9}$$

where the approximation sign indicates that, even in the leading semiclassical approximation, this formula is valid only for large a. It is not known whether the quantum soliton is a fermion or a boson.

(This last remark probably requires amplification. The spin-statistics theorem is useless, because there is no spin in two dimensions, since there is no little group. Of course, even in two dimensions there is a connection between the statistics of a particle and the Lorentz-transformation properties of the local field that creates the particle from the vacuum, but since the soliton is not created by a local field, this is also of no help. The semiclassical methods are in principle capable of answering the question (for example, by studying the symmetry of soliton-soliton scattering), but the necessary computations have not yet been done.)

I conjecture that the quantum soliton is a fermion, and is, in fact, the fundamental fermion of the massive Thirring model. Here are my reasons:

(1) A similar time-independent solution of finite energy exists for classical two-dimensional φ^4 theory. Goldstone and Jackiw[3] have shown that this particle is a fermion. Unfortunately, their argument depends on the identification of particle and antiparticle, possible for φ^4 but not for the sine-Gordon equation.

(2) Equation (1.10) is an exact result. If we apply it to a widely separated soliton-antisoliton pair, we find

$$\langle j_0(x)\rangle = \frac{1}{2\pi}\langle\partial_1\varphi'(x)\rangle$$
$$= \frac{1}{2\pi}\frac{d}{dx}[f(x-a) - f(x+a)] \ . \tag{5.10}$$

The right-hand side of this expression is peaked about the points $-a$ and a, the locations of the soliton and antisoliton. The integrated charge over the soliton peak is $+1$; that over the antisoliton peak is -1. Since the charge in question is fermion number for the Thirring model, this is evidence for the conjecture.

(3) In the massive Thirring model, we have defined the coupling constant such that positive g

corresponds to an attractive force between fermions and antifermions. In the nonrelativistic limit, the attractive interaction is a δ-function potential; in one spatial dimension, such a potential produces a bound state for arbitrarily small g. As we increase g, we would expect the mass of this bound state to decrease, becoming as small as possible (zero) as g goes to infinity. This bound state is the obvious candidate for the fundamental meson of the sine-Gordon equation.

Thus, I am led to conjecture a form of duality, or nuclear democracy in the sense of Chew, for this two-dimensional theory. A single theory has two equally valid descriptions in terms of Lagrangian field theory: the massive Thirring model and the quantum sine-Gordon equation. The particles which are fundamental in one description are composite in the other: In the Thirring model, the fermion is fundamental and the boson a fermion-antifermion bound state; in the sine-Gordon equation, the boson is fundamental and the fermion a coherent bound state. In the Thirring-model description, as the coupling constant goes to infinity, the bound-state mass goes to zero; in the sine-Gordon description, as the coupling constant vanishes, the coherent-bound-state mass goes to infinity; these are just two ways of describing the behavior of a single mass ratio in a single limit.

Speculation on extending these ideas to four dimensions is left as an excercise for the reader.

ACKNOWLEDGMENTS

I am indebted to Jeffrey Goldstone, Roman Jackiw, and Andre Neveu for discussion of their work in general and its applications to the sine-Gordon equation in particular, to Howard Georgi for guiding me through the literature of the Thirring model, and to Konrad Osterwalder for reassuring me of my sanity. This work was begun while I was a visitor at the Laboratoire de Physique Théorique et Hautes Energies, Orsay, and the Service de Physique Théorique of Centre d'Études Nucléaires de Saclay. I thank both of these groups for their very generous hospitality.

APPENDIX (ADDED IN PROOF)

Since this paper was accepted for publication, there have been several new developments:

1. I have learned from Schroer[13] that he has independently obtained many of the results in this paper. Schroer has also pointed out that many of the results obtained here are in close correspondence with the results of the studies of one-dimensional electron gasses by Luther and collaborators.[14] Luther and I are in total agreement with Schroer on this point; we are also united in our embarrassment that we were incapable of reaching this conclusion unprompted. (Our offices are on the same corridor.)

2. Mandelstam[15] has been able to construct the Fermi field operators as (nonlocal) functions of the canonical Bose field, following the methods of Dell'Antonio *et al.*[16]

3. Fröhlich[17] has been able to rigorously prove that the sine-Gordon Hamiltonian defines a physically sensible theory for $\beta^2 < 16/\pi$. (The restriction on β is connected only with certain technical details of the proof, not with any obvious pathology of the physics.) He has also been able to establish similar results for the sine-Gordon Hamiltonian with a mass term (the massive Thirring-Schwinger model).

---

*Work supported in part by the National Science Foundation under Grant No. MPS73-05038 A01.

[1]Notation: $g_{00} = -g_{11} = 1$, $\epsilon_{01} = -\epsilon_{10} = 1$. The symbol x is used for both a space-time point and for the associated spatial coordinate x^1; which is meant should always be clear from the context.

[2]For a review, see A. Scott, F. Chu, and D. McLaughlin, Proc. IEEE 61, 1443 (1973).

[3]J. Goldstone and R. Jackiw, Phys. Rev. D 11, 1486 (1975); R. Dashen, B. Hasslacher, and A. Neveu, Phys. Rev. D 10, 4114 (1974); 10, 4130 (1974); 10, 4138 (1974); K. Cahill, Ecole Polytechnique report (unpublished).

[4]This result is trivial and is well known to workers in the field.

[5]I thank Andre Neveu for pointing out to me that this result is prefigured in two papers by T. H. R. Skyrme [Proc. R. Soc. A247, 260 (1958); A262, 237 (1961)]. Skyrme argued that the soliton modes of the sine-Gordon equation were fermions, and that the interaction between the fermions was of Thirring-model type. This is in agreement with my results; however, others of Skyrme's conclusions contradict mine: He argued that the identification with fermions could only be made for $\beta^2 = 4\pi$ (this is Skyrme's condition $\epsilon = \frac{1}{2}\hbar c$), he found that a fermion interaction existed in this case, and he identified the fermion current with a very different operator than I do. I regret that I do not understand Skyrme's methods well enough to know whether an improved version of them, taking more careful account of renormalization effects, would yield results that agree with mine.

[6]The Thirring model was proposed by W. Thirring, Ann. Phys. (N.Y.) 3, 91 (1958). A formal operator solution was found by V. Glaser, Nuovo Cimento 9, 990 (1958). A more rigorous solution was found by K. Johnson, Nuovo Cimento 20, 773 (1961). The model was studied in more detail by C. Sommerfield, Ann. Phys. (N.Y.)

26, 1 (1963), and by B. Klaiber, in *Lectures in Theoretical Physics*, lectures delivered at the Summer Institute for Theoretical Physics, University of Colorado, Boulder, 1967, edited by A. Barut and W. Brittin (Gordon and Breach, New York, 1968), Vol. X, part A.

[7]This normalization of the current is that recommended by Sommerfield (Ref. 6). Johnson (Ref. 6) uses a different normalization, which leads to a different definition of the coupling constant. If we denote Johnson's coupling constant by g', then

$$g' = 2\pi g/(2\pi + g).$$

The allowed range of g' is $2\pi > g' > -2\pi$.

[8]See the discussion of A. Wightman, in *High Energy Electromagnetic Interactions and Field Theory*, edited by M. Lévy (Gordon and Breach, New York, 1967).

[9]S. Coleman, Commun. Math. Phys. 31, 259 (1973).

[10]A very similar transformation of fermions into bosons takes place in two-dimensional quantum electrodynamics of massless fermions (the Schwinger model). This aspect of the model has recently been clarified and emphasized by A. Casher, J. Kogut, and L. Susskind, Phys. Rev. D 10, 732 (1974). I am indebted to Leonard Susskind and Curtis Callan for discussions of this phenomenon.

[11]I thank R. Jackiw for discussions of this point.

[12]I do not mean to imply that these two methods are equivalent. They are not, any more than in atomic physics perturbation theory for the T matrix is equivalent to perturbation theory for the K matrix. However, in atomic physics, these are both small-coupling approximations, although one is a better small-coupling approximation than the other in certain circumstances (for example, resonance scattering). Likewise, there may be circumstances where one of the two methods mentioned above is a better small-coupling approximation than the other, although we do not have enough insight into either method at the moment to know what these circumstances are.

[13]B. Schroer, private communication.

[14]A. Luther and I. Peschel, Phys. Rev. B 9, 2911 (1974); A. Luther and V. Emery, Phys. Rev. Lett. 33, 598 (1974).

[15]S. Mandelstam, Phys. Rev. D (to be published).

[16]G. Dell'Antonio, Y. Frishman, and D. Zwanziger, Phys. Rev. D 6, 988 (1972).

[17]J. Fröhlich, private communication.

Soliton operators for the quantized sine-Gordon equation*

S. Mandelstam
Department of Physics, University of California, Berkeley, California 94720
(Received 24 February 1975)

Operators for the creation and annihilation of quantum sine-Gordon solitons are constructed. The operators satisfy the anticommutation relations and field equations of the massive Thirring model. The results of Coleman are thus reestablished without the use of perturbation theory. It is hoped that the method is more generally applicable to a quantum-mechanical treatment of extended solutions of field theories.

I. INTRODUCTION

The two-dimensional classical sine-Gordon field is probably the simplest nonlinear field which possesses extended solutions of the type currently under investigation, the so-called soliton solutions.[1-3] Coleman[4] has extended the results to the quantized theory by relating the sine-Gordon field to the massive Thirring model, i.e., to a two-dimensional self-coupled Fermi field with vector interaction. It is the purpose of this note to construct operators for the creation and annihilation of bare solitons; we shall thereby obtain a simple rederivation of Coleman's results. Operators analogous to ours have been obtained for the massless Thirring model by Dell'Antonio, Frishman, and Zwanziger,[5] but they do not possess a similar physical significance. The treatment of the massive model is in some respects simpler than that of the massless model as there are no infrared divergences. We shall therefore attempt to keep our treatment self-contained, at the risk of repeating some of the analysis of Ref. 5. Our work will also be logically independent of that of Coleman, though we shall be motivated by some of his results.

The sine-Gordon field satisfies the equation

$$\ddot{\phi}(x,t)-\phi''(x,t)+(\mu^2/\beta):\sin[\beta\phi(x,t)]:=0\ . \qquad (1.1)$$

The equation is invariant under the transformation

$$\phi\to\phi+2\pi n\beta^{-1}\ , \qquad (1.2)$$

so that the vacuum possesses a discrete degeneracy, characterized by an index n which can assume any integral value (positive, negative, or zero). Solitons are solutions of the field equations where the vacuum well to the left of the disturbance is different from the vacuum well to the right. We shall define a soliton and an antisoliton by the boundary conditions

$$\begin{aligned}&\langle\phi(x)\rangle\to 0, \quad x\to+\infty\ ,\\ &\langle\phi(x)\rangle\to\mp 2\pi\beta^{-1}, \quad x\to-\infty\ .\end{aligned} \qquad (1.3)$$

The sign is negative for a soliton, positive for an antisoliton.

Solitons may be compared with certain types of extended solutions of classical equations in three or four dimensions, where the degenerate vacuum is characterized by a continuous parameter which varies with the direction in which one recedes from the disturbance. The two-dimensional model has only two asymptotic directions and a discretely degenerate vacuum, but it nevertheless possesses the essential features of systems such as vortices or monopoles.

Most treatments of extended solutions have been classical or semiclassical, but the work of Coleman, referred to above, is fully quantum-mechanical. Coleman showed that the sine-Gordon field is equivalent to the massive Thirring field, the solitons corresponding to the states with a fermion number of unity. He found the following relation between the constant β in (1.1) and the coupling constant q of the Thirring model:

$$g/\pi = 1 - 4\pi\beta^{-2} . \quad (1.4)$$

For $\beta^2 > 4\pi$ the coupling between a soliton and an antisoliton is attractive, and sine-Gordon particles probably appear as soliton-antisoliton bound states. For $\beta^2 > 4\pi$ the coupling is repulsive, and stable sine-Gordon particles probably do not exist. For $\beta^2 = 4\pi$, we are led to the remarkable result that the sine-Gordon model is equivalent to a free Fermi field.

The results of Ref. 4 were obtained by comparing two rather unconventional perturbation series, and it is doubtful whether the methods could be extended to four dimensions. We wish to show that "bare" soliton creation and annihilation operators can be constructed fairly simply from sine-Gordon operators; the construction is motivated by the physical characteristics of solitons discussed above. The operators will be shown to satisfy the commutation relations and field equations of the massive Thirring model. The relations between sine-Gordon operators and bilinear functions of the Fermi operators will agree with those found by Coleman; they are generalizations of relations for the massless case which have been used to solve the Thirring model.[6] The correspondences have been listed in a recent paper by Kogut and Susskind,[7] who suggest applications to massive quantum electrodynamics.

Since our operators are local, the bare solitons which they create will be point particles. Physical solitons become spread out by the interaction in the usual way. It is, of course, not guaranteed that there is any relation (other than that of fermion number conservation) between soliton operators and actual particles, except for the interaction-free case $\beta^2 = 4\pi$. Nevertheless, it is very plausible that such a relation exists, at any rate for a range of β around this value.

II. CONSTRUCTION OF SOLITON OPERATORS

The quantized sine-Gordon system is described by Eq. (1.1), with the ϕ's satisfying the canonical commutation relations. Since all renormalization constants are finite, we can eliminate the infinities by normal ordering with respect to bare-particle creation and annihilation operators. We therefore write

$$\phi(x, t) = \phi^+(x, t) + \phi^-(x, t) , \quad (2.1)$$

where ϕ^+ and ϕ^- satisfy the commutation relations

$$[\phi^+(x, t+dt), \phi^-(y, t)] = \Delta_+((x-y)^2 - (dt + i\epsilon)^2) . \quad (2.2)$$

For small separations

$$\Delta_+ = -(4\pi)^{-1}\ln\{c^2\mu^2[x^2 - (dt + i\epsilon)^2]\} + O(x^2). \quad (2.3)$$

The value of the constant c need not concern us. It will be convenient to use the quantity $c\mu$ as our unit of mass in defining dimensionless quantities.

An operator $\psi(x)$ which annihilates a soliton at a point x must increase the value of ϕ by $2\pi\beta^{-1}$ in regions well to the left of x, but it must have no effect in regions well to the right of x. We therefore expect the operator to satisfy the commutation relations

$$[\phi(y), \psi(x)] = 2\pi\beta^{-1}\psi(x) \quad (y < x) , \quad (2.4a)$$

$$[\phi(y), \psi(x)] = 0 \quad (y > x) . \quad (2.4b)$$

Such commutation relations will hold if ψ has the form

$$\psi(x) = :A(x)\exp\left[-2\pi i\beta^{-1}\int_{-\infty}^{x} d\xi\, \dot\phi(\xi)\right]: , \quad (2.5)$$

where the operator A is yet to be determined. At the moment we leave it open whether ψ is a boson or fermion operator.

We cannot take $A(x)$ equal to unity, since ψ would then be a boson operator, and the commutator between $\psi(x)$ and $\dot\psi(y)$ would not be simple when $x = y$. Taking A to be a polynomial in ϕ, $\dot\phi$, or ϕ' merely complicates matters. We therefore try modifying the exponent, and the simplest dimensionless term we can add is $C\phi(x)$. Thus,

$$\psi(x) = :\exp\left[-2\pi i\beta^{-1}\int_{-\infty}^{x} d\xi\, \dot\phi(\xi) + C\phi(x)\right]: . \quad (2.6)$$

From the formula

$$e^A e^B = e^{[A,B]} e^B e^A \quad ([A, B] \text{ a } c \text{ number}) , \quad (2.7)$$

it follows that $\psi(x)$ and $\psi(y)$, defined by (2.6), do not in general commute or anticommute if $x \neq y$. However, if $C = Ni\beta$ they commute, while if $C = (N + \frac{1}{2})i\beta$ they anticommute. The simplest possibility is $C = \frac{1}{2}i\beta$ so that, with hindsight from the results of Refs. 4 and 7, we are led to suggest the operators

$$\psi_1(x) = (c\mu/2\pi)^{1/2} \times e^{\mu/8\epsilon} :\exp\left[-2\pi i\beta^{-1}\int_{-\infty}^{x} d\xi\, \dot\phi(\xi) - \tfrac{1}{2}i\beta\phi(x)\right]: , \quad (2.8a)$$

$$\psi_2(x) = -i(c\mu/2\pi)^{1/2} \times e^{\mu/8\epsilon} :\exp\left[-2\pi i\beta^{-1}\int_{-\infty}^{x} d\xi\, \dot\phi(\xi) + \tfrac{1}{2}i\beta\phi(x)\right]: . \quad (2.8b)$$

An adiabatic cutoff $e^{-\epsilon\xi}$ is implied in the integration. The constant factors have been inserted for convenience in our later work; in particular, the phase factor $-i$ in (2.8b) will be necessary if our operators are to correspond to the canonical γ matrices

$$\gamma^0=\sigma^1,\quad \gamma^1=i\sigma_2,\quad \gamma^5=\gamma^0\gamma^1=-\sigma_3\,. \tag{2.9}$$

We note that the reflection operator interchanges ψ_1 and ψ_2 and, at the same time, changes the vacuum index n by one unit.

It remains to find the commutation relations and field equations satisfied by the ψ's, and to show that they correspond to those of the massive Thirring model.

III. COMMUTATION RELATIONS AND CURRENT DENSITIES

We have already shown that two ψ's anticommute when $x\neq y$. When $x=y$ a more detailed investigation is necessary, both because the commutator in (2.7) is not well defined and because the product of two ψ's becomes singular as their arguments approach one another. First let us decide how to formulate the commutation relations between renormalized ψ's when Z is infinite. Formally we wish to show that

$$\{\psi_i(x),\psi_j^\dagger(y)\}=Z\delta(x-y)\,. \tag{3.1}$$

Alternatively, we may define

$$j^\mu(x)=Z^{-1}\bar\psi(x)\gamma^\mu\psi(x)+\text{const.} \tag{3.2}$$

We then have to verify the commutation relations

$$[j^\mu(x),\psi(y)]=-(g^{\mu 0}+\epsilon^{\mu 0}\gamma^5)\psi(x)\delta(x-y)\,. \tag{3.3}$$

Our procedure will be to replace (3.2) by the equation

$$\tilde j^\mu(x)=\lim_{y\to x}\,[\,|c\mu(x-y)|^\sigma\bar\psi(x)\gamma^\mu\psi(y)+F(x-y)]\,, \tag{3.4}$$

where the constant σ and the c-number function F are chosen so that the right-hand side approaches a finite limit as $y\to\infty$. The function $|x-y|^\sigma$ replaces the constant Z. We shall then verify the commutation relations (3.3), with j replaced by $\tilde j$.

We evaluate the product $\bar\psi(x)\psi(y)\,(x\simeq y)$ using the formula

$$:e^A::e^B:=e^{[A^+,B^-]}:e^{A+B}:\,, \tag{3.5}$$

which is true if $[A^+,B^-]$ is a c number. It follows by straightforward calculation from (2.2), (2.3), (2.8), and (3.5) that

$$\psi_\alpha^\dagger(x)\psi_\alpha(y)=\mp i[2\pi(x-y)]^{-1}|c\mu(x-y)|^{-\beta^2g^2/(2\pi)^3}\times{:}\exp\left\{-2\pi i\beta^{-1}\int_x^y d\xi\,\dot\phi(\xi)\mp\tfrac{1}{2}i\beta[\phi(y)-\phi(x)]+O(x-y)^2\right\}{:}\quad(\text{no sum over }\alpha)\,, \tag{3.6}$$

where the $\mp$ sign is $-$ for $\alpha=1$, $+$ for $\alpha=2$. The exponential in (3.6) may be expanded up to terms linear in $x-y$, and the result compared with (3.4). We find that

$$\sigma=\beta^2g^2(2\pi)^{-3}\,,$$
$$\tilde j_0(x)=(2\pi)^{-1}\beta\phi'(x)\,, \tag{3.7}$$
$$\tilde j_1(x)=-2\beta^{-1}\dot\phi(x)\,.$$

On inserting (2.8) and (3.7) in (3.3), we confirm that that our operators satisfy the required commutation relations.

Equation (3.7) shows that the two operators $\tilde j^0$ and $\tilde j^1$ are not components of a vector and that they do not satisfy the equation of continuity. We therefore replace (3.4), (3.3), and (3.7) by the equations

$$j^\mu(x)=\lim_{y\to x}\{[\delta_0^\mu+(4\pi)^{-1}\beta^2\delta_1^\mu]|c\mu(x-y)|^\sigma\bar\psi(x)\gamma^\mu\psi(y)+F(x-y)\}\,, \tag{3.8}$$

$$[j^\mu(x),\psi(y)]=-[g^{\mu 0}+(4\pi)^{-1}\beta^2\epsilon^{\mu 0}\gamma^5]\psi(x)\delta(x-y)\,, \tag{3.9}$$

$$j^\mu=-(2\pi)^{-1}\beta\epsilon^{\mu\nu}\partial_\nu\phi\,. \tag{3.10}$$

It is well known that a correct treatment of the infinities in the Thirring model requires the extra factor $(4\pi)^{-1}\beta^2$ in (3.8) and (3.9).

IV. FIELD EQUATIONS

Rather than expressing the Hamiltonian in terms of the ψ's we shall show that our operators do satisfy the field equations of the massive Thirring model. We thereby reduce the appearance of infinite quantities to a minimum.

We wish to establish the equations

$$(-i\gamma^\mu\partial_\mu-m_0)\psi(x)=\lim_{\delta x\to 0}\tfrac{1}{2}g\gamma^\mu[j_\mu(x+\delta x)+j_\mu(x-\delta x)]\psi(x)\,. \tag{4.1}$$

The limiting procedure on the right-hand side of (4.1) is familiar in the Thirring model, and we shall find that no singular terms appear when δx approaches zero.

The only infinity in (4.1) is that associated with mass renormalization, and it will be treated in the same way as before. We write

$$-m_0\psi\sim[S,\gamma^0\psi]\,, \tag{4.2}$$

where, formally,

$$S = Zm_0 \int_{-\infty}^{\infty} dx\,\bar{\psi}(x)\psi(x)\ . \qquad (4.3)$$

We then replace (4.3) by the equation

$$S = \int_{-\infty}^{\infty} dx \lim_{y\to x} |c\mu(x-y)|^{-\delta} m\bar{\psi}(x)\psi(y)\ , \qquad (4.4)$$

where δ is chosen so that the limit is finite. The constant m is a finite mass.

From (2.8a), we find that the time derivative of ψ_1 is given by the equation

$$\dot{\psi}_1(x) = :\left\{\left[-2\pi i\beta^{-1}\int_{-\infty}^{x} d\xi\ \ddot{\phi}(\xi) - \tfrac{1}{2}i\beta\dot{\phi}(x)\right], \psi_1(x)\right\}:$$

$$= N'\left\{\left[-2\pi i\beta^{-1}\phi'(x) - \tfrac{1}{2}i\beta\dot{\phi}(x) + 2i\pi\mu^2\beta^{-2}\int_{-\infty}^{x} d\xi\ :\sin\beta\phi(\xi):\right], \psi_1(x)\right\} \qquad (4.5)$$

by (1.1). Normal ordering is always understood to be with respect to sine-Gordon operators, not Thirring operators. The symbol N' indicates that the terms in the expansion of $\sin\beta\phi$ are treated as units; their individual factors are not normal ordered with respect to those of ψ. The term involving $\sin\beta\phi$ may then be written as follows:

$$N'\left\{2i\pi\mu^2\beta^{-2}\int_{-\infty}^{x} d\xi\ \sin\beta\phi(\xi), \psi_1(x)\right\} = -i\mu^2\beta^{-2}\left[\int d\xi\ :\cos\beta\phi(\xi):, \psi_1(x)\right]. \qquad (4.6)$$

Equation (2.8a) now shows that, apart from factors due to normal ordering, the operator $\cos\beta\phi$ is just $2\pi\bar{\psi}\psi$. Taking into account the normal ordering, we find that

$$\psi_2^\dagger(x)\psi_1(y) \simeq (c\mu/2\pi)|c\mu(x-y)|^{\delta} :e^{-i\beta\phi}:,\quad x\simeq y\ , \qquad (4.7a)$$

$$\psi_1^\dagger(x)\psi_2(y) \simeq (c\mu/2\pi)|c\mu(x-y)|^{\delta} :e^{i\beta\phi}:,\quad x\simeq y\ , \qquad (4.7b)$$

where

$$\delta = \frac{g}{2\pi}\left(1+\frac{\beta^2}{4\pi}\right)\ . \qquad (4.7c)$$

Comparing this equation with (4.4), we may write

$$S = c\mu m\pi^{-1}\int_{-\infty}^{\infty} d\xi\ :\cos\beta\phi(\xi):\ .$$

Equation (4.6) thus becomes

$$N'\left\{2i\pi\mu^2\beta^{-2}\int_{-\infty}^{x} d\xi\ \sin\beta\phi(\xi), \psi_1\right\} = -i[S, \gamma^0\psi_2]\ , \qquad (4.8)$$

with

$$m = \mu\pi/(c\beta^2)\ . \qquad (4.9)$$

As we are aiming at Eq. (4.1), we calculate ψ_1' as well as $\dot{\psi}_1$:

$$\psi_1'(x) = :\{[-2\pi i\beta^{-1}\dot{\phi}(x) - \tfrac{1}{2}i\beta\phi'(x)], \psi_1(x)\}:\ . \qquad (4.10)$$

From (4.5), (4.10), (4.8), and (1.4) we obtain the equation

$$\dot{\psi}_1(x) + \psi_1'(x) + i[S, \gamma^0\psi_2(x)] = -\frac{i\beta}{2\pi}g:\{[\phi'(x) - \dot{\phi}(x)], \psi_1(x)\}:\ . \qquad (4.11)$$

The operators ϕ' and $\dot{\phi}$ may be taken outside the normal-ordering signs by using the formula

$$A:e^B: = :\{A + [A^+, B^-]\}e^B:\ , \qquad (4.12)$$

valid when $[A^+, B^-]$ is a c number. If $A = \phi'(y) - \dot{\phi}(y)$ and $e^B = \psi_1(x)$, we find from (2.2), (2.3), and (2.8) that the commutator $[A^+, B^-]$ is an odd function of $x-y$, so that it gives no contribution to (4.12) if we take the average of terms with $y = x + \delta x$, $y = x - \delta x$. Thus, expressing the operators ϕ' and $\dot{\phi}$ in terms of current densities by (3.10), we obtain the final result

$$\dot{\psi}_1 + \psi_1' + i[S, \gamma^0\psi_2] = \lim_{\delta x\to 0} \tfrac{1}{2}ig[j^0(x+\delta x) + j^0(x-\delta x) + j^1(x+\delta x) + j^1(x-\delta x)]\psi_1(x)\ . \qquad (4.13)$$

Equation (4.13) is precisely the second component of Eq. (4.1a). The first component can be obtained in a similar way, and all required properties of our soliton operators are established.

V. CONCLUDING REMARKS

It is hoped that the methods presented here can be applied to extended solutions of four-dimensional field theories. For example, one might attempt to construct operators which create bare Nielsen-Olesen vortices. Just as our soliton operators create point particles, vortex operators would create infinitely thin strings. Interactions between strings would give the vortices a finite thickness. Vortex operators would depend on the shape of an entire string rather than on a single coordinate, and one might hope to identify them with the operators of the second-quantized dual model.[8]

It may therefore be possible to establish a "duality" between quantized Nielsen-Olesen systems and dual models, analogous to the duality between the sine-Gordon field and the massive Thirring model. The Higgs scalars may cause difficulty, since one cannot construct models of strings interacting with "elementary" particles. It will probably be necessary to start from a non-Abelian Nielsen-Olesen model without a Higgs field.

The relationship between the sine-Gordon field and the massive Thirring model does not in itself provide a practical approach to quantized solitons except when β^2 is approximately equal to 4π. Nevertheless, it suggests that we might approximate a physical soliton by suitably spreading out the operators in the exponents of (2.8). A similar approximation may be possible for Nielsen-Olesen vortices, even if the limit of taking an infinitely narrow bare vortex cannot be carried through consistently.

ACKNOWLEDGMENTS

Discussions with M. B. Halpern and A. Neveu are gratefully acknowledged. I should like to thank M. Bander for pointing out an error in the original draft.

---

*Work supported by the National Science Foundation under the Grant No. GP-42249X.

[1]A. C. Scott, F. Y. F. Chu, and D. W. McLaughlin, Proc. IEEE 61, 1443 (1973).

[2]A. Barone, F. Esposito, C. J. Magee, and A. C. Scott, Riv. Nuovo Cimento 1, 227 (1971).

[3]L. D. Faddeev and L. A. Takhtajan, Dubna report (unpublished).

[4]S. Coleman, Phys. Rev. D 11, 2088 (1975).

[5]G. F. Dell'Antonio, Y. Frishman, and D. Zwanziger, Phys. Rev. D 6, 988 (1972).

[6]See, for instance, B. Klaiber, in *Lectures in Theoretical Physics*, edited by A. O. Barut and W. E. Brittin (Gordon and Breach, New York, 1968), Vol. X-A, p. 141.

[7]J. Kogut and L. Susskind, Cornell report (unpublished).

[8]M. Kaku and M. Kikkawa, Phys. Rev. D 10, 1110 (1974).

PHYSICAL REVIEW D VOLUME 12, NUMBER 6 15 SEPTEMBER 1975

Quantum "solitons" which are SU(N) fermions*

M. B. Halpern
Department of Physics, University of California, Berkeley, California 94720
(Received 30 April 1975)

In two dimensions, we find a construction for an SU(N) quark field in terms of N real Bose fields. Hence, equivalence is shown between certain massive SU(N) Thirring models and systems of quantum sine-Gordon–type equations. From the point of view of the bosons, the "soliton-quark" SU(N) is topological. To minimize guesswork in the development of such correspondences, we employ a systematic blend of Mandelstam's operator approach with the interaction picture.

I. INTRODUCTION

Recently Coleman[1] established the surprising result that the quantum sine-Gordon equation is equivalent to the massive Thirring model, at least in the zero-fermion sector. He also gave simple arguments which imply that the Thirring fermion is the soliton of the sine-Gordon equation. Soon afterward, Mandelstam[2] gave a nonperturbative proof of this by constructing the fermion out of the Bose operators. Such results are of obvious significance to particle theorists trying to perceive the medium in which they are immersed. Extensions may even lead to a "derivation" of dual models as extended solutions of local field theory.

It is my purpose in this paper to extend such correspondence to SU(N). I have in mind finding representations of SU(N) quarks in terms of Bose fields, and hence finding the quantum sine-Gordon–like (SGL) equations that correspond to certain interacting models of these fermions. Thus, SU(N) quarks will be quantum "solitons." As pure fermion examples, one seeks correspondence with massive SU(N) Thirring models, but it would be nice to have the results in a form whereby other interactions, such as $A_\mu^\alpha \bar\psi \frac{1}{2}\lambda^\alpha \gamma^\mu \psi$ (gluon), can be explored. We also have a disadvantage in not knowing ahead of time the form of the SGL equations.

I have chosen an approach which handles all this, and which circumvents, in an orderly fashion, a great deal of guesswork. The approach is based on the following observation: *If* one knows *free-field correspondences*, it is generally very easy to add *many* interactions in the *interaction picture*. Getting back to Heisenberg equations of motion is quite simple (at least when there are no coupling constant renormalizations). One can always check one's results directly in the Heisenberg picture, either by calculating matrix elements, or, preferably, by Mandelstam's operator manipulations.

The plan of the paper is then as follows: In Sec. II I give a brief discussion of the interaction picture for certain (soft-chiral-breaking) SU(N) Thirring models. Section III is a rapid rederivation of Coleman's correspondence, starting from the free correspondences. In Sec. IV I guess generalizations of Mandelstam's operators for the case of free SU(N) fermions. The representation involves N bosons. In Sec. V I leap quickly, via the interaction-picture method, to the corresponding SGL equations. The intricate form of these equations compliments the method; the structure seems quite difficult to guess (but easy to check) in a more direct approach. From the point of view of the bosons, the SU(N) of their "soliton-quarks" is topological. For example, the diagonal component of isospin (say I_3) is a matter of asymptotic behavior of the soliton, just as was baryon number in the U(1) case. The scalars themselves have no simple transformations under SU(N), and the SGL systems no obvious invariance. In Appendix A, however, I discuss the sense in which the scalars form an extraordinary nonlocal representation of SU(N). This may be interesting in its own right, but the connection with the quark isospin is highly indirect. In Sec. VI I make some preliminary remarks about the more difficult case of *hard*-chiral-breaking interactions. There is another appendix, B, in which I note that many of the manipulations and correspondences within two-dimensional field theories are known in dual models.

II. INTERACTION PICTURE

I will briefly review[3] the interaction-picture formulation for a baryon-number current-current interaction, thus a simple SU(N) Thirring model. We will be interested later in mass terms and other interactions. We seek correspondence then with the family of Dirac equations[4]

$$i\partial\!\!\!/\psi = g_B :J\!\!\!/\psi:\,, \quad J\!\!\!/ = J_\mu \gamma^\mu \tag{2.1}$$

where ψ_a, $a=1,\ldots,N$ is an SU(N) quark, and

$$J^\mu = \sum_{a=1}^{N} \bar\psi_a \gamma^\mu \psi_a = \bar\psi\gamma^\mu\psi\,. \qquad (2.1)$$

SU(N) currents $J^\mu_\alpha = \bar\psi\gamma^\mu \frac{1}{2}\lambda^\alpha\psi$, $\mathrm{Tr}(\lambda^\alpha\lambda^\beta)=2\delta^{\alpha\beta}$, are interesting objects, but we shall not include these in the interaction until Sec. VI.

In the interaction picture, we deal only with free fields, and we expect to guarantee both Lorentz invariance and the usual Feynman series by studying $\theta^F_{0\mu}$ and $\theta^I_{0\mu}$, the free and interacting energy-momentum densities in the interaction picture (all functions of free fields, or free currents). We assume[5-7] for the free objects

$$\theta^F_{00} = -\frac{i}{2}\psi^\dagger_D\gamma_0\gamma^1\overleftrightarrow{\partial}_1\psi_D \equiv \theta^{FB}_{00}+\theta^{FV}_{00}\,,$$

$$\theta^{FB}_{00} = \frac{1}{2C_B}:J^2_{0D}+J^2_{1D}:\,,$$

$$\theta^{FV}_{00} = \frac{1}{2\tilde C_V}:J^\alpha_{0D}J^\alpha_{0D}+J^\alpha_{1D}J^\alpha_{1D}:\,, \qquad (2.2)$$

$$\theta^F_{01} = +\frac{i}{2}\psi^\dagger_D\overleftrightarrow{\partial}_1\psi_D \equiv \theta^{FB}_{01}+\theta^{FV}_{01}\,,$$

$$\theta^{FB}_{01} = \frac{1}{C_B}:J_{0D}J_{1D}:,\quad \theta^{FV}_{01} = \frac{1}{\tilde C_V}:J^\alpha_{0D}J^\alpha_{1D}:\,.$$

The subscript D denotes all free quantities, and $C_B=N/\pi$, $\tilde C_V=(N+1)/2\pi$. Now for the interaction. We assume a form, $H_I=\int dx\,\theta^I_{00}$,

$$\theta^I_{00} = \tfrac{1}{2}g_B(J_{0D})^2 + b(J_{1D})^2\,,\quad \theta^I_{01}=0\,, \qquad (2.3)$$

where b is to be determined and g_B will turn out to be that of Eq. (2.1). The form assumed is in general not a Lorentz scalar because there are Schwinger terms in the current algebra,[8]

$$[J_{0D}(x),J_{1D}(y)] = iC_B\partial_x\delta(x-y)\,,$$
$$[J^\alpha_{0D}(x),J^\beta_{1D}(y)] = if^{\alpha\beta\gamma}J^\gamma_{1D}(x)\delta(x-y) + i\delta^{\alpha\beta}C_V\partial_x\delta(x-y)\,, \qquad (2.4)$$

where $C_V=1/2\pi$. We will also need

$$[J_{0D}(x),\psi_D(y)] = -\psi_D(x)\delta(x-y)\,,$$
$$[J_{1D}(x),\psi_D(y)] = -\gamma_0\gamma_1\psi_D(x)\delta(x-y)\,,$$
$$[J^\alpha_{0D}(x),\psi_D(y)] = -\frac{\lambda^\alpha}{2}\psi_D(x)\delta(x-y)\,, \qquad (2.5)$$
$$[J^\alpha_{1D}(x),\psi_D(y)] = -\gamma_0\gamma_1\frac{\lambda^\alpha}{2}\psi_D(x)\delta(x-y)\,.$$

The way to determine b in terms of g_B is by requiring Lorentz invariance, that is, Schwinger's condition

$$[\theta_{00}(x),\theta_{00}(y)] = i[\theta_{01}(x)+\theta_{01}(y)]\partial_x\delta(x-y) \qquad (2.6)$$

on *both* the free and the total $\theta^T_{00}=\theta^F_{00}+\theta^I_{00}$. Since θ^F_{00} already satisfies this, you may pick any other θ_{00}, with the result that

$$b = -\frac{g_B}{2}G\,,\quad G=(1+C_Bg_B)^{-1}\,. \qquad (2.7)$$

As shown in Ref. 3, this guarantees a Lorentz-invariant S matrix (the Feynman series, if expanded in g_B).

We next turn our attention to the Lorentz-transformation properties of the currents. In the interaction picture (IP)

$$[\theta^F_{00}(x),J_{0D}(y)] = iJ_{1D}(x)\partial_x\delta(x-y)\,,$$
$$[\theta^F_{00}(x),J_{1D}(y)] = iJ_{0D}(x)\partial_x\delta(x-y)\,, \qquad (2.8)$$

and similarly for J^α_μ. These are useful in establishing that $J_{\mu D}$ is a two-vector in the IP.

$$M^F_{01} = \int dx[x_0\theta^F_{01}+x\theta^F_{00}]\,,$$
$$[M^F_{01},J_{0D}(x)] = i(t\partial_x+x\partial_t)J_{0D}(x) - iJ_{1D}(x)\,, \qquad (2.9)$$
$$[M^F_{01},J_{1D}(x)] = i(t\partial_x+x\partial_t)J_{1D}(x) - iJ_{0D}(x)\,,$$

and similarly for J^α_μ. However, because there are Schwinger terms in the commutator (θ^I_{00},J_μ), this will not persist in the Heisenberg picture (HP). Define HP fields as $\psi_H=U^\dagger\psi_DU$, $\dot U=-iH_IU$, etc., and the full Lorentz generator $M^T_{01}(\theta^T_{00},\theta^T_{01})$. Then applying $U^\dagger\cdots U$, one easily establishes that J^μ_H is not a two-vector in the HP. However, in the same calculation, one sees that

$$\tilde J_\mu \equiv (J_{0H},GJ_{1H}) \qquad (2.10)$$

is a vector. This result is well known in the Thirring model, and as noted above, it is a general phenomenon for any field which has Schwinger terms with θ^I_{00}.[3] $J^\alpha_{\mu H}$ remains a vector under this interaction.

We next make contact with the Thirring-Dirac equation. Applying $U^\dagger\cdots U$ to $i\partial\!\!\!/\psi_D=0$ and using (2.5) we obtain immediately that

$$i\partial\!\!\!/\psi_H = g_B:\tilde J\!\!\!/\psi_H:\,, \qquad (2.11)$$

completing the correspondence. We also list the obvious relations

$$[\tilde J_0(x),\psi_H(y)] = -\psi_H(x)\delta(x-y)\,,$$
$$[\tilde J_1(x),\psi_H(y)] = -G\gamma_0\gamma_1\psi_H\delta(x-y)\,, \qquad (2.12)$$
$$[\tilde J_0(x),\tilde J_1(x)] = iGC_B\partial_x\delta(x-y)\,,$$

all other commutators going over form-invariant to the HP.[9]

As a quick application of the method, we derive the Sugawara stress-tensor form for these theories.[5,6] Adding θ^F_{00} and θ^I_{00}, in either picture, and remembering to use $\tilde J_\mu$, we find

$$\theta_{\mu\nu}=\theta^{FV}_{\mu\nu}+\frac{1}{2C_BG}:2\tilde{J}_\mu\tilde{J}_\nu-g_{\mu\nu}\tilde{J}_\lambda\tilde{J}^\lambda:. \tag{2.13}$$

A word about renormalizations is in order here. The IP methods suffice to define the unique one-parameter (g_B) unrenormalized Dyson S matrix of the Thirring models. However, we have ignored wave-function renormalizations in our passage to the HP. With an adiabatic limit, one can show that $[\psi_H(x),\psi_H^\dagger(y)]_+=Z\delta(x-y)$. The finite currents that we want are then $J^\mu_H=Z^{-1}:\bar\psi_H\gamma^\mu\psi_H$, $J^{\mu\,\alpha}_H=Z^{-1}:\bar\psi_H\gamma^\mu\frac{1}{2}\lambda^\alpha\psi_H:$, and (still) $\tilde{J}_\mu=(J_{0H},GJ_{1H})$. We shall understand this identification in Eqs. (2.10)–(2.13) and so on. No renormalization is needed for g_B (see, however, Sec. VI).

As a warm-up for Secs. IV and V we will next apply the method to Coleman's correspondence.

III. A QUICK DERIVATION OF COLEMAN'S CORRESPONDENCE

Suppose we know the free-field correspondences[10] for U(1):

$$\begin{aligned}J^\mu_D&=-\frac{1}{\sqrt{\pi}}\epsilon^{\mu\nu}\partial_\nu\phi_D\\&=-\sqrt{C_B}\,\epsilon^{\mu\nu}\partial_\nu\phi_D,\quad C_B=\frac{1}{\pi}\\:\psi^\dagger_{1D}\psi_{2D}:&=-\frac{C\mu}{2\pi}N_\mu e^{i2\sqrt{\pi}\phi_D},\\:\bar\psi_D\psi_D:&=-\frac{C\mu}{\pi}N_\mu\cos(2\sqrt{\pi}\phi_D),\end{aligned}\tag{3.1}$$

where all fields are free (ψ = spinor field and ϕ = Bose field), and μ is the mass at which we choose to normal-order the ϕ field. For correspondence with a massless fermion, one should take $\mu\to0$ (*after* operator manipulation or calculation of matrix elements). Later, however, we will follow Mandelstam's *convention*, and identify μ with the Bose quantum mass in the sine-Gordon equation. In (3.1) all expressions are in fact independent of μ, but, in general, the limit must be smooth because the Fermi side is smooth. The constant C is the same one that Mandelstam defines in his Eq. (2.3).

Now the IP is just the place to use these. We rewrite the U(1) theory of Sec. II. Equations (2.3) and (3.1) give

$$\theta^I_{00}=\frac{g_BC_B}{2}[(\partial_x\phi_D)^2-G(\partial_0\phi_D)^2]. \tag{3.2}$$

This is quadratic, and when taken with $\theta^F_{00}=\frac{1}{2}[(\partial_0\phi)^2+(\partial_x\phi)^2]$, is easily seen to be only a finite wave-function renormalization on ϕ. I will first state the result and then prove it. The interaction (3.2) is equivalent to the HP Lagrangian

$$\begin{aligned}\mathcal{L}&=\tfrac{1}{2}(1+\kappa)(\partial_\lambda\phi_H)^2\\&=\frac{1}{2G}(\partial_\lambda\phi_H)^2,\quad\kappa=C_Bg_B\end{aligned}\tag{3.3}$$

that is, a misnormalized free field. The way to see this is the following. Starting from (3.3) with κ unknown, go to Hamiltonian formalism, and perturb in small κ. The correct breakup is $\theta_{00}=\theta^F_{00}+\theta^I_{00}$,

$$\begin{aligned}\theta^F_{00}&=\tfrac{1}{2}[\pi_H^2+(\partial_x\phi_H)^2],\\\theta^I_{00}&=-\frac{\kappa}{2(1+\kappa)}\pi_H^2+\frac{\kappa}{2}(\partial_1\phi_H)^2,\end{aligned}\tag{3.4}$$

where $\pi_H=\partial\mathcal{L}/\partial\dot\phi_H=(1+\kappa)\partial_0\phi_H$. Now go to the IP, $\phi_H=U^\dagger\phi_DU$. Because $U\pi_HU^\dagger=\partial_0\phi_D$, then θ^I_{00} in the IP is immediately compared with (3.2); it is the same when $\kappa=g_BC_B$.

The dynamics of (3.3) is entirely trivial. We write

$$\begin{aligned}U^\dagger\phi_DU&=\phi_H=\sqrt{G}\,\hat\phi,\\U^\dagger\dot\phi_DU&=G^{-1}\dot\phi_H=\frac{1}{\sqrt{G}}\dot{\hat\phi},\end{aligned}\tag{3.5}$$

where we have introduced the properly normalized free field $\hat\phi$. The second identity is immediately obtainable by differentiating the definition of ϕ_H in terms of ϕ_D. With the help of (3.5), we immediately transform our correspondences (3.1) to

$$\begin{aligned}:Z'\psi^\dagger_{1H}\psi_{2H}:&=-\frac{C\mu}{2\pi}N_\mu e^{i2(\pi G)^{1/2}\hat\phi},\\:Z'\bar\psi_H\psi_H:&=-\frac{C\mu}{\pi}N_\mu\cos[2(\pi G)^{1/2}\hat\phi],\\\tilde{J}^\mu&=(CG_B)^{1/2}\epsilon^{\mu\nu}\partial_\nu\hat\phi\\&=Z^{-1}:(\bar\psi_H\gamma^0\psi_H,G\bar\psi_H\gamma^1\psi_H):.\end{aligned}\tag{3.6}$$

Here, as promised in Sec. II, I have left room for a wave-function renormalization, and a (different) mass renormalization Z'. The fermion fields solve the massless Thirring model, and the Bose field is free and massless ($\mu\to0$ after calculation, as above).

The final step is the simplest: Add a mass term to the Thirring Lagrangian. In a "second" interaction picture, where the fields have the time dependence of the massless Thirring model, we use

$$\begin{aligned}+\theta^I_{00}&=+m':Z'\bar\psi\psi:\\&=-\frac{m'C\mu}{\pi}N_\mu\cos[2(\pi G)^{1/2}\hat\phi],\end{aligned}\tag{3.7}$$

where m' is finite.

This interaction commutes with itself at equal times, so we need do no work. Going immediately to the final HP, we obtain the sine-Gordon Lagrangian

$$\mathcal{L}=\tfrac{1}{2}(\partial_\lambda\phi)^2+\frac{m'C\mu}{\pi}N_\mu\cos[2(\pi G)^{1/2}\phi]. \qquad (3.8)$$

Comparing this with the standard form, $(\alpha/\beta^2)N_\mu\cos\beta\phi$, we read off

$$2(\pi G)^{1/2}=\beta\,,\quad \left(\frac{4\pi}{\beta^2}=1+\frac{g_B}{\pi}\right). \qquad (3.9)$$

Following Mandelstam's convention, we *choose* μ to be the (tree-approximation) Bose mass. Then $\alpha=\mu^2$, $m'=\mu\pi/C\beta^2$. The identities (3.6) are form-invariant under this last transformation, with the understanding that ψ solves the massive Thirring model and ϕ solves the sine-Gordon equation. Taken together with the commutators (2.12), this completes the correspondence.

Mandelstam's operators[2] can also be boosted by these transformations from free to interacting fields:

$$\psi_D^1=\left(\frac{\mu C}{2\pi}\right)^{1/2}N_\mu\exp\left\{-i\sqrt{\pi}\left[\int_{-\infty}^{x}\dot\phi_D(\xi)+\phi_D(x)\right]\right\}. \qquad (3.10)$$

Then $\psi_H^1=U^\dagger\psi_D^1U$ is, using (3.5), the same form, but with $\dot\phi_D\to(1/\sqrt{G})\dot\phi$, $\phi_D\to\sqrt{G}\,\phi$.

With my machinery well oiled, I proceed to guess free-field correspondences for SU(N).

IV. FREE-FIELD CORRESPONDENCES FOR SU(N)

Begin with the case of SU(2). We need two quarks ψ_a^r, $r=1,2$ (Lorentz index), $a=1,2$ (isospin index). Mandelstam's representation[2] is immediately generalized to

$$\begin{aligned}\chi_a^1&=\xi_1\psi_a^1\\&=\left(\frac{\mu C}{2\pi}\right)^{1/2}N_\mu\exp\left\{-i\sqrt{\pi}\left[\int_{-\infty}^{x}d\xi\,\dot\phi_a(\xi)+\phi_a(x)\right]\right\},\\ \chi_a^2&=\xi_1\psi_a^2\\&=+i\left(\frac{\mu C}{2\pi}\right)^{1/2}N_\mu\exp\left\{-i\sqrt{\pi}\left[\int_{-\infty}^{x}d\xi\,\dot\phi_a(\xi)-\phi_a(x)\right]\right\}.\end{aligned} \qquad (4.1)$$

Here[11] we have introduced two (Lorentz) scalar free Bose fields ϕ_a $(a=1,2)$. $\xi_1=(-1)^{N_1}$ is the Klein transformation operator, in terms of N_1, the number operator for the isospin-up quark. The fields χ_a anticommute when $a=b$ as they should, but commute when $a\neq b$. The Klein transformation corrects this for ψ_a^r; thus

$$[\psi_a^r(x),\psi_b^{s\dagger}(y)]_+=\delta_{rs}\delta_{ab}\delta(x-y)\,, \qquad (4.2)$$

etc. is easily verified following Mandelstam. There are many other equivalent Klein transformations. For example, the scheme

$$\begin{aligned}\chi_a^1&=\xi_1\cdots\xi_{a-1}\psi_a^1\\&=\left(\frac{\mu C}{2\pi}\right)^{1/2}N_\mu\exp\left\{i\sqrt{\pi}\left[\int_{-\infty}^{x}d\xi\,\dot\phi_a(\xi)+\phi_a(x)\right]\right\},\\ \chi_a^2&=\xi_1\cdots\xi_{a-1}\psi_a^2\\&=i\left(\frac{\mu C}{2\pi}\right)^{1/2}N_\mu\exp\left\{i\sqrt{\pi}\left[\int_{-\infty}^{x}d\xi\,\dot\phi_a(\xi)-\phi_a(x)\right]\right\}\end{aligned} \qquad (4.3)$$

generalizes to all SU(N). Here $\xi_j=(-1)^{N_j}$, $\xi_0=1$, and $a=1,\ldots,N$. I chose the one I did for SU(2) (a special case) because bilinears are so simple, $\chi_a^{\dagger r}\chi_b^s=\psi_a^{\dagger r}\psi_b^s$, and normal ordering with respect to ψ is the same as normal ordering with respect to χ. The normal-ordering equivalence persists for all SU(N), but, in general, the bilinears contain ξ's for $N\geq3$. "Diagonal" bilinears, such as the mass term, etc., and SU(N) symmetric quartics are ξ free. In any case, the ξ's cause no trouble with the correspondences. As a matter of taste, I would leave them associated with the ψ's; they amount only to certain phases in a final calculation of Fermi Green's functions.

I am going to concentrate on working out the case of SU(2) in detail, with occasional remarks on general features of SU(N). At the end of Sec. V I will give a general discussion for SU(N). Moreover, I am going to assume familiarity with Mandelstam's manipulation of these operators, except when a step is somewhat unusual.

A. Bilinears

The calculation of the bilinears follows Mandelstam closely. First, the "diagonal" currents (we will in general use : : for Fermi normal ordering),

$$\begin{aligned}J^\mu&=:\bar\psi\gamma^\mu\psi:\\&=-\frac{1}{\sqrt{\pi}}\epsilon^{\mu\nu}\partial_\nu(\phi_1+\phi_2)\\&=-\sqrt{C_B}\,\epsilon^{\mu\nu}\partial_\nu\phi_+,\\ J_3^\mu&=:\bar\psi\gamma^\mu\frac{\tau_3}{2}\psi:\\&=-\frac{1}{2\sqrt{\pi}}\epsilon^{\mu\nu}\partial_\nu(\phi_1-\phi_2)\\&=-\frac{1}{(2\pi)^{1/2}}\epsilon^{\mu\nu}\partial_\nu\phi_-,\end{aligned} \qquad (4.4)$$

where now $C_B=2/\pi$, and for later purposes, we have introduced the normalized combinations $\phi_\pm=(1/\sqrt{2})(\phi_1\pm\phi_2)$. For SU($N$), J^μ will be the sum over all ϕ_a, etc. Before going on, these forms are worth consideration. As in the Abelian model, the time components of these currents are total derivatives, so, e.g.,

$$I_3 = \int dx\, J_{03}$$

$$= \frac{1}{(2\pi)^{1/2}} \phi_- \Big|_{-\infty}^{+\infty} . \qquad (4.5)$$

Thus, ϕ_a, $\dot{\phi}_a$ do not transform under I_3 (or B). This tells us that when we get to the SGL equations whose solitons are these quarks, then, from the point of view of the ϕ's, the isospin of their quark solitons will be essentially topological (a conservation of asymptotic properties). The ϕ's themselves must have no simple isospin properties. Parenthetically, a group theorist would partially presage these remarks; he knows that SU(N) cannot be represented (linearly) on N real fields. These remarks are explored further in Appendix A.

Proceeding, we give the charged currents,

$$J_+(\tau_+) = J_+^\dagger(\tau_-)$$
$$= 2 :\psi_1^{1\dagger} \psi_2^1 :$$
$$= \frac{\mu C}{\pi} N_\mu$$
$$\times \exp\left\{ i(2\pi)^{1/2} \left[\int_{-\infty}^{x} d\xi\, \dot{\phi}_-(\xi) + \phi_-(x) \right] \right\}, \qquad (4.6)$$

$$J_-(\tau_+) = J_-^\dagger(\tau_-)$$
$$= 2 :\psi_1^{2\dagger} \psi_2^2 :$$
$$= \frac{\mu C}{\pi} N_\mu$$
$$\times \exp\left\{ i(2\pi)^{1/2} \left[\int_{-\infty}^{x} d\xi\, \dot{\phi}_-(\xi) - \phi_-(x) \right] \right\}.$$

Here $J_\pm = J_0 \pm J_1$ and $2\tau_\pm = \tau_1 \pm i\tau_2$. These appear spatially nonlocal, in distinction to the neutral currents. This may be an interesting effect in a theory probed by W mesons and photons, or a gluon interaction. Note, however, that in an SU(2) symmetric theory, any *particular* component of J^μ_α can be made local. I remark also that every time I have studied some property that all J^α_μ should have in common, it all works out. For example, one easily calculates directly by differentiation and $\Box^2 \phi_\pm = 0$ that

$$(\partial_0 \mp \partial_x) J_\pm = 0 ,$$
$$(\partial_0 \mp \partial_x) J_\pm^\alpha = 0 , \quad \alpha = 1, 2, 3 . \qquad (4.7)$$

The simple calculations, however, go along different lines for charged and neutral currents. Equation (4.7) is equivalent to $\partial_\mu J^\mu = \partial_\mu J_5^\mu = \partial_\mu J^\mu_\alpha = \partial_\mu J^\mu_{5\alpha} = 0$, and therefore *all* currents are proportional to gradients of free scalar fields, as expected. This was obvious for the neutrals (from our representation), but is extremely indirect for the charged currents. It appears that, despite appearances, all J^α_μ are on equal footing. I also remark that J^α_μ (J_μ) are made up entirely of ϕ_- (ϕ_+); hence they commute. Commutators of $\phi_\pm$ with the isospin generators are discussed in Appendix A.

The "mass" term has the form

$$- :\bar{\psi}\psi: = \frac{\mu C}{\pi} N_\mu [\cos(2\sqrt{\pi}\,\phi_1) + \cos(2\sqrt{\pi}\,\phi_2)] \qquad (4.8a)$$

$$= \frac{2\mu C}{\pi} N_\mu \cos[(2\pi)^{1/2} \phi_+] \cos[(2\pi)^{1/2} \phi_-] \qquad (4.8b)$$

$$= \frac{2C}{\pi} (\mu_+ \mu_-)^{1/2} N_+ \cos[(2\pi)^{1/2} \phi_+]$$
$$\times N_- \cos[(2\pi)^{1/2} \phi_-] . \qquad (4.8c)$$

In the last step, we have used Coleman's identity[1] $N_\mu(\cos\beta\phi) = (\mu'/\mu)^{\beta^2/4\pi} N_{\mu'}(\cos\beta\phi)$ to re-normal-order $\phi_\pm$ at different masses $\mu_\pm$. I shall have more to say about this later. For those who understood Secs. II and III, Eq. (4.8c) is very close to the interaction in the forthcoming SGL equations. For SU(N), this term is a sum of cosines for each ϕ_a.

For reference, I will simply state the other bilinears. Defining

$$P \equiv :\bar{\psi} \begin{pmatrix} 1 & 0 \\ 0 & -1 \end{pmatrix} \psi : , \quad S_\alpha \equiv :\bar{\psi} \tau_\alpha \psi : ,$$

and

$$P_\alpha \equiv :\bar{\psi} \tau_\alpha \begin{pmatrix} 1 & 0 \\ 0 & -1 \end{pmatrix} \psi : ,$$

we obtain

$$P = \frac{2iC}{\pi} (\mu_+ \mu_-)^{1/2} N_\pm \sin[(2\pi)^{1/2} \phi_+] \cos[(2\pi)^{1/2} \phi_-] ,$$

$$S_3 = \frac{2C}{\pi} (\mu_+ \mu_-)^{1/2} N_\pm \sin[(2\pi)^{1/2} \phi_+] \sin[(2\pi)^{1/2} \phi_-] ,$$

$$S_1 = -\frac{2C}{\pi} (\mu_+ \mu_-)^{1/2} N_\pm \sin[(2\pi)^{1/2} \phi_+]$$
$$\times \cos\left[(2\pi)^{1/2} \int_{-\infty}^{x} \dot{\phi}_-(\xi) d\xi \right] ,$$

$$S_2 = -\frac{2C}{\pi} (\mu_+ \mu_-)^{1/2} N_\pm \sin[(2\pi)^{1/2} \phi_+]$$
$$\times \sin\left[(2\pi)^{1/2} \int_{-\infty}^{x} \dot{\phi}_-(\xi) d\xi \right] , \qquad (4.9)$$

$$P_3=\frac{2iC}{\pi}(\mu_+\mu_-)^{1/2}N_\pm\cos[(2\pi)^{1/2}\phi_+]\sin[(2\pi)^{1/2}\phi_-],$$

$$P_1=-\frac{2iC}{\pi}(\mu_+\mu_-)^{1/2}N_\pm\cos[(2\pi)^{1/2}\phi_+]\times\cos\left[(2\pi)^{1/2}\int_{-\infty}^{x}\dot{\phi}_-(\xi)d\xi\right],$$

$$P_2=-\frac{2iC}{\pi}(\mu_+\mu_-)^{1/2}N_\pm\cos[(2\pi)^{1/2}\phi_+]\times\sin\left[(2\pi)^{1/2}\int_{-\infty}^{x}\dot{\phi}_-(\xi)d\xi\right].$$

Here, I am using the abbreviation $N_\pm$ for $N_{+,-}$. These equations all follow straightforwardly, with the observation that $\exp[i\beta\int_{-\infty}^{x}d\xi\,\dot{\phi}_\pm(\xi)]$ normal-orders just as $\exp(i\beta\phi_\pm)$. We now turn to some quartics. As examples, I will work these out for currents, and current algebra.

B. Some quartics and current algebra

Evidently

$$\begin{aligned}
&:J_0J_0:=N_\mu\frac{2}{\pi}(\partial_x\phi_+)^2,\\
&:J_1J_1:=N_\mu\frac{2}{\pi}(\partial_0\phi_+)^2,\\
&:J_\pm^3J_\pm^3:=N_\mu\frac{1}{\pi}[(\partial_0\pm\partial_x)\phi_-]^2,\\
&:J_+^3J_-^3:=-N_\mu\frac{1}{2\pi}\partial_\lambda\phi_-\partial^\lambda\phi_-;
\end{aligned}\tag{4.10}$$

one must, however, take more care with products of charged currents. Following Mandelstam, we obtain ($\dot{\phi}\equiv\partial_0\phi$)

$$J_+(\tau_+,x)J_+(\tau_-,y)\underset{x\to y}{\simeq}-\frac{1}{(x-y-i\epsilon)^2}\frac{1}{\pi^2}-\frac{1}{\pi^2}\frac{i(2\pi)^{1/2}}{x-y-i\epsilon}[\dot{\phi}_-(y)+\partial_y\phi_-(y)]$$
$$-\frac{1}{2\pi^2}\{i(2\pi)^{1/2}[\partial_y\dot{\phi}_-(y)+\partial_y^2\phi_-(y)]-2\pi[(\partial_0+\partial_y)\phi_-]^2\}+O(x-y),\tag{4.11a}$$

$$J_+(\tau_-,y)J_+(\tau_+,x)\underset{x\to y}{\simeq}-\frac{1}{\pi^2}\frac{1}{(x-y+i\epsilon)^2}-\frac{1}{\pi^2}\frac{i(2\pi)^{1/2}}{x-y+i\epsilon}[\dot{\phi}_-(x)+\partial_x\phi_-(x)]$$
$$-\frac{1}{2\pi^2}\{-i(2\pi)^{1/2}[\partial_x\dot{\phi}_-(x)+\partial_x^2\phi_-(x)]-2\pi[(\partial_0+\partial_x)\phi_-]^2\}+O(y-x)\tag{4.11b}$$

Here we were careful to expand the exponentials to second order. (4.11b) is just (4.11a) with $x\leftrightarrow y$ and $\phi_-\to-\phi_-$. It is crucial to note now that reexpanding a first-order term at x, around y, can change second-order terms. I choose to compare the two expressions at y. Thus, I reexpress (4.11b) as

$$J_+(\tau_-,y)J_+(\tau_+,x)\underset{x\to y}{\simeq}-\frac{1}{\pi^2}\frac{1}{(x-y+i\epsilon)^2}-\frac{1}{\pi^2}\frac{i(2\pi)^{1/2}}{x-y+i\epsilon}[\dot{\phi}_-(y)+\partial_y\phi_-(y)]$$
$$-\frac{1}{2\pi^2}\{+i(2\pi)^{1/2}[\partial_y\dot{\phi}_-(y)+\partial_y^2\phi_-(y)]-2\pi[(\partial_0+\partial_y)\phi_-]^2\}+O(y-x).\tag{4.11b'}$$

Note the sign change in the next-to-last term. Without this care, we would blunder into nonlocal current commutators. Now using $(z-i\epsilon)^{-1}=Pz^{-1}+i\pi\delta(z)$, together with (4.11a), (4.11b′), and (4.4), we obtain correctly

$$[J_+(\tau_+,x),J_+(\tau_-,y)]=4J_+^3(x)\delta(x-y)+i\frac{2}{\pi}\partial_x\delta(x-y).\tag{4.12}$$

The rest of current algebra follows smoothly: Commutators of charged currents with neutral currents are much easier.

With the help of (4.11a), and (4.11b′) we can also construct the point quartic

$$:J_+(\tau_+,x)J_+(\tau_-,x):=:\sum_{\alpha=1}^{2}J_+^\alpha(x)J_+^\alpha(x):.$$

This object is symmetric under $\tau_+\leftrightarrow\tau_-$, so we define

$$J_+(\tau_+,x)J_+(\tau_-,x)\equiv\lim_{x\to y}\tfrac{1}{2}[J_+(\tau_+,x)J_+(\tau_-,y)+J_+(\tau_-,x)J_+(\tau_+,y)].\tag{4.13}$$

You may also further symmetrize with respect to x and y, at no cost. After an ordinary *Fermi* normal ordering to remove the c-number vacuum expectation value, we obtain [just from the last term of (4.11a) and (4.11b′)]

$$:J_\pm(\tau_+,x)J_\pm(\tau_-,x):=N_\mu\frac{1}{\pi}[(\partial_0\pm\partial_x)\phi_-]^2. \quad (4.14)$$

We combine (4.10) and (4.14) into some more familiar forms:

$$\begin{aligned}\theta_{00}^{FB}&=\frac{1}{2C_B}:J_0J_0+J_1J_1:\\&=N_\mu\tfrac{1}{2}[(\dot\phi_+)^2+(\partial_x\phi_+)^2]\\&=\theta_{00}^F(\phi_+),\\\theta_{00}^{FV}&=\frac{1}{2\tilde C_V}:J_0^\alpha J_0^\alpha+J_1^\alpha J_1^\alpha:\\&=\frac{1}{4\tilde C_V}:J_+^\alpha J_+^\alpha+J_-^\alpha J_-^\alpha:\\&=N_\mu\tfrac{1}{2}[(\dot\phi_-)^2+(\partial_x\phi_-)^2]\\&=\theta_{00}^F(\phi_-).\end{aligned} \quad (4.15)$$

Remember that for SU(2), $C_B=2/\pi$, $\tilde C_V=3/2\pi$. We see that the free Fermi and Bose Hamiltonians are identical, a result which had to be true. It is also true for the entire free stress tensor. On the other hand, the forms (4.10) and (4.14) caused me some consternation. For example, I can write

$$:J_\pm^\alpha J_\pm^\alpha:=3:J_\pm^3J_\pm^3:, \quad (4.16)$$

where α is summed as usual from 1 to 3. This is a surprise, as the left-hand side is isoscalar and the right a tensor. Is there trouble, or is (4.16) true, plus many other relations (by isospin rotations, commutators, etc.)? The answer is that it *is* true, and it is a simple property of Fermi statistics. The reader is invited to write out (4.16) explicitly in terms of free fermions and see for himself. Such identities are also true in SU(N), where $J_\pm^\alpha J_\pm^\alpha$ equals a weighted sum of "diagonal" currents squared. Out of many identities of this sort, I mention also

$$:J_+^\alpha\tau^\alpha\psi^1:=3:J_+^3\tau_3\psi^1:, \quad (4.17)$$

and a similar equation for ψ^2 with J_-. These follow immediately on commutation of ψ with (4.16), and we will make mention of (4.16) and (4.17) again in Sec. VI.

None of this will directly help us simplify a $J_\mu^\alpha J_\alpha^\mu$ interaction, however; to construct this, we need to calculate

$$\tfrac{1}{2}:J_+(\tau_+)J_-(\tau_-)+J_+(\tau_-)J_-(\tau_+):$$
$$=-\left(\frac{C\mu}{\pi}\right)^2N_\mu\cos[2(2\pi)^{1/2}\phi_-], \quad (4.18)$$

or, with (4.9),

$$:J_\lambda^\alpha J_\alpha^\lambda:=:J_+^\alpha J_-^\alpha:$$
$$=-\left(\frac{C\mu}{\pi}\right)^2N_\mu\cos[2(2\pi)^{1/2}\phi_-]-\frac{1}{2\pi}N_\mu(\partial_\lambda\phi_-)^2. \quad (4.19)$$

We see manifest in this representation the basis of the statement that such an interaction breaks (naive) conformal invariance. Although the right-hand side of (4.19) is independent of μ, it still provides a mass scale. We will return to this interaction in Sec. VI. I will not write out the other quartics ($S^\alpha S^\alpha$, etc.). They are, however, quite local. I have also checked a representative selection of the quark algebra among the J^μ's, S's, and P's.

Before going on to introduce interaction, I want to use a little hindsight to introduce a slightly different free-field representation. We shall see that $\phi_\pm=(1/\sqrt2)(\phi_1\pm\phi_2)$ are the natural eigenstates of the theory, and they are not in general treated symmetrically. (The masses of the $\phi_\pm$ quanta are not the same.) It is therefore convenient to have also

$$\begin{aligned}\chi_1^1&=\xi_1\psi_1^1\\&=\left(\frac{C}{2\pi}\right)^{1/2}(\mu_+\mu_-)^{1/4}N_+\exp\left\{-i\left(\frac{\pi}{2}\right)^{1/2}\left[\int_{-\infty}^x d\xi\,\dot\phi_+(\xi)+\phi_+(x)\right]\right\}N_-\exp\left\{-i\left(\frac{\pi}{2}\right)^{1/2}\left[\int_{-\infty}^x d\xi\,\dot\phi_-(\xi)+\phi_-(x)\right]\right\},\\\chi_2^1&=\xi_1\psi_2^1\\&=\left(\frac{C}{2\pi}\right)^{1/2}(\mu_+\mu_-)^{1/4}N_+\exp\left\{-i\left(\frac{\pi}{2}\right)^{1/2}\left[\int_{-\infty}^x d\xi\,\dot\phi_+(\xi)+\phi_+(x)\right]\right\}N_-\exp\left\{+i\left(\frac{\pi}{2}\right)^{1/2}\left[\int_{-\infty}^x d\xi\,\dot\phi_-(\xi)+\phi_-(x)\right]\right\},\end{aligned} \quad (4.20)$$

and similarly for ψ_a^2 (with $\phi_\pm\to-\phi_\pm$, $\dot\phi_\pm\to\dot\phi_\pm$). What we have done here relative to (4.1) is to rewrite in terms of $\phi_\pm$, and normal-order each separately at $\mu_\pm$. I believe (4.1) and (4.20) are equivalent (certainly for the free theory), but the latter is more convenient for using Mandelstam's formalism, in which, in the end, one is normal ordering directly at the Bose quanta masses. I will therefore switch to (4.20) for the remainder of the paper.

For this representation, it is easily checked that there are only a few minor modifications in all our correspondences. All forms are the same, with the understanding that expressions involving pure ϕ_+ (or ϕ_-) are normal ordered at μ_+ (μ_-). Our mixed expressions, Eqs. (4.8) and (4.9), for S's and P's, come out directly as (4.8c) and (4.9). We are now ready for interaction.

V. INTERACTION AND THE SINE-GORDON-LIKE EQUATIONS

Armed with Secs. II and IV, introduction of interaction is mechanical. We first take the simple $J^\mu J_\mu$ interaction set up in Sec. II. Following Sec. III, we reexpress Eq. (2.3), using (4.10), as

$$\theta^I_{00} = \tfrac{1}{2}g_B(J_{0D}{}^2 - GJ_{1D}{}^2) = \tfrac{1}{2}g_B C_B[(\partial_x \phi_{+D})^2 - G(\partial_0\phi_{+D})^2], \tag{5.1}$$

where $G=(1+C_B g_B)^{-1}$, $C_B = 2/\pi$. Everything sweeps through as before, including $\kappa = C_B g_B$, $\bar{J}_\mu = (J_0, GJ_1)$, $U^\dagger \phi_{+D} U = \sqrt{G}\,\hat{\phi}_+$, $U^\dagger \dot{\phi}_{+D} U = (1/\sqrt{G})\dot{\hat{\phi}}_+$, where $\hat{\phi}_+$ is the correctly normalized free field. Thus (all in HP)

$$\begin{aligned}\bar{J}^\mu &= -(C_B G)^{1/2}\epsilon^{\mu\nu}\partial_\nu \phi_+ \\ &= :Z^{-1}(\bar{\psi}\gamma^0\psi, G\bar{\psi}\gamma^1\psi):,\end{aligned} \tag{5.2a}$$

$$\begin{aligned}\bar{J}^\mu_3 &= J^\mu_3 \\ &= -\frac{1}{(2\pi)^{1/2}}\epsilon^{\mu\nu}\partial_\nu\phi_- \\ &= :Z^{-1}\left(\bar{\psi}\gamma^0\frac{\tau_3}{2}\psi, \bar{\psi}\gamma^1\frac{\tau_3}{2}\psi\right):,\end{aligned}$$

$$\begin{aligned}&[\bar{J}_0(x), \psi(y)] = -\psi(x)\delta(x-y),\\ &[\bar{J}_1(x), \psi(y)] = -\gamma_0\gamma_1 G\psi(x)\delta(x-y),\\ &[J^\alpha_0(x), \psi(y)] = -\tfrac{1}{2}\lambda_\alpha\psi(x)\delta(x-y),\\ &[\bar{J}_0(x), \bar{J}_1(y)] = iC_B\partial_x\delta(x-y),\end{aligned} \tag{5.2b}$$

and so on. The isospin scale is not changed because ϕ_- is undisturbed. Indeed, all the isospin current relations of Sec. IV are completely unchanged for this reason. There are only the usual expected changes $\phi_+ \to \sqrt{G}\,\hat{\phi}_+$, $\dot{\phi}_+ \to (1/\sqrt{G})\dot{\hat{\phi}}_+$ everywhere. Thus, e.g., (4.20) is changed to

$$\begin{aligned}\chi^1_1 &= \xi_1\psi^1_1 \\ &= \left(\frac{C}{2\pi}\right)^{1/2}(\mu_+\mu_-)^{1/4}N_\pm \exp\left\{-i\left(\frac{\pi}{2}\right)^{1/2}\left[\int_{-\infty}^x d\xi \frac{1}{\sqrt{G}}\dot{\hat{\phi}}_+ + \sqrt{G}\,\hat{\phi}_+(x)\right]\right\}\exp\left\{-i\left(\frac{\pi}{2}\right)^{1/2}\left[\int_{-\infty}^x d\xi\,\dot{\phi}_- + \phi_-(x)\right]\right\},\end{aligned} \tag{5.3}$$

etc., and from (4.8c)

$$-:Z'\bar{\psi}\psi: = \frac{2C}{\pi}(\mu_+\mu_-)^{1/2}N_\pm \cos[(2\pi G)^{1/2}\hat{\phi}_+]\cos[(2\pi)^{1/2}\phi_-]. \tag{5.4}$$

Still following Sec. III, we next add in the mass term, $-m'{:}Z'\bar{\psi}\psi{:}$, obtaining the boson Lagrangian:

$$\mathcal{L} = \tfrac{1}{2}(\partial_\nu\phi_+)^2 + \tfrac{1}{2}(\partial_\nu\phi_-)^2 + \frac{2m'C}{\pi}(\mu_+\mu_-)^{1/2}N_\pm\cos[(2\pi G)^{1/2}\phi_+]\cos[(2\pi)^{1/2}\phi_-]. \tag{5.5}$$

This implies two coupled SGL equations for $\phi_\pm$. I will discuss the system presently.

Before that, however, it is wise to check our results directly. Starting from the SGL equations implied by (5.5), and our final fermions

$$\begin{aligned}\chi^1_1 &= \xi_1\psi^1_1 \\ &= \left(\frac{C}{2\pi}\right)^{1/2}(\mu_+\mu_-)^{1/4}N_\pm \exp\left\{-i\left(\frac{\pi}{2}\right)^{1/2}\left[\int_{-\infty}^x d\xi \frac{1}{\sqrt{G}}\dot{\phi}_+(\xi) + \sqrt{G}\,\phi_+(x)\right]\right\}\exp\left\{-i\left(\frac{\pi}{2}\right)^{1/2}\left[\int_{-\infty}^x d\xi\,\dot{\phi}_-(\xi) + \phi_-(x)\right]\right\},\\ \chi^1_2 &= \xi_1\psi^1_2 \\ &= \left(\frac{C}{2\pi}\right)^{1/2}(\mu_+\mu_-)^{1/4}N_\pm \exp\left\{-i\left(\frac{\pi}{2}\right)^{1/2}\left[\int_{-\infty}^x d\xi \frac{1}{\sqrt{G}}\dot{\phi}_+(\xi) + \sqrt{G}\,\phi_+(x)\right]\right\}\exp\left\{+i\left(\frac{\pi}{2}\right)^{1/2}\left[\int_{-\infty}^x d\xi\,\dot{\phi}_-(\xi) + \phi_-(x)\right]\right\},\\ \chi^2_1 &= \xi_1\psi^2_1 \\ &= i\left(\frac{C}{2\pi}\right)^{1/2}(\mu_+\mu_-)^{1/4}N_\pm \exp\left\{-i\left(\frac{\pi}{2}\right)^{1/2}\left[\int_{-\infty}^x d\xi \frac{1}{\sqrt{G}}\dot{\phi}_+(\xi) - \sqrt{G}\,\phi_+(x)\right]\right\}\exp\left\{-i\left(\frac{\pi}{2}\right)^{1/2}\left[\int_{-\infty}^x d\xi\,\dot{\phi}_-(\xi) - \phi_-(x)\right]\right\},\\ \chi^2_2 &= \xi_1\psi^2_2 \\ &= i\left(\frac{C}{2\pi}\right)^{1/2}(\mu_+\mu_-)^{1/4}N_\pm \exp\left\{-i\left(\frac{\pi}{2}\right)^{1/2}\left[\int_{-\infty}^x d\xi \frac{1}{\sqrt{G}}\dot{\phi}_+(\xi) - \sqrt{G}\,\phi_+(x)\right]\right\}\exp\left\{+i\left(\frac{\pi}{2}\right)^{1/2}\left[\int_{-\infty}^x d\xi\,\dot{\phi}_-(\xi) - \phi_-(x)\right]\right\},\end{aligned} \tag{5.6}$$

one can wend one's way back through all the identities to the massive-Thirring-model Dirac equation.

Indeed, following Mandelstam, we find the acknowledged wave-function renormalizations Z [in the form $(x-y)^{\circ}$] the same for baryon number and SU(2) currents. The renormalizations for all the S's and P's are the same among themselves, but different from the currents, hence Z'. Indeed, we recover *all* the results of the interaction picture, i.e., all[12] the results of Sec. IV, with the simple map $\phi_+ \to \sqrt{G}\,\phi_+$, $\dot{\phi}_+ \to (1/\sqrt{G})\dot{\phi}_+$. In going all the way back to the Thirring-Dirac equation, one calculates derivatives of (5.6). The useful identities analogous to Mandelstam's[2] Eq. (4.6) are of the form

$$\pi N'_\pm \int_{-\infty}^{x} d\xi(\{\cos[(2\pi G)^{1/2}\phi_+(\xi)]\sin[(2\pi)^{1/2}\phi_-(\xi)]+\sin[(2\pi G)^{1/2}\phi_+(\xi)]\cos[(2\pi)^{1/2}\phi_-(\xi)]\}\chi_1^1(x))$$

$$=\int_{-\infty}^{+\infty} d\xi[N_\pm\cos[(2\pi G)^{1/2}\phi_+(\xi)]\cos[(2\pi)^{1/2}\phi_-(\xi)],\chi_1^1(x)]\,, \quad (5.7)$$

where N' is Mandelstam's "block" normal ordering, and the second line is a commutator. Recognizing $\bar{\psi}\psi$ in the last line, one obtains the Dirac equation for χ. Multiplying in a ξ_1 from the left results in the expected equation for ψ. Everything goes through smoothly, so we turn out attention back to the SGL system (5.5).

We transform (5.5) into a "standard" form, say

$$\mathcal{L}_I=\frac{\alpha}{\beta_+^2}N_\pm\cos(\beta_+\phi_+)\cos[(2\pi)^{1/2}\phi_-]\,. \quad (5.8)$$

Again following Mandelstam's convention, we choose to set $\mu_\pm$ equal to the Bose quantum masses. Thus $\alpha=\mu_+^2$, $(m'2C/\pi)(\mu_+\mu_-)^{1/2}=\mu_+^2\beta_+^{-2}$, and

$$\beta_+=(2\pi G)^{1/2}\,,$$

$$\frac{2\pi}{\beta_+^2}=1+g_BC_B$$

$$=1+\frac{2g_B}{\pi}\,. \quad (5.9)$$

We also notice the curious fact that the "mass" μ_- of the ϕ_- quantum is *fixed* in terms of μ_+ (mass of ϕ_+),

$$\mu_-^2=\frac{2\pi}{\beta_+^2}\mu_+^2\,, \quad (5.10)$$

and thus $m'=(\mu_+/4C)G^{-3/4}=\mu_+\beta_+^{-3/2}(\pi/2C)(2\pi)^{-1/4}$. This fixed mass ratio is easily understood: The corresponding massive Thirring model has only one dimensional parameter m'. The fact that $\mu_+\neq\mu_-$ is also the last blow to any hope that the SGL system would exhibit a (linear) SU(2) symmetry. It apparently has no (linear) continuous symmetry at all. In Appendix A, however, we discuss the sense in which the sine-Gordon-like equations may be thought of as providing a (spatially) nonlocal representation of isospin. Of course, (5.10) is not a reliable prediction as large (calculable) higher-order corrections are expected from the interaction (even for small β_+).

What about soliton-like extended solutions of the classical SGL equations? I have not tried to find them analytically, but, by our very construction, they do exist [and will be free—no scattering—when $\beta_+=(2\pi)^{1/2}$]. The qualitative features of the solutions can be seen through the discrete symmetries of the system:

$$\phi_+\to\phi_++\frac{\pi n}{\beta_+}\ \textit{and}\ \phi_-\to\phi_-+\frac{\pi m}{(2\pi)^{1/2}}\ \text{(type I)}\,,$$

$$\phi_+\to\phi_++\frac{2\pi n}{\beta_+}\ \textit{or}\ \phi_-\to\phi_-+\frac{2\pi m}{(2\pi)^{1/2}}\ \text{(type II)}\,, \quad (5.11)$$

where n,m are odd for type I. Thus, we expect solutions with such (say one-sided) asymptotic behavior. Consulting Eqs. (5.2a), now in the form

$$\bar{J}^\mu=-\frac{\beta_+}{\pi}\epsilon^{\mu\nu}\partial_\nu\phi_+\,,\quad J_3^\mu=-\frac{1}{(2\pi)^{1/2}}\epsilon^{\mu\nu}\partial_\nu\phi_-\,, \quad (5.2\text{a}')$$

that is, $B=(\beta_+/\pi)\phi_+|_{-\infty}^{+\infty}$, $I_3=[1/(2\pi)^{1/2}]\phi_-|_{-\infty}^{+\infty}$, we easily correlate asymptotic behaviors with quark content. Assuming $\phi_\pm(-\infty)=0$, then a solution for which $(x\to+\infty)$

$$\phi_+\to\frac{\pi}{\beta_+}\,,\quad \phi_-\to\left(\frac{\pi}{2}\right)^{1/2} \quad (5.12)$$

is a quark with $I_3=+\frac{1}{2}$ (plus perhaps $qq\bar{q}$ in $I_3=\frac{1}{2}$, etc.). In general, states with asymptotic behavior of type I [Eq. (5.11)] have half-integer isospin correlated with odd quark number. States of type II have even quark number and integer isospin. It is not clear to me whether there is a simple way of reading *total* isospin from the classical solutions.

Finally, I will mention that, having repeated Coleman's[1] variational vacuum calculation on the Hamiltonian corresponding to (5.8), I find $\beta_+^2<6\pi$ ($g_B>-\frac{1}{3}\pi$), or else the energy is unbounded below. This is to be compared with the "obvious" bound $g_B>-\frac{1}{2}\pi$.

General remarks on SU(N). So much for SU(2). What can we say about SU(N)? In general, the isoscalar interaction

$$J^\mu = -\sqrt{C_B}\,\epsilon^{\mu\nu}\partial_\nu\phi_+, \quad \phi_+ \equiv \frac{1}{\sqrt{N}}\sum_a \phi_a, \quad C_B = \frac{N}{\pi} \tag{5.13}$$

$$\bar{J}^\mu = -(C_B G)^{1/2})\epsilon^{\mu\nu}\partial_\nu\phi_+, \quad G = (1+g_B C_B)^{-1}$$

scale shifts only ϕ_+, leaving undisturbed the other $N-1$ orthogonal combinations Φ_a, $a=2,\ldots,N$. It is, of course, a matter of taste how to choose the Φ_a. A choice for SU(3) might be

$$\begin{aligned}
\phi_+ &= \frac{1}{\sqrt{3}}(\phi_1+\phi_2+\phi_3),\\
\Phi_2 &= \frac{1}{\sqrt{2}}(\phi_1-\phi_2),\\
\Phi_3 &= \frac{1}{\sqrt{6}}(\phi_1+\phi_2-2\phi_3),\\
\phi_1 &= \frac{1}{\sqrt{3}}\phi_+ + \frac{1}{\sqrt{2}}\Phi_2 + \frac{1}{\sqrt{6}}\Phi_3,\\
\phi_2 &= \frac{1}{\sqrt{3}}\phi_+ - \frac{1}{\sqrt{2}}\Phi_2 + \frac{1}{\sqrt{6}}\Phi_3,\\
\phi_3 &= \frac{1}{\sqrt{3}}\phi_+ - \left(\frac{2}{3}\right)^{1/2}\Phi_3,
\end{aligned} \tag{5.14}$$

but $\Phi_{2,3}$ can be orthogonally mixed. In general

$$\phi_a = \frac{1}{\sqrt{N}}\phi_+ + D_{ab}\Phi_b, \quad \sum_a D_{ab} = 0,$$

$$\sum_a D_{ab}D_{ab'} = \delta_{bb'}.$$

We expect and will note later that the masses of the Φ_a are degenerate [a bosonic O($N-1$) symmetry?] under the isoscalar interaction, so the convenient quark representation [analogous to (4.19)] is constructed as follows: Start from (4.1); break ϕ_a up into ϕ_+, Φ_a, and shift $\dot{\phi}_+ \to (1/\sqrt{G})\dot{\phi}_+$, $\phi_+ \to \sqrt{G}\,\phi_+$. Normal-order ϕ_+ at μ_+ and Φ_a at μ_-. The breakup of the normalization factor is

$$\left(\frac{\mu C}{2\pi}\right)^{1/2} \to \left(\frac{\mu_+ C}{2\pi}\right)^{1/2N}\left(\frac{\mu_- C}{2\pi}\right)^{(1-1/N)/2}. \tag{5.15}$$

The SU(N) currents are functions of Φ_a only. The Fermi-mass term generates the SGL interaction

$$\begin{aligned}
\mathcal{L}_I &= -m' :Z'\bar{\psi}\psi:\\
&= 2m'\left(\frac{\mu_+ C}{2\pi}\right)^{1/N}\left(\frac{\mu_- C}{2\pi}\right)^{1-1/N}\\
&\quad \times \sum_a \cos\left[2\left(\frac{\pi G}{N}\right)^{1/2}\phi_+ + 2\sqrt{\pi}\,D_{ab}\Phi_b\right].
\end{aligned} \tag{5.16}$$

Expanding to second order in the fields, we obtain the boson mass matrix $\mathfrak{m}^2$,

$$\begin{aligned}
-\tfrac{1}{2}\mathfrak{m}^2 &= \mathcal{L}_{(2)}\\
&\cong -2m'\left(\frac{\mu_+ C}{2\pi}\right)^{1/N}\left(\frac{\mu_- C}{2\pi}\right)^{1-1/N}\\
&\quad \times\left(2\pi G\phi_+^2 + 2\pi\sum_a \Phi_a^2\right).
\end{aligned} \tag{5.17}$$

Thus $\mathfrak{m}^2$ has the promised O($N-1$) symmetry. Calling μ_+ the mass of ϕ_+ and μ_- all the others, we find $\mu_-^2 = G^{-1}\mu_+^2$, as above. Normalizing to μ_+, then $m' = (\mu_+/4C)G^{-1/2-1/2N}$. We also remark that as in Sec. IV this peculiar $\mu_\pm$ pattern in the mass term (5.20) can be obtained directly from the representation (4.1) by re-normal-ordering in the free theory before interaction, as in (4.8c).

Do we have a real bosonic (linear) O($N-1$)? The answer is no. The rest of the interaction completely ruins it. As a simple example of what is going on, consider a toy Lagrangian

$$\mathcal{L} = \tfrac{1}{2}(\partial\phi_1)^2 + \tfrac{1}{2}(\partial\phi_2)^2 + \mu^2(\cos\phi_1 + \cos\phi_2).$$

The mass and kinetic energy terms have a U(1) symmetry, but not the interaction. We could make a U(1) transformation on the Lagrangian, thus introducing a parameter, but it is not a symmetry transformation, only a change of variables. This is precisely the situation among the Φ_a in our SU(N) models, except that in the SGL equations the symmetry is *required* and then broken. Its breaking must then be *calculable*. [For that matter, the deviation of μ_-^2 from $\mu_+^2G^{-1}$ is also calculable in our SU(2) model.] It is certainly not surprising that there seems to be no direct connection between this O($N-1$) and the quark SU(N).

As a closing remark in this section, I mention that although we have worked in terms of an SU(N) quark representation for the fermions, one can as easily take quarks with color, octets (baryons), etc. As in Sec. IV, one simply takes one ϕ for each independent Fermi field.

VI. REMARKS ON HARD-CHIRAL-SYMMETRY-BREAKING INTERACTIONS

We have previously concerned ourselves only with interactions of the form $g_B J^\mu J_\mu$ (baryon number current-current interaction), and mass terms. Collectively, these imply at most a soft breaking of the chiral SU(N). Other hard-breaking interactions, such as $g_V J^\alpha_\mu J^\mu_\alpha$ are presently under investigation, and I will confine myself here to a few preliminary remarks.

In the first place, we are on much more treacherous grounds with these interactions: It is known[13] that g will require a renormalization, and it may even be necessary to include at least one other interaction (say $g_S\bar{\psi}\psi\bar{\psi}\psi$) for a consistent renormalization.[14] We expect our interaction picture approach to be less useful, but still suggestive.

Following Sec. II, we can write

$$\theta^I_{00}=\frac{g_V}{2}J^\alpha_0J^\alpha_0+b_VJ^\alpha_1J^\alpha_1+\frac{g_B}{2}J_0J_0+b_BJ_1J_1$$
$$=A_V\theta^{FV}_{00}+B_VJ^\alpha_+J^\alpha_-+A_B\theta^{FB}_{00}+B_BJ_+J_-, \qquad (6.1)$$

where $J_\pm=J_0\pm J_1$, $A_V=(\frac{1}{8}g_V+\frac{1}{4}b_V)4\bar{C}_V$, $B_V=\frac{1}{4}g_V-\frac{1}{2}b_V$, $A_B=(\frac{1}{8}g_B+\frac{1}{4}b_B)4C_B$, $B_B=\frac{1}{4}g_B-\frac{1}{2}b_B$, and we have used (2.2). Also $J_+J_-=J_\mu J^\mu$. These are convenient for evaluating Schwinger's condition. We obtain

$$A_V^2+2A_V=4B_V^2C_V\bar{C}_V,$$
$$A_B^2+2A_B=4B_B^2C_B^2. \qquad (6.2)$$

The second restriction is essentially that given in Sec. II; the vector restriction is new, and bears comment. Its solution has a two-sheeted structure, only one sheet of which is perturbative ($b_V\to 0$ as $g_V\to 0$). Presumably, this sheet defines ordinary formal unrenormalized perturbation theory for the vector interaction. If we further require the Dirac equation in terms of a simple rescaling $\tilde{J}^\alpha_\mu=(J^\alpha_0,\lambda J^\alpha_1)$, we need $g_V\lambda=-2b_V$. The requirement that $\tilde{J}_\mu$ be a two-vector in the Heisenberg picture is $\lambda-(A_V-2B_VC_V)=1$, $\lambda^{-1}-(A_V+2B_VC_V)=1$. The entire set of requirements is, in general, *inconsistent* (except at $g_V=0$), indicating that our scheme of passage to the Heisenberg picture is too naive: g_V needs a renormalization.

Curiously enough, there is one nontrivial value of the coupling, on the *second* sheet, for which all the equations can be satisfied. That value is

$$g_V=-\frac{4\pi}{n+1},$$

and $B_V=0$, $A_V=-2$, $\lambda=-1$. The value of g_V is the negative of the Dashen-Frishman[6] value, and $B_V=0$ means conserved axial-vector currents. $\lambda=-1$ is a change of sign in the axial-vector current commutators with the field. This was suggestive enough to pursue. For correspondence with the Dashen-Frishman equations, it turns out that one must map $\psi(-x,-t)|_{\text{here}}=\psi_{\text{DF}}(x,t)$, and similarly for all the currents. This changes the sign of g_V (but reverses g_B). We obtain, in the notation of Dashen and Frishman, precisely their Dirac equations, and $\delta=-1$, $C_1=1/2\pi$, $\tilde{C}_1=(n+1)/2\pi$, $\bar{a}=[1-g_B(n/\pi)]^{-1}$, $a=1$, $C_0=-(n/\pi)[1-g_B(n/\pi)]^{-1}$. Unfortunately, plugging into their Eq. (17), we get -1, i.e., an "antispinor" (with an interchange of ψ^1 and ψ^2). Presumably then, this is *not* their solution. Indeed, a glance at their equations (15) shows that a solution for $\delta=-1$ is the $\psi^1\leftrightarrow\psi^2$ interchange of the $\delta=+1$ solution.

We turn now to a few remarks about the $\phi_\pm$ representation. Using (4.19), we rewrite the vector part of (6.1) as

$$\theta^I_{00}=\tfrac{1}{2}(\dot{\phi}_-)^2\left(A_V-\frac{B_V}{\pi}\right)+\tfrac{1}{2}(\partial_x\phi_-)^2\left(A_V+\frac{B_V}{\pi}\right)$$
$$-B_V\left(\frac{C\mu_-}{\pi}\right)^2N\cos[2(2\pi)^{1/2}\phi_-]. \qquad (6.3)$$

We see immediately that, in general, g_V renormalization will be tangled with normal-ordering. Note that we can sweep out the quadratic form into a covariant wave-function renormalization (as in Sec. III),

$$\mathcal{L}=\tfrac{1}{2}(1+\kappa)(\partial_\mu\phi_-)^2-B_V\left(\frac{C\mu_-}{\pi}\right)^2N\cos[2(2\pi)^{1/2}\phi_-], \qquad (6.4)$$

if $\kappa=A_V+B_V/\pi$ and $-\kappa/(1+\kappa)=A_V-B_V/\pi$. This form is suggestive of the requisite SGL equation, but I do not trust it. These conditions on κ are inconsistent with (6.2) (being instead the conditions on λ above). I think the point is that the cosine interaction is quite bizarre. Remember that it represents $\sum_{\alpha=1}^2J^\alpha_\mu J^\mu_\alpha$ which certainly does *not* commute with itself at equal time. As a matter of fact, the same problem exists in the Abelian Thirring model, where $[\bar{\psi}(1+\gamma_5)\psi,\bar{\psi}(1-\gamma_5)\psi]\sim[e^{+i\beta\phi},e^{-i\beta\phi}]$ appears to be zero, and yet, from the ψ's, must be proportional to $\delta(x-y)J_1(x)\sim\delta(x-y)\partial_0\phi$. To see such things, one must do (at least) a smearing in time.

Our final remark in this section concerns the explicit $\phi_\pm$ construction of our solution to the Dashen-Frishman Dirac equation. One can work through the interaction picture, as described above, but I will just state the result directly. Take their equations in the form

$$i(\partial_0+\partial_x)\psi^2_{\text{DF}}=:g_B\tilde{J}^B_+\psi^2_{\text{DF}}+g_V\left[\tfrac{1}{4}\int_{-\infty}^{+\infty}dy\,\tilde{J}^\alpha_+(y)\tilde{J}^\alpha_+(y),\psi^2_{\text{DF}}(x)\right]:, \qquad (6.5)$$

where the last term is a commutator, and similarly for ψ^1_{DF}. By $\tilde{J}$ I mean Heisenberg currents that transform as two-vectors. When $g_V=4\pi/n+1$, this is their case $\delta=-1$. Now for the solution: Consider our $g_V=0$ solutions (Secs. IV and V), e.g.,

$$\chi_1^1 = \xi_1\psi_1^1$$
$$= \left(\frac{C}{2\pi}\right)^{1/2}(\mu_+\mu_-)^{1/4}N\exp\left\{-i\left(\frac{\pi}{2}\right)^{1/2}\left[\int_{-\infty}^{x} d\xi \frac{1}{\sqrt{\mathcal{G}}}\dot{\phi}_+ + \sqrt{\mathcal{G}}\,\phi_+(x)\right]\right\}N\exp\left\{-i\left(\frac{\pi}{2}\right)^{1/2}\left[\int_{-\infty}^{x} d\xi\,\dot{\phi}_-(\xi)+\phi_-(x)\right]\right\}, \tag{6.6}$$

and so on. Leave $\mathcal{G}$ undetermined, however, for the moment. Construct the Dashen-Frishman currents out of these, *as usual* [$\tilde{J}^\mu = -(2\mathcal{G}/\pi)^{1/2}\epsilon^{\mu\nu}\partial_\nu\phi_+$, $J_3^\mu = J_3^\mu = -(1/2\pi)^{1/2})\epsilon^{\mu\nu}\partial_\nu\phi_-$, etc.], *but* take the Dashen-Frishman fields to be $\psi_{DF}^{1,2} = \psi^{2,1}$. Thus in terms of *our* fields (6.6), Eq. (6.5) reads

$$i(\partial_0+\partial_x)\psi^1 = :g_B\tilde{J}_+^B\psi^1 + g_V\left[\tfrac{1}{4}\int_{-\infty}^{+\infty} dy\, J_+^\alpha J_+^\alpha,\ \psi^1(x)\right]:. \tag{6.7}$$

Recall our identity (4.16). Since ϕ_- is not excited, it is true for $\mathcal{G}\neq 1$ and the last term in (6.7) is merely

$$:g_V 3\frac{T_3}{2}\,J_+^3(x)\psi^1:.$$

Now, it is a simple matter to differentiate our ψ's. We obtain, e.g.,

$$i(\partial_0+\partial_1)\psi_1^1 = N_\mu\left\{\left[\frac{\pi}{2}\left(1+\frac{1}{\mathcal{G}}\right)J_+ + 2\pi J_+^3\right]\psi_1^1\right\}$$

and hence $g_B = \frac{1}{2}\pi(1+1/\mathcal{G})$, $g_V = \frac{4}{3}\pi$, as required. All this is as I reasoned it from the interaction picture, and it is no surprise that our solution ψ_{DF} is an "antispinor."

NOTE ADDED IN PROOF

By a simple extension of the approach in Sec. VI, I have been able also to construct the "correct" Dashen-Frishman (DF) solution. The solution has the form

$$(\chi_1^2)_{DF} = \xi_1(\psi_1^2)_{DF} = \chi_1^1 = \xi_1\psi_1^1 = \kappa N\exp\left\{-i\left[\alpha\int_{-\infty}^{x} d\xi\,\dot{\phi}_+(\xi)+\beta\phi_+(x)\right]\right\}$$
$$\times\exp\left\{-i(\pi/2)^{1/2}\left[\int_{-\infty}^{x} d\xi\,\dot{\phi}_-(\xi)+\phi_-(x)\right]\right\},$$

$$(\chi_2^2)_{DF} = \xi_1(\psi_2^2)_{DF} = \chi_2^1 = \xi_1\psi_2^1 = \kappa N\exp\left\{-i\left[\int_{-\infty}^{x} d\xi\,\dot{\phi}_+(\xi)+\beta\phi_+(x)\right]\right\}$$
$$\times\exp\left\{+i(\pi/2)^{1/2}\left[\int_{-\infty}^{x} d\xi\,\dot{\phi}_-(\xi)+\phi_-(x)\right]\right\},$$

$$(\chi_1^1)_{DF} = \xi_1(\psi_1^1)_{DF} = \chi_1^2 = \xi_1\psi_1^2 = i\kappa N\exp\left\{-i\left[\alpha\int_{-\infty}^{x} d\xi\,\dot{\phi}_+(\xi)-\beta\phi_+(x)\right]\right\}$$
$$\times\exp\left\{-i(\pi/2)^{1/2}\left[\int_{-\infty}^{x} d\xi\,\dot{\phi}_-(\xi)-\phi_-(x)\right]\right\},$$

$$(\chi_2^1)_{DF} = \xi_1(\psi_2^1)_{DF} = \chi_2^2 = \xi_1\psi_2^2 = i\kappa N\exp\left\{-i\left[\int_{-\infty}^{x} d\xi\,\dot{\phi}_+(\xi)-\beta\phi_+(x)\right]\right\}$$
$$\times\exp\left\{+i(\pi/2)^{1/2}\left[\int_{-\infty}^{x} d\xi\,\phi_-(\xi)-\phi_-(x)\right]\right\}.$$

Here $\kappa=(C\mu/2\pi)^{1/2}$, and we have generalized (6.6), leaving two parameters α,β; this will allow the proper "spin." Following Sec. VI, we take the isospin currents as in (4.4) and (4.6). Indeed, we do not tamper with the ϕ_- (isospin) structure at all, so current algebra follows, with $C_1 = 1/2\pi$, $\tilde{C}_1 = 3/2\pi$. Because we maintain the $\psi^1 \leftrightarrow \psi^2$ interchange of the text, $\delta=-1$, and the equation to be solved is still (6.7), with its companion remarks.

Anticommutativity of the Fermi fields requires $\alpha\beta/\pi = \frac{1}{2}+2j$ with j an integer ($j=0$ is the solution of the text). The baryon-number current (normalized to $a=1$) is easily seen to be $\bar{J}_\mu = -\alpha^{-1}\epsilon_{\mu\nu}\partial^\nu\phi_+$; then $\bar{a}=-\beta\alpha^{-1}$ and $C_0=\alpha^{-2}$. Differentiating $\psi(\Box^2\phi_\pm=0)$, and comparing with (6.7) we identify $g_V=4\pi/3$, $g_B=\alpha(\alpha+\beta)$. Finally, one calculates the spin [either directly via the Lorentz generator in terms of $\phi_\pm$, or just by substituting into DF(17)]. The result is $s=-\frac{1}{2}[2j+1]$. The solution of the text is $j=0$, $s=-\frac{1}{2}$; now we choose $j=-1$, $s=+\frac{1}{2}$. The final algebra is trivial, and we record ($s=+\frac{1}{2}$)

$$\bar{a}=\left(1+\frac{2g_B}{3\pi}\right)^{-1},\quad C_0=\left(g_B+\frac{3\pi}{2}\right)^{-1},$$

$$\alpha^2=g_B+\frac{3\pi}{2},\quad \beta=-\frac{3\pi}{2\alpha},$$

thus completing the DF solution.

It is easy to add a Fermi mass term. Following the method of the text, we calculate $[(\bar{\psi}\psi)_{\rm DI}=\bar{\psi}\psi]$ the equivalent SGL system,

$$\mathcal{L}=\tfrac{1}{2}(\partial\phi_+)^2+\tfrac{1}{2}(\partial\phi_-)^2 - \frac{2C\mu m'}{\pi}\cos(2\beta\phi_+)\cos[(2\pi)^{1/2}\phi_-]$$

with $\beta=-(3\pi/2)(g_B+3\pi/2)^{-1/2}$. The over-all sign of the interaction can be changed via a redefinition $\psi\to\gamma_5\psi$. Comparing with (5.8) we see that the boson systems for $\delta=\pm1$ are the same, with the identification $g_B^-=3\pi+9g_B^+$ ($\pm$ are $\delta=\pm1$). In this sense, the $\delta=\pm1$ solutions are themselves equivalent. This does not, however, imply equality of particular Fermi Green's functions.

I wish to thank M. Kaku for bringing to my attention recent similar work on this model by Dashen and Frishman, and by Bhattacharya and Roy.

ACKNOWLEDGMENTS

I would like to thank W. Siegel, I. Bars, and especially S. Mandelstam for helpful conversations.

APPENDIX A: THE SCALARS AS A (SPATIALLY) NONLOCAL REPRESENTATION OF ISOSPIN

As discussed in Sec. IV it is formally true that

$$[I_3,\phi_\pm]=[B,\phi_\pm]=[I_3,\dot\phi_\pm]=[B,\dot\phi_\pm]=0,\quad [I_\pm,\phi_+]=[I_\pm,\dot\phi_+]=0. \tag{A1}$$

Here, $I_\pm$ are isospin-raising and -lowering operators. Thus, although one may say that ϕ_+ is an isoscalar, the isospin transformation of ϕ_- cannot be simple. Indeed, as discussed in Sec. V, the quark isospin arises as a topological property of ϕ_-, so there can be no *direct* connection between quark isospin and ϕ_- transformation properties. Yet, there is a curious sense in which the scalar system provides an eldritch (spatially) nonlocal representation of isospin; I will now sketch how this goes.

Using Eqs. (4.6), it is not hard to show that

$$\begin{aligned}&[J_\pm(\tau_\pm,x),\phi_-(y)]=\pm(2\pi)^{1/2}\theta(x-y)J_\pm(\tau_\pm,x),\\ &[J_+(\tau_\pm,x),\dot\phi_-(y)]=\mp(2\pi)^{1/2}\delta(x-y)J_+(\tau_\pm,x),\\ &[J_-(\tau_\pm,x),\dot\phi_-(y)]=\pm(2\pi)^{1/2}\delta(x-y)J_-(\tau_\pm,x),\end{aligned} \tag{A2}$$

where the $\pm$ in front of $(2\pi)^{1/2}$ always goes with $\tau_\pm$. Note that the first equation of (A2) is nonlocal. Such commutators are easy to find in these constructions, even in the original U(1) case. I do not think any noncausality is implied. Defining $I_\pm=\int_{-\infty}^{+\infty}dx\,J_0(\tau_\pm,x)$, we obtain the isospin transformation properties of ϕ_-:

$$\begin{aligned}&[I_\pm,\phi_-(x)]=\pm(2\pi)^{1/2}\int_x^\infty dx\,J_0(\tau_\pm,x),\\ &[I_\pm,\dot\phi_-(x)]=\mp(2\pi)^{1/2}J_1(\tau_\pm,x),\end{aligned} \tag{A3}$$

The right-hand sides of (A3) are of course wild (spatially) nonlocal functions of ϕ_-. I have checked the Jacobi identities among (A1) and (A3) and they are satisfied. ϕ_- appears to be a legitimate representation.

Let us see how such transformations can be invariances of the SGL equations (5.5), and (5.8),

$$\Box^2\phi_+=\frac{\alpha}{\beta_+}\sin(\beta_+\phi_+)\cos[(2\pi)^{1/2}\phi_-], \tag{A4a}$$

$$\Box^2\phi_-=\frac{\alpha(2\pi)^{1/2}}{\beta_+^{\,2}}\cos(\beta_+\phi_+)\sin[(2\pi)^{1/2}\phi_-]. \tag{A4b}$$

I will concentrate on the left-hand sides, and sketch the result for the right-hand sides. Define infinitesimal transformations $\delta_\pm\phi_\pm$, $\delta_\pm\dot\phi_\pm$ by the commutators (A1), (A3). Here $\delta_\pm$ means the change due to $I_\pm$. One shows directly from (A3) that $\delta_\pm\dot\phi_\pm=(d/dt)\delta_\pm\phi_\pm$, and thus

$$\delta_\pm\Box^2\phi_+=0, \tag{A5a}$$

$$\delta_\pm\Box^2\phi_-=\mp(2\pi)^{1/2}[\partial_0J_1(\tau_\pm)-\partial_xJ_0(\tau_\pm)]. \tag{A5b}$$

The right-hand side of (A5b) is proportional to the divergence of the charged axial-vector currents. *Using the equations of motion,* these vanish for the free theory and are proportional to $P(\tau_\pm)$ in the interacting theory. Thus, for the free theory, we already see that our transformation is an invariance. I find it intriguing that a single free massless Bose field can formally support an isospin in thus way. With interaction, we need also transform the right-hand sides of (A4). For brevity, I will ignore β_+, $(2\pi)^{1/2}$'s and normal-ordering in sketching the transformation for (A4b):

$$\cos\phi_+\delta_\pm \sin\phi_- \sim \cos\phi_+(x)\cos\phi_-(x)\int_x^\infty d\xi\, J_0(\tau_\pm,\xi)$$

$$\sim \int_{-\infty}^{+\infty} dy[\cos\phi_+(x)\sin\phi_-(x), J_0(\tau_\pm, y)]. \quad \text{(A6)}$$

The commutator is proportional [see Eqs. (4.9)] to $[P_3, J_0(\tau_\pm)]$, which is proportional to $P(\tau_\pm)$. This exactly parallels the change (A5b) in the left-hand side of (A4b), and again the transformation is an invariance. For the right-hand side of Eq. (A4a), a chain of identities parallel to (A6) leads to $[P, J_0(\tau_\pm)] \sim 0$, so this equation is also invariant.

Is this invariance an observable symmetry in, say, a perturbative approach to the Bose system? I do not think so; our manipulations, though formal, suggest why. Notice that $I_\pm$, though well defined with respect to the fermions, fail to annihilate the boson vacuum. In fact, $I_\pm$ are quite poorly defined with respect to that vacuum (infinite expectation value, for example.) From the point of view of the bosons then, the isospin is (something like) spontaneously broken; it is thus not observable in any ordinary sense until the soliton-fermions are obtained.

APPENDIX B: CONNECTIONS WITH DUAL MODELS

As I was going through these two-dimensional theories, and the relevant correspondences, I noticed that much of the field-theoretic work has a direct map onto past work in dual models. I will begin by discussing free-field connections, and return to interactions later. Consider a free SU(N) Fermi field in two dimensions. Introduce $u = t + x$, $v = t - x$; then $\psi_a^1(u), \psi_a^2(v)$ form two independent spaces. I will focus on $\psi_a^1(u)$ alone, remembering that the complete system is a doubling. The claim is that $\psi_a^1(u)$ is the Bardakci-Halpern[8] dual quark field[15] $\psi_a(\theta)$.

Towards this result, an appropriate map is the projective transformation

$$e^{i\theta} = \frac{u-i}{u+i}, \quad 0 < \theta < 2\pi, \quad -\infty < u < \infty. \quad \text{(B1)}$$

Useful identities are $d\theta/du = 2/(u^2+1)$, and

$$\delta(u-u') = 2\sin^2(\tfrac{1}{2}\theta)\delta(\theta-\theta'),$$

$$\frac{\partial}{\partial u}\delta(u-u') = [2\sin^2(\tfrac{1}{2}\theta)][2\sin^2(\tfrac{1}{2}\theta')]\partial_\theta\delta(\theta-\theta'). \quad \text{(B2)}$$

Then it is easy to see from

$$[\psi_a^1(u), \psi_b^{1\dagger}(u')]_+ = \delta_{ab}\delta(u-u'),$$

$$[\psi_a(\theta), \psi_b^\dagger(\theta')]_+ = 2\pi\delta_{ab}\delta(\theta-\theta') \quad \text{(B3)}$$

that

$$\frac{1}{\sqrt{\pi}}\sin(\tfrac{1}{2}\theta)\psi_a(\theta) = \psi_a^1(-u). \quad \text{(B4)}$$

Actually the $(-u)$ is only a convention to keep dual-model creation (and annihilation) operators in correspondence with field-theory creation (and annihilation) operators—with $(+u)$, they are anticorrelated. Let us see how this goes. In the dual model one expands [drop SU(N) labels, with immediate generalization]

$$\psi(\theta) = \sum_{h=1}^{\infty} (e^{i(n+1/2)\theta} b_{1/2+n} + e^{-i(n+1/2)\theta} d_{1/2+n}^\dagger), \quad \text{(B5)}$$

whereas

$$\psi^1(u) = \frac{1}{(2\pi)^{1/2}}\int_{-\infty}^0 dp[d^\dagger(p)e^{ip_0u} + b(p)e^{-ip_0u}]. \quad \text{(B6)}$$

We can calculate, say $b_{n+1/2}$ in terms of the field-theoretic operators,

$$\begin{aligned} b_{1/2+n} &= \frac{1}{2\pi}\int_0^{2\pi} d\theta\, e^{-i\theta(n+1/2)}\psi(\theta) \\ &= \frac{1}{2\sqrt{\pi}}\int_{-\infty}^{+\infty} du \frac{2}{1+u^2}(1+u^2)^{1/2}\left(\frac{u+i}{u-i}\right)^{n+1/2} \times \psi^1(-u) \\ &= \frac{1}{\pi\sqrt{2}}\int_{-\infty}^0 b(p)\int_{-\infty}^{+\infty}\frac{du}{u-i}\left(\frac{u+i}{u-i}\right)^n e^{ip_0u} \\ &= \sqrt{2}\, i\int_{-\infty}^0 dp\, b(p)e^{-p_0}L_n(2p_0). \end{aligned} \quad \text{(B7)}$$

In the last step, we have recognized the Laguerre polynomials $L_n(Z) = e^Z d^n/dZ^n(Z^n e^{-Z})$. A similar calculation gives

$$b_{n+1/2}^\dagger = -i\sqrt{2}\int_{-\infty}^0 dp\, b^\dagger(p)e^{-p_0}L_n(2p_0).$$

With the usual orthonormality properties of L_n, one easily verifies $[b_{n+1/2}, b_{m+1/2}^\dagger]_+ = \delta_{n,m}$ from $[b(p), b^\dagger(\kappa)]_+ = \delta(p-\kappa)$. Evidently, 2 conformal spin $\frac{1}{2}$ dual quarks correspond to one Lorentz spinor.

We can go much further. We turn for a moment to scalar fields. It is known that if $\Box^2\phi = 0$, then $\phi = f(u) + g(v)$, and, for example,

$$f(u) = \int_{-\infty}^0 \frac{dk}{(2\pi 2k_0)^{1/2}}[e^{-iuk_0}a(k) + e^{+iuk_0}a^\dagger(k)], \quad \text{(B8a)}$$

$$[\partial_u f(u), f(u')] = -\frac{i}{2}\delta(u-u'). \quad \text{(B8b)}$$

Comparing with the dual model (fifth dimension, no four-vector index) conformal scalar-vector system,[15]

$$[Q_5(\theta), \pi_5(\theta')] = -2\pi\delta(\theta-\theta'),\qquad \pi_5(\theta) = i\partial_\theta Q_5(\theta) \quad \text{(B9)}$$

we identify

$$Q_5(\theta) = -i2\sqrt{\pi} f(-u),\qquad \pi_5(\theta) = \frac{\sqrt{\pi}\,\partial_{-u} f(-u)}{\sin^2(\frac{1}{2}\theta)}. \quad \text{(B10)}$$

The troubles in normal ordering the exponential of a free massless scalar field are thus the same as those of the dual vertex. In dual models, it was known[15] how to construct a $\pi_5(\theta)$ from the quark field,

$$i\partial_\theta Q_5 = \pi_5(\theta) = :\psi^\dagger(\theta)\psi(\theta):. \quad \text{(B11)}$$

This dual identity is then the familiar field-theoretic statement that vector currents can be written as gradients of free massless scalar fields. Something that was *not* realized in dual models is the analog of Mandelstam's representation in the field theory. Using that fact that for free fields $\int_{-\infty}^x d\xi\, \dot\phi(\xi) = f(u) - g(v)$, we easily show the inverse of (B11), namely that

$$\psi(\theta) \sim :e^{iQ_5(\theta)}: \sim :e^{\sqrt{2}k_5 Q_5(\theta)}: \quad \text{(B12)}$$

when $k_5^2 = -\frac{1}{2}$. That is, the spinor field can be expressed in terms of the scalar when k_5 is chosen so that the exponential is a conformal spinor.

We can go still further. The field-theoretic identity[5,6] [back to SU(N), $\theta_\pm = \theta_{00} \pm \theta_{01}$]

$$\theta_+(u) = i\psi^{1\dagger}\overleftrightarrow{\partial}_u\psi^1 = :\frac{1}{2\tilde C_V}(J_+^\alpha J_+^\alpha) + \frac{1}{2C_B}J_+J_+: \quad \text{(B13)}$$

[$\tilde C_V = (n+1)/2\pi$, $C_B = n/\pi$] maps directly onto[15]

$$\mathcal{L}(\theta) = -\frac{i}{2}\psi^+\overleftrightarrow{\partial}_\theta\psi = :\frac{1}{n+1}J^\alpha(\theta)J^\alpha(\theta) + \frac{1}{2n}J(\theta)J(\theta):. \quad \text{(B14)}$$

Here the dual quantities need some comment for nonexperts. $\mathcal{L}(\theta)$ is the conformal density, i.e., the object from which we build the conformal algebra $L_m = (1/2\pi)\int_0^{2\pi} e^{-im\theta}\mathcal{L}(\theta)$. The currents $J(\theta), J^\alpha(\theta)$ are formed as

$$J^\alpha(\theta) = :\psi^\dagger\frac{\lambda^\alpha}{2}\psi:,\qquad J(\theta) = :\psi^\dagger\psi: \quad \text{(B15)}$$

and satisfy

$$[J^\alpha(\theta), J^\beta(\theta')] = 2\pi i f^{\alpha\beta\gamma}J^\alpha(\theta)\delta(\theta-\theta') - \pi i\delta^{\alpha\beta}\delta_\theta(\theta-\theta'),$$
$$[J(\theta), J(\theta')] = -2\pi i n\partial_\theta\delta(\theta-\theta'). \quad \text{(B16)}$$

Thus, comparing with Refs. 5 and 6, we identify

$$J_+^\alpha(-u) = \frac{2}{\pi}\sin^2(\tfrac{1}{2}\theta)J^\alpha(\theta),\qquad J_+(-u) = \frac{2}{\pi}\sin^2(\tfrac{1}{2}\theta)J(\theta). \quad \text{(B17)}$$

Identity (B14) appears explicitly in Ref. 15 for the case of SU(3). (Caution: Here I have normalized the currents with $\frac{1}{2}\lambda^\alpha$. In Ref. 15, they are λ^α.) It is clear then that the dual method of defining normal-ordered quartics $[(J)^2]$ is the same as the field-theoretic method.

Going on toward interaction, we note some more general correspondences. The fact that $\{\psi(\theta)$ is a conformal spinor$\} \rightarrow \{\psi(x,t)$ is a Lorentz spinor$\}$. The fact that $\{\pi(\theta)$ is a conformal vector$\} \rightarrow \{\bar\psi\gamma^\mu\psi$ is a Lorentz vector$\}$, and so on. The most important of these is the fact that under the correspondence

$$\theta_+(-u) = \frac{4}{\pi}\sin^4(\tfrac{1}{2}\theta)\mathcal{L}(\theta) \quad \text{(B18)}$$

then, the "conformal-Schwinger condition" [Eq. (11) of Ref. 6] is equivalent to requiring that $\mathcal{L}$ is indeed a conformal density,

$$(\mathcal{L}(\theta), \mathcal{L}(\theta')) = -2\pi i[\partial_\theta\mathcal{L}(\theta)\delta(\theta-\theta') + 2\mathcal{L}(\theta)\partial_\theta\delta(\theta-\theta')]. \quad \text{(B19)}$$

I believe these correspondences provide the opportunity for further flow between the two fields. E.g., any conformally invariant two-dimensional field theory [hence a $\theta_+(u)$] provides us with the conformal algebra of a dual model: Just construct $\mathcal{L}(\theta)$ via (B18). Conversely, any dual $\mathcal{L}(\theta)$ provides us with a θ_+. From the dual-model viewpoint, the Dashen-Frishman phenomenon corresponds to reversing the sign of the Schwinger term in $J^\alpha(\theta)$. This corresponds to taking a $b|0\rangle = 0$ vacuum for J, but a $b^+|0\rangle = 0$ vacuum for J^α. Hence the difficulty in constructing a proper ψ. Much work has been done on $\mathcal{L}$ in dual models, most of which has involved four-Lorentz indices which would be superimposed above the two-Lorentz indices of the field theories. One application for which this feature will not appear is *discretely broken* SU(N). Such an application has been discussed[15] in the dual models, and hence one may be able to do the same in a conformal field theory. Quantized couplings occur all the time in dual $\mathcal{L}$'s, and, in general, I believe this is the same as the

Dashen-Frishman phenomenon: Conformal invariance quantizes couplings.

In general, however, because there is a doubling of two dual operators to one "Thirring operator," or, more simply, a u and v, the field theories correspond more closely to *Virasoro-Shapiro models*. Indeed it takes only a moment to show that the Thirring model Green's functions [for $\bar{\psi}(1 \pm \gamma_5)\psi$ at $\beta^2 = 8\pi$] are the Virasoro-Shapiro N-point functions, with a fifth dimension to bring the ground state up to $k^\mu k_\mu = 0$, all evaluated at $k_\mu = 0$. The sum of these functions, $Z(J)$ for the massive Thirring model, is then the function $W(J)$ useful in the dual-model spontaneous-breakdown approach of Bardakci and Halpern.[16]

I finally note that the work of the present paper is equivalent to constructing an SU(N) out of dual fifth-, sixth- (etc.) dimension orbital operators, even though the orbital operators have no simple transformation properties under SU(N). In the dual model, of course, $N \leqslant 22$ for orbital models and $N \leqslant 6$ for models with spin.

---

*This research is supported by the National Science Foundation under Grant No. MPS 74-08175-A01.

[1]S. Coleman, Phys. Rev. D 11, 2088 (1975).

[2]S. Mandelstam, Phys. Rev. D 11, 3026 (1975).

[3]D. Gross and M. B. Halpern, Phys. Rev. 179, 1436 (1969); also Univ. of California, Berkeley, report, 1969 (unpublished); C. Garwin, Report No. UCRL-20697 (unpublished).

[4]Our notation is that of B. Klaiber, in *1967 Boulder Lectures in Physics, Vol. XA: Quantum Theory and Statistical Theory,* edited by W. E. Brittin *et al.* (Gordon and Breach, New York, 1968), p. 141.

[5]G. F. Dell'Antonio, Y. Frishman, and D. Zwanziger, Phys. Rev. D 6, 988 (1972).

[6]R. Dashen and Y. Frishman, Phys. Lett. 46B, 439 (1973).

[7]See also Appendix B.

[8]The formal procedure here is the same in four dimensions. However, in two dimensions, the Schwinger terms are finite and calculable. For such reasons, a two-dimensional interaction picture has greater reliability than a four-dimensional one.

[9]It is a curious fact that, with our methods, we calculate zero anomalous dimension for ψ_H: Using as dilation operator the standard form $D = \int dx (t\theta_{00} + x\theta_{01})$, and assuming no anomalous dimension in the IP, the result follows immediately because $\theta^I_{01} = 0$. I checked through Refs. 5 and 6, and discovered that their generators are slightly different. Translating back from their (u, v) language, these references use $D \mp M = \int dx\, x\theta_\mp (0, \mp x)$ and $H \mp P = \int dx\, \theta_\mp (0, \mp x)$, whereas "standard" forms are $D \mp M = \mp\int dx\, x\theta_\mp (0, x)$ and $H \mp P = \int dx \theta_\mp (0, x)$. There are evidently certain advantages in these unconventional forms. (Caution: These are *only* for conformal models, whereas the "standard" forms work more generally.) Notice that with these unconventional forms, P has an interacting part $H_I(t) = \frac{1}{2}\int dx [\theta^I_{00}(x) + \theta^I_{00}(-x)]$, $P_I(t) = \frac{1}{2}\int dx [\theta^I_{00}(x) - \theta^I_{00}(-x)]$. The path is clear to the intriguing development of a "conformal interaction picture" in both H_I, P_I.

[10]The "current" identities among (3.1) are extremely well known. See also Appendix B. According to J. Kogut and L. Susskind, Phys. Rev. D 11, 3594 (1975), the others were known as well.

[11]See the remarks about $\mu \to 0$ of Sec. III. I have also suppressed the subscripts D on all the free fields of Sec. III.

[12]Except of course (4.7). It is not hard to show by direct differentiation that the interacting currents have the expected divergences. As in (4.7), however, the charged and neutral calculations go different routes, the charged calculation being somewhat tricky. For example, one finds

$$\partial \cdot J_5(\tau_+) \sim N' \left[\int_{-\infty}^{x} d\xi \cos[(2\pi G)^{1/2}\phi_+(\xi)] \times \sin[(2\pi)^{1/2}\phi_-(\xi)] J_1(\tau_+, x) \right]$$
$$\sim \int_{-\infty}^{+\infty} d\xi [S(\xi), J_1(\tau_+, x)]$$
$$\sim P(\tau_+).$$

For $\partial \cdot J(\tau_+)$, $J_1 \to J_0$, and the vector currents are still conserved.

[13]A. Mueller and L. Trueman, Phys. Rev. D 4, 1635 (1971); D. Gross and A. Neveu, *ibid.* 10, 3235 (1974).

[14]P. K. Mitter and P. H. Weisz, Phys. Rev. D 8, 4410 (1973).

[15]K. Bardakci and M. B. Halpern, Phys. Rev. D 3, 2493 (1971).

[16]K. Bardakci, Nucl. Phys. B68, 331 (1974); B70, 397 (1974); K. Bardakci and M. B. Halpern, *ibid.* B73, 295 (1974); Phys. Rev. D 10, 4230 (1974).

Nuclear Physics B108 (1976) 119–129

BOSONIZATION OF THE SU(N) THIRRING MODELS*

T. BANKS, D. HORN and H. NEUBERGER
Department of Physics and Astronomy, Tel Aviv University, Tel Aviv, Israel

Received 2 December 1975

Bosonization is applied to the SU(N) Thirring models, and interesting relations between various two-dimensional field theories arise. In particular, we show that the SU(2) model is equivalent to a version of the Sine-Gordon equation plus a free massless field.

1. Introduction

The kinematical constraints on field theories in two space-time dimensions lead to many unusual effects which have no analogue in higher dimensions. Perhaps the strangest of these is the equivalence between large classes of Bose and Fermi field theories. This correspondence, which has been dubbed bosonization, provides a transparent unified method for solving two dimensional Fermi theories [1**, 2–4].

In this paper we will apply bosonization to the non-abelian SU(N) Thirring models [5,6,7+]. Although we will not solve these models, we will find several amusing relations between them and other two-dimensional theories. Our most striking result is for the case $N = 2$: we show that the SU(2) Thirring model is equivalent to the theory of a free massless scalar field and a Sine-Gordon field. The bare coupling constant of the Sine-Gordon theory is fixed at $\beta^2 = 8\pi$, the value which makes the model exactly renormalizable. According to the results of Coleman [2], this implies a correspondence between the SU(2) model and another fermion model: a free massless Fermi field and a massive abelian Thirring model.

This work is divided into three parts. In sect. 2 we set up a precise operator scheme which implements the bosonization of free massless Fermi fields. Our formulas are very similar to those proposed by Dell'Antonio et al. [1] and by Mandelstam [3]. In sect. 3 we apply our formalism to the SU(N) models in the interaction picture and give a heuristic derivation of the equivalent boson Lagrangians for these models. Sect. 4 is devoted to a proof of the equivalence for the SU(2) case

* Supported in part by the Israel Commission on Basic Research.
** We use the γ-matrix conventions of Klaiber which imply $\psi_{L,R} = \psi_{1,2}$.
+ This author has arrived at some of the results of the present paper, in particular the Lagrangian (26).

based on the method and results of Coleman [2]. Finally we discuss the rather unusual picture of renormalization that emerges from our results and point out areas in which further study is necessary.

2. Bosonization of massless free fermion fields

We are going to write an explicit expression for a massless free fermion field as a non-local function of a free massless pseudo-scalar field. As is well-known, the latter needs infrared regularization in two dimensions; we will supply this regularization by working in a spatial box. We have found that we cannot write the fermion in terms of the boson field alone. We will have to introduce two discrete Fermi degrees of freedom. This is in accord with the results of Schroer [1].

The left- and right-handed components of the massless pseudo-scalar field are defined in a box of length L with periodic boundary conditions in the following explicit fashion:

$$\varphi_{\mathrm{L,R}}(x \pm t) = \frac{1}{\sqrt{L}} \sum_{n \neq 0} [a_n e^{ik_n(x \pm t)} + \text{h.c.}] \frac{\theta(\mp n)}{\sqrt{2|k_n|}} \; ; \; k_n = 2\pi n/L \,, \quad n = \pm \text{ integer} \,. \tag{1}$$

This field φ does not have the zero momentum mode. The latter is treated separately in terms of the charge (q) and axial charge ($\widetilde{q}$) operators and their conjugate momenta (p and $\widetilde{p}$):

$$\hat{\varphi}_{\mathrm{L,R}}(x \pm t) = \frac{1}{2}\left\{\frac{1}{\sqrt{\pi}}(\widetilde{p} \pm p) + \sqrt{\pi}\,(q \mp \widetilde{q})\frac{x \pm t}{L}\right\},$$

$$[q,p] = [\widetilde{q},\widetilde{p}] = i\,, \qquad [q,\widetilde{q}] = [q,\widetilde{p}] = [\widetilde{q},p] = [\widetilde{p},p] = 0\,. \tag{2}$$

The operators p and $\widetilde{p}$ are assumed to be angle variables so that q and $\widetilde{q}$ have integer eigenvalues. We will designate henceforth the Hilbert space on which q and $\widetilde{q}$ operate by $H_{q\widetilde{q}}$. This is distinct from the Fock space H_{B} on which the field φ operates. We will also use the pseudo-scalar field

$$\Phi = \varphi_{\mathrm{L}} + \varphi_{\mathrm{R}} + \hat{\varphi}_{\mathrm{L}} + \hat{\varphi}_{\mathrm{R}}\,. \tag{3}$$

It obeys the massless Klein-Gordon equation and operates on the Hilbert space $H_{\mathrm{B}} \otimes H_{q\widetilde{q}}$.

We now construct the massless spinor field

$$\psi_{\mathrm{L,R}} = \frac{1}{\sqrt{L}} : e^{\mp 2i\sqrt{\pi}\varphi_{\mathrm{L,R}}(x \pm t)} : e^{\mp 2i\sqrt{\pi}\hat{\varphi}_{\mathrm{L,R}}(x \pm t)} \chi_{\mathrm{L,R}}\,. \tag{4}$$

The normal ordering is used on the fields $\varphi_{L,R}$. Here we introduced two additional fermionic degrees of freedom

$$\chi_\alpha = A_\alpha + A_\alpha^+ , \qquad \{A_\alpha, A_\beta^+\} = \delta_{\alpha,\beta}, \qquad \{A_\alpha, A_\beta\} = 0, \qquad \alpha, \beta = \mathrm{L,R} , \tag{5}$$

which operate on the Hilbert space H_χ spanned by the operation of A_α^+ and $A_L^+ A_R^+$ on the vacuum. Altogether ψ is defined on $H_B \otimes H_{q\tilde{q}} \otimes H_\chi$. It is a straightforward exercise to show that ψ satisfies the Dirac equation and the correct Fermi anticommutation relations. We construct explicitly the Wightman functions of these ψ fields:

$$\langle 0|\psi_1(x_1) \ldots \psi_1(x_n)\psi_2(x_{n+1}) \ldots \psi_2(x_{n+m})\psi_1^+(y_1) \ldots \psi_1^+(y_n)\psi_2^+(y_{n+1}) \ldots \psi_2^+(y_{n+m})|0\rangle$$

$$= \frac{(-1)^{m(n+1)} \prod\limits_{1\leqslant j<k\leqslant n} (L/\pi) \sin [(\pi/L)(u_j - u_k)] \, (L/\pi) \sin [(\pi/L)(U_j - U_k)]}{(2\pi i)^{n+m} \prod\limits_{j,k=1}^{n} \{(L/\pi) \sin [(\pi/L)(u_j - U_k)] - i\epsilon\}}$$

$$\times \frac{\prod\limits_{1\leqslant i<l\leqslant m} (L/\pi) \sin[(\pi/L)(v_i - v_l)] \, (L/\pi) \sin[(\pi/L)(V_i - V_l)]}{\prod\limits_{i,l=1}^{m} \{(L/\pi) \sin[(\pi/L)(v_i - V_l)] + i\epsilon\}} ,$$

$$u_k = x_k^0 + x_k^1 , \qquad U_k = y_k^0 + y_k^1 , \qquad k = 1, \ldots n ,$$

$$v_{k-n} = x_k^1 - x_k^0, \quad V_{k-n} = y_k^1 - y_k^0, \qquad k = n+1, \ldots n+m . \tag{6}$$

In the limit $L \to \infty$ these expressions converge to the correct forms for a free massless spinor field in two dimensions (see, e.g. Klaiber, ref. [1]).

By introducing fermion normal ordering *via* point splitting, one can prove that the fermion number current is given by

$$j^\mu = N_F(\bar{\psi}\gamma^\mu\psi) = -\frac{1}{\sqrt{\pi}}\epsilon^{\mu\nu}\partial_\nu\Phi . \tag{7}$$

The appropriate current field algebra is now easily obtained. With some appropriate modifications, one can view eq. (4) as a construction of the fermion field in terms of the currents in this theory, as advocated by Dell-Antonio et al. [1].

Using the bosonization formulas one is able to easily solve a large class of fermion models by writing everything in the interaction picture. We have checked the validity of this procedure in the following soluble cases:

(a) The Schwinger model in the Coulomb gauge [4];

(b) The derivative coupling model [8];

(c) The abelian SU(N) Thirring models [1,7].

The solutions of these models obtained by bosonization are completely consistent with known solutions.

In the following we will apply this method to field theoretical models that have not been solved yet. Throughout this paper we will use an interaction picture approach which is based on the free field construction described in this section. Our approach is therefore different from that of Mandelstam [3], although the formulas may look similar.

Halpern [7] has also used an interaction picture formalism to discuss bosonization of the SU(N) models. He works with T products and enforces covariance via the Dirac-Schwinger commutation relations for the energy density. Instead, we will use T* products and Mathews' theorem, a procedure which simplifies the derivations.

3. Bosonization of the SU(N) Thirring model in the interaction picture

We begin the bosonization of the SU(N) Thirring model by defining the N two-component spinors:

$$\psi^a_{\mathrm{L,R}}(x,t) = \frac{1}{\sqrt{L}} : \mathrm{e}^{\mp 2i\sqrt{\pi}\Phi^a_{\mathrm{L,R}}(x,t)} : \chi^a_{\mathrm{L,R}} , \quad a = 1, ..., N . \tag{8}$$

Here Bose normal ordering should be understood to act only on the H_{B} part of the field Φ. The U(N) currents are defined by

$$J^\mu = \sum_a \bar{\psi}^a \gamma^\mu \psi^a , \qquad J^{(i)\mu} = \sum_{a,b} \bar{\psi}^a \gamma^\mu \tfrac{1}{2} \lambda^{(i)}_{ab} \psi^b , \tag{9}$$

where the $N \times N$ matrices $\lambda^{(i)}$, $i = 1, ..., N$ form the regular representation of SU(N). Eq. (9) has to be understood as a point-split definition so that the currents are regular. In terms of the boson field they turn out to be

$$J^0 = \frac{1}{\sqrt{\pi}} \sum_{a=1}^{N} \partial_x \Phi^a , \qquad J^1 = -\frac{1}{\sqrt{\pi}} \sum_{a=1}^{N} \partial_t \Phi^a ,$$

$$J^{(i)0} = \sum_{a \neq b}^{N} \frac{\lambda^{(i)}_{ab}}{2L} \{\chi^a_1 \chi^b_1 : \mathrm{e}^{2i\sqrt{\pi}(\Phi^a_{\mathrm{L}} - \Phi^b_{\mathrm{L}})} : + \chi^a_2 \chi^b_2 : \mathrm{e}^{-2i\sqrt{\pi}(\Phi^a_{\mathrm{R}} - \Phi^b_{\mathrm{R}})} :\}$$

$$+ \frac{1}{\sqrt{\pi}} \sum_{a=1}^{N} \tfrac{1}{2} \lambda^{(i)}_{aa} \partial_x \Phi^a , \qquad i = 1, ..., N^2 - 1 ;$$

$$J^{(i)1} = \sum_{a \neq b}^{N} \frac{\lambda^{(i)}_{ab}}{2L} \{-\chi^a_1 \chi^b_1 : \mathrm{e}^{2i\sqrt{\pi}(\Phi^a_{\mathrm{L}} - \Phi^b_{L})} : + \chi^a_2 \chi^b_2 : \mathrm{e}^{-2i\sqrt{\pi}(\Phi^a_{\mathrm{R}} - \Phi^b_{\mathrm{R}})} :\}$$

$$- \frac{1}{\sqrt{\pi}} \sum_{a=1}^{N} \tfrac{1}{2} \lambda^{(i)}_{aa} \partial_t \Phi^a , \qquad i = 1, ..., N^2 - 1 . \tag{10}$$

Notice that the diagonal currents are simple functions of the boson fields. It is a straightforward exercise to check that these expressions satisfy the non-abelian current algebra as given by Dashen and Frishman [5].

In order to perform the bosonization of the model it is necessary to use a point split definition of Bose normal ordering for arbitrary solutions of the Klein-Gordon equation. In ref. [9] it was shown that this is the way to get the right quantum expression for the energy-momentum tensor in the Sugawara form. The prescription is:

$$J_1(x)\,J_2(y) \to \lim_{y\to x}\ \{\tfrac{1}{2}(J_1(x)\,J_2(y) + J_2(y)\,J_1(x)) - \text{V.E.V.}\}\,. \tag{11}$$

We now start from the Lagrangian

$$\mathcal{L} = i\bar{\psi}\partial\!\!\!/\psi - \tfrac{1}{2}g_B\, J^\mu J_\mu - \sum_i \tfrac{1}{2}g_V\, J^{\mu(i)}J^{(i)}_\mu\,, \tag{12}$$

and use the Gell-Mann-Low formula to express time ordered products of Heisenberg operators in terms of interaction-picture fields.

$$_{\rm H}\langle 0|\mathrm{T}\prod_{i=1}^{n} O_{\rm H}(x_i)\ |0\rangle_{\rm H} = \frac{_{\rm I}\langle 0|\mathrm{T}^*[\prod_{i=1}^{n} O_{\rm I}(x_i)\exp(i\int \mathrm{d}^2y\ \mathcal{L}_{\rm I}(\psi_{\rm I}(y)))]\,|0\rangle_{\rm I}}{_{\rm I}\langle 0|\mathrm{T}^*[\exp(i\int \mathrm{d}^2y\ \mathcal{L}_{\rm I}(\psi_{\rm I}(y)))]\,|0\rangle_{\rm I}}\,. \tag{13}$$

We must use a T* product in eq. (13) because of the Schwinger terms in the current algebra.

Using eqs. (10) and (11) we can rewrite (12) in boson language. The effective boson interaction Lagrangian is

$$\mathcal{L}_{\rm I} = \frac{g_B N}{2\pi}\ :\partial_\mu\Phi'^1\partial^\mu\Phi'^1: + \frac{g_V}{4\pi}\sum_{a\geq 2}\ :\partial_\mu\Phi'^a\partial^\mu\Phi'^a:$$

$$-\frac{g_V}{L^2}\sum_{a\neq b}^{N} \chi_1^a\chi_2^a\,\chi_1^b\chi_2^b\ :\cos[2\sqrt{\pi}\sum_{c\geq 2}(C^{ca}-C^{cb})\,\Phi'^c]: \tag{14}$$

C is an orthogonal real $N\times N$ matrix that satisfies

$$C^{1a} = \frac{1}{\sqrt{N}}\qquad \text{for } a = 1,\dots,N;\quad \sum_{bd} C^{ab}C^{cd} = N\delta^{a1}\delta^{c1};\quad \sum_{b=1}^{N} C^{ab}C^{cb} = \delta^{ac}\,, \tag{15}$$

and the Φ' are related to the original fields Φ by

$$\Phi'^a = \sum_b C^{ab}\Phi^b\,. \tag{16}$$

We will now assume that eq. (13) can be implemented in the boson language by the use of Mathews' theorem [10], i.e. we use the standard recipe of regarding T* as a T product which commutes with time derivatives. The fermion Lagrangian, on the other hand, contains no derivative interactions. Formal evaluation of a T product of Fermi interaction Lagrangians *via* Wick's theorem leads to a covariant but divergent expression. When we define this expression by means of a covariant regularization and subtraction procedure we effectively convert it into a T* product. The crucial assumption of this section is that these two definitions coincide. This conjecture is actually justified in all of the soluble models mentioned above, and we will assume that it remains valid here. We are then led to a theory which can be represented by the Lagrangian

$$\mathcal{L} = \frac{\pi + g_B N}{2\pi} \partial_\mu \Phi'^1 \partial^\mu \Phi'^1 + \frac{\pi + \frac{1}{2} g_V}{2\pi} \sum_{a=2}^{N} \partial_\mu \Phi'^a \partial^\mu \Phi'^a$$
$$- \frac{g_V}{L^2} \sum_{a \neq b} \chi_1^a \chi_2^a \chi_1^b \chi_2^b : \cos[2\sqrt{\pi} \sum_{c=2}^{N} (C^{ca} - C^{cb}) \Phi'^c] : . \tag{17}$$

Let us perform a finite wave function renormalization on the fields Φ':

$$\theta^1 = \sqrt{1 + \frac{g_B N}{2\pi}}\, \Phi'^1 , \qquad \theta^a = \sqrt{1 + \frac{g_V}{2\pi}}\, \Phi'^a , \qquad a = 2, \ldots, N . \tag{18}$$

The Lagrangian (17) then becomes

$$\mathcal{L} = \tfrac{1}{2} \sum_{a=1}^{N} \partial_\mu \theta^a \partial^\mu \theta^a - \frac{g_V}{L^2} f(L^2 \Lambda^2) \sum_{a \neq b} \chi_1^a \chi_2^a \chi_1^b \chi_2^b : \cos[2\sqrt{\pi} \left(1 + \frac{g_V}{2\pi}\right)^{-1/2}$$
$$\times \sum_{c=2}^{N} (C^{ca} - C^{cb}) \theta^c] :; \qquad f(L^2 \Lambda^2) = \left(\frac{L^2 \Lambda^2}{4\pi^2}\right)^{g_V/(2\pi + g_V)} , \tag{19}$$

where Λ^2 is an ultra-violet cutoff. The cutoff-dependent factor in front of the cosine comes from re-normal ordering the cosine in terms of the fields θ.

One now sees immediately how to solve the problem for $g_V = 0$. The ground state of the Heisenberg Hamiltonian is the Fock vacuum of the fields θ. We can get a Fermi field which has finite Green functions in this ground state by re-normal ordering our field (8) with respect to the θ's. The finite wave function renormalization between Φ' and θ induces an infinite wave function renormalization of ψ and we pick up the correct anomalous dimension for the Fermi field.

$$\psi_r^{(a)} = \left(\frac{L\Lambda}{2\pi}\right)^{N g_B^2 / 4\pi(\pi + N g_B)} \psi^{(a)} . \tag{20}$$

The full $g_V \neq 0$ Lagrangian (19) becomes particularly simple for the case of SU(2).

$$\mathcal{L} = \tfrac{1}{2} \partial_\mu \theta^1 \partial^\mu \theta^1 + \tfrac{1}{2} \partial_\mu \theta^2 \partial^\mu \theta^2 - \frac{2 g_V}{L^2} f(L^2 \Lambda^2)\, \chi_1^1 \chi_2^1 \chi_1^2 \chi_2^2 : \cos\left[\sqrt{8\pi} \left(1 + \frac{g_V}{2\pi}\right)^{-1/2} \theta^2\right] : \tag{21}$$

Thus, the SU(2) Thirring model is equivalent to a massless free Bose field plus a Sine-Gordon field. The correspondence between couplings is (we use Coleman's notation) [2]

$$\frac{\alpha}{\beta^2} \leftrightarrow \frac{2g_V}{L^2} f(L^2\Lambda^2)\,, \qquad \beta \leftrightarrow \left(1+\frac{g_V}{2\pi}\right)^{-\frac{1}{2}} \sqrt{8\pi}\,. \tag{22}$$

Let us summarise this section. By using an interaction picture formalism based on eq. (8) we wrote the interaction Lagrangian in terms of the scalar fields. It separated automatically into two parts which can be simply expressed in terms of ϕ'^a which are linear combinations of ϕ^a. This field theory of bosons can be represented by the Lagrangian (17). Performing a finite wave function renormalization on ϕ', one is led to a new Lagrangian which is equivalent (in the SU(2) case) to one free Bose field plus one Sine-Gordon field with a fixed relation between the two Sine-Gordon couplings α and β. This is a rather surprising result and the skeptical reader may be suspicious of our cavalier treatment of T* products. We therefore check the result using a different method in sect. 4.

4. Proof of equivalence between the Bose and Fermi Lagrangians for the case of SU(2)

In this section we will use Coleman's results to prove the correspondences suggested in sect. 3. We will therefore give up our box normalization in favour of Coleman's infrared regularization. We will also set $g_B = \frac{1}{4}g_V$. This simplifies many formulas and implies no essential loss of generality because we already know that bosonization works for the Abelian part of the model.

After applying a Fierz transformation, we can write the fermion Lagrangian as

$$\begin{aligned}\mathcal{L} &= i\bar{\psi}\partial\!\!\!/\psi - \tfrac{1}{4}g_V\,(\bar{\psi}^1\gamma^\mu\psi^1)^2 - \tfrac{1}{4}g_V(\bar{\psi}^2\gamma^\mu\psi^2)^2 \\ &\quad + g_V\,[(\psi_1^{1+}\psi_2^1)(\psi_2^{2+}\psi_1^2) + (\psi_1^{2+}\psi_2^2)(\psi_2^{1+}\psi_1^1)]\,.\end{aligned} \tag{23}$$

Defining

$$\mathcal{L}^0(\psi) = \sum_{a=1}^{2} [i\bar{\psi}^a\partial\!\!\!/\psi^a - \tfrac{1}{4}g_V(\bar{\psi}^a\gamma^\mu\psi^a)^2]\,,$$

$$\sigma_\pm^{1,2}(x) \propto \lim_{(x-y)^2\to 0} (x-y)^{2\delta}\,\psi_{1,2}^{+1,2}(x)\,\psi_{2,1}^{1,2}(y)\,; \qquad \delta = \frac{g_V}{8\pi}\,\frac{4\pi+g_V}{2\pi+g_V}\,, \tag{24}$$

we get (following Coleman [2])

$$\langle 0|\mathrm{T}\prod_{i=1}^{I}\sigma_+^1(\xi_i)\prod_{k=1}^{K}\sigma_-^1(\eta_k)\prod_{j=1}^{J}\sigma_+^2(\tau_j)\prod_{l=1}^{L}\sigma_-^2(\lambda_l)\prod_{\alpha=1}^{n}\sigma_+^1(x_\alpha)\,\sigma_-^2(x_\alpha)$$

$$\times \prod_{\beta=1}^{m} \sigma_+^2(y_\beta)\, \sigma_-^1(y_\beta)|0\rangle \propto (\tfrac{1}{2})^{4(I+J+n+m)} \left\{ \prod_{i>j}^{I} (\xi_i - \xi_j)^2 \right. \tag{25}$$

$$\times \prod_{\alpha>\beta}^{n} (x_\alpha - x_\beta)^2 \prod_{i=1}^{I} \prod_{\alpha=1}^{n} (x_\alpha - \xi_i)^2 \prod_{i>j}^{K} (\eta_i - \eta_j)^2 \prod_{\alpha>\beta}^{m} (y_\alpha - y_\beta)^2$$

$$\times \prod_{i=1}^{K} \prod_{\beta=1}^{m} (\eta_i - y_\beta)^2 \Big[\prod_{i=1}^{I} \prod_{k=1}^{K} \prod_{\alpha=1}^{m} \prod_{\beta=1}^{n} (\xi_i - \eta_k)^2 (\xi_i - y_\alpha)^2 (x_\beta - \eta_k)^2$$

$$\times (x_\beta - y_\alpha)^2 \Big]^{-1} \times \text{[same expression with } I \to J; K \to L; \xi \to \tau; \eta \to \lambda;$$

$$\left. \times\, n \leftrightarrow m; x \leftrightarrow y] \right\}^{1/(1+g_V/2\pi)} .$$

if $I + n = K + m$ and $J + m = L + n$ and zero otherwise.

We write the boson Lagrangian as [7]

$$\mathcal{L} = \tfrac{1}{2}\partial_\mu\theta^1\partial^\mu\theta^1 + \tfrac{1}{2}\partial_\mu\theta^2\partial^\mu\theta^2 - \tfrac{1}{2}\mu^2[(\theta^1)^2 + (\theta^2)^2]$$

$$- g(x)\, G m^2 N_m \cos[\gamma\sqrt{8\pi}\,\theta^2] , \tag{26}$$

$$\gamma = \left(1 + \frac{g_V}{2\pi}\right)^{-\frac{1}{2}} .$$

The function g has compact support in space-time and is supposed to be taken to 1 everywhere after the perturbation series is summed. μ^2 is a regulator mass which allows us to make calculations with massless scalar fields in two dimensions. It should be taken to zero order by order in the perturbation theory. Note that the factors χ_i which appear in eq. (21) are absent from eq. (26). The reason is that they do not play any role in the matrix elements of eq. (25) since they will always appear squared and therefore lead to no observable factor in the boson sector of the theory. Therefore, although we will be able to show the equivalence using Coleman's method, we will not get the explicit representation of the spinor from which we started.

Now, consider the following expression:

$$\lim_{\mu^2 \to 0} \langle 0,\mu| \mathrm{T} \prod_{i=1}^{I} N_m \mathrm{e}^{-i\gamma\sqrt{2\pi}(\theta^1+\theta^2)(\xi_i)} \prod_{k=1}^{K} \mathrm{e}^{i\gamma\sqrt{2\pi}(\theta^1+\theta^2)(\eta_k)}$$

$$\times \prod_{j=1}^{J} N_m \mathrm{e}^{-i\gamma\sqrt{2\pi}(\theta^1-\theta^2)(\tau_j)} \prod_{l=1}^{L} N_m\, \mathrm{e}^{i\gamma\sqrt{2\pi}(\theta^1-\theta^2)(\lambda_l)} \tag{27}$$

$$\times \prod_{\alpha=1}^{n} N_m \mathrm{e}^{-i\gamma\sqrt{2\pi}\theta^2(x_\alpha)} \prod_{\beta=1}^{m} N_m \mathrm{e}^{i\gamma\sqrt{8\pi}\theta^2(y_\beta)} |0, \mu\rangle .$$

The dependence on μ^2 is

$$(\mu^2)^{\frac{1}{4}[(-L+K+J-I+2m-2n)^2+(L+K-I-J)^2]}\ , \tag{28}$$

so a contribution is non-zero as $\mu^2 \to 0$ if $J + m = L + n$ and $K + m = I + n$. When these conditions are satisfied it is easy to see (using Coleman's formulae again) that the result is proportional to eq. (25).

So, we are led to deduce the correspondences

$$\sigma^1_\pm \propto mN_m\, e^{\mp i\gamma\sqrt{2\pi}(\theta^1+\theta^2)}\ ,$$
$$\sigma^2_\pm \propto mN_m\, e^{\mp i\gamma\sqrt{2\pi}(\theta^1-\theta^2)}\ . \tag{29}$$

These are exactly the consequences that one would anticipate on the basis of the discussion in sect. 3. Thus we have proven the equivalence conjectured in the previous section, at least for a class of matrix elements. We have not checked the equivalence of other matrix elements of the two theories but we are confident that they are, in fact, equal.

5. Discussion

Let us discuss the renormalization procedure for the Lagrangian (21) of section 3 keeping the box length L fixed. We therefore ask what has to be done to the coupling constant g_V and the field normalization in order to make all Green functions Λ independent as $\Lambda \to \infty$. For $\Lambda < \infty$ the terms in the Lagrangian (and all of their powers) are finite operators on Fock space (we always work in a finite volume). The coefficient of the cosine term blows up as $\Lambda \to \infty$ so we must obviously let $g_V \to 0$ to obtain anything finite in this limit. The crucial point now is that as we let $g_V \to 0$ the coefficient inside the cosine goes to $\sqrt{8\pi}$. This is the very edge of the positivity domain which Coleman [2] has obtained for the Sine-Gordon equation. Moreover, for our purpose it is important to note that $\beta = \sqrt{8\pi}$ is exactly the value for which $\cos \beta\phi$ becomes an operator of scale dimension 2. Thus, the cosine interaction, which for $\beta^2 < 8\pi$ was superrenormalizable, becomes exactly renormalizable in this limit. It is easy to verify that new divergences appear in the Sine-Gordon perturbation series for this value of β^2.

The upshot of this is that if we take $g_V \to 0$ in such a way that

$$g_V(\Lambda) \left(\frac{L^2\Lambda^2}{4\pi^2}\right)^{g_V(\Lambda)/(2\pi+g_V(\Lambda))} \underset{\Lambda\to\infty}{\to} g_V^0 = \text{finite}\ , \tag{29}$$

then the higher powers of the resulting Hamiltonian will not be well defined. In more mathematical terms, the Hamiltonian will not converge to an essentially self-adjoint operator on Fock space. We know that if we take g_V to zero fast enough as $\Lambda \to \infty$, then we get a well-defined operator. For example, if we simply set $g_V = 0$ for some finite Λ, then we get back the abelian Thirring model. Since we have shown that the massive Thirring model is equivalent to the SU(2) model (which we know can be renormalized), it is now reasonable to conjecture that there is a way to choose the bare couplings

$$g_V \to 0\,, \qquad \beta = \sqrt{8\pi}\left(1 + \frac{g_V}{2\pi}\right)^{-1/2}, \tag{30}$$

in the Sine-Gordon equation and obtain finite Green functions for the theory which are essentially different from the $g_V(\Lambda) \equiv 0$ case.

In other words, the limit (27) leads to an equivalence between the SU(2) model and the Sine-Gordon model with parameters

$$\frac{\alpha}{\beta} = \frac{2g_V^0}{L^2}\,, \qquad \beta = \sqrt{8\pi}\,. \tag{31}$$

For finite g_V^0 this theory is not yet renormalized. However, our knowledge that the SU(2) model is renormalizable by coupling constant and Fermi wave function renormalization [6]* leads us to expect that this Sine-Gordon field theory is also renormalizable. We may expect that an additional finite renormalization of the boson field has to be carried out. This is because the anomalous dimension of the Fermi field in the non-abelian SU(2) model is not the same as that in the abelian model [5,11]. (The Bose renormalization must be finite because the curl of the Bose field is an SU(2) current which we know is not renormalized.) Thus, we expect that a finite version of the Sine-Gordon equation can be constructed even when $\beta^2 = 8\pi$. For larger values of β Coleman has shown [2] that the model does not exist.

The conjecture presented above is a bit disquieting when we think of it in terms of renormalization group trajectories. The standard way of computing renormalization constants by looking at the divergent parts of integrals leads us to believe that for $\beta^2 < 8\pi$ the trajectories for the Sine-Gordon equation are all β = constant, i.e. straight lines in the α, β plane. We seem to have found a new trajectory which moves in to the point $(\alpha = 0, \beta^2 = 8\pi)$ like a square root. We believe that the apparent contradiction lies in the fact that the relation between the bare and renormalized β is a completely finite one and is missed by the usual method of only taking infinite renormalizations into account. This point deserves further study.

We would like to emphasize that our analysis does not reveal any hint of the second scale invariant solution of the SU(N) model found by Dashen and Frishman.

* These authors discuss a slightly more general SU(N) model than ours. One can use their results to show that the U(1) chiral symmetry may be consistently implemented. This implies that the only renormalizations needed in our model are the wave function and coupling constant renormalizations.

The existence of the $g_V = 4\pi/(N+1)$ Dashen Frishman solution is a consequence of anomalies which make the SU(N) axial currents conserved for this value of the coupling. In previous applications of bosonization, fermion anomalies always appeared as canonical results in the boson language, and one might naively expect the same to happen for the SU(N) models. The difference, of course, is that in all previous models the non-quadratic part of the equivalent boson Lagrangian was super-renormalizable, while for the SU(N) models it is renormalizable. Thus the boson interaction will have anomalies of its own and we will not see anomalies arise as canonical results.

Nonetheless, there is a way that we should see the Dashen-Frishman solution in our formalism. Since the diagonal SU(N) axial currents are conserved in the Dashen-Frishman model, our Bose fields (from which we get the axial currents by taking derivatives) should be free. The only way we can see to make this happen is to take the bare coupling constant g_V to zero so fast that the interaction vanishes even in the infinite cutoff limit. In this case the Fermi fields (as defined by us) would also be free. The Dashen-Frishman Fermi fields for $g_V = 4\pi/(N+1)$ can also be written in terms of free boson fields, but we see no limit in which our fields can be identified with theirs. It may be that the Dashen-Frishman model can be obtained by other limiting procedures and not by a quasi-perturbative scheme such as our own.

We thank Professors L. Susskind and Y. Aharonov for many stimulating discussions.

References

[1] R. Klaiber, Lectures in theoretical physics ed. A. Barut and W. Britten (Gordon and Breach, 1968);
G.F. Dell'Antonio. Y. Frishman and D. Zwanziger, Phys. Rev. D6 (1972) 988;
B. Schroer, Freie Universität Berlin preprint (1975).
[2] S. Coleman, Phys. Rev. D11 (1975) 2088.
[3] S. Mandelstam, Phys. Rev. D11 (1975) 3026.
[4] J. Kogut and L. Susskind, Phys. Rev. D11 (1975) 3594.
[5] R. Dashen and Y. Frishman, Phys. Rev. D11 (1975) 2781.
[6] P.K. Mitter and P.H. Weisz, Phys. Rev. D8 (1973) 4410.
[7] M. Halpern, Berkeley preprint (1975).
[8] K.D. Rothe and I.O. Stamatescu, Heidelberg preprint (1975).
[9] S. Coleman, R.W. Jackiw and D.J. Gross, Phys. Rev. 180 (1969) 1359.
[10] P.T. Mathews, Phys. Rev. 76 (1949) 684L (erratum: 76 (1949) 1489 (L)).
[11] H. Romer, S. Yankielowicz and H.R. Rubinstein, Nucl. Phys. B79 (1974) 285.

J. Phys. C: Solid State Phys., **14** (1981) 2585–2609. Printed in Great Britain

'Luttinger liquid theory' of one-dimensional quantum fluids: I. Properties of the Luttinger model and their extension to the general 1D interacting spinless Fermi gas

F D M Haldane†
Institut Laue–Langevin, 156X, 38042 Grenoble Cedex, France

Received 6 October 1980, in final form 26 January 1981

Abstract. The explicitly soluble Luttinger model is used as a basis for the description of the general interacting Fermi gas in one dimension, which will be called 'Luttinger liquid theory', by analogy with Fermi liquid theory. The excitation spectrum of the Luttinger model is described by density-wave, charge and current excitations; its spectral properties determine a characteristic parameter that controls the correlation function exponents. These relations are shown to survive in non-soluble generalisations of the model with a non-linear fermion dispersion. It is proposed that this low-energy structure is universal to a wide class of 1D systems with conducting or fluid properties, including spin chains.

1. Introduction

This paper is the first in a series that will present a general description of the low-energy properties of a wide class of one-dimensional quantum many-body systems, which I will call 'Luttinger liquids'. The work to be described was originally motivated by the search for a replacement for Fermi liquid theory in one dimension, where it fails because of the infrared divergence of certain vertices it assumes to remain finite; these divergences make an approach based on conventional fermion many-body perturbation theory useless. However, there is a certain model of an interacting one-dimensional spinless fermion system, the Luttinger model (Luttinger 1963), which has been explicitly solved (Mattis and Lieb 1965). This solution, by a Bogoliubov transformation, in effect resums all the divergences encountered in perturbation theory. The excitation spectrum of the diagonalised model is described in terms of *non-interacting* boson collective modes.

The feature of the Luttinger model that allows its solution is its exactly linear fermion dispersion. What will be demonstrated in this paper is that correction terms representing non-linearity of the fermion dispersion can be added to the model, and give rise to non-linear boson couplings between the collective modes. A *boson* many-body perturbation expansion in these terms is shown to be completely regular, so the Bogoliubov transformation technique that solves the Luttinger model is shown to provide a general method for resumming *all* the infrared divergences present, at least in the spinless

† Address from September 1981: Department of Physics, University of Southern California, Los Angeles, CA90007, U.S.A.

0022-3719/81/192585 + 25 $01.50

Fermi gas. The name 'Luttinger liquid' has been chosen to reflect the idea that such systems have a low-energy excitation spectrum similar to the Luttinger model spectrum, but with *interactions* between the elementary excitations. This resembles the relation between the Fermi liquid theory and the soluble model on which it is based, the free Fermi gas.

This paper is perhaps the most technical of the planned series. It sets up the essential machinery for working with the Luttinger model and its generalisations, and uses it to discuss the effects of a non-linear fermion dispersion. The previous treatments of the model in the literature are often ambiguous on certain points, and there has been a certain amount of confusion, particularly associated with the role of cut-offs. I have therefore aimed to present a completely self-contained and precise treatment of the original Luttinger model, in particular emphasing the key role played by charge and current excitations (as opposed to collective density wave modes) which have in general been neglected in previous treatments. It was attention to these details that allowed the identification of a key part of the underlying structure of the solution that proved to remain valid in the 'Luttinger liquid' generalisation, with applications to be described in future papers.

The characteristic properties of a 'Luttinger liquid' that have emerged are: (i) a conserved charge; (ii) a characteristic 'Kohn anomaly' wavevector '$2k_F$', varying linearly with charge density; (iii) persistent currents at low temperatures, quantised in units that carry momentum $2k_F$; (iv) a spectrum of collective density wave elementary excitations, with a dispersion linear in $|q|$ at long wavelengths that defines a sound velocity v_S; (v) two additional velocities, v_N and v_J, associated with charge and current excitations, obeying $v_S = (v_N v_J)^{1/2}$; (vi) power-law decay of correlation functions at $T = 0$, with coupling-strength-dependent exponents that depend only on $\exp(-2\varphi)$, where $v_N = v_S \exp(-2\varphi)$ and $v_J = v_S \exp(2\varphi)$. It should be emphasised that this means that $\exp(-2\varphi)$ is a measure of the essential renormalised coupling constant, and can thus be obtained from knowledge of v_S and the change of ground state energy with charge, which gives v_N.

1D systems with this Luttinger liquid structure so far identified include: (*a*) interacting spinless fermions; (*b*) interacting spin-$\frac{1}{2}$ fermions (and those with higher internal symmetries; (*c*) the Bose fluid (including systems with internal symmetries); (*d*) the finite-density gas of solitons of the Sine–Gordon theory; (*e*) uniaxially anisotropic spin systems (the 'charge' here is azimuthal spin)—antiferromagnets (only in the case of finite azimuthal magnetisation in the easy-axis case) and ferromagnets (easy-plane only). For many of these classes there exist models exactly soluble by the Bethe *ansatz* (Bethe 1930), and the Luttinger liquid structure can then be explicitly tested and verified (Haldane 1981). Subsequent papers will present such 'case studies' (see also Haldane 1980).

The Luttinger liquid has a characteristic instability if a multiple of its fundamental momentum $2k_F$ is equal to a reciprocal lattice vector reflecting an underlying periodicity. For large enough values of the parameter $\exp(-2\varphi)$, a gap opens in the spectrum, and the system becomes insulating. This instability can be studied in detail using the precise operator machinery set up in this paper, and will be the subject of paper II in this series. A universal description of the behaviour of the strongly renormalised Luttinger liquid near this instability emerges. The precise agreement between the predictions of this description and the features found in many of the 'test case' models solvable by the Bethe *ansatz* will provide strong evidence for the universality of the 'Luttinger liquid' description.

The organisation of this paper is as follows. To avoid confusion, it deals only with the *spinless* form of the model. The necessary generalisation to spin-$\frac{1}{2}$ fermions, and from these to the Bose fluid and spin systems, will be dealt with in subsequent articles. In § 2 there is a brief introduction to specifically one-dimensional features of the Fermi gas. Section 3 contains the bulk of the technical development, and describes the structure of the *non-interacting* Luttinger model. The machinery set up in § 3 leads quickly to the solution of the *interacting*-fermion Luttinger model in § 4. Section 5 uses the machinery to discuss the effects of a non-linear fermion dispersion. Finally, § 6 summarises the results, and formulates the hypothesis that the 'Luttinger liquid' structure is universal to conducting spinless fermion systems in 1D.

2. The one-dimensional Fermi gas

Fermion systems in one dimension have features quite distinct from those in higher dimensions. This is because the one-dimensional Fermi surface consists of two discrete points, while in higher dimensions it is continuous. The special spectral structure resulting from this can be seen by examining the full spectrum of excited states above a ground state with Fermi wavevector k_F. Figures 1(*a*) and (*b*) show the single-particle dispersion (and ground state occupancy) and the particle–hole pair spectrum of the (spinless) 1D Fermi gas with periodic boundary conditions on a length L. The distinctive one-dimensional feature of the pair spectrum is the non-existence of low-energy pairs for $0 < |k| < 2k_F$; in higher dimensions this region of 'missing' states is filled in. The full spectrum of excited states with zero excited charge (with respect to a ground state with odd charge $N_0 = k_F L/\pi$) is obtained by using figure 1(*b*) to determine the allowed energies of multiple-pair states, and is shown in figure 1(*c*).

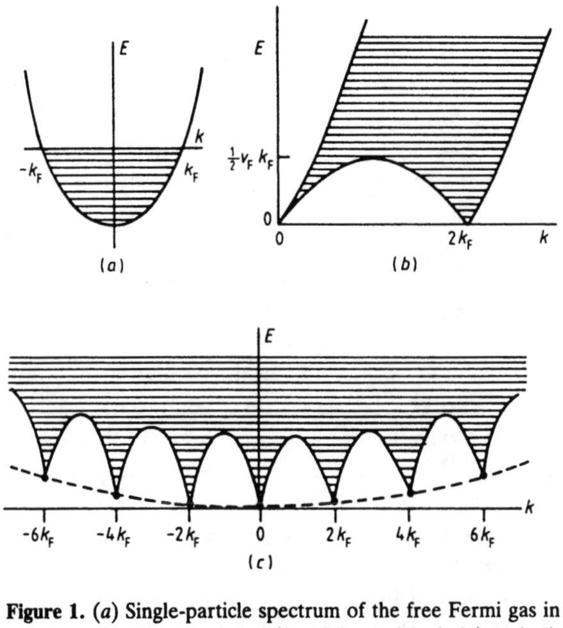

Figure 1. (*a*) Single-particle spectrum of the free Fermi gas in 1D; (*b*) Particle–hole pair spectrum; (*c*) full zero-charge (multiple particle–hole) excitation spectrum (energy differences $E(n) = 2\pi v_F n^2/L$ of extremal states at $k = 2nk_F$ greatly exaggerated).

At low energies $E \ll v_F k_F$, where v_F is the Fermi velocity $d\varepsilon(k_F)/dk$, the spectrum splits up into separate sectors that can be labelled by an even integer J, and can be described as excitations about a set of extremal states with momentum $k_F J$. These states have an energy $\frac{1}{2}(\pi/L)v_F J^2$, and this quadratic energy dependence (valid for energies $\ll v_F k_F$) is shown on a greatly exaggerated scale in figure 1(c). The spectrum of excitations with non-zero but even values of extra charge $(N - N_0)$ is similar, but with k_F replaced by $(k_F + \pi(N - N_0)/L)$, and the addition of a term $\frac{1}{2}(\pi/L)v_F(N - N_0)^2$ to the energy. The spectrum of excitations with odd values of $(N - N_0)$ differs only in that extremal states correspond to odd values of J. The form of the excitation spectrum suggests that at low energies it can be described by linear boson ('sound wave') excitations about the extremal states labelled by integers N and J. Such a classification breaks down at higher energies not only because of the non-linearity of the electron dispersion, but also because there is no longer any unambiguous operational way of assigning the quantum number J.

These observations can be summarised by the hypothesis that the *low-energy* spectrum can be represented by the form (where $b_q^\dagger$ are boson creation operators)

$$H = v_S \sum_q |q|\, b_q^\dagger b_q + \tfrac{1}{2}(\pi/L)[v_N (N - N_0)^2 + v_J J^2] \tag{2.1}$$

$$P = [k_F + (\pi/L)(N - N_0)]J + \sum_q q b_q^\dagger b_q; \qquad k_F = \pi N_0/L, \tag{2.2}$$

where $qL/2\pi = \pm 1, \pm 2, \ldots$; N and J are integers, subject to the selection rule (for periodic boundary conditions)

$$(-1)^J = -(-1)^N. \tag{2.3}$$

The parameters are identified as

$$v_S = v_N = v_J = v_F, \tag{2.4}$$

where v_F is the Fermi velocity. Such a spectral form is obviously compatible with the above discussion, but only has the status of a plausible hypothesis until it has been verified that it gives the correct multiplicity of states. This can in fact be verified, as is shown in the next section, by examination of the *non-interacting Luttinger model*, for which the spectral form (2.1)–(2.3) *holds exactly at all energies.*

Though the three parameters v_S, v_N and v_J are all equal to the Fermi velocity in the case of the non-interacting Fermi gas, they describe quite distinct properties of the spectrum. It is thus natural to wonder whether in fact *interacting* gapless fermion systems also have a low-energy spectrum described by (2.1)–(2.3), but with renormalised and unequal values of the three velocity parameters v_S, v_N and v_J. This can in fact be confirmed by the study of the *interacting Luttinger model* (described in § 4, which is explicitly soluble. Though it has the feature that v_S, v_N and v_J are no longer equal, they are not independent, and their ratios are determined by a parameter that characterises the essential interaction strength and low-energy physical properties such as the asymptotic forms of the various correlation functions. I will argue that these relations, together with the spectral form (2.1)–(2.3) are universally valid for the description of the low-energy properties of gapless interacting one-dimensional spinless fermion systems. The assignment of different values to the parameters v_S, v_N and v_J in this 'Luttinger liquid theory' will be analogous to the assignment of different effective masses to the quasi-particles for the characterisation of different physical properties in Fermi liquid theory.

3. The Luttinger model and its solution: I. The non-interacting limit

3.1. Historical development

The Luttinger (1963) model is an exactly soluble model of interacting fermions in one dimension with the following key features:

(i) its elementary excitations are non-interacting bosons;
(ii) the mean fermion current j is a good quantum number;
(iii) all its correlation functions can be explicitly evaluated.

The complete solubility of this model only emerged over a period of a decade; because the resolution of certain ambiguities in versions of the solution developed in the literature over this period turned out to be a key step in the work reported in this series of papers, I will merely cite some of the key papers in the literature, and then present a detailed version of the solution without further reference to its historical development.

The model was proposed by Luttinger (1963), but this first step in its correct solution was taken by Mattis and Lieb (1965), who discovered the free boson elementary excitations. Soon after, Overhauser (1965) pointed out that these bosons could be used to construct a complete set of eigenstates. Theumann (1967) and Dover (1968) gave early calculations of the single-particle correlation function, but the systematic calculation of correlation functions became trivial after the simultaneous discovery of the existence of a simple representation of the fermion operators in terms of the boson fields by Mattis (1974) and Luther and Peschel (1974). In fact, these fields are not on their own sufficient for the full construction of fermion operators in the diagonal basis, and both these early forms have problems associated with the characterisation of $q = 0$ modes. In particular, Luther and Peschel (1974) introduced a certain cut-off parameter α in their version, with the stipulation that it only became an exact operator identity in the limit $\alpha \to 0$. The necessity for any such limiting procedure has been entirely eliminated in the exact formulation reviewed below. The first completely precise formulation in the solid-state literature (though from a field-theory viewpoint) was given by Heidenreich *et al* (1975), though there has been an entirely parallel development in the field-theoretical literature on the related 'massless Thirring model' which I will not review here. The first construction of the important unitary charge-raising operators in terms of the bare fermions was apparently given by Haldane (1979). An important paper essentially parallel to, but not part of, the above developments is that of Dzyaloshinskii and Larkin (1973) who studied the spin-$\frac{1}{2}$ version of the (originally spinless-fermion) model, and provided an interpretation of the Mattis–Lieb solution from the point of view of conventional many-body diagrammatic perturbation theory. Similarly, Everts and Schulz (1974) have shown how the power-law character of the correlation functions can be simply recovered by the standard equation-of-motion techniques. Below, I give a description of the spinless fermion form of the model; the simple extension to the spin-$\frac{1}{2}$ case will be discussed elsewhere.

3.2. The fermion description

It is useful to begin a discussion of the Luttinger model by characterising its Hilbert space: this is *not* the usual electron Hilbert space, but has been expanded to include a branch of *'positron'* states as well. This second, unphysical set of fermions will require high energies for their excitation, so will not qualitatively affect low-energy properties, but are absolutely necessary for the construction of the new basis of eigenstates given here. Note that definition of the Hilbert space does *not* require any precise specification of the electron and positron dispersions, only that these energies are bounded below, and increase without limit as the momentum $|k| \to \infty$. The model is defined on a finite ring of length L; only periodic fermion boundary conditions will be considered. It is then useful in developing the formalism to take the ground-state charge $N_0 = k_F L/\pi$ to be *odd*, so the ground state is non-degenerate (this restriction is eventually dropped). The Hilbert space worked in is spanned by the set of finite-energy eigenstates of the free Luttinger model, measured from a ground state with electron states from $-k_F$ to k_F filled, and all positron states empty.

A correct definition of the Hilbert space is required before any operator acting in it, such as the Hamiltonian, is defined. It gives meaning to 'operator identities' such as $\hat{A} = \hat{B}$, shorthand for $\langle\alpha|\hat{A}|\beta\rangle = \langle\alpha|\hat{B}|\beta\rangle$ for all $|\alpha\rangle$, $|\beta\rangle$ forming a set that spans the Hilbert space. Operators are only well defined if $\langle\alpha|\hat{A}|\beta\rangle$ is finite for all α, β; the problem of ill-defined operators does not arise in finite-dimensional Hilbert spaces such as in lattice systems, but problems can arise with infinite-dimensional spaces arising from continuum problems. This type of problem flawed Luttinger's original solution of the model. One standard way to ensure all operators worked with are finite is to consider only quantities that are normal-ordered in a set of creation operators that create excited states out of the ground state.

Instead of working directly with charge $+1$ electron states, and charge -1 positron states, it is useful to describe the Luttinger model in terms of charge $+1$ 'right-' and 'left-moving' fermions labelled by $p = \pm 1$ (note that this label should *not* be confused with a momentum label, for which k is used here). The kinetic part of the Luttinger Hamiltonian is then given (using units where $\hbar = 1$) by

$$H^0 = v_F \sum_{kp} (pk - k_F)(n_{kp} - \langle n_{kp}\rangle_0) \qquad \langle n_{kp}\rangle_0 = \theta(k_F - pk) \tag{3.1}$$

$$= v_F \int_0^L \mathrm{d}x \sum_p : \psi_p^\dagger(x)(\mathrm{i}p\nabla - k_F)\psi_p(x): \tag{3.2}$$

where $:(\ldots):$ means fermion normal-ordering with respect to the ground state of (3.1). The term k_F is essentially a chemical potential to fix the ground state charge. The spectral diagrams corresponding to figure 1 for the non-interacting Luttinger model (3.1) are shown in figure 2. The fermion field $\psi_p^\dagger(x)$ is given by

$$\psi_p^\dagger(x) = \lim_{\varepsilon \to 0^+} \left[L^{-1/2} \sum_k \mathrm{e}^{\mathrm{i}kx} \exp(-\varepsilon|kL/2\pi|)\, c_{kp}^\dagger \right]. \tag{3.3}$$

The limiting procedure $\varepsilon \to 0^+$ is usually left implicit, but has been explicitly included here to emphasise the similarity with a analogous construction that will appear later. It is necessary for the definition of the periodic delta function generated by the anticommutation relations:

$$\{\psi_p^\dagger(x), \psi_{p'}(x')\} = \delta_{pp'} \left[\lim_{\varepsilon \to 0^+} \left(L^{-1} \sum_k \exp[-k(x - x')] \exp(-\varepsilon|kL/\pi|) \right) \right]$$

$$= L^{-1}\delta_{pp'} \lim_{\varepsilon \to 0^+} [(1 - z\mathrm{e}^{-2\varepsilon})^{-1} + (1 - z^*\mathrm{e}^{-2\varepsilon})^{-1} - 1]$$

$$= \delta_{pp'} \sum_{n=-\infty}^{\infty} \delta(x - x' - nL). \tag{3.4}$$

where $z = \exp(2\pi\mathrm{i}x/L)$, and allowed k-values in the sum satisfy $\exp(\mathrm{i}kL) = 1$.

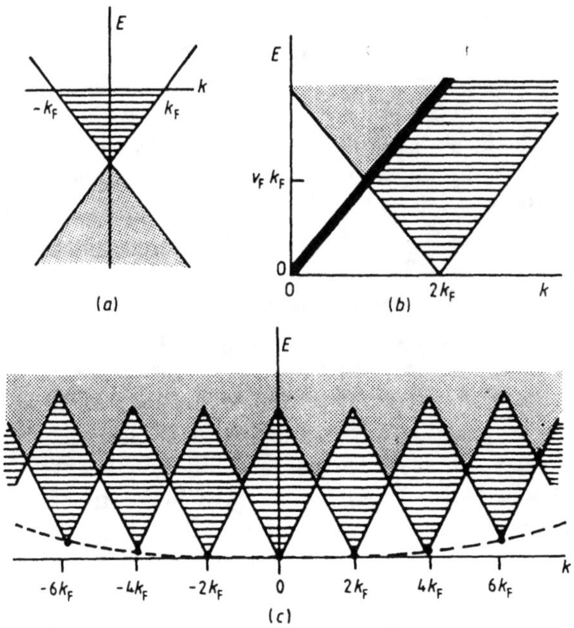

Figure 2. Diagrams corresponding to figure 1, this time for the spectrum of the non-interacting Luttinger model (3.1). Dotted areas indicate the presence of 'unphysical' states involving excited 'positrons'.

Note that the quantity ε appears as a dimensionless infinitesimal quantity necessary for controlling the sums over the infinite range of values of k, and in no way plays the role of a 'cut-off length'.

The fundamental electron and positron fields are related to the fields $\psi_p^\dagger(x)$ in a *non-local* way: in terms of the c_{kp}, they are given by

$$\psi^\dagger(x) = L^{-1/2} \sum_{kp} \theta(kp)\, \mathrm{e}^{\mathrm{i}kx}\, c_{kp}^\dagger$$
$$\bar{\psi}^\dagger(x) = L^{-1/2} \sum_{kp} \theta(-kp)\, \mathrm{e}^{-\mathrm{i}kx}\, c_{kp}. \tag{3.5}$$

When expressed in terms of $\psi_p^\dagger(x)$, the electron field $\psi^\dagger(x)$ is given by

$$\psi^\dagger(x) = \frac{1}{2\pi\mathrm{i}} \sum_p p \int_{-\frac{1}{2}L}^{\frac{1}{2}L} \mathrm{d}y\, K(y)\, \psi_p^\dagger(x+y), \tag{3.6}$$

where $K(y) = (\pi/L)[\tan(\pi y/L)]^{-1}$.

3.3. *The boson description: construction from fermion operators*

A central role in the theory of the Luttinger model is played by the density operator for type-p fermions:

$$\rho_{qp} = \sum_k c_{k+qp}^\dagger c_{kp} \qquad (q \neq 0)$$
$$\equiv N_p = \sum_k n_{kp} - \langle n_{kp} \rangle_0 \qquad (q = 0). \tag{3.7}$$

Note that the subtraction of the (infinite) ground state density of type-p fermions means that the $q = 0$ component of ρ_{qp} is a well defined operator in the sense discussed above: this procedure is equivalent to normal-ordering in the fermion variables. The commutation relations of the ρ_{qp} are

$$[\rho_{qp}, \rho_{-q'p'}] = \delta_{pp'}\delta_{qq'}(Lpq/2\pi). \tag{3.8}$$

This is easily established by direct evaluation of the commutator, then writing pair operators $c^\dagger_{kp}c_{k'p}$ as $(c^\dagger_{kp}c_{k'p} - \langle c^\dagger_{kp}c_{k'p}\rangle_0) + \langle c^\dagger_{kp}c_{k'p}\rangle_0$, where $\langle c^\dagger_{kp}c_{k'p}\rangle_0 \equiv \delta_{kk'}\langle n_{kp}\rangle_0$: this guarantees that operator quantities are effectively normal-ordered, and hence well defined, so they can be manipulated safely. The commutator (3.8) trivially vanishes when $p \neq p'$; for equal indices p, it is given by

$$\hat{O}_{pqq'} - \hat{O}_{pqq'} + \delta_{qq'} \sum_k (\langle n_{k+qp}\rangle_0 - \langle n_{kp}\rangle_0);$$

$$\hat{O}_{pqq'} \equiv \sum_k (c^\dagger_{k+q-q'p}c_{kp} - \langle c^\dagger_{k+q-q'p}c_{kp}\rangle_0).$$

Being well defined, the operators $\hat{O}_{pqq'}$ can safely be cancelled, and the remaining c-number term gives the RHS of (3.8). A second important commutation relation is that with the fermion fields:

$$[\rho_{qp}, \psi^\dagger_{p'}(x)] = \delta_{pp'}\, e^{-iqx}\psi^\dagger_p(x). \tag{3.9}$$

It will prove useful to define the following partial Fourier transforms:

$$\rho_p^{(\pm)}(x) = L^{-1} \sum_q \theta(\pm pq)\, e^{iqx}\, \rho_{qp}. \tag{3.10}$$

These operators have the property that $(\rho_p^{(-)}(x))^\dagger = \rho_p^{(+)}(x)$, and that $\rho_p^{(-)}(x)$ annihilates the vacuum state of (3.1). They satisfy periodic boundary conditions, $\rho_p^{(\pm)}(x+L) = \rho_p^{(\pm)}(x)$. The local density of p-type fermions, with respect to the ground state density, is

$$\lim_{a\to 0} [\psi^\dagger_p(x+a)\psi_p(x) - \langle \psi^\dagger_p(x+a)\psi_p(x)\rangle_0] \equiv \rho_p(x) = [\rho_p^{(+)}(x) + \rho_p^{(-)}(x)]. \tag{3.11}$$

The limiting procedure $a \to 0$ is to avoid direct reference to the infinite quantity $\langle \psi^\dagger_p(x)\psi_p(x)\rangle_0$.

The commutation algebra (3.8) immediately suggests the construction of boson operators. For $q \neq 0$,

$$a^\dagger_q = (2\pi/L|q|)^{1/2} \sum_p \theta(pq)\rho_{qp} \qquad (q \neq 0; q = 2\pi n/L, n = \pm 1, \pm 2, \ldots). \tag{3.12}$$

These obey exact boson commutation relations, and have the property that a_q annihilates the ground state of (3.1). There is no $q = 0$ boson mode (indeed, the form (3.12) is undefined at $q = 0$); the $q = 0$ mode is represented by the number operator N_p, which commutes with the bosons $a^\dagger_q$. The density operators are then expressed by

$$\rho_{qp} = N_p\delta_{p0} + (L|q|/2\pi)^{1/2}\{\theta(pq)a^\dagger_q + \theta(-pq)a_{-q}\}. \tag{3.13}$$

The algebra of the operators $(a^\dagger_q, a_q, N_p)$ that have been constructed so far is incomplete: it lacks a ladder operator U_p that raises the fermion charge N_p in unit steps, while commuting with the bosons a_q. The ladder of allowed values of N_p has no upper or lower limit, so (in contrast to the case of boson or finite spin ladder algebras) the number operator cannot be expressed in terms of the raising operator and its conjugate lowering operator. The raising operator can be chosen to be *unitary*, $U_p^{-1} = (U_p)^\dagger$. Finally, the fermion nature of the ladder operators means that U_p will *anti*commute with U_{-p} and U_{-p}^{-1}.

It is useful to study the construction of the ladder operators U_p in detail. It is important that these operators be given in a well defined form. A heuristic understanding of their form can be gained from the following argument. A special subset of eigenstates of (3.1) are those with occupations $n_{kp} = \theta(k_F + (2\pi N_p/L) - pk)$; these states $|\{N_p\}\rangle$ include the vacuum, and share with it the property that they are annihilated by a_q. The ladder operator U_p must have the property that

$$U_p\,|N_p, N_{-p}\rangle = \eta(p, N_p, N_{-p})\,|N_p + 1, N_{-p}\rangle, \quad \eta = \pm 1.\dagger$$

A construction with this property is

$$\sum_k c_{kp}^\dagger \delta(pk - [k_F + (2N_p + 1)\pi/L]).$$

Writing this in a more symmetrical form, using an integral representation of the Kronecker delta function, this becomes:

$$U_p = L^{-1/2}\int_0^L \mathrm{d}x\, \exp(-\mathrm{i}pk_F x)\exp[-\mathrm{i}\phi_p^\dagger(x)]\psi_p^\dagger(x)\exp[-\mathrm{i}\phi_p(x)];$$

$$\phi_p(x) = p(\pi x/L)N_p. \tag{3.14}$$

The ladder operator U_p must also have the property that it commutes with the boson operators a_q, or equivalently with the density operators ρ_{qp}, when $q \neq 0$. The commutation relation (3.9) means that the trial form above does *not* have this property. However, it would if the operators $\phi_p(x)$ were modified so that

$$[\rho_{qp}, \phi_p(x)] = -\mathrm{i}\delta_{pp'}\theta(pq)\,\mathrm{e}^{-\mathrm{i}qx}(1 - \delta_{q0}). \tag{3.15}$$

Because this commutator is a c-number,

$$[\rho_{qp}, \exp(-\mathrm{i}\phi_{p'}(x))] = -\theta(pq)(\delta_{pp'}\,\mathrm{e}^{-\mathrm{i}qx})\exp\{-\mathrm{i}\phi_p(x)\}(1 - \delta_{q0}),$$
$$[\rho_{pq}, \exp\{-\mathrm{i}\phi_{p'}^\dagger(x)\}] = -\theta(-pq)(\delta_{pp'}\,\mathrm{e}^{-\mathrm{i}qx})\exp\{-\mathrm{i}\phi_p^\dagger(x)\}(1 - \delta_{q0}). \tag{3.16}$$

When $q \neq 0$, these terms exactly counterbalance the commutator (3.9), so when U_p is given by the form (3.14) with a $\phi_p(x)$ that satisfies (3.15),

$$[\rho_{qp}, U_{p'}] = \delta_{pp'}\delta_{q0}U_p. \tag{3.17}$$

The explicit construction of the quantities $\phi_p(x)$ is now easily given; using the property (3.8):

$$\phi_p(x) = (2\pi p/L)\left(\tfrac{1}{2}xN_p + \mathrm{i}\sum_{q=0}\theta(-pq)(\mathrm{e}^{-\mathrm{i}qx}/q)\rho_{qp}\right). \tag{3.18}$$

The first term in this is just the $q = 0$ component of the sum, with the limit $q \to 0$ properly taken, so this can be re-expressed as

$$\phi_p(x) = \lim_{\varepsilon\to 0^+}\left((2\pi\mathrm{i}p/L)\sum_q \theta(-pq)\,\mathrm{e}^{-\mathrm{i}qx}/q)\exp[-\varepsilon(|q|L/2\pi)]\,\rho_{qp}\right). \tag{3.19}$$

The limiting procedure $\varepsilon \to 0^+$ has been included in order to properly define the sum in much the same spirit as in equation (3.3), and it will be needed to properly define the periodic delta function. ε is a positive dimensionless infinitesimal, and in no way should

† η depends on the ordering convention used in constructing $|\{N_p\}\rangle$.

be interpreted as a cut-off length, despite the formal similarity to the length parameter α introduced by Luther and Peschel (1974). The field $\phi_p(x)$ is easily found to have the following properties:

$$\nabla\phi_p(x) = 2\pi p\rho_p^{(-)}(x) \tag{3.20}$$

$$[\phi_p(x), \phi_{p'}(x')] = [\phi_p^\dagger(x), \phi_{p'}^\dagger(x')] = 0;$$

$$[\phi_p(x), \phi_{p'}^\dagger(x')] = \lim_{\varepsilon\to 0^+} \{\delta_{pp'}(-\ln(1 - e^{-2\varepsilon}z^{-2p}))\} \tag{3.21}$$

where $z \equiv \exp(i\pi(x - x')/L)$. $\phi_p(x)$ has the property that it annihilates the vacuum state of (3.1); this is easily seen when $\phi_p(x)$ is written in terms of N_p and a_q:

$$\phi_p(x) = p(\pi x/L)N_p + i\sum_{q\neq 0} \theta(pq)(2\pi/L|q|)^{1/2} e^{-iqx}a_q. \tag{3.22}$$

This means that the operator U_p defined by (3.14) plus (3.19) is normal-ordered in the boson operators that annihilate the vacuum state of (3.1). As will shortly be seen, this guarantees that this construction of U_p defines a well defined operator.

It is useful to give a representation of the unitary operators U_p in terms of Hermitian phase variables $\overline{\theta}_p = \overline{\theta}_p^\dagger$, conjugate to N_p:

$$U_p = (-1)^{(\frac{1}{2}pN_{-p})}\overline{U}_p, \qquad \overline{U}_p = \exp(i\overline{\theta}_p); \tag{3.23}$$

$$[N_p, \overline{\theta}_{p'}] = i\delta_{pp'}, \qquad [\overline{\theta}_p, \overline{\theta}_{p'}] = [N_p, N_{p'}] = 0. \tag{3.24}$$

The prefactor $(-1)^{(\frac{1}{2}pN_{-p})}$ ensures that U_p and U_{-p} anticommute, so that the unitary operators $\bar{U}_p$ and $\bar{U}_{-p}$ can commute. This choice of anticommutation factor corresponds to a particular ordering convention in constructing states $|\{N_p\}\rangle$ in terms of fermion operators; other choices are possible.

For some purposes, it is useful to introduce a *local* phase field $\theta_p(x)$:

$$\theta_p(x) = \overline{\theta}_p + \phi_p(x) + \phi_p^\dagger(x). \tag{3.25}$$

This has the property

$$\nabla\theta_p(x) = 2\pi p\rho_p(x). \tag{3.26}$$

It can easily be seen that $\rho_p(x)$ and $\theta_p(x)$ are canonical conjugate fields:

$$[\rho_p(x), \theta_{p'}(x')] = i\delta_{pp'}\sum_n \delta(x - x' + nL). \tag{3.27}$$

3.4. Boson form of the Hamiltonian

So far, no use whatsoever has been made of the fact that the Hamiltonian (3.1) has a linear fermion dispersion, and the above discussion has only depended on identification of its vacuum state and the structure of the associated Hilbert space. The linear spectral property will now be used to construct a new basis of eigenstates that will be shown to be complete and hence to span the Hilbert space defined by all finite-energy eigenstates of (3.1). The operators N_p commute with (3.1), but $a_q^\dagger$ and U_p do not: first note the commutator of the density operator with the Hamiltonian:

$$[H^0, \rho_{qp}] = v_F pq\rho_{qp}. \tag{3.28}$$

With this, the definition (3.12) of the boson operators a_q leads to

$$[H^0, a_q^\dagger] = v_F |q| a_q^\dagger. \tag{3.29}$$

Instead of attempting to directly evaluate the commutator $[H^0, U_p]$, I use the following argument: the special set of states $|\{N_p\}\rangle$ can be constructed from the vacuum by acting on it with U_p:

$$|\{N_p\}\rangle = \pm \prod_p (U_p)^{N_p}|0\rangle; \tag{3.30}$$

this is verified by explicitly showing that U_p as constructed indeed has the property $U_p|N_p, N_{-p}\rangle = \eta|N_p + 1, N_p\rangle$. Because a_q annihilates the states $|\{N_p\}\rangle$, (3.22) implies that

$$U_p|\{N_p\}\rangle = L^{-1}\sum_k \int_0^L \mathrm{d}x \exp[\mathrm{i}(k - pk_F)x] \exp[-\mathrm{i}\phi_p^\dagger(x)]c_{kp}^\dagger \exp[-\mathrm{i}p(\pi x/L)N_p]|\{N_p\}\rangle$$

$$= \prod_{q\neq 0}\left(1 + \theta(pq)\sum_{n_q=1}^{\infty}(2\pi/L|q|)^{\frac{1}{2}n_q}(n_q!)^{-1}(a_q^\dagger)^{n_q}\right)$$

$$\times \sum_k c_{kp}^\dagger \delta\left(pk - [k_F + (2N_p + 1)\pi/L] + \sum_q |q| n_q\right)|\{N_p\}\rangle.$$

The Kronecker delta ensures that no states containing boson excitations survive in this sum, and (3.30) is verified.

The energies of the eigenstates $|\{N_p\}\rangle$ are easily obtained by examining their construction: $E(\{N_p\}) = v_F(\pi/L)\,\Sigma_p(N_p)^2$. A larger set of eigenstates is obtained by acting these with the boson operators:

$$|\{N_p\}, \{n_q\}\rangle = \prod_{q\neq 0}\left(\frac{(a_q^\dagger)^{n_q}}{(n_q!)^{1/2}}\right)\prod_p (U_p)^{N_p}|0\rangle. \tag{3.31}$$

In its action on *these* eigenstates, the Hamiltonian is given by

$$H^0 = v_F\left[\sum_q |q| a_q^\dagger a_q + (\pi/L)\sum_p (N_p)^2\right]. \tag{3.32}$$

This can be written in terms of the phase fields $\theta_p(x)$ as

$$H^0 = v_F\frac{1}{\pi}\int_0^L \mathrm{d}x \sum_p :(\nabla\theta_p(x))^2:, \tag{3.33}$$

where *boson* normal-ordering is implied. The momentum operator is similarly given by

$$P = \sum_p p[k_F + (\pi/L)N_p]N_p + \sum_q q a_q^\dagger a_q \tag{3.34}$$

$$= \frac{1}{\pi}\int_0^L \mathrm{d}x \sum_p p[k_F\nabla\theta_p(x) + :(\nabla\theta_p(x))^2:]. \tag{3.35}$$

The question arises: *are the eigenstates* (3.31) *a complete set?* If so, U_p is proved to be a well defined operator, and (3.32) and (3.34) have the status of identities in the full Hilbert space based on the vacuum of (3.1). The possibility of two such different sets of eigenstates of the free Luttinger Hamiltonian (3.1) arises because of the high degree of degeneracy of the spectrum due to the linear dispersion: all eigenstates with even (odd) fermion charge $N - N_0$ have energies that are even (odd) multiples of $\pi v_F/L$ with respect to the ground state. One way to check the completeness of the set (3.31) is to directly investigate the degeneracy of states at a given energy. An equivalent, more elegant, way is to compute the grand partition sum of the Hamiltonian at arbitrary inverse temperature β, first using the 'obvious' set of fermion excitation states, then the set (3.31). This is a sum over positive definite quantities, so if any states were missing from (3.31), the result of the second calculation would be less than the first.

Defining $w = \exp(-\beta\pi v_F/L)$, the direct evaluation of the partition function using the free fermion basis gives

$$Z(w) = \left(\prod_{n=1}^{\infty} (1 + w^{2n-1})^2\right)^2. \tag{3.36a}$$

Using the set (3.31), one obtains

$$Z(w) = \left(\prod_{n=1}^{\infty} (1 - w^{2n})^{-1}\right)^2 \left(\sum_{m=-\infty}^{\infty} w^{(n^2)}\right)^2. \tag{3.36b}$$

These apparently different expressions are in fact both equal, since the elliptic theta function $\vartheta_3(0; w)$ (Gradsteyn and Ryzhik 1965, p 921) has both a series and a product representation:

$$\vartheta_3(0; w) = \sum_{n=-\infty}^{\infty} w^{(n^2)} = \prod_{n=1}^{\infty} (1 + w^{2n-1})^2(1 - w^{2n}).$$

The set (3.31) is thus complete, and spans the full Hilbert space.

3.5. *Boson form of fermion operators*

With the completeness of the set of eigenstates (3.31) established, the remaining task is to construct the representation of the fermion operators $\psi_p^\dagger(x)$ in this basis. The ground work has been laid: $\psi_p^\dagger(x)$ is trivially obtained by inverting the expression (3.14) for U_p:

$$\begin{aligned}\psi_p^\dagger(x) &= L^{-1/2} \exp(ipk_F x)\{\exp[i\phi_p^\dagger(x)]U_p \exp[i\phi_p(x)]\} \\ &= (-1)^{(\frac{1}{2}pN_{-p})} L^{-1/2} \exp(ipk_F x)\{\exp[i\phi_p^\dagger(x)] \exp(i\bar{\theta}_p) \exp[i\phi_p(x)]\},\end{aligned} \tag{3.37}$$

where $\phi_p(x)$ is now defined directly by (3.22). This is an explicitly well defined operator, since it is normal-ordered in terms of the bosons a_q. The anticommutation relations can be explicitly verified; here the limiting procedure defined in (3.19) and (3.21) is required: the procedure is to construct the anticommutators, and then re-normal-order the resulting products in terms of the bosons, so they become explicitly well defined operators that can be manipulated and cancelled. The anticommutation of fields with different labels p is trivially assured by the anticommuting properties of U_p; for equal p, the anticommutator $\{\psi_p^\dagger(x), \psi_p^\dagger(x')\}$ is given by

$$\begin{aligned}&L^{-1} \exp[ipk_F(x + x')] \exp[i\hat{O}_1^\dagger(x, x')]U_p^2 \exp[i\hat{O}_1(x, x')]F_1(x, x') \\ &\hat{O}_1(x, x') = \phi_p(x) + \phi_p(x') \\ &F_1(x, x') = G_1(x - x') + G_1(x' - x) \\ &G_1(x - x') = \exp[i\pi p(x - x')/L] \exp\{-[\phi_p(x), \phi_p^\dagger(x')]\}.\end{aligned} \tag{3.38}$$

The c-number function $F_1(x, x')$ is multiplying a well defined (i.e. normal-ordered) operator expression. $F_1(x, x')$ can be evaluated using equation (3.21): setting $z = \exp[i\pi(x - x')/L]$,

$$F_1 = \lim_{\varepsilon\to 0^+} \{z^p(1 - e^{-2\varepsilon}z^{-2p}) + z^{-p}(1 - e^{-2\varepsilon}z^{2p})\} = 0. \tag{3.39}$$

The anticommutator $\{\psi_p^\dagger(x), \psi_p^\dagger(x')\}$ thus vanishes correctly. The anticommutator $\{\psi_p(x), \psi_p^\dagger(x')\}$ is given by

$$L^{-1}\exp[ipk_F(x' - x)]\exp[i\hat{O}_2^\dagger(x, x')]\exp[i\hat{O}_2(x, x')]F_2(x, x')$$

$$\hat{O}_2^\dagger(x, x') = \phi_p(x) - \phi_p(x')$$

$$F_2(x, x') = G_2(x - x') + G_2(x' - x)$$

$$G_2(x - x') = \exp[i\pi p(x - x')/L]\exp\{+[\phi_p(x), \phi_p^\dagger(x')]\}. \tag{3.40}$$

Again this is a normal-ordered operator expression, times a c-number function $F_2(x, x')$. Again using (3.21),

$$F_2 = \lim_{\varepsilon\to 0^+} \{z^p(1 - e^{-2\varepsilon}z^{-2p})^{-1} + z^{-p}(1 - e^{-2\varepsilon}z^{2p})^{-1}\}$$

$$= L\sum_{n=-\infty}^{\infty} (-1)^n\delta(x - x' + nL). \tag{3.41}$$

when $x - x' = nL$, the operator-valued expression that multiplies F_2 in (3.41) takes the simple c-number values $L^{-1}\exp(-inpk_FL) = L^{-1}(-1)^n$. The anticommutator is thus correctly given by the periodic delta function as in (3.3).

This completes the derivation of the operator algebra needed to describe the model using the alternative basis set of eigenstates (3.31). This algebra is a precise tool, and I now use it to recover the expressions (3.32) and (3.34) for H^0 and P directly from the fermion representation (3.37). Consider the quantity

$$\int_0^L dx\exp(-ipk_Fa)\,\psi_p^\dagger(x + \tfrac{1}{2}a)\psi_p(x - \tfrac{1}{2}a). \tag{3.42}$$

Using the expression (3.3) for $\psi_p(x)$, this is easily found to be

$$L(2\pi ipa')^{-1} + \sum_k \exp[i(k - pk_F)a](n_{kp} - \langle n_{kp}\rangle_0) \tag{3.43}$$

where $a' = (L/\pi)\sin(\pi a/L)$. Using the alternative expression (3.37), and then normal-ordering, it is found to be

$$(2\pi ipa')^{-1}\left(L + \int_0^L dx\{\exp[2\pi ip(a/L)N_p]\exp[i\Phi_p^\dagger(x)]\exp[i\Phi_p(x)] - 1\}\right)$$

$$\Phi_p(x) = i\sum_q (2\pi/L\,|q|)^{1/2} e^{-iqx}\theta(qp)\,2\sin(\tfrac{1}{2}qa)a_q. \tag{3.44}$$

Cancelling the divergent term $L(2\pi ipa')^{-1}$ and comparing the term $O(a)$ in the expansions of the two expressions, one directly obtains

$$\sum_k (pk - k_F)(n_{kp} - \langle n_{kp}\rangle_0) = (\pi/L)N_p^2 + \sum_q pq\theta(qp)a_q^\dagger a_q. \tag{3.45}$$

The expressions for H^0 and P are now trivial to obtain.

3.6. The charge and current formalism

So far, the formalism has been developed in terms of operators labelled by $p = \pm 1$, corresponding to the right- and left-going fermions. It is convenient for some purposes to introduce the symmetric and antisymmetric combinations, labelled by N and J respectively, which will be related to charge and current variables. It is also useful to include the ground state electronic charge (number of electrons minus number of positrons) $N_0 = k_F L/\pi$ in the charge variables. The following combinations are defined:

$$N = N_0 + \sum_p N_p \qquad J = \sum_p pN_p \tag{3.46}$$

$$\overline{\theta}_N = \sum_p \overline{\theta}_p \qquad \overline{\theta}_J = \sum_p p\overline{\theta}_p \tag{3.47}$$

$$\rho_N(x) = (N_0/L) + \sum_p \rho_p(x) \qquad \rho_J(x) = \sum_p p\rho_p(x) \tag{3.48}$$

$$\phi_N(x) = \sum_p \phi_p(x) \qquad \phi_J(x) = \pi(N_0/L)x + \sum_p p\phi_p(x). \tag{3.49}$$

Phase fields $\theta_N(x)$ and $\theta_J(x)$ are then defined by, e.g.,

$$\theta_N(x) = \overline{\theta}_N + \phi_N(x) + \phi_N^\dagger(x). \tag{3.50}$$

The following relations are found:

$$\nabla\theta_N(x) = 2\pi\rho_J(x) \qquad \nabla\theta_J(x) = 2\pi\rho_N(x) \tag{3.51}$$

$$[\rho_N(x), \theta_N(x')] = [\rho_J(x), \theta_J(x')] = \mathrm{i}\sum_n \delta(x - x' + nL); \tag{3.52}$$

$$[\rho_N(x), \theta_J(x')] = [\rho_J(x), \theta_N(x')] = 0. \tag{3.53}$$

The fields $(\rho_N(x), \theta_N(x))$ and $(\rho_J(x), \theta_J(x))$ are canonically conjugate pairs. Note however that $[\rho_N(x), \rho_J(x')]$ and $[\theta_J(x), \theta_N(x')]$ do *not* vanish, except at equal positions, $x = x'$. On the other hand, $[\theta_N(x), \theta_N(x')]$, $[\rho_N(x), \rho_N(x')]$, etc, *do* vanish.

When the quantities $\rho_N(x)$, $\phi_N(x)$, etc are expressed in terms of boson variables, they are explicitly given by

$$\rho_N(x) = (N/L) + \sum_q (|q|/2\pi L)^{1/2}\, \mathrm{e}^{\mathrm{i}qx}\, (a_q^\dagger + a_{-q})$$

$$\phi_N(x) = \pi(J/L)x + \mathrm{i}\sum_q (2\pi/L\,|q|)^{1/2}\, \mathrm{e}^{-\mathrm{i}qx} a_q$$

$$\rho_J(x) = (J/L) + \sum_q (|q|/2\pi L)^{1/2}\, \mathrm{sgn}(q)\, \mathrm{e}^{\mathrm{i}qx}(a_q^\dagger - a_{-q})$$

$$\phi_J(x) = \pi(N/L)x + \mathrm{i}\sum_q (2\pi/L\,|q|)^{1/2}\, \mathrm{sgn}(q)\, \mathrm{e}^{-\mathrm{i}qx} a_q. \tag{3.54}$$

Note how $\mathrm{sgn}(q)$ characteristically appears in the boson part of J-labelled quantities.

The commuting unitary operators $\overline{U}_N = \exp(\mathrm{i}\overline{\theta}_N)$ and $\overline{U}_J = \exp(\mathrm{i}\overline{\theta}_J)$ respectively raise N and J by one. In this basis, the fermion field operator $\psi_p^\dagger(x)$ becomes

$$\psi_p^\dagger(x) = L^{-\frac{1}{2}}(-1)^{\frac{1}{2}(pJ-N)}\{\exp[\tfrac{1}{2}\mathrm{i}(p\phi_J^\dagger + \phi_N^\dagger)]\, \exp[\tfrac{1}{2}\mathrm{i}(p\overline{\theta}_J + \overline{\theta}_N)]$$
$$\times \exp[\tfrac{1}{2}\mathrm{i}(p\phi_J + \phi_N)]\}. \tag{3.55}$$

The dependence on k_F has been absorbed into the definition of $\phi_J(x)$.

The Hamiltonian takes the form

$$H^0 = v_F\left(\sum_q |q|\, a_q^\dagger a_q + \tfrac{1}{2}(\pi/L)((N-N_0)^2 + J^2)\right) \qquad (-1)^J = -(-1)^N; \tag{3.56}$$

$$P = [k_F + \pi(N-N_0)/L]\,J + \sum_q q a_q^\dagger a_q \qquad k_F = \pi N_0/L. \tag{3.57}$$

This is just the form postulated in § 2. The selection rule linking allowed values of J and N arises because N_p and N_{-p} are both integral.

In the phase field variables, the Hamiltonian can be written

$$H = v_F \frac{1}{\pi}\int_0^L \mathrm{d}x\, :(\nabla\theta_N(x))^2 + (\nabla\theta_J(x))^2:$$

$$P = \frac{1}{\pi}\int_0^L \mathrm{d}x\, :\nabla\theta_N(x)\nabla\theta_J(x): + \text{HC}. \tag{3.58}$$

Since $\rho_N(x) = (2\pi)^{-1}\nabla\theta_J(x)$ is the canonical conjugate to $\theta_N(x)$, and $\rho_J(x)$ to $\theta_J(x)$, the Hamiltonian (3.58) can be written as a Klein–Gordon field Hamiltonian in either the N or the J variables. The periodic fermion boundary conditions that must be satisfied by (3.55) imply that $\exp[i\theta_N(x+L)] = \exp[i\theta_N(x)]$, etc, so $\theta_N(x+L) = 2\pi J + \theta_N(x+L)$, and $\theta_J(x+L) = 2\pi N + \theta_J(x)$. The quantum numbers N and J thus can be related to *topological* excitations of the phase fields $\theta_N(x)$ and $\theta_J(x)$, while the bosons relate to their small fluctuations.

The physical interpretation of the quantum number N is simple: it is just the total electronic charge (electrons minus positrons). Similarly, it will now be shown that J is proportional to the mean *current*. It would be tempting to identify $\rho_N(x)$ with the *local* charge density operator $\rho(x)$; unfortunately, this is not correct, due to the non-local relation between the electron field and $\psi_p^\dagger(x)$. The fundamental definition of the local electronic density in terms of the electrons and positrons leads to

$$\rho(x) = \rho_N(x) + \tau(x),$$

$$\tau(x) = \sum_{kk'pp'} \exp[i(k-k')x](-pp')\theta(-kk')c_{kp}^\dagger c_{k'p'}. \tag{3.59}$$

At low energies, the extra term $\tau(x)$ involves only fluctuations with $q \sim 2k_F$. The fundamental definition of the current $j(x)$ is through the continuity equation for local charge:

$$\frac{\mathrm{d}}{\mathrm{d}t}\rho(x) \equiv i[H, \rho(x)] = \nabla j(x). \tag{3.60}$$

The mean current j is then given by

$$j \equiv L^{-1}\int_0^L \mathrm{d}x\, j(x) = \lim_{q\to 0}\{(qL)^{-1}[H, \rho_q]\}, \tag{3.61}$$

ρ_q being the Fourier transform of $\rho(x)$. In a low-energy subspace, and provided k_F is finite, the contribution from $\tau(x)$ can be neglected, and ρ_{Nq} substituted for ρ_q in (3.61). Then it is easily found that

$$j = v_F(J/L). \tag{3.62}$$

Actually, this is exact in the case of the free Luttinger model, but in a more general model extra terms will be present, and a linear relation like (3.62) will only be valid in a low-energy subspace, where the presence of long-wavelength boson excitations does not affect the current.

Finally, I note a useful low-energy, finite k_F *approximation* for $\tau(x)$:

$$\tau(x) \sim \sum_p \psi_p^\dagger(x)\psi_{-p}(x). \tag{3.63}$$

4. The Luttinger model and its solution: II. The interacting model

The full Luttinger model is obtained by taking the kinetic term (3.1), (3.32) and adding the fermion two-particle interaction:

$$H^1 = (\pi/L) \sum_{pp'q} (V_{1q}\delta_{pp'} + V_{2q}\delta_{p,-p'})\rho_{qp}\rho_{-qp'}. \tag{4.1}$$

The density operators ρ_{qp} are defined by (3.7) and (3.13). The coupling constants $V_{1q} \equiv V_1(|q|R)$ and $V_{2q} \equiv V_2(|q|R)$ have dimensions of velocity. They will be required to satisfy the following conditions:

(i) $V_1(0)$, $V_2(0)$ are finite;
(ii) $V_{2q}/(v_F + V_{1q}) \to 0$ as $|q| \to \infty$, faster than $|q|^{-\frac{1}{2}}$;
(iii) $|V_{2q}| < (v_F + V_{1q})$ for all q.

Conditions (i) could be relaxed somewhat, but this would alter the physics of the model. Conditions (ii) and (iii) are necessary to ensure that the Hilbert space of the model $H^0 + H^1$ remains the same as that of H^0. The conditions (i) and (ii) imply the existence of some length scale R that controls the crossover from the small-q to large-q regimes. The inclusion of this length scale in (4.1) means that V_1 and V_2 can be written as functions with a dimensionless argument. R is an effective range of the interaction in real space.

Using the phase-field formalism of § 3.6, the *low-energy* ($E \ll v_F/R$) form of the Hamiltonian can be written

$$H \simeq \frac{1}{\pi}\int_0^L \mathrm{d}x :v_N(\nabla\theta_N(x))^2 + v_J(\nabla\theta_J(x))^2: \tag{4.2}$$

where

$$v_N = v_F + V_1(0) + V_2(0), \qquad v_J = v_F + V_1(0) - V_2(0). \tag{4.3}$$

It is also useful to define the quantities

$$\omega_q = |(v_F + V_{1q})^2 - (V_{2q})^2|^{1/2}|q| \tag{4.4}$$

$$\tanh(2\varphi_q) = -V_{2q}/(v_F + V_{1q}). \tag{4.5}$$

Then the quantities v_S and φ are defined by

$$v_S = \lim_{q\to 0} (\omega_q/|q|); \qquad \varphi = \lim_{q\to 0} (\varphi_q). \tag{4.6}$$

The definitions (4.2)–(4.5) imply the relations

$$v_N = v_S \exp(-2\varphi); \qquad v_J = v_S \exp(2\varphi). \tag{4.7}$$

It will be convenient to represent φ_g as $\varphi g(|q|R)$, where the function $g(y)$ has the properties

$$g(0) = 1; \qquad y^{1/2}g(y) \to 0 \quad \text{as} \quad y \to \infty. \tag{4.8}$$

The conditions (i)–(iii) assure this, and also that φ_q is finite, ω_q is positive definite (except at $q = 0$) and v_N, v_J are positive definite. The model is fully parametrised by L, k_F, ω_q, and φ_q (or φ, R and $g(y)$). R has not been defined up to a multiplicative factor: it should be chosen so the crossover in $g(y)$ is around $y \sim 1$ (a unique definition might be provided by demanding that $(g(y))^2$ is normalised, for example).

When the full Hamiltonian $H^0 + H^1$ is written out in terms of a_q and the number operators N, J, it takes the simple bilinear form

$$H = -\tfrac{1}{2}\Big(\sum_q v_F |q|\Big) + \tfrac{1}{2}(\pi/L)(v_N N^2 + v_J J^2) + \tfrac{1}{2}\sum_q |q|\big[(v_F + V_{1q})(a_q^\dagger a_q + a_q a_q^\dagger) + V_{2q}(a_q^\dagger a_{-q}^\dagger + a_q a_{-q})\big]. \tag{4.9}$$

This is trivially diagonalised by a Bogoliubov transformation. The new ground state is given by

$$|\mathrm{GS}\rangle = \exp[-(A^2 L/R)] \exp\Big(\sum_{q>0} \tanh(\varphi_q) a_q^\dagger a_{-q}^\dagger\Big) |0, 0\rangle$$

$$A^2(\varphi g(y)) = \frac{1}{2\pi}\int_0^\infty dy\, \ln[\cosh(\varphi g(y))]. \tag{4.10}$$

For the ground state to belong to the Hilbert space of H^0, the normalisation constant must be finite; this means that the limit $R \to 0$ cannot be taken. The condition (4.8) assures that the constant A^2 is finite.

The Hamiltonian is diagonal in terms of the new boson operators

$$b_q^\dagger = \cosh(\varphi_q) a_q^\dagger - \sinh(\varphi_q) a_{-q} = \sum_p \alpha(pq, -\varphi_q)\,\rho_{qp} \qquad (q \neq 0); \tag{4.11}$$

$$\alpha(q, \varphi_q) = (2\pi/L\,|q|)^{1/2}[\theta(q)\cosh(\varphi_q) + \theta(-q)\sinh(\varphi_q)]. \tag{4.12}$$

The diagonalised Hamiltonian is given by substituting these into (4.9):

$$H = E_0 + \sum_q \omega_q b_q^\dagger b_q + \tfrac{1}{2}(\pi/L)(v_N N^2 + v_J J^2),$$

$$E_0 = \tfrac{1}{2}\sum_q (\omega_q - v_F|q|). \tag{4.13}$$

The ground state energy shift E_0 may well be divergent if ω_q does not tend to $v_F|q|$ fast enough as $|q| \to \infty$; however, in contrast to the case of a divergence of the ground state normalisation parameter A^2, this divergence is subtractable, and causes no problems. The form of the momentum operator remains essentially unchanged:

$$P = [k_F + \pi(N/L)]J + \sum_q q b_q^\dagger b_q. \tag{4.14}$$

The relation between the 'true' Fermi momentum $[k_F + \pi(N/L)]$ and the total charge N is unaffected by the interactions.

In addition to the total charge N with respect to the ground state remaining a good quantum number, the current quantum number J is also conserved. This reflects an invariance of the Hamiltonian, under which it is unchanged by independent global gauge transformations of the 'right-' and 'left-moving' ('clockwise' and 'anticlockwise') fermion fields $\psi_p(x)$, $p = \pm 1$. The density operators ρ_{qp} are given by

$$\rho_{qp} = \tfrac{1}{2}(N + pJ)\,\delta_{q0} + (L|q|/2\pi)[\alpha(pq, \varphi_q) b_q^\dagger + \alpha(-qp, \varphi_q) b_{-q}],$$

$$[H, \rho_{qp}] = p\,\mathrm{sgn}(q)\omega_q[\cosh(2\varphi_q)\rho_{qp} - \sinh(2\varphi_q)\rho_{q\,-p})]. \tag{4.15}$$

Following the arguments of (3.60)–(3.62), the mean current j is given by

$$j \simeq \lim_{q\to 0}\left\{(qL)^{-1}\sum_p [H, \rho_{qp}]\right\} = \lim_{q\to o}\left(L^{-1}(\omega_q/|q|)\, e^{2\varphi_q}\sum_p p\rho_{qp}\right). \tag{4.16}$$

From (4.7) the mean current is found to be $j = v_J(J/L)$; v_J thus plays the role of the renormalised Fermi velocity for fermion currents, as well as controlling the energy of $2k_F$ excitations. Note that it is somewhat unphysical for V_1 and V_2 to differ: if they are set equal, as would be the case if the model was derived as an effective Hamiltonian for a model where only the total charge density was coupled, v_J remains equal to the bare value v_F due to the kinetic term, and is not renormalised.

It is now necessary to transcribe the fermion field $\psi_p(x)$ (3.37) into a form normal-ordered in the new basis. First the definition of the quantities $\phi_p(x)$ (3.22) must be generalised:

$$\phi_p(x, \varphi_q) = p(\pi x/L)N_p + \mathrm{i}\sum_{q\neq 0}\alpha(pq, -\varphi_q)\, e^{-\mathrm{i}qx} b_q; \tag{4.17}$$

note that the phase field $\theta_p(x)$ is still given by

$$\theta_p(x) = \bar\theta_p + \phi_p(x, \varphi_q) + \phi_p^\dagger(x, \varphi_q) \tag{4.18}$$

independent of φ_q. Then the fermion field is given by

$$\psi_p(x) = e^{-\frac{1}{4}A} L^{-\nu} R^{\nu-1/2} \exp(\mathrm{i}pk_F x)\exp[\mathrm{i}\phi_p^\dagger(x, \varphi_p)]U_p \exp[\mathrm{i}\phi_p(x, \varphi_q)] \tag{4.19}$$

where $\nu = \frac{1}{2}\cosh(2\varphi)$, and the cut-off-dependent constant $\bar A(\varphi, g(y))$ and a similar quantity $\bar B$ are given by

$$\bar A = \lim_{\varepsilon\to 0^+}\left[2\sinh^2(\varphi)\left(C + \ln(\varepsilon/2\pi) + \int_\varepsilon^\infty \mathrm{d}y\, y^{-1} 2\sinh^2(\varphi g(y))\right)\right],$$

$$\bar B = \lim_{\varepsilon\to 0^+}\left[-\sinh(2\varphi)\left(C + \ln(\varepsilon/2\pi) - \int_\varepsilon^\infty \mathrm{d}y\, y^{-1}\sinh(2\varphi g(y))\right)\right]; \tag{4.20}$$

(C here is Euler's constant).

It is also useful to define two cut-off-dependent and φ-dependent functions $A_1(u) \equiv A_1(u; \varphi g(y'))$ and $B_1(u)$:

$$A_1(u) = \int_0^\infty \mathrm{d}y\, y^{-1}\sinh^2(\varphi g(y))[2\sin(\tfrac{1}{2}uy)]^2,$$

$$B_1(u) = -\tfrac{1}{2}\int_0^\infty \mathrm{d}y\, y^{-1}\sinh(2\varphi g(y))[2\sin(\tfrac{1}{2}uy)]^2. \tag{4.21}$$

The even functions $A_1(u)$ and $B_1(u)$ vanish as $u\to 0$; for large $|u|$, they behave as

$$A_1(u) \sim \bar A + 2\sinh^2(\varphi)\ln(2\pi|u|) + O(|u|^{-1}),$$

$$B_1(u) \sim \bar B - \sinh(2\varphi)\ln(2\pi|u|) + O(|u|^{-1}). \tag{4.22}$$

Together with $\bar A$ and $\bar B$, they vanish in the non-interacting limit $\varphi\to 0$.

These quantities characterise the commutation algebra of the quantities $\phi_p(x, \varphi_q)$, which I henceforth write as $\phi_p(x)$, suppressing the explicit dependence on φ_q: in the limit $L \gg R$,

$$[\phi_p(x), \phi_{p'}(x')] = [\phi_p^\dagger(x), \phi_{p'}^\dagger(x')] = 0;$$

$$[\phi_p(x), \phi_p^\dagger(x')] = \lim_{\varepsilon\to 0^+} \{-\ln[1 - e^{-2\varepsilon}\exp(-2\pi i p(x-x')/L)]\}$$
$$-A_1(d(x-x')/R) + \bar{A} - 2\sinh^2(\varphi)\ln(R/L),$$

$$[\phi_p(x), \phi_{-p}^\dagger(x')] = -B_1(d(x-x')/R) + \bar{B} + \sinh(2\varphi)\ln(R/L). \quad (4.23)$$

Here $d(x) \equiv (L/\pi)|\sin(\pi x/L)|$ is the *chord* distance between points with separation x along the circumference of the ring of length L.

The necessary mechanism for calculation of correlation functions has now been established: the desired quantity must be constructed in terms of the fermion operators (4.19), and then manipulated into normal-ordered form in boson variables. The limit $L \to \infty$ can then be taken. As an example, the *electron* single-particle correlation function is easily constructed (using (3.5) to construct the electron field in terms of $\psi_p(x)$); the finite-temperature terms are easily evaluated using the familiar property that $\langle \exp(\alpha b^\dagger) \exp(\alpha' b)\rangle = \exp(\alpha\alpha'\langle b^\dagger b\rangle)$ if $H = \omega b^\dagger b$:

$$\langle \psi^\dagger(x)\psi(0)\rangle_{T=0} = (k_F/\pi)[\sin(k_F x)/(k_F x)]\exp[-A_1(|x|/R)]; \quad (4.24)$$

$$\langle \psi^\dagger(x), \psi(0)\rangle = \langle \psi^\dagger(x)\psi(0)\rangle_{T=0}\exp[-F(|x|)],$$

$$F(x) = \int_0^\infty \mathrm{d}q\, q^{-1}[\exp(\beta\omega_q) - 1]^{-1}\cosh(2\varphi_q)[2\sin(\tfrac{1}{2}qx)]^2. \quad (4.25)$$

At $T = 0$ the familiar free-electron result is reduced at large separations by a factor $\exp(-\bar{A})|2\pi x/R|^{1-2\nu}$. At low but finite temperatures $T \ll v_S/R$, it is further reduced at separations $|x| \gg v_S/T \gg R$ by a factor $\exp(-2\nu|x|\xi)$, where $\xi = (v_S\pi T)$. Note that when models with the same sound velocity v_S are compared, the single-particle correlation function of the interacting model is always reduced below that of the free model.

The recipe for such calculations of correlation functions was first given by Luther and Peschel (1974). The calculation is easily extended to give the dynamic correlation functions, as shown by these authors. In table 1 I summarise the low-energy properties of the spinless fermion Luttinger model, and list the static single-particle, density, and pair correlation functions. In the Luttinger model itself, the linear relation (3.6) between the electron field $\psi(x)$ and the fields $\psi_p(x)$ means that the single-particle correlation only has a k_F oscillatory component, while the density and pair correlations only have 0 and $2k_F$ components, just as in the case of the free Fermi gas. However, in a more general model where J is not strictly conserved, interaction effects will give rise to additional periodic components with extra multiples of $2k_F$ in the period. For example, in addition to the two components $\psi_p^\dagger(x)$ and $\psi_{-p}^\dagger(x)$ making up the operator representing the electron field $\psi^\dagger(x)$, there will be admixture of terms like $\psi_p^\dagger(x)\,\nabla\psi_p^\dagger(x)\,\psi_{-p}(x)$ which adds a $3k_F$ oscillatory term to the single-particle correlation function. Charge conservation allows terms with periodicity $(2m+1)k_F$ in the single-particle correlation function, and $2mk_F$ in the density and pair correlation functions, and the relevant terms are listed in table 1.

To conclude the discussion of the Luttinger model solution, I note that the low-energy properties of the diagonalised model depend on five distinct parameters: v_S, v_N and v_J parametrise the Hamiltonian, k_F the momentum operator, and φ the fermion field operator. A fundamental result is the relations $v_N = v_S\exp(-2\varphi)$, $v_J = v_S\exp(2\varphi)$, which were deduced from the structure of the solution. The question arises: are these relations fundamental, in that they can be deduced solely from the low-energy structure

Table 1. Summary of 'Luttinger liquid' properties of the spinless 1D Fermi gas. $[\psi_p(x)]^m$ means

$$\lim_{a\to 0}[a^{-\frac{1}{2}m(m-1)}\psi_p(x)\psi_p(x+a)\ldots\psi_p(x+(m-1)a)].$$

Higher harmonics of $2k_F$ allowed by charge conservation, and likely to be present in a more general model, are also included in the list of correlation functions. The phase (cos or sin) of the asymptotic oscillations is also indicated.

1. Interaction parameter (> 1 for repulsive forces): $\exp(-2\varphi)$
2. Relation of Fermi vector k_F to charge density $\rho = N/L$: $k_F = \pi\rho$
3. Density fluctuation sound velocity: v_S
4. Change of chemical potential with Fermi vector: $v_N \equiv d\mu/dk_F = v_S e^{-2\varphi}$
5. Fermi velocity (for currents): $v_J = v_S e^{2\varphi}$
6. Asymptotic form of low-temperature correlation functions:

$$\langle A^\dagger(x)A(x')\rangle \sim \sum_i C_i \begin{Bmatrix}\cos(n_i k_F(|x-x'|)\\ \sin(n_i k_F|x-x'|)\end{Bmatrix}[|x-x'|^{-1}\exp(-\pi T|x-x'|/v_S)]^{\eta_i}.$$

Correlation	$A^\dagger(x)$	Luttinger model form	n	η
Single-particle (sin)	$\psi^\dagger(x)$	$\psi_p^\dagger(x)$ $[\psi_p^\dagger(x)]^{m+1}[\psi_{-p}(x)]^m$	1 $(2m+1)$	$\frac{1}{2}e^{-2\varphi}+\frac{1}{2}e^{2\varphi}$ $\frac{1}{2}e^{-2\varphi}+2(m+\frac{1}{2})^2e^{2\varphi}$
Density (cos)	$[\psi^\dagger(x)\psi(x)-\rho]$	$\rho_p(x)$ $[\psi_p^\dagger(x)]^m[\psi_{-p}(x)]^m$	0 $2m(\geq 2)$	2 $2m^2e^{2\varphi}$
Pair (cos)	$\psi^\dagger(x)\nabla\psi^\dagger(x)$	$\psi_p^\dagger(x)\psi_{-p}^\dagger(x)$ $[\psi_p^\dagger(x)]^{m+1}[\psi_{-p}(x)]^{m-1}$	0 $2m$	$2e^{-2\varphi}$ $2e^{-2\varphi}+2m^2e^{2\varphi}$

of the diagonalised form of the Hamiltonian, without reference to the 'bare' form of the model? The answer is yes: the relations (4.7) can be obtained by considering the static response functions of the density components $\Sigma_p\rho_{pq}$ and $\Sigma_p p\rho_{pq}$; when $q \neq 0$, the calculation only involves the boson variables, and v_S and $\exp(-2\varphi)$. In the limit $q \to 0$, the results must go over into the results $1/2\pi v_N$ and $1/2\pi v_J$ calculated when $q = 0$, and the relations (4.7) are recovered.

In addition to the above five characteristic parameters, various multiplicative factors appear in the asymptotic form of the various correlation functions. These depend only on the length scale R, and the two constants $\bar{A}$ and $\bar{B}$; however, in contrast to (4.7), the relation between these various multiplicative factors is likely to be a model-dependent feature of the Luttinger model, as R, $\bar{A}$ and $\bar{B}$ depend on the high-energy structure of the model (i.e., the cut-off function $g(y)$).

5. Generalisation to non-soluble models: the 'Luttinger liquid' concept

The complete solubility of the Luttinger model makes it a fascinating example of an interacting one-dimensional system. Nevertheless, its solubility rests on quite specific properties that are lost if the model is modified. However, I will argue that its low-energy structure still provides a model of the most important features of more general, non-soluble models. As an example I consider a generalisation of the Luttinger model that incorporates a non-linear fermion dispersion relation:

$$\varepsilon(kp) = v_F(kp - k_F) + (1/2m)(kp - k_F)^2 + \lambda(1/12m^2v_F)(kp - k_F)^3. \qquad (5.1)$$

For stability reasons, it is necessary to include the cubic term: the ground state of the non-interacting model is altered unless $\lambda > \frac{3}{4}$, when $\mathrm{sgn}(\varepsilon(kp)) = \mathrm{sgn}(kp)$; $\varepsilon(kp)$ increases monotonically if $\lambda > 1$. In general, the interacting model will remain stable for λ greater than some positive limit λ_c. This modification of the model retains the feature that J is a good quantum number; though the non-linear dispersion means that the mean current operator j is no longer simply proportional to J, it remains so in a low-energy subspace.

The procedure for translating this generalised Luttinger model into normal-ordered boson form is extremely simple. An expansion technique as in equations (3.42)–(3.45) can be used to transcribe the non-linear fermion dispersion terms. The general fermion representation (4.19) should be used, with arbitrary parameter φ_q. The final result is a boson normal-ordered Hamiltonian with quadratic boson terms that depend on N and J, plus new cubic and quartic boson interaction terms. The parameter $\varphi_q(N/L, J/L)$ is then chosen to diagonalise the quadratic boson terms, giving a Luttinger model with N- and J-dependent parameters, *plus* irreducible boson interaction terms. The dependence of the Luttinger model parameters on N and J merely reflects the change in Fermi velocity for non-zero N and J, so in order to show up more clearly the other new feature (the boson–boson interaction), I give the new Hamiltonian only in the subspace $N = J = 0$; when $J \neq 0$, the structure of the boson spectrum is slightly altered in that $\varphi_q(N/L, J/L)$ and $\omega_q(N/L, J/L)$ are no longer even functions of q because the right- and left-travelling fermions then have different Fermi velocities. The boson part of the Hamiltonian has the form

$$H(N, J=0) = \sum_q \omega_q b_q^\dagger b_q + \sum_p \frac{1}{2\pi} \int_0^L \mathrm{d}x [:(1/6m)(\Phi_p(x))^3 + (\lambda/48m^2 v_F)(\Phi_p(x))^4:]$$

$$\Phi_p(x) = \sum_p pq\alpha(pq, -\varphi_q)(\mathrm{e}^{\mathrm{i}qx} b_q^\dagger + \mathrm{e}^{-\mathrm{i}qx} b_q). \tag{5.2}$$

The colons :(...): mean boson normal-ordering. The parameters ω_q and φ_q are now given by modified versions of the expressions (4.4) and (4.5), where v_F has been replaced by $\tilde{v}_{Fq} = v_F + (\lambda/4m^2 v_F)(c_1 + \frac{1}{6}q^2)$; the equation for φ_q must be solved self-consistently, since the constant term c_1 itself depends on φ_q:

$$c_1 = \frac{2\pi}{L} \sum_q |q| \sinh^2(\varphi_q) \equiv A_1''(0)/R^2.$$

The constant c_1 exists provided the large-q behaviour of the fermion interaction matrix elements is sufficiently good for $yg(y)$ to vanish as $y \to \infty$. In fact, as will be seen, the requirement that the renormalisation of the ground state of the quadratic part of (5.2) by the boson interactions be finite imposes the stronger requirement $y^3 g(y) \to 0$ as $y \to \infty$. Assuming $V_1(q)$ does not diverge as $q \to \infty$, this implies the condition $qV_2(q) \to 0$ as $q \to \infty$, a slightly stronger condition than in the absence of a non-linear dispersion ($q^{1/2} V_2(q) \to 0$).

With the explicit construction (5.2) of the boson–boson interaction terms induced by a non-linear fermion dispersion, it is possible to construct an expansion in m^{-1} for the changes in the model properties due to the modification. This is particularly interesting in the case of the correlation functions: it allows the rigorous proof, at least for this type of generalised model, that the relations (4.7) between the spectral parameters v_S, v_N, and v_J and the parameter φ, and the relation between φ and the correlation exponents, remain unchanged from those found in the unmodified Luttinger model. This provides evidence in favour of the universal nature of these relations which will be proposed in this paper.

The relation between the spectral parameters is easiest to demonstrate; I give the form of the Hamiltonian in the subspace where no boson modes are excited:

$$H(n_q=0)=\tfrac{1}{2}(\pi/L)(v_N N^2+v_J J^2)+(1/6m)(\pi/L)^2(N^3+3NJ^2)$$
$$+(\lambda/48m^2v_F)(\pi/L)^3(N^4+6N^2J^2+J^4);$$
$$v_N=\bar{v}_{F0}+V_1(0)+V_2(0);\qquad v_J=\bar{v}_{F0}+V_1(0)-V_2(0).\tag{5.3}$$

The relations (4.7) between v_N, v_J, v_S and φ are clearly unchanged. The stability condition giving the lower bound λ_c to allowed values of λ is clearly obtained by demanding that the ground state of (5.3) has $N=J=0$. A necessary condition is that v_N and v_J are positive definite, i.e., that $|V_2(0)|<v_F+V_1(0)+(\lambda/4m^2v_F)c_1(\lambda)$; since $c_1(\lambda)$ is positive, this is a less restrictive condition than that in the original Luttinger model with $m^{-1}=0$. The condition $\lambda/v_F>\max(3/4v_N,\ 1/v_J)$ ensures (5.2) has no stationary points other than $N=J=0$, and is sufficient to guarantee stability.

The effect of the non-linear dispersion on the correlation functions will now be discussed. I study the single-electron correlation function $\langle\psi^\dagger(x)\psi(0)\rangle_{T=0}$ (4.24) discussed earlier, as an example. Following that discussion, this is given (after a little manipulation) by

$$\langle\psi^\dagger(x)\psi(0)\rangle_{T=0}=(k_F/\pi)(k_Fx)^{-1}\exp[-A_1(x/R)]\frac{1}{2\mathrm{i}}\sum_p p\exp(\mathrm{i}pk_Fx)$$
$$\times\langle\exp[\mathrm{i}\chi_p^\dagger(x)]\exp[\mathrm{i}\chi_p(x)]\rangle$$
$$\chi_p(x)=\sum_q\alpha(pq,-\varphi_q)\,2\sin(\tfrac{1}{2}qx)b_q.\tag{5.4}$$

$\langle\psi^\dagger\psi\rangle=(k_F/\pi)$, so the relation between k_F and electron density is unaffected by the non-linear dispersion. The expectation value is of course taken in the ground state of the interacting boson system (5.2), and hence differs from unity when m^{-1} is non-zero. A perturbation expansion in m^{-1} can be developed; the ground state expansion is

$$|\mathrm{GS}\rangle=\mathcal{N}\left(1+\frac{1}{6mv_S}\sum_{q_1+q_2+q_3=0}f(q_1,q_2,q_3)b^\dagger_{q_1}b^\dagger_{q_2}b^\dagger_{q_3}+\mathrm{O}(m^{-2})\right)|0\rangle$$

$$f(q_1,q_2,q_3)=\frac{L}{2\pi}\frac{q_1q_2q_3}{|q_1|+|q_2|+|q_3|}\sum_p p\prod_{i=1}^{3}\alpha(pq_i,-\varphi_{q_i}).\tag{5.5}$$

The normalisation constant $\mathcal{N}$ is given by

$$\mathcal{N}=1-\tfrac{1}{2}(1/mv_SR)^2(L/2\pi R)c_2[\varphi g(y)]+\mathrm{O}(m^{-4});$$
$$c_2=\frac{R^3}{6}\frac{2\pi}{L}\sum_{q_1+q_2+q_3=0}f(q_1,q_2,q_3)^2\tag{5.6}$$

$c_2[\varphi g(y)]$ is a positive dimensionless constant that is finite provided $y^3g(y)\to0$ as $y\to\infty$, as mentioned earlier:

$$c_2=\int_0^\infty\mathrm{d}x\int_0^x\mathrm{d}y\,x^{-1}(x^2-y^2)\,h(x,y)^2$$

$$h(x, y) = \tfrac{1}{2}[cg(x + y)cg(x - y)sg(2x) + sg(x + y)sg(x - y)cg(2x)] \tag{5.7}$$

where $cg(x)$ and $sg(x)$ are $\cosh(\varphi g(y))$ and $\sinh(\varphi g(y))$. Note that c_2 vanishes in the absence of fermion interactions ($\varphi = 0$), when there is no renormalisation of the ground state by the boson interactions.

I now calculate the single-electron correlation function to $O(m^{-1})$. From (5.4) and (5.5), this is given by

$$(k_F/\pi) \exp[-A_1(|x|/R)](k_F x)^{-1}\left[\sin(k_F x) - \cos(k_F x)\left(\tfrac{1}{6}\left\langle\sum_p p(\chi_p(x))^3\right\rangle + \text{HC}\right) + O(m^{-2})\right];$$

$$\left\langle\sum_p p(\chi_p(x))^3\right\rangle = -(1/mv_S x)F(|x|/R) + O(m^{-2});$$

$$F(u) = 4\int_0^\infty dx \int_0^x dy\, x^{-1}\sin x\,(\cos x - \cos y)h(x/u, y/u)^2.$$

The function $F(u)$ vanishes at $u = 0$, and remains bounded as $u \to \infty$; the corrections to the correlation function thus do not affect the asymptotic behaviour of the correlation functions. Physically, this is because the factors $|q_i|^{1/2}$ in the interaction matrix elements of (5.2) kill the effects of the boson interactions at long wavelengths. The relation between the various correlation exponents and the parameter φ is thus identical to that in the original Luttinger model; the *value* of the parameter φ, on the other hand, *is* affected by the interaction terms, and varies with the ground state charge density.

6. Discussion: the Luttinger liquid concept

To summarise the results of this paper: it has been shown that the low-energy excitation of the soluble Luttinger model of interacting fermions in one dimension consists of three parts: the well known collective density fluctuation boson modes, plus charge and current excitations, which have not previously been emphasised. Associated with these three types of excitations are three velocities, v_S, v_N and v_J, which obey the relation $v_S = (v_N v_J)^{1/2}$. The current of the Luttinger model is a good quantum number, and is quantised in units $2v_J/L$, each unit carrying momentum $2k_F$. $v_N = d\mu/dk_F$ describes the rate of change of chemical potential with the Fermi vector, which is unrenormalised by interactions, and given by the charge density, $k_F = \pi(N/L)$; v_S is the density excitation sound velocity. The relation between the three velocities defines a parameter φ: $v_N = v_S \exp(-2\varphi)$, $v_J = v_S \exp(2\varphi)$. This parameter φ is the intrinsic renormalised coupling constant of the model, and determines the non-integer power laws characterising the asymptotic behaviour of the correlation functions. The elementary excitations of the Luttinger model are non-interacting, which explains why it can be explicitly solved. An important tool for working with the model and its generalisation is the representation of the fermion fields in terms of the elementary excitations: this is given here in a fully precise form.

A generalisation of the Luttinger model with a non-linear fermion dispersion, but where the current quantum number J is still conserved, was considered here. It was shown that the characteristic low-energy structure of the Luttinger model was preserved, including the relations between its velocities and correlation exponents, but that its renormalised parameters now depend on the position of the Fermi level, and non-linear couplings appear between the elementary excitations.

On the basis of this demonstration that this structure remains valid in a much wider class of models than the Luttinger model itself, I will propose that it is generally valid for conducting spinless fermion systems in one dimension. For full generality, it is necessary to consider models where the current quantum number J is no longer a good quantum number: this will be done in the next paper in this series. What emerges is that unless a multiple of the fundamental wavevector $2k_F$ is some multiple of a reciprocal lattice vector reflecting an underlying periodicity of the system, momentum conservation eventually inactivates a non-J-conserving term at low energies (though such terms will give rise to renormalisations of the low-energy spectral parameters), and the low-energy structure is again of the form described here. If $2k_F = (n/m)G$, this remains valid provided $\exp(-2\varphi)$ is less than a critical value $\frac{1}{2}m^2$, above which an instability against an insulating pinned charge-density-wave state occurs. If such Umklapp processes are present, but $\exp(-2\varphi) < \frac{1}{2}m^2$, there is a characteristic non-analytic scaling dependence of the renormalised $\exp(-2\varphi)$ on $|2k_F - (n/m)G|$, reflecting the power laws of the correlation functions.

A very important test of the universality of the Luttinger model structure is provided by the class of models exactly soluble by the Bethe *ansatz*, mentioned in the Introduction. For these models, v_S, v_N, and v_J can be explicitly calculated, though their correlation functions have not as yet been obtained. As described in Haldane (1981), the relation $v_S = (v_J v_N)^{1/2}$ can be explicitly verified, and the parameter $\exp(-2\varphi)$ obtained from these velocities shows the characteristic behaviour due to Umklapp processes when $2k_F \sim (n/m)G$ mentioned above, providing additional confirmation that the relation between $\exp(-2\varphi)$ and the correlation exponents is valid (Haldane 1980).

It is obviously possible to generalise the discussion to the case of spin-$\frac{1}{2}$ fermions; the spin-$\frac{1}{2}$ Fermi gas has a characteristic instability against a gap opening in the spin excitation spectrum in zero magnetic field, if $2k_F$ exchange (backscattering) processes are attractive (Luther and Emery 1974); the resulting state is the one-dimensional analogue of superconductivity, though no long-range order is involved, and can be related to the 1D Bose fluid. Similarly, when Umklapp processes open up a gap in the charge density excitation spectrum, leaving gapless low-energy spin-wave modes (Emery *et al* 1976), the resulting system models the antiferromagnetic chain. This in turn can be related to a ferromagnetic chain by a sublattice rotation. In this way, the apparently diverse collection of systems mentioned in the Introduction can be brought into the framework of what I propose to call 'Luttinger liquid theory', which can be tested on those models soluble by the Bethe *ansatz*. Of course, this description of these models is only valid in those regimes where they have a gapless linear density wave excitation, and are conductors of a locally conserved charge, with associated quantised persistent currents at $T = 0$. This underlying unity explains the rather bizarre fact that spin systems and Bose fluids in one dimension have the fermion-like property of a characteristic momentum $2k_F$, as seen in the equivalence of the $S = \frac{1}{2}XY$ spin chain and hard core bose lattice gas to a spinless fermion system (Lieb *et al* 1961, Matsubara and Matsuda 1956). These generalisations will be discussed in detail in subsequent papers.

The emphasis here has been on spectral properties and correlation functions. As a final comment, I note that the approach introduced here could be used as the basis of a theory of transport processes in 'Luttinger liquids'; for example, in the Luttinger model itself, transport of energy by the boson modes would be purely ballistic, since they are non-interacting. The boson interactions due to a non-linear fermion dispersion would introduce lifetime effects and dissipative behaviour.

References

Bethe H A 1930 *Z. Phys.* **71** 205
Dover C B 1968 *Ann. Phys., NY* **50** 500
Dzyaloshinskii I E and Larkin A I 1973 *Zh. Eksp. Teor. Fiz.* **65** 411 (Engl. Transl. 1974 *Sov. Phys.–JETP* **38** 202)
Emery V J, Luther A I and Peschel I 1976 *Phys. Rev.* B **13** 1272
Everts H V and Schulz H 1974 *Solid State Commun.* **15** 1413
Gradsteyn I S and Ryzhik I M 1965 *Tables of Integrals, Series, and Products* (New York: Academic Press)
Haldane F D M 1979 *J. Phys. C: Solid State Phys.* **12** 4791
—— 1980 *Phys. Rev. Lett.* **45** 1358
—— 1981 *Phys. Lett.* **81**A 153
Heidenreich R, Schroer B, Seiler R and Uhlenbrock D 1975 *Phys. Lett.* **54**A 119
Lieb E H, Schultz T and Mattis D C 1961 *Ann. Phys., NY* **16** 407
Luther A and Emery V J 1974 *Phys. Rev. Lett.* **33** 389
Luther A and Peschel I 1974 *Phys. Rev.* **B9** 2911
Luttinger J M 1963 *J. Math. Phys.* **15** 609
Matsubara T and Matsuda H 1956 *Prog. Theor. Phys.* **16** 569
Mattis D C 1974 *J. Math. Phys.* **15** 609
Mattis D C and Lieb E H 1965 *J. Math. Phys.* **6** 304
Overhauser A W 1965 *Physics* **1** 307
Theumann A 1967 *J. Math. Phys.* **8** 2460

Volume 131B, number 1,2,3 PHYSICS LETTERS 10 November 1983

THEORY OF NONABELIAN GOLDSTONE BOSONS IN TWO DIMENSIONS

A. POLYAKOV and P.B. WIEGMANN
Landau Institute of Theoretical Physics, Moscow, USSR

Received 20 May 1983
Revised manuscript received 12 July 1983

We present an exact theory of an O(4)-σ-model based on its relation to a certain fermionic model. The S-matrix and the vacuum energy in a constant external field are computed.

One of the most typical phenomena of contemporary physics is spontaneous symmetry breaking. When such a thing happens the system acquires massless particles in its spectrum (provided the original symmetry was continuous). These particles represent slowly varying fluctuations of the order parameter.The low energy interaction of these Goldstone particles are uniquely determined by the pattern of symmetry breaking and described by the so called non-linear σ-models.

In two-dimensional space–time and in the nonabelian case this interaction creates a severe infrared problem. As has been found in ref. [1] the theory becomes asymptotically free at small distances and, most probably, the Goldstone bosons acquire a mass through the dimensional transmutation, thus indicating complete restoration of symmetry.

One might ask here why any four-dimensional physicist should be interested in this theory? The answer is that four-dimensional gauge bosons behave in many respects similarly to the two-dimensional Goldstone bosons. Apart from the already mentioned asymptotic freedom, let us recall that in many cases both systems have similar instanton structure [2], that they are identical under Migdal–Kadanoff recursion computations [3], and that the gauge fields can be formulated as a nonlinear σ-model in loop space [4]. There are also numerous applications in solid state physics [5].

After these justifications let us recall the theoretical status of the model. In most cases of symmetry breaking it has an infinite number of quantum conserved currents [6,7] and if one guesses that the mass gap really appears in the theory and assumes the most natural quantum numbers for the elementary excitations then the S-matrix can be explicitly found [8].

However up to now the exact solution of the model has not been presented, though all experts agreed that exists.

In this paper we present such a solution. It is obtained by first reducing the model to another model of the four fermion interaction and by solving the latter by means of the Bethe-Ansatz technique. We shall be dealing mostly with the case of the principal chiral field for the SU(2) group but our methods are applicable in other interesting cases as well. The principal chiral field is described by an SU(2) matrix $\Omega(x)$ and the lagrangian

$$\mathscr{L} = (1/2\lambda_0)\,\mathrm{tr}(\partial_\mu \Omega^{-1} \partial_\mu \Omega) \qquad (1)$$

(λ_0 is a bare coupling constant).

Let us compare this theory with another one:

$$\mathscr{L} = \sum_{a=1}^{N} \mathrm{i}\bar{\psi}_a \gamma_\mu \partial_\mu \psi_a + \lambda_0 \Big(\sum_{a=1}^{N} \bar{\psi}_a \gamma_\mu \tau^i \psi_a \Big)^2 \qquad (2)$$

(τ^i are Pauli matrices).

We shall prove that the theories (1) and (2) are equivalent at $N = \infty$. It is convenient to introduce an auxiliary field A^i_μ and to rewrite the lagrangian as:

$$\mathcal{L} = \sum_{a=1}^{N} \bar{\psi}_a \gamma_\mu (\mathrm{i}\partial_\mu + A_\mu)\psi_a + \frac{1}{2\lambda_0}(A_\mu)^2 . \qquad (3)$$

After standard integration over fermions we obtain an effective action:

$$S_{\text{eff}} = \int \mathrm{d}^2x\{(1/2\lambda_0)(A_\mu)^2 + NW[A_\mu]\} , \qquad (4)$$

here

$$A_\mu = A_\mu^i \tau^i, \quad W[A] = \ln\det(\gamma_\mu(\mathrm{i}\partial_\mu + A_\mu)) .$$

We shall compute W explicitly in a moment, but first let us explain the connection between the theories (2) and (1) on a qualitative level. The basic observation is that (4) is gauge invariant, being unchanged under transformations

$$A_\mu(x) \to \Omega^{-1}(x)A_\mu(x)\Omega(x) + \Omega^{-1}(x)\partial_\mu\Omega(x) \qquad (5)$$

and it depends effectively on the field strength

$$F_{\mu\nu} = \partial_\mu A_\nu - \partial_\nu A_\mu + [A_\mu, A_\nu] . \qquad (6)$$

Therefore if $N \to \infty$ the fluctuations of $F_{\mu\nu}$ become negligible:

$$F_{\mu\nu} \approx 0 .$$

But the langragian

$$\mathcal{L} = (1/2\lambda_0)\,\mathrm{tr}(A_\mu^2) ,$$

with the constraint

$$F_{\mu\nu} = 0 ,$$

is precisely the chiral field (1).

We expect therefore that as $N \to \infty$, the models (1) and (2) become equivalent.

To make the argument rigorous, let us compute $W[A]$. The computation is based on the axial anomaly, following the method found by Schwinger and Johnson. However some unusual features arise in the nonabelian case.

Let us define:

$$J_\mu = \delta W / \delta A_\mu . \qquad (7)$$

Then the standard Johnson [9]-like argument leads to the relations:

$$\partial_\mu J_\mu + [A_\mu, J_\mu] = 0 ,$$
$$\epsilon_{\mu\nu}(\partial_\mu J_\nu + [A_\mu, J_\nu]) = (1/2\pi)\epsilon_{\mu\nu}F_{\mu\nu} . \qquad (8)$$

The system (8) determines in principle J_μ and therefore W. What is most interesting, however, is that its solution can be obtained in an explicit form. To find this let us use coordinates $x_\pm = x_0 \pm x_1$, and represent the external field as

$$A_+ = g^{-1}\partial_+ g, \quad A_- = h^{-1}\partial_+ h .$$

(Notice that it is not a pure gauge for $h \neq g$.) Then an easy computation shows that

$$J_+ = g^{-1}\partial_+ g - h^{-1}\partial_+ h ,$$
$$J_- = h^{-1}\partial_- h - g^{-1}\partial_- g , \qquad (9)$$

are solutions of (8). The further calculations are simplified in the axial gauge $A_- = 0$, $h = I$. In this case:

$$\delta W = \int \mathrm{d}^2x\, \mathrm{tr}(J_- \delta A_+)$$
$$= \int \mathrm{d}^2x\, \mathrm{tr}(\partial_-(g^{-1}\partial_+ g)\delta g g^{-1}) . \qquad (10)$$

(We have used an identity:

$$\nabla_+(g^{-1}\partial_- g) - \partial_-(g^{-1}\partial_+ g) = 0 ,$$

where $\nabla_+ f = (\partial_+ f + [g^{-1}\partial_+ g, f])$ is the covariant derivative.) We have now to find a functional W the variation of which is equal to (10). It is quite surprising that this problem leads us to the multivalued action functionals for chiral fields considered recently by Novikov [10] and Witten [11] in a quite different setting ‡1. A single valued W does not exist, or more accurately, does not exist globally. The answer has the following form. It is convenient to consider x-space to be a sphere S^2. Introduce a three-dimensional sphere with coordinates ξ^A and consider its hemisphere bounded by our S^2. Then W is given by

‡1 In the local form this lagrangian appeared in the paper by Wess and Zumino [12]. Constructions for the Yang–Mills case, which are close in spirit, can be found in ref. [13].

$$W = \frac{1}{2}\int_{S^2} d^2x \, \mathrm{tr}(\partial_\mu g^{-1}\partial_\mu g)$$

$$+ \frac{i}{8\pi^2}\int_{Q} d^3\xi \, \epsilon^{ABC} \, \mathrm{tr}(g^{-1}\partial_A g g^{-1}\partial_B g g^{-1}\partial_C g)$$

$$(\partial Q = S^2) . \qquad (11)$$

Another choice of hemisphere would lead to the action $\bar{W}$, such that

$$\bar{W} - W = \frac{i}{8\pi^2}\oint d^3\xi\epsilon^{ABC} \, \mathrm{tr}(g^{-1}\partial_A g g^{-1}\partial_B g g^{-1}\partial_C g) ,$$

which is an integer, since the homotopy gauge $\Pi_3(SU(2)) = Z$. The proof of (11) is easily obtained by taking its variation, using Stokes theorem and comparing with (10). Locally, (11) can be rewritten in many equivalent ways directly in x-space. For example if $g = e^{i\phi}$ (ϕ being an antihermitean matrix) then integration of (10) gives:

$$W = \int_0^1 d\alpha (1-\alpha) \int d^2x \, \mathrm{tr}(\partial_- \phi \, e^{-\alpha\phi} \partial_+ \phi \, e^{\alpha\phi}) , \qquad (12)$$

The multivaluedness of W is harmless under the condition that the number of flavours N in (3) is integer. It is remarkable that our lagrangian remembers that it was obtained from an integral number of fermions.

The last part of the proof of the equivalence is easy. Pass from the axial gauge to the general one by a gauge rotation Ω; then (3) is rewritten as

$$\mathscr{L} = (1/2\lambda_0) \, \mathrm{tr}(\Omega^{-1}\partial_-\Omega)\{\Omega^{-1}\partial_+\Omega + \Omega^{-1}(g^{-1}\partial_+ g)\Omega\}$$
$$+ NW(g) . \qquad (13)$$

If we look at the perturbation expansion in (13) for small λ_0 and large N, we find a theory of the chiral Goldstones (π-mesons) interacting with another massless field ϕ-mesons. Combining (12, 13) we have

$$\mathscr{L} = (1/2\lambda_0) \, \mathrm{tr}\{(\Omega^{-1}\partial_\mu\Omega)^2$$
$$+ \partial_\mu\Omega\Omega^{-1}(\partial_\mu\phi + [\partial_\mu\phi, \phi] + \cdots)\}$$
$$+ N \, \mathrm{tr}\{\tfrac{1}{2}(\partial_\mu\phi)^2 + i\epsilon_{\mu\nu}[\partial_\mu\phi, \partial_\nu\phi] + \cdots\} . \qquad (14)$$

This theory is renormalizable and has two coupling constants λ_0 and N^{-1}. An important conclusion which we derive from the multivaluedness of the action is that one of them (i.e. N^{-1}) is not renormalized. Since any amplitude with radiation of ϕ-particles contains the factor $1/N$ we conclude that as $N \to \infty$ only π-mesons remain with us. In other words the fluctuations of the ϕ-field are precisely those of $F_{\mu\nu}$, this result proves our assumption that fluctuations of F disappear from the physical degrees of freedom, so that A_μ in (6) becomes a pure gauge field as $N \to \infty$.

It is perfectly possible to solve exactly the N-flavour fermionic model (2) via a hierarchy of Bethe-Ansätze. The direct application of the Bethe-Ansatz technique to this model encounters a difficulty connected with the proper treatment of the axial anomaly. To avoid this difficulty it is necessary to consider a non-relativistic integrable version of this model. Such a model does exist but its solution is rather cumbersome. For this reason we take another route. Namely, there exists a much simpler model (from the point of view of Bethe-Ansatz application) which appears to be equivalent to the N-flavour model (2). This is the completely integrable theory of the single fermion field but with arbitrary isotopic spin S. The interaction lagrangian is given by

$$\mathscr{L}_{\mathrm{int}} = \sum_{n=0}^{2S} P_n(\lambda_0)(\chi^+ \gamma_\mu S^{a_1} S^{a_2} \cdots S^{a_n} \chi)^2 . \qquad (15)$$

Here χ is a $2S+1$ component isospinor, $S^a = (S^x, S^y, S^z)$ are SU(2) generators for the spin S representation, λ_0 is a coupling constant and polynomials $P_n(\lambda_0) \sim \lambda_0^n$ at small λ_0 are determined from the condition of complete integrability of the model and are given by

$$\sum_{n=0}^{2S} P_n(\lambda) z^n = i^{-1} \ln R(\lambda; z) ,$$

$$R(\lambda; z) = -\sum_{l=0}^{2S} \prod_{k=1}^{l} \frac{1-k\lambda}{1+k\lambda} \pi^l(z) ,$$

$$\pi_l(z) = \prod_{\substack{l=0 \\ l \neq l'}}^{2S} \frac{z - z_{l'}}{z_l - z_{l'}} ,$$

$$z_l = \tfrac{1}{2}[l(l+1) - 2s(s+1)] , \qquad (16)$$

Here $R(\lambda, z)$ is a scattering matrix for the particles of spin S, satisfying Yang–Baxter equations with λ_0^{-1} being the spectral parameter and $z = \bar{S}_1\bar{S}_2$; $\pi_l(z)$ are projections onto the state $|l\rangle$ with the given spin l ($\bar{S}_1\bar{S}_2|l\rangle = Z_l|l\rangle$). This factorized S-matrix has been discovered in refs. [14,15]. Integrability of the fermion lagrangian (15) follows from factorizability of the R-matrix. This model is equivalent to the N-flavoured one in the non-Goldstone sector provided that $N = 2S$. This can be proved by direct comparison of the corresponding exact solutions and perhaps by more intelligent methods. Since we do not intend to derive here the Bethe-Ansatz solution of the N-flavour model the proof of equivalence is beyond the scope of the present letter and will be published elsewhere. For physical applications we consider the chiral field in the external "magnetic" field by adding the terms

$$\mathcal{L}_{\rm H} = H \sum_{a=1}^{N} (\bar{\psi}_a \gamma_0 \tau^3 \psi_a) ,$$

or

$$\mathcal{L}_{\rm H} = H(\bar{\chi}\gamma_0 S^3 \chi) .$$

This is equivalent to the following extra term in the chiral lagrangian (1):

$$\mathcal{L}_{\rm H}^{(\rm chiral)} = H \, {\rm tr}(\tau^3 \Omega^{-1} \partial_0 \Omega) .$$

Application of the Bethe-Ansatz technique to the S-spin fermionic model (15) begins from the diagonalization of the transfer-matrix obtained from the corresponding R-matrix (16) [16,17]. This problem has already been solved in the context of a generalized Heisenberg model in refs. [14,18,19]. Adapted to our case, these results lead to the following Bethe-Ansatz equations

$$\exp({\rm i}p_j^{\mp}L) = \prod_{\alpha=1}^{M} \left(\frac{\theta_\alpha + {\rm i}S \mp \lambda_0^{-1}}{\theta_\alpha - {\rm i}S \mp \lambda_0^{-1}}\right) ,$$

$$\prod_{(\pm)} \left(\frac{\theta_\alpha \mp \lambda_0^{-1} + {\rm i}S}{\theta_\alpha \mp \lambda_0^{-1} - {\rm i}S}\right)^{N_\pm} = \prod_{\beta=1}^{M} \left(\frac{\theta_\alpha - \beta + {\rm i}}{\theta_\alpha - \theta_\beta - {\rm i}}\right) . \qquad (17)$$

Here $p_j^{\mp}$, $N_\pm$ are momenta and numbers of particles of positive and negative chiralities and $\{\theta_\alpha\}$ is a set of "rapidities" which parametrize the eigenvalues of the transfer-matrix. The number M is the number of reversed spins. The projection of spin and the energy of states are given by

$$S^z = (N_+ + N_-)S - M, \quad E = \sum_{j=1}^{N_+} p_j^+ - \sum_{j=1}^{N_-} p_j^- . \qquad (18)$$

The analyses of (17), (18) at finite $S = N/2$ proceed in a standard fashion. In the thermodynamic limit all solutions of (17) group in the so-called "strings" of complex rapidities $\theta_{\alpha k}^{(n)} = \theta_\alpha^{(n)} + {\rm i}k$ $[k = -(n-1), \ldots, (n-1)]$ and are described by the distributions $\rho_n(\theta)$ of the real centres $\theta_\alpha^{(n)}$ of the strings. It is necessary to introduce the density of holes in the θ-distribution, $\tilde{\rho}_n(\theta)$. The eqs. (17) give the following relations between the distributions for the given state

$$\tilde{\rho}_n(\theta) + \int A_{n,m}(\theta - \theta')\rho_m(\theta')\,{\rm d}\theta'$$

$$= \int A_{n,2S}(\theta - \theta') \cdot s_+(\theta')\,{\rm d}\theta' ,$$

$$S^z = S - \int \rho_1(\theta)\,{\rm d}\theta ,$$

$$s_\pm = \tfrac{1}{4}[{\rm sech}\,\tfrac{1}{2}\pi(\theta - \lambda_0^{-1}) \pm {\rm sech}\,\tfrac{1}{2}\pi(\theta + \lambda_0^{-1})] ,$$

$$E = \int A_{n,2S}(\theta) \cdot s_-(\theta)\,{\rm d}\theta \qquad (19)$$

(we omit here the Goldstone part of the energy which is decoupled from the massive part of the spectrum). Here A_{nm} is connected with the scattering phase $\Phi_{nm}(\theta)$ of the bare strings n and m by the relation

$$A_{n,m} = \delta_{nm}\delta(\theta) + (1/2\pi)\,{\rm d}\Phi_{nm}(\theta)/{\rm d}\theta .$$

The explicit form of this matrix is:

$$A_{n,m}(\theta) = \int_0^\infty 2\coth\omega\,\sinh(\min(n,m)\omega)$$

$$\times \exp[-\max(n,m)\omega]\cos\omega\theta\,{\rm d}\omega/\pi . \qquad (20)$$

In order to determine the excitation spectrum, the S-matrix of the model and the magnetic field dependence of the vacuum energy we shall choose the most straightforward method due to Yang et al. [19]. In this method spectral integral equations are written directly for the energies of the ground state and the elementary excitation spectrum. Let $(-\epsilon_n^{(-)}(\theta)) < 0$ and $\epsilon_n^{(+)}(\theta) > 0$ be the energies for particles and holes in the n-string. Then according to eqs. (19) the Yang equations take the form

$$\epsilon_n^{(-)}(\theta) + (A^{-1})_{nm} * \epsilon_m^{(+)}(\theta) = (H - m \cosh \tfrac{1}{2}\pi\theta)\delta_{n,2S}. \tag{21}$$

Here (*) means convolution and $m \sim \Lambda \exp(-\pi/\lambda_0)$ is a gap in the physical spectrum. The continuous function

$$\epsilon_n(\theta) = \epsilon_n^{(+)}(\theta), \quad \text{if } \epsilon_n > 0,$$
$$\text{or } \epsilon_n(\theta) = \epsilon_n^{(-)}(\theta), \quad \text{if } \epsilon_n < 0,$$

which is completely determined by eq. (21), is the energy of the elementary excitation in the presence of a finite field H and the vacuum energy is

$$E(H) - E(0) = -\int m \cosh(\tfrac{1}{2}\pi\theta)\epsilon_{2S}^{(+)}(\theta)\, d\theta. \tag{22}$$

It is easy to show that $\epsilon_n^{(-)} = 0$ for $n \neq 2S$ indicating that, in a manner similar to ref. [20], the ground state is formed by the sea of "strings" of length $2S$. Using this fact one derives

$$\epsilon_{2S}^{(-)}(\theta) + A_{2S,2S}^{-1} * \epsilon_{2S}^{(+)}(\theta) = H - m \cosh(\tfrac{1}{2}\pi\theta), \tag{23}$$

This can be rewritten as a single integral equation for the function ϵ_{2S}:

$$\epsilon_{2S}(\theta) + \int_{-F}^{F} \mathcal{K}^{(S)}(\theta - \theta')\epsilon_{2S}(\theta')$$
$$= H - m \cosh(\tfrac{1}{2}\pi\theta). \tag{24}$$

The Fourier transform of the kernel $\mathcal{K}$ here is given by:

$$\mathcal{K}^{(S)}(\omega) = A_{2S,2S}^{-1}(\omega) - 1$$
$$= \text{th}|\omega|/(1 - e^{-2S|\omega|}) - 1, \tag{25}$$

and the Fermi-rapidity F is defined by the condition

$$\epsilon_{2S}(\pm F) = 0. \tag{26}$$

We conclude with the description of results obtained from (21)–(26). First of all, analogously to the S-spin Heisenberg model [21] the spectrum consists of isotopic spin 1/2 fermions irrespectively of S. But here this spectrum is doubly degenerate.

The four eigenstates of the scattering matrix – two triplet and two singlet states – are represented as the Bethe states as follows:

Triplet (tt): two holes in the $2S$-string. The S-matrix for this state is expressed through the kernel of the integral equation [22]

$$S_{tt}^{(S)}(\theta) = \exp\left(i \int_{-\infty}^{+\infty} \frac{\mathcal{K}^{(S)}(\omega)}{\omega} \sin(\omega\theta/\pi) \frac{d\omega}{\pi}\right). \tag{27a}$$

Triplet (ts): two $2S$-strings plus $(2S-1)$-string:

$$S_{ts}^{(S)}(\theta) = S_{tt}(\theta)$$
$$\times \exp\left(i \int_{-\infty}^{\infty} \frac{A_{2S-1,2S}(\omega)}{A_{2S,2S}(\omega)} \sin(\omega\theta/\pi) \frac{d\omega}{\omega\pi}\right)$$
$$= S_{tt}(\theta) \frac{\sin(i\theta - \pi)/4S}{\sin(i\theta + \pi)/4S}. \tag{27b}$$

Singlet (st): two $2S$-strings plus $(2S+1)$

$$S_{st}^{(S)}(\theta) = S_{tt}(\theta)$$
$$\times \exp\left(i \int_{-\infty}^{\infty} \frac{A_{2S+1,2S}(\omega)}{A_{2S,2S}(\omega)} \sin(\omega\theta/\pi) \frac{d\omega}{\omega\pi}\right)$$
$$= S_{tt}(\theta) \frac{i\theta - \pi}{i\theta + \pi}. \tag{27c}$$

Singlet (ss): two $2S$-strings plus $(2S+1)$-string plus $(2S-1)$-string:

$$S_{ss}^{(S)}(\theta) = S_{st}(\theta) S_{tt}^{-1}(\theta) S_{ts}(\theta). \tag{27d}$$

When $S \to \infty \mathcal{K}^{(S)}(\theta) = 2\mathcal{K}^{(1/2)}(\theta)$ and the prefactors in (27b,c) become equal. Therefore the scattering matrix of the chiral SU(2) field is:

$$S^{(\infty)}(\theta) = S^{(1/2)}(\theta) \otimes S^{(1/2)}(\theta) \tag{28}$$

and coincides with the S-matrix (28) for the SU(2)⊗SU(2) group for the $(\frac{1}{2}, \frac{1}{2})$ representation obtained previously by Zamolodchikov et al. [8] for the O(4) – σ-model by the S-matrix bootstrap method. Here

$$S^{(1/2)}(\theta) = i \frac{i\theta + \pi\hat{P}}{i\theta + \pi}$$
$$\times \frac{\Gamma(1 + i\theta/2\pi)\Gamma(1/2 - i\theta/2\pi)}{\Gamma(1 - i\theta/2\pi)\Gamma(1/2 + i\theta/2\pi)} \tag{29}$$

is the factorized S-matrix for SU(2) group for the fundamental representation. The situation is quite remarkable. Namely, the fermionic model had only one SU(2) symmetry for any finite S, and only this continuous symmetry was present in S-matrix. In the limit of $2S \equiv N = \infty$ another SU(2) appears. Thus, our analysis has confirmed a long standing conjecture that the excitations in the $SU(2) \otimes SU(2) = O(4)$ model are massive and transform according to the vector representation of O(4).

Investigation of eq. (24) permits us to determine the energy dependence on the magnetic field and compare the high-field expansion with the perturbation theory results. At finite S the analyses of (24), (25) proceed in a standard fashion (for instance see ref. [23]). The result is:

$$H/m \sim e^{\pi F/2}(1 + O(S/F)), \quad (30)$$

$$E(H) \sim -H^2 \sum_{n=0}^{\infty} c_n/F_*^n, \quad (31)$$

where

$$F_* - S \ln F_* = F.$$

Combining these formulas one gets the renormalization group prediction for the $N = 2S$-flavour model up to two loop order. However the behaviour (30), (31) is reached only in the region $\ln(H/m) \gg N$ i.e. at finite N. An interesting feature arises in the opposite limit $\ln(H/m) =$ fixed, $N \to \infty$ which corresponds to the chiral field and also should have a renormalization group structure, but a different one. The analysis of the eqs. (24), (26) at $S = \infty$ gives the result

$$H/m \sim (e^{\pi F/2}/\sqrt{F})(1 + O(1/F)), \quad (32)$$

$$E(H) \sim -H^2 \sum_{n=0}^{\infty} d_n/F^n, \quad (33)$$

which coincides with the two loop approximations for the chiral model [5,24].

The method described above can be clearly generalized for the $SU(N) \otimes SU(N)$ theories. The $O(3) - \sigma$-model can also be described by replacing $(A^i)^2$ by $\lambda_{\parallel}^{-1}(A^3)^2 + \lambda_{\perp}^{-1}[(A^1)^2 + (A^2)^2]$ in (3) or (6) with subsequent limit $\lambda_{\parallel} \to \infty$. However in this case the limiting procedure might be very subtle.

Recently there was an interesting suggestion concerning the solution of the O(3)-σ-model [25]. Its connection with our methods remains obscure to us.

We are grateful to A.A. Belavin, L.D. Faddeev, V.A. Fateev, A.M. Tsvelick and A.B. Zamolodchikov for many useful and interesting discussions. One of us (A.M.P.) would like to acknowledge an extremely stimulating discussion of this project with Alan Luther in 1976, and very helpful conversations with Ed Witten about anomalies.

References

[1] A.M. Polyakov, Phys. Lett. 59B (1975) 79.
[2] A.A. Belavin and A.M. Polyakov, Sov. Phys. JETP Lett. 47 (1975) 12.
[3] A.A. Migdal, Sov. Phys. JETP 42 (1976) 413.
[4] A.M. Polyakov, Phys. Lett. 82B (1979) 247.
[5] V.L. Pokrovsky, Adv. Phys. 28 (1979) 595.
[6] M. Lusher, Nucl. Phys. B135 (1978) 1.
[7] A.M. Polyakov, Phys. Lett. 72B (1977) 224.
[8] A.B. Zamolodchikov and Al.B. Zamolodchikov, Ann. Phys. 120 (1979) 253.
[9] K. Johnson, Phys. Lett. 5 (1963) 253.
[10] S.P. Novikov, to be published.
[11] E. Witten, Princeton preprint (1983).
[12] J. Wess and B. Zumino, Phys. Lett. 37B (1971) 25.
[13] S. Deser, R. Jackiw and N. Templeton, Phys. Rev. D23 (1981) 229.
[14] V.A. Fateev, unpublished.
[15] P.P. Kulish, N.Yu. Reshetichin and F.K. Sklyanin, Lett. Math. Phys. 5 (1981) 393.
[16] A.A. Belavin, Phys. Lett. 87B (1979) 117.
[17] N. Andrei and I.H. Lowenstein, Phys. Rev. Lett. 43 (1979) 698.
[18] P. Kulish, Proc. Intern. Conf. on Statistical mechanics, Vol. 2 (Dubna, 1982).
[19] C.N. Yang and C.P. Yang, J. Math. Phys. 16 (1969) 19.
[20] G.M. Babudjian, Nucl. Phys. B215 (FS7) (1983) 317.
[21] L.A. Takhtadjan, Phys. Lett. 87A (1982) 479.
[22] V.E. Korepin, Theor. Math. Phys. 41 (1979) 169.
[23] G.I. Japaridze and A.A. Nersesyan, Phys. Lett. 85A (1981) 23.
[24] Z. Brezin and J. Zinn-Justin, Phys. Rev. B14 (1976) 240.
[25] L.D. Faddeev and L.A. Takhtajan, preprint, LOMI (1983).

Commun. Math. Phys. 92, 455–472 (1984)

Communications in
Mathematical
Physics

Non-Abelian Bosonization in Two Dimensions

Edward Witten*

Joseph Henry Laboratories, Princeton University, Princeton, NJ 08544, USA

Abstract. A non-abelian generalization of the usual formulas for bosonization of fermions in $1+1$ dimensions is presented. Any fermi theory in $1+1$ dimensions is equivalent to a local bose theory which manifestly possesses all the symmetries of the fermi theory.

One of the most startling aspects of mathematical physics in $1+1$ dimensions is the existence of a (non-local) transformation from local fermi fields to local bose fields. Thus, consider the theory of a massless Dirac fermion:

$$\mathscr{L}_D = \bar{\psi} i \partial\!\!\!/ \psi . \tag{1}$$

This theory is equivalent [1] to the theory of a free massless scalar field:

$$\mathscr{L}_S = \tfrac{1}{2}\partial_\mu \phi \partial^\mu \phi . \tag{2}$$

The fermi field ψ has a relatively complicated and non-local expression [2] in terms of ϕ. However, fermion bilinears such as $\bar{\psi}\gamma_\mu\psi$ or $\bar{\psi}\psi$ take a simple form in the bose language. For example, the current $J_\mu = \bar{\psi}\gamma_\mu\psi$ becomes in terms of ϕ

$$J_\mu = \frac{1}{\sqrt{\pi}} \varepsilon_{\mu\nu} \partial^\nu \phi . \tag{3}$$

Similarly the chiral densities $\mathcal{O}_\pm = \bar{\psi}(1 \pm \gamma_5)\psi$ become

$$\mathcal{O}_\pm = M \exp \pm i \sqrt{4\pi}\phi , \tag{4}$$

where the value of the mass M depends on the precise normal ordering prescription that is used to define the exponential in (4).

By means of formulas like (3) and (4), the equivalence between the free Dirac theory and the free scalar theory can be extended to interacting theories. A perturbation of the free Dirac Lagrangian can be translated, via (3) and (4), into an equivalent perturbation of the free scalar theory. This procedure is remarkably

* Supported in part by NSF Grant PHY-80-19754

useful for elucidating the properties of 1+1 dimensional theories. Many phenomena that are difficult to understand in the fermi language have simple, semiclassical explanations in the bose language. A major limitation of the usual bosonization procedure, however, is that in the case of fermi theories with non-abelian symmetries, these symmetries are not preserved by the bosonization. For instance, a theory with N free Dirac fields has a $\mathrm{U}(N)\times\mathrm{U}(N)$ chiral symmetry [actually $\mathrm{O}(2N)\times\mathrm{O}(2N)$, as we will see later]. Upon bosonization, this becomes a theory with N free scalar fields. The *diagonal* fermi currents can be bosonized conveniently, as in Eq. (3), but the *off-diagonal* currents are complicated and non-local in the bose theory. [Although the free scalar theory with N fields has an $\mathrm{O}(N)$ symmetry, this $\mathrm{O}(N)$ does not correspond to any subgroup of the fermion symmetry group.] For this reason, it is rather difficult [3] to bosonize non-abelian theories by the usual procedure. It is also sometimes difficult to understand via bosonization the realization of non-abelian global symmetries.

In this paper, an alternative bosonization procedure will be described which generalizes the usual one and can be used to bosonize *any* theory in a local way, while manifestly preserving *all* of the original symmetries. Unfortunately, the resulting bose theories are somewhat complicated.

First, we rewrite Eq. (3) for the currents in a way susceptible of generalization.

We define an element U of the U(1) or O(2) group by $U=\exp i\sqrt{4\pi}\phi$. Then (3) can be written

$$J_\mu = -\frac{i}{2\pi}\varepsilon_{\mu\nu}U^{-1}\partial^\nu U = -\frac{i}{2\pi}\varepsilon_{\mu\nu}(\partial^\nu U)\cdot U^{-1}\,. \tag{5}$$

We have emphasized in (5) that the ordering of factors does not matter, because the group U(1) is abelian. In generalizing (5) we will have to be careful about factor ordering.

It is convenient to rewrite (5) in light cone coordinates. Let $x^\pm=(x^0\pm x^1)/\sqrt{2}$. In these coordinates the Lorentz invariant inner product is $A_\mu B^\mu = A^+B^- + A^-B^+ = A_+B_- + A_-B_+$; the components of a vector obey $A_+=A^-$, $A_-=A^+$. If we normalize the Levi-Civita symbol so that $\varepsilon_{01}=+1=-\varepsilon_{+-}$, then (3) and (5) become

$$\begin{aligned} J_+ &= -\frac{1}{\sqrt{\pi}}\partial_+\phi = \frac{i}{2\pi}U^{-1}\partial_+U\,,\\ J_- &= +\frac{1}{\sqrt{\pi}}\partial_-\phi = -\frac{i}{2\pi}(\partial_-U)U^{-1}\,. \end{aligned} \tag{6}$$

Of course, the ordering of factors in (6) is still arbitrary.

For the massless Dirac particle, the vector and axial vector currents $\bar\psi\gamma^\mu\psi$ and $\bar\psi\gamma^\mu\gamma_5\psi$ are both conserved.[1] But in 1+1 dimensions $\bar\psi\gamma^\mu\gamma_5\psi=\varepsilon^{\mu\nu}\bar\psi\gamma_\nu\psi$. So the current conservation equations are $0=\partial_\mu J^\mu=\varepsilon^{\mu\nu}\partial_\mu J_\nu$. In light cone coordinates this means $0=\partial_-J_+=\partial_+J_-$. The bosonization formula (6) is compatible with that strong condition because the free massless ϕ field obeys $0=\nabla^2\phi=2\partial_+\partial_-\phi$.

1 As usual, we define $\{\gamma_\mu,\gamma_\nu\}=2\eta_{\mu\nu}$, $\gamma_5=\gamma^0\gamma^1$ (so $\gamma_5^2=+1$), and $\bar\psi=\psi^*\gamma^0$. A convenient basis is $\gamma^0=\begin{pmatrix}0&1\\1&0\end{pmatrix}$, $\gamma^1=\begin{pmatrix}0&-1\\1&0\end{pmatrix}$, $\gamma_5=\begin{pmatrix}1&0\\0&-1\end{pmatrix}$. We define light cone components $\psi_\pm$ of ψ by requiring $\gamma_5\psi_-=\psi_-$, $\gamma_5\psi_+=-\psi_+$. (The sign convention may seem odd but is useful.) Thus $\psi=\begin{pmatrix}\psi_+\\\psi_-\end{pmatrix}$. ψ_+ and ψ_- are left movers and right movers, respectively, as one may see from Eq. (8) later

We wish to generalize this to fermion theories with non-abelian symmetries. As we wish to be general, we will consider a theory with N Majorana fermions ψ^i, $i=1\dots N$. [If one prefers, one can choose N even and consider this to be a theory of $N/2$ Dirac fields. If so, in much of the subsequent discussion one can consider the chiral group $U(N/2)\times U(N/2)$ instead of $O(N)\times O(N)$.] The conventional Lagrangian for free Majorana fields is

$$\mathscr{L}=\int d^2x\tfrac{1}{2}\bar{\psi}_k i\partial\!\!\!/\psi^k\,. \tag{7}$$

The conserved vector currents are $V_\mu^a=\bar{\psi}\gamma_\mu T^a\psi$, T^a being any generator of $O(N)$. The axial currents are $A_\mu^a=\varepsilon_{\mu\nu}V^{\nu a}=\bar{\psi}\gamma_\mu\gamma_5 T^a\psi$. These currents generate chiral $O(N)\times O(N)$.

Since $\bar{\psi}\gamma^\mu\partial_\mu\psi=\psi^T(\partial_0+\gamma^0\gamma^1\partial_1)\psi$, the free Lagrangian, in terms of the light cone components of ψ, is

$$\mathscr{L}=\tfrac{1}{2}i\int d^2x\left[\psi_-^k\left(\frac{\partial}{\partial t}+\frac{\partial}{\partial x}\right)\psi_-^k+\psi_+^k\left(\frac{\partial}{\partial t}-\frac{\partial}{\partial x}\right)\psi_+^k\right]. \tag{8}$$

Instead of vector and axial vector currents, it is more useful to work with chiral components. We define $J_+^{ij}(x,t)=-i\psi_+^i\psi_+^j(x,t)$ and $J_-^{ij}(x,t)=-i\psi_-^i\psi_-^j(x,t)$. Note that $J_\pm^{ij}$ are hermitian and that by fermi statistics they obey $J_+^{ij}=-J_+^{ji}$, $J_-^{ij}=-J_-^{ji}$. J_+^{ij} and J_-^{ij} generate chiral $O(N)_R$ and $O(N)_L$, respectively. [By $O(N)_R$ and $O(N)_L$ we mean $O(N)$ transformations for right-moving and left-moving fermions.] The conservation laws for $J_\pm$ are very simple

$$\partial_-J_+^{ij}=\partial_+J_-^{ij}=0\,. \tag{9}$$

Thus, J_+ is a function only of x^+, and J_- is a function only of x^-.

We wish to find an ansatz writing J_+ and J_- in terms of suitable bose fields. In the usual bosonization procedure, one considers a current $\bar{\psi}\gamma_\mu\psi$ that generates an abelian or U(1) symmetry; it is written [Eq. (5)] in terms of a field that takes values in the U(1) group. Now we are dealing with currents J_-^{ij} and J_+^{ij} that generate $O(N)_L\times O(N)_R$, and it is natural to try to express these currents in terms of a suitable field g that takes values in the $O(N)$ group. $O(N)_L\times O(N)_R$ will act on g by $g\to AgB^{-1}$, $A, B\in O(N)$.

What is a suitable expression for the currents in terms of g? One is tempted to try $J_+\sim g^{-1}\partial_+g$, $J_-\sim g^{-1}\partial_-g$. However, this is incompatible with (9) because in a *non-abelian* group the equations $0=\partial_-(g^{-1}\partial_+g)$ and $0=\partial_+(g^{-1}\partial_-g)$ are inconsistent. Instead, we generalize the factor ordering of Eq. (6) and write

$$J_+=\frac{i}{2\pi}g^{-1}\partial_+g\,,\qquad J_-=-\frac{i}{2\pi}(\partial_-g)g^{-1}\,. \tag{10}$$

[The ij indices are suppressed, it being understood that J_+ and J_- are elements of the $O(N)$ Lie algebra.] Notice that the equations $0=\partial_-(g^{-1}\partial_+g)$ and $0=\partial_+((\partial_-g)g^{-1})$ are compatible and in fact equivalent.

What Lagrangian will govern g? The obvious guess is

$$\mathscr{L}=\frac{1}{4\lambda^2}\mathrm{Tr}\,\partial_\mu g\partial_\mu g^{-1}\,. \tag{11}$$

This is the unique renormalizable and manifestly chirally invariant Lagrangian for g. However, for many reasons, (11) is wrong.

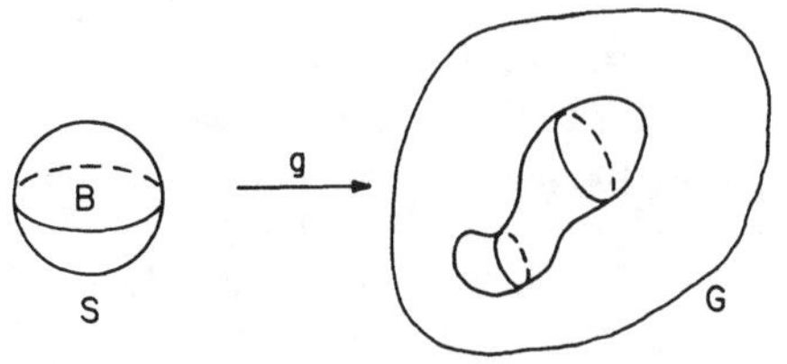

Fig. 1. A mapping g from a two sphere S (representing space-time) into a group manifold G. Since $\pi_2(G)=0$, any mapping of the *surface* S into G can be extended to a mapping into G of the solid sphere B (S and its interior)

First of all, (11) describes an asymptotically free theory with interactions that become strong in the infrared. It is certainly not equivalent to the conformally invariant free massless fermi field theory. Second, (11) leads to the equation of motion $0=\partial_\mu(g^{-1}\partial_\mu g)$ rather than the desired $0=\partial_-(g^{-1}\partial_+ g)=\partial_+((\partial_- g)g^{-1})$. Third, by analogy with similar considerations in QCD current algebra, [8] it may be shown that (11) has *more* discrete symmetries than the free, massless fermi theory.

Although (11) is the only renormalizable interaction for the non-linear sigma model that is *manifestly* chirally invariant, there is another one that is chirally invariant but not manifestly so. This is the two-dimensional analogue of the Wess-Zumino term [4], which has figured in various recent discussions of two dimensional models [5–7].

The two dimensional Wess-Zumino term can be constructed by analogy [8] with a similar treatment in four dimensions. Working in Euclidean space, we imagine space time to be a large two sphere S^2. Since π_2 $(\mathrm{O}(N))=0$, a mapping g from S into the $\mathrm{O}(N)$ manifold can be extended to a mapping $\bar{g}$ of a solid ball B whose boundary is S into $\mathrm{O}(N)$ (Fig. 1). If y_1, y_2, and y_3 are coordinates for B, the Wess-Zumino functional is

$$\Gamma=\frac{1}{24\pi}\int_B d^3y\varepsilon^{ijk}\,\mathrm{Tr}\,\bar{g}^{-1}\frac{\partial\bar{g}}{\partial y^i}\bar{g}^{-1}\frac{\partial\bar{g}}{\partial y^j}\bar{g}^{-1}\frac{\partial\bar{g}}{\partial y^k}. \tag{12}$$

As in four dimensions, the Wess-Zumino functional has a very essential property [8, 9]: it is well-defined only modulo a constant. Equation (12) has been normalized so that if g is a matrix in the fundamental representation of $\mathrm{O}(N)$, (12) is well-defined modulo $\Gamma\to\Gamma+2\pi$. The ambiguity in Γ arises because of the existence of topologically inequivalent ways to extend g into a mapping from B into $\mathrm{O}(N)$; the topologically distinct possibilities are classified by π_3 $(\mathrm{O}(N))\simeq Z$.

In what sense is Γ an ordinary Lagrangian – an integral over space-time? This question is answered in the appendix, where it is shown that (locally in field space) Γ can be written as the integral over space-time of an ordinary but not manifestly chirally invariant Lagrangian which under a chiral transformation changes by a total divergence.

2 In Minkowski space, we instead consider *space* to be compact. We then consider finite time transition amplitudes between specified initial and final states of the g field. This "ties down" the fields at the boundary of space-time and leads to a similar quantization argument for Γ

Making use of Γ, we can consider a more general action for the field g:

$$I=\frac{1}{4\lambda^2}\int d^2x\,\mathrm{Tr}\,\partial_\mu g\partial^\mu g^{-1}+n\Gamma\,. \tag{13}$$

Here n must be an integer [19], since Γ is well-defined only modulo 2π. The theory (13) is renormalizable, since the new coupling constant is a dimensionless integer. Perhaps it should be stressed that (13) is not invariant under naive parity $x\to -x$, but is invariant under $x\to -x$, $g\to g^{-1}$.

We wish to ask whether for some values of λ and n this theory might be equivalent to the free massless fermi theory.

The first step is to calculate the equations of motion from (13). As has been discussed previously [6, 8], the variation of Γ is a simple, local functional. We find from (13) that the change of I under $g\to g+\delta g$ is

$$\delta I=\frac{1}{2\lambda^2}\int d^2\,\mathrm{Tr}\,g^{-1}\delta g\partial_\mu(g^{-1}\partial_\mu g)-\frac{n}{8\pi}\int d^2x\,\mathrm{Tr}\,g^{-1}\delta g\varepsilon^{\mu\nu}\partial_\mu(g^{-1}\partial_\nu g)\,. \tag{14}$$

The variational equations are therefore

$$\begin{aligned}0&=\frac{1}{2\lambda^2}\partial_\mu(g^{-1}\partial_\mu g)-\frac{n}{8\pi}\varepsilon^{\mu\nu}\partial_\mu(g^{-1}\partial_\nu g)\\&=\left(\frac{1}{2\lambda^2}+\frac{n}{8\pi}\right)\partial_-(g^{-1}\partial_+g)+\left(\frac{1}{2\lambda^2}-\frac{n}{8\pi}\right)\partial_+(g^{-1}\partial_-g)\,.\end{aligned} \tag{15}$$

We see therefore that if $\lambda^2=\frac{4\pi}{n}$ the equation is as desired, $0=\partial_-(g^{-1}\partial_+g)$. Of course, λ^2 must be positive for stability, so this is only possible for $n>0$. For $n<0$ the parity conjugate equation $0=\partial_+(g^{-1}\partial_-g)$ arises at $\lambda^2=-\frac{4\pi}{n}$.

At $\lambda^2=\frac{4\pi}{n}$ the equations of motion of the theory can easily be solved in closed form. The general solution of $0=\partial_-(g^{-1}\partial_+g)$ is

$$g(x^+,x^-)=A(x^-)B(x^+)\,, \tag{16}$$

where $A(x^-)$ and $B(x^+)$ are arbitrary $O(N)$ valued functions of one coordinate. [At $\lambda^2=-4\pi/n$ the factorization is instead $g(x^+,x^-)=B(x^+)A(x^-)$.] Equation (16) means that left-moving and right-moving waves pass through each other without any interference. This property is strongly reminiscent of the fermion free field theory, in which the left- and right-moving waves are the γ_5 eigenstates. Combining this analogy with the fact that at $\lambda^2=4\pi/n$ the equation of motion for g reproduces the behavior of the fermion currents, we are led to conjecture that at $\lambda^2=4\pi/n$ and some value of n the non-linear sigma model is equivalent to the fermion free field theory.

What are the renormalization group properties of the theory with action (13)? Being an integer, n must not be subject to renormalization. This can be established in the background field method; in that method the counter-terms are local and *manifestly chinally invariant* functionals of the background field, so there is no

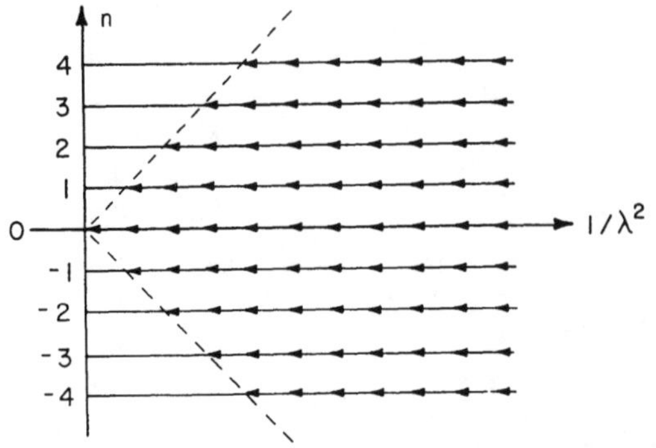

Fig. 2. Renormalization group flows. Plotted is the behavior of $1/\lambda^2$ in coming from high-energy to low-energy. Weak coupling is at the right, strong coupling is at the left. Assuming there are no non-trivial fixed points apart from the one found in the text, the $n=0$ theory flows to strong coupling at long distances, while for $n \neq 0$ the theory flows to $\lambda^2 = |4\pi/n|$. The behaviour of the non-asymptotically free theory with bare coupling bigger than $|4\pi/n|$ is not considered

counterterm proportional to Γ [3]. We will illustrate this shortly at the one loop level.

The theory therefore requires only renormalization of λ. However, the renormalization of λ depends on both λ and n. For any n, the theory is asymptotically free, just as at $n=0$. This is so because for λ so small that $\frac{1}{\lambda^2} \gg n$, (13) is dominated by the first term, and the renormalization group calculation coincides with the standard calculation at $n=0$. However, as λ becomes large, the effects of the Wess-Zumino term can become important. We will argue that the beta function always has a zero at $\lambda^2 = \pm\frac{4\pi}{n}$. We will first illustrate this point with a one loop calculation; then we will establish the point by showing that the theory at $\lambda^2 = \pm 4\pi/n$ is equivalent to a known exactly soluble, conformally invariant theory. The existence of a non-trivial zero of the beta function at $\lambda^2 = \frac{4\pi}{n}$ means that the physical content of the weakly coupled theory with $n \neq 0$ is dramatically different from what it is for $n=0$. Instead of flowing in the infrared to strong coupling, the coupling constant flows (Fig. 2) to $\sqrt{4\pi/n}$ (or perhaps to another zero of the beta function closer to the origin).

Let us now calculate the one loop beta function of the theory. We will use the background field and expand around an arbitrary solution g_0 of the classical field equations. We write $g = g_0 \exp i\lambda T^a \pi^a$, where the T^a (normalized so $\mathrm{Tr}\, T^a T^b = 2\delta^{ab}$) are the generators of O(N) and π^a are the small fluctuation fields. The action becomes

$$I = \int d^2x \left[\frac{1}{4\lambda^2} \mathrm{Tr}\, \partial_\mu g_0 \partial_\mu g_0^{-1} + \frac{1}{2} \sum_a (\partial_\mu \pi^a)^2 + \frac{\eta^{\mu\nu}}{4} \mathrm{Tr}\, g_0^{-1} \partial_\mu g_0 [T\cdot\pi, \partial_\nu T\cdot\pi] \right.$$
$$\left. - \frac{1}{4}\frac{\lambda^2 n}{4\pi} \varepsilon^{\mu\nu} \mathrm{Tr}\, g_0^{-1} \partial_\mu g_0 [T\cdot\pi, \partial_\nu T\cdot\pi] \right] \qquad (17)$$

3 Similar reasoning has been given in discussion of the θ angle in four dimensions by Novikov, Shifman, Vainshtain, and Zakharov (private communication)

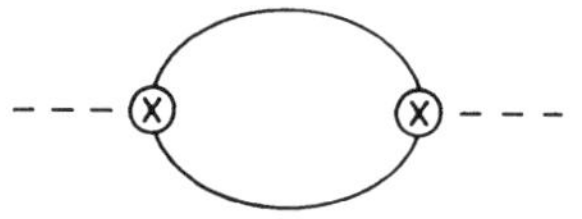

Fig. 3. The one loop renormalization calculation in the nonlinear sigma model. The dotted line is the background field; the solid line represents quantum fluctuations. The divergence arises only if the two vertices contain both $\eta^{\mu\nu}$ or both $\varepsilon^{\mu\nu}$

up to terms cubic or higher order in π. By power counting, a one loop divergence will be quadratic in $g_0^{-1}\partial_\mu g_0$. The only possible quadratic term is $\mathrm{Tr}(g_0^{-1}\partial_\mu g_0)^2 = -\mathrm{Tr}\,\partial_\mu g_0 \partial_\mu g_0^{-1}$, since $\varepsilon^{\mu\nu}\mathrm{Tr}(g_0^{-1}\partial_\mu g_0)(g_0^{-1}\partial_\nu g_0)=0$. This shows at the one loop level that only renormalization of λ is necessary – as was asserted earlier.

Since the sought for counterterm $\mathrm{Tr}\,\partial_\mu g_0 \partial_\mu g_0^{-1}$ is even under naive parity $x \to -x$, $g \to g$, the divergent one loop diagrams (Fig. 3) have two vertices both proportional to $\eta^{\mu\nu}$ or both proportional to $\varepsilon^{\mu\nu}$. The $\eta^{\mu\nu}$ vertex is the usual one that give asymptotic freedom [10]. The $\varepsilon^{\mu\nu}$ vertex is known from an old calculation in a different model [11] to give a positive contribution to the beta function. Actually even without evaluating the diagrams it is easy to see that they cancel if $\lambda^2 = \pm\frac{4\pi}{n}$. Apart from a factor of $\left(\frac{\lambda^2 n}{4\pi}\right)^2$, they differ in that one diagram has a factor of $\eta^\alpha_\mu \eta_{\alpha\nu} = \eta_{\mu\nu}$ while the other has $\varepsilon^\alpha_\mu \varepsilon_{\nu\alpha} = -\eta_{\mu\nu}$. Actual evaluation of the diagrams of Fig. 3 is not difficult. The divergent term in the effective action is

$$\int \frac{i(N-2)}{16\pi}\mathrm{Tr}\,\partial_\mu g_0 \partial_\mu g_0^{-1} \ln\left(\frac{\Lambda^2}{\mu^2}\right) d^2x, \tag{18}$$

where Λ is a momentum space cut-off and μ is a renormalization mass. From this we read off the one loop beta function

$$\beta(\lambda, n) = -\frac{\lambda^2(N-2)}{4\pi}\left[1-\left(\frac{\lambda^2 n}{4\pi}\right)^2\right] \tag{19}$$

which, as claimed, vanishes for $\lambda^2 = \left|\frac{4\pi}{n}\right|$.

If n is very large, say $n = 10^{10}$, this perturbative calculation reliably shows the existence of a zero of the beta function, since the computed zero is at a very small coupling for which higher order terms are negligible. Of course, this reasoning does not show that the zero of the beta function is *precisely* at $|4\pi/n|$; and for n of order one the lowest order calculation does not reliably show even the *existence* of a zero. To show that the beta function vanishes for $\lambda^2 = |4\pi/n|$ and that the theory at the zero is exactly soluble requires more information.

Let us return to the fermion currents, $J^{ij}_+ = -i\psi^i_+\psi^j_+$, $J^{ij}_- = -i\psi^i_-\psi^j_-$, and to our hypothesis that these currents can be equated with suitable expressions constructed from g. What commutation relations do the fermion currents obey? The canonical anticommutation relations for the fermi fields are $\{\psi^i_+(x), \psi^j_+(y)\} = \{\psi^i_-(x), \psi^j_-(y)\} = \delta^{ij}\delta(x-y)$, $\{\psi^i_-(x), \psi^j_-(y)\} = 0$. Using these equations one can readily work out the canonical commutation rules for $J_\pm$. These canonical

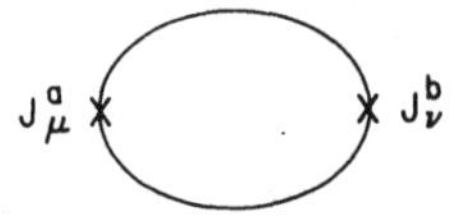

Fig. 4. The one loop diagram that yields the Schwinger anomaly in 1+1 dimensions

relations, however, are not valid quantum mechanically. The proper quantum mechanical formulas contain a c-number anomaly term, the Schwinger term. It arises [12] from diagram (4)[4]. The quantum mechanical commutation relations can be compactly written

$$\begin{aligned}&[\mathrm{Tr}\,AJ_-(x), \mathrm{Tr}\,BJ_-(y)] = 2i\delta(x-y)\,\mathrm{Tr}[A,B]J_-(x) + \frac{i}{\pi}\delta'(x-y)\,\mathrm{Tr}\,AB\,,\\ &[\mathrm{Tr}\,AJ_+(x), \mathrm{Tr}\,BJ_+(y)] = 2i\delta(x-y)\,\mathrm{Tr}[A,B]J_+(x) - \frac{i}{\pi}\delta'(x-y)\,\mathrm{Tr}\,AB\,,\\ &[J^{ij}_-, J^{kl}_+] = 0\,,\end{aligned} \tag{20}$$

where A and B are arbitrary antisymmetric matrices [generators of O(N)]. The terms proportional to $\delta'(x-y)$ originate from the anomaly.

Consider the following generalization of the first line of Eq. (20):

$$[\mathrm{Tr}\,AJ_-(x), \mathrm{Tr}\,BJ_-(y)] = 2i\delta(x-y)\,\mathrm{Tr}[A,B]J_-(x) + k\frac{i}{\pi}\delta'(x-y)\,\mathrm{Tr}\,AB\,. \tag{21}$$

Here we allow the coefficient of the anomaly to be rescaled by an arbitrary constant k. This algebra is known in the mathematical literature as the Kac-Moody algebra with a central extension, the central extension being $k \neq 0$ [13]. In the mathematical literature it is shown that this algebra has well-behaved unitary representations if and only if k is an integer. Actually, in quantum field theory one can easily find a system in which the anomaly has an arbitrary integer strength k. Consider a theory with k "flavors" and N "colors" of fermions ψ^{ia}, $a=1\ldots k$, $i=1\ldots N$, and define $J^{ij}_- = -i\sum_{a=1}^{k}\psi^{ia}_-\psi^{jk}_-$. The anomaly is then k times as large, coming from a sum over the flavor index in Fig. 4. [This gives an arbitrary positive integer k in (21); if a negative integer is desired, one may consider J_+ instead.] The fact that the Kac-Moody representation theory is well behaved only for integral k is another aspect of the a priori quantization of anomalies, a phenomenon that can also be seen from instanton physics [14] or from the multivaluedness of the Wess-Zumino term [8].

Our one flavor theory obeys (20) with $k = \pm 1$. The following very important facts are known about the Kac-Moody algebra. The unitary irreducible representation for $k = \pm 1$ is *essentially unique*. For $k > 1$, there are a finite number of irreducible representations, obtained by taking tensor products of the $k=1$ representation with different symmetry or antisymmetry conditions. To prove the equivalence of a boson theory to the one flavor fermion theory it is sufficient to show that the boson theory gives currents that obey a Kac-Moody algebra with $k = \pm 1$.

4 The evaluation of the anomaly is standard. One derivation of this formula is described in detail by Coleman et al. [12, Eqs. (3.4), (4.19), and (4.28)]. They use, however, a notation based on Dirac fermions

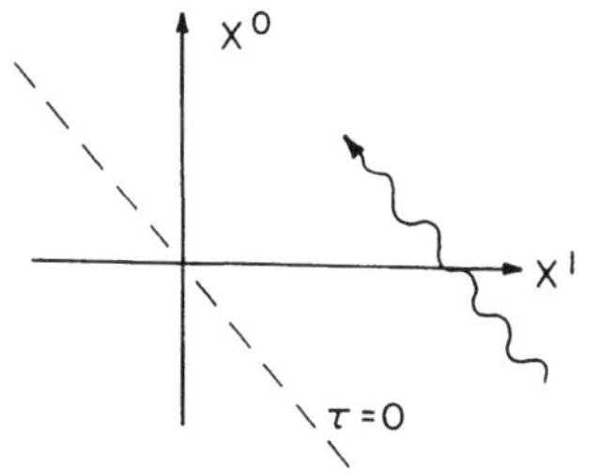

Fig. 5. This diagram is meant to illustrate the limitations of a canonical formalism based on light cone coordinates in $1+1$ dimensions. The dotted line is an "initial value surface", $\tau=0$ ($\tau=(x^0+x^1)/\sqrt{2}$). The wave line is a massless particle traveling to the left at the speed of light. Its world path never meets $\tau=0$, so its existence cannot be predicted from initial data at $\tau=0$

We are thus led to try to calculate the canonical commutation relations of the currents $g^{-1}\partial_+ g$ and $(\partial_- g)g^{-1}$. Actually, we will calculate the purely classical Poisson bracket (PB). This calculation may appear formidable because of the complexity of the Wess-Zumino term but in fact at the critical coupling $\lambda^2=\left|\frac{4\pi}{n}\right|$ it can be carried out more or less simply.

First of all, since $g^{-1}\partial_+ g$ is only a function of x^+ and $(\partial_- g)g^{-1}$ is only a function of x^-, their Poisson bracket $[(g^{-1}\partial_+ g)_{ij}, ((\partial_- g)g^{-1})_{kl}]_{\mathrm{PB}}$ vanishes. It is enough to calculate the Poisson bracket of $(\partial_- g)g^{-1}$ with itself; the Poisson bracket of $g^{-1}\partial_+ g$ can be deduced from the symmetry under $x \leftrightarrow -x$, $g \leftrightarrow g^{-1}$.

We will carry out the canonical analysis in a "light-cone frame." This means that we will regard $\sigma=x^-=\dfrac{x^0-x^1}{\sqrt{2}}$ as "space" while regarding $\tau=x^+=\dfrac{x^0+x^1}{\sqrt{2}}$ as "time." Actually, in $1+1$ dimensions the light cone framework has a drawback. A left-moving massless degree of freedom may be unpredictable on the basis of initial data at $\tau=0$ (Fig. 5). For this reason, the light cone treatment fails to give the Poisson bracket of operators like $g^{-1}\partial_+ g$ that contain τ derivatives. (Their Poisson brackets can be obtained from an opposite light cone treatment in which τ is regarded as space and σ as time.) But the light cone framework yields straightforwardly the Poisson brackets of operators like $(\partial_- g)\cdot g^{-1}$ that do not contain τ derivatives.

With $\lambda^2=4\pi/n$, the action, in light cone coordinates, is

$$I=\frac{n}{16\pi}\int d\sigma d\tau\, \mathrm{Tr}\,\partial_\tau g \partial_\sigma g^{-1}+n\Gamma. \tag{22}$$

Γ is rather complicated but it has one simple property: it is first order in time derivatives. Therefore, the whole action (22) has this property.

To introduce a canonical formalism it is necessary to formulate a theory with an action that is first order in time derivatives. Usually this requires introducing momenta that are independent of the coordinates, passing for instance from $\frac{1}{2}\dot{q}^2-V$ to $p\dot{q}-\frac{1}{2}p^2-V$. The case at hand is an exception. Equation (22) is *already* in Hamiltonian form; that is, it is already of first order in time derivatives.

In one other way, (22) differs from usual experience. In the light cone non-linear sigma model it is not convenient to split the dynamical variables into coordinates and momenta. Let us discuss, therefore, how Poisson brackets may in general be computed without making an explicit choice of p's and q's.

Consider a theory with dynamical variables ϕ^i and an arbitrary action that is first order in time derivatives:

$$I=\int dt A_i(\phi)\frac{d\phi^i}{dt}. \tag{23}$$

(The action may also contain terms independent of time derivatives; such terms are ignored in computing Poisson brackets.) We calculate the change in I under an arbitrary infinitessimal variation $\phi^i \to \phi^i + \delta\phi^i$:

$$\begin{aligned}\delta I &= \int dt\left(\frac{\partial A_i}{\partial\phi^j}\delta\phi^j\frac{d\phi^i}{dt}+A_i\frac{d}{dt}\delta\phi^i\right)\\ &= \int dt\left(\frac{\partial A_i}{\partial\phi^j}\delta\phi^j\frac{d\phi^i}{dt}-\frac{dA_i}{dt}\delta\phi^i\right)\\ &= \int dt\left(\frac{\partial}{\partial\phi^i}A_j-\frac{\partial}{\partial\phi^j}A_i\right)\delta\phi^i\frac{d\phi^j}{dt}.\end{aligned} \tag{24}$$

Define a matrix $F_{ij}=\partial_i A_j-\partial_j A_i$ as the coefficient of $\delta\phi^i\frac{d\phi^j}{dt}$. Notice that F_{ij} is always antisymmetric. Let F^{jk} be the universe matrix of F_{ij} (so $F^{jk}F_{ki}=\delta^j_i$).[5] Then the Poisson bracket of any two functions on phase space X and Y is defined by

$$[X,Y]_{\mathrm{PB}}=\sum_{i,j}F^{ij}\frac{\partial X}{\partial\phi^i}\frac{\partial Y}{\partial\phi^j}. \tag{25}$$

In the simple case in which the ϕ^i *are* decomposed into coordinates and momenta q^i and p^i, and in which the part of the action containing time derivatives is $\int dt \sum_i p^i\frac{dq^i}{dt}$, (25) agrees with the usual definition of Poisson brackets.

In this calculation it is unnecessary to choose an explicit set of coordinates ϕ^i for the classical phase space. (Such a choice would be very awkward in the non-linear sigma model becuase of the nonlinearity of the phase space.) It is enough to have a basis of tangent vectors to the phase space (analogous to the tetrad in general relativity). The matrices F_{ij} and F^{jk} may be constructed relative to any such basis. In the non-linear sigma model a very convenient basis of tangents to the phase space are the matrices $g^{-1}\delta g(\sigma)$. The matrix F must act both on the Lie algebra index of $g^{-1}\delta g(\sigma)$ and on σ.

In this basis, it is very easy to calculate the matrix F in the non-linear sigma model. From (14), with $\lambda^2=4\pi/n$, the variation of the action is

$$\delta I=\frac{n}{4\pi}\int d\sigma d\tau\, \mathrm{Tr}\, g^{-1}\delta g\frac{d}{d\sigma}g^{-1}\frac{dg}{dt}. \tag{26}$$

---

5 If this inverse does not exist, one must introduce "constraints.") This does not occur in the case at hand

From (26) we see that F is $1\otimes\frac{n}{4\pi}\frac{d}{d\sigma}$, where "1" acts on the Lie algebra index and $\frac{n}{4\pi}\frac{d}{d\sigma}$ acts on the spatial coordinate. The inverse matrix is, of course $1\otimes\frac{4\pi}{n}\left(\frac{d}{d\sigma}\right)^{-1}$.

We now wish to apply definition (25) of the Poisson bracket with $X=\operatorname{Tr}A\frac{dg}{d\sigma}g^{-1}(\sigma)$, $Y=\operatorname{Tr}B\frac{dg}{d\sigma}g^{-1}(\sigma')$. Note that (25) can be understood as follows. First calculate $\delta X\delta Y=\frac{\partial X}{\partial\phi^i}\frac{\partial Y}{\partial\phi^j}\delta\phi^i\delta\phi^j$; then replace $\delta\phi^i\delta\phi^j$ by F^{ij}. So we calculate δX:

$$\begin{aligned}\delta X&=\operatorname{Tr}A\left(\frac{d}{d\sigma}\delta g\right)g^{-1}(\sigma)+\operatorname{Tr}A\frac{dg}{d\sigma}\delta g^{-1}\\&=\operatorname{Tr}A\left(\frac{d}{d\sigma}\delta g\right)g^{-1}-\operatorname{Tr}A\frac{dg}{d\sigma}g^{-1}\delta g g^{-1}\\&=\operatorname{Tr}g^{-1}(\sigma)Ag(\sigma)\frac{d}{d\sigma}(g^{-1}\delta g(\sigma)).\end{aligned}\tag{27}$$

δY is evaluated similarly, so

$$\delta X\delta Y=\operatorname{Tr}g^{-1}(\sigma)Ag(\sigma)\frac{d}{d\sigma}(g^{-1}\delta g(\sigma))\cdot\operatorname{Tr}g^{-1}(\sigma')Bg(\sigma')\frac{d}{d\sigma'}(g^{-1}\delta g(\sigma')).\tag{28}$$

After evaluating $\delta X\delta Y=\frac{\partial X}{\partial\phi^i}\frac{\partial Y}{\partial\phi^j}\delta\phi^i\delta\phi^j$, the next step is to replace $\delta\phi^i\delta\phi^j$ with F^{ij}. In our problem the role of $\delta\phi^i$ and $\delta\phi^j$ is played by $(g^{-1}\delta g(\sigma))^a$ and $(g^{-1}\delta g(\sigma'))^b$ (here we explicitly exhibit – temporarily – the Lie algebra indices a and b carried by these matrices). In view of our previous determination of F_{ij} and F^{ij}, we are to replace $(g^{-1}\delta g(\sigma))^a(g^{-1}\delta g(\sigma))^b$ by $\delta^{ab}\frac{4\pi}{n}\theta(\sigma,\sigma')$ where $\theta(\sigma,\sigma')$ is an inverse of $\frac{d}{d\sigma}$. Hence $\frac{d}{d\sigma}(g^{-1}\delta g(\sigma))^a\cdot\frac{d}{d\sigma'}(g^{-1}\delta g(\sigma'))^b$ is replaced by $\delta^{ab}\frac{4\pi}{n}\frac{d}{d\sigma}\cdot\frac{d}{d\sigma'}\theta(\sigma-\sigma')=-\delta^{ab}\frac{4\pi}{n}\delta'(\sigma-\sigma')$. For the Poisson bracket of X and Y we get therefore

$$\begin{aligned}[X,Y]_{\mathrm{PB}}&=-\frac{4\pi}{n}\delta'(\sigma-\sigma')\operatorname{Tr}g^{-1}(\sigma)Ag(\sigma)g^{-1}(\sigma')Bg(\sigma')\\&=-\frac{4\pi}{n}\delta(\sigma-\sigma')\operatorname{Tr}[A,B]\frac{dg}{d\sigma}g^{-1}-\frac{4\pi}{n}\delta'(\sigma-\sigma')\operatorname{Tr}AB.\end{aligned}\tag{29}$$

Bearing in mind the definition of X and Y and the relation between Poisson brackets and quantum mechanical commutation relations, this corresponds to the commutation relations

$$\left[\operatorname{Tr} A \frac{dg}{d\sigma} g^{-1}(\sigma), \operatorname{Tr} B \frac{dg}{d\sigma} g^{-1}(\sigma')\right]$$

$$= \frac{4\pi}{n} i\delta(\sigma-\sigma') \operatorname{Tr}[A,B] \frac{dg}{d\sigma} g^{-1} + \frac{4\pi}{n} i\delta'(\sigma-\sigma') \operatorname{Tr} AB . \tag{30}$$

Now, let us compare this to the Kac-Moody algebra (21). We see that they coincide if $k=n$ and if J_- is identified with $\frac{n}{2\pi} \frac{dg}{d\sigma} g^{-1}$.

Now, that conclusions can we draw? In the nonlinear sigma model and in Eq. (30), n is an integer because of the multivaluedness of the Wess-Zumino coupling. In (21) k is an integer because only then does the Kac-Moody algebra have well-behaved unitary representations. We see that single valuedness of e^{iI}, required in quantum mechanics for mathematical consistency, leads to a Kac-Moody algebra with properly normalized central charge.

Second, the theory at $\lambda^2 = 4\pi/n$ really does have a vanishing β function, because it is known [15] that the irreducible representation of the Kac-Moody algebra is conformally invariant (can be extended to the semi-direct product of the Kac-Moody algebra with the conformal algebra).

Third, and most important, it follows from Eq. (30) that the non-linear sigma model with $n=1$ and $\lambda^2 = 4\pi$ is equivalent to the free field theory of N massless Majorana fermions. For, with the identifications

$$\begin{aligned} J_-^{ij} &= i\psi_-^i \psi_-^j = \frac{1}{2\pi}\left(\frac{dg}{d\sigma} g^{-1}\right)^{ij}, \\ J_+^{ij} &= i\psi_+^i \psi_+^j = \frac{1}{2\pi}\left(g^{-1}\frac{dg}{d\tau}\right)^{ij}, \end{aligned} \tag{31}$$

the currents of these theories obey the same algebra [Eqs. (20) and (30)], and, as has been mentioned, this algebra has an essentially unique irreducible representation. (The Hilbert spaces of the quantum theories in question furnish irreducible representations of the current algebras because no operators commute with all the currents. This has been proved [13] in the fermi case and also [16] in the bose case.[6]) The Kac-Moody representation is not quite unique, but the non-uniqueness just refers to superselection rules and boundary conditions in the quantum field theory.

Moreover, the Hamiltonian H and the momentum operator P of the free fermi theory coincide with those of the non-linear sigma model at the special values of

6 Our discussion of the non-linear sigma model is closely related to the discussion of the Kac-Moody representations in [16]. The phase space of our theory in the light cone frame is a complex manifold, the loop space Z of $O(N)$. (Actually, this is only half the phase space of the theory, since it omits left-moving waves. The full phase space is $Z \times Z$.) The operator F_{ij} that we have constructed represents the first Chern class of a holomorphic line bundle E over Z. The Hilbert space of the theory is the space of holomorphic sections of E. This construction generalizes some classical theorems about representations of finite dimensional Lie groups to the Kac-Moody system. The novelty of our discussion of the non-linear sigma model is to show that the construction of Kac-Moody representations just mentioned can be realized by canonical quantization of a quantum field theory

the couplings under discussion. This can be seen in various ways. Because the Hilbert spaces form irreducible representations of the current algebras, H and P are uniquely determined in each case by their commutation relations with the currents. These commutation relations are suitable; H and P generate translations in both the bose and fermi theories. A more explicit argument is the following. In the fermi theory one can show $H+P=\text{const}\lim_{\varepsilon\to 0}\int dx J_+^{ij}(x+\varepsilon)J_+^{ij}(x)$, and a similar formula for $H-P$ in terms of J_-. (This is not a canonical equation. One must study the short distance behavior of the product of currents, and subtract an infinite c-number.) On the other hand, in the bose theory one can calculate canonically an equivalent formula

$$H+P=\text{const}\int dx\,\mathrm{Tr}\left(g^{-1}\frac{dg}{d\tau}\right)^2$$

(and similarly for $H-P$). In view of (31) these relations show that H and P of the bose theory equal those of the fermi theory.

By introducing several fermion flavors, it is possible to make a bose-fermi translation also for $n\neq 1$. Consider a theory with N "colors" and n "flavors" of Majorana fermions ψ^{ia}, $i=1\ldots N$ and $a=1\ldots n$. We can define currents $\tilde{J}_\pm^{ij}=-i\sum_a \psi_\pm^{ia}\psi_\pm^{ja}$, $\tilde{J}_\pm^{ab}=-i\sum_i \psi_\pm^{ia}\psi_\pm^{ib}$. These currents generate $\mathrm{O}(N)_L\times\mathrm{O}(N)_R\times\mathrm{O}(n)_L\times\mathrm{O}(n)_R$. The $\mathrm{O}(N)_L\times\mathrm{O}(N)_R$ current commutators have anomalies of strength n. The $\mathrm{O}(n)_L\times\mathrm{O}(n)_R$ current commutators have anomalies of strength N. The free field theory Hilbert space is an irreducible representation [17] of the combined current algebra.

An equivalent bose theory is a theory with two fields g and h; g takes values in $\mathrm{O}(N)$ and h in $\mathrm{O}(n)$. For g we take a Wess-Zumino coupling n and $\lambda^2=4\pi/n$; for h we take a Wess-Zumino coupling N and $\lambda^2=4\pi/N$. g and h are decoupled, corresponding to the fact that at the fermion level amplitudes with a product of $\mathrm{O}(N)_L\times\mathrm{O}(N)_R\times\mathrm{O}(n)_L\times\mathrm{O}(n)_R$ currents factorize [18] as a product of $\mathrm{O}(N)_L\times\mathrm{O}(N)_R$ amplitudes and $\mathrm{O}(n)_L\times\mathrm{O}(n)_R$ amplitudes. The Hamiltonian and momentum operators of ψ^{ia} can be identified with the sum of those constructed from g and h.

As we have discussed, Eq. (31) generalizes to the non-abelian case the equation $\bar\psi\gamma_\mu\psi=\frac{1}{\sqrt{\pi}}\varepsilon_{\mu\nu}\partial^\nu\phi$ of conventional bosonization. In conventional bosonization, there also are formulas $\bar\psi\psi=M\cos\sqrt{4\pi}\,\phi$, $\bar\psi i\gamma_5\psi=M\sin\sqrt{4\pi}\,\phi$ (M is a renormalization mass). What are the analogues of those formulas here?

Let $Q_k^i=-i\psi_-^i\psi_{+k}$. We would like to translate Q_k^i into the bose language. The commutators of Q^ik with the fermion currents are as follows:

$$\begin{aligned}
[J_{ij}^-(x),Q_l^k(y)]&=-i\delta(x-y)(\delta_{jk}Q_l^i(y)-\delta_{ik}Q_l^j(y)],\\
[J_{ij}^+(x),Q_l^k(y)]&=-i\delta(x-y)(\delta_{jl}Q_i^k(y)-\delta_{il}Q_j^k(y)).
\end{aligned}\tag{32}$$

We must therefore find in the bose theory operators obeying the algebra (32). One need not look far. The matrix elements $g_j^i(x)$ of the matrix g are the required operators. Canonically,

$$\left[\frac{1}{2\pi}\left(\frac{dg}{d\sigma}g^{-1}(x)\right)^{ij}, g_l^k(y)\right] = -i\delta(x-y)(\delta^{jk}g_l^i(y)-\delta^{ik}g_l^j(y)),$$
$$\left[\frac{1}{2\pi}\left(g^{-1}\frac{dg}{d\tau}(x)\right)^{ij}, g_i^k(y)\right] = -i\delta(x-y)(\delta^{jl}g_i^k(y)-\delta^{il}g_j^k(y)). \tag{33}$$

[The evaluation of (33) is simple for the following reason. From (30) we know that $\int dx\frac{dg}{d\sigma}g^{-1}$ and $\int dxg^{-1}\frac{dg}{d\tau}$ generate chiral $O(N)\times O(N)$. Therefore, (33) holds up to total derivatives that vanish after integrating over x. By virtue of locality and dimensional analysis – g is dimensionless – there are no such possible terms. For the same reason, there can be no anomaly in (33) quantum mechanically.]

At least heuristically, it appears that $Q_k^i(x)$ and $g_k^i(x)$ are uniquely characterized (up to normalization) by the relations (32) and (33). For instance, in the free fermi theory one cannot find another operator that transforms like $Q_k^i(x)$, so we are led to conclude

$$Q_j^i(x) = -i\psi^i_-\psi_{j+}(x) = Mg_j^i(x), \tag{34}$$

where (as in the conventional bosonization) M is a mass that depends on the renormalization procedure for the bosonic operator. It should be noted that – while $-i\psi^i_-\psi^j_+$ has canonical dimension one – g_j^i is dimensionless classically. So Eq. (34) is possible only if g_j^i has anomalous dimension one in the fixed point theory with $n=1$.

Equation (34) is a generalization of the conventional bosonization formulas, in which $\bar\psi\psi$ and $\bar\psi i\gamma_5\psi$ are identified with matrix elements of the O(2) matrix $\begin{pmatrix}\cos\sqrt{4\pi}\,\phi & \sin\sqrt{4\pi}\,\phi\\ -\sin\sqrt{4\pi}\,\phi & \cos\sqrt{4\pi}\,\phi\end{pmatrix}$. Of course, in the free bose field theory, it is easy to see that $\cos\sqrt{4\pi}\,\phi$ and $\sin\sqrt{4\pi}\,\phi$ do have anomalous dimension one.

Equation (34) can be tested in the following way. We have identified $i\psi_i^-\psi_j^-$ with $\frac{1}{2\pi}\left(\frac{dg}{dx^-}g^{-1}\right)^i_j = \frac{1}{2\pi}\sum_k\frac{d}{dx^-}g_k^i(g^{-1})^k_j$. Since g is orthogonal, $(g^{-1})^k_j = g^j_k$. So $i\psi_i^-\psi_j^- = \frac{1}{2\pi}\sum_k\frac{dg_k^i}{dx^-}g_k^j$. If we identify g_j^i with $\frac{-i}{M}\psi^i_-\psi_{j+}$, we are led to require

$$i\psi_i^-\psi_j^-(x) = -\frac{1}{2\pi M^2}\sum_k\left(\frac{d}{dx^-}(\psi_i^-\psi_k^+(x))\right)(\psi_j^-\psi_k^+(x)). \tag{35}$$

At first sight Eq. (35) looks preposterous. It relates an operator quadratic in spinors to one quartic in spinors. However, to understand the right-hand side of (35), we must study the small Δ behavior of

$$\frac{1}{\Delta^2}\int_{\substack{|y^0-x^0|<\Delta/2\\|y^1-x^1|<\Delta/2}} d^2y\frac{\partial}{\partial y^-}T(\psi_i^-\psi_k^+(y)\psi_j^-\psi_k^+(x)). \tag{36}$$

Since $\sum_{k=1}^{N} \frac{\partial}{\partial y^-} T(\psi_k^+(y)\psi_k^+(x)) = N\delta^2(x-y)$, (36) has a piece $-\frac{N}{\Delta^2}\psi_i^-\psi_j^-(x)$. This is the most singular part of (36), as $\Delta \to 0$. Equation (35) must be understood to mean that the operator on the left-hand side equals the most singular part of the operator on the right-hand side. Note that while Δ is cut-off dependent, the mass M appearing in $-i\psi_i^-\psi_j^+(x) = Mg_j^i(x)$ is also cutoff dependent [since $-i\psi_i^-\psi_j^+$ has no anomalous dimension, but as already noted g_j^i must have anomalous dimension one in the equivalent bose theory, if the relation $-i\psi_i^-\psi_j^+(x) = Mg_j^i(x)$ holds]. Evidently, in view of (35) and (36), the product $M\Delta$ is cut-off independent. Equation (35) holds in the limit as $M \to \infty$ and $\Delta \to 0$ with $M\Delta$ fixed. While these manipulations are somewhat bizarre, the same bizarre manipulations are needed in conventional bosonization to show the consistency of the relations

$$\bar{\psi}\gamma^\mu\psi = \frac{1}{\sqrt{\pi}}\varepsilon^{\mu\nu}\partial_\nu\phi\,, \qquad \bar{\psi}\psi = M\cdot\cos\sqrt{4\pi}\phi\,, \qquad \bar{\psi}i\gamma_5\psi = M\sin\sqrt{4\pi}\phi\,.$$

This completes the dictionary of bosonization of fermi bilinears. Given the dictionary, it should be clear that the bosonization can be carried out also for arbitrary massive or interacting fermi theories. For instance, a fermion bare mass $m\bar{\psi}\psi = mi\sum_{k=1}^{N}\psi_-^k\psi_+^k$ can be included by adding to the Lagrangian a term proportional to $\sum_{i=1}^{N} g_i^i = \mathrm{Tr}\,g$. A $(\bar{\psi}\psi)^2$ coupling becomes $(\mathrm{Tr}\,g)^2$. One can likewise study gauge theories in this way. For instance, in the fermion language one may choose to gauge an arbitrary anomaly free subgroup H of chiral $O(N)\times O(N)$. The corresponding theory can be studied in the bose language by gauging the same subgroup H of the symmetry group of the nonlinear sigma model. One will be limited to anomaly free groups H – as one should be – because only for anomaly free groups does the Wess-Zumino term have a gauge invariant generalization [8].

Various applications of the present work can be imagined, but will not be explored here. The non-abelian bosonization may help in understanding 1+1 dimensional field theories. It may be helpful in understanding the Callan-Rubakov effect, which is described by an effective 1+1 dimensional s-wave field theory. And the conformally invariant theory with $\lambda^2 = 4\pi/n$ may provide a starting point for constructing generalizations of the usual string theories.

Appendix: Explicit Form of the Wess-Zumino Functional

In this appendix we will work out an explicit formula for the Wess-Zumino functional in the simplest case of an SU(2) non-linear sigma model in two space-time dimensions. We will use an index free notation, so antisymmetric tensors ω_{ijk} are denoted simply as ω, and the curl of ω, $(\partial_i\omega_{jkl} \pm$ cyclic permutations) is denoted $d\omega$. Differentials are considered to anticommute, so $dxdy = -dydx$ and $(dx)^2 = 0$, if x and y are functions.

First let us write the Wess-Zumino functional in an abstract form. On the group manifold of any simple, non-abelian group G, there is a $G\times G$ invariant third rank tensor field ω. ω obeys $d\omega=0$, and locally but not globally $\omega=d\lambda$ for some second rank tensor field λ. ω may be normalized so that its integral over any three sphere in G is an integral multiple of 2π.

Let B be a three-dimensional ball whose boundary, the two sphere S, is identified with space time. Given a mapping g from S into G, which has been extended to a mapping (also denoted g) from B into G, the Wess-Zumino functional is defined as

$$\Gamma=\int_B g^*\cdot\omega=\int_B g^*\cdot d\lambda=\int_{\partial B} g^*\cdot\lambda=\int_S g^*\cdot\lambda\,. \tag{37}$$

Here g^* is the "pull-back" of differential forms. In the third step of (37) $\partial B=S$ is the boundary of B; Stokes' theorem has been used. The last formula in (37) exhibits Γ as the integral of an ordinary two-dimensional Lagrangian. In concrete terms, the meaning of this formula as follows. Let ϕ^i be a set of coordinates for the group manifold G. Let λ_{ij} be the components of the anti-symmetric tensor λ. Then the mapping $g:S\to G$ can be described by means of functions ϕ^i, and

$$\Gamma=\int d^2x\varepsilon^{\mu\nu}\lambda_{ij}(\phi^k(x))\partial_\mu\phi^i\partial_\nu\phi^j\,. \tag{38}$$

Equation (38) has "Dirac string" type singularities, because the defining equation of λ, $\omega=d\lambda$, can be solved only locally on the group manifold.

Equation (38) is $G\times G$ invariant, although not manifestly so. Under a $G\times G$ transformation λ transforms as $\lambda_{ij}\to\lambda_{ij}+\dfrac{\partial\beta_j}{\partial\phi^i}-\dfrac{\partial\beta_i}{\partial\phi^j}$ for some $\beta_i(\phi^k)$. So

$$\delta\Gamma=\int d^2x\varepsilon^{\mu\nu}(\partial_i\beta_j-\partial_j\beta_i)\partial_\mu\phi^i\partial_\nu\phi^j=2\int d^2x\frac{\partial}{\partial x^\mu}(\varepsilon^{\mu\nu}\beta_j\partial_\nu\phi^j)\,, \tag{39}$$

and Γ changes by a total divergence whose integral vanishes.

Now let us construct explicit formulas for the simplest non-abelian group SU(2). The SU(2) manifold is a three sphere; it can be described by polar angles ψ, θ, ϕ with line element $ds^2=d\psi^2+\sin^2\psi(d\theta^2+\sin^2\theta d\phi^2)$. The only $SU(2)\times SU(2)$ invariant third rank antisymmetric tensor is the Levi-Civita tensor or volume form, so

$$\omega=\frac{1}{\pi}\sin^2\psi\sin\theta\, d\psi d\theta d\phi\,. \tag{40}$$

Note the normalization of (40). Since the volume of the SU(2) manifold is $2\pi^2$, (40) is chosen so that the integral of ω over the whole manifold is 2π.

The equation $\omega=d\lambda$ can be solved in many ways, for instance

$$\lambda=\frac{1}{\pi}\phi\sin^2\psi\sin\theta\, d\psi d\theta\,. \tag{41}$$

[Recall $(d\psi)^2=(d\theta)^2=0$.] So given a mapping of space-time into SU(2), the properly normalized Wess-Zumino interaction is

$$\Gamma=\frac{1}{\pi}\int d^2x\phi(x)\sin^2\psi(x)\sin\theta(x)\varepsilon^{\mu\nu}\partial_\mu\psi(x)\partial_\nu\theta(x)\,. \tag{42}$$

[In this parametrization, the Dirac-string type singularities occur at $\theta(x)=0$ or π. For at $\theta=0$ or π, everything should be independent of ϕ. This is not true in Eq. (42). In the case of the group SU(2), it is possible to choose another parametrization that is singular only at a single point on the group manifold.]

Formulas similar to (42) can be constructed for other groups and also for the Wess-Zumino interaction in four dimensions. These formulas are not very enlightening, however.

Acknowledgements. I wish to thank A. Feingold and I. Frenkel for discussions about Kac-Moody algebras, and D. J. Gross for suggesting a test of Eq. (34). I also wish to acknowledge discussions of bosonization and related matters with S. Coleman and D. Olive.

References

1. Coleman, S.: Quantum sine-Gordon equation as the massive Thirring model. Phys. Rev. D **11**, 2088 (1975)
2. Mandelstam, S.: Soliton operators for the quantized sine-Gordon equation. Phys. Rev. D **11**, 3026 (1975)
3. Baluni, V.: The Bose form of two-dimensional quantum chromodynamics. Phys. Lett. **90** B, 407 (1980)
   Steinhardt, P.J.: Baryons and baryonium in QCD_2. Nucl. Phys. B **176**, 100 (1980)
   Amati, D., Rabinovici, E.: On chiral realizations of confining theories. Phys. Lett. **101** B, 407 (1981)
4. Wess, J., Zumino, B.: Consequences of anomalous word identities. Phys. Lett. **37** B, 95 (1971)
5. D'Adda, A., Davis, A.C., DiVecchia, P.: Effective actions in non-abelian theories. Phys. Lett. **121** B, 335 (1983)
6. Polyakov, A.M., Wiegmann, P.B.: Landau Institute preprint (1983)
7. Alvarez, O.: Berkeley preprint (1983)
8. Witten, E.: Global aspects of current algebra. Nucl. Phys. B (to appear)
9. Novikov, S.P.: Landau Institute preprint (1982)
10. Polyakov, A.M.: Interaction of Goldstone particles in two dimensions. Applications to ferromagnets and massive Yang-Mills fields. Phys. Lett. **59** B, 79 (1975)
    Belavin, A.A., Polyakov, A.M.: Metastable states of two-dimensional isotropic ferromagnets. JETP Lett. **22**, 245 (1975)
11. Nappi, C.R.: Some properties of an analog of the chiral model. Phys. Rev. D **21**, 418 (1980)
12. Goto, T., Imamura, I.: Note on the non-perturbation-approach to quantum field theory. Prog. Theor. Phys. **14**, 396 (1955)
    Schwinger, J.: Field-theory commutators. Phys. Rev. Lett. **3**, 296 (1959)
    Jackiw, R.: In: Lectures on current algebra and its applications, Treiman S.B., et al. (eds.): Princeton, NJ: Princeton University Press 1972
    Coleman, S., Gross, D., Jackiw, R.: Fermion avatars of the Sugawara model. Phys. Rev. **180**, 1359 (1969)
13. Kac, V.G.: J. Funct. Anal. Appl. **8**, 68 (1974)
    Lepowsky, J., Wilson, R.L.: Construction of the affine Lie algebra. $A_1(1)$. Commun. Math. Phys. **62**, 43 (1978)
    Frenkel, I.B.: Spinor representations of affine Lie algebras. Proc. Natl. Acad. Sci. USA **77**, 6303 (1980); J. Funct. Anal. **44**, 259 (1981)
    Feingold, A.J., Frenkel, I.B.: IAS preprint (1983)
14. Belavin, A.M., Polyakov, A.M., Schwar, A.S., Tyupkin, Yu.S.: Pseudoparticle solutions of the Yang-Mills equations. Phys. Lett. **59** B, 85 (1975)
    't Hooft, G.: Symmetry breaking through Bell-Jackiw anomalies. Phys. Rev. Lett. **37**, 8 (1976); Computation of the quantum effects due to a four-dimensional pseudoparticle. Phys. Rev. D **14**, 3432 (1976)
    Callan, C.G., Jr., Dashen, R., Gross, D.J.: The structure of the gauge theory vacuum. Phys. Lett. **63** B, 334 (1976)
    Jackiw, R., Rebbi, C.: Vacuum periodicity in a Yang-Mills quantum theory. Phys. Rev. Lett. **37**, 172 (1976)

15. Segal, G.: Unitary representations of some infinite-dimensional groups. Commun. Math. Phys. **80**, 301 (1981)
Frenkel, I., Kac, V.G.: Basic representations. Invent Math. **62**, 23 (1980)
16. Kac, V.G., Peterson, D.H.: Spin and wedge representations of infinite-dimensional Lie algebras and groups. Proc. Natl. Acad. Sci. USA **78**, 3308 (1981)
17. Frenkel, I.: Private communication
18. Frishman, Y.: Quark trapping in a model field theory. Mexico City 1973. Berlin, Heidelberg, New York: Springer 1975
19. Deser, S., Jackiw, R., Templeton, S.: Three-dimensional massive gauge theories. Phys. Rev. Lett. **48**, 975 (1982); Topologically massive gauge theories. Ann. Phys. (NY) **140**, 372 (1982)

Communicated by A. Jaffe

Received September 29, 1983

ON THE EQUIVALENCE BETWEEN THE WESS–ZUMINO ACTION AND THE FREE FERMI THEORY IN TWO DIMENSIONS

P. Di VECCHIA and P. ROSSI
CERN, Geneva, Switzerland

Received 24 February 1984

Using the functional technique we prove the bosonization rules of Witten for the currents in a non-abelian two-dimensional theory with a particular regularization of the Fermi theory.

Recently, Witten [1] has derived bosonization rules for non-abelian theories in two dimensions by using arguments based on the uniqueness of the representation of certain Kac–Moody algebras and showing the equivalence between a free Fermi theory and a bosonic chiral theory described by the Wess–Zumino lagrangian in two dimensions [2].

The bosonization rules for the currents have been also derived [3] by using directly the functional integral technique, but some limitations in the application of these rules have been noticed.

In this paper, a study of the properties of the bosonic chiral model is presented and by using its invariance under the Kac–Moody algebra a certain operator identity is shown. This implies that this theory is equivalent to a free Fermi theory provided that it is regularized in a very specific way in which the vector current relative to the SU(N) part of U(N) is not conserved. The bosonization rules for the currents proposed in ref. [1] follow then trivially.

Let us consider the two-dimensional Wess–Zumino action in a two-dimensional Minkowski space–time [1,2]:

$$S(U)=\frac{1}{8\pi}\int d^2x\,\mathrm{Tr}(\partial_\mu U\partial_\mu U^{-1})+\frac{1}{12\pi}\int_Q d^3\xi\,\epsilon^{ijk}\,\mathrm{Tr}\,[U^{-1}\partial_i U\cdot U^{-1}\partial_j U\cdot U^{-1}\partial_k U]\,, \tag{1}$$

where Q is a three-dimensional hemisphere with compactified two-dimensional space as boundary. The field U_{ij} is taken to be an element of the U(N) group, such that $U^+ = U^{-1}$. Later, we shall consider the extension of our result to O(N) groups.

We define the following currents:

$$J_\mu=(\mathrm{i}/8\pi)\{[U^{-1}\partial_\mu U+U\partial_\mu U^{-1}]-\epsilon_{\mu\nu}[U^{-1}\partial_\nu U-U\partial_\nu U^{-1}]\}\,,\quad J_+=(\mathrm{i}/4\pi)U^{-1}\partial_+U,\quad J_-=(\mathrm{i}/4\pi)U\partial_-U^{-1}\,, \tag{2}$$

with the notations

$$J_\pm=J_0\pm J_1\,,\quad \partial_\pm=\partial_0\pm\partial_1\,. \tag{3}$$

The action (1) possesses a local U(N) × U(N) symmetry corresponding to its invariance under the transformation

$$U\to a(x^-)Ub^{-1}(x^+)\,, \tag{4}$$

where $a(x^-)$ and $b(x^+)$ are arbitrary unitary matrices depending only on one of the two light-cone coordinates. A special case of this symmetry is the global U(N) × U(N) invariance whose generators $J_\pm$ are, however, not covariant under the transformations (4):

$$J_+ \to b(x^+)J_-b^-(x^+) + (i/4\pi)b(x^+)\partial_+ b^{-1}(x^+), \quad J_- \to a(x^-)J_-a^{-1}(x^-) + (i/4\pi)a(x^-)\partial_- a^{-1}(x^-) . \tag{5}$$

The local invariance [eq. (4)] generates the manifold of the classical solutions (local minima of the classical action)

$$U_{\rm cl}(x) = a(x^-)\, b^{-1}(x^+) \tag{6}$$

such that $S(U_{\rm cl}) = 0$ and the classical conservation equations

$$\partial_+ J_-^{\rm cl} = \partial_- J_+^{\rm cl} = 0 , \tag{7}$$

are satisfied.

We are interested in the generating function for the Green functions of the currents $J_\pm$. For this purpose, we couple them to external sources $A_\pm$, that we define to be hermitean matrices belonging to the algebra of U(N) and we parametrize in terms of unitary matrices A, B, by means of

$$iA_+ = A^{-1}\partial_+ A , \quad iA_- = B^{-1}\partial_- B . \tag{8}$$

Eqs. (8) have a trivial reparametrization invariance

$$A \to \alpha(x^-)A , \quad B \to \beta(x^+)B . \tag{9}$$

We now can define

$$\exp[iW(A_+, A_-)] \equiv \left\langle \exp\left(i\int d^2x\, {\rm Tr}[A_+J_- + A_-J_+]\right)\right\rangle$$

$$\equiv \left[\int \exp\left(iS(U) + i\int d^2x\, {\rm Tr}[A_+J_- + A_-J_+]\right) DU\right] \left(\int \exp[iS(U)]\, DU\right)^{-1} . \tag{10}$$

We shall compute exactly $W(A_+, A_-)$. For this purpose, we must recall the following relation [3] that is a fundamental property of the Wess–Zumino action $S(U)$:

$$S(AUB^{-1}) = S(U) + S(AB^{-1}) + \int d^2x\, {\rm Tr}[A_+J_- + A_-J_+] + \frac{1}{4\pi}\int d^2x\, {\rm Tr}[A_+A_- - A_+UA_-U^{-1}] . \tag{11}$$

This equation allows us to rewrite (10) in the form:

$$\exp[iW(A_+, A_-)] = \left[\int DU \exp\left(iS(AUB^{-1}) + \frac{i}{4\pi}\int d^2x\, {\rm Tr}(A_+UA_-U^{-1})\right) \exp\{-i[S(A) + S(B^{-1})]\}\right]$$

$$\times \left(\int DU \exp[iS(U)]\right)^{-1} . \tag{12}$$

We can now make a change of variable in the functional integral by exploiting the local U(N) × U(N) invariance of the Haar measure and we get

$$\exp[iW(A_+, A_-)] = \left\langle \exp\left(\frac{i}{4\pi}\int d^2x\, {\rm Tr}[(AA_+A^{-1})U(BA_-B^{-1})U^+]\right)\right\rangle \exp[-iS(A) - iS(B^{-1})] . \tag{13}$$

In order to compute the expectation value indicated in (13), we can make use of the invariance under the transformation (4) and write

$$\left\langle \exp\left(\frac{i}{4\pi}\int d^2x\, {\rm Tr}[A'_+UA'_-U^{-1}]\right)\right\rangle = \left\langle \exp\left(\frac{i}{4\pi}\int d^2x\, {\rm Tr}[(a^{-1}A'_+a)\, U(b^{-1}A'_-b)U^+]\right)\right\rangle$$

$$= \left\langle \int da(x^{-1}) \int db(x^+) \exp\left(\frac{i}{4\pi}\int d^2x\, {\rm Tr}[(a^{-1}A'_+a)U(b^{-1}A'_-b)U^+]\right)\right\rangle , \tag{14}$$

where we have defined for simplicity of notations

$$A'_+ = AA_+A^{-1}, \quad A'_- = BA_-B^{-1}, \tag{15}$$

and the integration over a, b is intended to be normalized

$$\int da(x^-) = \int db(x^+) = 1. \tag{16}$$

In the derivation of eq. (14), we have assumed that the physical vacuum is invariant under the symmetry described by eq. (4). This symmetry generates any classical solution starting from the trivial $U = 1$. Therefore, our assumption implies that the *invariant* vacuum to vacuum amplitudes can be computed taking any classical solution as boundary condition at $t = \pm\infty$ or equivalently averaging over all solutions. The last step of eq. (14) can be consistently interpreted in this way.

Amazingly enough, we are able to perform explicitly this integration by means of a lattice regularization where $\int d^2x \to a^2 \Sigma_{x^+x^-}$, where a is the lattice spacing. The main ingredient of our calculation is a result due to Itzykson and Zuber [4], who performed a U(N) group integration dV for any given couple of hermitean matrices M_1, M_2:

$$\int dV \exp[\beta \operatorname{Tr}(M_1 V M_2 V^+)] = \sum_{\{n\}} \left(\beta^{|n|} \prod_{k=0}^{N-1} k! \Big/ \prod_{k=0}^{N-1} (k+n_k)! \right) \chi_{\{n\}}(M_1) \chi_{\{n\}}(M_2), \tag{17}$$

where $\{n\}$ is a set of N non-negative integers characterizing a polynomial representation of U(N), $|n| = \Sigma_{k=0}^{N-1} n_k$ and $\chi_{\{n\}}$ is the corresponding character of U(N) analytically continued to an arbitrary $N \times N$ matrix.

After little thought, it is easy to convince oneself that:

$$\int \prod_{x^+x^-} da(x^-)\, db(x^+) \exp\left(\frac{ia^2}{4\pi} \sum_{x_+x_-} \operatorname{Tr}[a^{-1}(x^-) A'_+ a(x^-) U b^{-1}(x^+) A'_- b(x^+) U^+] \right)$$

$$= \exp\left\{ \sum_{x^+x^-} \log\left[\sum_{\{n\}} \left(\frac{ia^2}{4\pi} \right)^{|n|} \left(\prod_{k=0}^{N-1} k! \Big/ \prod_{k=0}^{N-1} (k+n_k)! \right) \chi_{\{n\}}(A'_+(x^+,x^-)) \chi_{\{n\}}(A'_-(x^+,x^-)) \right] \right\}. \tag{18}$$

In the continuum limit, only the term proportional to a^2 in the exponent is non-vanishing and therefore

$$\int da(x^-)\, db(x^+) \exp\left(\frac{i}{4\pi} \int d^2x \operatorname{Tr}[(a^{-1}A'_+a) U (b^{-1}A'_-b) U^{-1}] \right)$$

$$= \exp\left(\frac{i}{4\pi} \int d^2x \frac{1}{N} \operatorname{Tr}[A'_+(x)] \operatorname{Tr}[A'_-(x)] \right) = \exp\left(\frac{i}{4\pi N} \int d^2x \operatorname{Tr}(A_+) \operatorname{Tr}(A_-) \right). \tag{19}$$

Notice that the dependence on U has already disappeared at the level of eq. (18). Therefore, taking the expectation value in (14) is a trivial operation and we reach the following final result:

$$W(A_+, A_-) = -S(A) - S(B^{-1}) + \frac{1}{4\pi N} \int d^2x \operatorname{Tr}(A_+) \operatorname{Tr}(A_-). \tag{20}$$

We can also reinterpret our result as indicating that the proper normal ordering prescription needed to define the operator $U_{ij}(x) U_{hk}^{-1}(x)$, appearing in (13), would lead to the operator identity:

$$: U_{ij}(x) U_{hk}^{-1}(x) : = \frac{1}{N} \delta_{ik} \delta_{jh} = \int dU\, U_{ij} U_{hk}^+. \tag{21}$$

It is interesting to rewrite our main result, eq. (20), by considering the natural decomposition of the U(N) group element in terms of an abelian factor belonging to U(1) and an element of the SU(N) group $\tilde{U}$:

$$U_{ij} = \exp(i\varphi/\sqrt{N}) \tilde{U}_{ij}, \quad \det \tilde{U} = 1. \tag{22}$$

Under this decomposition, the action $S(U)$ splits into separate factors:

$$S(U_1\tilde{U}) = S(U_1) + S(\tilde{U}), \quad S(U_1) = \frac{1}{8\pi}\int d^2x\, \partial_\mu\varphi\, \partial_\mu\varphi . \tag{23}$$

When we apply the same procedure to the currents and sources, we find that they decompose into irreducible tensors

$$J_\mu^{ij} = \tilde{J}_\mu^{ij} + (\epsilon_{\mu\nu}\partial^\nu\varphi/8\pi\sqrt{N})\delta^{ij}; \quad J_\pm^{ij} = \tilde{J}_\pm^{ij} \mp (1/4\pi\sqrt{N})\, \partial_\pm\varphi\delta^{ij}, \quad \mathrm{Tr}(\tilde{J}_\pm) = 0 , \tag{24}$$

and

$$A_+ = \tilde{A}_+ + (1/\sqrt{N})\partial_+\alpha , \quad A_- = \tilde{A}_- + (1/\sqrt{N})\partial_-\beta ; \quad \mathrm{Tr}(\tilde{A}_\pm) = 0 , \tag{25}$$

where we have defined $A_1 = \exp(i\alpha/\sqrt{N})$ and $B_1 = \exp(i\beta/\sqrt{N})$. Inserting these expressions in (20), we find that it also decomposes according to

$$W(A_+, A_-) = -S(\tilde{A}) - S(\tilde{B}^{-1}) - S(A_1 B_1^{-1}) . \tag{26}$$

This formulation makes it evident that there is an invariance of the generating functional, corresponding to the transformation:

$$A \to A a_1(x), \quad B \to B a_1(x); \quad a_1^{ij} = \exp[i\alpha(x)/\sqrt{N}]\, \delta^{ij} . \tag{27}$$

This local U(1) invariance is a reflection of the fact that the abelian current $(1/8\pi\sqrt{N})\, \epsilon_{\mu\nu}\partial^\nu\phi\delta^{ij}$, introduced in eq. (24), is automatically conserved and this feature must survive in the quantized version of the model.

We conclude the discussion of the bosonic theory described by (1) with an observation concerning the structure of the quantum effective action of the model. In order to define this quantity, we must introduce the "classical" currents

$$j_+ = \delta W/\delta A_- = (i/4\pi)(B^{-1}\partial_+ B) + (1/4\pi N)\mathrm{Tr}(A_+)\mathbf{1} , \quad j_- = \delta W/\delta A_+ = (i/4\pi)(A^{-1}\partial_- A) + (1/4\pi N)\mathrm{Tr}(A_-)\mathbf{1} . \tag{28}$$

The effective action is defined to be

$$\Gamma(j_+, j_-) = W(A_+, A_-) - \int d^2x\, \mathrm{Tr}(A_+ j_- + A_- j_+) = S(\tilde{A}^{-1}) + S(\tilde{B}) + S(B_1 A_1^{-1}) . \tag{29}$$

Notice that the relationship between the "classical" currents $\tilde{j}_\pm$ and their respective contributions to the effective action Γ is the same as that of the "quantum" currents $\tilde{J}_\pm$ and the action $S(U)$ (where U is thought as a function of $\tilde{J}_\pm$), but the two contributions to Γ decouple, in contrast with what happens in $S(U)$. The abelian factor behaves as a free field theory, as expected.

In the last part of this work, we want to show that the generating functional for the currents, obtained in the bosonic theory described by the action $S(U)$, is equal to the corresponding one in the free Fermi theory provided that we regularize this theory in a very specific way.

The generating functional $W_F(A_\mu)$ for the Green functions involving currents of the free Fermi theory with Dirac fermions can be written as

$$\exp[iW_F(A_+, A_-)] = \int D(\bar{\psi}, \psi) \exp\left(i \int d^2x\, \bar{\psi} i \not{D} \psi \right), \tag{30}$$

where $D_\mu = \partial_\mu + iA_\mu$.

The fermion determinant appearing in (30) has been exactly computed and the result is

$$W_F(A_+, A_-) = -S(AB^{-1}) , \tag{31}$$

where $S(AB^{-1})$ is the same expression defined in (1) and the sources A_+, A_- have been parametrized as in (8).

The result (31) is based on the fact that one regularizes the Fermi theory preserving the vector symmetry.

However, one could also regularize it in such a way that the Green functions involving right- and left-handed currents are factorized in a product of a Green function with only the right currents and a Green function with only the left currents. In this case, instead of (31), one would get the following expression for the determinant:

$$W_F(A_+, A_-) = -S(A) - S(B^{-1}) = -S(AB^{-1}) - \frac{1}{4\pi}\int d^2x\, \mathrm{Tr}(A_+ A_-)\,. \tag{32}$$

Finally, one could also regularize it in such a way that only the vector current, corresponding to the U(1) factor of U(N), is conserved. In this case, one would get [‡1]

$$W_F(A_+, A_-) = -S(A) - S(B^{-1}) + \frac{1}{4\pi N}\int d^2x\, \mathrm{Tr}(A_+)\mathrm{Tr}(A_-) = -S(\tilde{A}) - S(\tilde{B}^{-1}) - S(A_1 B_1^{-1})\,, \tag{33}$$

that is exactly equal to the corresponding expression in the case of the bosonic theory described by the action in (1).

This equality implies the following bosonization rules for the currents in a non-abelian Fermi theory:

$$-\bar{\psi}\gamma_+\psi \to (i/2\pi)U^{-1}\partial_+ U\,, \quad -\bar{\psi}\gamma_-\psi \to (i/2\pi)U\partial_- U^{-1}\,, \tag{34}$$

that are the same proposed by Witten in ref. [1]. However, it must be stressed that the equivalence is exactly valid only if the fermion theory is regularized in the way implied by (33). If we regularize the Fermi theory as in (31), the bosonization rules (34) are not exactly valid anymore and some caution must be used in bosonizing an interacting theory as, for instance, the non-abelian Thirring model. Notice also that the regularization implied by (33) destroys vector gauge invariance and therefore it is not allowed in a locally invariant theory as QCD_2. This is reflected in the fact that the generating functional (10) is not vector gauge invariant [3].

Finally, the equivalence shown in the case of U(N) is also true for an O(N) group that was considered by Witten in ref. [1]. The only differences are that the action (1) must have an additional factor $\frac{1}{2}$ to be equivalent to a Fermi theory with Majorana fermions and that for $N > 2$ there is no abelian factor group and therefore the last term in (26) is absent. Notice also that for $N = 2$, the theory is completely equivalent to the U(1) case.

We thank B. Durhuus and J.L. Petersen for their criticism on some points of the paper by K. Yoshida, G.C. Rossi and M. Testa for a useful discussion.

‡1 Notice that the three results, that we find for the fermion determinant, differ by a local polynomial quadratic in the external field A_μ. This is precisely the arbitrariness due to the infinities of the theory, as discussed by Gasser and Leutwyler in Appendix A of ref. [5].

References

[1] E. Witten, Non-abelian bosonization in two dimensions, Princeton University preprint (1983).
[2] A. D'Adda, A. Davis and P. Di Vecchia, Phys. Lett. 121B (1982) 335;
O. Alvarez, Fermion determinants, chiral symmetry and the Wess–Zumino anomaly, Berkeley preprint (1983);
A. Polyakov and A.S. Wiegman, Phys. Lett. 131B (1983) 121.
[3] P. Di Vecchia, B. Durhuus, J.L. Petersen, The Wess–Zumino action in two dimensions and non-abelian bosonization, preprint NBI-HE-84-02 (1984).
[4] C. Itzykson and J.B. Zuber, J. Math. Phys. 21 (1980) 411.
[5] J. Gasser and H. Leutwyler, CERN preprint TH.3689 (1983).

THE WESS–ZUMINO ACTION IN TWO DIMENSIONS AND NON-ABELIAN BOSONIZATION

P. DI VECCHIA [1], B. DURHUUS and J.L. PETERSEN
The Niels Bohr Institute, University of Copenhagen, Blegdamsvej 17, DK-2100 Copenhagen Ø, Denmark

Received 10 May 1984

Using recent results on fermionic determinants in two-dimensional non-abelian background fields we give a very simple path integral demonstration of the equivalence between the free Fermi theory in this background and a corresponding chiral Bose theory with Wess–Zumino action. The result is compared to previously proposed bosonization rules and certain limitations to the general validity of these are found.

1. Introduction. QCD is a non-abelian gauge theory of quarks and gluons. Its relation to the physical hadron states remains obscure, however. Some information about the behaviour of low energy mesonic amplitudes may be obtained using effective lagrangian techniques. Although the effective theory is unknown it must satisfy the same ordinary and anomalous Ward identities as QCD itself, when subjected to an external flavour field.

As argued by Wess and Zumino [1] a long time ago this information may be used to obtain low energy behaviour of amplitudes involving Goldstone bosons of spontaneously broken symmetry.

Recently Witten [2] has given an explicit form of the Wess–Zumino action in four dimensions and discussed its topological significance. For alternative approaches to the explicit construction of the Wess–Zumino action in four dimensions see ref. [3].

The Wess–Zumino action may be defined in the following way. We start with a free massless Fermi theory with a U(N) × U(N) chiral flavour symmetry. Coupling the fermions to an external gauge field A_μ, the chiral invariance is broken by the axial anomaly [4]. Integrating over the Fermi fields an effective action

$$w(A_\mu) = \mathrm{Tr}\log \not{D}\,, \tag{1}$$

where $\not{D} = i\gamma^\mu(\partial_\mu + A_\mu)$, is obtained. Here A_μ denotes a combined vector and axial vector field, $v_\mu + \gamma_5 A_\mu$. Notice that in two dimensions there are altogehter two degrees of freedom only.

Letting A_μ^{L} and A_μ^{R} denote the left- and right-handed components of A_μ, these transform under a general U(N) × U(N) transformation as

$$A_\mu^{\mathrm{L}} \to g_{\mathrm{L}}^{-1}(\partial_\mu + A_\mu^{\mathrm{L}})g_{\mathrm{L}}\,, \tag{2}$$

$$A_\mu^{\mathrm{R}} \to g_{\mathrm{R}}^{-1}(\partial_\mu + A_\mu^{\mathrm{R}})g_{\mathrm{R}}\,, \tag{3}$$

where $(g_{\mathrm{L}}, g_{\mathrm{R}}) \in \mathrm{U}(N)\times\mathrm{U}(N)$.

The determinant in eq. (1) is defined so that it is invariant under a pure vector-gauge transformation, i.e. for $g_{\mathrm{R}} = g_{\mathrm{L}}$.

Because of the anomaly it is not, however, invariant under pure axial (or chiral) transformations, i.e. for $g_{\mathrm{R}} = g_{\mathrm{L}}^{-1} \equiv g$. Instead

$$w(A_\mu^g) = w(A_\mu) + \mathrm{WZ}(g^2, A_\mu)\,, \tag{4}$$

where A_μ^g denotes the chirally transformed A_μ. This defines the Wess–Zumino action WZ.

Any effective action for the mesons (in the background A_μ) must transform in the same way under a chiral transformation. As observed by Wess and Zumino, one candidate with this transformation property is (minus) the Wess–Zumino action WZ itself, where the chiral transformation matrix g is replaced by the Goldstone-boson field.

In four dimensions there is an ambiguity in the determination of the effective action by this procedure

[1] Permanent address: Gesamthochschule Wuppertal, D-5600 Wuppertal 1, Fed. Rep. Germany.

since not all A_μ can be obtained by applying transformations of the form (2) and (3) to some fixed A_μ. In two dimensions, on the other hand, there is no such ambiguity. In the rest of this paper we present a very simple analysis of the situation in two dimensions. We show that (independently of the number of colours) the Wess–Zumino prescription may be interpreted in an exact way: the free quantum Fermi theory in the background A_μ is equivalent to a certain quantum Bose theory in the same background, and this theory is given precisely by the Wess–Zumino action. This finding is similar to the results on bosonization discussed at length by many authors [5] in the abelian case and recently extended to the non-abelian case by Witten [6]. Indeed our analysis may be viewed as a simple direct proof of (some of) these results. On the other hand, we also find some limitations in the application of the bosonization rules in the non-abelian case, at least when interpreted in a straightforward manner in terms of functional integrals.

2. Two-dimensional fermion determinant. In two dimensions the gauge potential A_μ contains two matrix degrees of freedom and may be parametrized by

$$A_+ \equiv A_0 + \mathrm{i}A_1 = A(x)^{-1}\partial_+ A(x)\,, \tag{5}$$

$$A_- \equiv A_0 - \mathrm{i}A_1 = B(x)^{-1}\partial_- B(x)\,, \tag{6}$$

where $\partial_\pm = \partial_0 \pm \mathrm{i}\partial_1$. We employ a euclidean notation. Thus $A(x), B(x)$ belong to the group whose Lie algebra is the complexified Lie algebra of $\mathrm{U}(N)$, i.e. $A(x), B(x)$ belong to the group of complex invertible matrices, and since $A_+ = -A_-^\dagger$ we have

$$A(x)^{-1} = B(x)^\dagger\,. \tag{7}$$

(In Minkowski space we would leave out the i in the definition of $A_\pm$ and $\partial_\pm$, and then $A(x)$ and $B(x)$ would be independent matrix functions with values in $\mathrm{U}(N)$.)

The evaluation of the two-dimensional fermionic determinant corresponding to (1) has been discussed recently by D'Adda et al. [7], by Alvarez [8] and by Polyakov and Wiegman [9], whose result may be written as

$$\mathrm{Tr}\log \not{D} = S(AB^{-1})\,, \tag{8}$$

where

$$S(G) = \frac{1}{8\pi}\int \mathrm{d}^2x\, \mathrm{tr}(\partial_\mu G\, \partial_\mu G^{-1}) - \frac{\mathrm{i}}{12\pi}\int_Q \mathrm{d}^3x\, \epsilon^{ABC}\, \mathrm{tr}(G^{-1}\partial_A G G^{-1}\partial_B G G^{-1}\partial_C G). \tag{9}$$

Here Q is a three-dimensional hemisphere with compactified two-dimensional space as boundary. As discussed by Polyakov and Wiegman [9] and by Witten [2] the last integral taken over S^3 is a topological number, the value of which is $2\pi n$, where n is an integer.

Under a vector gauge transformation we have $A \to Ag$ and $B \to Bg$, where g is unitary (both in euclidean and in Minkowski space). Clearly the effective action $S(AB^{-1})$ is vector gauge invariant.

Under a purely chiral transformation, on the other hand,

$$A \to Ag,\quad B \to Bg^{-1}\,, \tag{10}$$

where g is hermitean in euclidean space and unitary in Minkowski space. Thus under a chiral transformation we get for the Wess–Zumino relation, eq. (4),

$$w(A_\mu^g) = S(Ag^2B^{-1}) = S(AB^{-1}) + \mathrm{WZ}(g^2, A_\mu), \tag{11}$$

or

$$\mathrm{WZ}(U, A_\mu) = S(AUB^{-1}) - S(AB^{-1})\,, \tag{12}$$

where A_μ is given by A, B as in eqs. (5) and (6).

Under a chiral transformation defined by

$$U \to g^{-1}Ug^{-1},\quad A \to Ag,\quad B \to Bg^{-1} \tag{13}$$

we have

$$\begin{aligned} -\mathrm{WZ}(U, A_\mu) &\to -\mathrm{WZ}(U^g, A_\mu^g) \\ &= S(Ag^2B^{-1}) - S(AUB^{-1}) \\ &= -\mathrm{WZ}(U, A_\mu) + \mathrm{WZ}(g^2, A_\mu)\,, \end{aligned} \tag{14}$$

showing that $-\mathrm{WZ}(U, A_\mu)$ indeed satisfies the Wess–Zumino transformation. In section 3 we show that the theory described by the action $\mathrm{WZ}(U, A_\mu)$ is indeed equivalent to the original Fermi theory in the background field A_μ.

3. Bosonization. The action for the Fermi theory is

$$S_F(\psi, \bar{\psi}, A_\mu) = \int d^2x\, \bar{\psi} \not{D} \psi$$

$$= \int d^2x [\bar{\psi} i \not{\partial} \psi + \mathrm{tr}(J_+ A_- + J_- A_+)] , \qquad (15)$$

where $J_\pm$ are the chiral currents. Thus we want to prove the functional integral identity

$$\int D(\psi, \bar{\psi}) \exp[-S_F(\psi, \bar{\psi}, A_\mu)]$$

$$= \mathrm{const.} \times \int DU \exp[-\mathrm{WZ}(U, A_\mu)] , \qquad (16)$$

where the functional measure DU is (formally) a product of Haar measures on U(N). Unlike the Fermi theory, the Bose theory is assumed not to have any anomalies. Instead the lack of chiral invariance is explicit in the bosonic action, whereas the quantum measure DU is taken chirally invariant. Note that the field $U(x)$ is now unitary and not hermitean as in (14). The relation (16) follows from the identity

$$\mathrm{const.} = \int DU \exp[-S(U)] = \int DU \exp[-S(AUB^{-1})]$$

$$= \exp[-S(AB^{-1})] \int DU \exp[-\mathrm{WZ}(U, A_\mu)] .$$

This identity, in turn, follows from the invariance of the Haar measure under $U \to AUB^{-1}$ in the case that A and B are unitary. By analytic continuation it is then extended to the case where $A^\dagger = B^{-1}, A \in \mathrm{Gl}(N, \mathbf{C})$. (In Minkowski space this analytic continuation would not be necessary since in that case A and B are unitary.)

A straightforward calculation shows that

$$\mathrm{WZ}(U, A_\mu) = S(U) + \frac{1}{4\pi} \int d^2x\, \mathrm{tr}[A_+ U \partial_- U^{-1}$$

$$+ A_- U^{-1} \partial_+ U + A_+ U A_- U^{-1} - A_+ A_-] . \qquad (17)$$

It is possible to show that this result agrees (locally in field space) with the following expression in ref. [7]

$$\mathrm{WZ}(U, A_\mu) = \frac{1}{2\pi} \int d^2x\, (D_\mu \theta)_i$$

$$\times \left[\frac{1 - \cos(T\theta)}{(T\theta)^2} \delta_{\mu\nu} + \frac{\sin(T\theta) - T\theta}{(T\theta)^2} \epsilon_{\mu\nu} \right]_{ij} (D_\nu \theta)_j$$

$$+ \frac{i}{4\pi} \int d^2x\, \epsilon_{\mu\nu} F_{\mu\nu} \theta ,$$

where the index i refers to the adjoint representation of U(N) with generators $\{T\}$ and where

$$U = \exp(-i\tau\theta) , \quad D_\mu = \partial_\mu + \tfrac{1}{2} T A_\mu ,$$

and $\{\tau\}$ are the generators of U(N) in the fundamental representation.

This result, eq. (17), admits an interpretation in terms of bosonization. First let us consider the abelian case and take $U = \exp(i\varphi)$, φ being a scalar field. Then $DU = D\varphi$ and eqs. (16), (17) read

$$\int D(\psi, \bar{\psi}) \exp\left(\int d^2x (\bar{\psi} i \not{\partial} \psi + J_+ A_- + J_- A_+) \right)$$

$$= \mathrm{const.} \int D\varphi \exp\left(- \int d^2x [(1/8\pi) \partial_\mu \varphi\, \partial_\mu \varphi \right.$$

$$\left. + (i/4\pi)(A_+ \partial_- \varphi + A_- \partial_+ \varphi)] \right) . \qquad (18)$$

By taking functional derivatives with respect to A_+ and A_- we see that

$$\langle J_+(x_-) \ldots J_+(x_n) J_-(y_1) \ldots J_-(y_m) \rangle_F$$

$$= \langle (-i/4\pi) \partial_+ \varphi(x_1) \ldots (-i/4\pi) \partial_+ \varphi(x_n)$$

$$\times (-i/4\pi) \partial_- \varphi(y_1) \ldots (-i/4\pi) \partial_- \varphi(y_m) \rangle_B , \qquad (19)$$

where $\langle\ \rangle_F$ and $\langle\ \rangle_B$ denote expectation values in the fermionic respectively bosonic theory, thereby expressing the usual result of the bosonization prescription [5].

Since eq. (19) is true even with external sources A_+, A_- present, it is straightforward to extend it to any interacting theory with an action of the form

$$S_F(\psi, \bar{\psi}, A_\mu) = \int d^2x [\bar{\psi} i \not{\partial} \psi + \mathcal{L}(J_+, J_-, A_\mu)] \qquad (20)$$

simply by applying a Fourier transformation to the interaction term and using (19) or, alternatively, by using perturbation theory.

It is also possible to derive the bosonization rules for $\bar{\psi}\psi$ and $\bar{\psi}\gamma_5\psi$ in a functional integral context, although in a less trivial way compared to the above derivation for the currents (see ref. [10]).

We now proceed to the non-abelian case where the last term in eq. (17) is non-vanishing. It is, however, quadratic in the sources A_+, A_-. Hence by inserting eq. (17) into (16), differentiating with respect to A_- and setting $A_- = A_+ = 0$ we get

$$\langle J_+^{i_1j_1}(x_1) \ldots J_+^{i_nj_n}(x_n)\rangle_F$$
$$= \langle(-1/4\pi)(U^{-1}\partial_+U(x_1))^{i_1j_1} \ldots$$
$$\times (-1/4\pi)(U^{-1}\partial_+U(x_n))^{i_nj_n}\rangle_B , \qquad (21)$$

where $\langle \ \rangle_F$ now denotes the expectation value in the free Dirac theory and $\langle \ \rangle_B$ is the expectation value in the model whose action $S(U)$ is given by eq. (9).

Similarly by differentiating with respect to A_+

$$\langle J_-^{i_1j_1}(x_1) \ldots J_-^{i_nj_n}(x_n)\rangle_F$$
$$= \langle(-1/4\pi)(U\partial_-U^{-1}(x_1))^{i_1j_1} \ldots$$
$$\times (-1/4\pi)(U\partial_-U^{-1}(x_n))^{i_nj_n}\rangle_B . \qquad (22)$$

Thus we have found that in the non-abelian case the bosonization rules for the chiral currents J_+ and J_- are given by

$$J_+^{ij} \to -(1/4\pi)(U^{-1}\partial_+U)^{ij} , \qquad (23)$$

$$J_-^{ij} \to -(1/4\pi)(U\partial_-U^{-1})^{ij} . \qquad (24)$$

We remark that these are identical to those found by Witten [6] for the bosonic action given by eq. (9).

Expectation values of products of currents of *both* chiralities, on the other hand, are different in the two theories due to the presence of the last two terms in eq. (17). These differences, however, have support in coinciding points.

We expect that the collection of Green functions $\langle J_+(x_1) \ldots J_+(x_n)J_-(y_1) \ldots J_-(y_m)\rangle$ restricted to those arguments which are pairwise *non-coincident*, uniquely determine the operators $J_+(x)$ and $J_-(x)$. We therefore conclude that the bosonization rules (23) and (24) are valid for vanishing external fields, as far as the operators are concerned.

A generalization to the case of interacting theories (consider for example the Thirring interaction $\int d^2x\, \lambda_{ij,kl} J_+^{ij}(x) J_-^{kl}(x))$ would have been immediate if the last two terms in eq. (17) were not present. However, for the gauge invariant regularization of the Fermi theory, which we employ, these terms cannot be avoided and seem to generate different interactions than the ones obtained by a direct application of the bosonization rules (23) and (24).

An equivalent way of viewing the same point is to observe, that our result eq. (17) implies a *field dependent* bosonization prescription,

$$J_+ \to -(1/4\pi)(U^{-1}\partial_+U + U^{-1}A_+U - A_+) ,$$

$$J_- \to -(1/4\pi)(U\partial_-U^{-1} + UA_-U^{-1} - A_-) .$$

It is the field dependence of this result which prevents a generalization to the case of interacting theories. Notice that in the abelian case the bosonization rules are field independent.

Let us present a simple example to indicate the difference between mixed Green functions in the two theories. For the two-point function $\langle J_+(x)J_-(y)\rangle$ we find

$$\langle J_+^{ij}(x)\, J_-^{kl}(y)\rangle_F$$
$$= \langle(-1/4\pi)\,[U^{-1}\partial_+U(x)]^{ij}$$
$$\times (-1/4\pi)[U\partial_-U^{-1}(y)]^{kl}\rangle_B$$
$$-(1/4\pi)\langle U^{-1}(x)^{il}U(x)^{kj} - \delta^{il}\delta^{kj}\rangle_B\,\delta^2(x-y) . \qquad (25)$$

In eq. (25) we assume that $\langle U^{-1}(x)^{il}U(x)^{kj}\rangle_B$ is defined by a normal ordering prescription which respects the global invariance $U(x) \to U_1U(x)U_2$, $U_1, U_2 \in U(N)$. Using that $U(x)$ is unitary this implies that

$$\langle U^{-1}(x)^{il}U(x)^{kj}\rangle = N^{-1}\delta^{ij}\delta^{kl} . \qquad (26)$$

Inserting eq. (26) in eq. (25) we get

$$\langle J_+^{ij}(x)J_-^{kl}(y)\rangle_F$$
$$= \langle(-1/4\pi)(U^{-1}\partial_+U(x))^{ij}$$
$$\times (-1/4\pi)(U\partial_-U^{-1}(y))^{kl}\rangle_B$$
$$+ (1/4\pi)(\delta^{il}\delta^{kj} - N^{-1}\delta^{ij}\delta^{kl})\delta^2(x-y) . \qquad (27)$$

The lhs of eq. (27) can be explicitly computed from eqs. (8) and (9). One gets

$$\langle J_+^{ij}(x) J_-^{kl}(y)\rangle_F = (1/4\pi)\delta^{il}\delta^{kj}\delta^2(x-y)\,. \tag{28}$$

If we insert this expression in eq. (27) we find the following bosonic two-point function

$$\langle(-1/4\pi)(U^{-1}\partial_+ U(x))^{ij}(-1/4\pi)(U\partial_- U^{-1}(y))^{kl}\rangle_B$$

$$= (1/4\pi N)\delta^{ij}\delta^{kl}\delta^2(x-y)\,. \tag{29}$$

It is interesting to notice the different index structure that one gets in the fermionic and bosonic theory.

Remark. We would like to point out that if eq. (26) was valid at the operator level (as suggested in ref. [11]), i.e.

$$U^{-1}(x)^{il}U(x)^{kj} = N^{-1}\delta^{ij}\delta^{lk}$$

we would have that $\mathrm{tr}(A_+ U A_- U^{-1}) = N^{-1}\,\mathrm{tr}A_+\,\mathrm{tr}A_-$, and hence the last two terms in eq. (17) can be replaced by quadratic terms in A_+, A_-, which are independent of U. In this case it is easy to see that the bosonization rules (23) and (24) can still be maintained in the interacting case, if a suitable shift of the coupling constant matrix $\lambda_{ij,kl}$ is also performed.

We have benefited from discussions with H. Aratyn, N.K. Nielsen, P. Rossi and B. Schroer. One of us (PDV) is grateful to the Danish research council for partial support.

References

[1] J. Wess and B. Zumino, Phys. Lett. 37B (1971) 95.
[2] E. Witten, Princeton Univ. preprint (1983).
[3] Chou Kuang-chao, Guo Han-ying, Wu Ke and Song Xing-chang, Beijing preprints AS-ITP-83-033; AS-ITP-83-032;
B. Zumino, Wu Yong-shi and A. Zee: Chiral anomalies, higher dimensions and differential geometry, preprint 40048-18, p. 3;
G.C. Rossi, M. Testa and K. Yoshida, Phys. Lett. 134B (1984) 78.
[4] S. Adler, Phys. Rev. 177 (1969) 2426;
J. Bell and R. Jackiw, Nuovo Cimento 60A (1969) 47.
[5] S. Coleman, Phys. Rev. D11 (1975) 2088;
S. Mandelstam, Phys. Rev. D11 (1975) 3026;
A. Luther and I. Peschel, Phys. Rev. B9 (1974) 2911.
[6] E. Witten, Princeton Univ. preprint (1983).
[7] A. D'Adda, A.C. Davis and P. Di Vecchia, Phys. Lett. 121B (1983) 335.
[8] O. Alvarez, Berkeley preprint (1983).
[9] A.M. Polyakov and P.B. Wiegman, Phys. Lett. 131B (1983) 121.
[10] A.V. Kulikov, Serpukov preprint IHEP 82-23 (1982).
[11] E. Abdalla and M.C.B. Abdalla, Niels Bohr Institute preprint, NBI-HE-84-11.

PHYSICAL REVIEW D VOLUME 31, NUMBER 8 15 APRIL 1985

Abelian and non-Abelian bosonization in the path-integral framework

C. M. Naón
Departamento de Física, Universidad Nacional de La Plata, Argentina
(Received 31 July 1984)

Abelian and non-Abelian bosonization of two-dimensional models is discussed within the path-integral framework. Concerning the Abelian case, the equivalence between the massive Thirring and the sine-Gordon models is rederived in a very simple way by making a chiral change in the fermionic path-integral variables. The massive Schwinger model is also studied using the same technique. The extension of this bosonization approach to the solution of non-Abelian models is performed in a very natural way, showing the appearance of the Wess-Zumino functional through the Jacobian associated with the non-Abelian chiral change of variables. Relevant features of massless two-dimensional QCD are discussed in this context.

I. INTRODUCTION

Two-dimensional field theories have been widely explored in the last ten years and various phenomena such as dynamical mass generation, asymptotic freedom, and quark confinement, relevant in more realistic models, have been tested. The startling property which was exploited in these studies is related to the possibility of transforming Fermi fields into Bose fields. The existence of such a transformation, called bosonization, provided a powerful tool to obtain nonperturbative information of two-dimensional field theories.

Bosonization has its historical roots in Klaiber's[1] work on the massless Thirring model and Lowenstein and Swieca's[2] investigations on the massless Schwinger[3] model. It found a remarkable application in Coleman's equivalence proof[4] between the massive Thirring and the sine-Gordon theories.

For Abelian models the bosonization prescription is by now very well understood and quite rigorously established.[5–7] On the other hand, when Fermi fields belong to a multiplet transforming under a non-Abelian group, the usual bosonization procedure becomes rather difficult.[8–10] It is only very recently that non-Abelian bosonization in the operator approach has begun to be understood after the works of Polyakov and Wiegman[11] and Witten.[12] (See also Refs. 13–16).

There is an alternative approach to bosonization recently developed using the path-integral framework[17] which has been shown to be very appropriate for non-Abelian theories.[18–20] Basically, this approach parallels, in the path-integral framework, the operator fit of Lowenstein and Swieca,[2] through the use of a chiral change in the fermionic variables. Fujikawa's observation[21] on the noninvariance of the path-integral measure under γ_5 transformations is crucial for this method. In particular, it is the chiral Jacobian which gives rise to the Wess-Zumino term in the study of non-Abelian models.[21–23]

It is the purpose of this work to present a detailed discussion of the path-integral approach to bosonization, both in the Abelian and the non-Abelian cases. To this end we give in Sec. II the path-integral version of Coleman's proof of the equivalence between the massive Thirring and sine-Gordon models.[4] It is important to stress that in our method we first introduce an auxiliary vector field A_μ which is then decoupled from the fermions, exploiting the peculiarities of the algebra of $d=2$ Dirac matrices, as is done in the solution of the massless Schwinger model.[3,2,17] It is precisely this connection which allows us to extend our procedure in a very natural way to the massive Schwinger model. This is done in Sec. III, where the equivalence between this last model and a massive sine-Gordon theory is derived. The results are in complete agreement with those obtained using the operational approach.[24,25,6]

The non-Abelian extension of our bosonization method is discussed in Sec. IV. We show how the Wess-Zumino functional arises after carefully computing the chiral Jacobian associated with the decoupling change in the fermionic variables. Taking as an example massless two-dimensional QCD we compute the fermion determinant and establish the equivalence between this model and a boson theory (related to a certain chiral model). It is worthwhile to note that the role of the (non-Abelian) chiral anomaly becomes apparent in our treatment, showing its relevance in non-Abelian bosonization, exactly as it happens in the Abelian case. We also give a qualitative discussion of the principal features of this non-Abelian theory and indicate the steps to follow in order to complete its solution. A brief summary of our results and conclusions is given at the end of Sec. IV.

II. ABELIAN BOSONIZATION: MASSIVE THIRRING AND SINE-GORDON MODELS

Exactly as in the operator approach, where bosonization was derived by analyzing the equivalence between the massive Thirring and sine-Gordon models,[4] we shall show in this section how an analogous derivation can be performed very simply in the path-integral framework. We work in Euclidean $(1+1)$-dimensional space-time with γ_μ matrices chosen in the form

$$\gamma_0=\begin{bmatrix}0 & 1\\ 1 & 0\end{bmatrix}, \quad \gamma_1=\begin{bmatrix}0 & i\\ -i & 0\end{bmatrix}, \quad \gamma_5=i\gamma_0\gamma_1 . \tag{2.1}$$

The following relations hold:

$$\{\gamma_\mu,\gamma_\nu\}=2\delta_{\mu\nu} , \tag{2.2}$$

$$\gamma_\mu\gamma_5=i\epsilon_{\mu\nu}\gamma_\nu , \tag{2.3}$$

with $\epsilon_{01}=-\epsilon_{10}=1$.

The dynamics of the massive Thirring model is determined by the Lagrangian density

$$\mathcal{L}_T=-i\bar\psi\partial\!\!\!/\psi-\tfrac{1}{2}g^2(\bar\psi\gamma_\mu\psi)^2+izm\bar\psi\psi , \tag{2.4}$$

where z is a cutoff-dependent constant. In the path-integral approach the generating functional is given by

$$Z_T=\mathcal{N}\int \mathcal{D}\bar\psi\,\mathcal{D}\psi \exp\left[-\int d^2x\, \mathcal{L}_T\right] . \tag{2.5}$$

Following Ref. 19 we now use the identity

$$\exp\left[\frac{g^2}{2}\int(\bar\psi\gamma_\mu\psi)^2 d^2x\right] =\int \mathcal{D}A_\mu \exp\left[-\int d^2x(\tfrac{1}{2}A_\mu A^\mu-g\bar\psi A\!\!\!/\psi)\right] \tag{2.6}$$

in order to eliminate the quartic interaction. Here A_μ is a two-component vector field which (in two dimensions) can be written as

$$A_\mu=-\frac{1}{g}(\epsilon_{\mu\nu}\partial_\nu\phi-\partial_\mu\eta) . \tag{2.7}$$

Using Eqs. (2.5)–(2.7), we get

$$Z_T=\mathcal{N}\int \mathcal{D}\bar\psi\,\mathcal{D}\psi\,\mathcal{D}\phi\,\mathcal{D}\eta \exp\left[-\int d^2x\, \mathcal{L}_{\rm eff}\right] , \tag{2.8}$$

where

$$\mathcal{L}_{\rm eff}=-\bar\psi[i\partial\!\!\!/-\gamma_\mu(\epsilon_{\mu\nu}\partial_\nu\phi-\partial_\mu\eta)] +izm\bar\psi\psi+\frac{1}{2g^2}[(\partial_\mu\phi)^2+(\partial_\mu\eta)^2] . \tag{2.9}$$

At this point we perform a change in the fermion variables which corresponds, in the path-integral framework, to the bosonization realization in the operator approach.[2,17] The change of variables takes the form

$$\psi(x)=\exp[\gamma_5\phi(x)+i\eta(x)]\chi(x) ,$$
$$\bar\psi(x)=\bar\chi(x)\exp[\gamma_5\phi(x)-i\eta(x)] \tag{2.10}$$

and it has been chosen so as to cancel the coupling between scalars and fermions in the kinetic term of $Z_{\rm eff}$. Indeed, using Eq. (2.10) the Lagrangian (2.9) can be written in the form

$$\mathcal{L}_{\rm eff}=-\bar\chi i\partial\!\!\!/\chi+izm\bar\chi e^{2\gamma_5\phi}\chi +\frac{1}{2g^2}[(\partial_\mu\phi)^2+(\partial_\mu\eta)^2] . \tag{2.11}$$

As we see, the scalar field η is completely decoupled from the rest and this fact remains valid at the quantum level. We can write the generating functional in terms of the new variables provided the corresponding Jacobians are taken into account,

$$\mathcal{D}\bar\psi\,\mathcal{D}\psi=J_F\mathcal{D}\bar\chi\,\mathcal{D}\chi , \tag{2.12a}$$

$$\mathcal{D}A_\mu=J_A\mathcal{D}\phi\,\mathcal{D}\eta . \tag{2.12b}$$

The fermion Jacobian is nontrivial due to the noninvariance of the measure under chiral changes (this being related to the axial anomaly). It can be computed following Fujikawa's procedure.[21] We sketch the calculation in an appendix and only state here the final result,

$$J_F=\exp\left[-\frac{1}{2\pi}\int d^2x(\partial_\mu\phi)^2\right] . \tag{2.13}$$

Concerning the change (2.12b), it trivially yields

$$J_A=\frac{1}{g^2}\det\nabla^2 \tag{2.14}$$

which can be absorbed in the normalization constant. We then have

$$Z_T=\mathcal{N}\int \mathcal{D}\bar\chi\,\mathcal{D}\chi\,\mathcal{D}\phi\,\mathcal{D}\eta \times\exp\left\{-\int d^2x\left[-\bar\chi i\partial\!\!\!/\chi+izm\bar\chi e^{2\gamma_5\phi}\chi +\left[\frac{1}{2g^2}+\frac{1}{2\pi}\right](\partial_\mu\phi)^2 +\frac{1}{2g^2}(\partial_\mu\eta)^2\right]\right\} . \tag{2.15}$$

The addition of a source term

$$\mathcal{L}_{\rm source}=\bar\theta e^{\gamma_5\phi+i\eta}\chi+\bar\chi e^{\gamma_5\phi-i\eta}\theta \tag{2.16}$$

allows the computation of any Green's function in terms of the new fields.[19] However, this is not necessary in order to prove the equivalence between this model and the sine-Gordon theory. This can be done just by making a perturbative expansion (in the mass) in the generating functional

$$Z_T=\mathcal{N}\int \mathcal{D}\phi\,\mathcal{D}\bar\chi\,\mathcal{D}\chi \times\exp\left\{-\int d^2x\left[-\bar\chi i\partial\!\!\!/\chi+\frac{1}{2\lambda^2}(\partial_\mu\phi)^2\right]\right\} \times\sum_{n=0}^{\infty}\frac{(-izm)^n}{n!}\prod_{j=1}^{n}\int d^2x_j\bar\chi(x_j) \times e^{2\gamma_5\phi(x_j)}\chi(x_j) , \tag{2.17}$$

where we have defined

$$\lambda^2=\frac{g^2}{1+g^2/\pi} . \tag{2.18}$$

We then get

$$Z_T=\sum_{n=0}^{\infty}\frac{(-izm)^n}{n!}\left\langle\prod_{j=1}^{n}\int d^2x_j\bar\chi(x_j)e^{2\gamma_5\phi(x_j)}\chi(x_j)\right\rangle_0 , \tag{2.19}$$

where $\langle\ \rangle_0$ means the vacuum expectation value (VEV) in a theory of free fermions and massless free scalars. Note that owing to the presence of λ in the scalar Lagrangian, the scalar propagator is defined through the identity

$$\frac{1}{\lambda^2}\Box\Delta_F(x)=-\delta^2(x)\ , \tag{2.20a}$$

$$\Delta_F(x)=-\frac{\lambda^2}{2\pi}\ln(\mu x)\ . \tag{2.20b}$$

In order to avoid infrared divergences, we shall follow the usual procedure[4] of adding a small mass μ to Eq. (2.20a) then getting for the scalar propagator, instead of (2.20b),

$$\Delta_F(x)=\frac{\lambda^2}{2\pi}K_0(\mu x)\ . \tag{2.21}$$

Of course, we shall take $\mu^2\to 0$ at the end of our computations. On the other hand, the fermion propagator is just the free fermion Green's function

$$G_F(x)=\frac{i}{2\pi}\frac{\gamma_\mu x^\mu}{x^2}\ . \tag{2.22}$$

In order to compute (2.19) one just separates the boson factor from the free fermionic part by writing

$$\bar{\chi}e^{2\gamma_5\phi}\chi=e^{2\phi}\bar{\chi}\frac{1+\gamma_5}{2}\chi+e^{-2\phi}\bar{\chi}\frac{1-\gamma_5}{2}\chi \tag{2.23}$$

and uses the well-known identity

$$\left\langle \exp\left[i\sum_i \beta_i\phi(x_i)\right]\right\rangle_{0\ \text{bosonic}}=\exp\left\{-\tfrac{1}{2}\sum_{i,j}\beta_i\beta_j[\Delta_F(\mu,x_i-x_j)-\Delta_F(\Lambda,x_i-x_j)]\right\}$$
$$=\left[\frac{\mu}{\rho}\right]^{(\lambda^2/4\pi)\left[\sum_i\beta_i\right]^2}\left[\frac{\rho}{\Lambda}\right]^{(\lambda^2/4\pi)\sum_i\beta_i^2}\prod_{i>j}(\rho c\,|x_i-x_j|)^{(\lambda^2/2\pi)\beta_i\beta_j}\ , \tag{2.24}$$

where Λ is a large mass introduced to cut off the theory and the arbitrary mass ρ is included as a normal-ordering mass.

We have also considered the restricted region

$$\Lambda\,|x_i-x_j|\gg 1\ , \tag{2.25}$$

$$\mu\,|x_i-x_j|\ll 1\ . \tag{2.26}$$

This last condition is required in order to circumvent the trivial infrared problems which arise when one performs a mass perturbation expansion about a massless theory.[4]

Note that if $\sum_i\beta_i\neq 0$, then (2.24) vanishes in the limit $\mu\to 0$. We shall then restrict our analysis to the case

$$\sum_i\beta_i=0\ . \tag{2.27}$$

We then obtain

$$Z_T=\sum_{k=0}^{\infty}\frac{(-izm)^{2k}}{(k!)^2}\int\left[\prod_{i=1}^{k}d^2x_id^2y_i\right]\left\langle\exp\left[2\sum_i[\phi(x_i)-\phi(y_i)]\right]\right\rangle_{0\ \text{bosonic}}$$
$$\times\left\langle\prod_{i=1}^{k}\bar{\chi}(x_i)\frac{1+\gamma_5}{2}\chi(x_i)\bar{\chi}(y_i)\frac{1-\gamma_5}{2}\chi(y_i)\right\rangle_{0\ \text{fermionic}}\ . \tag{2.28}$$

The fermionic part is readily computed by writing

$$\bar{\chi}\frac{1+\gamma_5}{2}\chi=\bar{\chi}_1\chi_1\ , \tag{2.29}$$

$$\bar{\chi}\frac{1-\gamma_5}{2}\chi=\bar{\chi}_2\chi_2\ , \tag{2.30}$$

where

$$\chi=\begin{bmatrix}\chi_1\\ \chi_2\end{bmatrix},\quad \bar{\chi}=(\bar{\chi}_1,\bar{\chi}_2) \tag{2.31}$$

and using now Eq. (2.24) for the boson part we get

$$Z_T=\sum_{k=0}^{\infty}\frac{m^{2k}}{k!^2}\int\frac{\prod_{i>j}^{k}(\rho^2c^2\,|x_i-x_j||y_i-y_j|\,)^{2-2\lambda^2/\pi}\prod_{i=1}^{k}d^2x_id^2y_i}{\prod_{i,j}^{k}(\rho c\,|x_i-y_j|\,)^{2-2\lambda^2/\pi}}\ . \tag{2.32}$$

We shall compare Eq. (2.32) with the corresponding one arising in the sine-Gordon case. The generating functional for this last model is given by

$$Z_{SG}=\mathcal{N}\int \mathcal{D}\varphi \exp\left[-\int d^2x\left[\tfrac{1}{2}(\partial_\mu\varphi)^2-\frac{\alpha_0}{\beta^2}\cos\beta\varphi+\gamma_0\right]\right]. \tag{2.33}$$

Performing a perturbative expansion in α_0 we get

$$Z_{SG}=\sum_{k=0}^{\infty}\left[\frac{1}{k!}\right]^2\left\langle\left[\frac{\alpha_0}{\beta^2}\right]^{2k}\int\prod_{i=1}^{k}e^{i\beta\varphi(x_i)}e^{-i\beta\varphi(y_i)}d^2x_id^2y_i\right\rangle_0, \tag{2.34}$$

where use has been made of Eq. (2.27). We then proceed exactly as for the boson part of the Thirring model obtaining

$$Z_{SG}=\sum_{k=0}^{\infty}\left[\frac{1}{k!}\right]^2\left[\frac{\alpha}{\beta^2}\right]^{2k}\int\frac{\prod_{i>j}^{k}(c^2M^2|x_i-x_j||y_i-y_j|)^{\beta^2/2\pi}\prod_{i=1}^{k}d^2x_id^2y_i}{\prod_{i,j}^{k}(cM|x_i-y_j|)^{\beta^2/2\pi}}, \tag{2.35}$$

where we have defined the renormalized constant

$$\alpha=\frac{\alpha_0}{2}\left[\frac{M}{\Lambda}\right]^{\beta^2/4\pi}. \tag{2.36}$$

Here Λ is, as above, a certain cutoff, and M is an arbitrary mass used to normal-order the scalar theory.

We can easily see that both generating functionals (2.32) and (2.35) are identical provided we make the identifications

$$\frac{\beta^2}{4\pi}=\frac{1}{1+g^2/\pi}, \tag{2.37}$$

$$\frac{\alpha}{\beta^2}=m, \tag{2.38}$$

$$M=\rho, \tag{2.39}$$

which are, of course, the ones obtained by Coleman in his original work.[4]

Note that Eq. (2.37) does not depend on the way one has performed the renormalization in both theories [Eq. (2.39)]. In contrast, Eq. (2.38) does depend on the renormalization convention and so it has no independent meaning.

We conclude that it is possible to study the massive Thirring model in terms of a bosonic Lagrangian by considering the usual bosonization relations:

$$-i\bar\psi\partial\!\!\!/\psi=\tfrac{1}{2}(\partial_\mu\varphi)^2, \tag{2.40}$$

$$\bar\psi\gamma_\mu\psi=\frac{i}{\sqrt{\pi}}\epsilon_{\mu\nu}\partial_\nu\varphi, \tag{2.41}$$

$$imz\bar\psi\psi=-\frac{\alpha_0}{\beta^2}\cos\beta\varphi. \tag{2.42}$$

Thus we see that the corresponding sine-Gordon Lagrangian is

$$\mathcal{L}_{SG}=\tfrac{1}{2}(\partial_\mu\varphi)^2\left[1+\frac{g^2}{\pi}\right]-\frac{\alpha_0}{\beta^2}\cos\beta\varphi+\gamma_0. \tag{2.43}$$

Note that one can always define a new scalar field φ' so that

$$\varphi'=\left[1+\frac{g^2}{\pi}\right]^{1/2}\varphi \tag{2.44}$$

and then

$$\mathcal{L}_{SG}=\tfrac{1}{2}(\partial_\mu\varphi')^2-\frac{\alpha_0}{\beta^2}\cos\left[\frac{\beta^2\varphi'}{2\sqrt{\pi}}\right]+\gamma_0. \tag{2.45}$$

Using now the free value $\beta=2\sqrt{\pi}$ ($g=0$), the usual bosonization identification is obtained:

$$\mathcal{L}_{SG}=\tfrac{1}{2}(\partial_\mu\varphi)^2-\frac{\alpha_0}{\beta^2}\cos(2\sqrt{\pi}\varphi)+\gamma_0. \tag{2.46}$$

III. ABELIAN BOSONIZATION: MASSIVE SCHWINGER AND SINE-GORDON MODELS

Let us now consider the path-integral bosonization for massive quantum electrodynamics in two space-time dimensions,

$$\mathcal{L}_{SM}=-\bar\psi(i\partial\!\!\!/+eA\!\!\!/)\psi+\tfrac{1}{4}F_{\mu\nu}F^{\mu\nu}+im_0\bar\psi\psi. \tag{3.1}$$

We start from the generating functional

$$Z_{SM}=\mathcal{N}\int\mathcal{D}A_\mu\mathcal{D}\bar\psi\mathcal{D}\psi\exp\left[-\int d^2x\mathcal{L}_{SM}\right] \tag{3.2}$$

and in analogy with the Thirring model case we perform a decoupling change of variables

$$\psi=e^{\gamma_5\phi}\chi, \tag{3.3a}$$

$$\bar\psi=\bar\chi e^{\gamma_5\phi}, \tag{3.3b}$$

$$A_\mu=-\frac{1}{e}\epsilon_{\mu\nu}\partial_\nu\phi, \tag{3.3c}$$

where we are working in the Lorentz gauge. One then gets

$$Z_{SM}=\mathcal{N}\int\mathcal{D}\phi\mathcal{D}\bar\chi\mathcal{D}\chi\exp\left[-\int d^2x\,\mathcal{L}_{\rm eff}\right], \tag{3.4}$$

where

$$\mathcal{L}_{\rm eff}=-i\bar\chi\partial\!\!\!/\chi+\frac{1}{2e^2}\phi\Box\Box\phi-\frac{1}{2\pi}\phi\Box\phi+im_0\bar\chi e^{2\gamma_5\phi}\chi \tag{3.5}$$

and the nontrivial Jacobian J_F has been computed as in the Thirring case, adding a term

$$\mathscr{L}_m = -\frac{1}{2\pi}\phi\Box\phi$$

in the effective Lagrangian.

We can now follow the procedure of the preceding section defining the scalar propagator

$$\left[\frac{1}{e^2}\Box\Box - \frac{1}{\pi}\Box\right]\widetilde{\Delta}_F(x) = \delta^2(x) , \quad (3.6)$$

$$\widetilde{\Delta}_F(x) = -\tfrac{1}{2}\left[K_0\left[\frac{e}{\sqrt{\pi}}x\right] + \ln\left[\frac{ecx}{\sqrt{\pi}}\right]\right]$$

$$\equiv -\pi\left[\Delta_F\left[\frac{e}{\sqrt{\pi}},x\right] - \Delta_F(0,x)\right] .$$

As we see $\widetilde{\Delta}_F$ corresponds to the propagator of a free scalar field of mass $e/\sqrt{\pi}$ and a massless free field. This last is a manifestation in the path-integral approach of the zero-mass gauge excitation which appears in the Lowenstein-Swieca solution for the massless Schwinger model.[2]

Note that due to the particular form of the propagator (3.6) there are no ultraviolet divergences. On the other hand, one has to introduce, as in the preceding section, a mass μ^2 in order to avoid infrared divergences and take $\mu^2 \to 0$ at the end of the computation. We shall not repeat this part of the computations since they are in complete analogy with those detailed in Sec. II.

The boson part in Eq. (3.4) can be separated using again an identity of the form

$$\left\langle \exp\left[-i\sum_i \beta_i \phi(x_i)\right]\right\rangle_0 = \exp\left[-\tfrac{1}{2}\sum_{i,j}\beta_i\beta_j\widetilde{\Delta}_F(x_i - x_j)\right] . \quad (3.7)$$

Concerning the fermion part it is computed exactly as in the Thirring model case. We then get for the generating functional of the massive Schwinger model the following expression:

$$Z_{\rm SM} = \sum_{k=0}^{\infty}\left[\frac{1}{k!}\right]^2\left[\frac{m_0}{2\pi}\right]^{2k}\int B_k \frac{\prod_{i>j}^{k}|x_i - x_j|^2|y_i - y_j|^2\left[\prod_{i=1}^{k} d^2x_i d^2y_i\right]}{\prod_{i,j}^{k}|x_i - y_j|^2} , \quad (3.8)$$

where B_k is the scalar contribution to $Z_{\rm SM}$,

$$B_k = \left[\frac{ec}{\sqrt{\pi}}\right]^{2k}\frac{\prod_{i>j}^{k}|x_i - x_j|^{-2}|y_i - y_j|^{-2}}{\prod_{i,j}^{k}|x_i - y_j|^{-2}}$$

$$\times\exp\left\{-2\sum_{i>j}\left[K_0\left[\frac{e}{\sqrt{\pi}}|x_i - x_j|\right] + K_0\left[\frac{e}{\sqrt{\pi}}|y_i - y_j|\right] - K_0\left[\frac{e}{\sqrt{\pi}}|x_i - y_j|\right]\right]\right\} . \quad (3.9)$$

We then see that the massless excitation [see Eq. (3.6)] cancels out the free-fermion contribution to $Z_{\rm SM}$. We then get

$$Z_{\rm SM} = \sum_{k=0}^{\infty}\left[\frac{mec}{\sqrt{\pi}}\right]^{2k}\left[\frac{1}{k!}\right]^2\int\left[\prod_{i=1}^{k} d^2x_i d^2y_i\right]\exp\left\{-2\sum_{i>j}\left[K_0\left[\frac{e}{\sqrt{\pi}}|x_i - x_j|\right] + K_0\left[\frac{e}{\sqrt{\pi}}|y_i - y_j|\right] - K_0\left[\frac{e}{\sqrt{\pi}}|x_i - y_j|\right]\right]\right\} , \quad (3.10)$$

where $m = m_0/2\pi$.

One can now proceed to the bosonization identification. It is evident that the generating functional (3.10) coincides with the one for a massive sine-Gordon model with Lagrangian density

$$\mathscr{L}_{\rm SG} = \tfrac{1}{2}(\partial_\mu\varphi)^2 + \frac{e^2}{2\pi}\varphi^2 - \frac{\alpha}{\beta^2}\cos\beta\varphi + \gamma \quad (3.11)$$

provided the following identifications hold:

$$\frac{\alpha}{\beta^2} = m\frac{ec}{\sqrt{\pi}} , \quad (3.12)$$

$$\beta^2 = 4\pi . \quad (3.13)$$

Of course, the analysis of the massless Schwinger model can be performed by taking $m = 0$ in Eq. (3.12) and hence the isomorphism between this model and a free massive (with mass $e/\sqrt{\pi}$) scalar theory becomes apparent. [There is also the massless gauge excitation; see the discussion in Refs. (2), (18) and (19)].

We shall end this section with a brief discussion on the way the θ vacuum can be very easily incorporated in our approach to bosonization. As it is well known, the θ vacuum can be studied by adding to the generating functional of the Schwinger model, Eq. (3.2), a θ term

$$Z[\theta]=\mathcal{N}\int \mathcal{D}A_\mu \mathcal{D}\bar{\psi}\mathcal{D}\psi \times\exp\left[-\int d^2x\left[\mathcal{L}_0+\frac{e\theta}{4\pi}\epsilon_{\mu\nu}F_{\mu\nu}\right]\right]. \tag{3.14}$$

In principle, for the massless theory any reference to θ can be eliminated from Z by making a finite chiral rotation of fermions,

$$\psi=e^{\gamma_5\alpha}\chi\,,\qquad \bar{\psi}=\bar{\chi}e^{\gamma_5\alpha}\,, \tag{3.15}$$

since it gives rise to a Jacobian (see the Appendix)

$$J=\exp\left[\frac{\alpha e}{2\pi}\int \epsilon_{\mu\nu}F_{\mu\nu}d^2x\right] \tag{3.16}$$

and hence the choice $\alpha=\theta/2$ cancels the θ term in (3.14), thus giving

$$Z_{m=0}[\theta]=Z_{m=0}[0]\,. \tag{3.17}$$

However, as is discussed in Ref. 28, quantities involving chiral nonsinglet operators keep trace of the θ term. For example, in order to compute the quantity

$$\langle\bar{\psi}(x)\psi(x)\bar{\psi}(0)\psi(0)\rangle \tag{3.18}$$

one has to add a source term to the Lagrangian

$$Z_{m=0}[\theta,j]=\int \mathcal{D}A_\mu\mathcal{D}\bar{\psi}\mathcal{D}\psi \times\exp\left[-\int d^2x\left[\mathcal{L}_0+\frac{e\theta}{4\pi}F_{\mu\nu}\epsilon_{\mu\nu}+j\bar{\psi}\psi\right]\right] \tag{3.19}$$

and this source term keeps trace of the θ angle when the rotation (3.15) is performed. This aspect is discussed at length in Ref. 29.

It is now evident what happens in the massive theory. The chiral rotation (3.15) eliminates the θ term from $Z_m[\theta]$ through the Jacobian, but it changes the mass term

$$im_0\bar{\psi}(x)\psi(x)\rightarrow im_0\bar{\psi}(x)e^{\gamma_5\theta}\psi(x)\,. \tag{3.20}$$

At this point one has to proceed as in the $\theta=0$ case, performing transformations (3.3) and then making an expansion in the mass. The only change in the effective Lagrangian (3.5) is the presence of a mass term of the form

$$\mathcal{L}_{\text{mass}}=im_0\bar{\chi}e^{2\gamma_5[\phi(x)+\theta/2]}\chi\,. \tag{3.21}$$

All the analysis then follows exactly as before, except that now the cosine term in the equivalent sine-Gordon theory is changed [cf. Eq. (3.11)]:

$$\mathcal{L}_{\text{SG}}=\tfrac{1}{2}(\partial_\mu\varphi)^2+\frac{e^2}{2\pi}\varphi^2-\frac{\alpha}{\beta^2}\cos\beta(\varphi-\theta/2)+\gamma\,. \tag{3.22}$$

Equation (3.22) makes contact with the usual treatment of the θ-vacuum Schwinger model.[25]

IV. NON-ABELIAN BOSONIZATION

We shall show in this section how the path-integral bosonization approach developed above can be naturally extended to the non-Abelian case.

As we stated in the introduction the usual bosonization procedure turns out to be very awkward in the case of fermion theories with non-Abelian symmetries.[8–10] An alternative bosonization procedure has been proposed by Witten,[12] and from his work and those of other authors[11–16] interesting connections with the Wess-Zumino functional—originally constructed as an effective action for chiral anomalies—were discovered. In the path-integral framework it has been pointed out by Gamboa Saraví, Schaposnik, and Solomin[20] how this relation emerges in a very natural way by extending the chiral change of variables (2.10) or (3.3) to the non-Abelian case.

Consider for simplicity the case of massless QCD in two dimensions. The Euclidean Lagrangian for this model is

$$\mathcal{L}=-\bar{\psi}\not{D}\psi+\tfrac{1}{4}F_{\mu\nu}F_{\mu\nu}+\text{gauge-fixing terms}\,, \tag{4.1}$$

where $\not{D}=i\not{\partial}+e\not{A}$, and A_μ takes values in the Lie algebra of SU(2) [the extension to SU(N) is trivial]. The massless fermions are taken in the fundamental representation of SU(2).

As it was stressed in Ref. 18 there exists a non-Abelian analog of the decoupling change of variables (2.10) or (3.3). Indeed, if one makes the following transformation,

$$\psi=U_5\chi\,,\qquad \bar{\psi}=\bar{\chi}U_5\,, \tag{4.2}$$

where

$$U_5=e^{\gamma_5\phi} \tag{4.3}$$

and $\phi=\phi^a t^a$ [takes values in the Lie algebra of SU(2), generated by the t^a's], it is straightforward to check that the fermion Lagrangian decouples completely from gauge fields, i.e.,

$$\mathcal{L}_F=-\bar{\psi}\not{D}\psi=-\bar{\chi}i\not{\partial}\chi\,. \tag{4.4}$$

Although Eq. (4.4) holds in an arbitrary gauge it is simpler and more instructive to work in the decoupling gauge, defined by the relation

$$\not{A}=-\frac{i}{e}(\not{\partial}U_5)U_5^{-1} \tag{4.5}$$

[Eq. (4.5) becomes $A_\mu=(1/e)\epsilon_{\nu\mu}\partial_\nu\phi$ in the Abelian case, and hence the decoupling gauge coincides with the Lorentz gauge for the Schwinger model].

That the choice of the decoupling gauge is possible can be proved following Roskies's work[26] by considering the $j=i\gamma_5$ complexification of SU(2), SL(2,c). Indeed, U_5 can be taken as an element of the form

$$U_5=e^{-ij\phi}\,, \tag{4.6}$$

that is, a positive-definite Hermitian matrix of determinant one. We shall call G_5 the set of all such elements, $G_5 \subset SL(2,c)$.

Defining

$$x_\pm = x_1 \pm x_0 \tag{4.7}$$

Eq. (4.5) can be written as

$$A_+ = -2i(\partial_- U_5)U_5^{-1} . \tag{4.8}$$

Note that in our way to bosonization the role of fermion currents, which in the operator approach are written in terms of scalar fields [for example, $J_\mu = (i/\pi)\epsilon_{\mu\nu}\delta_\nu\phi$ in the Abelian case, see Eq. (2.41)], is played now by the gauge field. We then see that in our approach to non-Abelian bosonization we have naturally arrived to the analog of currents $J_\pm$ introduced in Refs. 11 and 12, except for the fact that $x_\pm$ are defined by Eq. (4.7) and are not then true light-cone coordinates. (Precisely this difference simplifies considerably our treatment.)

In order to write the generating functional in terms of the new variables we have to take into account as before the change in the fermionic measure under transformation (4.2):

$$\mathscr{D}\bar\psi\,\mathscr{D}\psi = J_F \mathscr{D}\bar\chi\,\mathscr{D}\chi . \tag{4.9}$$

It is interesting to note that the Jacobian J_F coincides with the fermionic determinant

$$\det \not D = \int \mathscr{D}\bar\psi\,\mathscr{D}\psi\, e^{-\int \bar\psi \not D \psi d^2x} = J_F \int \mathscr{D}\bar\chi\,\mathscr{D}\chi\, e^{-\int \bar\chi i\not\partial\chi d^2x} = J_F \det i\not\partial . \tag{4.10}$$

As we shall see, this relation allows one to make contact with the Polyakov-Wiegman solution of the nonlinear σ model.[11]

In computing J_F we consider an extended U_5 transformation depending on a parameter t ($t \in [0,1]$):

$$U_5(x,t) = e^{(1-t)\gamma_5\phi(x)} . \tag{4.11}$$

The whole transformation (4.3) is then built up by iteration from the infinitesimal one, varying t from 0 to 1.

The method described in the appendix for the Abelian Jacobian can be extended to the non-Abelian case following Refs. 18—20. One then gets

$$\ln J_F = -\frac{e^2}{2\pi}\int d^2x\, \mathrm{tr}\left[\tfrac{1}{2}\not A \not A + \int_0^1 dt\, \gamma_5 \not A_t \phi \not A_t\right] , \tag{4.12}$$

where tr means trace both in Lorentz and SU(2) indices and

$$\not A_t = -\frac{i}{e}[\not\partial U_5(x,t)]U_5^{-1}(x,t) . \tag{4.13}$$

One can then infer from the first term in (4.12) that the Schwinger mechanism[3] also takes place in QCD_2, giving a mass m ($m^2 = e^2/2\pi$) to the scalar fields (see the discussion below). Concerning the second term, it is related to the Wess-Zumino functional in two dimensions. Indeed, using Eq. (4.13), the second term in Eq. (4.12) can be written as

$$-\frac{e^2}{2\pi}\mathrm{tr}\int \gamma_5 \not A_t \phi \not A_t dt\, d^2x - \frac{e^2}{2\pi}\int d^2x\, \mathrm{tr} \not A \not A = \frac{i}{\pi}\int_0^1 dt \int d^2x\, \mathrm{tr}^c\left[\frac{-i}{4}\partial_\mu U \partial_\mu U^{-1} + [(\partial_t U)U^{-1}(\partial_\mu U)U^{-1}(\partial_\nu U)U^{-1}\epsilon_{\mu\nu}]\right] \tag{4.14}$$

with $U(x,t) = e^{t\phi(x)}$, and tr^c indicates an SU(2) trace. The relation between these results and the Wess-Zumino functional can be better discussed by considering the analytic continuation of U to an SU(2) element.

Let us now rewrite the Jacobian (4.12) in the form

$$\ln J_F = \frac{1}{\pi}\mathrm{tr}\left[\tfrac{1}{4}\int d^2x[\partial_\mu U(x,1)][\partial_\mu U^{-1}(x,1)] + \frac{i}{2}\epsilon_{\mu\nu}\int_0^t dt \int d^2x[\partial_t U(x,t)]U^{-1}(x,t)[\partial_\mu U(x,t)]U^{-1}(x,t)[\partial_\nu U(x,t)]U^{-1}(x,t)\right] . \tag{4.15}$$

By considering the analytic continuation of U to an element $U_c = e^{it\phi} \in SU(2)$ one can discover the relation between the second term in the Jacobian (4.15) and the Wess-Zumino functional.[22] This term is endowed with deep topological meaning since it corresponds to the Chern-Simons secondary invariant in differential geometry. Again, the chiral transformation makes manifest the role of topology in bosonization [for details see Ref. (20)].

In order to establish the rigorous equivalence between QCD_2 and a certain bosonic model one now has to follow the steps described in the preceding section for Abelian theories, i.e., to write an effective Lagrangian constructed from the original one in terms of fields U and $\chi, \bar\chi$, including the Jacobian (4.14),

$$\int \mathscr{L}_{\mathrm{eff}} d^2x = \int d^2x[-\bar\chi i\not\partial\chi + \tfrac{1}{4}F_{\mu\nu}(U)F_{\mu\nu}(U)] + \ln J_F . \tag{4.16}$$

We leave for a forthcoming paper[27] the detailed discussion of the effective action (4.16). However, one can give a qualitative picture by making a perturbative expansion of the form

$$U = 1 + 2\phi^a t^a + O(\phi^2) . \tag{4.17}$$

In this approximation the Jacobian reads

$$\ln J_F = -\frac{4}{\pi}\mathrm{tr}\int d^2x[(\partial_\mu\phi)^2 + \tfrac{1}{6}\epsilon_{\mu\nu}\phi(\partial_\mu\phi)(\partial_\nu\phi) + \text{higher-order terms}] . \tag{4.18}$$

This Jacobian resembles the effective Lagrangian discussed by Witten[23] in order to describe low-energy hadron phenomenology. However, in the present case the effective Lagrangian contains the $F_{\mu\nu}{}^2$ term and reads

$$\mathcal{L}_{\rm eff}=\frac{1}{e^2}\,{\rm tr}\left[\phi\left[\Box\Box+\frac{e^2}{2\pi}\Box\right]\phi+\frac{e^2}{6\pi}\phi(\partial_\mu\phi)(\partial_\nu\phi)\epsilon_{\mu\nu}\right.$$
$$\left.+2\phi\partial_\mu\Box\phi\partial_\nu\phi\epsilon_{\mu\nu}\right]. \tag{4.19}$$

As in the Abelian case we have gotten an effective Lagrangian with high-order derivative terms. The free Lagrangian $\mathcal{L}_0$,

$$\mathcal{L}_0=\frac{1}{e^2}\,{\rm tr}\left[\phi\left[\Box\Box+\frac{e^2}{2\pi}\Box\right]\phi\right], \tag{4.20}$$

corresponds to N^2-1 [$2^2-1=3$ for SU(2)] massive scalars (with mass $m=e/2\sqrt{\pi}$) and N^2-1 massless gauge excitations [the propagator associated to the Lagrangian (4.20) takes the form (3.6)].

In contrast with the Abelian cases described previously (where the massive scalars were free), here a self-interaction (given by the Wess-Zumino functional and the nonquadratic part of the $F_{\mu\nu}{}^2$ term) is present.

Because of the Wess-Zumino term, the Lagrangian violates both naive parity operation ($P_0\equiv X_0\rightarrow X_0$, $X_1\rightarrow -X_1$, $U\rightarrow U$) and (modulo 2) boson number N_B conservation [$(-1)^{N_B}\equiv U\rightarrow U^{-1}$ or $\phi^a\rightarrow-\phi^a$] but it is invariant under the product $P_0(-1)^{N_B}$.

As was pointed out above, the fermion determinant (4.14) coincides with the one computed by Polyakov and Wiegman[11] in their solution of the nonlinear σ model. However, their effective Lagrangian has no $F_{\mu\nu}{}^2$ term and hence bosons remain massless. Similarly to the σ-model case, the solution of the chiral Gross-Neveu model leads, in our approach, to a theory of fermions interacting with an effective vector field (see Ref. 19) and after decoupling the bosons remain massless.

In summary, we have been able to develop a path-integral approach to bosonization which parallels, in the Abelian case, the operator prescription established by Coleman[4] and Madelstam.[5] For the massive Thirring model we reobtained Coleman's results in a very simple way. It is interesting to note that it is the contribution of the chiral Jacobian that makes our results meaningful: were the Jacobian absent the theory would become free, since in that case $\beta^2/4\pi=1$ [see Eq. (2.37)]. Our approach reveals then the importance of the chiral anomaly in two-dimensional bosonization, making then apparent the role of topology, usually hidden in the operator procedure.

It is important to stress that the decoupling change of variables which allowed us to establish the equivalence between fermion and boson models in the Abelian case showed how the non-Abelian extension has to be performed. Indeed, in analogy with the Abelian case we were able to decouple fermions from bosons by making a non-Abelian chiral transformation

$$\psi=e^{\gamma_5\phi^a t^a}\chi\,,$$
$$\bar\psi=\bar\chi e^{\gamma_5\phi^a t^a}\,. \tag{4.21}$$

The decoupling was possible for the case we analyzed (massless QCD_2) due to the fact that the gauge field can be written as in Eq. (4.5). This in terms of complexified variables $X_\pm$ takes the form

$$A_+=-2i(\partial_-U_5)U_5{}^{-1}\,. \tag{4.22}$$

As we stressed before this equation is the path-integral version of the usual identification allowing to write the fermionic current in terms of boson fields. It makes natural the introduction of currents J_+ and J_- used by Polyakov and Wiegman[11] and Witten.[12] Again the role of the anomaly becomes apparent in the non-Abelian case: it originates the Wess-Zumino term [Eq. (4.14)], establishing a link with other two-dimensional models. In particular, by using the non-Abelian analog of the Gaussian identity (2.6) one can easily extend our treatment to the SU(N) Thirring model. One has to introduce a non-Abelian vector field

$$\exp\left[\frac{g^2}{2}\int d^2x(\bar\psi\gamma_\mu\lambda^a\psi)^2\right]$$
$$=\int\mathcal{D}A_\mu^a\exp\left[-\int d^2x(\tfrac{1}{2}A_\mu^aA_\mu^a-g\bar\psi A^a\lambda^a\psi)\right],$$

where λ^a are the SU(N) generators and the fermions are taken in the fundamental representation of SU(N). One can then decouple the fermions from the vector field A_μ by making a non-Abelian chiral transformation. The corresponding Jacobian originates also in this case a Wess-Zumino term.

Due to the fact that the SU(N) chiral Gross-Neveu interaction Lagrangian

$$\mathcal{L}=-\frac{g^2}{4N}[(\bar\psi\psi)^2-(\bar\psi\gamma_5\psi)^2]$$

can be written using a Fierz-type transformation in the form

$$\mathcal{L}_{\rm int}=-\frac{g^2}{2N}(\bar\psi\gamma_\mu t^a\psi)^2$$

[with t^a being the generators of U(N)] this model can also be solved using our method.

All relevant features of the Abelian models discussed in Secs. II and III can be inferred very simply from our treatment. Concerning the non-Abelian model (QCD_2) we were able to show that scalars become massive and self-interacting due in particular to the Wess-Zumino term. From their decoupling one can easily show that at short distances fermions become free. The long-distance behavior of fermion Green's functions as well as the analysis of bosonic correlation functions necessitates a more thorough investigation of the complete effective Lagrangian (4.16). Work on this aspect will be reported elsewhere.

ACKNOWLEDGMENTS

I am grateful to Fidel Schaposnik for calling my attention to this problem and for his advice and encouragements. I am also indebted to H. Falomir and R. E. Gamboa Saraví for useful comments and H. de Vega and E. Fradkin for a fruitful discussion. C. M. Naón was financially supported by Consejo Nacional de Investigaciones Cientificas y Técnicas, Argentina.

Note added in proof. After this paper was submitted for publication we discovered the paper of D. Gonzales and A. N. Redlich, Phys. Lett. **147B**, 150 (1984) where similar results are presented.

APPENDIX

We shall now compute, for the Abelian case, the Jacobian J_F associated with the fermion change of variables

$$\psi(x)=U_5(x)\chi(x)\ ,\quad \bar\psi(x)=\bar\chi(x)U_5(x)\ ,\quad U_5(x)=e^{\gamma_5\phi(x)}\ . \tag{A1}$$

Let us now consider the normalized eigenvectors of the Hermitian operator $\not{D}=i\not\partial+e\not{A}$:

$$\not{D}\varphi_n(x)=\lambda_n\varphi_n(x)\ . \tag{A2}$$

The classical fields $\psi(x)$ and $\bar\psi(x)$ can be expanded as

$$\psi(x)=\sum_n a_n\varphi_n(x)\ ,\quad \bar\psi(x)=\sum_n \varphi_n^\dagger(x)\bar b_n\ , \tag{A3}$$

where a_n and $\bar b_n$ are elements of a Grassmann algebra. The fermionic part of the functional-integral measure is then defined by

$$\mathscr{D}\bar\psi\,\mathscr{D}\psi=\prod_{n,m} d\bar b_n da_m\ . \tag{A4}$$

The new fields $\chi(x)$ and $\bar\chi(x)$ can also be expanded in terms of the φ_n's,

$$\chi(x)=\sum_n a_n'\varphi_n(x)\ , \tag{A5}$$

where the coefficients of the expansions (A3) and (A5) are related by

$$a_m'=\sum_n c_{mn}a_n \tag{A6}$$

with

$$c_{mn}=\langle\varphi_m\mid U_5^{-1}\mid\varphi_n\rangle\ . \tag{A7}$$

One then easily obtains

$$J_F=(\det c_{mn})^2\ . \tag{A8}$$

In order to compute the determinant of the c matrix defined in Eq. (A8) we shall consider the one-parameter-dependent γ_5 transformation

$$\psi=U_5(\alpha)\chi\ ,\quad \bar\psi=\bar\chi U_5(\alpha)\ , \tag{A9}$$

where $U_5(\alpha)=e^{\alpha\gamma_5\phi}$ and α is a real parameter to be varied from 0 to 1, allowing us to build up the whole transformation (A1) by iteration of the infinitesimal one,

$$1+\gamma_5\phi(x)\delta\alpha\ . \tag{A10}$$

In order to follow Fujikawa's procedure[21] at each state of the transformation one must take the eigenvectors which correspond to that stage:

$$\tilde{\not{D}}(\alpha)\tilde\varphi_n(x,\alpha)=\tilde\lambda_n(\alpha)\tilde\varphi_n(x,\alpha)\ , \tag{A11}$$

where

$$\not{D}(\alpha)=U_5(\alpha)\not{D}U_5(\alpha)=i\not\partial+e\tilde{\not{A}}(\alpha) \tag{A12}$$

with

$$\tilde{\not{A}}(\alpha)=U_5(\alpha)\not{A}U_5(\alpha)+\frac{i}{e}U_5(\alpha)\not\partial U_5(\alpha)\ . \tag{A13}$$

This α-dependent field satisfies

$$\tilde{A}(\alpha)\Big|_{\alpha=0}=\not{A}\ ,\quad \tilde{\not{A}}(\alpha)\Big|_{\alpha=1}=\not{A}'\ ,\quad \frac{\partial\tilde{\not{A}}(\alpha)}{\partial\alpha}=-\gamma_5\tilde{\not{D}}(\alpha)U_5(\alpha)\ . \tag{A14}$$

Let us now consider the matrix

$$B_{pn}(\beta;\alpha)=\langle\tilde\varphi_p(x,\beta+\alpha)\mid U_5^{-1}(\alpha)\mid\tilde\varphi_n(x,\beta)\rangle \tag{A15}$$

which satisfies

$$B(\beta;\alpha+\delta)=B(\beta+\alpha;\delta)B(\beta;\alpha)\ . \tag{A16}$$

Taking $\beta=0$ and $\delta=\delta\alpha$, we obtain

$$\frac{d}{d\alpha}[\ln\det B(0,\alpha)]=\frac{\ln\det[B(\alpha,\delta\alpha)]}{\delta\alpha}\ . \tag{A17}$$

Using the relation

$$\mid\tilde\varphi_n(\alpha+\delta\alpha)\rangle\simeq\mid\tilde\varphi_n(\alpha)\rangle+\sum_p b_{np}\mid\tilde\varphi_p(\alpha)\rangle\delta\alpha\ , \tag{A18}$$

where $b_{np}=0$ if $n=p$, we can rewrite (A17) in the form

$$\frac{d}{d\alpha}\ln\det B(0,\alpha)=-\sum_n\langle\tilde\varphi_n(\alpha\mid\gamma_5\phi\mid\tilde\varphi_n(\alpha)\rangle=-w(\alpha)\ . \tag{A19}$$

Taking into account that

$$B_{nm}(0,0)=\langle\varphi_n\mid\varphi_m\rangle=\delta_{nm}\ ,\quad B_{nm}(0,1)=\langle{\varphi_n}^1\mid U_5^{-1}\mid\varphi_m\rangle\ , \tag{A20}$$

we can easily integrate Eq. (A19) obtaining

$$\det\langle\varphi_p^1\mid U_5^{-1}\mid\varphi_m\rangle=e^{-\int_0^1 d\alpha\, w(\alpha)}\ . \tag{A21}$$

We can write from Eq. (A7)

$$\det c_{mn}=\det\langle\varphi_m\mid{\varphi_p}^1\rangle\det\langle{\varphi_p}^1\mid U_5^{-1}\mid\varphi_n\rangle\ . \tag{A22}$$

One now uses Eq. (A18) to show that

$$\det\langle\varphi_m\mid\varphi_p^1\rangle=1\ . \tag{A23}$$

Inserting Eq. (A21) in Eq. (A8) we obtain

$$J_F = e^{-2\int_0^1 w(\alpha)d\alpha}, \quad \text{(A24)}$$

where

$$w(\alpha) = \int d^2x\, \phi(x) \sum_n \tilde{\varphi}_n^\dagger(x,\alpha)\gamma_5\tilde{\varphi}_n(x,\alpha)\,. \quad \text{(A25)}$$

The summation appearing in this integrand is an ill-defined quantity and we evaluate it by using a well-known regularization procedure

$$\sum_n \tilde{\varphi}_n^\dagger(x,\alpha)\gamma_5\tilde{\varphi}_n(x,\alpha) = \lim_{M\to\infty} \sum_n \tilde{\varphi}_n^\dagger(x,\alpha)\gamma_5 e^{-\tilde{D\!\!\!/}^2(\alpha)/M^2}\tilde{\varphi}_n(x,\alpha) \quad \text{(A26)}$$

getting

$$w(\alpha) = -\frac{ie}{8\pi}\int d^2x\, \mathrm{tr}[\gamma_5\gamma_\mu\gamma_\nu\phi(x)\tilde{F}_{\mu\nu}(x,\alpha)]\,, \quad \text{(A27)}$$

where

$$\tilde{F}_{\mu\nu}(x,\alpha) = \partial_\mu\tilde{A}_\nu(x,\alpha) - \partial_\nu\tilde{A}_\mu(x,\alpha)\,. \quad \text{(A28)}$$

Now one can use γ properties and the antisymmetry of $F_{\mu\nu}$ in order to write

$$w(\alpha) = \frac{e}{2\pi}\int d^2x\,(1-\alpha)A_\nu(x)\epsilon_{\mu\nu}\partial_\mu\phi(x)\,, \quad \text{(A29)}$$

where we have used Eq. (A13), for the Abelian case, and neglected a surface term. The fermion Jacobian then reads

$$J_F = \exp\left[-\frac{e}{\pi}\int d^2x \int_0^1 d\alpha(1-\alpha)A_\nu(x)\epsilon_{\mu\nu}\partial_\mu\phi(x)\right]. \quad \text{(A30)}$$

It is straightforward to integrate in α, and for Abelian vector fields decomposed in transverse and longitudinal parts,

$$A_\nu = -\frac{1}{e}\epsilon_{\nu\mu}\partial_\mu\phi + \frac{a}{e}\partial_\nu\eta\,, \quad \text{(A31)}$$

we finally obtain

$$J_F = e^{-(1/2\pi)\int d^2x(\partial_\mu\phi)^2}\,. \quad \text{(A32)}$$

The non-Abelian case can be treated in a completely analogous way (see Refs. 18 and 19).

[1]B. Klaiber, in *Lectures in Theoretical Physics, Boulder, 1967* (Gordon and Breach, New York, 1968), p. 141.
[2]J. H. Lowenstein and J. A. Swieca, Ann. Phys. (N.Y.) 68, 172 (1971).
[3]J. S. Schwinger, Phys. Rev. **128**, 2425 (1962).
[4]S. Coleman, Phys. Rev. D **11**, 2088 (1975).
[5]S. Mandelstam, Phys. Rev. D **11**, 3026 (1975).
[6]J. A. Swieca, Fortschr. Phys. **25**, 303 (1977).
[7]J. Frohlich, Phys. Rev. Lett. **34**, 833 (1975).
[8]I. V. Belvedere, J. A. Swieca, K. D. Rothe, and B. Schroer, Nucl. Phys. **B153**, 112 (1979).
[9]B. Baluni, Phys. Lett. **90B**, 407 (1980).
[10]P. J. Steinhardt, Nucl. Phys. **B106**, 100 (1980).
[11]A. M. Polyakov and P. Wiegman, Phys. Lett. 131B, 121 (1983).
[12]E. Witten, Commun. Math. Phys. **92**, 455 (1984).
[13]P. Di Vecchia and P. Rossi, Phys. Lett. **140B**, 344 (1984).
[14]P. Di Vecchia, B. Durhuus, and J. L. Petersen, Phys. Lett. **114B**, 245 (1984).
[15]E. Abdalla and M. C. B. Abdalla, Niels Bohr Report No. NBI-HE-84-11, 1984 (unpublished).
[16]G. Bhattacharya and S. Rajeev, Nucl. Phys. **B246**, 157 (1984).
[17]R. Roskies and F. A. Schaposnik, Phys. Rev. D **23**, 558 (1981).
[18]R. E. Gamboa Saraví, F. A. Schaposnik, and J. E. Solomin, Nucl. Phys. **B185**, 239 (1981).
[19]K. Furuya, R. E. Gamboa Saraví, and F. A. Schaposnik, Nucl. Phys. **B208**, 159 (1982).
[20]R. E. Gamboa Saraví, F. A. Schaposnik, and J. E. Solomin, Phys. Rev. D **30**, 1353 (1984).
[21]K. Fujikawa, Phys. Rev. Lett. **42**, 1195 (1979); Phys. Rev. D **21**, 2848 (1980).
[22]J. Wess and B. Zumino, Phys. Lett. **37B**, 95 (1971).
[23]E. Witten, Nucl. Phys. **B233**, 422 (1983); **B233**, 433 (1983).
[24]S. Coleman, R. Jackiw, and L. Susskind, Ann. Phys. (N.Y.) **93**, 267 (1975).
[25]S. Coleman, Ann. Phys. (N.Y.) **101**, 239 (1976).
[26]R. Roskies, Festschrift for Feza Gursey's 60th Birthday (unpublished).
[27]C. M. Naón and F. A. Schaposnik (unpublished).
[28]R. E. Gamboa Saraví, M. A. Muschietti, F. A. Schaposnik, and J. E. Solomin, Phys. Lett. **183B**, 145 (1984).
[29]R. E. Gamboa Saraví, M. A. Muschietti, F. A. Schaposnik, and J. E. Solomin, Ann. Phys. (N.Y.) **157**, 360 (1984).

CALCULATION OF A PROPAGATOR ON A RIEMANN SURFACE ☆

Hidenori SONODA

Lawrence Berkeley Laboratory, University of California, Berkeley, CA 94720, USA

Received 18 July 1986

A field theory is considered on a Riemann surface of genus p, where the field is a section of a holomorphic line bundle ξ with Chern number $p-1$. When ξ has no holomorphic section, one can calculate the propagator explicitly. As an application the expectation value of the energy–momentum tensor is computed.

Recently field theory on a Riemann surface has attracted much attention, especially in the context of string theory [1–3].

One novel feature of this class of field theory is the implication of the complex structure of the Riemann surface. In this letter we discuss one of such implications, i.e., the calculation of a certain propagator. As we will see later, for the fields we discuss, the propagator becomes analytic with respect to the complex coordinate of the Riemann surface. Then the propagator can be deduced only from the analytic properties of the Riemann surface; the propagator is written in terms of holomorphic one-forms and ϑ-functions.

Let M be a Riemann surface of genus p. We introduce a Riemann metric on M, which has only a $g_{z\bar{z}}$ component in the local complex coordinate system z.

We consider a spin bundle s. There are 2^{2p} number of spin structures. We especially consider s which has only one holomorphic section $h(z)$ [4]. The bundle ξ, which we are interested in, can be obtained from s by changing the transition functions of s by multiplicative constants. We call the bundles with constant transition functions flat bundles. Thus, ξ is a product of s and a flat bundle ζ:

$$\xi = s \cdot \zeta. \tag{1}$$

The transition functions of ζ are constants. The sections of ζ become multi-valued on M when continued analytically. The bundle ζ can be characterized by the multi-valuedness of its sections: Let A_i, B_i $(i = 1,\ldots,p)$ be the nontrivial loops on M. The loop (or period) A_i intersects only with B_i but not with any other loop. (For the A_i, B_i periods, see ref. [4] for more details.) The sections of ζ are multiplied by $\exp(-2\pi i a_i)$ $(a_i \in \mathbf{R})$ around the A_i period, and by $\exp(2\pi i b_i)$ $(b_i \in \mathbf{R})$ around the B_i period.

The bundle ξ has Chern number $p-1$, where Chern number is the difference between the number of zeros and the number of poles of any meromorphic section of ξ [5].

Now, we are ready to start discussing field theory. We consider the following action:

$$I = \int_{\mathrm{M}} \mathrm{d}^2 z \sqrt{g}\, \chi \nabla^z \psi, \tag{2}$$

where the field ψ is a section of ξ, and the field χ is a section of $K \cdot \xi^{-1}$ (k is the canonical bundle whose local basis is $\mathrm{d}z$). The covariant derivative ∇^z is defined by

☆ This work was supported by the Director, Office of Energy Research, Office of High Energy and Nuclear Physics, Division of High Energy Physics of the US Department of Energy under Contract DE-AC03-7600098.

0370-2693/86/\$ 03.50 © Elsevier Science Publishers B.V. (North-Holland Physics Publishing Division)

$$\nabla^z \equiv g^{z\bar{z}}\partial_{\bar{z}}. \tag{3}$$

We define the covariant derivative ∇_z acting on sections of $K^{-1}\cdot\xi$ by

$$\nabla_z \equiv \partial_z + \tfrac{1}{2} g^{z\bar{z}}\partial_z g_{z\bar{z}}. \tag{4}$$

Then the laplacian is defined by $-\nabla_z\nabla^z$ for ξ. Let $\{\phi_n(z,\bar{z})\}$ be the eigenmodes of the laplacian:

$$-\nabla_z\nabla^z\phi_n = \lambda_n\phi_n. \tag{5}$$

Here we assume that ξ has no holomorphic section. Then, by the Riemann–Roch theorem [5], $K\cdot\xi^{-1}$ has no holomorphic section. Under this assumption, all the eigenvalues of the laplacian are strictly positive:

$$\lambda_n > 0. \tag{6}$$

We normalize the eigenmodes by

$$\int_M d^2z\,\sqrt{g}\sqrt{g^{z\bar{z}}}\,\phi_n^*\phi_m = \delta_{nm}. \tag{7}$$

The propagator is now defined as follows:

$$F_z(z,w) \equiv \langle\chi(z,\bar{z})\,\psi(z,\bar{z})\rangle = -2\pi\sum_n \frac{1}{\lambda_n}\sqrt{g^{z\bar{z}}}\,\partial_z\phi_n^*(z,\bar{z})\cdot\phi_n(w,\bar{w}). \tag{8}$$

We see that this is the correct definition from the following equations:

$$\nabla^z F_z(z,w) = 2\pi(1/\sqrt{g})\,\delta^{(2)}(z,w), \quad \nabla^w F_z(z,w) = -2\pi(1/\sqrt{g})\,\delta^{(2)}(z,w). \tag{9, 10}$$

We get (9), since $\{\phi_n\}$ is an orthonormal basis of ξ. Likewise we get (10), since $\{(1/\sqrt{\lambda_n})\,\nabla^z\phi_n\}$ is an orthonormal basis of $K^{-1}\cdot\xi$.

Eqs. (9), (10) imply that $F_z(z,w)$ is meromorphic in z and w with the only singularity at $z=w$:

$$F_z(z,w) \to 1/(z-w) \quad \text{as } z\to w. \tag{11}$$

With respect to z, $F_z(z,w)$ is a section of $K\cdot\xi^{-1}$, and with respect to w, it is a section of ξ.

Here we should make a remark on the importance of the assumption that ξ has no holomorphic section. Suppose $G_z(z,w)$ is a meromorphic section of $K\cdot\xi^{-1}(\xi)$ with respect to $z\,(w)$ with the singularity (11). The $F_z(z,w) - G_z(z,w)$ is a holomorphic section of $K\cdot\xi^{-1}(\xi)$ with respect to $z\,(w)$. This has to vanish due to the assumption. Thus, the problem of finding the propagator reduces to constructing the meromorphic section of $K\cdot\xi^{-1}(\xi)$ with respect to $z\,(w)$ with the singularity (11). Such a section can be found entirely by an analytic method as follows:

We consider

$$m(z,w) \equiv F_z(z,w)/h(z)\,h(w), \tag{12}$$

where $h(z)$ is the holomorphic section of the spin bundle s. Then, $m(z,w)$ is a meromorphic section of $\zeta^{-1}(\zeta)$ with respect to $z\,(w)$. Let $P_1,\ldots,P_{p-1}$ be the zeros of $h(z)$. Then, with respect to z, $m(z,w)$ has poles at $z=w$, $P_1,\ldots,P_{p-1}$. Likewise with respect to w, the poles are at $w=z, P_1,\ldots,P_{p-1}$. Since $m(z,w)$ is a section of $\zeta^{-1}(\zeta)$ with respect to $z\,(w)$, $m(z,w)$ is multiplied by $\exp(\pm 2\pi i a_i)$ when we move $z\,(w)$ around the A_i period, and by $\exp(\mp 2\pi i b_i)$ around the B_i period. We note that up to a multiplicative constant, $m(z,w)$ can be completely characterized by its singularities and multi-valuedness, for if $m'(z,w)$ has the same properties, then $m'(z,w)\,h(z)\,h(w)$ has a pole only at $z=w$ and must be a constant multiple of $F_z(z,w)$.

The meromorphic section $m(z,w)$ can be readily constructed using ϑ-functions [4]:

$$m(z,w) = \vartheta\begin{bmatrix}a_0+a\\ b_0+b\end{bmatrix}\left(\int_w^z\omega\right)\Big/\vartheta\begin{bmatrix}a_0\\ b_0\end{bmatrix}\left(\int_w^z\omega\right), \tag{13}$$

where a_{0i} and b_{0i} $(i = 1,\ldots,p)$ are half-integers and are given by

$$\Omega_{ij} a_{0j} + b_{0i} \equiv -\Delta_i + \sum_{k=1}^{p-1} \int_Q^{P_k} \omega_i. \tag{14}$$

Here $\{\omega_i\ (i = 1,\ldots,p)\}$ is a basis of the holomorphic one-forms on **M**, which is normalized by

$$\oint_{A_i} \omega_j = \delta_{ij}. \tag{15}$$

The period matrix Ω_{ij} is defined by

$$\Omega_{ij} \equiv \oint_{B_i} \omega_j. \tag{16}$$

The Riemann constant Δ_i corresponds to a spin bundle which has a meromorphic section with its only pole at Q [4]. The denominator of (13) has simple zeros at $z = w, P_1,\ldots,P_{p-1}$, and $w = z, P_1,\ldots,P_{p-1}$. The multi-valuedness of (13) can be easily checked from

$$\vartheta\begin{bmatrix} a \\ b \end{bmatrix}(u + m + \Omega n) = \exp(2\pi i a m - 2\pi i b n - \pi i n \Omega n - 2\pi i n u)\, \vartheta\begin{bmatrix} a \\ b \end{bmatrix}(u), \tag{17}$$

where m_i, n_i are integers [4].

From the uniqueness of $m(z, w)$, we can find the propagator:

$$F_z(z, w) = h(z)\, h(w)\, \vartheta\begin{bmatrix} a_0 + a \\ b_0 + b \end{bmatrix}\left(\int_w^z \omega\right) \Big/ \vartheta\begin{bmatrix} a_0 \\ b_0 \end{bmatrix}\left(\int_w^z \omega\right). \tag{18}$$

Due to (11), the normalization of $h(z)$ is given by

$$h^2(z) = \omega_{iz} \frac{\partial}{\partial u_i} \vartheta\begin{bmatrix} a_0 \\ b_0 \end{bmatrix}(u = 0) \Big/ \vartheta\begin{bmatrix} a_0 + a \\ b_0 + b \end{bmatrix}(0). \tag{19}$$

This is the main result of this paper.

Using the propagator (18), we can compute any correlation function of the field operators ψ and χ. It is particularly interesting to compute the expectation value of the energy–momentum tensor, which is defined by

$$T_{zz} \equiv \lim_{w \to z} [\tfrac{1}{2}(\partial_z - \partial_w)\chi(z, \bar{z})\, \psi(w, \bar{w}) + 1/(z - w)^2]. \tag{20}$$

This is not quite a rank-two tensor. The covariant tensor can be obtained from (20) by adding local counter terms:

$$\tilde{T}_{zz} \equiv T_{zz} + \tfrac{1}{24}[(g^{z\bar{z}}\partial_z g_{z\bar{z}})^2 - 2\partial_z(g^{z\bar{z}}\partial_z g_{z\bar{z}})]. \tag{21}$$

The expectation value of $\tilde{T}_{zz}$ gives a change of the functional determinant of the laplacian $-\nabla_z \nabla^z$ under a change of the metric δg^{zz}:

$$\frac{\delta}{\delta g^{zz}} \ln \det -\nabla_z \nabla^z = \frac{1}{4\pi}\langle \tilde{T}_{zz} \rangle. \tag{22}$$

It is straightforward to calculate the expectation value of T_{zz} from (18); the result is

$$\langle T_{zz} \rangle = \tfrac{1}{2}\omega_{iz}\omega_{jz} \frac{\partial^2}{\partial u_i \partial u_j} \ln \vartheta\begin{bmatrix} a_0 + a \\ b_0 + b \end{bmatrix}(u = 0) + \frac{1}{2}\left(\omega_{iz} \frac{\partial}{\partial u_i} \ln \vartheta\begin{bmatrix} a_0 + a \\ b_0 + b \end{bmatrix}(u = 0)\right)^2 + \tfrac{1}{2}\pi\, \omega_{iz} (\mathrm{Im}\, \Omega)^{-1}_{ij} \omega_{jz} + \langle T^x_{zz} \rangle. \tag{23}$$

Here T^x_{zz} is the energy–momentum tensor of the scalar field x, which is given by

$$\langle T^x_{zz}\rangle \equiv -\tfrac{1}{2} \lim_{w\to z} [\langle \partial_z x \partial_w x\rangle + 1/(z-w)^2]$$

$$= -\tfrac{1}{2}\pi\, \omega_{iz} (\mathrm{Im}\, \Omega)^{-1}_{ij} \omega_{jz} - \tfrac{1}{8}(C^{-1}\partial_z C)^2 + \tfrac{1}{12} C^{-1}\partial_z^2 C - \tfrac{1}{6} C^{-1} \frac{\partial^3}{\partial u_i \partial u_j \partial u_k} \vartheta\begin{bmatrix} a_0 \\ b_0 \end{bmatrix} (u=0)\, \omega_{iz}\omega_{jz}\omega_{kz}, \qquad (24)$$

where

$$C \equiv \omega_{iz} \frac{\partial}{\partial u_i} \vartheta \begin{bmatrix} a_0 \\ b_0 \end{bmatrix} (u=0).$$

The energy–momentum tensor (24) is not covariant, either. By adding the same counter terms as for T_{zz}, we can construct the covariant tensor [6]

$$\widetilde{T}^x_{zz} \equiv T^x_{zz} + \tfrac{1}{24} [(g^{z\bar z}\partial_z g_{z\bar z})^2 - 2\partial_z (g^{z\bar z}\partial_z g_{z\bar z})]. \qquad (25)$$

This gives a change of the determinant of the ordinary laplacian $-g^{z\bar z}\partial_{\bar z}\partial_z$:

$$\frac{1}{4\pi}\langle \widetilde{T}^x_{zz}\rangle = \frac{\delta}{\delta g^{zz}} \ln \left(\frac{\det' - g^{z\bar z}\partial_{\bar z}\partial_z}{\int d^2 z \sqrt{g}} \right)^{-1/2}, \qquad (26)$$

where the prime means the omission of the constant mode.

Both (23) and (24) are holomorphic, therefore the contributions to the $\bar z$-derivatives of the covariant energy–momentum tensors (21), (25) solely come from the counter terms

$$\nabla^z \langle \widetilde{T}_{zz}\rangle = \nabla^z \langle \widetilde{T}^x_{zz}\rangle = \tfrac{1}{24}\partial_z R, \qquad (27)$$

where R is the Ricci scalar. From the coordinate invariance, (27) implies

$$\langle \widetilde{T}_{z\bar z}\rangle = \langle \widetilde{T}^x_{z\bar z}\rangle = -\tfrac{1}{24} R g_{z\bar z}. \qquad (28)$$

The expression of the energy–momentum tensor (23) gives a relation between the determinants of the two laplacians $-\nabla_z\nabla^z$ and $-g^{z\bar z}\partial_z\partial_{\bar z}$. Since

$$\delta\Omega_{ij}/\delta g^{zz} = \tfrac{1}{2}\mathrm{i}\, \omega_{iz}\omega_{jz}, \qquad (29)$$

we find

$$4\pi \frac{\delta}{\delta g^{zz}} \ln \left((\det \mathrm{Im}\, \Omega_{ij})^{1/2} \left| \vartheta \begin{bmatrix} a_0 + a \\ b_0 + b \end{bmatrix} (0) \right|^2 \right)$$

$$= \tfrac{1}{2}\omega_{iz}\omega_{jz} \frac{\partial^2}{\partial u_i \partial u_j} \ln \vartheta \begin{bmatrix} a_0 + a \\ b_0 + b \end{bmatrix} (u=0) + \frac{1}{2}\left(\omega_{iz} \frac{\partial}{\partial u_i} \ln \vartheta \begin{bmatrix} a_0 + a \\ b_0 + b \end{bmatrix} (u=0) \right)^2 + \tfrac{1}{2}\pi\, \omega_{iz} (\mathrm{Im}\, \Omega)^{-1}_{ij}\omega_{jz}, \qquad (30)$$

which is the first three terms of (23).

Therefore, from (23), (30), and (28), we obtain

$$\det - \nabla_z \nabla^z = \text{const.} \times \left(\frac{\det' - g^{z\bar z}\partial_{\bar z}\partial_z}{\int d^2 z \sqrt{g} \det \mathrm{Im}\, \Omega} \right)^{-1/2} \left| \vartheta \begin{bmatrix} a_0 + a \\ b_0 + b \end{bmatrix} (0) \right|^2. \qquad (31)$$

The constant is independent of the metric tensor $g_{z\bar z}$. This relation (31) has already been found in refs. [2,7] from the consideration of the determinant bundles over the moduli space.

I thank O. Alvarez, M. Dugan, N. Marcus, and P. Windey for discussions.

References

[1] D. Friedan, E. Martinec and S. Shenker, Conformal invariance, supersymmetry, and string theory, Chicago preprint (1985).
[2] L. Alvarez-Gaumé, G. Moore and C. Vafa, Harvard preprint HUTP-86/A017.
[3] E. Martinec, Conformal field theory on a (super) Riemann surface, Princeton preprint (1986).
[4] D. Mumford, Tata Lectures on Theta I, II (Birkhäuser, Basel, 1983).
[5] R.C. Gunning, Lectures on Riemann surfaces (Princeton U.P., Princeton, NJ, 1966).
[6] H. Sonoda, LBL preprint LBL-21363.
[7] J.B. Bost and P. Nelson, Harvard preprint HUTP-86/A039.

Nuclear Physics B288 (1987) 357–396
North-Holland, Amsterdam

CHIRAL BOSONIZATION, DETERMINANTS AND THE STRING PARTITION FUNCTION

Erik VERLINDE and Herman VERLINDE

Institute for Theoretical Physics, Princetonplein 5, P.O. Box 80.006, 3508 TA Utrecht, The Netherlands

Received 17 November 1986

We study the bosonization of chiral fermion theories on arbitrary compact Riemann surfaces. We express the fermionic and bosonic correlation functions in terms of theta functions and prove their equality. This is used to obtain explicit expressions for a class of chiral determinants relevant to string theory. The anomaly structure of these determinants and their behaviour on degenerate Riemann surfaces is analysed. We apply these results to multi-loop calculations of the bosonic string.

1. Introduction and summary

In the path integral approach to string perturbation theory the g-loop contribution to a scattering amplitude is given by the functional integral over all geometries of a two-dimensional surface of genus g and over quantum fields living on this surface [1,2]. Depending on the type of geometry and the field content one distinguishes between bosonic, fermionic and heterotic strings [3,4]. The 2d conformal field theories corresponding to these string theories in the critical dimension are all free of both local and global anomalies. Due to this fact, g-loop amplitudes can be reduced to a finite dimensional integral over moduli space $\mathcal{M}_g$, the space of conformally inequivalent surfaces of genus g [5,6]. The integrand is build up from Green functions and determinants of differential operators. To understand the physics of string loop corrections, one needs to have a good description of this integrand on moduli space.

An important step in this direction has been made by Belavin and Knizhnik [7]. They showed that string amplitudes are essentially the product of an analytic and an anti-analytic function on $\mathcal{M}_g$. This factorization has a direct correspondence with the decomposition of the string modes into left and right movers. The analytic building units of the string integrand are the determinants of chiral operators. The main purpose of this paper is to study these chiral determinants. We are especially interested in their behaviour near the boundary of moduli space, i.e. on singular surfaces with nodes, because this is of direct relevance to the analysis of possible

0550-3213/87/$03.50 © Elsevier Science Publishers B.V.
(North-Holland Physics Publishing Division)

divergences in string amplitudes [7–9]. Another feature of the chiral determinants that is important for string theory is the behaviour under modular transformations.

The determinants of differential operators have a well-known representation in terms of a fermionic path integral. The fermion theories corresponding to the chiral operators that are relevant to string theory are all of the same type: they describe two conjugate chiral fermions with conformal spin λ and $1-\lambda$. (The relevant values for λ are $\frac{1}{2}$, 1, $\frac{3}{2}$ and 2.) The local structure of these theories has been analysed by Friedan, Martinec and Shenker [10], using techniques of 2d conformal field theory [11, 12]. In particular they showed that the operator algebra of these Fermi theories is identical to that of a bosonic theory of a free scalar field taking values on $\mathbb{R}/2\pi\mathbb{Z}$ and coupled to a background charge $Q=2\lambda-1$. This background charge is the manifestation of the anomaly of the fermion number current. Assuming this local equivalence extends to a complete identification of both theories, this gives a second representation of the chiral determinants in terms of a bosonic partition function.

It was first realized by Alvarez-Gaumé, Moore and Vafa [13], that Bose-Fermi equivalence provides a powerful method for calculating the determinants. They showed that it gives a relation between the Dirac determinant, the Riemann theta function, and the determinant of the scalar laplacian. The theta function represents the partition sum over the soliton sectors of the Bose theory. This relation was used in ref. [13] to analyse the modular transformation properties of the Dirac determinant. Motivated by this result, we study in this paper the generalization of this idea to fermions with arbitrary spin.

Our starting point is the correlation functions on a fixed surface Σ. In the chiral fermion theory these are analytic functions of the positions of the fields, which can be studied using complex analysis on Σ. The main technical difficulty in showing that these fermionic correlation functions can be reproduced by bosonic theories is that a free scalar field is not a chiral field. Locally it is possible to define a chiral projection, but there can be global obstructions due to the constant zero mode. As we will show, this obstruction cancels after including the contributions of all soliton sectors.

Theta functions are very useful for obtaining explicit expressions for the correlation functions, because the position of their zeroes and poles are well described by the so-called Riemann vanishing theorem, and because of their relation with spinor bundles on Σ [15–17]. In particular, all fermionic propagators and also the scalar Green function can be expressed in terms of theta functions. Rather surprisingly it then turns out that the identities stating the equivalence of the fermionic and the bosonic expressions for the correlation functions are already known to mathematicians as so-called "addition theorems" [15], the basic idea of which can be traced back to Riemann himself!

These results for the correlation functions on Σ can be translated to statements about the partition function on moduli space by integrating the expectation value of the stress-energy tensor with respect to the moduli [18,19]. We will perform this integration explicitly, using the bosonized results for the expectation values. Via this procedure the chiral determinant can be expressed in terms of theta functions. This expression is useful for studying the behaviour of the chiral determinants on moduli space, because the properties of the theta function on $\mathbb{M}_g$ (their transformation rule under the modular group, behaviour near the boundary of $\mathbb{M}_g$ etc.) are well-known [15–17].

The organization of this paper is as follows. In sect. 2 we show, using some basic facts about divisor classes of complex line bundles, that the chiral fermion correlation functions are uniquely determined by the local operator products. In sect. 3 we review some aspects of the theory of theta functions, and describe the so-called prime form. This is used in sects. 4 and 5 to construct the fermion propagator and the scalar Green function. In sect. 6 we compute the correlation functions of the vertex operators and show how they are related to the fermionic correlation functions. The construction of the chiral determinants is discussed in sect. 7. We give explicit expressions in terms of theta functions, and describe their anomaly structure. We also discuss the determinants of the laplacians. In sect. 8 we study the modular transformation properties of the determinants and in sect. 9 their behaviour on degenerate surfaces. Finally, in sect. 10 these results are applied to the partition function of the closed bosonic string.

After our calculations were completed, we received a preprint by Alvarez-Gaumé, Bost, Moore, Nelson, and Vafa in which they discuss the bosonization of general non-chiral fermion systems on higher genus surfaces and report results similar to ours for the determinants of laplacians [14].

2. Chiral fermions on a Riemann surface

We consider the quantum theory of two conjugate chiral fermions, b and c, with conformal spin λ and $1-\lambda$, living on a compact Riemann surface Σ of genus g. The spin λ may be integer or half integer. In sect. 7 we will use the partition function of these theories to represent the chiral determinants $\det \nabla^{\bar{z}}_{1-\lambda}$, where $\nabla^{\bar{z}}_{1-\lambda}$ is the adjoint of the covariant derivative $\nabla_z^{-\lambda}$ on the space of $(-\lambda)$-differentials (= tensors of rank $-\lambda$). In local complex coordinates, in which the metric looks like $ds^2 = \rho(z,\bar{z})\,dz\,d\bar{z}$, these derivatives are given by:

$$\nabla_z^{-\lambda} = \rho^{-\lambda}\partial_z\rho^{\lambda}, \qquad \nabla^{\bar{z}}_{1-\lambda} = \rho^{-1}\partial_{\bar{z}}. \tag{2.1}$$

In this section we discuss the correlation functions of the theory. The action is given by:

$$S[b,c] = \frac{1}{2\pi}\int d^2z\,\sqrt{g(z)}\,b(z)\nabla^{\bar{z}}_{1-\lambda}c(z). \tag{2.2}$$

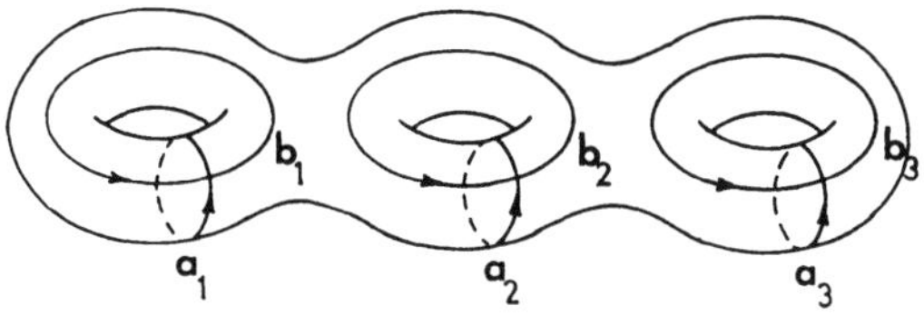

Fig. 1. A canonical homology basis a_i, b_i $i = 1,2,3$ for a surface with genus $g = 3$.

To fix the boundary conditions on the fields, we choose a basis of homology cycles a_i and b_i $(i = 1,2,\dots, g)$ as drawn in fig. 1. Such a basis is called a canonical basis. For integer λ we will only consider periodic boundary conditions. If λ is half-integer we impose boundary conditions corresponding to some spin structure $(\alpha, \beta) \in (\frac{1}{2}\mathbb{Z}/\mathbb{Z})^{2g}$, which means that (relative to the spin structure (0,0)) around a_i the fields are multiplied with $\exp(2\pi i \alpha_i)$, and around b_i with $\exp(2\pi i \beta_i)$.

The theory is invariant under the large group of conformal transformations, which consist of all analytic coordinate changes $z \to z'$. These transformations are generated by the stress-energy tensor T_{ab}, which can be found by varying the action with respect to the metric:

$$T_{zz} = -\lambda b(\nabla_z c) + (1-\lambda)(\nabla_z b)c. \tag{2.3}$$

The other components of T_{ab} vanish classically, which reflects the invariance of the action under Weyl transformations and anti-holomorphic variations of the complex structure.

Turning to the correlation functions of the quantum theory, our first concern are the fermion zero modes. By the Riemann-Roch theorem we know that the number of b zero modes ($= \dim(\ker \nabla_\lambda^z)$) minus the number of c zero modes ($= \dim(\ker \nabla_{1-\lambda}^z)$) is equal to $I = (2\lambda - 1)(g-1)$. This result can be also derived by integrating the anomaly of the fermion number current $j_z = -bc$:

$$\partial_{\bar z} j_z = -\tfrac{1}{8}(2\lambda - 1)\sqrt{g}\,R, \tag{2.4}$$

yielding a total violation of fermion number of I. So correlation functions of N b's with M c's are only non-vanishing if $N - M = I$. Consider such a correlation function:

$$A(z_1,\dots, w_M) = \left\langle \prod_1^{M+I} b(z_i) \prod_1^{M} c(w_j) \right\rangle. \tag{2.5}$$

We can use the techniques of two dimensional conformal field theory [11] to analyse the local properties of A. $A(z_1,\dots, w_M)$ depends analytically on the z's and w's, due to the equation of motion. The behaviour of the correlation function A under conformal transformations is determined by the operator product expansions of the b and c fields with the stress-energy tensor. From the fundamental operator products, for $z \sim w$:

$$b(z)c(w) \sim \frac{1}{z-w},$$

$$b(z)b(w) \sim O(z-w),$$

$$c(z)c(w) \sim O(z-w), \tag{2.6}$$

one finds [10]:

$$T_{zz}(z)b(w) \sim \left(\frac{\lambda}{(z-w)^2} + \frac{1}{(z-w)}\partial_w\right)b(w),$$

$$T_{zz}(z)c(w) \sim \left(\frac{1-\lambda}{(z-w)^2} + \frac{1}{(z-w)}\partial_w\right)c(w). \tag{2.7}$$

Under an infinitesimal conformal transformation $z \to z + \varepsilon(z)$ the variation of any conformal field $\varphi(z)$ is:

$$\delta_\varepsilon\varphi(z) = \frac{1}{2\pi i}\oint_C \mathrm{d}w\, \varepsilon(w) T_{ww}(w)\varphi(z),$$

where the contour C surrounds z. So from (2.7) it follows that the quantum fields b and c transform (as expected) as their classical counterparts:

$$A(z_1',\dots,w_M') = \prod_1^{M+I}\left(\frac{\mathrm{d}z}{\mathrm{d}z'}\right)^{\lambda}\prod_1^{M}\left(\frac{\mathrm{d}w}{\mathrm{d}w'}\right)^{1-\lambda} A(z_1,\dots,w_M).$$

Further we know from (2.6) that $A(z_1,\dots,w_m)$ has zeroes for $z_i = z_j$ or $w_i = w_j$, $i \neq j$, and (in general) poles for $z_i = w_j$. The residues at these poles are equal to the correlators of the remaining b's and c's. We will now show that these facts uniquely determine the correlation function $A(z_1,\dots,w_M)$, using some results from the theory of complex line bundles.

Considered as a function of e.g. z_1, A is a meromorphic λ-differential. It is clear that $A(z_1)$ is determined up to a constant factor independent of z_1 by its divisor D_A, i.e. the position of all its zeroes and poles. So what we need is a characterization of the possible divisors of a meromorphic λ-differential.

For the case that $A(z_1)$ is a function, i.e. $\lambda = 0$, it is well-known that (because Σ is compact) the degree of D_A, defined as the number of zeroes minus the number of poles of $A(z_1)$, is zero. Since the quotient of two arbitrary meromorphic λ-differentials A_1 and A_2 is a meromorphic function, the difference $D_1 - D_2$ of their divisors is the divisor of a meromorphic function. Two divisors related in this way are called equivalent. Thus the divisors of all meromorphic λ-differentials form an equivalence class of divisors, or shortly, a divisor class, $\mathbb{D}_\lambda$. In particular, they all have the same degree. Next, because the line bundle of λ-differentials for integer λ is equal to the λth tensor power of the bundle of 1-differentials (or 1-forms), it follows that their divisor classes are related by $\mathbb{D}_\lambda = \lambda\mathbb{D}_1$. This class $\mathbb{D}_1$ is called the canonical class, and is commonly denoted by K. The degree of K is equal to $2(g-1)$. For $\frac{1}{2}$-integer λ we have to distinguish the spin structures. Let D_α denote the divisor class of the bundle S_α of $\frac{1}{2}$-forms with spin structure α. Then for general $\frac{1}{2}$-integer λ we have: $\mathbb{D}_{\lambda,\alpha} = (\lambda - \frac{1}{2})\mathrm{K} + \mathrm{D}_\alpha$. All D_α satisfy $2\mathrm{D}_\alpha = \mathrm{K}$ and are of degree $g-1$.

Applying all this to the correlation function $A(z_1)$ we conclude that the number of zeroes minus the number of poles of $A(z_1)$ is equal to $2\lambda(g-1)$. Therefore, besides the "physical" zeroes z_i $(i = 2,\dots, M+I)$ and poles w_j $(j = 1,\dots, M)$ which are determined by the operator products, there must be g additional zeroes p_k. The positions of these zeroes are constrained by the condition that the divisor of $A(z_1)$ is in the class $\mathbb{D}_{\lambda,\alpha}$:

$$\left[\sum_2^{M+I} z_i - \sum_1^{M} w_j + \sum_1^{g} p_k\right] = \mathbb{D}_{\lambda,\alpha}, \tag{2.8}$$

where $[D]$ denotes the divisor class of D. Now it is a well known mathematical result, called the Jacobi inversion theorem [16], that for almost every divisor D of degree g (= genus of Σ) there is a *unique* set of points $p_1,\dots, p_g$ on Σ such that: $[D] = [\Sigma p_i]$. So for generic z_i and w_j the condition (2.8) indeed uniquely determines the points p_k. (For completeness we note that there are some special cases for which residue of one of the poles for $z_1 = w_j$ vanishes. In these cases the counting is somewhat different, but the conclusion that all the zeroes of A are determined remains unaltered.)

Thus the correlation function A is determined up to a constant factor. To fix this factor we only have to normalize one correlator, since the relative normalization between the correlators is fixed by the operator products. The condition (2.8) and the fact that it determines the positions of all the zeroes p_k, will be important for the proof of the bosonization formulas.

3. Theta functions and spin structures

In this section we introduce the theta function and discuss its relation with spin structures. For more details we refer to [15–17].

Let Σ again be a compact Riemann surface, with a complex structure defined on it. As before, we choose on Σ a canonical basis of homology cycles a_i, b_i. Then there exists a normalized basis of holomorphic 1-forms ω_i, $i = 1,\dots, g$, such that:

$$\oint_{a_i} \omega_j = \delta_{ij}, \qquad \oint_{b_i} \omega_j = \tau_{ij}. \tag{3.1}$$

τ is called the period matrix of Σ and is a complex symmetric $g \times g$ matrix with positive definite imaginary part. The space of such matrices is called the Siegel upper half plane $\mathbb{H}_g$. The basis ω_i and the period matrix τ depend on the choice of the canonical homology basis a_i, b_i. For two different choices the ω's and τ's are related by a modular transformation (see sect. 8). Given a base point p_0 we can now associate to every point p on Σ a complex g-component vector z by the Jacobi map, defined as:

$$\mathbb{I}: \quad p \to z_i(p) = \int_{p_0}^{p} \omega_i .$$

This vector is unique up to the periods (3.1). So z is an element of the complex torus $J(\Sigma) = \mathbb{C}^g/(\mathbb{Z}^g + \tau\mathbb{Z}^g)$, called the jacobian variety of Σ. The Jacobi map can be generalized to any degree zero divisor $D = B - A$, with A and B positive divisors:

$$\mathbb{I}: \quad D \to \int_A^B \omega \in J(\Sigma). \tag{3.2}$$

$\mathbb{I}$ has the important property, called Abel's theorem, that two divisors are mapped onto the same point in $J(\Sigma)$ if and only if they are equivalent. So $\mathbb{I}$ acts injectively on the space of divisor classes. From now on we will mostly just write D for $\mathbb{I}(D)$.

The Riemann theta function is defined for $z \in J(\Sigma)$ by the infinite sum:

$$\theta(z, \tau) = \sum_{n \in \mathbb{Z}^g} \exp(i\pi n_i \tau_{ij} n_j + 2\pi i n_i z_i). \tag{3.3}$$

The basic property of $\theta(z, \tau)$ is that it has a simple transformation law if we shift z by the lattice $\mathbb{Z}^g + \tau\mathbb{Z}^g$:

$$\theta(z + \tau n + m, \tau) = \exp(-i\pi n\tau n - 2\pi i n z)\theta(z, \tau). \tag{3.4}$$

Via the Jacobi map $\mathbb{I}$ one can define the theta function on degree-zero divisor classes D on Σ. Most applications of theta functions are based on the Riemann vanishing theorem, describing the set of zeroes of $\theta(z, \tau)$. It states that there exists a divisor class Δ of degree $g-1$, such that $\theta(z, \tau) = 0$ if and only if there are $g-1$ points $p_1, \dots, p_{g-1}$ on Σ such that:

$$z = \Delta - \sum_1^{g-1} p_i .$$

Δ is called the Riemann class. This result can be proved by a straightforward application of the Green theorem, see ref. [16].

One of the corrolaries of this theorem is in fact the Jacobi inversion theorem, used in the previous section. Namely, the Riemann vanishing theorem can be reformulated as follows: There exists a degree $g-1$ divisor class Δ such that for any divisor class $\mathbb{D}$ of degree g the function $f(x)$ defined by $f(x)=\theta(x+\Delta-\mathbb{D})$, $x\in\Sigma$, either vanishes identically, or has exactly g zeroes p_i, which are uniquely determined by the condition $[\Sigma_1^g p_i]=\mathbb{D}$. The latter situation is generic: the exceptions form a set of codimension 2 and are of the form: $\mathbb{D}=2\Delta-[\Sigma_1^{g-2}q_i]$.

The close relation between theta functions and the bundles of $\frac{1}{2}$-differentials on Σ is based on the fact that the Riemann class Δ is related to the canonical class K, the divisor class of the bundle of 1-forms, by $2\Delta=\mathrm{K}$. This implies that Δ is the divisor class D_0 of a spin structure S_0. The divisor classes D_α of the other spin structures S_α are given by

$$D_\alpha=\Delta-\tau\alpha_1-\alpha_2,\qquad \alpha=(\alpha_1,\alpha_2)\in\left(\tfrac{1}{2}\mathbb{Z}/\mathbb{Z}\right)^{2g}. \tag{3.5}$$

In the same way S_0 is related to the Riemann theta function, we can associate to every spin structure S_α a theta function with characteristics (α_1,α_2), defined by:

$$\theta[\alpha](z,\tau)=\exp\left(i\pi\alpha_1\tau\alpha_1+2\pi i\alpha_1(z+\alpha_2)\right)\theta(z+\tau\alpha_1+\alpha_2,\tau). \tag{3.6}$$

For the characteristics $\alpha\in(\frac{1}{2}\mathbb{Z}/\mathbb{Z})^{2g}$ the theta function $\theta[\alpha](z)$ is either an even or an odd function of z. One easily shows:

$$\theta[\alpha](-z,\tau)=(-1)^{4\alpha_1\cdot\alpha_2}\theta[\alpha](z,\tau). \tag{3.7}$$

Spin structures are called even or odd, depending on whether the corresponding theta function is even or odd. So there are $2^{g-1}(2^g-1)$ odd and $2^{g-1}(2^g+1)$ even spin structures.

For odd characteristics α we have $\theta[\alpha](0,\tau)=0$. By Riemann's theorem, this implies that there are $g-1$ points r_i on Σ such that $\mathrm{D}_\alpha=[\Sigma r_i]$. This suggests that S_α has a holomorphic section $h_\alpha(z)$ having $g-1$ zeroes for $z=r_i$. Indeed, this section can be explicitly constructed as follows. Consider the function:

$$f_\alpha(z,w)=\theta[\alpha]\left(\int_w^z\omega\right),\qquad z,w\in\Sigma. \tag{3.8}$$

It follows again from the Riemann vanishing theorem that $f_\alpha(z,w)$ has single zeroes for $z=w$, $z=r_i$ or $w=r_i$. So for both z and w in the neighbourhood of one of the r $f_\alpha(z,w)$ looks like: $f_\alpha(z,w)\sim\mathrm{const}\cdot(z-w)(z-r_i)(w-r_i)$. Differentiating this w.r.t. w at $w=z$ one finds that the 1-form:

$$g_\alpha(z)=\sum_i\partial_i\theta[\alpha](0)\omega_i(z) \tag{3.9}$$

has $g-1$ *double* zeroes for $z=r_i$. Note that, since the divisor of g_α is in K, this shows that indeed $2D_\alpha = K$. Furthermore, this implies that there exists a holomorphic $\frac{1}{2}$-differential h_α such that:

$$g_\alpha(z) = h_\alpha^2(z). \tag{3.10}$$

This $h_\alpha(z)$ is the holomorphic section of S_α.

As we will see in the next section, all correlation functions of the spin-$\frac{1}{2}$ fermion theories can be expressed in terms of $f_\alpha(z, w)$'s and $h_\alpha(z)$'s. Particularly useful is the so-called prime form $E(z, w)$ [15]. This form is defined as the holomorphic differential form on $\Sigma \times \Sigma$ of weight $(-\frac{1}{2},0) \times (-\frac{1}{2},0)$:

$$E(z, w) = \frac{f_\alpha(z, w)}{h_\alpha(z) h_\alpha(w)} \qquad (\alpha \text{ odd}). \tag{3.11}$$

$E(z, w)$ is independent of the choice of the odd spin structure α. Its key property is that it is only zero for $z = w$. For $z \sim w$ it behaves as:

$$E(z, w) \sim z - w, \qquad z \sim w.$$

$E(z, w)$ is not single-valued around the b-cycles: if z is moved around the cycle b_i to z', then:

$$E(z', w) = -\exp\left(-i\pi\tau_{ii} - 2\pi i \int_z^w \omega_i\right) E(z, w). \tag{3.12}$$

The prime form $E(z, w)$ is the generalization of the function $z - w$ on the Riemann sphere; it can be used to factorize arbitrary meromorphic functions on Σ [15–16].

4. Fermionic construction of the correlation functions

We now discuss the fermionic construction of the correlation functions: Wick's theorem is used to express all correlation functions in terms of the zero modes and the fermion propagator, and we use theta functions to construct the propagator.

When applied to the correlation functions (2.5), Wick's theorem can be formulated as follows. A b-field is either used to absorb a zero mode or it is contracted with a c-field, meaning that the pair $b(z_i)c(w_j)$ is replaced by the propagator $P(z_i, w_j)$. The c-zero modes, if they are present, are absorbed by the remaining c-fields. The complete correlation function is obtained by antisymmetrization over all permutations of the b- and c-fields.

The propagator $P(z, w)$ is usually defined as the two-point function $\langle b(z)c(w)\rangle$ in the system with the zero modes projected out. This propagator $P(z, w)$ is a $(\lambda, 1-\lambda)$ differential with a single pole for $z = w$, and is the Green function of the differential operator $\partial_{\bar{z}}$ on the space orthogonal to the zero modes. It satisfies:

$$\partial_{\bar{z}} P(z, w) = \pi\delta(z-w) - \pi\sum_{1}^{m} \nu_k^*(z)\nu_k(w)\rho^{\lambda}(z),$$

$$\partial_{\bar{w}} P(z, w) = -\pi\delta(z-w) + \pi\sum_{1}^{n} \mu_k(z)\mu_k^*(w)\rho^{1-\lambda}(w), \tag{4.1}$$

where $\mu_1, \dots, \mu_n$, and $\nu_1, \dots, \nu_m$, $(n-m=I)$ are orthonormal bases for the b resp. c zero modes, with respect to the usual inner products:

$$\langle\mu_1|\mu_2\rangle = \int d^2z\, \rho^{1-\lambda}\mu_1^*\mu_2, \qquad \langle\nu_1|\nu_2\rangle = \int d^2z\, \rho^{\lambda}\nu_1^*\nu_2. \tag{4.2}$$

From (4.1) we see that $P(z, w)$ is in general not analytic in z or w. Correlation functions built from $P(z, w)$ are analytic, because terms in $P(z, w)$ proportional to the b or c zero modes drop out, due to the anti-symmetrisation in all z's and w's. For the same reason we can, instead of $P(z, w)$, also use as the propagator any meromorphic differential $p(z, w)$ with a single pole for $z = w$, satisfying the correct boundary conditions up to terms proportional to the zero modes. The Green function $P(z, w)$ is obtained from $p(z, w)$ by a projection onto the space orthogonal to the zero modes. We will now give expressions for the differential $p(z, w)$ for the cases $\lambda = \frac{1}{2}$ and $\lambda = 1$.

For spin-$\frac{1}{2}$ fermions there are for even spin structures generically no zero modes. (There are certain values of the period matrix τ for which $\theta[\alpha](0, \tau) = 0$, in which case the Dirac operator has two or more zero modes. We will not consider these special situations here.) So for even α the propagator $P_\alpha(z, w) = \langle\bar{\psi}(z)\psi(w)\rangle_\alpha$ is a meromorphic $\frac{1}{2}$-differential in z and w, with one single pole with residue 1 for $z = w$. There exists only one such $\frac{1}{2}$-differential for every even spin structure, called the Szego kernel [15, 20]. It is given in terms of the theta function and the prime form by:

$$P_\alpha(z, w) = \frac{1}{E(z, w)} \frac{\theta[\alpha](z-w)}{\theta[\alpha](0)}. \tag{4.3}$$

The correlation functions follow from Wick's theorem:

$$\left\langle \prod_{1}^{M} \bar{\psi}(z_i) \prod_{1}^{M} \psi(w_j) \right\rangle_\alpha = \det_{ij}\left(P_\alpha(z_i, w_j)\right). \tag{4.4}$$

For odd spin structures the situation is different. Both $\bar{\psi}$ and ψ have a zero mode given by the holomorphic $\frac{1}{2}$-differential $h_\alpha(z)$ introduced in the previous section. For the function $p_\alpha(z, w)$ we choose a simple modification of (4.3)

$$p_\alpha(z, w) = \frac{1}{E(z, w)} \frac{\Sigma\partial_i\theta[\alpha](z-w)\omega_i(y)}{\Sigma\partial_i\theta[\alpha](0)\omega_i(y)}, \tag{4.5}$$

where y is an arbitrary point on Σ, for which $h_\alpha(y) \neq 0$. The correlation functions are:

$$\left\langle \prod_1^M \bar{\psi}(z_i) \prod_1^M \psi(w_j) \right\rangle_\alpha = \sum_{\substack{\text{pairs} \\ k,l}} (-1)^{k+l} h_\alpha(z_k) h_\alpha(w_l) \det_{\substack{i\neq k \\ j\neq l}} (p_\alpha(z_i, w_j)). \quad (4.6)$$

The $\frac{1}{2}$-differential $p_\alpha(z, w)$ depends on the point y, but the correlation functions (4.6) do not.

Next consider $\lambda = 1$. In this case there are g b- zero modes and one c- zero mode, given by resp. the holomorphic 1-forms ω_i, and the constant function. For the meromorphic differential $p(z, w)$ one can take $p(z, w) = \partial_z \log(E(z, w))$. One then finds for the correlation functions:

$$\left\langle \prod_1^{M+g-1} b(z_i) \prod_1^M c(w_j) \right\rangle = \det_{kl}(\varphi_k(z_l)), \quad (4.7)$$

where

$$\varphi_k(z) = \omega_k(z), \qquad k = 1, \dots, g,$$

$$\varphi_{g+l}(z) = \omega_{w_l - w_{l+1}}(z), \qquad l = 1, \dots, M-1,$$

where $\omega_{a-b}(z)$ is the so-called third abelian differential defined by:

$$\omega_{a-b}(z) = \partial_z \log\left(\frac{E(z, a)}{E(z, b)} \right). \quad (4.8)$$

A similar analysis can be done for $\lambda > 1$. Then there are $(2\lambda - 1)$ b-zero modes and no c-zero modes. A difficulty is that there does not exist a convenient representation for these zero modes. Therefore the fermionic expressions for the correlation functions for $\lambda > 1$ are not as useful as those for $\lambda = \frac{1}{2}$ and $\lambda = 1$.

Until now we have only considered correlation functions of b- and c-fields at different points. These do not yet contain all information of the theory. To define the theory we have to choose a regularization procedure for defining correlation functions containing composite operators such as the fermion current and the stress-energy tensor. The chiral fermion theories discussed here all have a gravitational anomaly: there does not exist a covariant regularization. We will choose a Weyl invariant regularization procedure. Together with the correlation functions discussed in this section, this uniquely determines the partition function as a function of the geometry of the surface, via the expectation value of the stress-energy tensor (see sect. 7). For proving bosonization it is sufficient to consider the correlation functions with all $z_i \neq w_j$, since if these are the same for the Bose and the Fermi theory, then both theories also have the same partition function and anomaly structure, provided we subtract the infinities in the same way.

5. Green function of a free scalar field

As a preparation for the discussion of the bosonization of the fermion theories we first consider the theory of a free scalar field $\varphi(z, \bar{z})$ on a Riemann surface Σ. So the action is:

$$S[\varphi] = \frac{1}{4\pi} \int d^2 z \sqrt{g}\, g^{z\bar{z}}\, \partial_z \varphi\, \partial_{\bar{z}} \varphi , \tag{5.1}$$

Classically this theory is invariant under Weyl rescalings: $g \to e^{\sigma} g$. This symmetry is broken in the quantum theory. Although this fact is well-known, let us briefly discuss the reasons. We define correlation functions using the path integral:

$$\langle \varphi(z_1)\varphi(z_2)\dots\varphi(z_n)\rangle = \frac{1}{Z_0} \int D\varphi\, \varphi(z_1)\dots\varphi(z_n) e^{-S[\varphi]},$$

where Z_0 is the partition function: $Z_0 = \int D\varphi\, e^{-S[\varphi]}$. (Here and from now on we just write $\varphi(z)$, meaning $\varphi(z, \bar{z})$.) The field φ is expanded in eigenfunctions of the scalar laplacian Δ_0. This laplacian has one normalized zero mode: $\varphi = 1/\sqrt{A}$. where A is the area of the surface: $A = \int d^2 z \sqrt{g}$. By Wick's theorem, all correlation functions can be expressed in terms of the two-point function $f(z, w) = \langle \varphi(z)\varphi(w)\rangle$. which is the Green function for Δ_0. From the representation of this Green function in terms of eigen functions, it follows that $f(z, w)$ satisfies:

$$\int d^2 z \sqrt{g} f(z, w) = 0,$$

$$\partial_z \partial_{\bar{z}} f(z, w) = -\pi \delta(z - w) + \frac{\pi}{A} \sqrt{g}(z), \tag{5.2}$$

$$\partial_z \partial_{\bar{w}} f(z, w) = \pi \delta(z - w) - \pi \sum_{i,j} \omega_i(z) (\operatorname{Im} \tau)^{-1}_{ij} \bar{\omega}_j(w), \tag{5.3}$$

where $\omega_i = \omega_i(z) dz$ is the basis of holomorphic 1-forms and τ is the period matrix. Note these equations are related to eq. (4.1) for the propagator $P(z, w)$ for the *b-c* system with $\lambda = 1$, via the identification $P(z, w) = -\partial_z f(z, w)$. The properties (5.2), reflecting the presence of the constant zero mode, are not Weyl invariant.

Instead we find that for $g_{ab} = e^{\sigma}\hat{g}_{ab}$, $f(z, w)$ transforms as:

$$f(z,w) = \hat{f}(z,w) - \frac{1}{A}\int d^2y \sqrt{g}(y)(\hat{f}(z,y) + \hat{f}(y,w))$$

$$+ \frac{1}{A^2}\int\int d^2x\, d^2y \sqrt{g}(x)\sqrt{g}(y)\hat{f}(x,y). \qquad (5.4)$$

To define the two-point function for coincident points $z = w$, we have to introduce a covariant cut-off ε, and remove the divergence $\log(\varepsilon)$. As is well-known, this procedure introduces an additional dependence on the conformal factor $\sigma(z)$ [1]:

$$f_R(z,z) = \hat{f}_R(z,z) + \sigma(z) - \frac{2}{A}\int d^2y \sqrt{g}(y)\hat{f}(z,y)$$

$$+ \frac{1}{A^2}\int\int d^2x\, d^2y \sqrt{g}(x)\sqrt{g}(y)\hat{f}(x,y). \qquad (5.5)$$

So we can see that, in general, correlation functions are not Weyl invariant. The reason for this trouble is that $\varphi(z)$ is not a proper conformal field. Derivatives of φ are good conformal operators: their correlation functions are Weyl invariant. Other examples of conformal fields are vertex operators:

$$V_q(z) = \rho^{q^2/2}(z) e^{iq\varphi(z)}, \qquad (\rho = 2g_{z\bar{z}}).$$

Their two-point function is given by:

$$\langle V_q(z) V_{q'}(w)\rangle = (F(z,w))^{-q^2}\delta(q+q'), \qquad (5.6)$$

where

$$F(z,w) = (\rho(z)\rho(w))^{-1/2}\exp(-f(z,w) + \tfrac{1}{2}f_R(z,z) + \tfrac{1}{2}f_R(w,w)) \qquad (5.7)$$

is indeed a Weyl invariant function, as can be seen from the Weyl transformation properties of $f(z, w)$ and $f_R(z, z)$.

Let us examine this function $F(z, w)$ somewhat closer. From its definition it follows that $F(z, w)$ is a single valued real symmetric function. Under conformal transformations $z \to z'(z)$ it transforms as an $(-\frac{1}{2}, -\frac{1}{2}) \times (-\frac{1}{2}, -\frac{1}{2})$ differential:

$$F(z',w') = \left|\frac{\partial z'}{\partial z}\right|\left|\frac{\partial w'}{\partial w}\right| F(z,w). \qquad (5.8)$$

Further, a short computation using eqs. (5.2)–(5.5) gives:

$$\partial_z\partial_{\bar{z}} \log F(z,w) = \pi\delta(z-w) - \pi\sum_{i,j}\omega_i(z)(\operatorname{Im}\tau)^{-1}_{ij}\bar{\omega}_j(z). \qquad (5.9)$$

These properties uniquely determine $F(z, w)$:

$$F(z,w) = \exp\left[-2\pi \operatorname{Im}\int_w^z \omega (\operatorname{Im}\tau)^{-1} \operatorname{Im}\int_w^z \omega\right] |E(z,w)|^2. \tag{5.10}$$

It follows from the properties of the prime form that the r.h.s is a real symmetric single-valued $(-\frac{1}{2}, -\frac{1}{2})$-form in z and w, satisfying the differential equation (5.9).

The Green function $f(z, w)$ can be expressed in terms of $F(z, w)$ by inverting the relation (5.7). The result can be written as eq. (5.4) with on the r.h.s. $\hat{f}$ replaced by $-\log F$.

6. Chiral bosonization

6.1. INTRODUCTION

The bosonization of fermions with arbitrary spin λ was first considered by Friedan, Martinec and Shenker [10]. They showed that, locally, the fermionic systems of the previous sections can be represented by a bosonic scalar field coupled to a background charge $Q = 2\lambda - 1$, which is described by the action:

$$S_Q[\varphi] = \frac{1}{2\pi}\int \mathrm{d}^2 z \left(\partial_z\varphi \partial_{\bar z}\varphi - \tfrac{1}{4} iQ\sqrt{g}\, R\cdot\varphi\right). \tag{6.1}$$

Their key observation is that the current algebra of the fermion number current j_z and of $-i\partial_z\varphi$ are identical. Also the anomaly in j_z is reproduced by the equation of motion:

$$\partial_z\partial_{\bar z}\varphi = \tfrac{1}{8} iQ\sqrt{g}\, R. \tag{6.2}$$

The background charge changes the weight of the vertex operators $\mathrm{e}^{iq\varphi}$ under Weyl transformations from $\frac{1}{2}q^2$ to $\frac{1}{2}q(q+Q)$. Multiplied with the appropriate power of $\rho = 2g_{z\bar z}$ to make them Weyl invariant, they transform under conformal transformations as conformal fields of dimension $(\frac{1}{2}q(q+Q), \frac{1}{2}q(q+Q))$. In the conformal quantum theory this follows from the operator products of the vertex operators with the stress-energy tensor:

$$T_{zz} = -\tfrac{1}{2}(\partial_z\varphi)^2 - \tfrac{1}{2} iQ\partial_z^2\varphi,$$

$$T_{zz}(z)\mathrm{e}^{iq\varphi(w)} \sim \left(\frac{\frac{1}{2}q(q+Q)}{(z-w)^2} + \frac{1}{z-w}\partial_w\right)\mathrm{e}^{iq\varphi(w)}. \tag{6.3}$$

A second effect of the background charge is that only those correlation functions are non-zero, for which the total background charge is cancelled by the external charges:

$$\Sigma q_i = -\frac{Q}{8\pi}\int d^2z\sqrt{g}R(z) = Q(g-1).$$

This is the translation onto bosonic language of the Reimann-Roch theorem, counting the b- and c-zero modes in the fermion theory.

The vertex operators with $q = \pm 1$ are identified with the fermi bilinears $b\bar{b}$ and $c\bar{c}$. Locally, we can perform a "chiral projection" by taking the holomorphic square root of the vertex operators. The transformation properties and operator products then coincide with those of the Fermi fields b and c. However, if we want to relate the correlation functions of the Fermi and Bose systems, we must be able to take the holomorphic square root of the bosonic correlation functions in a globally well-defined way. We will see that this is possible if the scalar field takes its values on a circle with radius one: $\varphi \in \mathbb{R}/2\pi\mathbb{Z}$. Because of this compactification one can have soliton configurations on Riemann surfaces with genus $g > 0$. This means that for nontrivial homology cycles the winding- (or soliton-) number: $N = (1/2\pi)\oint d\varphi$ may be non-zero. Solitons are created and destroyed by the vertex operators $e^{i\varphi}$ and $e^{-i\varphi}$, so the fermions are the solitons of the bosonic theory.

We will now explicitly calculate the correlation functions of the vertex operators, and show how they correspond to the fermionic amplitudes. We first consider the case with vanishing background charge, i.e. the bosonized spin-$\frac{1}{2}$ theory.

6.2. SPIN-$\frac{1}{2}$ BOSONIZATION

We start with the action of the classical solitons. The soliton sectors can be labeled by the winding numbers for the canonical homology basis a_i, b_i. The soliton solution with winding numbers n_i, m_i is given by:

$$\varphi_{nm}(z) = i\pi(m+\bar{\tau}n)(\operatorname{Im}\tau)^{-1}\int_{p_0}^{z}\omega + \text{c.c.}. \tag{6.4}$$

Note that the r.h.s. indeed satisfies the equation of motion, since it is the sum of a holomorphic and an anti-holomorphic function of z. For the action of this soliton one finds, using that $\int\omega_i \wedge \bar{\omega}_j = 2i\operatorname{Im}\tau_{ij}$:

$$S_{nm} = \tfrac{1}{2}\pi(m+\bar{\tau}n)(\operatorname{Im}\tau)^{-1}(m+\tau n). \tag{6.5}$$

This result will be used below.

Now let us compute the correlation functions of the vertex operators:

$$A(z_1,\dots,w_M) = \left\langle \prod_{i=1}^{M}\rho^{1/2}(z_i)e^{i\varphi(z_i)}\prod_{j=1}^{M}\rho^{1/2}(w_j)e^{-i\varphi(w_j)}\right\rangle.$$

In a given soliton sector (n, m) the correlation functions can be computed in the same way as in sect. 5.

$$A_{nm}(z_1, \dots, w_M) = \prod_i e^{i\varphi_{nm}(z_i)} \prod_j e^{-i\varphi_{nm}(w_j)} \frac{\prod_{i<j} F(z_i, z_j) \prod_{i<j} F(w_i, w_j)}{\prod_{i,j} F(z_i, w_j)}. \tag{6.6}$$

The correlation function for the full theory is obtained by summing over all soliton sectors (n, m). To evaluate this sum we write, following ref. [13], $m = 2(m' + \alpha)$ with $m' \in \mathbb{Z}^g$, $\alpha \in (\frac{1}{2}\mathbb{Z}/\mathbb{Z})^g$, and apply the Poisson summation formula to the sum over m'. The result can be expressed in terms of theta functions:

$$\begin{aligned} A_{\text{sol}}(z_1, \dots, w_M) &= \sum_{n,m} e^{-S_{nm}} \prod_i e^{i\varphi_{nm}(z_i)} \prod_j e^{-i\varphi_{nm}(w_j)} \\ &= (\det \operatorname{Im} \tau)^{1/2} \sum_{\alpha \in (\frac{1}{2}\mathbb{Z}/\mathbb{Z})^{2g}} \|\theta[\alpha](\Sigma z_i - \Sigma w_j)\|^2, \end{aligned} \tag{6.7}$$

where:

$$\|\theta[\alpha](z)\|^2 = \exp\big(2\pi \operatorname{Im} z (\operatorname{Im} \tau)^{-1} \operatorname{Im} z\big) |\theta[\alpha](z)|^2.$$

This indicates that the bosonic theory corresponds to the sum over all spin structures of the Fermi theory. We can define the correlation functions for a particular even spin structure α by:

$$\begin{aligned} \langle \dots \rangle_\alpha &= \frac{1}{Z_{\text{sol}}^{(\alpha)}} \sum_{n,m}{}^{(\alpha)} e^{-S_{nm}} \langle \dots \rangle_{nm}, \\ Z_{\text{sol}}^{(\alpha)} &= (\det \operatorname{Im} \tau)^{1/2} |\theta[\alpha](0)|^2, \end{aligned} \tag{6.8}$$

where the superscript (α) indicates that, after the Poisson resummation, one only takes the term corresponding to the spin structure α. Combining (6.6)–(6.8) and substituting for $F(z, w)$ the expression (5.10) one finds that all anholomorphic parts cancel. So A_α is the modulus squared of a meromorphic function for all z_i and w_j. This is precisely what we need, since now we can perform a chiral projection by taking the meromorphic square root. This gives the following result for the chiral amplitude:

$$A_\alpha^{\text{chiral}}(z_1, \dots, w_m) = \frac{\theta[\alpha](\Sigma z_i - \Sigma w_j)}{\theta[\alpha](0)} \frac{\prod_{i<j} E(z_i, z_j) \prod_{i<j} E(w_i, w_j)}{\prod_{i,j} E(z_i, w_j)}. \tag{6.9}$$

According to the bosonization prescription the r.h.s. of this equation should be equal to the correlation functions for the spin-$\frac{1}{2}$ fermions with spin structure α, given in (4.4). The correlation function of two vertex operators indeed gives the fermion propagator (4.3). Equating the fermionic and bosonic expression for $M=2$ (i.e. two z's and two w's) gives the following identity:

$$\theta[\alpha](z_1-w_1)\theta[\alpha](z_2-w_2)E(z_2,w_1)E(z_1,w_2)-(z_1\leftrightarrow z_2)$$
$$=\theta[\alpha](z_1+z_2-w_1-w_2)\theta[\alpha](0)E(z_1,z_2)E(w_1,w_2). \qquad (6.10)$$

This is indeed a well-known mathematical result, called the trisecant identity [16], and was first written down by Fay [15]. It holds for arbitrary characteristics $\alpha\in(\mathbb{R}/\mathbb{Z})^{2g}$. This identity is very special, because it is only valid for theta functions associated with Riemann surfaces. (In ref. [16] a detailed proof of (6.10) is given.) The bosonization formulas for general M can also be found in the book of Fay, as an "addition theorem" for abelian functions (p. 33). They can be derived from the trisecant identity. At the end of this section we will sketch a direct proof of a generalized version of this addition theorem.

The definition (6.8) has to be modified for odd spin structures α, because then $\theta[\alpha](0)=0$, which corresponds to the presence of the zero mode $h_\alpha(z)$ in the fermion theory. If for this case we simply take $Z^{(\alpha)}$ to be equal to 1 then the bosonic theory again reproduces the fermion correlation functions given in (4.6) (see ref. [15]).

6.3. BOSONIZATION OF FERMIONS WITH ARBITRARY SPIN

Let us now switch on the background charge Q. First, consider the soliton action. Here we meet some trouble with the $R\varphi$-term because the field φ is not single valued. To give the action a proper definition, one has to specify how to integrate over the surface. To do so, we cut the surface Σ in the standard way along $2g$ curves all going through one base point p_0 as shown in fig. 2 for a surface with $g=2$. The

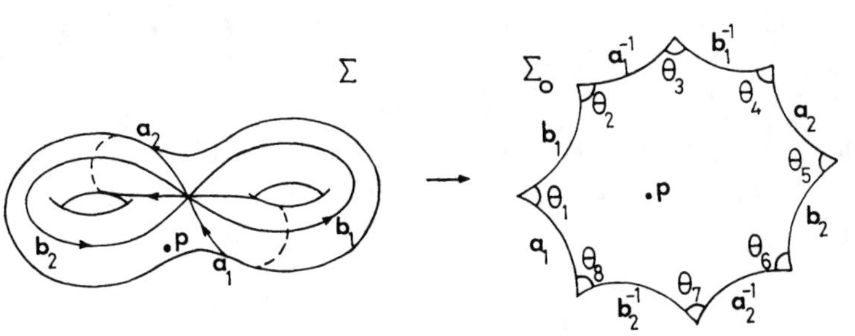

Fig. 2. But cutting a genus two surface along the homology cycles a_i, b_j one obtains a octagon with angles θ_i satisfying $\Sigma\theta_i=2\pi$.

resulting polygon is called Σ_0. We now define:

$$\frac{1}{8\pi}\int_\Sigma d^2z\sqrt{g}\,R\varphi=\frac{1}{8\pi}\int_{\Sigma_0}d^2z\sqrt{g}\,R\varphi+\frac{1}{4\pi}\int_{\partial\Sigma_0}ds\,k\varphi+\frac{1}{4\pi}\sum_{k=1}^{4g}\Lambda_k\varphi(p_k).\qquad(6.11)$$

Here k is the geodesic curvature and $\Lambda_k=1/2\pi-\theta_k-2\pi l_k$, where the θ_k are the inner angles of the polygon Σ_0 (see fig. 2), and the l_k are arbitrary integers satisfying $\Sigma l_k=g-1$. The correlation functions will turn out to be independent of the choice for these integers. It follows from the Gauss-Bonnet theorem that the definition (6.11) is invariant under variations of the base point p_0 and the curves a_i and b_i. This is in fact the reason for adding the boundary terms. Note that these boundary terms vanish if the field φ is single valued.

To find the soliton action we have to calculate the l.h.s. of (6.11) for $\varphi(z)=\int_p^z\omega_i$. Using that $\varphi(z)$ is holomorphic it is easily shown that the result is Weyl invariant. We use this to take as metric on Σ: $ds^2=|s_\alpha|^4dz\,d\bar{z}$ where s_α is a section of a twisted spin bundle S_α, i.e. s_α is meromorphic $\frac{1}{2}$-differential with periodicities $\exp(2\pi i\alpha_1)$ and $\exp(2\pi i\alpha_2)$ for the a- resp. b-cycles, $(\alpha_1,\alpha_2)\in(\mathbb{R}/\mathbb{Z})^{2g}$. Let $D_\alpha=\Sigma\varepsilon_i r_i$ ($\varepsilon_i=\pm1$, $\Sigma\varepsilon_i=g-1$) be the divisor of s_α. The curvature scalar is then given by:

$$\sqrt{g}\,R(z)=-8\,\partial_z\partial_{\bar{z}}\log|s_\alpha|^2=-8\pi\sum_i\varepsilon_i\,\delta(z-r_i).$$

Substituting this in (6.11) we obtain:

$$\frac{1}{8\pi}\int d^2z\sqrt{g}\,R\int_p^z\omega_i=-\int_{(g-1)p}^{D_\alpha}\omega_i+(\text{boundary terms})_i,\qquad(6.12)$$

$$(\text{b.t.})_i=\frac{1}{4\pi}\tau_{ij}\left[\oint_{a_j}ds\,k+\pi-\theta_{a_j}+2\pi N_j\right]+\frac{1}{4\pi}\left[\oint_{b_i}ds\,k+\pi-\theta_{b_i}+2\pi M_i\right].$$

Here θ_{a_j} and θ_{b_i} are the inner angles of the a- resp. b-cycles and are equal to the sum of two adjacent angles θ_k of the polygon Σ_0. The integers N_j and M_i are the sum of the corresponding l_k. The boundary terms in (6.12) can be rewritten by applying the Gauss-Bonnet theorem to the cycles a_j and b_i:

$$\oint_{a_j}ds\,k+\pi-\theta_{a_j}=2\pi(1-2\alpha_{1j}),$$

$$\oint_{b_i}ds\,k+\pi-\theta_{b_i}=2\pi(1-2\alpha_{2i}).$$

Inserting this in (6.12) and using the relation: $D_\alpha=\Delta-\tau\alpha_1-\alpha_2$ we find that the result is independent of α. The boundary terms cancel modulo $\mathbb{Z}^g + \tau\mathbb{Z}^g$ for some $\alpha \in (\frac{1}{2}\mathbb{Z}/\mathbb{Z})^{2g}$, i.e. if D_α is the divisor of some spin structure. Which spin structure this is, depends on the choice of the integers l_k in (6.11). We will see that for the correlation functions it does not matter which one we take. However, for definiteness we choose for D_α the Riemann class Δ. So we have:

$$\frac{1}{8\pi}\int d^2z\sqrt{g}\,R\int_p^z \omega_i = -\int_{(g-1)p}^{\Delta}\omega_i \in J(\Sigma). \tag{6.13}$$

Using this result we obtain for the action for the soliton configuration (6.5):

$$S[\varphi_{mn}] = S_{mn} + QS_b,$$

with

$$S_b = \pi(m+\bar{\tau}n)(\mathrm{Im}\,\tau)^{-1}\int_{(g-1)p}^{\Delta}\omega - \text{c.c.}. \tag{6.14}$$

Note that now the soliton is not a solution to the classical equation of motion. For $Q \neq 0$ there are no solutions to (6.2).

Next, we again turn to the correlation functions of the vertex operators:

$$A(z_1,\dots,w_m) = \left\langle \prod_{i=1}^{M+I}\rho^{\lambda}(z_i)e^{i\varphi(z_i)}\prod_{j=1}^{M}\rho^{1-\lambda}(w_j)e^{-i\varphi(w_j)}\right\rangle, \tag{6.15}$$

where we already took account of charge conservation. Here we should note that, because the identity operator has a vanishing vacuum expectation value, there is no canonical normalization of the correlation functions: only the ratio of two correlation has an invariant meaning. Keeping this in mind, let us calculate the amplitude (6.15). It is again given by the product of a soliton sum Λ_{sol} and the amplitude in the zero soliton sector A_{00}. The soliton sum can be evaluated with the same technique as for $Q=0$:

$$\Lambda_{\text{sol}}^{(Q)}(z_1,\dots,w_M) = (\det\mathrm{Im}\,\tau)^{1/2}\sum_{\alpha}\|\theta[\alpha](\Sigma z_i - \Sigma w_j - Q\Delta)\|^2. \tag{6.16}$$

Because of the sum over all spin structures, this result is independent of the choice of the divisor D_α in the classical soliton action. The amplitude in the zero soliton sector is:

$$A_{00}(z_1,\dots,w_M) = \frac{\prod_{i<j}F(z_i,z_j)\prod_{i<j}F(w_i,w_j)}{\prod_{i,j}F(z_i,w_j)}\,\frac{\prod_i H^Q(z_i)}{\prod_j H^Q(w_j)}$$

where:

$$H(z)=\rho^{1/2}(z)\exp\left\{\frac{1}{8\pi}\int d^2y\sqrt{g}R(y)\log(F(z,y))\right\} \tag{6.17}$$

represents the coupling of the vertex operators to the background charge. Because, if possible, we want to take the holomorphic square root of the total correlation function, we extract the an-holomorphic part of $H(z)$:

$$H(z)=\text{const}\cdot\exp\left\{\frac{-2\pi}{g-1}\operatorname{Im}\Delta_z(\operatorname{Im}\tau)^{-1}\operatorname{Im}\Delta_z\right\}|\sigma(z)|^2, \tag{6.18}$$

where we have introduced the notation $\Delta_z=\int_{(g-1)z}^{\Delta}\omega$. Using $\sqrt{g}R=-4\partial_z\partial_{\bar z}\log\rho$ one shows that $\sigma(z)$ is indeed holomorphic in z, and furthermore a section of a trivial line bundle (i.e. it has no zeroes or poles). Under analytic coordinate transformations it transforms as a tensor of rank $\frac{1}{2}g$. Further, σ is multivalued around the b-cycles:

$$\sigma(z')=\sigma(z)\exp\{-\pi i(g-1)\tau_{ii}+2\pi i\Delta_z\}. \tag{6.19}$$

These properties determine $\sigma(z)$ up to a constant factor independent of z. At this point, because of our previous remark, we only need to know $\sigma(z)$ up to a constant factor. In the next section we will also consider the dependence of $\sigma(z)$ on the moduli parameters, parametrizing the complex structures on Σ. This will fix the constant in (6.18). The quotient $\sigma(z)/\sigma(w)$ can be expressed in terms of theta functions as:

$$\frac{\sigma(z)}{\sigma(w)}=\frac{\theta(z-\Sigma p_i+\Delta)}{\theta(w-\Sigma p_i+\Delta)}\prod_i\frac{E(w,p_i)}{E(z,p_i)}, \tag{6.20}$$

where p_i, $i=1,\dots,g$, are arbitrary point on Σ.

Putting all ingredients together and projecting onto a single spin structure, we again find that the correlation function A is the modulus squared of a meromorphic, or chiral, amplitude:

$$A^{\text{chiral}}_{\alpha,Q}(z_1,\dots,w_M)$$
$$=\theta[\alpha](\Sigma z_i-\Sigma w_j-Q\Delta)\frac{\prod_{i<j}E(z_i,z_j)\prod_{i<j}E(w_i,w_j)}{\prod_{i,j}E(z_i,w_j)}\frac{\prod_i\sigma^Q(z_i)}{\prod_j\sigma^Q(w_j)}. \tag{6.21}$$

This is the final result for the correlation functions of the vertex operators.

Now we want to show that $A_{\alpha,Q}$ can be identified with the correlation functions of the fermionic b–c theory with spin $\lambda=\frac{1}{2}(Q+1)$ and spin structure α, which are discussed in sect. 2. We first note that the r.h.s. of (6.21) is indeed a λ- resp. $1-\lambda$ differential in z_i, resp w_j. Next we must check that all the zeroes and poles coincide. So consider $A_{\alpha,Q}$ again as a function of z_1. It is clear that $A_{\alpha,Q}$ has zeroes for $z_1 = z_i$, $i = 2, \dots, M+I$, and poles for $z_1 = w_j$, $j = 1, \dots, M$ due to the prime forms, and that the only other zeroes come from the theta function. To find these we use the Riemann vanishing theorem. It gives:

$$\operatorname{div}_{z_1}\theta[\alpha]\left(z_1 + \sum_2^{M+I} z_i - \sum_1^{M} w_j - Q\Delta\right) = \sum_1^{g} p_i ,$$

where $p_1, \dots, p_g$ is the unique g-tuple of points on Σ such that:

$$\left[\sum_1^{g} p_i\right] = \left[\sum_2^{M+I} z_i - \sum_1^{M} w_j\right] - Q\Delta - D_\alpha .$$

This is precisely the same condition as the one which determines the positions of the g additional zeroes of the fermion amplitude (see eq. (2.8)). So indeed, the positions of the poles and zeroes match. (An interesting special case to check is $\lambda = 1$ ($Q = 1$), $\alpha = 0$, since then there is also a c zero mode. So for $M = 1$ the fermion amplitude is regular for all $z_1, \dots, z_g$. This is also the case for the bosonic amplitude: the poles of the factor $\prod E(z_i, w)$ are cancelled by the zeroes of the θ-function.) Since further the periodicities around the a- and b-cycles coincide, we can conclude that the amplitudes are identical, and therefore that both theories are indeed equivalent.

7. Chiral determinants

7.1. DEFINITION OF THE CHIRAL DETERMINANTS

In this section we will use two-dimensional conformal field theory and the bosonization formulas to construct explicit expressions for the chiral determinant $\det \nabla^z_{1-\lambda}$ and the determinant of the corresponding laplacian Δ_λ. As these determinants are functions on the space of riemannian geometries of Σ, we start with a short description of this space.

The geometry of a Riemann surface is specified by the metric $ds^2 = g_{ab}\, d\xi^a d\xi^b$. Two metrics give the same geometry if they are related by a diffeomorphism from Σ to Σ. Note that this group of diffeomorphisms contains besides the transformations connected to the identity also topologically non-trivial transformations. As a coordinate system on the space of geometries one can take the reparametrisation invariant Weyl factor $\varphi(\xi)$, relating the metric $g_{ab}(\xi)$ to the constant curvature metric $\hat{g}_{ab}(\xi)$ on Σ with $\hat{R} = -1$ together with a set of parameters m_i, parametrizing the space of inequivalent constant metrics: $g_{ab}(\xi) = e^{\varphi(\xi)}\hat{g}_{ab}(m)(\xi)$. This space of inequivalent constant curvature metrics is isomorphic to the space of different complex structures on Σ, known as the moduli space $\mathbb{M}_g$. The complex dimension of $\mathbb{M}_g$ is equal to the dimension of the space of holomorphic quadratic differentials, which is $3(g-1)$. This relation can be understood by noting that infinitesimal changes in the complex structure correspond to symmetric traceless variations δg_{zz} of the metric, which are orthogonal to infinitesimal diffeomorphisms. These so-called Teichmüller deformations are solutions to the equation $\nabla^z \delta g_{zz} = 0$, so they are holomorphic quadratic differentials. The elements of the space dual to the space of quadratic differentials are called Beltrami differentials.

As described in sect. 2, we can for a given choice of the homology basis associate to every complex structure on Σ a period matrix τ, which is an element of the Siegel upper half plane $\mathbb{H}_g$. The set of period matrices associated to Riemann surfaces form a $3(g-1)$-dimensional subspace of $\mathbb{H}_g$. Moduli space is obtained from this subspace by dividing out the action of the modular group (i.e. the group of all diffeomorphisms, modulo the identity component). Thus $\mathbb{M}_g$ is not a manifold, but a so-called V-manifold or orbifold. $\mathbb{M}_g$ has a complex structure [21], which coincides with that on $\mathbb{H}_g$ meaning that holomorphic functions of the period matrix τ (e.g. theta functions) are also holomorphic sections of line bundles on $\mathbb{M}_g$. Functions on $\mathbb{M}_g$ correspond to modular invariant functions on $\mathbb{H}_g$.

Finally, by including in addition to the smooth regular surfaces, also singular surfaces with nodes the moduli space $\mathbb{M}_g$ is extended to a compact space $\overline{\mathbb{M}}_g$ [22]. The boundary $\mathbb{D} = \overline{\mathbb{M}}_g - \mathbb{M}_g$ is the union of $\frac{1}{2}g$ components $\mathbb{D}_k$, where the surface breaks down into two components of genus k, resp. $g-k$. All $\mathbb{D}_k$ have complex codimension 1.

Now that we know some facts about the space of geometries, let us turn to the chiral determinants as functions on this space. First we have to explain what we mean by the determinant of the covariant derivative $\nabla^z_{1-\lambda}$. This is a rather subtle matter, since $\nabla^z_{1-\lambda}$ is an operator between two different Hilbert spaces, and moreover, because it has zero modes. An important property of $\nabla^z_{1-\lambda}$, which can be used as guide line, is that it is invariant under anti-holomorphic variations δg_{zz} of the complex structure i.e. $\nabla^z_{1-\lambda}$ depends holomorphically on the moduli.

Let us first consider the corresponding finite dimensional situation. So we have a family of linear operators D_y, depending analytically on some complex parameter $y \in Y$, acting between two finite dimensional vector spaces V and W. If D_y is invertible for almost all y, then we can use as a definition for the "chiral" determinant det D_y the (formal) equation:

$$\frac{\mathrm{d}}{\mathrm{d}y}(\log \det D_y) = \mathrm{Tr}\left(D_y^{-1} \frac{\mathrm{d}}{\mathrm{d}y}(D_y)\right). \tag{7.1}$$

The r.h.s. is a well-defined analytic function of y, which can be integrated to give $\det D_y$ (modulo a constant factor). There is also a second, equivalent, definition of $\det D_y$, namely via the determinant of $D_y^\dagger D_y$. Because D_y varies holomorphically with y, $\det D_y^\dagger D_y$ is the square of an analytic function of y. It is therefore natural to define $\det D_y$ as the holomorphic square root of $\det D_y^\dagger D_y$:

$$|\det D_y|^2 = \det D_y^\dagger D_y . \tag{7.2}$$

This uniquely determines $\det D_y$, up to a y-independent phase factor. As we will see, both definitions have useful analogues in the infinite dimensional case.

The above definitions fail if D_y has a non-zero kernel or cokernel. The situation can then be saved as follows. Choose a basis $\{e_i, i = 1, \dots, n\}$ for $\ker D_y$, and a basis $\{f_j, j = 1, \dots, m\}$ for $\operatorname{coker} D_y$. With these bases $\ker D_y$ and $\operatorname{coker} D_y$ are identified with $\mathbb{C}^n$ resp. $\mathbb{C}^m$. Using this identification D_y can be naturally extended to a bijective map $\hat{D}_y(e, f)$ from $\mathrm{V} \times \mathbb{C}^m$ to $\mathrm{W} \times \mathbb{C}^n$. For this $D_y(e, f)$ we can define a non-zero determinant function $\det D_y(e, f)$, which is analytic in y if the bases $\{e_i\}$ and $\{f_j\}$ are chosen to vary holomorphically. Under changes of the bases $\{e_i\}$ and $\{f_j\}$ this function transforms as:

$$\det \hat{D}_y(\tilde{e}, \tilde{f}) = \left(\det\langle e_i|\tilde{e}_j\rangle\right)^{-1} \det\langle f_i|\tilde{f}_j\rangle \det \hat{D}_y(e, f) .$$

We now define $\det D_y$ as the element of $\lambda(\ker D_y) \otimes \lambda(\operatorname{coker} D_y)^*$ (where $\lambda(\mathrm{V})$ denotes the highest exterior power of V) given by:

$$\det D_y = \det \hat{D}_y(e, f)(e_1 \wedge \dots \wedge e_n) \otimes (f_1^* \wedge \dots \wedge f_m^*) . \tag{7.3}$$

This definition is independent of the choice of the bases $\{e_i\}$ and $\{f_j\}$. Together the $\det D_y$ define a holomorphic section of the line bundle $\lambda(\ker D_y) \otimes \lambda(\operatorname{coker} D_y)^*$ over the parameter space Y.

This construction can be generalized to determinants of chiral differential operators. Applied to our case it means that $\det \nabla^z_{1-\lambda}$ is defined as a holomorphic section of the line bundle $\mathbb{L}_\lambda$ over $\mathbb{M}_g$ with fibre generated by $(\nu_1 \wedge \dots \wedge \nu_n) \otimes (\mu_1 \wedge \mu_2 \wedge \dots \wedge \mu_{I+n})$, where $\{\nu_i\}$ is a basis of $\ker(\nabla^z_{1-\lambda})$ and $\{\mu_j\}$ a basis of $\ker(\nabla^z_\lambda)$. (For half-integer spin the line bundle $\mathbb{L}_\lambda$ in fact must be defined over the so-called spin covering $\mathbb{S}_g$ of $\mathbb{M}_g$, see next section.) Formally, the chiral determinant can be represented by a fermionic path integral. We can write (for $\lambda > 1$):

$$\det \nabla^z_{1-\lambda} = \frac{Z_\lambda(z_1, \dots, z_I)}{\det\left(\mu_i(z_j)\right)} \cdot \mu_1 \wedge \mu_2 \wedge \dots \wedge \mu_I ,$$

where

$$Z_\lambda(z_1, \dots, z_I) = \int \mathrm{D}b\, \mathrm{D}c\, b(z_1) \dots b(z_I) \exp\left(-\frac{1}{2\pi} \int \mathrm{d}^2 z \sqrt{g}\, b(z) \nabla^z_{1-\lambda} c(z)\right) \tag{7.4}$$

is the partition function of the b-c system, treated as a conformal field theory. This means that in defining Z_λ we use a Weyl invariant regularization scheme ensuring that det $\nabla^z_{1-\lambda}$ depends holomorphically on the moduli parameters. We will describe two ways to calculate Z_λ. The first uses the expectation value of stress-energy tensor and corresponds to eq. (7.1). The second is based on the relation between det $\nabla^z_{1-\lambda}$ and the determinant of the laplacian $\Delta_{1-\lambda} = \nabla_z^{-\lambda}\nabla^z_{1-\lambda}$ and generalizes the construction (7.2). The final result for the chiral determinants is the same for both methods, and is given by the following expression for $\lambda \neq 1$:

$$Z_{\alpha,\lambda}(z_1,\dots,z_I) = Z_1^{-1/2}\theta[\alpha]\Big(\sum_i z_i - Q\Delta\Big)\prod_{i<j}E(z_i,z_j)\prod_i \sigma^Q(z_i) \tag{7.5}$$

and for $\lambda = 1$:

$$Z_1(z_1,\dots,z_g,w) = Z_1^{-1/2}\theta\Big(\sum_i z_i - w - \Delta\Big)\frac{\prod_{i<j}E(z_i,z_j)\prod_i\sigma(z_i)}{\prod_i E(z_i,w)\sigma(w)}. \tag{7.6}$$

The factor $Z_1^{-1/2}$ can be considered as the partition function of a chiral scalar field and is related to $Z_1(z_1,\dots,z_g,w)$ by:

$$Z_1(z_1,\dots,z_g,w) = Z_1 \det \omega_i(z_j). \tag{7.7}$$

Together (7.5)–(7.7) form a closed set of equations from which the Z_λ's can be solved in terms of the theta function, the 1-forms ω_i, the prime form and the holomorphic $\frac{1}{2}g$-differential $\sigma(z)$, which we will describe in more detail below. These equations are the main result of this paper. After the discussion of their derivation, we will use these expressions in the next sections to examine some of the properties of the chiral determinants.

7.2. CHIRAL DETERMINANTS FROM THE STRESS-ENERGY TENSOR

In this subsection we follow the strategy advocated in [18,19] (see also [20]) by using the expectation value of the analytic stress-energy tensor of the chiral theory to define Z_λ:

$$\frac{4\pi}{\sqrt{g}}\frac{\delta}{\delta g^{ww}}\log(Z_\lambda(z_1,\dots,z_I)) = \frac{\langle T_{ww}(w)\Pi b(z_i)\rangle}{\langle \Pi b(z_i)\rangle}, \tag{7.8}$$

where T_{ww} is defined by:

$$T_{ww}(w) = \lim_{z\to w}\left(-\lambda\partial_z + (1-\lambda)\,\partial_w\right)\left\{b(z)c(w) - \frac{1}{z-w}\right\}. \tag{7.9}$$

Here, by choosing this Weyl invariant subtraction procedure, we have sacrificed coordinate invariance. As a consequence T_{ww} is not a proper quadratic differential. It transforms anomalously under conformal transformations:

$$T_{ww}(w) = \left(\frac{\mathrm{d}z}{\mathrm{d}w}\right)^2 T_{zz}(z) - \tfrac{1}{6}c_\lambda\{z, w\}, \tag{7.10}$$

where

$$\begin{aligned}\{z, w\} &= \frac{\mathrm{d}^3 z/\mathrm{d}w^3}{\mathrm{d}z/\mathrm{d}w} - \frac{3}{2}\left(\frac{\mathrm{d}^2 z/\mathrm{d}w^2}{\mathrm{d}z/\mathrm{d}w}\right)^2 \\ &= -\tfrac{1}{2}(\partial_w \log(\mathrm{d}z/\mathrm{d}w))^2 + \partial_w^2 \log(\mathrm{d}z/\mathrm{d}w)\end{aligned}$$

is the schwarzian derivative and $c_\lambda = 6\lambda^2 - 6\lambda + 1$. This transformation law of T_{ww} is well-known from conformal field theory, and is related to the presence of the central charge in the Virasoro algebra [11]. It implies that the chiral determinants have a gravitational anomaly: it depends on the coordinate system on Σ which local section of $\mathbb{L}_\lambda$ is identified with $\det \nabla^z_{1-\lambda}$.

The stress-energy tensor T_{ww} can also be represented in terms of the chiral fermion number current $j_w = -bc$:

$$T_{ww}(w) = \lim_{z \to w} \tfrac{1}{2}\left\{ j_z(z) j_w(w) - \frac{1}{(z-w)^2}\right\} - \tfrac{1}{2}Q\partial_w j_w(w).$$

Using this representation and the bosonic expressions for the b-c correlation functions given in sect. 6, it is straightforward to compute the expectation value of T_{ww}. One finds (for $\lambda \neq 1$):

$$\begin{aligned}\frac{\langle T_{ww}(w)\Pi b(z_i)\rangle}{\langle \Pi b(z_i)\rangle} &= T_{ww}^{(\varphi)}(w) + \tfrac{1}{2}\partial_i\partial_j \log\theta[\alpha]\Big(\sum_j z_j - Q\Delta\Big)\omega_i(w)\omega_j(w) \\ &\quad + \tfrac{1}{2}\langle j_w\rangle^2 - \tfrac{1}{2}Q\,\partial_w\langle j_w\rangle,\end{aligned} \tag{7.11}$$

where

$$\begin{aligned}T_{ww}^{(\varphi)} &= \lim_{z \to w} -\tfrac{1}{2}\left(\partial_z\partial_w \log E(z,w) - \frac{1}{(z-w)^2}\right) \\ &= \tfrac{1}{6}\left\{(\partial_w \log E(z,w))^2 + \partial_w^2 \log E(z,w)\right\}\Big|_{z=w}\end{aligned}$$

and $\langle j_w\rangle$ denotes the expectation value of the fermion current:

$$\begin{aligned}\langle j_w\rangle &= \frac{\langle j_w(w)\Pi b(z_i)\rangle}{\langle \Pi b(z_i)\rangle} \\ &= \partial_i \log\theta[\alpha]\Big(\sum_j z_j - Q\Delta\Big)\omega_i(w) + \partial_w \log\Big(\sigma^Q(w)\prod_j E(w, z_j)\Big).\end{aligned} \tag{7.12}$$

The result for $\lambda = 1$ is slightly different because of the c-zero mode. Note that both (7.11) and (7.12) are indeed holomorphic on moduli space.

We can also use the fermionic representations of the correlation functions, discussed in sect. 4, to calculate the expectation value of T_{ww}. For $\lambda = \frac{1}{2}$ the result is the same as (7.11); for $\lambda = 1$ we find from (4.7):

$$\frac{\langle T_{ww}(w)\Pi b(z_i)c(x)\rangle}{\langle \Pi b(z_i)c(x)\rangle}$$
$$= -2T_{ww}^{(\varphi)}(w) + \frac{1}{\det w_i(z_j)}\sum_{k,l}(-1)^{k+l}\partial_w\partial_{z_l}\log E(w,z_l)\,\omega_k(w)\det_{\substack{i\neq k\\ j\neq l}}\omega_i(z_j). \tag{7.13}$$

Now we will show that eqs. (7.11)–(7.13) can be integrated to (7.5)–(7.7). We do this by studying the variation of (7.5)–(7.7) and verify that it is given by (7.11)–(7.13). So let us examine the behaviour of the theta function and related objects under variations of the complex structure. We use the notation:

$$\delta_{ww} = \frac{4\pi}{\sqrt{g}}\frac{\delta}{\delta g^{ww}}.$$

Locally, any deformation of the metric is equivalent to an infinitesimal reparametrization, combined with a Weyl rescaling. Since all functions we consider have a Weyl invariant definition, their variations can be determined from their behaviour under reparametrizations. The variation $\delta_{ww}\Phi(z)$ of a holomorphic λ-differential $\Phi(z)$ behaves for w in the neighborhood of z as (cf. eq. (2.7)):

$$\delta_{ww}\Phi(z) \sim \left(\frac{\lambda}{(z-w)^2} + \frac{1}{z-w}\partial_z\right)\Phi(z). \tag{7.14}$$

Now consider the variation $\delta_{ww}\omega_i$ of the holomorphic 1-forms ω_i. From (7.14) we see that $\delta_{ww}\omega_i(z)$ is a meromorphic differential with a double pole for $z = w$: $\delta_{ww}\omega_i(z) \sim \omega_i(w)(z-w)^{-2}$. Further, since the ω's are normalized by their integral around the a-cycles, we have: $\oint_{a_i}\delta_{ww}\omega_j(z)\,\mathrm{d}z = 0$. This uniquely determines the variations of the ω_i:

$$\delta_{ww}\omega_i(z) = \omega_i(w)\,\partial_w\,\partial_z\log E(w,z). \tag{7.15}$$

All other variational formulas can be derived from this one. The variation of the period matrix is given by the b-period of the r.h.s.:

$$\delta_{ww}\tau_{ij} = \oint_{b_i}\delta_{ww}\omega_j(z)\,\mathrm{d}z = 2\pi i\omega_i(w)\,\omega_j(w). \tag{7.16}$$

The theta function $\theta(z,\tau)$ depends only through its arguments on the complex structure. Therefore its variation just follows from the chain rule. Note that derivatives of $\theta(z,\tau)$ w.r.t. τ can be replaced by z-derivatives using the "heat equation":

$$2\pi i \partial_{\tau_{ij}} \theta(z,\tau) = \tfrac{1}{2}\partial_i \partial_j \theta(z,\tau). \tag{7.17}$$

It requires a bit more work to obtain the variations of the prime form and the Riemann class. The results are:

$$\delta_{ww} \log E(z,x) = -\tfrac{1}{2}(\omega_{z-x}(w))^2, \tag{7.18}$$

$$\delta_{ww} \Delta_z = \tfrac{1}{2}\partial_w \omega(w) - \omega(w)\, \partial_w \log \psi(w,z), \tag{7.19}$$

where $\omega_{z-x}(w)$ is defined in (4.8) and

$$\psi(w,z) = \sigma(w) E(w,z)^{g-1}.$$

Now let us consider the variation of eqs. (7.5)–(7.7). First, comparing (7.7) with (7.13) and (7.15), it is clear that we can identify Z_1 as the integral of $-2T^{(\varphi)}_{ww}$:

$$\delta_{ww} \log Z_1 = -2T^{(\varphi)}_{ww}. \tag{7.20}$$

This also accounts for the first term in (7.11). Most of the other terms can be recognized as variations of the theta function $\theta[\alpha](\Sigma z - Q\Delta)$ and the prime forms $E(z_i, z_j)$. The remaining part should come from the holomorphic $\frac{1}{2}g$-differentials $\sigma(z_i)$. In the previous section we have specified $\sigma(z)$ only up to a z-independent factor. (Note that this was sufficient for calculating $\langle T_{ww}\rangle$.) We now fix this factor by giving $\sigma(z)$ the following variational formula:

$$\delta_{ww} \log \sigma(z) = \frac{1}{2g-z}\left\{(\partial_w \log \psi(w,z))^2 - \partial_w^2 \log \psi(w,z)\right\}. \tag{7.21}$$

This equation can be considered as the definition of $\sigma(z)$, and is consistent with our earlier description of $\sigma(z)$. It is easy to verify that with this definition the variation of (7.5) is indeed given by (7.11). Similarly, one can also prove (7.6).

The differential $\sigma(z)$ is the carrier of the gravitational anomaly. Since $\psi(w,z)$ is a $\frac{1}{2}$-form in w, the variation of $\sigma(z)$ transforms with a schwarzian derivative:

$$\delta_{ww} \log \sigma(x) = \left(\frac{dz}{dw}\right)^2 \delta_{zz} \log \sigma(x) + \frac{1}{2g-2}\{z, w\}.$$

Further, if we eliminate Z_1 from (7.5) one finds that the total number of σ's is indeed proportional to the anomaly in det $\nabla^z_{1-\lambda}$, namely $\frac{1}{3}c_\lambda(g-1)$.

7.3. CHIRAL DETERMINANTS FROM HOLOMORPHIC FACTORIZATION

There is a second way to construct the determinants of chiral differential operators, which is due to Quillen [23]. It makes use of the holomorphic structure of the ζ-regularized determinants of the corresponding laplacian. Roughly speaking, the chiral determinant is defined as the holomorphic square root of that of the laplacian. Quillen's work was continued in the context of string theory by Belavin and Knizhnik [7], who investigated the obstruction to holomorphic factorization of the determinant of $\Delta_{1-\lambda} = \nabla_z^{-\lambda}\nabla^z_{1-\lambda}$, the laplacian acting on $(1-\lambda)$-differentials on moduli space. They showed that this obstruction, the so-called analytic anomaly, is proportional to the Weyl anomaly. This result can be derived as follows.

The determinant of $\Delta_{1-\lambda}$ also has a fermionic path integral representation, namely as the partition function of the covariant non-chiral b-c field theory with the zero modes projected out, or equivalently:

$$|\det(\mu_i(z_j))|^2 \frac{\det'\Delta_{1-\lambda}}{\det\langle\mu_i|\mu_j\rangle}$$

$$= \int \mathrm{D}b\,\mathrm{D}\bar{b}\,\mathrm{D}c\,\mathrm{D}\bar{c} \prod_i b(z_i)\bar{b}(z_i) \exp\left(-\frac{1}{2\pi}\int \mathrm{d}^2z\sqrt{g}\,b(z)\,\nabla^z_{1-\lambda}c(z) + \text{c.c.}\right). \tag{7.22}$$

In this covariant theory, correlation functions of b- and c-fields at different points coincide with those of the chiral theory. However, correlation functions of composite operators are different in both theories. In particular the definition of the covariant stress-energy tensor differs from that of the chiral one by the addition of a local counterterm depending on the surface metric:

$$T^{\text{cov}}_{zz}(z) = T^{\text{chiral}}_{zz}(z) - \tfrac{1}{6}c_\lambda T^{(\rho)}_{zz}(z),$$

$$T^{(\rho)}_{zz}(z) = \left\{-\tfrac{1}{2}(\partial_z\log\rho(z))^2 + \partial_z^2\log\rho(z)\right\}. \tag{7.23}$$

The necessity of this extra term is the cause of the breakdown of Weyl invariance. Integrating this anomaly w.r.t. the metric gives the famous Liouville action:

$$\tfrac{1}{6}T^{(\rho)}_{zz}(z) = 4\pi\frac{\delta}{\delta g^{zz}}S_{\mathrm{L}}(\log\rho, g)|_{g_{ab}=\delta_{ab}},$$

$$S_{\mathrm{L}}(\log\rho, g) = \frac{1}{24\pi}\int \mathrm{d}^2z\sqrt{g}\left(g^{ab}\partial_a\log\rho\,\partial_b\log\rho - R(g)\log\rho\right). \tag{7.24}$$

Besides the Weyl anomaly this action $S_{\mathrm{L}}(\log\rho,\delta)$ also carries a gravitational anomaly. Under conformal transformations $z \to w$ it transforms as: $S_{\mathrm{L}}(\log\rho, \delta) \to S_{\mathrm{L}}(\log\rho, \delta) + S_{\mathrm{L}}(\log(|dz/dw|^2), \delta)$. Furthermore, since (as shown in the previous section) the stress-energy tensor of the chiral theory depends analytically on the moduli parameters, the analytic anomaly of the covariant partition function is entirely contained in $S_{\mathrm{L}}(\log\rho, \delta)$. Thus, we find that the determinant of $\Delta_{1-\lambda}$ has the following structure:

$$|\det(\mu_i(z_j))|^2 \frac{\det'\Delta_{1-\lambda}}{\det\langle\mu_i|\mu_j\rangle} = |Z_\lambda(z_1,\dots,z_I)|^2 \mathrm{e}^{-c_\lambda S_{\mathrm{L}}(\log\rho,\delta)}, \tag{7.25}$$

where $Z_\lambda(z_1,\dots,z_I)$ is holomorphic on $\mathbb{M}_g$ and has a gravitational anomaly which cancels that of the Liouville action. This equation can also be used as a definition of the chiral determinant. It is the basis of the second construction.

We now compute the l.h.s. of (7.25) using bosonization, and show that it indeed has the holomorphic structure of the r.h.s.. So we start by equating the non-chiral fermionic partition function with the corresponding bosonic one:

$$\int [\mathrm{D}b\,\mathrm{D}\bar{b}\,\mathrm{D}c\,\mathrm{D}\bar{c}]_\alpha \prod_i b(z_i)\bar{b}(z_i)\mathrm{e}^{-S[b,c]-\mathrm{c.c.}}$$

$$= \sum_{n,m}{}^{(\alpha)} \int [\mathrm{D}\varphi]_{nm} \prod_{i=1}^{I} \rho^\lambda(z_i)\mathrm{e}^{i\varphi(z_i)}\,\mathrm{e}^{-S_Q(\varphi)},$$

where α denotes the spin structure, and the summation symbol $\Sigma^{(\alpha)}$ is defined in sect. 6. Evaluating the path integrals on both sides we obtain the following equation (for $\lambda \neq 1$):

$$|\det(\mu_i(z_j))|^2 \frac{\det'_\alpha \Delta_{1-\lambda}}{\det\langle\mu_i|\mu_j\rangle}$$

$$= \left(\frac{\det'\Delta_0}{A\det(\mathrm{Im}\,\tau)}\right)^{-1/2} \|\theta[\alpha]\Big(\sum_i z_i - Q\Delta\Big)\|^2 \prod_{i<j} F(z_i, z_j) \prod_i H^Q(z_i)\,\mathrm{e}^{-3/2Q^2 U(g)}, \tag{7.26}$$

where $\|\theta\|$, F and H are defined in resp. (6.7), (5.10) and (6.18), and $U(g)$ is given by:

$$U(g) = \frac{1}{192\pi^2} \int\int \mathrm{d}^2x\,\mathrm{d}^2y\,\sqrt{g}R(x)\sqrt{g}R(y)\log(F(x,y)). \tag{7.27}$$

For spin $\lambda = \frac{1}{2}$ this result was first obtained by Alvarez-Gaumé, Moore, and Vafa

[13]. For $\lambda = 1$ one finds a similar equation, in which $\det'\Delta_0(=\det'\Delta_1)$ occurs on both sides. Thus $\det'\Delta_0$ can be solved in terms of θ, F, H and U. This equation for $\lambda = 1$ has also been found in a different context by Faltings [24].

In order to recover the results (7.5)–(7.7) for the chiral determinants, we must extract the an-holomorphic part from these expressions for the laplacians. So for $\|\theta\|$ and F we substitute their definitions in terms of θ and E. Further, we define $\sigma(z)$ as the unique holomorphic (both on Σ and on moduli space $\mathbb{M}_g$) differential. satisfying (c.f. eq. (6.18)):

$$H(z)\,\mathrm{e}^{-3/(2g-2)U(g)} = \exp\left\{\frac{-2\pi}{g-1}\operatorname{Im}\Delta_z(\operatorname{Im}\tau)^{-1}\operatorname{Im}\Delta_z\right\}|\sigma(z)|^2\mathrm{e}^{-[3/(2g-2)]S_L} \quad (7.28)$$

Using the variational formula of the prime form given in the previous subsection and the formula: $\delta R = \nabla^z\nabla^z\delta g_{zz} - \nabla_z\nabla_z\delta g^{zz}$, it is a straightforward calculation to show that the entire an-holomorphic part of the l.h.s. is contained in the first and last factor of the r.h.s., and furthermore that the variation of $\sigma(z)$ is given by (7.21). Inserting (7.28) into the expressions for the determinants of the laplacians, all an-holomorphic factors cancel except for the Liouville action. So the result can be written as eq. (7.25), where $Z_{\alpha,\lambda}$ is given in (7.5). The relation (7.7) between Z_1 and $Z_1(z_1,\dots,z_g,w)$ is obvious in this approach. This concludes the second derivation of the results (7.5)–(7.7).

8. Modular transformations

Moduli space $\mathbb{M}_g$ is obtained from its analytic covering space $\mathbb{T}_g$, called Teichmüller space, by dividing out the action of the mapping class group Γ_g: $\mathbb{M}_g = \mathbb{T}_g/\Gamma_g$. The elements of Γ_g are the non-trivial loops in $\mathbb{M}_g$ and correspond to diffeomorphisms on Σ which are not connected to the identity. These diffeomorphisms can act non-trivially on the homology of Σ. As a consequence, the holomorphic 1-forms ω_i and the period matrix τ transform under Γ_g according to:

$$\begin{aligned}\tilde{\omega}_i &= \omega_j(C\tau + D)^{-1}_{ji},\\ \tilde{\tau} &= (A\tau + B)(C\tau + D)^{-1},\end{aligned} \qquad \begin{pmatrix} D & C \\ B & A\end{pmatrix} \in \mathrm{Sp}(2g,\mathbb{Z}). \quad (8.1)$$

In this section we study the action of this modular group on the chiral determinants $\det\nabla^z_{1-\lambda}$. We have constructed these determinants by integrating the stress-energy tensor. It is therefore possible that around a non-trivial loop in $\mathbb{M}_g$ $\det\nabla^z_{1-\lambda}$ acquires a non-zero phase. If this is the case, the chiral determinant has a global gravitational anomaly [25].

What can be the origin of this phase? One might think that it occurs because the loop surrounds a boundary component of $\overline{\mathbb{M}}_g$ where the chiral determinant has a singularity (see next section). This is only partly true. $\mathbb{M}_g$ contains so-called orbifold points $\mathcal{O}$ that represent surfaces having a group of automorphisms $\Gamma_{\mathcal{O}}$ (= a subgroup of Γ_g). For a general orbifold point the corresponding surface $\Sigma_{\mathcal{O}}$ is non-singular and the chiral determinants det $\nabla^z_{1-\lambda}$ are completely finite at $\mathcal{O}$. Still det $\nabla^z_{1-\lambda}$ can be multiplied by a finite phase when transported around an infinitesimal loop surrounding $\mathcal{O}$.

To understand how this happens, let us consider as an example the modular behaviour of det ∇^z_0. After we have chosen a canonical basis of holomorphic 1-forms ω_i, det ∇^z_0 is given by:

$$\det \nabla^z_0 = Z_1 \omega_1 \wedge \ldots \wedge \omega_g \otimes e\,, \tag{8.2}$$

where Z_1 is defined in the previous section and e denotes the constant function 1 on Σ. From (8.1) we have that under Γ_g:

$$\tilde{\omega}_1 \wedge \ldots \wedge \tilde{\omega}_g = \det(C\tau + D)^{-1} \omega_1 \wedge \ldots \wedge \omega_g\,. \tag{8.3}$$

The factor Z_1, being the integral of the stress-energy tensor $T^{(\varphi)}_{zz}$, is a single-valued function on $\mathbb{T}_g$. Furthermore, Z_1 is regular on the interior of $\mathbb{T}_g$, because $T^{(\varphi)}_{zz}$ is regular on smooth surfaces. Z_1 is not invariant under the modular group Γ_g. From the relation between Z_1 and the (modular invariant) determinant of the scalar laplacian one deduces that the transformation law of Z_1 under Γ_g is:

$$\tilde{Z}_1 = \zeta \det(C\tau + D) Z_1\,, \tag{8.4}$$

where ζ is a yet undetermined phase factor. Combining (8.2)–(8.4) we see that ζ is the global anomaly in det ∇^z_0. Now consider Z_1 near one the originals $\hat{\mathcal{O}}$ in $\mathbb{T}_g$ of an orbifold point $\mathcal{O}$. So $\hat{\mathcal{O}}$ is left fixed under some finite subgroup of Γ_g, $\Gamma_{\mathcal{O}}$, the automorphism group of $\Sigma_{\mathcal{O}}$. This subgroup $\Gamma_{\mathcal{O}}$ also leaves the value of Z_1 at $\hat{\mathcal{O}}$ invariant. Thus the phase factor ζ_0 for an element γ_0 of $\Gamma_{\mathcal{O}}$ is determined: at $\mathcal{O}$ it compensates for the factor $\det(C_0\tau + D_0)$ in (8.4). (Note that $\det(C_0\tau + D_0)$ indeed has absolute value 1 at $\hat{\mathcal{O}}$.) Since this factor $\det(C_0\tau + D_0)$ at $\hat{\mathcal{O}}$ will in general be different from 1, we conclude that det ∇^z_0 indeed has a global anomaly, and that the presence of the global anomaly can be seen as a consequence of the topological structure of the line bundle $\mathbb{L}_1$ near the orbifold points.

This is the general situation for integer spin λ. For $\frac{1}{2}$-integer spin things are different, because the spin structures α transform under the modular group. The transformation rule is:

$$\alpha \to \alpha' + \delta\,,$$

$$\begin{pmatrix} \alpha'_1 \\ \alpha'_2 \end{pmatrix} = \begin{pmatrix} D & -C \\ -B & A \end{pmatrix} \begin{pmatrix} \alpha_1 \\ \alpha_2 \end{pmatrix}, \qquad \begin{pmatrix} \delta_1 \\ \delta_2 \end{pmatrix} = \frac{1}{2} \begin{pmatrix} \mathrm{diag}(CD^{t}) \\ \mathrm{diag}(AB^{t}) \end{pmatrix}. \tag{8.5}$$

Therefore, the chiral determinants $\det_\alpha \nabla^z_{1-\lambda}$ for $\lambda \in \mathbb{Z} + \frac{1}{2}$ have to be defined on the spin covering $\mathbb{S}_g$ of $\mathbb{M}_g$. Even and odd spin structures, as can be easily checked from (8.5), are permuted separately under Γ_g. So $\mathbb{S}_g$ is a 2^{2g} sheeted covering of $\mathbb{M}_g$, consisting of two components corresponding to the odd and even spin structures. There is a global anomaly in $\det_\alpha \nabla^z_{1-\lambda}$ if it is changed by a phase around a non-trivial loop on $\mathbb{S}_g$. Such loops correspond to modular transformations leaving the spin structure α fixed.

Now let us use the expressions (7.5)–(7.7) to investigate these phases for all $\det_\alpha \nabla^z_{1-\lambda}$. The action of the modular group on the theta functions is given by:

$$\theta[\alpha' + \delta](\tilde{z}, \tilde{\tau}) = \varepsilon\, e^{i\varphi(\alpha)} \det(C\tau + D)^{1/2} e^{i\pi z (C\tau + D)^{-1} C z} \theta[\alpha](z, \tau),$$

$$\tilde{z} = (\tau C^{t} + D^{t})^{-1} z, \qquad \varphi(\alpha) = \alpha_1' \cdot \alpha_2' - \alpha_1 \cdot \alpha_2 + 2\alpha_1' \cdot \delta_2 . \tag{8.6}$$

ε is a phase with a rather complicated dependence on A, B, C and D, satisfying:

$$\varepsilon^8 = 1 .$$

The z-dependent factor can be determined by checking the behaviour of both the l.h.s. and r.h.s. of (8.6) under the shift $z \to z + n$. Also the δ can be found in this way. Since the Riemann class Δ is defined in terms of the divisor of $\theta(z, \tau)$, it transforms as:

$$\tilde{\Delta} - \tilde{\tau}\delta_1 - \delta_2 = (\tau C^{t} + D^{t})^{-1} \Delta . \tag{8.7}$$

The modular behaviour of $E(z, w)$ and $\sigma(z)$ can be deduced from (8.6) and their definitions (3.11) resp. (7.28):

$$E(z, w) \to \exp\left\{ i\pi \int_z^w \omega (C\tau + D)^{-1} C \int_z^w \omega \right\} E(z, w), \tag{8.8}$$

$$\sigma(z) \to \exp\left\{ \frac{i\pi}{g-1} \Delta_z (C\tau + D)^{-1} C \Delta_z \right\} \sigma(z), \tag{8.9}$$

where we have restricted ourselves to those transformations for which $\delta = 0 \pmod{\mathbb{Z}}$. Eq. (7.28) fixes the transformation rule for $\sigma(z)$ only up to a constant phase. In principle, this phase can be determined by considering the behaviour of $\sigma(z)$ near an orbifold point, similarly as discussed above for Z_1. We have not explicitly done this analysis, so strictly (8.9) holds up to an unknown phase.

Combining all this we find for chiral determinants the following modular transformation rules (modulo a phase proportional to the local anomaly):

$$\det \nabla^z_{1-\lambda}(\tilde{\tau}) = \varepsilon^{2/3} \det \nabla^z_{1-\lambda}(\tau) \qquad (\lambda \in \mathbb{Z}), \tag{8.10}$$

$$\det_{\tilde{\alpha}} \nabla^z_{1-\lambda}(\tilde{\tau}) = \varepsilon^{2/3} e^{i\pi\varphi(\alpha, \lambda)} \det_\alpha \nabla^z_{1-\lambda}(\tau) \qquad (\lambda \in \mathbb{Z} + \tfrac{1}{2}),$$

$$\varphi(\alpha, \lambda) = \varphi(\alpha) + 2(2\lambda - 1)(\alpha_1' \cdot \delta_2 + \alpha_2' \cdot \delta_1) . \tag{8.11}$$

From these results one can construct combinations of chiral determinants which are free of local and global anomalies. Of particular interest is the combination $\det_{\alpha}\nabla^{z}_{-1/2}(\det_{\alpha}\nabla^{z}_{1/2})^{-5}(\det_{\beta}\nabla^{z}_{1/2})^{8}(\det_{\gamma}\nabla^{z}_{1/2})^{8}$, since it describes the chiral fields of the heterotic string. The fact that it has no global anomalies was first proved by Witten [25].

Finally we note that the expression (7.27) for the determinants of the laplacians $\det\Delta_{\lambda}$ and $\det_{\alpha}\Delta_{\lambda+1/2}(\lambda\in\mathbb{Z})$ are manifestly single valued on $\mathbb{M}_g$ resp. $\mathbb{S}_g$. The function $F(z,w)$, and hence also $H(z)$ and $U(g)$, are modular invariant, while the norm of the theta function transforms as:

$$(\det\operatorname{Im}\tilde{\tau})^{1/2}\|\theta[\tilde{\alpha}](\Sigma\tilde{z}-Q\tilde{\Delta},\tilde{\tau})\|^{2}=(\det\operatorname{Im}\tau)^{1/2}\|\theta[\alpha](\Sigma z-Q\Delta,\tau)\|^{2},\quad(8.12)$$

where $\tilde{\alpha}=\alpha',\ \alpha'+\delta$ for $\lambda\in\mathbb{Z}$ resp. $\mathbb{Z}+\frac{1}{2}$.

9. Chiral determinants on degenerate surfaces

On smooth Riemann surfaces the chiral determinants are well-behaved and finite. This changes when the surface degenerates into a singular surface with nodes. These divergences are of importance for string theories, because they can give rise to infinities in scattering amplitudes [7–9]. In this section we discuss how the expressions (7.5)–(7.7) can be used to analyse the behaviour of the chiral determinants on degenerate surfaces.

There are two basic types of degeneracies corresponding to either pinching a cycle dividing the surface in two components, or to pinching a handle. As an example we discuss the pinching of a dividing cycle, so we let the surface approach one of the boundary components $\mathbb{D}_{g_1}$ of $\mathbb{M}_g$ with $0<g_1<[\frac{1}{2}g]$. One can construct a family of degenerating surfaces near $\mathbb{D}_{g_1}$ over the unit disk $D=\{t\in\mathbb{C}\mid |t|<1\}$ as follows [15]. Take two surfaces Σ_1 and Σ_2 of genus g_1 and $g_2=g-g_1$. Choc.e on each surface a point p_i, $i=1,2$, and a coordinate neighbourhood $U_i=\{z_i|\,|z_i|<1\}$ near each p_i, such that $p_i=\{z_i=0\}$. Remove a small disk $|z_i|<|t|^{1/2}$ from both surfaces, and glue the remaining surfaces together by attaching the annulus $A_t=\{w|\,|t|^{1/2}<|w|<|t|^{-1/2}\}$ according to:

$$w=\begin{cases}\dfrac{t^{1/2}}{z_1} & \text{if } |t|^{1/2}<|w|<1\\[2ex] t^{-1/2}z_2 & \text{if } 1<|w|<|t|^{-1/2}\end{cases}.$$

We denote the resulting surface by Σ_t. The two components of Σ_t-A_t we call X_t resp. Y_t. The parameter t, the points p_1 and p_2, together with the moduli of Σ_1 and Σ_2 provide a parametrisation of $\mathbb{M}_g$ near $\mathbb{D}_{g_1}$. Furthermore, t is the correct analytical coordinate on $\mathbb{M}_{g_1}$) transversal to $\mathbb{D}_g$, for $g_1>1$. (Near $\mathbb{D}_1$ the analytic transversal coordinate is t^2. This is because the punctured surfaces at $\mathbb{D}_1$ have an automorphism of order two, which means that t and $-t$ correspond to the same point on $\mathbb{M}_g$.)

Variations of the parameter t correspond to infinitesimal variations of the metric on the annulus A_t. It is a straightforward calculation to show that for small t:

$$\frac{4\pi}{\sqrt{g}}\frac{\delta}{\delta g^{ww}}\Phi = \frac{t}{w^2}\frac{\partial}{\partial t}\Phi, \qquad w \in A_t, \tag{9.1}$$

where Φ denotes any function or differential, which does not explicitly depend on a point on the annulus A_t. This relation can be used to deduce from the variational formula of Φ the leading power of t in Φ as $t \to 0$.

We now give the limiting behaviour of the theta function and associated quantities for small t. For proofs see ref. [15].

The basis of holomorphic differentials ω_i on Σ_t approaches for $t \to 0$ just the sum of the bases on Σ_1 and Σ_2. Therefore, the period matrix becomes:

$$\tau(t) \to \begin{pmatrix} \tau_1 & 0 \\ 0 & \tau_2 \end{pmatrix}, \tag{9.2}$$

where τ_i is the period matrix for Σ_i. Next let D_t be a degree zero divisor on Σ_t and suppose it splits as $D_t = D_x + D_y + D_a$, with D_x, D_y, D_a divisors on X_t, Y_t, resp. A_t, of degree d_x, d_y, and d_a. Then the theta function with argument D_t factorizes as:

$$\theta(D_t, \tau(t)) \to \theta_1(D_x - d_x p_1, \tau_1)\theta_2(D_y - d_y p_2, \tau_2). \tag{9.3}$$

From this it follows that the Riemann class goes to: $\Delta \to \Delta_1 + \Delta_2 + p_{1,2}$ For the behaviour of the prime form $E(z, z')$ we have to distinguish several cases. If both points z and z' lie on $X_t(Y_t)$ then $E(z, z')$ goes to the prime form of $\Sigma_1(\Sigma_2)$. On A_t the prime form is just $E(w, w') = (w - w')$. The other cases are:

$$\begin{aligned} E(x, w) &\to -E_1(x, p_1)\, w\, t^{-1/4}, \\ E(y, w) &\to E_2(y, p_2)t^{-1/4}, \\ E(x, y) &\to E_1(x, p_1)E_2(p_2, y)t^{-1/2}, \end{aligned} \tag{9.4}$$

where $x \in X_t$, $y \in Y_t$ and $w \in A_t$. Note that these formulas are compatible at the boundary of A_t and with the variational formula for $E(x, y)$.

The factorization formula for $\sigma(x)$ can be most easily found with the help of the variational formulas given in the previous section and eqs. (6.20), (9.1) and (9.4). One finds for $x \in \Sigma_1$:

$$\sigma(x) \to \sigma_1(x)\frac{1}{E_1(x, p_1)^{g_2}}\sigma_1(p_1)^{a_1 - 1}\sigma_2(p_2)^{a_2}t^{b_2},$$

$$a_i = \frac{g_i - 1}{g - 1}; \qquad b_i = \tfrac{1}{2}g_i a_i. \tag{9.5}$$

The exponents $a_1 - 1$, a_2 and b_2 are determined from the requirement that the small t limit commutes with a variation of the metric on X_t,Y_t resp. A_t. In a similar way one finds for a point w on the annulus:

$$\sigma(w) \to w^{-g_1}\sigma_1(p_1)^{a_1}\sigma_2(p_2)^{a_2}t^{(b_1+b_2)/2}. \tag{9.6}$$

These equations for $\sigma(z)$ can also be derived from eq. (7.28), choosing some limiting behaviour for the surface metric on Σ_t. A useful metric is the $R = -1$ metric. Its behaviour for small t is described in ref. [27].

These formulas can now be used to analyse the partition functions $Z_{\alpha,\lambda}(z_1,\dots,z_I)$ near $\mathbb{D}_g$. The result depends on the way we divide the points z_i over the three sub-surfaces. Let us consider the situation that the points z_i split into two groups $x_1,\dots,x_{I_1+q_1}$ on Σ_1 and $y_1,\dots,y_{I_2+q_2}$ on Σ_2, where $I_i = Q(g_i - 1)$ and $q_1 + q_2 = Q$. So both Σ_1 and Σ_2 have a nett charge, given by q_1 resp. q_2. Applying the above results one finds that the partition function $Z_{\alpha,\lambda}(z_1,\dots,z_I)$ factorizes as:

$$Z_{\alpha,\lambda}(z_1,\dots,z_I)(t) \underset{t\to 0}{\longrightarrow} t^{-q_1q_2/2}Z_{\alpha_1\lambda}(x_k,p_1;q_1)Z_{\alpha_2\lambda}(y_l,p_2;q_2), \tag{9.7}$$

with

$$Z_{\alpha_1\lambda}(x_k,p_1;q_1) = (Z_1)_1^{-1/2}\theta_1[\alpha_1](\Sigma x_k - q_1p_1 - Q\Delta_1)$$

$$\times \frac{\Pi E_1(x_i,x_j)}{\Pi E_1(x_i,p_1)^{q_1}}\frac{\Pi\sigma_1(x_i)^Q}{\sigma_1(p_1)^{q_1Q}}$$

and an identical expression for $Z_{\alpha 2\lambda}(y_l,p_2;q_2)$. An intermediate result we have used here, is that the chiral scalar partition function $Z_1^{-1/2}$ behaves simply as:

$$Z_1^{-1/2}(t) \underset{t\to 0}{\longrightarrow} (Z_1)_1^{-1/2}(Z_1)_2^{-1/2} \tag{9.8}$$

Eq. (9.7) has a clear physical interpretation. It is the first term of the expansion of the partition function $Z_\lambda(z_1,\dots,z_I)$ which is obtained by inserting at both boundaries of A_t a complete set of states $|\varphi\rangle\langle\varphi|$ and by representing the evolution from one boundary to the other by $U(t) = \Sigma|\varphi_h\rangle t^h\langle\varphi_h|$ where h is the conformal weight or energy of φ_h, [18]. The leading term in this expansion corresponds to the state with the lowest weight having a non-zero matrix element with the density matrix representing the theory on the rest of the surface. In our case this last requirement means that the states must carry a charge $-q_1$ resp. $-q_2$. The conformal field with the lowest weight, carrying this charge, is the vertex operator: $e^{-iq_1\varphi(p_1)} = (e^{-iq_2\varphi(p_2)})^\dagger$. Its weight is $-\frac{1}{2}q_1q_2$. Returning to eq. (9.7), we indeed recognize in both factors the amplitudes of the above vertex operators with the

b-fields on their side of the node. Also the power of t matches with the conformal weight. It is of course clear that by expanding θ, E and σ to higher order in t one should recover also the contribution of the higher energy states.

The same analysis can be done for the pinching of a non-zero homology cycle. In that case one also finds that the leading term has a similar interpretation as eq. (9.7). A new feature is that, since one now can have non-trivial boundary conditions around the pinched cycle, one also finds the correlation functions of spin fields, i.e. vertex operators with half-integer charge mapping one spin structure to another. This is the reason why in string theories unitarity requires the sum over all spin structures [26].

Now let us proceed with the chiral determinants themselves. We first discuss the case $\lambda > 1$, $Q > 1$. On Σ_t there are $Q(g-1)$ b-zero modes. In the limit $t \to 0$ Q of these will disappear, because the total number of zero modes on Σ_1 and Σ_2 is $Q(g_1 - 1 + g_2 - 1) = Q(g-2)$. Disappear here means: become zero or develop a pole at the node. We can choose a basis $\{\mu_i\}$ for the zero modes, such that as $t \to 0$ it goes to the union of the bases of the zero modes on Σ_1 resp Σ_2, together with Q λ-differentials $\hat{\mu}_m$ which in local coordinates w on A_t are given by: $\hat{\mu}_m(w) = 1/w^m$ $(m = 1, \ldots, Q)$. It is now convenient to divide the points z_i where the b zero modes are absorbed into three groups $x_1, \ldots, x_{l1}$ on X_t, $y_1, \ldots, y_{l2}$ on Y_t, and $w_1, \ldots, w_Q$ on A_t. With this distribution of the external charges the three subsurfaces $\mathrm{X}_t, \mathrm{Y}_t$, and A_t each have total charge equal to zero. One therefore expects that for $t \to 0$ the three parts decouple. Indeed, from (9.2)–(9.6) one finds that the partition function of the b-c fields living on Σ_t factorizes into the product of the three partition functions of the fields on $\mathrm{X}_t, \mathrm{Y}_t$, resp. A_t:

$$Z_{\alpha,\lambda}(x_k, y_l, w_m) \to \big(Z_{\alpha_1\lambda}(x_k)\big)_1 \big(Z_{\alpha_2\lambda}(y_l)\big)_2 \prod_{m<n} (w_m - w_n) \prod_m w_m^{-Q}. \tag{9.9}$$

So we find the following behaviour for the chiral determinants for $\lambda > 1$:

$$\det{}_\alpha \nabla^z_{1-\lambda}(t) \xrightarrow[t \to 0]{} \big(\det{}_{\alpha_1} \nabla^z_{1-\lambda}\big)_1 \otimes \big(\det{}_{\alpha_2} \nabla^z_{1-\lambda}\big)_2 \otimes (\hat{\mu}_1 \wedge \hat{\mu}_2 \wedge \ldots \wedge \hat{\mu}_Q),$$

$$\hat{\mu}_m = w^{-m} dw^\lambda; \qquad m = 1, \ldots, Q, \qquad w \in \mathrm{A}_t \tag{9.10}$$

For $\lambda = 1$ the structure is again slightly different. There are no b-zero modes destroyed for $t \to 0$; instead an extra c-zero mode is created. Let e_t, e_1 and e_2 denote the constant function 1 on Σ_t, Σ_1 resp. Σ_2, then the behaviour of $\det \nabla^z_0$ is given by:

$$\det \nabla^z_0(t) \otimes e_t^{-1} \xrightarrow[t \to 0]{} (\det \nabla^z_0)_1 \otimes (\det \nabla^z_0)_2 \otimes (e_1 \otimes e_2)^{-1}. \tag{9.11}$$

Finally, the factorization of the Dirac determinants $\det{}_\alpha \nabla^z_{1/2}$ follows directly from

(9.8) and the leading behaviour of the theta function. We leave the analysis of this case to the reader.

10. Application to string amplitudes

In this final section, we discuss the application of our results to string multi-loop calculations. We first consider the vacuum amplitude of the closed bosonic string. In Polyakov's path-integral approach this vacuum amplitude is given by [1, 2]:

$$Z_{\text{str}} = \frac{1}{N} \int [\mathrm{D}g][\mathrm{D}X]\, \mathrm{e}^{-S[g,X]}$$

$$S[g,X] = \frac{1}{2\pi} \int \mathrm{d}^2\xi \sqrt{g}\, g^{ab} \partial_a X^\mu \partial_b X_\mu \qquad (\mu = 1, \ldots, d) \tag{10.1}$$

The theory is invariant under the group of reparametrisations and, in the critical dimension, under Weyl rescalings. The normalization constant N is needed to compensate for the volume of this symmetry group. Using the standard Faddeev-Popov procedure to factor out this volume the integral over all metrics g_{ab} reduces to a finite dimensional integral over moduli space. After integrating out the scalar fields X one obtains [5]:

$$Z_{\text{str}} = \int \prod_i \mathrm{d}m^{(i)} \mathrm{d}\overline{m}^{(i)} \frac{\det' \Delta_2}{\det\langle \mu_i | \mu_j \rangle} \left(\frac{\det' \Delta_0}{A} \right)^{-13}. \tag{10.2}$$

The $m^{(i)}$, $i = 1, \ldots, 3(g-1)$, are analytic coordinates on the moduli space corresponding to a variation of the metric: $\delta(\mathrm{d}s^2) = \rho \delta m^{(i)} \eta_i \mathrm{d}z^2$ where η_i is a set of Beltrami differentials dual to the zero modes μ_i. Using the results of sect. 7 we can rewrite Z_{str} as:

$$Z_{\text{str}} = \int W \wedge \overline{W} \det(\operatorname{Im} \tau(m))^{-13},$$

with

$$W = \mathrm{d}m^{(1)} \wedge \ldots \wedge \mathrm{d}m^{(3g-3)} \frac{Z_2(x_1, \ldots, x_{3g-3})}{\det(\mu_k(x_1))} Z_1^{-13} \tag{10.3}$$

We can insert eqs. (7.5)–(7.7) and express W in terms of theta functions and related objects. The resulting expression has an equal number of σ's in the numerator and in the denominator and therefore has no local anomaly. It simplifies somewhat if we choose the points on Σ where all the zero modes are absorbed to lie in the Riemann class Δ; it then resembles the result of Manin [28]. A considerable simplification is made if one absorbs all zero modes in one single point z. The only objects we then need to express W are theta functions and the wronskians of the bases of zero modes:

$$W = \mathrm{d}m^{(1)} \wedge \ldots \wedge \mathrm{d}m^{(3g-3)} \frac{\theta(3\Delta_z, \tau(m))}{\det\left(\partial_z^{i-1}\mu_j\right)} \left(\frac{\det\left(\partial_z^{k-1}\omega_l\right)}{\partial_z^g \theta(\Delta_z, \tau(m))} \right)^9, \tag{10.4}$$

with $\Delta_z = \Delta - (g-1)z$.

It is instructive to analyse the properties of this expression for W. The wronskian of holomorphic 1-forms has $g(g^2-1)$ zeroes on Σ, known as the Weierstrass points [29]. It is easy to show using the Riemann vanishing theorem that numerator $\partial_z^g\theta(\Delta_z, \tau)$ has exactly the same zeroes. Similarly, the zeroes of the wronskian of the quadratic differentials μ are cancelled by those of $\theta(3\Delta_z, \tau)$. Further, using that $\partial_z^j\theta(\Delta_z) = 0$ if $j < g$, one can verify that (10.4) is independent of the point z. Finally, as follows from (8.1) and (8.6), W is a modular form of weight -13:

$$W(\tilde{\tau}) = \left[\det(C\tau + D)\right]^{-13} W(\tau). \tag{10.5}$$

Since $\det(\operatorname{Im}\tilde{\tau}) = |\det(C\tau + D)|^{-2} \det(\operatorname{Im}\tau)$ this indeed makes the total integrand in (10.3) modular invariant.

Eq. (10.3) for the string partition function is still somewhat formal, because little is known about the region of integration. For the cases $g = 1$, 2 and 3 more is known because then the dimension of $\mathbb{M}_g$ coincides with that of the Siegel upper half-plane $\mathbb{H}_g$. This means that one can use the entries of the period matrix as a coordinate system on $\mathbb{M}_g$ [30, 31].

The vacuum amplitude (10.3) is infinite because W has a singularity near the boundary of the moduli space $\mathbb{M}_g$. To find this singularity, consider again the pinching described in previous section. Near the boundary we can split the integral over the moduli into: $\mathrm{d}m_1^{(i)} \wedge \mathrm{d}m_2^{(j)} \wedge \mathrm{d}p_1 \wedge \mathrm{d}p_2 \wedge \mathrm{d}t$ where m_k are moduli for Σ_k $(k = 1, 2)$. The quadratic differentials corresponding to $\mathrm{d}p_1$, $\mathrm{d}p_2$ and $\mathrm{d}t$ are $-t^{1/2}/w\,\mathrm{d}w^2$, $t^{1/2}/w^3\,\mathrm{d}w^2$ resp. $t/w^2\,\mathrm{d}w^2$. It then follows from the results of the preceding section that W near the boundary behaves as:

$$W \underset{t\to 0}{\sim} t^{-2} W_1 \wedge W_2 \wedge \mathrm{d}p_1 \wedge \mathrm{d}p_2 \wedge \mathrm{d}t. \tag{10.6}$$

This singularity has been found by several authors [7, 8, 18] and is related the presence of the tachyon. This relation can be made more explicit by studying the factorization of the g-loop contribution to the scattering amplitude of n tachyons with momenta $k_1, \ldots, k_n$:

$$A_g(k_1, \ldots, k_n) = \int W \wedge \overline{W} \det(\operatorname{Im}\tau)^{-13} G(k_1, \ldots, k_n),$$

where

$$\begin{aligned} G(k_1, \ldots, k_n) &= \left\langle \prod_1^n \int \mathrm{d}^2 z_j \sqrt{g}(z_j) \exp\left(ik_j \cdot X(z_j)\right) \right\rangle \\ &= \left(\prod_1^n \int \mathrm{d}^2 z_j \right) \prod_{i<j} \left(F(z_i, z_j) \right)^{k_i \cdot k_j} \delta(\Sigma k_i). \end{aligned} \tag{10.7}$$

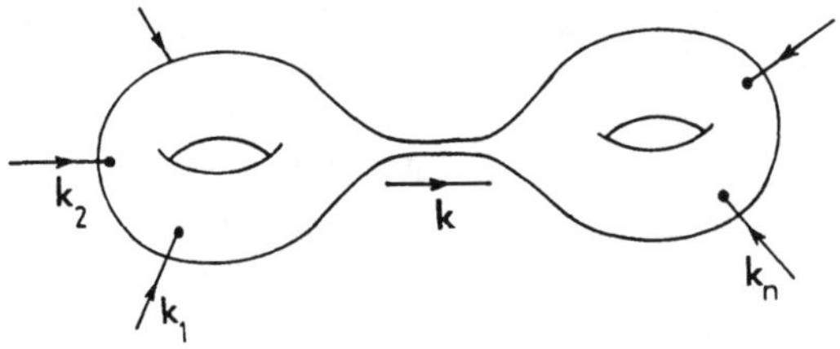

Fig. 3. Factorization of a two-loop tachyon amplitude.

We have chosen units such that for the tachyon $k^2 = 2$. Consider the situation that the surface Σ breaks up into Σ_1 and Σ_2 with m resp. $n - m$ tachyon lines attached to it, as is shown in fig. 3 for the case $g = 2$. Using eq. (9.3) and momentum conservation one finds the following behaviour for G as $t \to 0$:

$$\int d^2p_1\, d^2p_2 G(k_1,\dots,k_n) \underset{t\to 0}{\sim} G_1(k_1,\dots,k_m,-k)G_2(k,k_{m+1},\dots,k_n)|t|^{k^2}, \tag{10.8}$$

where k is the momentum going through the tube. Combined with (10.6) this gives for the tachyon amplitude:

$$A_g(k_1,\dots,k_n) \sim A_{g_1}(k_1,\dots,k_m,-k)\frac{1}{k^2-2}A_{g_2}(k,k_{m+1},\dots,k_n). \tag{10.9}$$

So indeed we see a tachyon pole arising from the singularity in the partition function W.

The partition functions for the fermionic and heterotic strings are much more complicated, because of the presence of supermoduli and the sum over spin structures [6]. It is expected that the vacuum amplitudes for both theories vanish, but until now only intuitive arguments have been given. In these arguments both 2d world-sheet supersymmetry as 10d space-time supersymmetry play a crucial role [32]. For studying the space-time supersymmetry it is convenient to bosonize the theory [10]. However 2d supersymmetry then becomes totally obscure. We believe that for the study of finiteness of superstring amplitudes it is important to have a better understanding of the role of bosonization (or some generalization there of) in supersymmetric conformal field theories. Work in this direction is in progress.

We thank B. de Wit for reading the manuscript, and B. van Geemen for helpful discussions on theta functions. This work is financially supported by the Stichting voor Fundamenteel Onderzoek der Materie (F.O.M.).

Note added in proof:

Just after this paper had been submitted for publication, we received ref. [33], in which part of our results are also reported.

References

[1] A.M. Polyakov, Phys. Lett 103B (1981) 207, 211
[2] D. Friedan, *in* Les Houches 1982, Recent advances in Field Theory and Statistical Physics, eds. J-B Zuber and R. Stora (North-Holland, 1984)
[3] Superstrings, the first fifteen years, eds. J. Schwarz (World Scientific, 1985)
[4] D. Gross, J. Harvey, E. Martinec and R. Rhom, Nucl. Phys. B256 (1985) 253: B267 (1986) 75
[5] O. Alvarez, Nucl. Phys. B216 (1983) 125;
E. D'Hoker, and D.H. Phong, Nucl. Phys. B269 (1986) 205;
G. Moore, and P. Nelson, Nucl. Phys B266 (1986) 58
[6] E. Martinec, Phys. Rev. D28 (1983) 2604;
J. Polchinsky, G. Moore, and P. Nelson, Phys. Lett. 169B (1986) 47;
S. Chaudhuri, H. Kawai and S-H Tye, preprint CLNS 86/723
[7] A.A. Belavin and V.G. Knizhnik, Phys. Lett. 168B (1986) 201
[8] E. Gava, R. Iengo, T. Jayaraman and R. Ramachandran, Phys. Lett. 168B (1986) 207
[9] R. Catenacci, M. Cornalba, M. Martellini and C. Reina, Phys. Lett. 172B (1986) 328
[10] D. Friedan, E. Martinec and S.H. Shenker, Nucl. Phys. B271 (1986) 93
[11] A.A. Belavin, A.M. Polyakov and A.B. Zamolodchikov, Nucl. Phys. B241 (1984) 333
[12] V.S. Dotsenko, and V. Fateev, Nucl. Phys. B240 (1984) 312
[13] L. Alvarez-Gaumé, G. Moore, and C. Vafa, Comm. Math. Phys. 106 (1986) 40
[14] L. Alvarez-Gaumé, J-B Bost, G. Moore, P. Nelson and C. Vafa, preprint HUTP 86/A039
[15] J.D. Fay, Theta functions on Riemann surfaces, Springer Notes in Mathematics 352 (Springer, 197[illegible]
[16] D. Mumford, Tata Lectures on Theta, vols I and II, Birkhauser (1983)
[17] J. Igusa, Theta functions (Springer, 1972)
[18] D. Friedan and S.H. Shenker, Nucl. Phys. B281 (1987) 509
[19] E. Martinec, Nucl. Phys. B281 (1987) 157
[20] M.A. Namazie, K.S. Narain and M.H. Sarmadi, Phys. Lett. 177B (1986) 329;
H. Sonoda, Phys. Lett. 178B (1986) 390
[21] L. Bers, Bull. Amer. Math. Soc. [NS] 5 (1981) 131
[22] P. Deligne and D. Mumford, Publ. IHES 36 (1969) 1
[23] D. Quillen, Funk. Anal. i Pril 19 (1985) 37;
J-M. Bismut and D.S. Freed, Comm. Math. Phys. 106 (1986) 159
[24] G. Faltings, Ann. Math. 119 (1984) 387
[25] E. Witten, *in* Geometry, anomalies and topology, ed. W.A. Bardeen, A.R. White (World Scientific, 1985)
[26] N. Seiberg and E. Witten, Nucl. Phys. B276 (1986) 272
[27] H. Masur, Duke Math. J. 43 (1976) 623
[28] Yu. Manin, Phys. Lett. 172B (1986) 184
[29] R.C. Gunning, Lectures on Riemann surfaces (Princeton University Press, 1966)
[30] A.A. Belavin, V.G. Knizhnik, A. Morozov and A. Perelemov, preprint ITEP 86-59;
A. Morozov, preprint ITEP 86-88;
[31] G. Moore, Phys. Lett. 176B (1986) 369
[32] E. Martinec, Phys. Lett. 171B (1986) 189
[33] V.G. Knizhnik, Phys. Lett. 180B (1986) 247

CHIRAL BOSONIZATION ON A RIEMANN SURFACE

Tohru EGUCHI [1]
Laboratoire de Physique Théorique de l'Ecole Normale Supérieure [2], *24 rue Lhomond, F-75231 Paris Cedex 05, France*

and

Hirosi OOGURI
Department of Physics, University of Tokyo, Tokyo, Japan

Received 23 December 1986

We point out that the basic addition theorem of θ-functions, Fay's identity, implies an equivalence between bosons and chiral fermions on Riemann surfaces with arbitrary genus. We present a rule for a bosonized calculation of correlation functions. We also discuss ghost systems of n and $(1-n)$ tensors and derive formulas for their chiral determinants.

1. Introduction. Fermion–boson equivalence is one of the most remarkable aspects of two-dimensional quantum field theories. Various physical quantities may be analysed either from a bosonic or a fermionic point of view and their equivalence provides a deep insight into the algebraic structure of the theory. In particular, bosonization or fermionization techniques have been essential in facilitating the construction of conformal and current algebras on a Riemann surface with genus $g=1$ [1]. Recently, there have been attempts at generalizing the fermion–boson equivalence to theories defined on a general Riemann surface [2]; however, the treatment has so far been limited to the case of non-chiral theories where right- and left-moving degrees of freedom are not treated separately.

In this article we would like to generalize this construction and discuss a chiral bosonization of spinor and ghost fields on a general Riemann surface. We first present a simple rule for bosonization of chiral fermions and point out that the consistency of our prescription is assured by the basic addition theorem of θ-functions (Fay's triscecant identity and its generalization [3,4]) on a general Riemann surface. We also show that we recover current algebra Ward identities [5] when a suitable limit of arguments is taken in the addition formula of θ-functions.

We then discuss the bosonic description of a ghost system of b and c fields which are n and $(1-n)$ differentials, respectively. In the case of ghosts, the U(1) current is no longer conserved and the scalar field is coupled to the scalar curvature of the Riemann surface. This would make it difficult to separate the system into left- and right-moving degrees of freedom. We shall show, however, by using a suitable (singular) metric on the surface, that we can still split the system into left and right components and describe each of them by a holomorphic (or anti-holomorphic) scalar field.

2. Bosonization of chiral fermions. Let us start with the formula for the Green function for a complex spinor field on a general Riemann surface [3,6]:

[1] On leave of absence from the University of Tokyo; Tokyo, Japan.
[2] Laboratoire Prope du Centre National de la Recherche Scientifique associé à l'Ecole Normale Supérieure et à l'Université de Paris-Sud.

0370-2693/87/$ 03.50 © Elsevier Science Publishers B.V.
(North-Holland Physics Publishing Division)

$$\langle\psi(z)\bar{\psi}(w)\rangle=\frac{\theta[^{\alpha}_{\beta}](\int_w^z \boldsymbol{\omega}(v)\,dv)}{\theta[^{\alpha}_{\beta}](0)E(z,w)}\,. \tag{1}$$

Here $E(z, w)$ is the prime form [3,4], a holomorphic $-1/2$ order differential in both z and w, with a unique zero at $z=w$ [$E(z,w)\approx z-w$ as $z\approx w$]. $\theta[^{\alpha}_{\beta}](u_1, ..., u_g)$ is the Riemann θ-function with the characteristic $\boldsymbol{\alpha}$, $\boldsymbol{\beta}\in\mathbb{Z}^g/(2\mathbb{Z})^g$,

$$\theta[^{\alpha}_{\beta}](\boldsymbol{u})=\sum_{\boldsymbol{n}\approx\boldsymbol{Z}^g}\exp\{\pi i(\boldsymbol{n}+\boldsymbol{\alpha})\Omega(\boldsymbol{n}+\boldsymbol{\alpha})+2\pi i(\boldsymbol{n}+\boldsymbol{\alpha})(\boldsymbol{u}+\boldsymbol{\beta})\}\,. \tag{2}$$

$\omega_1, ..., \omega_g$ is the basis of abelian differentials normalized as $\oint_{a_i}\omega_j(v)\,dv=\delta_{ij}$, $\oint_{b_i}\omega_j(v)\,dv=\Omega_{ij}$ where a_i, b_i ($i=1$, ..., g) is the canonical basis of $H_1(M; Z)$ of the Riemann surface M, and Ω is the period matrix. We consider an even characteristic $(-1)^{4\alpha\beta}=1$ in (1). The prime form $E(z, w)$ can be written explicitly in terms of the θ-function as

$$E(z,w)=\frac{\theta[^{\alpha_0}_{\beta_0}](\int_w^z \boldsymbol{\omega}(v)\,dv)}{h(z)h(w)}\,,\quad h(z)=\left(\sum_{i=1}^{g}\omega_i(z)\frac{\partial}{\partial u_i}\theta[^{\alpha_0}_{\beta_0}](\boldsymbol{u})\bigg|_{\boldsymbol{u}=0}\right)^{1/2}\,. \tag{3}$$

Here α_0, β_0 is an arbitrary odd characteristic $(-1)^{4\alpha_0\beta_0}=-1$ and $E(z, w)$ is independent of the choice of α_0, β_0. We see that the right-hand side of (1) transforms like a spinor (half-order differential) both at z and w and is analytic everywhere except at $z=w$ where there is a simple pole with unit residue. These are just the properties we want for a spinor Green function.

The spin structure of the spinor field $\psi(z)$ is specified by the characteristic $\boldsymbol{\alpha}, \boldsymbol{\beta}\cdot\theta[^{\alpha}_{\beta}](\int_w^z \boldsymbol{\omega}(v)\,dv)$ is multiplied by

$$\exp(2\pi i\alpha_j)\,,\quad \exp\left(-2\pi i\beta_j-\pi i\Omega_{jj}-2\pi i\int_w^z\omega_j(v)\,dv\right) \tag{4}$$

as z is moved around the a_j, b_j cycle, respectively. On the other hand, the prime form $E(z, w)$ has multipliers

$$1\,,\quad \exp\left(-\pi i\Omega_{jj}-2\pi i\int_w^z\omega_j(v)\,dv\right) \tag{5}$$

along the a_j and b_j cycles. Thus (1) describes a spinor field with the boundary condition

$$\begin{aligned}\psi(z)&\to\exp(2\pi i\alpha_j)\,\psi(z)\,,\quad\text{moved along the } a_j \text{ cycle}\,,\\ &\to\exp(-2\pi i\beta_j)\,\psi(z)\,,\quad\text{moved along the } b_j \text{ cycle}\,.\end{aligned} \tag{6}$$

Now let us introduce a free scalar field $\varphi(z)$ and express the spinor field $\psi(z)$ as

$$\psi(z)=\exp[i\varphi(z)]\,. \tag{7}$$

Let us suppose that $\varphi(z)$ is decomposed into its zero-mode and nonzero-mode components

$$\varphi(z)=2\pi\sum_{i=1}^{g}p_i\int^z\omega_i(v)\,dv+\hat{\varphi}(z)\,, \tag{8}$$

where p_i ($i=1$, ..., g) are zero-mode oscillators, and p_i and $\hat{\varphi}$ are assumed to be independent free fields (the lower end point of the integration is left arbitrary). Now our bosonization prescription is to postulate that

(A) zero-modes $\boldsymbol{p}$ have eigenvalues $\boldsymbol{n}+\boldsymbol{\alpha}$, $\boldsymbol{n}\in\mathbb{Z}^g$; (9)

(B) averaging over zero-modes is weighted by the action

$$\exp(\pi i \boldsymbol{p}\Omega\boldsymbol{p}+2\pi i\boldsymbol{p}\boldsymbol{\beta})\ ; \tag{10}$$

(C) averaging over a free field $\hat{\varphi}(z)$ is performed with the contraction function

$$\langle\hat{\varphi}(z)\hat{\varphi}(w)\rangle=-\log E(z,w)\ . \tag{11}$$

We then find

$$\langle\exp[i\varphi(z)]\exp[-i\varphi(w)]\rangle$$

$$=\sum_{\boldsymbol{n}\in Z^g}\exp\left[\pi i(\boldsymbol{n}+\boldsymbol{\alpha})\Omega(\boldsymbol{n}+\boldsymbol{\alpha})+2\pi i(\boldsymbol{n}+\boldsymbol{\alpha})\left(\int_w^z\boldsymbol{\omega}\,dv+\boldsymbol{\beta}\right)\right]\left(\theta[^{\alpha}_{\beta}](0)E(z,w)\right)^{-1}$$

$$=\theta[^{\alpha}_{\beta}]\left(\int_w^z\boldsymbol{\omega}\,dv\right)\Big/\left(\theta[^{\alpha}_{\beta}](0)E(z,w)\right). \tag{12}$$

(We have introduced a normalization factor $(\theta[^{\alpha}_{\beta}](0))^{-1}$ for the sum over the zero-mode.) We thus recover (1). The spin structure of the fermion is controlled by the spectrum and the weight of the zero-modes (A), (B). We note that $\exp(2\pi i\boldsymbol{p}\boldsymbol{\beta})$ in (10) corresponds to $(-1)^F$ (F is the fermion number) insertion in the case of the torus.

The U(1) current is defined by

$$j(z)=\lim_{w\to z}\left(\bar{\psi}(w)\psi(z)-\frac{1}{w-z}\right)=i\partial_z\varphi(z)=2\pi i\boldsymbol{p}\boldsymbol{\omega}+i\partial_z\bar{\varphi}(z)\ . \tag{13}$$

Using our rules (A)–(C) we may compute current correlation functions as

$$\langle j(z)\rangle=\sum_{i=1}^{g}\omega_i(z)\frac{\partial}{\partial u_i}\log\theta[^{\alpha}_{\beta}](\boldsymbol{u})|_{\boldsymbol{u}=0}\ , \tag{14}$$

$$\langle j(z)j(w)\rangle-\langle j(z)\rangle\langle j(w)\rangle=\sum_i\sum_j\omega_i(z)\omega_j(w)\frac{\partial}{\partial u_i}\frac{\partial}{\partial u_j}\log\theta[^{\alpha}_{\beta}](\boldsymbol{u})|_{\boldsymbol{u}=0}+\partial_z\partial_w\log E(z,w)\ . \tag{15}$$

Actually, the current one-point function vanishes for an even spin structure ($\theta[^{\alpha}_{\beta}](\boldsymbol{u})$ being an even function of $\boldsymbol{u}$). We may also compute other correlation functions using (A)–(C), for instance,

$$\langle j(z)\psi(w_1)\bar{\psi}(w_2)\rangle=\sum_i\omega_i(z)\frac{\partial}{\partial u_i}\log\theta[^{\alpha}_{\beta}]\left(\int_{w_2}^{w_1}\boldsymbol{\omega}(v)\,dv+\boldsymbol{u}\right)\Bigg|_{\boldsymbol{u}=0}\langle\psi(w_1)\bar{\psi}(w_2)\rangle$$

$$+\partial_z[\log E(z,w_1)-\log E(z,w_2)]\langle\psi(w_1)\bar{\psi}(w_2)\rangle$$

$$=\left[\sum_i\omega_i(z)\frac{\partial}{\partial u_i}\log\theta[^{\alpha}_{\beta}]\left(\int_{w_2}^{w_1}\boldsymbol{\omega}\,dv+\boldsymbol{u}\right)\Bigg|_{\boldsymbol{u}=0}+\partial_z\log\frac{E(z,w_1)}{E(z,w_2)}\right]\frac{\theta[^{\alpha}_{\beta}](\int_{w_2}^{w_1}\boldsymbol{\omega}\,dv)}{\theta[^{\alpha}_{\beta}](0)E(w_1,w_2)}\ . \tag{16}$$

We note that formulas (15), (16) have the form of a current algebra Ward identity [5].

On the other hand, these correlation functions may also be computed using the fermion representation $j(z)\approx\bar{\psi}(z)\psi(z)$ and the correlation function (1). We obtain $\langle j(z)\rangle=0$ and

$$\langle j(z)j(w)\rangle=-\left(\frac{\theta[^{\alpha}_{\beta}](\int_w^z\boldsymbol{\omega}\,dv)}{\theta[^{\alpha}_{\beta}](0)E(z,w)}\right)^2, \tag{17}$$

$$\langle j(z)\psi(w_1)\bar{\psi}(w_2)\rangle = \frac{\theta[^{\alpha}_{\beta}](\int_{w_2}^{z} \omega\, \mathrm{d}v)\, \theta[^{\alpha}_{\beta}](\int_{z}^{w_1} \omega\, \mathrm{d}v)}{\theta[^{\alpha}_{\beta}](0)E(z, w_2)\theta[^{\alpha}_{\beta}](0)E(w_1, z)} . \tag{18}$$

We wonder if there exist identities which show (15) and (17), (16) and (18) are in fact equal to each other. Actually (15), (17) is a limit of (16), (18) as w_1, w_2 approach w and the most general case we may consider is the fermion second-point function

$$\langle \psi(x_1)...\psi(x_n)\bar{\psi}(y_1)...\bar{\psi}(y_n)\rangle \tag{19}$$

from which current insertion amplitudes are obtained in the limit of coinciding arguments. By computing (19) either by bosonic (left-hand side) or fermionic (right-hand side) method we obtain,

$$\frac{\theta[^{\alpha}_{\beta}](\sum_{i=1}^{n}\int_{y_i}^{x_i} \omega\, \mathrm{d}\,v)}{\theta[^{\alpha}_{\beta}](0)} \frac{\prod_{i<j}E(x_i, x_j)\prod_{i<j}E(y_i, y_j)}{\prod_{i,j}E(x_i, y_j)} = (-1)^{n(n-1)/2} \det \frac{\theta[^{\alpha}_{\beta}](\int_{y_j}^{x_i} \omega\, \mathrm{d}v)}{\theta[^{\alpha}_{\beta}](0)E(x_i, y_j)} . \tag{20}$$

To our surprise, (20) is in fact a basic addition theorem of θ-functions (whose period matrix comes from the Riemann surface) discovered by Fay [3] and has been one of the cornerstones of the modern development of the theory of θ-functions [4]. (20) in the case $n=2$ is called the triscecant identity and the general case may be derived from it by mathematical induction. Thus our heuristic rule of bosonization (A)–(C) is in fact consistent and even reproduces deep results in complex function theory.

3. Bosonization of the ghost system. Let us now turn to the discussion of the ghost system, a pair of n and $(1-n)$ differentials, denoted as b and c fields, coupled by the $\bar{\partial}$ operator [7]. Their action and energy–momentum tensor is given by

$$S=\frac{1}{\pi}\int \mathrm{d}^2 z \sqrt{g}[b(z)\bar{\partial}c(z)+\text{c.c.}] , \quad T(z) = -[nb(z)\partial c(z)+(n-1)\partial b(z)c(z)] . \tag{21,22}$$

Although when n is an integer the b and c fields have integral spin, they are anti-commuting fields described by the first-order system (21), (22), and thus are regarded as fermionic degrees of freedom. On a Riemann surface of genus $g\geqslant 2$, b(n-differential) has $(2n-1)(g-1)$ zero-modes while c has none when $n\geqslant 2$. When $n=1$ there are g and 1 zero-mode(s) for b (vector) and c (scalar), respectively.

In order to fix our discussions let us consider the case $n=1$ first. In the presence of zero-modes it is necessary to make a suitable number of insertions of b, c fields to obtain meaningful results in path integration. The partition function is now defined by

$$Z(z_1, ..., z_g, w) = \int \mathrm{d}b\, \mathrm{d}c\, b(z_1)...b(z_g)c(w) \exp[-S(b, c)] \tag{23}$$

and the analogue of the two-point function is given by

$$\langle b(x)c(y)\rangle = Z(z_1, ..., z_g, w)^{-1} \int \mathrm{d}b\, \mathrm{d}c\, b(z_1)...b(z_g)c(w)b(x)c(y) \exp[-S(b, c)] . \tag{24}$$

An explicit formula can be obtained for (24). One first differentiates in $\bar{y}$ and uses an equation of motion to get

$$\partial_{\bar{y}}\langle b(x)c(y)\rangle = -\pi\delta(x-y)+\pi\sum_{j=1}^{g} \delta(z_j-y) \frac{\det\|\omega_i(z_1)...\omega_i(x)...\omega_i(z_g)\|}{\det\|\omega_i(z_1)...\omega_i(z_j)...\omega_i(z_g)\|} , \tag{25}$$

where in the second term of the right-hand side x replaces z_j. Then by integrating (25) we find

$$\langle b(x)c(y)\rangle = \frac{1}{\det\|\omega_i(z_j)\|}\det\begin{bmatrix} \partial_x \log\dfrac{E(x,y)}{E(x,w)} & \partial_{z_1}\log\dfrac{E(z_1,y)}{E(z_1,w)} & \cdots & \partial_{z_g}\log\dfrac{E(z_g,y)}{E(z_g,w)} \\ \omega_1(x) & \omega_1(z_1) & \cdots & \omega_1(z_g) \\ \omega_2(x) & \omega_2(z_1) & \cdots & \omega_2(z_g) \\ \vdots & \vdots & & \vdots \\ \omega_g(x) & \omega_g(z_1) & \cdots & \omega_g(z_g) \end{bmatrix} . \tag{26}$$

(26) is known as the Cauchy kernel in mathematical literature and has the property $f(y) = (1/2\pi i)\int_{\partial D} f(x)\langle b(x)c(y)\rangle\, dx$ for any function $f(x)$ holomorphic in D not containing z_i, w. (26) shows the fermionic nature of the ghost field in its determinantal structure. We need an identity analogous to (20) relating a determinant to a product of θ-functions and prime forms which allows a bosonic description. In fact there exists the identity [3]

$$\text{RHS of (26)} = \frac{\theta(\boldsymbol{I}(x-y-w+\sum z_i-\Delta))\theta(\boldsymbol{I}(\sum z_i-x-\Delta))E(y,w)}{\theta(\boldsymbol{I}(\sum z_i-w-\Delta))\theta(\boldsymbol{I}(\sum z_i-y-\Delta))E(x,y)E(x,w)} . \tag{27}$$

Here θ is $\theta[^{\alpha}_{\beta}]$ with $\alpha=\beta=0$ and the Jacobi map $\boldsymbol{I}$ is defined as

$$\boldsymbol{I}(\textstyle\sum_i a_i) = \sum_i \int^{a_i} \boldsymbol{\omega}(v)\, dv \tag{28}$$

for a divisor $\sum_i a_i$. The Riemann divisor class $\Delta = \sum r_j$ is defined by

$$I_k(\Delta) = \sum_j \int^{r_j} \omega_k\, dv = \pi i - \pi i \Omega_{kk} + \sum_{l\neq k} \oint_{a_l} \omega_l(u) \int^{u} \omega_k(v)\, dv\, du \quad (\text{mod. period}) . \tag{29}$$

Δ, degree $g-1$, is in general not a positive divisor, and $\sum r_j = r_1 + ... + r_g - r_0$. (27) may be checked by comparing the divisors of both sides of the equation and is actually a variant of the triscecant identity. We next introduce a σ-function defined by

$$\theta\left(\boldsymbol{I}\left(\sum_{i=1}^{g} z_i - x - \Delta\right)\right) = S(z_1, ..., z_g)\prod_{i=1}^{g} E(z_i, x)\sigma(x) . \tag{30}$$

Since the LHS of (30) vanishes only at $x = z_i$ $(i=1, ..., g)$, $\sigma(x)$ has no poles or zeros and hence a holomorphic section of a trivial bundle on M [$S(z_1, ..., z_g)$ is a holomorphic section of a line bundle of degree $g-1$ in each variable]. The explicit form of $\sigma(x)$ is given by [3]

$$\sigma(x) = \exp\left(-\sum_{j=1}^{g}\oint_{a_j} \omega_j(v) \log E(v,x)\, dv\right) . \tag{31}$$

We then rewrite (27) as

$$\langle b(x)c(y)\rangle = \frac{\theta(\boldsymbol{I}(x-y-w+\sum z_i-\Delta))\prod_{i=1}^{g} E(z_i,x)E(y,w)\sigma(x)}{\theta(\boldsymbol{I}(\sum z_i - w-\Delta))\prod_{i=1}^{g} E(z_i,y)E(x,w)E(x,y)\sigma(y)} . \tag{32}$$

We see in (32) again a coulombic structure, positive charges at z_i, x and negative charges at y, w, together with additional factors of θ and σ. Thus we again introduce a free scalar field $\varphi(z)$ and express ghost fields as

$$b(z) = \exp[i\varphi(z)] , \quad c(z) = \exp[-i\varphi(z)] . \tag{33}$$

$\varphi(z)$ is again written as a sum of nonzero- and zero-modes (8). We can then reproduce (32) with the following bosonization prescription:

(A′) the zero-modes $\boldsymbol{p}$ have eigenvalues $\boldsymbol{n}$, $\boldsymbol{n}\in\mathbb{Z}^g$; (34)

(B′) averaging over zero-modes is weighted by the action

$$\exp[\pi \mathrm{i}\boldsymbol{p}\Omega\boldsymbol{p}-2\pi\mathrm{i}\boldsymbol{p}\cdot\boldsymbol{I}(\Delta)] \; ; \tag{35}$$

(C′) averaging over the free field $\hat{\varphi}(z)$ is done with an extra insertion of an operator

$$(\mathrm{C}'1) \quad \exp\left(-\mathrm{i}\sum_{j=1}^{g}\oint_{a_j}\omega_j(v)\hat{\varphi}(v)\,\mathrm{d}v\right) \tag{36}$$

and by the contraction function

$$(\mathrm{C}'2) \quad \langle\hat{\varphi}(z)\hat{\varphi}(w)\rangle=-\log E(z,w)\,. \tag{37}$$

Note that the insertion of (36) generates the factor $\sigma(x)/\sigma(y)$ in (32). We would like to point out that this extra insertion (36) and the linear term, $-2\pi\mathrm{i}\boldsymbol{p}\cdot\boldsymbol{I}(\Delta)$, in (35) are due to the presence of an anomaly in the ghost system.

If we bosonize the ghost energy–momentum tensor (22), we find

$$T(z)=\tfrac{1}{2}[\partial_z\varphi(z)]^2+\mathrm{i}(n-1/2)\nabla_z^2\varphi(z) \tag{38}$$

which is derived from the action

$$S=\frac{1}{4\pi}\int \mathrm{d}^2z\,\sqrt{g}\,[\tfrac{1}{2}\partial_\mu\varphi\partial_\mu\varphi-\tfrac{1}{2}\mathrm{i}(2n-1)\varphi R]\,. \tag{39}$$

(39) involves a coupling to the scalar curvature of the surface and this induces an anomaly in the current conservation law,

$$j(z)=\lim_{w\to z}\left(b(z)c(w)-\frac{1}{z-w}\right)=\mathrm{i}\partial_z\varphi(z)\,,\quad \partial_{\bar{z}}j(z)=\tfrac{1}{8}(2n-1)\sqrt{g}R\,. \tag{40,41}$$

While the ghost action written in the fermionic form (22) does not involve explicit reference to the metric, the bosonized action, somewhat surprisingly, has an explicit dependence on the metric. Thus it is necessary to specify the metric properties of the Riemann surface if one tries to bosonize ghost fields. Now we shall show that our bosonization procedure (34)–(37) is associated with a (singular) metric whose curvature is concentrated on the divisor of the Riemann class Δ.

Let us introduce the fermion wave function $\psi(z)$ which is holomorphic except for a pole at $z=r_0$ and has zeros at $z=r_i$ ($i=1, \ldots, g$) (the meromorphic section of a spin bundle corresponding to the Riemann divisor class Δ). $\psi(z)$ obeys the boundary condition (6) with $\alpha=\beta=0$. We consider the metric

$$g_{z\bar{z}}(z,\bar{z})=c_z(z)\overline{c_z(z)}\,,\quad c_z(z)=[\psi(z)]^2\,. \tag{42,43}$$

Then the Ricci tensor $R_{z\bar{z}}(z,\bar{z})=-\partial_{\bar{z}}\partial_z\log g_{z\bar{z}}(z,\bar{z})$ is given by

$$R_{z\bar{z}}(z,\bar{z})=-4\pi\left(\sum_{i=1}^{g}\delta(z-r_i)-\delta(z-r_0)\right). \tag{44}$$

Then, in the case $n=1$, the curvature term in the action (39) contributes a factor

$$\exp[-2\pi\mathrm{i}\boldsymbol{p}\cdot\boldsymbol{I}(\Delta)] \tag{45}$$

to the zero-mode sector. (45) explains the origin of our prescription (B′). On the other hand, the curvature term gives rise to an operator insertion

$$\exp\left[\mathrm{i}\left(-\sum_{i=1}^{g}\hat{\varphi}(r_i)+\hat{\varphi}(r_0)\right)\right] \tag{46}$$

in the nonzero-mode sector. We now show that the effect of the insertion (46) is equivalent to that of (C′1) when we use the contraction function (C′2).

This is seen in the following way; we first note that the insertion (46) would generate a factor

$$\frac{E(u,r_0)}{\prod_{i=1}^{g}E(u,r_i)}\quad\left[\frac{\prod_{i=1}^{2}E(v,r_i)}{E(v,r_0)}\right] \tag{47}$$

for each $b(u)$ $[c(v)]$ field in the correlation functions. The factors (47), however, do not have a correct normalization and must be multiplied by $e_z(z)^{1/2}$ $[e_z(z)^{-1/2})]$ (these factors may have been included in the bosonization formula as $b(z)=e_z(z)^{1/2}\exp[\mathrm{i}\varphi(z)]$, $c(z)=e_z(z)^{-1/2}\exp[-\mathrm{i}\varphi(z)]$. We then observe

$$e_u(u)^{1/2}\frac{E(u,r_0)}{\prod_{i=1}^{g}E(u,r_i)}=f(r_0,\ldots,r_g)\sigma(u)\ . \tag{48}$$

In fact poles and zero in u cancel on the LHS, and both sides are differentials of order $g/2$ in the variable u. Furthermore, both sides are multiplied by

$$1\ ,\quad \exp\left(\pi\mathrm{i}(g-1)\Omega_{jj}-2\pi\mathrm{i}\int_{(g-1)u}^{\Delta}\omega_j\,\mathrm{d}v\right), \tag{49}$$

when u moves along the a_j, b_j cycle, respectively. This proves (48). Thus the factor (47) is essentially $\sigma(u)$ which is generated from our operator insertion (C′1).

We have seen that our modified rule reflects the presence of an anomaly in the ghost system. While we have discussed the case $n=1$ so far, generalization to $n\geqslant 2$ is straightforward. We simply replace the rules (B′1), (C′1) by

$$(\mathrm{B}'')\quad \exp[\pi\mathrm{i}\boldsymbol{p}\Omega\boldsymbol{p}-2\pi\mathrm{i}(2n-1)\boldsymbol{p}\cdot\boldsymbol{I}(\Delta)]\ , \tag{50}$$

$$(\mathrm{C}''1)\quad \exp\left(-\mathrm{i}(2n-1)\sum_{j=1}^{g}\oint_{a_j}\omega_j(v)\hat{\varphi}(v)\,\mathrm{d}v\right). \tag{51}$$

(50), (51) are in accord with (39) where the scalar curvature has a coefficient $(2n-1)$ in front.

Using our bosonization rule we may obtain various identities for functional determinants and correlation functions. In particular, if we compute the partition function (23) in both fermionic and bosonic methods, we find

$$Z(z_1,\ldots,z_g,w)=\det\|\omega_i(z_j)\|\,\det{}'\bar{\partial}$$

$$=\theta(\boldsymbol{I}(\textstyle\sum z_i-w-\Delta))\frac{\prod_{i<j}E(z_i,z_j)}{\prod_i E(z_i,w)}\frac{\prod_i\sigma(z_i)}{\sigma(w)}\times\text{chiral component of}\left(\frac{\det{}'-\Delta}{\int \mathrm{d}^2z\sqrt{g}\det\operatorname{Im}\Omega}\right)^{-1/2}. \tag{52}$$

Similarly in the case $n\geqslant 2$ we have

$$Z(z_1,\ldots,z_N)=\det\|h_i(z_j)\|\,\det{}'\bar{\partial}$$

$$=\theta(\boldsymbol{I}(\textstyle\sum z_i-(2n-1)\Delta))\prod_{i<j}E(z_i,z_j)\prod_i\sigma(z_i)^{2n-1}\times\text{chiral component of}\left(\frac{\det{}'-\Delta}{\int \mathrm{d}^2z\sqrt{g}\det\operatorname{Im}\Omega}\right)^{-1/2}, \tag{53}$$

where $N=(2n-1)(g-1)$ and $\bar{\partial}$ is the Cauchy–Riemann operator acting on $(1-n)$-differentials. h_i $(i=1, ..., N)$ are holomorphic n-differentials. (52), (53) are a "chiral version" of formulas obtained in ref. [2].

While in the case of non-chiral bosonization Arakelov's metric, $g_{z\bar{z}} \approx \sum \omega_i(z)(\mathrm{Im}\ \Omega)_{ij}^{-1}\overline{\omega_j(z)}$, has been used [2], use of a singular metric is necessary in order to keep the holomorphy property in our bosonization procedure. A similar use of a singular metric is discussed in ref. [8]. Our procedure is somewhat reminiscent of Witten's treatment of ghosts in string field theory where they are coupled to the defect angles of the surface [9]. Metric properties of the Riemann surface are relevant to the description of ghost systems and these aspects may be worth further study.

Since neither lagrangian nor hamiltonian descriptions are available for the chiral Bose field on a general Riemann surface, our bosonization procedure has necessarily been heuristic and non-rigorous. We hope, however, with more efforts, that in the near future, the chiral bosonization method will be put on a rigorous mathematical footing.

After completion of this work we have learned from Dr. B. Julia of a preprint by E. and H.L. Verlinde [10] where many of our results are obtained independently using a somewhat different line of argument.

T.E. would like to thank Dr. C. Bouchiat and Dr. J.-L. Gervais and members of the Theoretical Physics Division of Ecole Normale Supérieure for their kind hospitality and encouraging discussions. He is also grateful to Dr. B. Julia for his reading the manuscript.

References

[1] I. Frenkel and V. Kac, Invent. Math. 62 (1980) 23;
G. Segal, Commun. Math. Phys. 80 (1981) 301;
P. Goddard and D. Olive, Nucl. Phys. B 257 (1985) 226.

[2] L. Alvarez-Gaumé, J.B. Bost, G. Moore, P. Nelson and C. Vafa, Phys. Lett. B 178 (1986) 41;
J.B. Bost and P. Nelson, Phys. Rev. Lett. 57 (1986) 795.

[3] J.D. Fay, Theta functions on Riemann surfaces, Lectures Notes in Mathematics, Vol. 352 (Springer, Berlin, 1973).

[4] D. Munford, Tata lectures on theta, Vols. I and II (Birkhäuser, Basel, 1983).

[5] T. Eguchi and H. Ooguri, University of Tokyo preprint 491 (1986), Nucl. Phys. B, to be published.

[6] M.A. Namazie, K.S. Narain and M.H. Sarmadi, Phys. Lett. B 178 (1986) 329;
H. Sonoda, Phys. Lett. B 178 (1986) 390.

[7] D. Friedan, E. Martinec and S. Shenker, Nucl. Phys. B 271 (1986) 93.

[8] V.G. Knizhnik, Landau Institute preprint (1986).

[9] E. Witten, Nucl. Phys. B 268 (1986) 253.

[10] E. Verlinde and H.L. Verlinde, Utrecht preprint (1986).

Commun. Math. Phys. 106, 1–40 (1986)

Communications in
Mathematical
Physics

Theta Functions, Modular Invariance, and Strings

Luis Alvarez-Gaumé, Gregory Moore, and Cumrun Vafa

Department of Physics, Harvard University, Cambridge, MA 02138, USA

Abstract. We use Quillen's theorem and algebraic geometry to investigate the modular transformation properties of some quantities of interest in string theory. In particular, we show that the spin structure dependence of the chiral Dirac determinant on a Riemann surface is given by Riemann's theta function. We use this result to investigate the modular invariance of multiloop heterotic string amplitudes.

1. Introduction

Two-dimensional quantum field theories have served as toy models in attempts to understand more complicated four-dimensional theories. The two dimensional theories capture many essential features of higher dimensions, without sharing the complexities of higher dimensions. Certain features of $2d$ QFT's such as Bose-Fermi equivalence have led to a large number of exactly solvable theories.

In string theories, two-dimensional conformal QFT plays an even more important role [1–3]. The string sweeps out a surface as it moves through space-time, and therefore, the first quantized theory corresponds to a two dimensional QFT. The string can sweep out a surface with any number of handles. Whereas in most well known results one considers the underlying space to be R^2 (or $R \times S^1$, or $S^1 \times S^1$ corresponding to periodic boundary conditions in space or time), for string theories one must consider the space to be an arbitrary Riemann surface. Thus, understanding multiloop string amplitudes requires an understanding of QFT on a Riemann surface. In the general case, few explicit facts are known.

Many questions remain unanswered in string theory. For instance, how can we prove the vanishing of the cosmological constant in superstring theories? This has been shown explicitly at 1-loop. For higher loops, even though there is an indirect argument for the vanishing of the cosmological constant [3, 4], one would like to show this important fact more directly.

There are other issues: What happens to the string amplitude in the limit of a degenerating Riemann surface? Are superstring theories finite? Is the perturbation expansion valid? Can one develop an operator formalism for surfaces with more than one handle? How does the bosonization work at higher loops? etc.

Motivated by these questions, we address below some issues regarding the determinant of some operators, in particular the chiral Dirac operator, defined on a Riemann surface. We will discuss in some detail the spin structure dependence of the determinant of the Dirac operator. Our main tools are Quillen's holomorphic anomaly, algebraic geometry, and bosonization. We feel that the algebraic geometric techniques are very powerful and well suited for applications to a better understanding of string theories. This viewpoint has also been advocated recently by several groups.

In Sect. 2 we illustrate some of the issues which will be discussed in this paper, in the context of the simple case of the torus. In Sect. 3 we discuss the mathematical concepts which will be used. Many of them, for example, modular transformations, generalized theta functions and spin structures appear as natural generalizations of the concepts already appearing in the case of the torus. Yet others, such as divisors of theta functions and their relation to spin structures, although important for higher loops, do not play a central role in the genus one case.

In Sect. 4 we discuss Quillen's holomorphic anomaly, and a slight generalization of it. This anomaly is the obstruction to expressing as a holomorphic function the determinant of a family of operators depending holomorphically on some complex parameter.

In Sect. 5 we show that the spin dependence of the determinant of the chiral Dirac operator is given by theta functions. In Sect. 6 we discuss these results from the viewpoint of bosonization. In Sect. 7 we apply the results obtained to address some questions related to the modular invariance of the measure for heterotic strings [5]. Finally, in Sect. 8 we present our conclusions.

2. Fermions on a Torus

In this section we will consider fermions on a torus. We label the points of the torus by the complex quantity $\sigma_1+\tau\sigma_2$, where σ_1 and σ_2 are periodic variables with period 2π. This means that

$$x \sim x+2\pi \sim x+2\pi\tau\,.$$

For bosonic fields defined on the torus one requires the periodic boundary conditions:

$$X(\sigma_1+2\pi,\sigma_2)=X(\sigma_1,\sigma_2+2\pi)=X(\sigma_1,\sigma_2)\,.$$

For fermions, we have the option of choosing periodic or antiperiodic boundary conditions in each direction. Therefore we have four different possible boundary conditions: (A,A), (P,A), (A,P), and (P,P), where $P(A)$ stands for periodic (antiperiodic). These are called the spin structures. So for genus 1 we have four different spin structures.

For the torus we have chosen 2π and $2\pi\tau$ to generate the lattice defining it. We could have chosen a different basis for the same lattice, thereby obtaining the same torus. For example the tori defined by $2\pi(1,\tau)$ and $2\pi(1,\tau+1)$ differ from one another by a global diffeomorphism. To see this we simply cut the torus along the

σ_1 cycle and rotate the two boundary circles relative to one another by 2π and glue them back together. (This is called a Dehn twist, and its general form will be discussed in Sect. 3.)

Different choices of the lattice defining the torus could be obtained by means of 2×2, invertible (unit determinant) and integral matrices acting on a basis. This is the group $SL(2,Z)$ which is referred to as the modular group. Under an element of $SL(2,Z)$, with entries a, b, c, d,

$$\tau\to\frac{a\tau+b}{c\tau+d}.$$

The Dehn twists about the σ_1 and σ_2 cycles correspond respectively to

$$\begin{pmatrix}1 & 1\\ 0 & 1\end{pmatrix}\quad\begin{pmatrix}1 & 0\\ 1 & 1\end{pmatrix}$$

and they generate $SL(2,Z)$.

Under $SL(2,Z)$ it is easy to see that three spin structures mix, in the sense that

$$(P,A)_\tau=(A,A)_{\tau+1}=(A,P)_{-1/\tau}.$$

The (P,P) spin structure is invariant under the action of $SL(2,Z)$. So in a sense we have only two inequivalent spin structures. There is a simple way to characterize each class. We consider the Dirac operator for each of the spin structures which we label by α. Let $n(\alpha)$ denote the number of zero modes of the Dirac operator for spin structure α. Since the Dirac operator is real, there is no chiral index in two dimensions and therefore $n(\alpha)$ is even. We call a spin structure even (odd) if $\frac{1}{2}n(\alpha)$ is even (odd). It is easy to see in our example of the torus that there are three even spin structures [with $n(\alpha)=0$] (A,P), (P,A), (A,A) and one odd spin structure [with $n(\alpha)=2$] (P,P). Under the action of global diffeomorphisms the two different classes of spin structures cannot mix because they have different number of zero modes for the Dirac operator. In fact more is true: All the spin structures in each class mix under the action of global diffeomorphisms. This is true in the above example and, as we will discuss, it continues to be true for higher genus Riemann surfaces.

Modular transformations are important for string theories. In a string theory we have to divide the path integral measure by the action of diffeomorphisms, and this includes global diffeomorphisms which cannot be reached from identity. Therefore it is necessary for the path integral measure to be invariant under global diffeomorphisms, i.e., modular transformations (For each spin structure it should be invariant under diffeomorphisms preserving that spin structure.)

Note that each spin structure is simply related to a particular one by changing boundary conditions by ± 1. We can in fact continuously interpolate between various spin structures if we consider arbitrary twistings of the boundary conditions by a phase. (For this we should allow the spinor to be complex.) So we get a two parameter family of complex spinors on the torus defined by θ,ϕ:

$$\psi(\sigma_1+2\pi,\sigma_2)=-e^{2\pi i\theta}\psi(\sigma_1,\sigma_2),$$
$$\psi(\sigma_1,\sigma_2+2\pi)=-e^{-2\pi i\phi}\psi(\sigma_1,\sigma_2).$$

The four spin structures (A, A), (P, A), (A, P) and (P, P) correspond respectively to the twistings $(\theta, \phi) = (0, 0)$, $(\frac{1}{2}, 0)$, $(0, \frac{1}{2})$ and $(\frac{1}{2}, \frac{1}{2})$. (Even though we have used the twistings of boundary conditions as a formal technique to interpolate between various spin structures, they in fact appear naturally when considering string amplitudes on orbifolds [6].)

Let us denote by $\mathrm{Det}(\theta, \phi)$ the determinant of the chiral Dirac operator with the above twisted boundary conditions. The Hamiltonian for the twisted fermion can be expressed as

$$H = \sum_{-\infty}^{+\infty} \left(n + \theta - \frac{1}{2}\right) : b^{\dagger}_{n+\theta-1/2} b_{n+\theta-1/2} : + \left(\frac{\theta^2}{2} - \frac{1}{24}\right),$$

where the constant term comes from normal ordering, and

$$\{b^{\dagger}_{\eta_1}, b_{\eta_2}\} = \delta_{\eta_1, \eta_2},$$
$$\{b^{\dagger}_{\eta_1}, b^{\dagger}_{\eta_2}\} = \{b_{\eta_1}, b_{\eta_2}\} = 0.$$

We define g (up to a phase) by $g b_{n+\theta-1/2} g^{-1} = -e^{-2\pi i\phi} b_{n+\theta-1/2}$, and also let $q = e^{2\pi i\tau}$, then the determinant can be computed up to a phase using the partition function

$$\mathrm{Det}(\theta, \phi) = \mathrm{Tr}\, g q^H = e^{2\pi i\theta\phi} q^{\frac{\theta^2}{2} - \frac{1}{24}} \prod_{n=1}^{\infty} (1 + q^{n+\theta-\frac{1}{2}} e^{2\pi i\phi})(1 + q^{n-\theta-\frac{1}{2}} e^{-2\pi i\phi}). \quad (2.1)$$

(The phase in front has been chosen for convenience.) The argument for the equality between the determinant and partition function is the usual one based on the equivalence between path-integral formulation of quantum field theories and the operator formulation, when τ is purely imaginary. One simply views $2\pi\tau_2$ as the time variable. In the case when τ also has a real part $\tau = \tau_1 + i\tau_2$, the torus is slightly skewed and so the trace should be taken after multiplying by the operator which takes $\sigma_1 \to \sigma_1 + 2\pi\tau_1$. The operator which accomplishes this is $e^{2\pi i\tau_1(H - \bar{H})}$, where $\bar{H}$ denotes the Hamiltonian for the other chirality. So altogether, we get $e^{2\pi i\tau_1(H-\bar{H}) - 2\pi\tau_2(H+\bar{H})} = q^H \bar{q}^{\bar{H}}$. Keeping one of the chiralities, we obtain the above result (multiplied by the g twist operator).

In the above example, rather than changing the boundary conditions, we could fix the boundary conditions and instead introduce flat gauge fields which couple to the fermions. If we take the gauge field $A = \theta d\sigma_1 - \phi d\sigma_2$, the chiral Dirac operator depends on the twistings only through the combination $\phi + \tau\theta$. So, defining the complex quantity u by $u = \phi + \tau\theta$, we naively expect the determinant to be a function only of u and not of its complex conjugate $\bar{u}$. In fact the explicit expression (2.1) for the $\mathrm{Det}(\theta, \phi)$ is almost of that form. The terms in the infinite product are only functions of u, but the quadratic term in the exponent of the q term in front prevents us from writing the determinant solely as a function of u:

$$\theta^2 = \left(\frac{1}{2i\,\mathrm{Im}\,\tau}(u - \bar{u})\right)^2.$$

So it seems that there is some sort of anomaly. In fact this is an example of a holomorphic anomaly which will be discussed in detail in Sect. 4.

It is possible to write the (θ,ϕ) dependence of $\mathrm{Det}(\theta,\phi)$ in a way which generalizes to higher genus. For this purpose consider the function:

$$\vartheta\begin{bmatrix}\theta\\ \phi\end{bmatrix}(0|\tau)=\sum_n e^{i\pi(n+\theta)^2\tau+2i\pi(n+\theta)\phi}=\sum_n q^{\frac{(n+\theta)^2}{2}}e^{2\pi i(n+\theta)\phi}. \tag{2.2}$$

Using the product representation of theta functions (2.1) becomes

$$\mathrm{Det}(\theta,\phi)=\frac{\vartheta\begin{bmatrix}\theta\\ \phi\end{bmatrix}(0|\tau)}{\eta(\tau)}, \tag{2.3}$$

where $\eta(\tau)$, is the Dedekind eta function:

$$\eta(\tau)=q^{\frac{1}{24}}\prod_n(1-q^n). \tag{2.4}$$

Therefore the twist dependence of the determinant is given by a theta function. Theta functions arise naturally in string theories as sums over winding sectors and internal momenta. The same is true here, and the right-hand side of (2.3) is in fact the twisted partition function of a single boson allowed to have momenta on a shifted lattice. The above identity is a statement about bosonization. To see this, note that the chiral fermionic current is replaced by the translation operator in the bosonized theory. The twist operator g turns out to be given by $g=e^{2\pi i\phi P}$, where P is the translation operator, and we should require the bosonic momenta to lie on the shifted lattice $(n+\theta)$, where n is any integer. The partition function $\mathrm{Tr}\,gq^H$, where $H=\frac{P^2}{2}+\sum na_n^\dagger a_n$ with obvious commutation relations, is easily seen to be the expression (2.3). We will explore the analog of this argument for the higher genus case in Sect. 6.

In Sect. 7 we will discuss the transformation properties of theta functions and the modular invariance of multiloop string amplitudes. These matters are easily understood for the torus. It is straightforward to check, for example, that under modular transformations, $\vartheta\begin{bmatrix}0\\ 0\end{bmatrix}$, $\vartheta\begin{bmatrix}0\\ \frac{1}{2}\end{bmatrix}$, and $\vartheta\begin{bmatrix}\frac{1}{2}\\ 0\end{bmatrix}$ mix with each other, corresponding to the fact that all the even spin structures mix under global diffeomorphisms. $\vartheta\begin{bmatrix}\frac{1}{2}\\ \frac{1}{2}\end{bmatrix}$ is zero, corresponding to the fact that the Dirac operator has zero modes. (If we delete the zero mode, the odd theta function is seen to transform to itself under modular transformations.)

The one-loop vacuum to vacuum amplitudes for superstrings are well-known. The contribution of the right-moving spinors is proportional to:

$$\vartheta^4\begin{bmatrix}0\\ 0\end{bmatrix}-\vartheta^4\begin{bmatrix}\frac{1}{2}\\ 0\end{bmatrix}-\vartheta^4\begin{bmatrix}0\\ \frac{1}{2}\end{bmatrix}.$$

We have five complex spin $\frac{1}{2}$ fermions and one complex spin $\frac{3}{2}$ ghost, which cancels the determinant of one of the fermions, to leave us with the power 4 for the theta function. This expression is modular covariant, i.e., under modular transformations it transforms to itself, up to a prefactor. In fact the above combination is equal to zero, and it implies that the cosmological constant vanishes, at one-loop order.

This concludes our discussion of fermions on a torus. We will see that many of these concepts are easily generalized to higher loops.

3. Mathematical Background

In this section we summarize some aspects of the theory of Riemann surfaces which are useful in the computation of higher loop string amplitudes. Since we only consider closed strings we will be concerned with compact Riemann surfaces. We will consider the topology, differential geometry, line bundles, and theta functions associated with a surface.

a) The Mapping Class Group

Topologically, orientable two-dimensional surfaces Σ are completely classified by the Euler number $\chi(\Sigma)=2-2g$, where g, the genus of Σ, is the number of handles (see Fig. 1). We describe first the homology of Σ.

When Σ is compact the homology groups are free groups with dimensions

$$\dim H_0(\Sigma)=1\,, \qquad \dim H_1(\Sigma)=2g\,, \qquad \dim H_2(\Sigma)=1\,.$$

We can identify a canonical homology basis a_i, b_i, $1\leqq i\leqq g$ for $H_1(\Sigma)$ as in Fig. 1. Then any closed curve on Σ generates a homology class which can be uniquely decomposed in terms of the classes generated by a_i, b_i. The reason for calling a_i, b_i a *canonical* basis is the following. If we define the intersection number $J(\gamma,\gamma')$ between two curves γ and γ', as the number of points at which they intersect countihg orientation, then, since the number $J(\gamma,\gamma')$ only depends on the homology classes generated by γ and γ', J defines a quadratic form on $H_1(\Sigma)$. In terms of the a_i, b_i cycles, J takes the canonical form

$$J(a_i,a_j)=J(b_i,b_j)=0\,, \qquad J(a_i,b_j)=-J(b_i,a_j)=\delta_{ij}\,, \tag{3.1}$$

or, as a matrix,

$$J=\begin{pmatrix} 0 & 1\\ -1 & 0\end{pmatrix}. \tag{3.2}$$

Once we have chosen a canonical homology basis, we can represent Σ by a $4g$-sided polygon with appropriate identifications on the boundary. To do this, choose a point on Σ, and cut the surface along $2g$ curves homologous to the canonical basis (this is depicted in Fig. 2 for the case of $g=2$). If in Fig. 2 we glue together the sides $a_ia_i^{-1}$ and $b_ib_i^{-1}$ we get back the original surface. Thus each handle is represented by the symbol $a_ib_ia_i^{-1}b_i^{-1}$.

We will often use the basis dual to the canonical homology basis. In terms of differential forms, we may use the Hodge-De Rham theory to set up a one to one

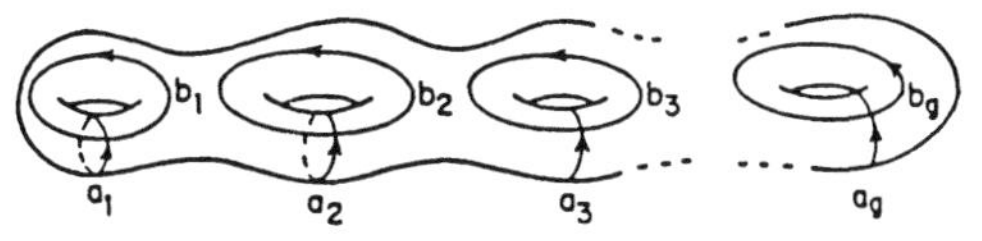

Fig. 1

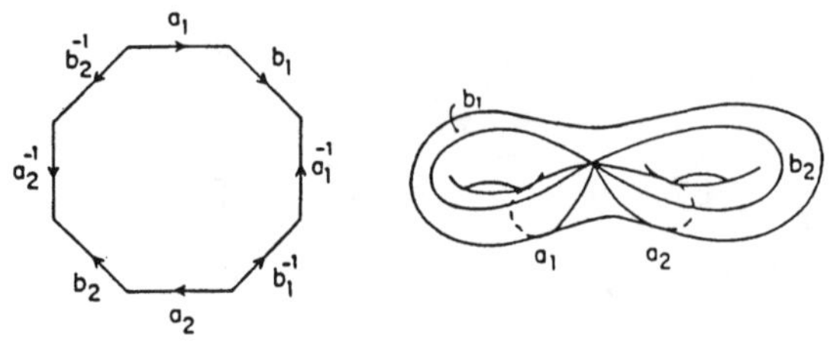

Fig. 2

correspondence between elements in $H_1(\Sigma)$ and harmonic 1-forms on Σ. Thus there are $2g$ harmonic 1-forms α_i, β_i, $1 \leqq i \leqq g$ normalized so that

$$\int_{a_i} \alpha_j = \delta_{ij}, \quad \int_{a_i} \beta_j = 0, \qquad \int_{b_i} \alpha_j = 0, \quad \int_{b_i} \beta_j = \delta_{ij}. \tag{3.3}$$

The representation of Σ in terms of a polygon is very helpful in proving some useful identities. For example, if θ, η are closed 1-forms then one can easily show [7]

$$\int_\Sigma \theta \wedge \eta = \sum_{i=1}^{g} \left[\int_{a_i} \theta \int_{b_i} \eta - \int_{b_i} \theta \int_{a_i} \eta \right]. \tag{3.4}$$

This completes our discussion of the homology and cohomology of Σ.

When we discuss global anomalies on the string world sheet and modular invariance of the string path integral we will need some facts about the nontrivial diffeomorphisms of Σ which we now outline [8]. Even though the proofs of the statements which follow are nontrivial, the results themselves are easily described. Let $\mathrm{Diff}(\Sigma)$ be the group of diffeomorphisms of Σ, and let $\mathrm{Diff}_0(\Sigma)$ be the normal subgroup of diffeomorphisms homotopic to the identity. Then the mapping class group is defined by

$$\Omega(\Sigma) = \mathrm{Diff}(\Sigma)/\mathrm{Diff}_0(\Sigma).$$

We first describe the generators of $\Omega(\Sigma)$. These can be taken to be Dehn twists around closed curves γ. In general, a Dehn twist is defined by excising a small tubular neighborhood of γ, twisting one boundary of the tube by 2π, and glueing the tube back into the surface. This sequence of operations defines an active transformation of the surface to itself (see Fig. 3). A useful result is that a set of

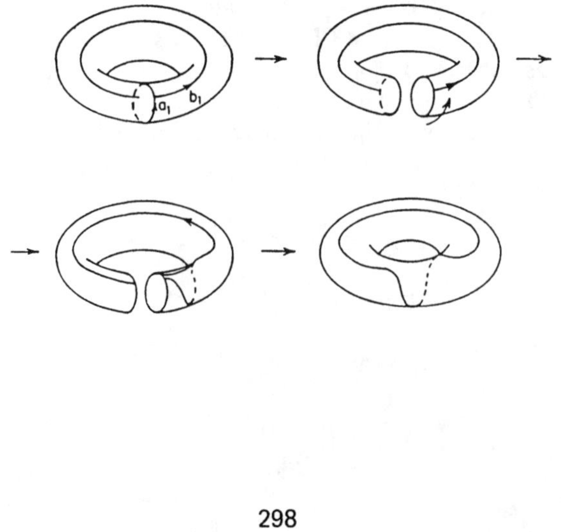

Fig. 3

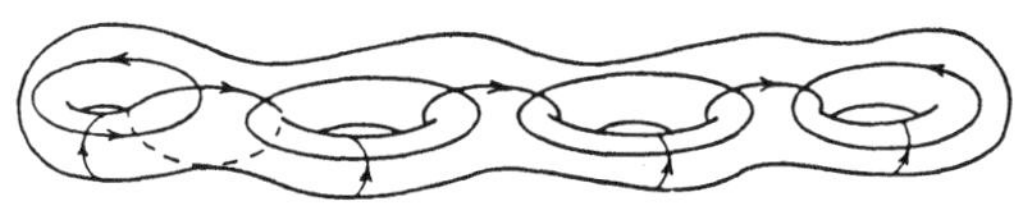

Fig. 4

generators of $\Omega(\Sigma)$ is provided by the Dehn twists about the curves illustrated in Fig. 4. (This is not, in fact, the minimal set.) Thus we have two generators for each handle (as on the torus) and a generator for the curve linking the holes of two consecutive handles. Thus, using the labeling of the homology basis in Fig. 1, the generators of $\Omega(\Sigma)$ are twists around

$$a_1, b_1, a_1^{-1}a_2, a_2, b_2, a_2^{-1}a_3, \ldots, a_g, b_g .$$

Thus it suffices to check modular invariance of the string integrand by checking modular invariance under these diffeomorphisms. We will apply this remark to obtain nontrivial restrictions on heterotic string theories in Sect. 7.

A useful representation of nontrivial diffeomorphisms is provided by their action on the homology of Σ. If the curve γ generates a nontrivial homology class, then a Dehn twist around γ acts nontrivially on the homology basis. For instance, in Fig. 3 the Dehn twist around a_1 induces the following transformation on the homology: $a_1 \to a_1$, $b_1 \to b_1 + a_1$. Let D_γ be the diffeomorphism defined by the twist around γ. The intersection matrix is manifestly invariant under diffeomorphisms, so the action of $\Omega(\Sigma)$ on the homology group $H_1(\Sigma, Z)$ must necessarily preserve (3.2). Thus, the matrix $M(D_\gamma)$ representing the action of D_γ on $H_1(\Sigma, Z)$ is a nonsingular $2g \times 2g$ matrix with integer entries, leaving the symplectic form (3.2) invariant, i.e. $M(D_\gamma)$ is an element of $Sp(2g, Z)$, the group of integer symplectic matrices, also known as the symplectic modular group. An important result is that the set of matrices $M(D_\gamma)$ in fact generate all of $Sp(2g, Z)$ [9].

As an example we construct an explicit representation of the generators of $\Omega(\Sigma)$ for $g=2$. They are given by the following 4×4 matrices:

$$D_{a_1} = \begin{pmatrix} 1 & 0 & 0 & 0 \\ 0 & 1 & 0 & 0 \\ 1 & 0 & 1 & 0 \\ 0 & 0 & 0 & 1 \end{pmatrix}, \quad D_{b_1} = \begin{pmatrix} 1 & 0 & 1 & 0 \\ 0 & 1 & 0 & 0 \\ 0 & 0 & 1 & 0 \\ 0 & 0 & 0 & 1 \end{pmatrix},$$

$$D_{a_2} = \begin{pmatrix} 1 & 0 & 0 & 0 \\ 0 & 1 & 0 & 0 \\ 0 & 0 & 1 & 0 \\ 0 & 1 & 0 & 1 \end{pmatrix}, \quad D_{b_2} = \begin{pmatrix} 1 & 0 & 0 & 0 \\ 0 & 1 & 0 & 1 \\ 0 & 0 & 1 & 0 \\ 0 & 0 & 0 & 1 \end{pmatrix},$$

$$D_{a_1^{-1}a_2} = \begin{pmatrix} 1 & 0 & 0 & 0 \\ 0 & 1 & 0 & 0 \\ -1 & 1 & 1 & 0 \\ 1 & -1 & 0 & 1 \end{pmatrix}.$$

An explicit construction of the action of $D_{a_1^{-1}a_2}$ on the homology is illustrated in Fig. 5.

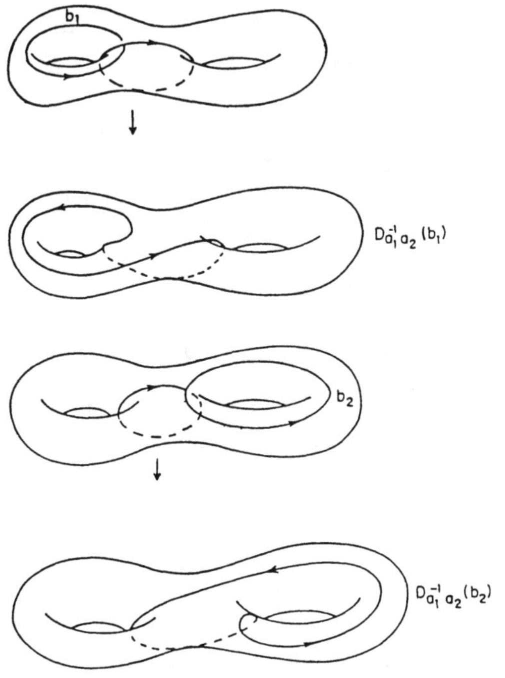

Fig. 5

b) Differential Geometry and Spin Structures

In this section we will define spin structures and review some of the elementary differential geometry of Riemann surfaces. Thus far we have considered Σ as a topological surface. The surfaces relevant to string theory are also endowed with a complex structure, which always exists if Σ is orientable. We will view the complex structure of Σ as that defined by a Riemannian metric g. Locally, we can always write any metric on Σ in the form

$$ds^2 = e^{2\phi}[(dx^0)^2 + (dx^1)^2] = 2g_{w\bar{w}}dw\,d\bar{w} \tag{3.5}$$

with $w = x^0 + ix^1$, $\bar{w} = x^0 - ix^1$. When we cover Σ with such complex coordinate patches $\mathscr{U}$, the transition functions on the overlaps are holomorphic, and thus define a complex structure. Using this complex structure we can divide the 1-forms

$$T^*\Sigma = T^{*(1,0)}\Sigma \oplus T^{*(0,1)}\Sigma$$

according to whether they are locally of the form $f(w,\bar{w})dw$ or $f(w,\bar{w})d\bar{w}$. In particular, there are $g(1,0)$ – forms ω_i which are holomorphic, i.e., are locally of the form $f_i(w)dw$ where f_i is holomorphic. A convenient way of normalizing the ω_i (also known as abelian differentials of the first kind) is to require

$$\int_{a_i} \omega_j = \delta_{ij}. \tag{3.6}$$

This completely specifies the ω_i [7]. One then finds for the b_i cycles

$$\int_{b_i} \omega_j = \Omega_{ij}. \tag{3.7}$$

Using (3.4) with $\theta=\omega_i$, $\eta=\omega_j$, it follows that $\Omega_{ij}=\Omega_{ji}$, and if one uses again (3.4) with $\theta=\bar{\eta}$, and η holomorphic, then

$$\frac{1}{2i}\int\bar{\theta}\wedge\theta=\operatorname{Im}\sum_i\overline{\int_{a_i}\theta}\int_{b_i}\theta>0\,, \tag{3.8}$$

and we see that $\operatorname{Im}\Omega>0$. The matrix Ω is thus a $g\times g$ complex symmetric matrix with positive imaginary part, known as the period matrix of the Riemann surface. The space of all matrices satisfying these conditions is known as the Siegel upper half plane $\mathscr{H}_g$ (in analogy with the $g=1$ case where Ω is the standard modular parameter of the torus).

Tensor algebra and analysis are particularly simple on a two-dimensional surface. By raising and lowering indices with the metric (3.5) any tensor can be decomposed into higher order differentials $\psi(dw)^n$, $n\in Z$. The transformation properties of ψ in going from a patch with complex coordinate w to another with coordinate u are obtained by requiring

$$\psi'(w)\,(dw)^n=\psi(u)\,(du)^n\;\Rightarrow\;\psi'(w)=\left(\frac{du}{dw}\right)^n\psi(u)\,.$$

In order to introduce spinors we choose local frames

$$ds^2=\delta_{ab}e^a\otimes e^b\,.$$

If $\mathscr{U}_\alpha$, $\mathscr{U}_\beta$ are two overlapping coordinate patches, the frames $e_{(\alpha)}$, $e_{(\beta)}$ are related by a local $SO(2)$ rotation:

$$e^a_{(\alpha)}=R^{ab}(\theta)e^b_{(\beta)}\,,$$

then the spinor bundles have transition functions $\tilde{R}(\theta/2)$, so that $(\tilde{R}(\theta/2))^2=R(\theta)$. If we consider three overlapping patches $\mathscr{U}_\alpha$, $\mathscr{U}_\beta$, $\mathscr{U}_\gamma$ the cocycle condition $R_{\alpha\beta}R_{\beta\gamma}R_{\gamma\alpha}=1$, imposes a nontrivial constraint on $\tilde{R}$, namely,

$$\tilde{R}_{\alpha\beta}\tilde{R}_{\beta\gamma}\tilde{R}_{\gamma\alpha}=w_2(\alpha,\beta,\gamma)$$

is a two-cocyle with Z_2 coefficients (i.e. for any three overlapping patches α, β, γ, $w_2=\pm 1$), w_2 is the second Stieffel-Whitney class. If w_2 is not cohomologous to zero, then the surface does not admit a spin structure. Since w_2 is the reduction modulo two of the Euler character [10], which is even for a compact Riemann surface, we know that w_2 is a coboundary: $w_2=\delta\eta$. Hence spinor structures always exist. The number of inequivalent spin structures is then the number of solutions to the equation $\delta(\eta-\eta')=0$, i.e., the number of Z_2-1-cocycles. These span the space $H^1(\Sigma, Z_2)$. Thus, even though describing a spin structure explicitly may in general be awkward, the *difference* between two spin structures is simply a question of assigning plus or minus signs to the generators of the homology of Σ. This implies in particular that Σ has 2^{2g} inequivalent spin structures.

One of the pleasant features of Riemann surfaces is that we can describe spinors in terms of half-order differentials. This is most easily done by choosing frames:

$$e^z=e^0+ie^1=e^\phi dw \quad \text{for} \quad T^{*(1,0)}\,,$$

$$e^{\bar{z}}=e^0-ie^1=e^\phi d\bar{w} \quad \text{for} \quad T^{*(0,1)}\,,$$

where ϕ is the factor in (3.5). Then across patches $\mathcal{U}_\alpha$, $\mathcal{U}_\beta$ with coordinates w_α, w_β

$$e^{2\phi_\alpha}|dw_\alpha|^2 = e^{2\phi_\beta}|dw_\beta|^2\,,$$

which implies

$$2\phi_\alpha = 2\phi_\beta + \log\left|\frac{dw_\beta}{dw_\alpha}\right|^2,$$

so we have

$$e^z_{(\alpha)} = e^{i\theta} e^z_{(\beta)}\,, \qquad e^{i\theta} = \frac{dw_\alpha}{dw_\beta}\left|\frac{dw_\beta}{dw_\alpha}\right|,$$

and the left and right spinors $\psi \in S^\pm$ transform as

$$\psi_{\pm\alpha} = e^{\pm i\theta/2}\eta_{\alpha\beta}\psi_{\pm\beta}\,.$$

Here again the cocycle η gives the spin structure relative to one particular choice of square roots of $e^{i\theta}$. When we refer the spinors $\psi_\pm$ to the frames $(e^z)^{1/2}$, $(e^{\bar z})^{1/2}$ the transition functions are one-by-one unitary matrices. It is occasionally more convenient to consider the bundles $S^\pm$ as *holomorphic* line bundles. These will have transition functions $\eta_{\alpha\beta}(dw_\alpha/dw_\beta)^{1/2}$ for S^+ and $\eta_{\alpha\beta}(dw_\alpha/dw_\beta)^{-1/2}$ for S^-. In the holomorphic category more appropriate local sections are the holomorphic half-order differentials $(dw_\alpha)^{1/2}$. The relation between the standard and holomorphic descriptions of spinors is given by

$$\psi_+(e^z)^{1/2} = \psi_\theta(dw)^{1/2}\,.$$

The holomorphic line bundle defined by S^+ will be denoted by L. As suggested by the notation, this bundle can be interpreted as a holomorphic square root of the bundle of (1,0)-forms:

$$T^{*(1,0)}\Sigma = K = L_\alpha^2\,, \tag{3.9}$$

here α labels the spin structure. Since there are 2^{2g} spin structures K will admit 2^{2g} inequivalent holomorphic square roots.

Once we have introduced the spinor bundle L_α, we can define tensor powers L_α^n corresponding to differentials $\psi(dw_\alpha)^{n/2}$. In the local coordinates (3.5) the covariant derivative for fields in L_α^n is:

$$\nabla^n_w : L_\alpha^n \to L_\alpha^{n+2}\,,$$

where

$$\nabla^n_w\psi = (g_{w\bar w})^{n/2}\frac{\partial}{\partial w}(g_{w\bar w})^{-n/2}\psi\,. \tag{3.10}$$

We can introduce a scalar product in L_α^n:

$$\langle\phi|\psi\rangle = \int d^2\sigma\sqrt{g}(g^{w\bar w})^{n/2}\phi^*\psi\,. \tag{3.11}$$

The operator ∇^n_w is just the unique holomorphic connection on L_α^n compatible with the metric (3.11). With respect to (3.11) the adjoint of (3.10) is

$$\begin{aligned} \nabla^w_{n+2} &= (\nabla^n_w)^\dagger : L_\alpha^{n+2} \to L_\alpha^n\,, \\ \nabla^w_{n+2} &= -g^{w\bar w}\frac{\partial}{\partial\bar w}\psi\,. \end{aligned} \tag{3.12}$$

With (3.10) and (3.12) we can construct two Laplacians:

$$\Delta_n^{(+)} = \nabla^w_{n+2}\nabla^n_w ,$$
$$\Delta_n^{(-)} = \nabla^{n-2}_w\nabla^w_n .$$

It is easy to show that $\Delta_n^{(+)}$ and $\Delta_{n+2}^{(-)}$ have the same spectrum of non-vanishing eigenvalues. For the zero-modes, the Riemann-Roch theorem states

$$\operatorname{ind}\nabla^n_w \equiv \dim\ker\nabla^n_w - \dim\ker\nabla^w_{n+2} = -(n+1)(g-1) .$$

When $g>1$ any Riemann surface admits a natural metric with constant negative curvature. Since

$$\Delta_n^{(+)} = \Delta_n^{(-)} - \frac{n}{4}R ,$$

where

$$R = -2e^{-2\phi}\partial_w\partial_{\bar w}\phi ,$$

and since $\Delta_n^{(\pm)}$ are both positive definite operators, we conclude that for $R<0$, and $n\geqq 1$, the dimension of $\ker\nabla^n_u$ is zero, and

$$\dim\ker\nabla^w_{n+2} = (n+1)(g-1) .$$

c) Divisors and Line Bundles

We now describe some aspects of the general theory of holomorphic line bundles on Σ. In the next section we will outline their relationship to Riemann's theta function. A holomorphic line bundle can be defined by its transition functions: $\psi_\alpha = g_{\alpha\beta}\psi_\beta$. (The transition functions $g_{\alpha\beta}$ are nowhere vanishing holomorphic functions on the overlaps $\mathcal{U}_\alpha \cap \mathcal{U}_\beta$.) There is an alternative description of L in terms of so-called divisors. Given a meromorphic section, ψ, of L we can consider the set of points where ψ vanishes or blows up: $\{P_i\}$. Since n_i, the order of the zero or pole of ψ at P_i is independent of the trivialization of L, we can define the divisor of ψ as the formal sum $\operatorname{div}(\psi) = \sum n_i P_i$. Another section of the same line bundle L is obtained if we multiply ψ by any meromorphic function on Σ (recall that a meromorphic function f has the same number of zeroes and poles, and that f is determined up to a constant by its divisor). Given this ambiguity in the construction of a section, we define an equivalence relation between two divisors D_1 and D_2 by $D_1 \sim D_2$ if $D_1 - D_2$ is the divisor of a meromorphic function. Thus, to a given line bundle L, we can associate a divisor class (L). Conversely, given a divisor class (D), we can construct a line bundle L, as follows. Choose a representative divisor D of the class (D). On any patch $\mathcal{U}_\alpha$, we can find a meromorphic function f_α whose divisor on $\mathcal{U}_\alpha$ coincides with the restriction of D to $\mathcal{U}_\alpha$. Then, on the overlap $\mathcal{U}_\alpha \cap \mathcal{U}_\beta$, f_α and f_β represent the same divisor, so $g_{\alpha\beta} = f_\alpha / f_\beta$ is a nowhere vanishing holomorphic function, and it trivially satisfies the cocycle condition. Hence D defines a line bundle L. This construction shows that there is a one-to-one correspondence between divisor classes and holomorphic line bundles on a Riemann surface. The degree of a divisor $D = \sum n_i P_i$ is defined as $\deg D = \sum n_i$. The degree only depends on the divisor class, and coincides with the first Chern class of L. For example, the first Chern class of the spinor bundle L_α can be found by noting that $L_\alpha^2=K$, and recalling that the first Chern class of K is minus the Euler class. Hence $c_1(L_\alpha)=g-1$.

The concept of a divisor generalizes to higher dimensions. A divisor on a complex manifold is a formal sum $\sum n_i V_i$, where V_i are complex codimension one analytic subvarieties. (A variety is a generalization of a manifold which allows certain singularities.) The equivalence between divisor classes and line bundles described above continues to hold in the higher dimensional case. Technical details can be found in standard texts [11]. This concludes our introduction to divisors.

Differentiable complex line bundles are completely classified by their first Chern class, but the space of holomorphic line bundles has more structure. We will now briefly outline the classification of holomorphic complex line bundles [12]. The difference $L_1\otimes L_2^{-1}$ of two line bundles on a Riemann surface with the same Chern class is a holomorphic line bundle which admits a flat connection. It can be shown that the group of flat line bundles is a torus, known as the Picard variety of Σ, and denoted by Pic(Σ). We can describe this space intuitively as follows. We cut Σ along a basis a_i, b_i for $H_1(\Sigma, Z)$ as in Fig. 2. The sections of a flat bundle are then characterized by their transition functions around a_i and b_i, which can be taken to be constant phases. Thus we identify the section ψ along a_i with $e^{-2\pi i\varphi_i}$ times ψ along a_i^{-1}, and ψ along b_i with $e^{2\pi i\theta_i}$ times ψ along b_i^{-1}. In other words, flat holomorphic line bundles are completely classified by their twists on the homology, and these twists are parametrized by the torus R^{2g}/Z^{2g}. Therefore a holomorphic line bundle is completely characterized by an integer (its first Chern class) and a point on the Picard variety which determines the twists on the homology. For example, if we denote by $V(\boldsymbol{\theta}, \boldsymbol{\phi})$ the flat line bundle with twists $e^{2\pi i\theta_k}$, $e^{-2\pi i\varphi_k}$, then the line bundles of degree $1-g$ can be parametrized by $L_\alpha^{-1}\otimes V(\boldsymbol{\theta}, \boldsymbol{\phi})$, where L_α^{-1} is some fixed spin structure. From (3.9) we see that twice the difference of spin bundles is trivial. Thus, in this parametrization of degree $1-g$ line bundles the points corresponding to spin structures are the half-points of the torus, that is, the points where $(\boldsymbol{\theta}, \boldsymbol{\phi})\in(\frac{1}{2}Z/Z)^{2g}$.

In Sect. 5 we will compute the spin-structure dependence of the determinant of the Dirac operator. One important ingredient will be a trick which we now discuss. Recall that the chiral Dirac operator is the holomorphic connection on L_α^{-1} compatible with the metric (3.11). Therefore, we consider the holomorphic connections on $L_\alpha^{-1}\otimes V(\boldsymbol{\theta}, \boldsymbol{\phi})$. Again considering Σ cut along its homology basis, we see that eigenfunctions of $\not{D}$ in $L_\alpha^{-1}\otimes V(\boldsymbol{\theta}, \boldsymbol{\phi})$ are related by the unitary transformation

$$\mathcal{U}(P)=\exp\left(\int_{P_0}^{P} A\right)$$

to eigenfunctions of the coupled Dirac operator $\not{D}_A$ in L_α^{-1}, where the gauge field A can be constructed in terms of the harmonic one-forms on Σ:

$$A=2\pi i\sum_1^g \theta^i\alpha_i-2\pi i\sum_1^g \phi^i\beta_i\,. \tag{3.13}$$

Locally, $A(\boldsymbol{\theta},\boldsymbol{\phi})$ is pure gauge, but it cannot be gauged away because it has nontrivial holonomy. The advantage of introducing the gauge field (3.13) is that if we compute $\det \not{D}_A$ as a function of A for a given spin structure, then evaluating the answer for the points $(\boldsymbol{\theta},\boldsymbol{\phi})$ in $(\frac{1}{2}Z/Z)^{2g}$ we have effectively computed the spin-structure dependence of $\det \not{D}$. Thus there are two equivalent procedures for computing the determinant of the Dirac operator on all line bundles of degree $1-g$. The first is to keep the operator ∇_w^{-1} "fixed" and vary the line bundle $L_\alpha^{-1}\otimes V(\boldsymbol{\theta},\boldsymbol{\phi})$, and the second is to keep the line bundle L_α^{-1} fixed and vary the operator by including the flat $U(1)$ gauge field A. We will usually adopt the second point of view.

We can write A in a more useful way in terms of the abelian differentials. It is easy to show that the α_i, β_i and the ω_i, $\bar{\omega}_i$ are related by

$$\begin{aligned}\alpha_i &= A_{ij}\omega_j + \text{cplx. conj.},\\ \beta_i &= B_{ij}\omega_j + \text{cplx. conj.},\end{aligned} \tag{3.14}$$

where

$$A = -\bar{\Omega}(\Omega-\bar{\Omega})^{-1}, \qquad B = (\Omega-\bar{\Omega})^{-1},$$

so that

$$A = 2\pi i(\boldsymbol{\phi}+\Omega\boldsymbol{\theta})\cdot(\Omega-\bar{\Omega})^{-1}\cdot\bar{\omega} + \text{h.c.}. \tag{3.15}$$

Thus the chiral Dirac operator $\nabla_w^{-1}\colon L_\alpha^{-1}\to L_\alpha$ when coupled to A becomes

$$D(\mathbf{u}) = \nabla_w^{-1} - 2\pi i(\boldsymbol{\phi}+\bar{\Omega}\boldsymbol{\theta})\cdot(\Omega-\bar{\Omega})^{-1}\cdot\omega\,, \tag{3.16}$$

where ∇_w^{-1} is the operator introduced in (3.10). Thus we have a family of Dirac operators parametrized by $\mathbf{u}=\boldsymbol{\phi}+\bar{\Omega}\boldsymbol{\theta}$. Since $\boldsymbol{\phi}\to\boldsymbol{\phi}+\mathbf{m}$, $\boldsymbol{\theta}\to\boldsymbol{\theta}+\mathbf{n}$ for $\mathbf{n},\mathbf{m}$ integral vectors defines the same point on $\mathrm{Pic}(\Sigma)$, the family is parametrized by another complex torus known as the Jacobian variety of Σ, denoted by $J(\Sigma)$, and defined by $J(\Sigma)=C^g/L_\Omega$, where L_Ω is the lattice generated by $Z^g+\Omega Z^g$. Since Ω (the period matrix) is a non-singular $g\times g$ complex matrix, there is a $1-1$ correspondence between $\mathrm{Pic}(\Sigma)$ and $J(\Sigma)$.

d) Theta Functions

In this section we define the theta functions and explain their relation to spin structures. We follow the notation and treatment of Mumford [13]. Another excellent reference is the book of Fay [14]. In analogy with the genus one case, once we have introduced the Jacobian $J(\Sigma)$, we can introduce the Riemann theta function by the series expansion:

$$\vartheta(\mathbf{z}|\Omega) = \sum_{\mathbf{n}\in Z^g} \exp(i\pi\mathbf{n}\cdot\Omega\cdot\mathbf{n} + 2\pi i\mathbf{n}\cdot\mathbf{z})\,.$$

We also define theta functions with characteristics $\mathbf{a}$, $\mathbf{b}$ for $\mathbf{a},\mathbf{b}\in R^g$ by the sum

$$\begin{aligned}\vartheta\begin{bmatrix}\mathbf{a}\\ \mathbf{b}\end{bmatrix}(\mathbf{z}|\Omega) &= \sum_{\mathbf{n}\in Z^g}\exp(i\pi(\mathbf{n}+\mathbf{a})\cdot\Omega\cdot(\mathbf{n}+\mathbf{a}) + 2\pi i(\mathbf{n}+\mathbf{a})\cdot(\mathbf{z}+\mathbf{b}))\\ &= e^{i\pi a\cdot\Omega\cdot a + 2\pi i a\cdot(z+b)}\,\vartheta(\mathbf{z}+\Omega\mathbf{a}+\mathbf{b}|\Omega)\end{aligned} \tag{3.17}$$

if we shift $\mathbf{z}$ by the lattice L_Ω, ϑ transforms by

$$\vartheta\begin{bmatrix}\mathbf{a}\\ \mathbf{b}\end{bmatrix}(\mathbf{z}+\Omega\mathbf{n}+\mathbf{m}|\Omega)=e^{-i\pi n\cdot\Omega\cdot n-2\pi i n\cdot(z+b)}e^{2\pi i a\cdot m}\vartheta\begin{bmatrix}\mathbf{a}\\ \mathbf{b}\end{bmatrix}(\mathbf{z}|\Omega)\,. \tag{3.18}$$

Another useful transformation law is:

$$\vartheta\begin{bmatrix}\mathbf{a}+\mathbf{n}\\ \mathbf{b}+\mathbf{m}\end{bmatrix}(\mathbf{z}|\Omega)=e^{2\pi i a\cdot m}\vartheta\begin{bmatrix}\mathbf{a}\\ \mathbf{b}\end{bmatrix}(\mathbf{z}|\Omega)\,. \tag{3.19}$$

Thus the theta function is not strictly a function on $J(\Sigma)$; rather, it is a section of a holomorphic line bundle on $J(\Sigma)$ called the ϑ – line bundle, $\mathscr{L}$. From the transition functions (3.18) one can compute the first Chern class of $\mathscr{L}$. Alternatively we can define the hermitian norm on sections of $\mathscr{L}$ by

$$\|s\|^2=e^{i\pi(u-\bar{u})\cdot(\Omega-\bar{\Omega})^{-1}\cdot(u-\bar{u})}|s|^2 \tag{3.20}$$

and compute the curvature

$$c_1(\mathscr{L})=\frac{i}{2\pi}\partial\bar{\partial}\log\|s\|^2=\sum_1^g d\phi^i\wedge d\theta^i\,. \tag{3.21}$$

We can now apply the Kodaira vanishing theorem [11] and the index theorem for the $\bar{\partial}$-complex coupled to $\mathscr{L}$ to show that $\mathscr{L}$ admits only one holomorphic section which is represented by the theta function.[1]

The relation between spinor bundles and theta functions is based on Riemann's vanishing theorem. To state this fundamental theorem we first introduce the Jacobian map of Σ into $J(\Sigma)$. Given a canonical homology basis, and a point P_0 on Σ, we can associate to any point P on Σ a point in $J(\Sigma)$ by

$$\Phi_i(P)=\int_{P_0}^{P}\omega_i\,, \tag{3.22}$$

where the ω_i are the holomorphic differentials. As a map from Σ into C^g, Φ is multivalued, but, considered as a map into $J(\Sigma)$ it is single-valued since if we move P around one of the cycles a_i or b_i, then (3.22) changes by an element of the Jacobian lattice. Similarly, the function on Σ defined by

$$f(P)=\vartheta\left(\mathbf{z}+\int_{P_0}^{P}\boldsymbol{\omega}|\Omega\right) \tag{3.23}$$

for $\mathbf{z}$ in C^g is multivalued. However, f is a well-defined function on the interior of the region in Fig. 2. By repeatedly applying Green's theorem to the one-form df/f on this region one can prove the

Riemann Vanishing Theorem. *The function $f(P)$ either vanishes identically for all $P\in\Sigma$ or $f(P)$ has exactly g zeroes $P_1,\dots,P_g$ on Σ. Furthermore, in the latter case, there exists a vector $\boldsymbol{\Delta}$ depending only on P_0 and the canonical homology basis so that the points $\{P_i\}$ satisfy*

$$\mathbf{z}+\sum_{i=1}^{g}\int_{P_0}^{P_i}\boldsymbol{\omega}=\boldsymbol{\Delta}\,. \tag{3.24}$$

[1] We thank Dan Freed for explaining this to us

Conversely, for all $P_1, \ldots, P_g \in \Sigma$, *if we define* $\mathbf{z}$ *according to* (3.24) *then* $f(P_i)=0$ *for all* $i=1, \ldots, g$.

Δ is known as the vector of Riemann constants. Since we will not need the explicit form of Δ we refer to [13, 14] for details, and the proof of this result.

The set of points $\mathbf{z} \in J(\Sigma)$, where $\vartheta(\mathbf{z}|\Omega)=0$, known as the Θ-divisor, is a variety of complex codimension one. Θ will be important to us in Sect. 5. The Riemann vanishing theorem leads to the following description of Θ:

Corollary. $\vartheta(\mathbf{e}|\Omega)=0$ iff *there exist* $g-1$ *points* $P_1, \ldots, P_{g-1} \in \Sigma$ *so that*

$$\mathbf{e}=\Delta-\sum_1^{g-1} \int_{P_0}^{P_i} \omega . \tag{3.25}$$

One direction of the proof of the corollary is simple. If $\mathbf{e}$ is given by (3.25) then we choose any point $P_g \in \Sigma$ and set

$$\mathbf{z}=\Delta-\sum_1^{g} \int_{P_0}^{P_i} \omega .$$

Then the Riemann vanishing theorem implies:

$$0=f(P_g)=\vartheta\left(\mathbf{z}+\int_{P_0}^{P_g} \omega|\Omega\right)=\vartheta(\mathbf{e}|\Omega) .$$

With a little more work one can show the converse [13].

The sums which arise in the above theorems are clearly related to divisors. In particular, the divisor $\sum_1^{g-1} P_i$ defines a (possibly twisted) spin bundle. Thus the connection with spinors begins to emerge. The relation can be made more precise by generalizing the Jacobian map (3.22) to a map I from divisors of degree zero to $J(\Sigma)$ as follows. If D is a degree zero divisor, we choose a 1-cycle σ on Σ whose boundary is D, i.e. $\partial\sigma=D$. Then the Jacobian map is just

$$I(D)=\int_\sigma \omega \in J(\Sigma) . \tag{3.26}$$

Note that the choice of cycle does not matter in (3.26). Two important properties of I are,

i) **Abel's Theorem.** $I(D)=0$ *if and only if* D *is the divisor of a meromorphic function. Thus* I *is defined on divisor classes.*

ii) **Jacobi Inversion Theorem.** *The set of points* $I[\sum_1^g P_i-gP_0]$ *for* $P_1, \ldots, P_g \in \Sigma$ *is all of* $J(\Sigma)$.

We are finally in a position to relate spin structures to Θ. We begin with the remark that, by the index theorem, a degree $g-1$ bundle E (that is, a twisted spin bundle) will have a holomorphic section if and only if the bundle $K^{-1}\otimes E$ has a zero-mode of ∇_w. We can take the complex conjugate of the zero-mode equation [considering $K^{-1}\otimes E$ as a $U(1)$ bundle] and translate back into holomorphic language to see that E has a holomorphic section if and only if $K\otimes E^{-1}$ has a holomorphic section. That is, E has a divisor of the form $P_1+\ldots P_{g-1}$ if and only if $K\otimes E^{-1}$ has a divisor of the form $Q_1+\ldots Q_{g-1}$. Since a spin structure L_α satisfies

$L_\alpha^2 = K$, one can show that the set of points

$$S_\alpha = \{I[P_1 + \dots P_{g-1} - D_\alpha] | P_1, \dots, P_{g-1} \in \Sigma\}, \tag{3.27}$$

where D_α is a divisor of L_α, is a symmetric subset of $J(\Sigma)$. However, from the Riemann vanishing theorem we see that the set S_α is just a translate of Θ by $\Delta - I[D_\alpha - (g-1)P_0]$. Thus, to any spin structure we may associate a symmetric translate of Θ. That is, if $\mathbf{e}$ is in S_α, then so is $-\mathbf{e}$.

On the other hand, since ϑ is an even function of $\mathbf{z}$, Θ is itself symmetric. Therefore, a translate $\Theta + \mathbf{e}$ of Θ is symmetric if $\Theta + 2\mathbf{e} = \Theta$. An application of Liouville's theorem to the function

$$\frac{\vartheta(\mathbf{z} + 2\mathbf{e})}{\vartheta(\mathbf{z})}$$

on C^g shows that $2\mathbf{e} \in L_\Omega$. Thus the symmetric translates of Θ are the translates by points of order two.

Finally, recall that the difference of two spin structures $D_1 - D_2$ is a point of order two in the Picard variety, hence $I[D_1 - D_2]$ is of order two in $J(\Sigma)$. Thus each of the 2^{2g} spin structures corresponds to one of the 2^{2g} half-points in $J(\Sigma)$. In particular, some spin structure must correspond to the symmetric translate of Θ given by Θ itself. We call this spin structure D_0, i.e.

$$\Delta = I[D_0 - (g-1)P_0]. \tag{3.28}$$

The vector of Riemann constants depends on P_0, but D_0 does not. Both Δ and D_0 depend on the choice of canonical homology basis.[2]

The correspondences we have just outlined allow a division of spin structures into two classes known as even and odd spin structures. Note that if $(\varepsilon_1, \varepsilon_2) \in (\frac{1}{2}Z/Z)^{2g}$ is a half-point then the divisor of $\vartheta\begin{bmatrix}\varepsilon_1\\ \varepsilon_2\end{bmatrix}(\mathbf{z}|\Omega)$ is a symmetric translate of Θ. Also, it is straightforward to show that

$$\vartheta\begin{bmatrix}\varepsilon_1\\ \varepsilon_2\end{bmatrix}(-\mathbf{z}|\Omega) = (-1)^{4\varepsilon_1 \cdot \varepsilon_2}\vartheta\begin{bmatrix}\varepsilon_1\\ \varepsilon_2\end{bmatrix}(\mathbf{z}|\Omega).$$

Therefore, we call a spin structure with characteristics $[\varepsilon_1\, \varepsilon_2]$ even or odd depending on whether $4\varepsilon_1 \cdot \varepsilon_2$ is even or odd. As we will see in Sect. 5, the parity of a spin structure is related to the existence of zero-modes of the Dirac operator. A simple induction argument shows that there are $2^{g-1}(2^g - 1)$ odd spin structures and $2^{g-1}(2^g + 1)$ even ones. For example, in the one loop case, there is a single odd spin structure corresponding to $\vartheta\begin{bmatrix}\frac{1}{2}\\ \frac{1}{2}\end{bmatrix}(0|\Omega)$, and three even spin structures with associated theta functions $\vartheta\begin{bmatrix}0\\ 0\end{bmatrix}(0|\Omega)$, $\vartheta\begin{bmatrix}\frac{1}{2}\\ 0\end{bmatrix}(0|\Omega)$, $\vartheta\begin{bmatrix}0\\ \frac{1}{2}\end{bmatrix}(0|\Omega)$. In the two-loop

[2] Under a change of homology basis Δ shifts by an even half point, defined below

case there are six odd spin structures associated to the theta functions

$$\begin{gathered}\vartheta\begin{bmatrix}\frac{1}{2} & 0\\ \frac{1}{2} & 0\end{bmatrix}(0|\Omega)\ \vartheta\begin{bmatrix}0 & \frac{1}{2}\\ 0 & \frac{1}{2}\end{bmatrix}(0|\Omega)\ \vartheta\begin{bmatrix}\frac{1}{2} & 0\\ \frac{1}{2} & \frac{1}{2}\end{bmatrix}(0|\Omega),\\ \vartheta\begin{bmatrix}\frac{1}{2} & \frac{1}{2}\\ \frac{1}{2} & 0\end{bmatrix}(0|\Omega)\ \vartheta\begin{bmatrix}\frac{1}{2} & \frac{1}{2}\\ 0 & \frac{1}{2}\end{bmatrix}(0|\Omega)\ \vartheta\begin{bmatrix}0 & \frac{1}{2}\\ \frac{1}{2} & \frac{1}{2}\end{bmatrix}(0|\Omega),\end{gathered} \tag{3.29}$$

and ten even ones with theta functions

$$\begin{gathered}\vartheta\begin{bmatrix}\frac{1}{2} & 0\\ 0 & 0\end{bmatrix}(0|\Omega)\ \vartheta\begin{bmatrix}0 & 0\\ \frac{1}{2} & 0\end{bmatrix}(0|\Omega)\ \vartheta\begin{bmatrix}0 & \frac{1}{2}\\ 0 & 0\end{bmatrix}(0|\Omega)\\ \vartheta\begin{bmatrix}0 & 0\\ 0 & \frac{1}{2}\end{bmatrix}(0|\Omega)\ \vartheta\begin{bmatrix}\frac{1}{2} & 0\\ 0 & \frac{1}{2}\end{bmatrix}(0|\Omega)\ \vartheta\begin{bmatrix}0 & \frac{1}{2}\\ \frac{1}{2} & 0\end{bmatrix}(0|\Omega)\\ \vartheta\begin{bmatrix}\frac{1}{2} & \frac{1}{2}\\ 0 & 0\end{bmatrix}(0|\Omega)\ \vartheta\begin{bmatrix}0 & 0\\ \frac{1}{2} & \frac{1}{2}\end{bmatrix}(0|\Omega)\ \vartheta\begin{bmatrix}0 & 0\\ 0 & 0\end{bmatrix}(0|\Omega)\ \vartheta\begin{bmatrix}\frac{1}{2} & \frac{1}{2}\\ \frac{1}{2} & \frac{1}{2}\end{bmatrix}(0|\Omega).\end{gathered} \tag{3.30}$$

Finally, we will see in Sect. 7 that modular transformations permute all the even and odd spin structures separately among themselves. This rather mathematical section will prove to be quite useful in our discussion of the Dirac determinant and of modular invariance.

4. Holomorphic Factorization

Quillen [15] has recently pointed out that functional determinants of operators on Riemann surfaces have interesting holomorphic properties. We will use these properties to investigate the Dirac determinant. However, holomorphic factorization has other important applications. For example, as pointed out in [16, 17, 18] the very existence of chiral string theories relies on the existence of such holomorphic square roots.

In this section we will review the statement of Quillen's theorem and its relation to the geometry of determinant line bundles. We first give an heuristic Feynman-diagram proof of the theorem, and then proceed to the heat kernel proof of a slight generalization of the theorem. This generalization can also be understood in the context of determinant line bundles. In a beautiful recent paper [18] Belavin and Knizhnik have used similar methods to investigate the holomorphic properties of functional determinants on Teichmüller space. For completeness we extend the heuristic discussion of Quillen's theorem to rederive their results at the end of this section.

Abstractly, we will consider families of operators $D_y: L^n \to L^{n+2}$. If the parameter space Y is a complex manifold, the D_y can vary holomorphically with y. Quillen showed that in this case the determinant line bundle $\mathscr{L} \to Y$ can be given a holomorphic structure, and, if D_y has no index, then the ζ-regulated determinant satisfies

$$\det{}_\zeta D^\dagger D = e^{-q} |\det D_y|^2 . \tag{4.1}$$

Here $\det D_y$ is a holomorphic function on Y. The "counterterm" q is defined by choosing an operator D_0 to define the origin on Y and setting

$$q = \frac{i}{2\pi} \int_\Sigma \operatorname{tr}(D - D_0)^\dagger \wedge (D - D_0) . \tag{4.2}$$

The notation means that the family of operators is chosen so that any two differ by a one form. The trace is a finite dimensional trace over the fiber indices. Note that $\det D$ also depends on the choice of D_0.

More concretely, we can give a simple proof of Quillen's theorem using Feynman diagrams. Locally, D_y looks like $\bar\partial + A$, so we can interpret A as a gauge field. Since the obstruction to holomorphic factorization is local in A we should be able to rederive (4.2) from perturbation theory. In perturbation theory

$$W = \log\det{}_\zeta D^\dagger D \tag{4.3}$$

is simply given by a sum of Feynman diagrams

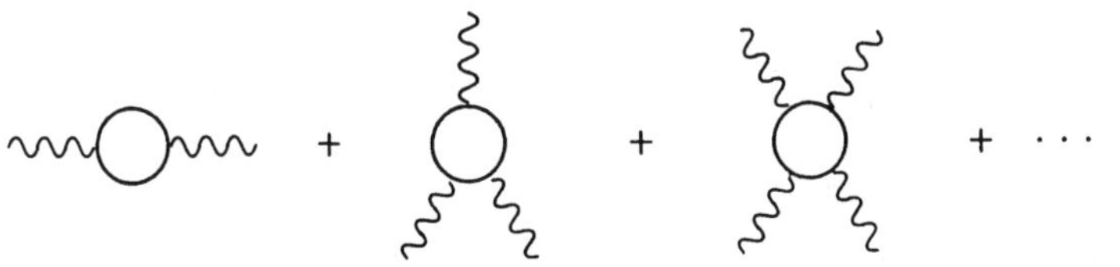

All diagrams but the first are convergent without regularization. Since the fields are massless their sum is a functional of the form $f[A_z] + \overline{f[A_z]}$. Thus their contribution to the determinant is gauge invariant and factorizes holomorphically as a functional of A_z. The situation is rather different for the vacuum polarization graph, which must be regulated. One can use Pauli-Villars regularization to maintain gauge invariance. Then as the PV mass goes to infinity one finds the nonholomorphic residue

$$q = \frac{1}{2\pi}\int_\Sigma d^2z A_z A_{\bar z}\,. \tag{4.4}$$

It is more illuminating, however, to follow the two-dimensional calculation in [19]. From the axial anomaly we know that to order A^2 the contribution of a Weyl fermion to the vacuum functional is (in momentum space)

$$\frac{1}{4\pi}\int\frac{d^2p}{(2\pi)^2}\frac{p_z}{p_{\bar z}}A_{\bar z}(p)A_{\bar z}(-p)\,. \tag{4.5}$$

If we are computing $\det D^\dagger D$ we must add the other chirality. The result

$$\frac{1}{4\pi}\int\frac{d^2p}{(2\pi)^2}\frac{p_z}{p_{\bar z}}A_{\bar z}(p)A_{\bar z}(-p) + \frac{1}{4\pi}\int\frac{d^2p}{(2\pi)^2}\frac{p_{\bar z}}{p_z}A_z(p)A_z(-p) \tag{4.6}$$

continues to satisfy holomorphic factorization, but is not gauge invariant. We can restore gauge invariance – at the cost of factorization – by adding the counterterm

$$\frac{1}{2\pi}\int\frac{d^2p}{(2\pi)^2}A_z(p)A_{\bar z}(-p)\,, \tag{4.7}$$

which is Quillen's counterterm [20].

In applications to string theory we will need to generalize (4.1)–(4.2) to the case when D has an index, in order to handle the ghost operators. Therefore, we now give a heat kernel proof of the Quillen theorem in the case that D_y has a nonzero index. We will simplify our calculation by assuming that $\ker D_y = 0$ so that the cokernel has constant dimension. This is true for the ghost operators on higher loop Riemann surfaces. In any case the following analysis is easily generalized to the case when D has a kernel [21].

We begin with the heat kernel definition of the determinant

$$\log\det D^\dagger D = -\int_\varepsilon^\infty \frac{dt}{t}\operatorname{Tr} e^{-tD^\dagger D}, \tag{4.8}$$

and, following Quillen, we take the second variation:

$$\delta\bar\delta \det D^\dagger D = \delta\int_\varepsilon^\infty \operatorname{Tr}\bar\delta D^\dagger D e^{-tD^\dagger D}. \tag{4.9}$$

Since D has no kernel the right-hand side becomes simply

$$= \delta \operatorname{Tr}\left(\bar\delta D^\dagger D \frac{1}{D^\dagger D} e^{-\varepsilon D^\dagger D}\right) \tag{4.10}$$

which is

$$\begin{aligned} &= \operatorname{Tr}\bar\delta D^\dagger\left(1 - D\frac{1}{D^\dagger D}D^\dagger\right)\delta D\frac{1}{D^\dagger D}e^{-\varepsilon D^\dagger D} \\ &\quad - \varepsilon\int_0^1 ds \operatorname{Tr}\bar\delta D^\dagger D\frac{1}{D^\dagger D}e^{-s\varepsilon D^\dagger D}D^\dagger\delta D e^{-(1-s)\varepsilon D^\dagger D}. \end{aligned} \tag{4.11}$$

Note that

$$P = 1 - D\frac{1}{D^\dagger D}D^\dagger \tag{4.12}$$

is the projector onto the kernel of $D^\dagger$. Thus the first trace in (4.11) is finite dimensional and we can take the limit $\varepsilon\to 0$. We can get a better understanding of this trace if we take a basis $\chi_i(y)$ for $\ker D_y^\dagger$ which varies antiholomorphically with y. (Such a basis exists since $D_y^\dagger$ varies antiholomorphically.) Then, a simple application of perturbation theory shows that

$$\delta\bar\delta \log\det\langle\chi_i|\chi_j\rangle = \operatorname{Tr} P\delta D\frac{1}{D^\dagger D}\bar\delta D^\dagger. \tag{4.13}$$

We can simplify the second trace by using

$$e^{-s\varepsilon D^\dagger D}D^\dagger = D^\dagger e^{-s\varepsilon D D^\dagger} \tag{4.14}$$

and applying (4.12). Since a finite dimensional trace is killed by the $\varepsilon\to 0$ limit, (4.11) becomes

$$\delta_y\bar\delta_y \log\frac{\det D^\dagger D}{\det\langle\chi_i|\chi_j\rangle} = -\varepsilon\int_0^1 ds \operatorname{Tr}\bar\delta D^\dagger e^{-s\varepsilon D D^\dagger}\delta D e^{-(1-s)\varepsilon D^\dagger D}. \tag{4.15}$$

If D_y has zero-modes we choose a basis ϕ_i varying holomorphically with y and multiply $\det\langle\chi_i|\chi_j\rangle$ by $\det\langle\phi_i|\phi_j\rangle$ [21]. In the case considered by Quillen $\delta D = \delta A$ does not involve derivatives and Eq. (4.15) reduces to the simpler statement

$$\delta\bar\delta \det D_y^\dagger D_y = -\varepsilon \operatorname{Tr}\bar\delta D^\dagger \delta D e^{-\varepsilon D^\dagger D}. \tag{4.16}$$

Then, using the heat kernel expansion one recovers (4.1), (4.2).

There is a third useful way to understand Quillen's theorem in terms of the geometry of determinant line bundles. If D_y has no index then one can define a holomorphic section σ of $\mathscr{L}$. The Hermitian norm

$$\|\sigma\|^2 = \det_\zeta D^\dagger D \tag{4.17}$$

has a unique holomorphic compatible connection, whose curvature is just

$$\frac{i}{2\pi}\partial\bar{\partial}\log\|\sigma\|^2 . \tag{4.18}$$

On the other hand, Bismut and Freed [22] have computed this curvature in terms of the family index density. Using their formula one easily recovers (4.1) and (4.2).

The holomorphic factorization of determinants of operators with index can also be understood from the third point of view. We consider again a family with no kernel so that the index bundle

$$\mathscr{L} = (\Lambda^{\max}\ker D_y)^* \otimes (\Lambda^{\max}\operatorname{cok} D_y) \tag{4.19}$$

is well-defined as it stands. We can define a section of $\mathscr{L}$ which is holomorphic in the holomorphic structure defined by Quillen. Choose a point y in some neighborhood of y_0 and take

$$\sigma(y) = \frac{\det\langle\chi_i(y)|\chi_j(y_0)\rangle}{\det\langle\chi_i(y)|\chi_j(y)\rangle}\,\chi_1(y)\wedge\ldots\wedge\chi_k(y) . \tag{4.20}$$

We define the neighborhood on which (4.20) holds to be the set of points y such that

$$\det\langle\chi_i(y)|\chi_j(y_0)\rangle \neq 0 .$$

We can extend σ to the rest of Y by taking $\sigma = 0$ outside the neighborhood. This is a holomorphic section of $\mathscr{L}$ [22]. The Quillen norm of this section is just

$$\|\sigma\|^2 = |\det\langle\chi_i(y)|\chi_j(y_0)\rangle|^2\left(\frac{\det D^\dagger D}{\det\langle\chi_i(y)|\chi_j(y)\rangle}\right). \tag{4.21}$$

On the other hand, the curvature of (4.7) has been computed by Bismut and Freed in terms of the family index density. Thus if we choose the basis $\chi_i(y)$ to vary antiholomorphically, then the obstruction to the holomorphic factorization of

$$\frac{\det D_y^\dagger D_y}{\det\langle\chi_i(y)|\chi_j(y)\rangle} \tag{4.22}$$

is again just given by the family index density.

In the following section we will apply Quillen's theorem to families of operators over the Picard variety. For completeness we describe here a recent result of Belavin and Knizhnik who considered families of operators over moduli space. In [18] a heat kernel or index bundle approach was used. We show here how the result can be understood as the gravitational analog of the Feynman diagram argument we used before. We begin by defining

$$W[\hat{g}] \equiv \log\left(\frac{\det' D^\dagger D}{\det\langle\chi_i|\chi_j\rangle\det\langle\phi_i|\phi_j\rangle}\right), \tag{4.23}$$

where D is now the operator ∇_w^n of (3.10), and e^W is the partition function for a pair of spin $\pm\frac{n}{2}$ particles on the worldsheet. Following [18], we consider metric deformations $g+\delta g$ which look like

$$e^{2\phi}|dz+\eta d\bar{z}|^2\,, \tag{4.24}$$

where η is a (small) Beltrami differential. From the family index theorem or from the general heat kernel formula (4.15), we know that the obstruction to holomorphic factorization (in η and $\bar{\eta}$, i.e. on Teichmüller space) is local so it suffices to consider Beltrami differentials with support in a coordinate patch, $\mathscr{U}$. These will not correspond to Teichmüller deformations, but the final result will apply to such deformations. Using the conformal anomaly we can write

$$\begin{aligned} W[g+\delta g] &= W[\hat{g}] + \frac{3(n+1)^2-1}{24\pi} S_L[\phi,\hat{g}] \\ &= W[\hat{g}] + \frac{3(n+1)^2-1}{24\pi} \int_\Sigma d^2\sigma \sqrt{\hat{g}}(\hat{g}^{ab}\partial_a\phi\partial_b\phi - \hat{R}\phi)\,, \end{aligned} \tag{4.25}$$

where $\hat{g}$ is just $|dz+\eta d\bar{z}|^2$ in $\mathscr{U}$, and (4.25) defines the Liouville action S_L. Thus $\hat{g}$ is a small deviation from a flat metric in $\mathscr{U}$, and we can use perturbation theory to find the part in $W[\hat{g}]$ of second order in η. We now use Orlando Alvarez's trick [16] to write the appropriate perturbation theory by considering the operator $D: L^n \to L^{n+2}$ as a chiral Dirac operator coupled to the vector bundle L^{n+1}. Thus the effective action W is, to second order in η, just given by the gravitational contribution computed in [19] and the gauge contribution of (4.6) with $A=\frac{n+1}{2}\omega$, where ω is the Riemannian spin connection. That is

$$\begin{aligned} W^{(2)}[\hat{g}] = &-\frac{1}{192\pi}\int\frac{d^2p}{(2\pi)^2}\frac{p_z^3}{p_{\bar{z}}}\hat{g}_{\bar{z}\bar{z}}(p)\hat{g}_{\bar{z}\bar{z}}(-p) - \frac{1}{192\pi}\int\frac{d^2p}{(2\pi)^2}\frac{p_{\bar{z}}^3}{p_z}\hat{g}_{zz}(p)\hat{g}_{zz}(-p) \\ &+\frac{1}{4\pi}\int\frac{d^2p}{(2\pi)^2}\frac{p_z}{p_{\bar{z}}}A_{\bar{z}}(p)A_{\bar{z}}(-p) + \frac{1}{4\pi}\int\frac{d^2p}{(2\pi)^2}\frac{p_{\bar{z}}}{p_z}A_z(p)A_z(-p)\,. \end{aligned} \tag{4.26}$$

Substituting for the gauge field and adding the counterterms to restore coordinate invariance (as described in [19]) we obtain the famous result

$$W^{(2)}[\hat{g}] = \frac{3(n+1)^2-1}{192\pi}\int\frac{d^2p}{(2\pi)^2}\frac{\hat{R}(p)\hat{R}(-p)}{p_z p_{\bar{z}}}\,. \tag{4.27}$$

The coefficient in front of the integral is just that of the usual conformal anomaly in our conventions. Finally, one computes the curvature for the metric $|dz+\eta d\bar{z}|^2$

$$\hat{R} = -\partial_z^2\bar{\eta} - \partial_{\bar{z}}^2\eta + 2\partial_z\partial_{\bar{z}}(\eta\bar{\eta}) + \tfrac{1}{2}[\partial_z\eta\partial_{\bar{z}}\bar{\eta} - \partial_z\bar{\eta}\partial_{\bar{z}}\eta]\,. \tag{4.28}$$

Combining (4.25), (4.27), and (4.28) we find

$$\begin{aligned} \delta_\eta\delta_{\bar{\eta}}W = \frac{3(n+1)^2-1}{96\pi}\int d^2\sigma[&\partial_z\eta\partial_{\bar{z}}\bar{\eta} + 2(\partial_z\eta\bar{\eta}\partial_z\phi + \partial_z\bar{\eta}\eta\partial_{\bar{z}}\phi) \\ &+4\eta\bar{\eta}\partial_{\bar{z}}\phi\partial_z\phi - 6\eta\bar{\eta}\partial_z\partial_{\bar{z}}\phi]\,, \end{aligned} \tag{4.29}$$

and in terms of the original metric deformations δg_{ww} this is simply

$$\frac{3(n+1)^2-1}{96\pi}\left[\int d^2\sigma\sqrt{g}(g^{z\bar{z}})^3\nabla_z^{-2}\delta g_{\bar{z}\bar{z}}\nabla_2^z\delta g_{zz}-\tfrac{3}{2}\int d^2\sigma\sqrt{g}(g^{z\bar{z}})^2\delta g_{zz}\delta g_{\bar{z}\bar{z}}R\right], \tag{4.30}$$

reproducing the result in [18].

The Belavin-Knizhnik obstruction can be rewritten in terms of the Weil-Petersson Kähler potential on Teichmüller space [23]. Recall that Teichmüller space T can be represented as the space of constant curvature metrics: namely, for any metric g there is Weyl transformation so that $\hat{g}=e^{-2\phi}g$ is a constant curvature metric.[3] It therefore suffices to consider the holomorphic obstruction on T. In the standard Bers embedding of T as a bounded domain in C^{3g-3} the holomorphic cotangent space of T is identified with the integrable holomorphic quadratic differentials [24]. For these deformations the first term in (4.30) is zero so

$$\delta\bar{\delta}W=-\frac{3(n+1)^2-1}{64\pi}\langle\delta g,\delta g\rangle_{\mathrm{WP}} \tag{4.31}$$

where $\langle\ ,\ \rangle_{\mathrm{WP}}$ is the Weil-Petersson metric. Since the Weil-Petersson metric is Kähler [24] and the Bers embedding provides globally defined analytic coordinates on T, there must be a globally defined Kähler potential U_{WP}. It follows that the complete obstruction to holomorphic factorization of (4.22) is given by

$$\exp\frac{3(n+1)^2-1}{24\pi}\left(S_L[\phi,\hat{g}]-\tfrac{3}{8}U_{\mathrm{WP}}\right). \tag{4.32}$$

Note that S_L and U_{WP} themselves make no reference to the spin n. Thus the product of a collection of determinants of spins n_i will have a total obstruction proportional to $\sum_i[3(n_i+1)^2-1]$. Therefore, as Belavin and Knizhnik observed, when the conformal anomaly cancels the string integrand is the square of a holomorphic function on Teichmüller space (even though the individual determinants are not). We believe this result is very powerful and should allow one to apply arguments similar to those used in the next section to obtain useful expressions for the functional determinants appearing in string theory in terms of "canonical" or "natural" functions associated to a Riemann surface.

5. Chiral Dirac Determinants

In this section we will apply the Quillen theorem to a family of Dirac operators over the Picard variety, which is essentially the original context considered by Quillen. Since the Dirac operator has no index the theorem will be especially easy to apply. We will begin by considering the family of operators in (3.16) with $\mathbf{u}\in C^g$ (or, if we wish to consider a twisted family of operators we may take $\mathbf{u}\in J(\Sigma)$). A simple argument using the Quillen theorem and gauge invariance shows that $\det D(\mathbf{u})$ is proportional to a theta function, but the argument is not strong enough

[3] In general g will also have to be pulled back by a diffeomorphism, but this complication is irrelevant for our purposes. Note too that for the torus we can take $R=0$. That is why explicit calculations on the torus do not encounter this obstruction

to determine the characteristics. Thus we give a second independent argument using the methods of Sect. 3d. The idea is simple: Since the Dirac operator has no index, $D(\mathbf{u})$ generically has no zero-modes. In other words, $\det D(\mathbf{u})$ will vanish in $J(\Sigma)$ on a subset of complex codimension 1. By choosing the spin structure judiciously, we will show that the locus of zeroes of $\det D(\mathbf{u})$ is the Θ-divisor described in Sect. 3. At the end of this section we comment on the ghost operators.

If we apply (4.1) and (4.2) to the $U(1)$ gauge field (3.15) parametrized by the Jacobian, then, using (3.7), (3.8), (3.15) and making a trivial redefinition of q in (4.2), we find:

$$\det D(\mathbf{u})^{\dagger} D(\mathbf{u}) = e^{i\pi(u-\bar{u})\cdot(\Omega-\bar{\Omega})^{-1}\cdot(u-\bar{u})}|g(\mathbf{u})|^2 . \tag{5.1}$$

Notice now that shifts of $\mathbf{u}$ by the Jacobian lattice $\mathbf{u}\to\mathbf{u}+\mathbf{n}+\Omega\mathbf{m}$ are equivalent to well-defined $U(1)$ gauge transformations on Σ given by

$$\mathscr{U}(P) = \exp - 2\pi i \left(\mathbf{m}\cdot \int_{P_0}^{P} \boldsymbol{\alpha} - \mathbf{n}\cdot \int_{P_0}^{P} \boldsymbol{\beta} \right). \tag{5.2}$$

Since the determinant (5.1) was computed by means of a gauge-invariant ζ-function regulator it must necessarily be gauge invariant under these gauge transformations. Since the exponent in (5.1) is not gauge invariant we can deduce the transformation properties of $g(\mathbf{u})$ under lattice shifts,

$$\begin{aligned} g(\mathbf{u}+\mathbf{n}) &= e^{i\phi(\mathbf{n})} g(\mathbf{u}) , \\ g(\mathbf{u}+\Omega\mathbf{m}) &= e^{i\psi(\mathbf{m})} e^{-i\pi\mathbf{m}\cdot\Omega\cdot\mathbf{m}-2\pi i\mathbf{m}\cdot\mathbf{u}} g(\mathbf{u}) , \end{aligned} \tag{5.3}$$

and since the phases $e^{i\phi(\mathbf{n})}$ and $e^{i\psi(\mathbf{m})}$ provide a unitary representation of Z^g, they must be of the form $e^{i\phi(\mathbf{n})} = e^{2\pi i\mathbf{n}\cdot\mathbf{a}}$, $e^{i\psi(\mathbf{m})} = e^{-2\pi i\mathbf{m}\cdot\mathbf{b}}$. Using a holomorphic Fourier decomposition of $g(\mathbf{u})$ we see that (5.3) uniquely defines a ϑ-function up to a constant. Hence, by (3.17),

$$g(\mathbf{u}) = \text{const}\,\vartheta \begin{bmatrix} \mathbf{a} \\ \mathbf{b} \end{bmatrix} (\mathbf{u}|\Omega) \tag{5.4}$$

for some real characteristics $\mathbf{a}, \mathbf{b}$.

We have not used any information about the spin structure to derive (5.4), so it is not unreasonable that the characteristics $\mathbf{a}, \mathbf{b}$ remain undetermined. We now use the methods of Sect. 3d to compute $\mathbf{a}, \mathbf{b}$. Quillen's construction of the chiral Dirac determinant shows that it is a holomorphic section of a holomorphic line bundle $\mathscr{L}$ on $J(\Sigma)$ (note that $\det D^{\dagger}D$ does not blow up for any $\mathbf{u}$). From the family index theorem and (3.21) we know that $\mathscr{L}$ is the line bundle of a theta function, and therefore by the comments following (3.21) that bundle admits a single holomorphic section. In order to determine this section, we simply compute its divisor. The divisor of the determinant consists of those points $\mathbf{u}$ in $J(\Sigma)$ where $D(\mathbf{u})$ has a zero mode. By the arguments preceding (3.15), $D(\mathbf{u})$ has a zero mode whenever the line bundle $L_{\alpha}^{-1}\otimes V(\mathbf{u})$ has a holomorphic section. We can determine those bundles admitting a holomorphic section by studying the divisor of $L_{\alpha}^{-1}\otimes V(\mathbf{u})$. Let A_{α} be a divisor of L_{α}^{-1}. If $d_{\mathbf{u}}$ is the divisor of $V(\mathbf{u})$, then $L_{\alpha}^{-1}\otimes V(\mathbf{u})$ has divisor

$A_\alpha + d_{\mathbf{u}}$. We can write $d_{\mathbf{u}}$ explicitly in terms of $\mathbf{u}$ as follows. The function

$$f_{\mathbf{u}}(P) = \frac{\vartheta\begin{bmatrix}\theta\\ \phi\end{bmatrix}\left(\mathbf{z} + \int_{P_0}^{P} \omega | \Omega\right)}{\vartheta\begin{bmatrix}0\\ 0\end{bmatrix}\left(\mathbf{z} + \int_{P_0}^{P} \omega | \Omega\right)} \tag{5.5}$$

is meromorphic and has periodicities

$$\Delta_{n \cdot a} f_u = e^{2\pi i n \cdot \theta} f_u, \qquad \Delta_{m \cdot b} f_u = e^{-2\pi i m \cdot \phi} f_u$$

around the $\mathbf{a}$ and $\mathbf{b}$ cycles respectively. ($\mathbf{z}$ is fixed and arbitrary.) Thus $f_{\mathbf{u}}(P)$ is a meromorphic section of $V(\mathbf{u})$. By the Riemann vanishing theorem $f_{\mathbf{u}}(P)$ has poles at g points Q_i satisfying

$$\mathbf{z} + \sum_{i=1}^{g} \int_{P_0}^{Q_i} \omega = \Delta,$$

and zeros at g points R_i such that

$$\mathbf{z} + \mathbf{u} + \sum_{i=1}^{g} \int_{P_0}^{R_i} \omega = \Delta,$$

hence

$$\mathbf{u} = -\sum_{i=1}^{g} \left(\int_{P_0}^{R_i} \omega - \int_{P_0}^{Q_i} \omega \right) = -I\left[\sum_1^g R_i - \sum_1^g Q_i \right]. \tag{5.6}$$

Now recall that $L_\alpha^{-1} \otimes V$ has a holomorphic section if and only if it has a divisor of the form

$$A_\alpha + d_{\mathbf{u}} = \sum_{i=1}^{g-1} P_i. \tag{5.7}$$

Choosing an arbitrary point P_0 on Σ, (5.7) can be rewritten as

$$d_{\mathbf{u}} = (g-1)P_0 - A_\alpha + \left(\sum_{i=1}^{g-1} P_i - (g-1)P_0 \right), \tag{5.8}$$

and using (5.6) we have:

$$-\mathbf{u} = I[(g-1)P_0 - A_\alpha] + I\left[\sum_1^{g-1} P_i - (g-1)P_0 \right]. \tag{5.9}$$

From (3.28) there is a spin structure D_0 such that the first term on the right-hand side of (5.9) is just the vector of Riemann constants. Thus

$$\mathbf{u} = \Delta - \sum_{i=1}^{g-1} \int_{P_0}^{P_i} \omega. \tag{5.10}$$

Using the characterization (3.25) of the Θ-divisor, we conclude that for the spin structure D_0, the characteristics $\mathbf{a}$, $\mathbf{b}$ in (5.4) are equal to zero. It follows that the determinant is given by

$$\det D(\mathbf{u})^\dagger D(\mathbf{u}) = |c|^2 e^{i\pi(u-\bar{u})\cdot(\Omega-\bar{\Omega})^{-1}\cdot(u-\bar{u})} |\vartheta(\mathbf{u}|\Omega)|^2. \tag{5.11}$$

The overall positive constant $|c|^2$ is a function only of the metric, not of the twisting.

In order to define the chiral determinant we must take the square root of (5.11). There are many ways to define chiral square roots, and we must choose a definition suitable to the physical problem under study. In the context of string theories we would like to separate the contributions of left-movers from right-movers. In Sect. 2 we found that the left- and right-movers contribute holomorphic and antiholomorphic functions of the moduli. Furthermore, from the formula for variation of a covariant derivative under a traceless deformation of the metric [25],

$$\delta\nabla^n_w = -\tfrac{1}{2}\delta g_{ww}\nabla^w + \frac{n}{2}\nabla^w(\delta g_{ww}),$$

we see that naively we expect $\det \nabla^n_w$ to be a holomorphic function on Teichmüller space [26]. Therefore, we will proceed with the hypothesis that on higher genus surfaces the contributions of left- and right-movers can be distinguished by holomorphy in the moduli. In this context the work of Belavin and Knizhnik shows that the quantity in (4.32) is the obstruction to the decoupling of left-movers from right-movers. Since the $|c|^2$ is, up to the obstruction (4.32), the square of a holomorphic function on moduli space, we sacrifice holomorphy on the Picard variety and write:

$$u-\bar{u}-(\Omega-\bar{\Omega})\boldsymbol{\theta},$$

so that

$$\det D^\dagger(\mathbf{u})D(\mathbf{u}) = |c|^2\left|e^{i\pi\theta\cdot\Omega\cdot\theta}\vartheta\begin{bmatrix}0\\0\end{bmatrix}(\mathbf{u}|\Omega)\right|^2 = |c|^2\left|\vartheta\begin{bmatrix}\boldsymbol{\theta}\\\boldsymbol{\phi}\end{bmatrix}(0|\Omega)e^{-2\pi i\theta\cdot\phi}\right|^2. \quad (5.12)$$

Therefore, we have

$$\det D(\mathbf{u}) = c\vartheta\begin{bmatrix}\boldsymbol{\theta}\\\boldsymbol{\phi}\end{bmatrix}(0|\Omega). \quad (5.13)$$

Actually the left-hand side of (5.13) is not purely a function of the moduli because of conformal and diffeomorphism anomalies. If we represent the metrics by $g=f^*e^{2\phi}\hat{g}$, then the modulus and phase of c will depend on ϕ and f respectively. In applications to string theory we may be cavalier about the anomalies since we know that the combination of determinants in the string path integral is anomaly free. Thus we can consider c as a function only of the moduli.

As a simple application, (5.12) makes transparent some well-known theorems on the existence of harmonic spinors. For example from (5.12) we learn that all odd spin structures have at least one zero-mode while the even spin structures generically have no zero modes. These results are derived in [27, 28]. Furthermore, it is known that for genus two the even theta functions do not vanish at $z=0$ except when Ω is diagonal [13]. (This corresponds to a degenerate curve for which the handles have been pulled off to infinity.) For genus three and larger there are period matrices and even characteristics such that $\vartheta(\mathbf{z}=0)=0$ [13], so that the dimension of the space of harmonic spinors is constant for $g=2$ and varies with the complex structure for $g\geqq 3$. These results were derived by Hitchin in 1974 [29].

We can also apply the Quillen theorem to the fermionic ghost operator of string theory, namely

$$\nabla_w^{+1}: L_\alpha \otimes V \to L_\alpha^3 \otimes V. \tag{5.14}$$

One finds that if χ_i is a basis of $\ker \nabla_3^w$ (that is, a basis of supermoduli) varying holomorphically with $\bar{u}$ then

$$\frac{\det(\nabla(\mathbf{u}))^\dagger \nabla(\mathbf{u})}{\det\langle\chi_i|\chi_j\rangle} = |c'|^2\, |e^{i\pi\theta\cdot\Omega\cdot\theta}|^2\, |g(\mathbf{u})|^2\,, \tag{5.15}$$

where $|c'|^2$ is independent of $\mathbf{u}$, and $g(\mathbf{u})$ is holomorphic. Unfortunately, there are many choices of holomorphically varying bases, so gauge invariance is not as powerful in this case. Therefore we will simply make a few remarks. First, note that there is no holomorphically varying basis χ_i which changes by a unitary transformation under a shift $\mathbf{u} \to \mathbf{u} + \mathbf{m} + \Omega\mathbf{n}$. For, if there were, gauge invariance would imply that $g(\mathbf{u})$ is a theta function, which must vanish on a subvariety of $J(\Sigma)$. On the other hand, the left-hand side of (5.15) never vanishes. In fact, we could choose our basis so that $g(\mathbf{u}) = 1$. Alternatively, using the Riemann vanishing theorem, it is possible to write a choice of basis χ_i for which the $\mathbf{u}$-dependence is explicit. One finds in this case that $g(\mathbf{u})$ can be expressed in terms of theta functions of weight $l = 2g - 2$ ($=$ the number of supermoduli). However the theta functions of weight l form a vector space of dimension l^g so that we do not learn much. Although the combination (5.15) is basis-dependent, the superstring path-integral measure is in fact basis-independent (see Sect. 7). An astute choice of basis should lead to important simplifications.

6. Bosonization

Bosonization is an important aspect of two-dimensional field theory. Although the local theory of bosonization on higher genus Riemann surfaces has been worked out [3], no complete treatment yet exists. We now show that the results of Sect. 5 can be used to give a prescription for the lattice sum needed in bosonization of spin-$\frac{1}{2}$ particles. In particular we use bosonization to evaluate $|c|^2$, where c is the function of the metric defined in (5.13). Furthermore, we will see that non-chiral bosonization involves an interesting subtlety. Since, roughly speaking, a Bose theory doesn't "know" about a spin structure, it will turn out that a Bose theory corresponds to an average over spin structures of Fermi theories. We consider first nonchiral bosonization to avoid subtleties peculiar to left-moving bosons. This is the reason we only evaluate the absolute square $|c|^2$. Then using the remarks of Sect. 5, we can separate out the contribution of the left-movers and give a prescription for the lattice contribution in chiral bosonization.

As we saw in Sect. 2, bosonization on compact spaces involves a sum over soliton sectors. If we formulate the $SO(n)$ current algebra on the cylinder we will have n scalar fields which can have shifts by the integral or half-integral vectors in the weight lattice of $SO(n)$. Furthermore, on the torus or the cylinder we see that twisted fermions correspond under bosonization to scalars which shift by a *translate* of the $SO(n)$ weight lattice. We now generalize these notions to higher loops, taking $n=2$ for simplicity. Therefore, consider a real Bose field x on a Riemann surface Σ. We can introduce soliton sectors into the theory if we allow x to have shifts around the nontrivial homology cycles. Therefore we can define the $(n^1, \ldots, n^g, m^1, \ldots, m^g)$-soliton sector by requiring

$$\begin{aligned} x &\to x+(n^i+\theta^i) \text{ around } a_i, \\ x &\to x+(m^i+\phi^i) \text{ around } b_i, \end{aligned} \tag{6.1}$$

where $0 \leqq \theta^i, \phi^i \leqq 1$ and $n^i, m^i \in \frac{1}{2}Z$, $i=1, \ldots, g$. We can simplify the derivation by first summing over soliton sectors with n, m integral, and then summing over half-integral shifts in θ, ϕ. The soliton solution is given by

$$dx_{n,m} = \sum_1^g (n+\theta)^i \alpha_i + \sum_1^g (m+\phi)^i \beta_i, \tag{6.2}$$

where α_i, β_i are the harmonic forms of Sect. 3. We can express the action for the soliton:

$$S[x] = 2\pi \int dx \wedge * dx, \tag{6.3}$$

in terms of the matrices

$$A_{ij} = \int \alpha^i \wedge * \alpha^j, \quad B_{ij} = \int \beta^i \wedge * \beta^j, \quad C_{ij} = \int \alpha^i \wedge * \beta^j. \tag{6.4}$$

Using Sect. 3 one finds that the period matrix is simply given in terms of these matrices by

$$\Omega = B^{-1} C^T + i B^{-1}. \tag{6.5}$$

With the soliton action one can then write the partition function for the theory:

$$Z = \left(\frac{\det' - \nabla^2}{\int \sqrt{g}} \right)^{-1/2} \sum_{\mathbf{n}, \mathbf{m} \in Z^g} e^{-S_{n,m}}, \tag{6.6}$$

where $S_{n,m}$ is the action for the soliton (6.2). We will now relate this sum to theta functions, thereby establishing a link with fermionic partition functions.

Holding $\mathbf{n}$ fixed, we use (6.3)–(6.5) and apply the Poisson summation formula to the sum on $\mathbf{m}$ to obtain:

$$Z = \left(\frac{\det' - \nabla^2}{\int \sqrt{g} \det \operatorname{Im} \Omega} \right)^{-1/2} \frac{1}{2^g} \sum_{n,m} e^{\mathscr{A}_{n,m}} e^{\overline{\mathscr{A}_{n,m}}}, \tag{6.7}$$

where $\mathscr{A}_{n,m}$ is a complex number. Explicitly one finds, using (6.2), (6.4), (6.5), and (6.7),

$$\begin{aligned} \mathscr{A}_{n,m} &= i\pi(\mathbf{n}+\boldsymbol{\theta}+\tfrac{1}{2}\mathbf{m}) \cdot \Omega \cdot (\mathbf{n}+\boldsymbol{\theta}+\tfrac{1}{2}\mathbf{m}) + 2\pi i(\mathbf{n}+\boldsymbol{\theta}+\tfrac{1}{2}\mathbf{m}) \cdot \boldsymbol{\phi}, \\ \bar{\mathscr{A}}_{n,m} &= -i\pi(\mathbf{n}+\boldsymbol{\theta}-\tfrac{1}{2}\mathbf{m}) \cdot \bar{\Omega} \cdot (\mathbf{n}+\boldsymbol{\theta}-\tfrac{1}{2}\mathbf{m}) - 2\pi i(\mathbf{n}+\boldsymbol{\theta}-\tfrac{1}{2}\mathbf{m}) \cdot \boldsymbol{\phi}. \end{aligned} \tag{6.8}$$

We can express the sum (6.7) in terms of theta functions. The result is:

$$Z = \left(\frac{\det' - \nabla^2}{\int \sqrt{g} \det \operatorname{Im} \Omega} \right)^{-1/2} \frac{1}{2^g} \sum_{\substack{\varepsilon_1 \in Z_2^g \\ \varepsilon_2 \in Z_2^g}}' \left| \vartheta \begin{bmatrix} \boldsymbol{\theta}+\boldsymbol{\varepsilon}_1 \\ -\boldsymbol{\phi}+\boldsymbol{\varepsilon}_2 \end{bmatrix} (0|\Omega) \right|^2 e^{4\pi i \varepsilon_1 \cdot \varepsilon_2}. \tag{6.9}$$

Now we include the full weight lattice, allowing soliton sectors with n^i, m^i integral and half-integral. The partition function then becomes

$$Z=|c|^2 \sum_{\varepsilon_1,\varepsilon_2 \in Z_2^g} \left| \vartheta \begin{bmatrix} \theta+\varepsilon_1 \\ \phi+\varepsilon_2 \end{bmatrix} (0|\Omega) \right|^2 . \tag{6.10}$$

Therefore, this Bose theory corresponds to a sum over all the 2^{2g} Fermi theories on Σ. Weighted sums over subsets of the lattice correspond to particular spin structures, suggesting the possibility of bosonizing a Fermi theory with a single spin structure. Comparing the overall normalization of the partition functions we learn that the function $|c|^2$ of the last section is given by

$$|c|^2=\mathrm{const}\left(\frac{\det'-\nabla^2}{\int\sqrt{g}\det\operatorname{Im}\Omega}\right)^{-1/2} \tag{6.11}$$

up to some *numerical* constant. This derivation is not rigorous, but (6.11) does pass two important consistency checks. First, both sides of the equation have the same conformal anomaly. Second, in the case of the torus one can easily compute both quantities directly. For the metric $|d\sigma_1+\tau d\sigma_2|^2$ one finds the right-hand side is given by

$$\frac{1}{2|\eta(\tau)|^2},$$

which agrees with (2.3) and (5.13). Thus (6.11) is also consistent with the naive factorization of determinants in the limit when a handle is pulled off to infinity.

We now turn to the case of chiral bosonization. We see from (6.7) that the contributions of the various soliton sectors are holomorphic squares. In the natural complex structure on Teichmüller space the period matrix is a holomorphic function of the moduli [24]. Therefore, according to the identification of Sect. 5, $\mathscr{A}_{n,m}$ should be thought of as the action of a right-moving soliton. Thus, the expression (6.10) allows a separation of the contributions of left-movers and right-movers, which we identify as holomorphic and anti-holomorphic functions of the moduli. If we simply drop the contribution of the left- or right-movers we should obtain the partition function for a chiral boson. Using bosonization arguments such as these one can verify that the contribution of the left moving $E_8\times E_8$ solitons to the string path integral is

$$\left(c^8 \sum_{\varepsilon_1,\varepsilon_2} \left(\vartheta \begin{bmatrix} \varepsilon_1 \\ \varepsilon_2 \end{bmatrix} (0|\Omega)\right)^8\right)^2 \tag{6.12}$$

as expected from the fermionic formulation of the current algebra. In string theory we also need to consider the ghosts. For these the bosonization is more subtle [3] and the sum over soliton sectors seems to yield rather awkward expressions.

7. Modular Invariance in the Heterotic String

In the previous sections we have been able to compute the spin-structure dependence of the determinant of the chiral Dirac operator on Riemann surfaces.

In this section, we will apply those results to discuss some aspects of the higher loop heterotic string amplitudes. In particular, as a first application, we will verify some results of Witten on global anomalies [30]. It has recently been pointed out that two-dimensional anomaly cancellation does not completely fix the heterotic string amplitudes [31, 32]. As a second application, we will discuss this ambiguity from the point of view of modular orbits of theta characteristics.

We begin by presenting an expression for the higher loop vacuum to vacuum amplitudes in the heterotic string. At g-loops this amplitude can be derived from a functional integral over two-dimensional supergeometries of a g-handled Riemann surface Σ, and over quantum matter fields living on this surface [2]. For the heterotic string a supergeometry on a Riemann surface is specified by a Riemannian metric (a background graviton field configuration) together with a section of L_α^{-3} (a background gravitino field configuration) [33]. The matter fields may be described as follows. There are 10 left- and right-moving bosonic coordinates describing the embedding of Σ onto the world surface of the string in 10-dimensional spacetime. Spin waves on the world surface are described by the $SO(10)$ current algebra constructed from the 10 right-moving spinor sections of L_α familiar from the NSR model.[4] In addition, heterotic strings have gauge currents. In the $O(32)$ string these currents can be constructed from 32 left-moving spinor sections of L_β^{-1}. The spin structures α, β are independent. In the $E_8 \times E_8$ string the 32 gauge fermions are split into two sets of 16, each with independent spin structures β and γ. In Minkowski space the spinors are Majorana-Weyl, but in Euclidean space they must be Weyl. The Euclidean path integral quantization of a Majorana-Weyl spinor introduces a Pfaffian of the complex chiral Dirac operator which results is an extra square root of the relevant functional determinant.

Gauge-fixing the symmetries of the Euclidean heterotic string action reduces the infinite-dimensional functional integral to a finite-dimensional integral over the moduli space of Σ which describes gauge-inequivalent graviton field configurations [34–37] together with a finite dimensional Grassmann integral over supermoduli space which describes gauge-inequivalent background gravitino field configurations [3, 17, 38]. The integration over moduli space is expressed as an integration over a fundamental domain F of the Teichmüller space T under the action of $\Omega(\Sigma)$ [39, 35]. In addition, in order to define a physically reasonable string theory one must include projections onto sectors with even fermion number [40]. This is accomplished in the path integral formulation by summing over the independent spin structures for the two sets of fermions[5].

The path-integral measure has been derived in [17] and in a slightly different language in [3] and in [38]. In [17] it was shown that one can find an acceptable integrand for the heterotic string given the holomorphic factorization of certain functional determinants. By the Belavin-Knizhnik theorem we know that the appropriate holomorphic square roots do, in fact, exist. Furthermore, it can be shown [41] that there is no obstruction to the holomorphic factorization in the supermoduli, and that after integration over the supermoduli the measure in [17]

[4] Two-dimensional supersymmetry requires that the NSR fermions and the background gravitino have the same spin structure

[5] This is very clearly explained in [31]

can be cast in the form

$$Z=\int_F \prod_1^{3g-3} dt^r dt^{r*} \frac{|\det\langle S^r|T^{r'}\rangle|^2}{(\det \operatorname{Im}\Omega)^5} \frac{\det \Delta_2^{(+)}}{\det\langle S^r|S^{r'}\rangle} \left(\frac{\det' - \nabla^2}{2\pi\int\sqrt{g}\det \operatorname{Im}\Omega}\right)^{-5}$$

$$\times \sum_{\alpha\beta\gamma} C_{\alpha\beta\gamma}(\det_\beta \nabla_w^{-1})^8 (\det_\gamma \nabla_w^{-1})^8 (\det_\alpha \nabla_1^{w})^5 \left(\frac{\det_\alpha\langle\chi_i|\chi_j\rangle}{\det \Delta_1^{(+)}}\right)^{1/2}$$

$$\times \frac{1}{\det_\alpha\langle\chi|\chi\rangle} \int \det_{i,j}\chi_i(w_j)\langle S_{z\theta}(w_1)\ldots S_{z\theta}(w_{2g-2})\rangle . \tag{7.1}$$

[We have chosen the case of $E_8\times E_8$ in (7.1).] Here we have represented Teichmüller space by a slice $\hat{g}(t)$ which has traceless symmetric tangent vectors T^r. The zero modes S^r and χ_i form a basis for $\ker \nabla^{-4}$ and $\ker \nabla^{-3}$, respectively. The quantity $S_{z\theta}$ can be most easily expressed in the language of [3] as the ghost supercurrent:[6]

$$S_{z\theta}=\gamma^\theta b_{zz}+\beta_{z\theta}\nabla_z^2 c^z ,$$

where $c^z\gamma^\theta$ are the reparametrization and super-reparametrization ghosts and $b_{zz}\beta_{z\theta}$ are their conjugate fields. Thus the expectation value of the ghost currents can be expressed via Wick's theorem in terms of the Green's functions for the operators ∇_z^1 and ∇_z^2. Finally, one must choose the (anti)holomorphic square-root of $\det\Delta_1^{(+)}$, which exists, by the Belavin-Knizhnik theorem.

As is well-known, (7.1) has no local anomalies. Witten [30] has shown that the expression also has no global anomalies. His argument proceeds in two steps. First, it is shown that the terms with $\alpha=\beta=\gamma$ have no phase change under diffeomorphisms which preserve the spin structure α.[7] We cannot explicitly verify this because we do not know the modular dependence of the Rarita-Schwinger determinant. The second step of the proof in [30] proceeds by considering ratios of Dirac determinants. In particular, it is shown that (7.1) is modular invariant if

$$\left(\frac{\det_\alpha \nabla_z^{-1}}{\det_\beta \nabla_z^{-1}}\right)^4 \tag{7.2}$$

is invariant under diffeomorphisms preserving spin structures α and β. We can explicitly verify this result since we know (7.2) as a function of the moduli, in particular, as a function of Ω.

We can find the transformation of Ω under a nontrivial diffeomorphism as follows. Recall that if $\kappa_i=(a_i, b_i)$ is a canonical homology basis then $f_*\kappa_i=T_{ij}\kappa_j$, where

$$T=\begin{pmatrix} D & C \\ B & A\end{pmatrix}\in Sp(2g,Z) .$$

---

[6] After an integration by parts

[7] Insertions of the ghost currents lead to Green's functions which at first sight vitiate the argument. However, the determinant over $\chi_i(w_j)$ suppresses the singular terms so that the correction factor is finite and does not need to be regularized. Since it needs no regularization, it should have no anomaly

To preserve the normalization condition (3.6) we must redefine the basis of holomorphic differentials

$$\tilde{\omega}_i = \omega_j (C\Omega + D)^{-1}_{ji}\,.$$

Thus

$$\Omega \to \tilde{\Omega}_{ij} = \int_{\tilde{b}_i} \tilde{\omega}_j$$

leads to the modular transformation

$$\Omega \to \tilde{\Omega} = (A\Omega + B)(C\Omega + D)^{-1}\,.$$

It can be shown [42] that the theta functions transform as

$$\vartheta\begin{bmatrix}\tilde{a}\\ \tilde{b}\end{bmatrix}(\tilde{\Omega}) = \varepsilon(\Lambda)\, e^{-i\pi\phi(a,b,\Omega)} \det(C\Omega + D)^{1/2}\, \vartheta\begin{bmatrix}a\\ b\end{bmatrix}(\Omega)\,, \tag{7.3}$$

where,

$$\Lambda = \begin{pmatrix} A & B\\ C & D\end{pmatrix},$$

$$\begin{bmatrix}\tilde{a}\\ \tilde{b}\end{bmatrix} = \begin{pmatrix} D & -C\\ -B & A\end{pmatrix}\begin{bmatrix}a\\ b\end{bmatrix} + \frac{1}{2}\begin{bmatrix}(CD^t)_d\\ (AB^t)_d\end{bmatrix},$$

$$\phi(a,b,\Omega) = [aD^tBa + bC^tAb] - [2aB^tCb + (aD^t - bC^t)(AB^t)_d]\,. \tag{7.4}$$

The phase ε is subtle. It is always an eighth root of unity and, if

$$\Lambda = I_{2g} \bmod 2\,,$$

then we have

$$\varepsilon^2(\Lambda) = e^{\frac{i\pi}{2}\operatorname{tr}(D-1)}\,.$$

In (7.4), the subscript d means that we are taking the diagonal elements of the matrix to form a column vector. The above formula for the transformation of ϑ is valid for arbitrary real g-dimensional vectors a and b. It can be easily verified for a set of generators of $Sp(2g, Z)$, although the extension to arbitrary symplectic transformations is more difficult.

We are now in a position to consider the transformation of (7.2). Let $[a_1, b_1]$ and $[a_2, b_2]$ be the half-integral characteristics corresponding to spin structures α and β respectively. The modular transformation fixes spin structures α and β if and only if for $i = 1, 2$

$$\begin{bmatrix}\tilde{a}_i\\ \tilde{b}_i\end{bmatrix} = \begin{bmatrix}a_i\\ b_i\end{bmatrix} \bmod 1\,. \tag{7.5}$$

Using the explicit dependence of determinants, we would therefore like to show that

$$\frac{\vartheta^4\begin{bmatrix}\tilde{a}_1\\ \tilde{b}_1\end{bmatrix}(\tilde{\Omega})}{\vartheta^4\begin{bmatrix}\tilde{a}_2\\ \tilde{b}_2\end{bmatrix}(\tilde{\Omega})} = \frac{\vartheta^4\begin{bmatrix}a_1\\ b_1\end{bmatrix}(\Omega)}{\vartheta^4\begin{bmatrix}a_2\\ b_2\end{bmatrix}(\Omega)}\,. \tag{7.6}$$

In the above expression we can shift the a's and b's by integral vectors, without changing the expression, because such a shift introduces a phase $e^{2\pi i a\cdot n}$ for the theta function, where n is an integral vector. This phase disappears when we raise the theta function to the fourth power. Using the modular transformation rule for the theta function, it is easy to see that the above equality holds provided

$$e^{-i4\pi[\phi(a_1,b_1,\Omega)-\phi(a_2,b_2,\Omega)]}=1\,.$$

Since a_i and b_i are half-integral vectors, and Λ is an integral matrix, it follows that a priori the above phase could be ± 1. To show that it is equal to 1, one has to show that

$$4(a_1D^tBa_1+b_1C^tAb_1-(1\to 2))=0 \bmod 2\,.$$

From now on for brevity, when we write an equality, we mean equality mod 2. Note that in such expressions $\pm$ signs play no role since $r=-r \bmod 2$ when r is integral. Let us define integral vectors c and d by

$$c=2a_1-2a_2\,,\qquad d=2b_1-2b_2\,.$$

The condition we will have to show becomes

$$Q=cD^tBc+dC^t\mathrm{A}d=0\,.$$

(We have used $D^tB=B^tD$ and $C^tA=A^tC$. This can be proved by writing the right-inverse of Λ and demanding that it be a left-inverse.) The condition that the spin structures α and β are preserved implies that (modulo two)

$$(1+D)c+Cd=0\,, \tag{7.7}$$

$$Bc+(1+A)d=0\,. \tag{7.8}$$

If we take the transpose of (7.7), multiply it by (7.8), and use $D^tA-B^tC=1$, we get

$$Q+c(D^t+A)d+cBc+dC^td=0\,. \tag{7.9}$$

Using (7.7) and (7.8) we see that

$$0=2c\cdot d=(c\ \ d)\begin{pmatrix} d\\ c\end{pmatrix}=(c\ \ d)\begin{pmatrix} D^t & B^t\\ C^t & A^t\end{pmatrix}\begin{pmatrix} d\\ c\end{pmatrix}.$$

And so Eq. (7.9) simplifies to $Q=0$, the desired result. Thus, (7.1) is free of global anomalies. As noted in [30], four is the smallest power for which (7.2) is invariant. This is important in constraining variations on the heterotic string, as we will see below.

We now study the constants $C_{\alpha\beta\gamma}$ in (7.1). Naively, one might expect these constants to factorize for the sums over right- and left-movers, an impression which might be reinforced by the observation that the gauge fermions would then contribute a modular covariant sum:

$$\sum_{a,b}\vartheta^8\begin{bmatrix} a\\ b\end{bmatrix}\to \det(C\Omega+D)^4\sum_{a,b}\vartheta^8\begin{bmatrix} a\\ b\end{bmatrix}.$$

In fact, the standard $E_8\times E_8$ theory corresponds to this choice, but it is not the most general choice. The correct procedure, which was outlined in [30], involves *modular orbits* which we now describe.

We have explained the one-to-one correspondence between spin structures and half-integral theta characteristics in Sect. 3. Using the transformation law (7.4) we can therefore find the orbit of a spin structure under the action of the modular group. As an example, we show that even and odd spin structures are separately permuted among themselves. The proof proceeds by induction. For $g=1$ the result is easily checked explicitly. In genus two we must use the nontrivial Dehn twist $D_{a_1^{-1}a_2}$ of Sect. 3a. The action of this twist on the half-integral characteristics is:[8]

$$\begin{bmatrix} 0 & 1 \\ a & b \end{bmatrix} \to \begin{bmatrix} 0 & 1 \\ a & b \end{bmatrix} \quad \begin{bmatrix} 1 & 0 \\ a & b \end{bmatrix} \to \begin{bmatrix} 1 & 0 \\ a & b \end{bmatrix}$$

$$\begin{bmatrix} 0 & 0 \\ a & b \end{bmatrix} \to \begin{bmatrix} 0 & 0 \\ a+1 & b+1 \end{bmatrix} \quad \begin{bmatrix} 1 & 1 \\ a & b \end{bmatrix} \to \begin{bmatrix} 1 & 1 \\ a+1 & b+1 \end{bmatrix}. \tag{7.10}$$

Using this transformation and the Dehn twists for the separate handles one can easily see that all the odd characcteristics of (3.29) can be transformed to

$$\begin{bmatrix} 0 & 1 \\ 0 & 1 \end{bmatrix}$$

while all the even characteristics of (3.30) can be transformed to

$$\begin{bmatrix} 0 & 0 \\ 0 & 0 \end{bmatrix}.$$

We now assume that at $g-1$ loops all odd characteristics can be transformed to

$$\begin{bmatrix} 0\dots 0 & 1 \\ 0\dots 0 & 1 \end{bmatrix}, \tag{7.11}$$

while all even characteristics can be transformed to

$$\begin{bmatrix} 0\dots 0 \\ 0\dots 0 \end{bmatrix}. \tag{7.12}$$

Consider any even characteristic at g loops. The first $g-1$ columns define an even or odd spin structure depending on whether the final column is even or odd. In the case that both are even we can use the inductive hypothesis and Dehn twists on the gth handle to bring the characteristic to the form (7.12). In the case that both are odd we again use the inductive hypothesis to transform the g-loop characteristic to

$$\begin{bmatrix} 0\dots 0 & 1 & 1 \\ 0\dots 0 & 1 & 1 \end{bmatrix}.$$

Applying the transformation (7.10) to the last two handles and using two further torus-Dehn-twists, we can finally bring the g-loop characteristic to the form (7.12). A similar argument for an arbitrary odd characteristic at g-loops completes the inductive step. Thus, the set of spin structures decomposes into two distinct modular orbits.

[8] For notational simplicity we have multiplied all the half-integral characteristics by two. They are therefore specified by zeros and ones, modulo two

When we consider more than one independent spin structure, the modular orbits become more intricate. For example, in the $E_8 \times E_8$ heterotic string we must consider triplets of spin structures $\alpha\beta\gamma$. Each such triplet has an associated modular orbit $\mathcal{O}$ obtained by simultaneously transforming all three spin structures. Denote by $S[\alpha\beta\gamma]$ the contribution of the functional determinants in (7.1) for the triplet of spin structures $\alpha\beta\gamma$. It is easy to check that if S is invariant under diffeomorphisms preserving the triplet of spin structures, then the sum

$$\sum_{[\alpha\beta\gamma]\in\mathcal{O}} S[\alpha\beta\gamma] \tag{7.13}$$

is modular invariant. This leaves an ambiguity in the definition of the amplitude (7.1), for if we write the integrand as

$$I_g = \sum_{\text{orbits}\,\mathcal{O}} C(\mathcal{O}) \sum_{[\alpha\beta\gamma]\in\mathcal{O}} S[\alpha\beta\gamma], \tag{7.14}$$

the coefficients C are not fixed by absence of global anomalies on the world sheet.

The importance of modular invariance and factorization in relating different terms in the path-integral measure for superstrings has been pointed out in [31]. Similar ideas have been applied in the context of orbifolds in [43] and in determining inequivalent phases for heterotic strings in [32]. In particular these ideas have led to new heterotic strings [44, 31, 32]. We will now discuss some aspects of the new theories using the transformation law (7.3), (7.4) for the theta characteristics. We will only investigate the possible modifications of the $O(32)$ and $E_8 \times E_8$ strings for simplicity. (One can also consider breaking up the sets of fermions further into sets in eight, or some multiples of eight, adding up to 32 fermions. This is the furthest one can go: since four is the smallest power for which (7.2) is invariant, allowing smaller sets of fermions would lead to a measure with global anomalies [30].)

An amplitude factorizes if, in the corner of Teichmüller space corresponding to the degeneration of a Riemann surface illustrated in Fig. 6 (that is, when the handles are pulled apart), the integrand factorizes into a product of g_1- and g_2-loop amplitudes: $I_g \to I_{g_1} I_{g_2}$. In this limit the period matrix becomes [14]

$$\Omega \to \begin{pmatrix} \Omega_1 & 0 \\ 0 & \Omega_2 \end{pmatrix} \tag{7.15}$$

so the theta functions also factorize into g_1- and g_2-loop theta functions:

$$\vartheta\begin{bmatrix} a \\ b \end{bmatrix}(0|\Omega) \to \vartheta\begin{bmatrix} a_1 \\ b_1 \end{bmatrix}(0|\Omega_1)\,\vartheta\begin{bmatrix} a_2 \\ b_2 \end{bmatrix}(0|\Omega_2).$$

We will assume that the supersymmetry ghost determinant and the supermoduli correction (which depend on spin structure) similarly factorize so that

$$I_{g_1+g_2} \to \sum_{\text{orbits}\,\mathcal{O}} C_{g_1+g_2}(\mathcal{O}) \sum S[\alpha_1\beta_1\gamma_1]\, S[\alpha_2\beta_2\gamma_2],$$

Fig. 6

where $[\alpha_i\beta_i\gamma_i]$ are any spin structures at genus g_i which combine to give a spin structure in orbit $\mathcal{O}$ at genus g. By our assumption that the S's factorize, the choice $C=1$ is consistent with factorization of the amplitude, but this is not the unique choice. In general the coefficients must satisfy

$$C_{g_1+g_2}(\mathcal{O}[\alpha\beta\gamma])=C_{g_1}(\mathcal{O}[\alpha_1\beta_1\gamma_1])C_{g_2}(\mathcal{O}[\alpha_2\beta_2\gamma_2]), \tag{7.16}$$

where $\mathcal{O}[\alpha\beta\gamma]$ denotes the modular orbit of a triplet of spin structures. Note that if two choices of coefficients satisfy (7.16) then so does their product. Thus the solutions of (7.16) form a group. The condition (7.16) is nontrivial because two triplets of spin structures, $\alpha\beta\gamma$ and $\mu\nu\varrho$, in the same modular orbit at (g_1+g_2)-loops can factorize into products of triplets of spin structures

$$[\alpha\beta\gamma]\rightarrow[\alpha_1\beta_1\gamma_1][\alpha_2\beta_2\gamma_2], \quad [\mu\nu\varrho]\rightarrow[\mu_1\nu_1\varrho_1][\mu_2\nu_2\varrho_2] \tag{7.17}$$

in *different* modular orbits at g_1- or g_2-loops. Thus we must have

$$C(\mathcal{O}[\alpha_1\beta_1\gamma_1])C(\mathcal{O}[\alpha_2\beta_2\gamma_2])=C(\mathcal{O}[\mu_1\nu_1\varrho_1])C(\mathcal{O}[\mu_2\nu_2\varrho_2]) \tag{7.18}$$

for all pairs of triplets factorizing from (g_1+g_2)-loop triplets which have the same modular orbit $\mathcal{O}$. From (7.16) we see that the g-loop coefficients can be written as products of one-loop coefficients. Since $\Omega(\Sigma)$ is generated by the Dehn twists about the curves in Fig. 4 we see that the only nontrivial conditions from (7.18) come from taking $g_1=g_2=1$, and if these are satisfied then (7.18) holds for all g_1, g_2, and all triplets of spin structures.

There are always some trivial solutions of (7.18) which we now describe. Consider either the first, second, or third set of fermions. It is easy to check that if we take $C=+1$ when this set has an even spin structure and $C=-1$ when the set has an odd spin structure, then (7.18) is satisfied. These solutions form the group $Z_2\times Z_2$ for the $O(32)$ string and $Z_2\times Z_2\times Z_2$ for the $E_8\times E_8$ string. By comparing the modified signs in the path integral with the corresponding expressions in the operator formalism one finds that these solutions correspond to physically irrelevant redefinitions of the chiralities of the massless ground state spinors in the spectrum of the theory.

We now search for nontrivial solutions of (7.18) in the specific example of the $O(32)$ string. In this case the modular orbits are generated by doublets of spin structures. There are five distinct one-loop modular orbits: Either both spin structures are even and identical, or both are even and distinct, or one is odd and the other is even, or both are odd. These may be denoted by *ee ee′ eo oe oo*. We may choose $C(ee)=1$ by convention. We can see that $C(ee')=1$ by considering the following example. The genus two doublet of spin structure with characteristics

$$\begin{bmatrix}0 & 0\\ 0 & 0\end{bmatrix}_1 \begin{bmatrix}0 & 1\\ 0 & 0\end{bmatrix}_2 \tag{7.19}$$

factorizes into the product of doublets of one-loop spin structures:

$$\left(\begin{bmatrix}0\\0\end{bmatrix}_1 \begin{bmatrix}0\\0\end{bmatrix}_2\right)\left(\begin{bmatrix}0\\0\end{bmatrix}_1 \begin{bmatrix}1\\0\end{bmatrix}_2\right) \tag{7.20}$$

which has coefficient $C(ee')$ in the product of one-loop amplitudes (since $C(ee)=1$). On the other hand the doublet of spin structures with characteristics

$$\begin{bmatrix} 0 & 0 \\ 1 & 1 \end{bmatrix}_1 \begin{bmatrix} 0 & 1 \\ 0 & 0 \end{bmatrix}_2 \tag{7.21}$$

factorizes into the product of doublets:

$$\left(\begin{bmatrix} 0 \\ 1 \end{bmatrix}_1 \begin{bmatrix} 0 \\ 0 \end{bmatrix}_2\right)\left(\begin{bmatrix} 0 \\ 1 \end{bmatrix}_1 \begin{bmatrix} 1 \\ 0 \end{bmatrix}_2\right) \tag{7.22}$$

which has coefficient $C(ee')^2$. Since the orbits of (7.19) and (7.21) are related by the Dehn twist $D_{a_1^{-1}a_2}$ we must have $C(ee')^2 = C(ee')$, which means $C(ee')=0$ or 1. The first choice leads to a theory with tachyons [31], so we investigate the second choice. We then have effectively only four distinct modular orbits corresponding to whether the members of the doublet of spin structures are even or odd. Since a spin structure with the characteristics $\begin{bmatrix} a \\ b \end{bmatrix}$ is even or odd depending on whether $ab=0$ or 1 mod 2, we can consider C as a function on $Z_2 \times Z_2$. Generalizing the above example of (7.19)–(7.22) to the genus two orbit of

$$\begin{bmatrix} a & a' \\ b & b' \end{bmatrix}_1 \begin{bmatrix} c & c' \\ d & d' \end{bmatrix}_2$$

we find the condition

$$C(ab, cd)\,C(a'b', c'd') = C(ab + aa', cd + cc')\,C(a'b' + aa', c'd' + cc')$$

which implies

$$C(0,1)^2 = C(1,0)^2 = 1\,, \qquad C(1,1) = C(0,1)\,C(1,0)\,,$$

which corresponds to the trivial solutions found above. A similar but more tedious analysis of the $E_8 \times E_8$ case shows that the solutions form the group $Z_2 \times Z_2 \times Z_2 \times Z_2$. The first three Z_2's correspond to the trivial solutions discussed above, but the fourth Z_2 is nontrivial, and leads to a heterotic string with $O(16)\times O(16)$ gauge symmetry [44, 32].

In general we can start with all spin structures identified [31, 32] and take the quotient by various groups to get the other theories. If we take the quotient with groups which are direct products of Z_2, we can get sums over up to five independent spin structures. For example, using a single Z_2 we can get the $O(32)$ heterotic string or a tachyonic $O(16)\times E_8$ string [44, 31], where the latter theory is obtained by identifying the spin structure of the right moving spacetime fermions with the spin structure of eight of the left moving gauge fermions. It seems that in all the known examples identification of the spin structures of some left and right movers leads to a theory with tachyons.

8. Conclusions

Much work remains to be done. We have seen that theta functions are useful for understanding spin one-half particles. Can the ghost determinants and zero modes also be usefully expressed in theta functions? We have also seen that some pieces of the string path integral extend as functions of the Siegel upper half plane. Can the entire string integrand be expressed as a function on this space [45]? If so, what are the constraints of symplectic modular invariance on physical theories? We have investigated one simple aspect of bosonization on higher genus surfaces, but a full treatment is still lacking.

The naive expectation that the determinant of the spin-$\frac{3}{2}$ operator should be given in its spin-structure-dependence by a theta function is misguided, as explained in Sect. 5. Furthermore, although there are identities (known as the Riemann identities) which are of the form $\sum C_\alpha \vartheta^4 = 0$ [13], they are not (formally) modular covariant. A direct proof of the vanishing of the cosmological constant will involve more sophisticated identities, and might point the way to new Schottky relations on the period matrix. We simply note here that a direct approach is likely to encounter subtleties, since there must be delicate cancellations between the contributions of the various spin structures. Indeed, merely changing a few of the coefficients $C_{\alpha\beta\gamma}$ leads to the $O(16) \times O(16)$ heterotic string which is modular invariant and unitary, and yet has a finite nonzero cosmological constant.

Acknowledgements. We would like to thank P. Nelson for suggesting many improvements of a preliminary manuscript and for many conversations on strings and Riemann surfaces. We would also like to thank Stephen Della Pietra, Vincent Della Pietra, D. Freed, P. Ginsparg, D. Mumford, J. Polchinski, and I. Singer for much help throughout the course of this work. This work was supported in part by NSF contract PHY-82-15249. G.M. and C.V. are also supported by the Harvard Society of Fellows, and L. A.-G. by the A.P. Sloan Foundation.

Similar results on the dependence of the Dirac determinant on spin structure have also been obtained by D. Freed from the viewpoint of the geometry of index bundles and by J. Fay using the Szegö kernel.

References

1. Schwarz, J.: Superstrings: The first fifteen years, Vol. 5, I, II. New York: World Scientific 1985
2. Polyakov, A.M.: Quantum geometry of bosonic strings. Phys. Lett. **103B**, 207 (1981); Quantum geometry of fermionic strings. Phys. Lett. **103B**, 211 (1981)
3. Friedan, D., Martinec, E., Shenker, S.: Conformal invariance, supersymmetry, and strings. Princeton preprint
4. Martinec, E.: Nonrenormalization theorems and fermionic string finiteness. Princeton preprint
5. Gross, D.J., Harvey, J.A., Martinec, E., Rohm, R.: Phys. Rev. Lett. **54**, 502 (1985); Nucl. Phys. B**256**, 253 (1985), and Nucl. Phys. B (to appear)
6. Dixon, L., Harvey, J., Vafa, C., Witten, E.: Strings on orbifolds I. Nucl. Phys. B**261**, 651 (1985). Strings on orbifolds II. Princeton preprint
7. Farkas, H.M., Kra, I.: Riemann surfaces. Berlin, Heidelberg, New York: Springer 1980
8. See, for example J. Birman in W.J. Harvey (ed.), Discrete groups and automorphic functions. New York: Academic Press 1977; L. Greenberg (ed.), Discontinuous groups and Riemann surfaces, Princeton, NJ: Princeton University Press 1974
9. Magnus, Karras, Solitar: Combinatorial group theory. New York: Interscience 1966
10. Milnor, J.W., Stasheff, J.D.: Characteristic classes. Ann. Math. Studies, Vol. 76. Princeton, NJ: Princeton University Press 1974

11. Griffiths, P., Harris, J.: Principles of algebraic geometry. New York: Wiley 1978
12. Gunning, R.: Introduction to Riemann surfaces princeton Math. Notes. Princeton, NJ: Princeton University Press 1966
13. Mumford, D.: Tata lectures on theta, Vols. I, II. Boston: Birkhäuser 1983
14. Fay, J.: Theta functions on Riemann surfaces Springer notes in math., Vol. 352. Berlin, Heidelberg, New York: Springer 1973
15. Quillen, D.: Determinants of Cauchy-Riemann operators on Riemann surfaces. Funkts. Anal. Prilozh. **19**, 37 (1985)
16. Alvarez, O.: Conformal anomalies and the index theorem. Berkeley preprint
17. Moore, G., Nelson, P., Polchinski, J.: Strings and supermoduli. Phys. Lett. **169**B,47 (1986)
18. Belavin, A.A., Knizhnik, V.G.: Algebraic geometry and the geometry of quantum strings. Landau Institute preprint; See also, Catenacci, R., Cornalba, M., Martellini, M., Reina, C.: Algebraic geometry and path integrals for closed strings; Bost, J.B., Jolicoeur, J.: A holomorphy property and critical dimension in string theory from an index theorem. Saclay PhT/86-28
19. Alvarez-Gaumé, L., Witten, E.: Gravitational anomalies. Nucl. Phys. B**234**, 269 (1984)
20. Actually, this particular example is well-known to physicists. See, for example, Jackiw, R.: Topological methods in field theory. Les Houches lectures 1983. What is new here is the better geometrical understanding in terms of holomorphic line bundles, and the idea that holomorphy is a powerful tool for understanding determinants
21. The generalization of Quillen's theorem has been independently derived by Belavin and Knizhnik. We thank Stephen Della Pietra for pointing out an error in an earlier version of Eq. (4.15). We also thank Phil Nelson and Joe Polchinski for discussions on the application of Eq. (4.15) to holomorphic factorization on moduli space, and on the important difference between Eq. (4.15) and Eq. (4.16)
22. Bismut, J.-M., Freed, D.S.: Geometry of elliptic families: Anomalies and determinants. M.I.T. preprint; The analysis of elliptic families: Metrics and connections on determinant bundles. Commun. Math. Phys. (in press); The analysis of elliptic families: Dirac operators, eta invariants, and the holonomy theorem. Commun. Math. Phys. (in press)
23. This can also be understood from the point of view of Wolpert, S.: On obtaining a positive line bundle from the Weil-Petersson class. Am. J. Math. **107**, 1485 (1985)
24. Bers, L.: Finite dimensional Teichmüller spaces and generalizations. Bull. Am. Math. Soc. **5**, 131 (1981)
25. Friedan, D.: Introduction to Polyakov's string theory. In: Zuber, J.-B., Stora, R. (eds.). Les Houches 1982. Recent advances in field theory and statistical mechanics. Amsterdam: Elsevier 1984
26. We owe this observation to Phil Nelson
27. Atiyah, M.: Riemann surfaces and spin structures. Ann. Sci. Ec. Norm. Super. **4** (1971)
28. In fact, more is true: the *number* of zero modes is given by the multiplicity of the zero of ϑ. See [14] and Mumford, D.: Theta characteristics of an algebraic curve. Ann. Sci. Éc. Norm. Supér. **4**, 24 (1971)
29. Hitchin, N.: Harmonic spinors. Adv. Math. **14**, 1–55 (1974)
30. Witten, E.: Global anomalies in string theory. In: Geometry anomalies topology. Bardeen, W.A., White, A.R. (eds.). New York: World Scientific 1985
31. Seiberg, N., Witten, E.: Spin structures in string theory. Princeton preprint
32. Alvarez-Gaumé, L., Ginsparg, P., Moore, G., Vafa, C.: An $O(16) \times O(16)$ heterotic string. Harvard preprint HUTP-86/AO13
33. For a pedagogical treatment and further references to the literature see Nelson, P., Moore, G.: Heterotic geometry. Harvard preprint HUTP-86/A014
34. Alvarez, O.: Theory of strings with boundary: Topology, fluctuations, geometry. Nucl. Phys. B**216**, 125 (1983)
35. Polchinski, J.: Evaluation of the one loop string path integral. Commun. Math. Phys. **104**, 37 (1986)

36. Moore, G., Nelson, P.: Measure for moduli: The Polyakov string has no nonlocal anomalies. Nucl. Phys. B**266**, 58 (1986)
37. D'Hoker, E., Phong, D.: Multiloop amplitudes for the bosonic string. Nucl. Phys. B **269**, 205 (1986)
38. D'Hoker, E., Phong, D.: Loop amplitudes for the fermionic string. CU-TP-340
39. Shapiro, J.: Loop graph in the dual-tube model. Phys. Rev. D**5**, 1945 (1972); Rohm, R.: Spontaneous supersymmetry breaking in supersymmetric string theories. Nucl. Phys. B**237**, 553 (1984)
40. Gliozzi, F., Scherk, J., Olive, D.: Supersymmetry, supergravity theories and the dual spinor model. Nucl. Phys. B**122**, 253 (1977)
41. In preparation
42. Igusa, J.: Theta functions. Berlin, Heidelberg, New York: Springer 1972
43. Vafa, C.: Modular invariance and discrete torsion on orbifolds
44. Dixon, L., Harvey, J.: String theories in ten dimensions without spacetime supersymmetry. Princeton preprint
45. After this work was completed we received a preprint in which this question is answered in the affirmative. See Manin, Yu.I.: The partition function of the Polyakov string can be expressed in terms of theta functions. Phys. Lett. (submitted)

Communicated by A. Jaffe

Received April 4, 1986

Commun. Math. Phys. 112, 503–552 (1987)

Communications in
**Mathematical
Physics**
© Springer-Verlag 1987

Bosonization on Higher Genus Riemann Surfaces[†]

Luis Alvarez-Gaumé[★][††], Jean-Benoît Bost[2], Gregory Moore[3][★★][†††],
Philip Nelson[3][★★][††], and Cumrun Vafa[3][★★]

[1] Theory Division, CERN, CH-1211 Genève 23, Switzerland
[2] Ecole Normale Supérieure, 45 Rue d'Ulm, F-75230 Paris Cedex 05, France
[3] Lyman Laboratory of Physics, Harvard University, Cambridge, MA 02138, USA

Abstract. We prove the equivalence between certain fermionic and bosonic theories in two spacetime dimensions. The theories have fields of arbitrary spin on compact surfaces with any number of handles. Global considerations require that we add new topological terms to the bosonic action. The proof that our prescription is correct relies on methods of complex algebraic geometry.

1. Introduction

Two-dimensional quantum field theory is very special. Many surprising and beautiful results turn out to be true only in two dimensions, including for example the exact solvability of certain models, the equivalence of fermionic and bosonic field theories, and so on. One way of describing the root cause for all these miracles is to note that in two dimensions the light cone is disconnected; it consists of a left moving and a right moving branch, and massless particles stay on one branch or the other[1].

This cleavage in turn comes from the fact that in two dimensions the scalar wave operator factorizes into the product of left and right moving derivatives. In euclidean space the analogous statement is

$$\nabla^2 = \bar\partial^\dagger \bar\partial \ , \tag{1.1}$$

where $\bar\partial$ is the Cauchy-Riemann operator. Thus in a sense we can say that $2d$ fields are special because for them complex analysis plays a key role.

In this paper we will see how complex analytic methods can extend our understanding of $2d$ fields from surfaces with the topology of the plane (or sphere)

[†] Work supported in part by NSF grant PHY-82-15249 and DOE contracts DE-FG02-84-ER-40164-A001 and DE-AC02-76ER02220
[★] Alfred P. Sloan Foundation Fellow
[††] Present address: Department of Physics, Boston University, Boston, MA 02215, USA
[★★] Harvard Society of Fellows
[†††] Present address: Institute for Advanced Study, Princeton, NJ 08540, USA
[1] See, e.g. the physical discussion in § V.B of [1]

to arbitrary compact euclidean spacetimes. Specifically we will study Fermi-Bose equivalence, or "bosonization". We give a prescription for bosonizing the correlation functions of a first order fermionic system with fields of any spin and any twisted spin structure. Our prescription generalizes that of [2, 3, 4]; in particular certain global terms must be added to the scalar action for nontrivial spacetimes. Most of these results were announced in [5] and build on [6] and [7].

Field theory on complicated surfaces, and in particular bosonization, has become an important tool in the study of string theory. For example, bosonization has been used in light-cone gauge to prove the equivalence of the Green-Schwarz and NSR superstring [8, 9]. Bosonization also plays a key role in understanding the gauge- and super-symmetry of the heterotic string [10] and in formulating the covariant fermion emission vertex [11, 12]. The methods we use however are quite general and we expect them to be of use in $2d$ field theory for problems other than bosonization. For instance we obtain some expressions for functional determinants in terms of the natural functions associated to a Riemann surface.

The key step in understanding $2d$ fields on compact surfaces is the observation that while the amplitudes are functionals on the large space of metric background fields, nevertheless most of this dependence is understood using the various known anomalies. The only interesting dependence is on the "moduli space" $\mathcal{M}_g$ of conformally-inequivalent surfaces with g handles. Similarly the dependence on flat background gauge fields boils down to one on the "jacobian variety" $J(\Sigma)$ of inequivalent bundles on a given surface Σ. The spaces $\mathcal{M}_g$ and $J(\Sigma)$ are both finite-dimensional. (Indeed both are *trivial* on the plane, corresponding to the well-known fact that fermion dynamics on the plane is completely given by the anomaly). Furthermore each is naturally a complex space, a consequence of the complex form of the wave operator (1.1). Thus as mentioned earlier, powerful complex analytic methods are available to study quantum amplitudes. This is why two dimensions is so special.

The link between field-theoretic and algebraic-geometric methods is provided by the theorems in [13, 14, 15] (see also [16, 17, 18]), which describe the determinant of a family of Cauchy-Riemann operators in terms of the complex structure of $\mathcal{M}_g$. The case originally studied by Quillen involves families of operators at one point of $\mathcal{M}_g$, parametrized by background gauge potentials on the given surface. These results were later generalized by Belavin and Knizhnik and by Bost and Jolicœur (using results of Bismut and Freed) to include families of Cauchy-Riemann operators parametrized by $\mathcal{M}_g$, which are of interest in string theory. The main conclusion is that the combinations of determinants appearing in the integrand over $\mathcal{M}_g$ in the bosonic and fermionic strings factorize into sections of flat holomorphic line bundles on $\mathcal{M}_g$. This factorization is useful for example when we study the infinities of string theory by allowing Riemann surfaces with nodes.

It has also been known for some time that Quillen's work is closely related to Falting's work on Arakelov geometry [19, 20]. It was suggested in [15] that a combination of Quillen's and Faltings' ideas would be of use in string theory. We will use just such a combination to prove our results on bosonization.

As mentioned, we will generalize the bosonization prescription given in [2, 3, 4] for anticommuting fields of any spin on the sphere. When we try to generalize to arbitrary compact surfaces, however, we face the problem that there is in general no euclidean "time" to use in a canonical formalism. Fortunately there is one case, the torus, where operator methods still work and yet the topology is interesting enough to show what happens in higher loops. We will use the canonical formalism to get the correct prescription in a simple case, then use modular invariance and factorization to guess the correct general prescription, in any number of loops. The prescription so obtained is unique.

To prove that our bosonization rules really do work, we will compute corresponding quantities in the bosonic and fermionic languages. Setting these equal gives a set of identities which express the content of Fermi-Bose equivalence. Finally we prove the identities using methods of algebraic geometry, thus establishing bosonization.

While the last steps get rather involved, we emphasize that the prescription itself is not too complicated. The reader may wish to turn immediately to Sect. 4 to see the statement of the bosonization rules.

Throughout this paper we will discuss only nonchiral theories. We restrict to this case because, as is well known, chiral determinants are problematic in $2d$ gravity due to anomalies. In the bosonic language this appears as a difficulty in defining chiral scalar fields in a path integral. There has been some progress in chiral bosonization in [21–24] and elsewhere, but a discussion is beyond the scope of this paper.

Also, in this paper we discuss bosonization physically in terms of path integrals and mathematically in terms of isometries of determinant bundles. Historically, another approach to bosonization in spin 1/2 has proceeded via the isomorphism between spinor and vertex operator representations of affine Lie algebras [25, 26]. It would be extremely interesting to unify and generalize these two approaches using a general operator formalism. Recent progress on this problem has been made along these lines in [27, 24, 28] (see also [29, 30]).

In Sect. 2 we describe various aspects of the theory of Riemann surfaces which we will need. In particular we discuss various ways to describe bundles, choices of homology basis, holonomy, and Arakelov metrics. References [6, 16, 31, 17] may be useful background for this section and for the whole paper. In Sect. 3 we arrive at the bosonization rules and in particular show that the scalar action is independent of various choices made in defining it. In Sect. 4 we give the complete set of rules, and work out the identities mentioned above. Section 5 contains the proofs of these identities.

We draw the reader's attention to several preprints on related topics which we received after this work was completed. These include [32, 22, 23, 33–36].

2. Foundations

This is a long introductory section in which we introduce some machinery[2]. In particular we describe bundles over Riemann surfaces in three different ways: via transition functions, via divisors, and via points in the jacobian mentioned ealier. We also briefly review theta functions and Arakelov metrics. The reader may wish to skip this section and refer to it as needed.

---

[2] We thank V. Dellapietra for many discussions on the material in this section, and also Sect. 3C

A. Surfaces. Let Σ be a smooth compact connected surface. We will always assume Σ is oriented, as for example in the heterotic string. We will soon need to give Σ more structure, but first we will briefly note some facts about its topology.

The homology of Σ is simple. $H_0(\Sigma;\mathbf{Z})$ has one generator since Σ is connected, while $H_2(\Sigma;\mathbf{Z})$ has one generator since Σ is compact, connected, and oriented [37]. The first homology group has $2g$ generators, where the genus g is an integer which completely specifies Σ topologically. See Fig. 1. By triangulating Σ one can see that the Euler number of Σ is $\chi=2-2g$; χ is also the Chern number of the tangent to Σ. The oriented intersection number of two 1-cycles is a signed integer, and

$$a\cdot a'=-a'\cdot a\ .$$

This pairing defines a quadratic form on $H_1(\Sigma)$.

We will want our amplitudes to depend only on intrinsic information. For example, the partition function for spin-1/2 fermions depends on a surface and a choice of spin bundle. In practice, however, we need coordinates to describe the intrinsic data, and this requires that we make some noninvariant choices. We then have to verify later that our answers are independent of the choices made. The most important such choice, which we will use throughout this paper, is that of a basis of $H_1(\Sigma;\mathbf{Z})$. While there is no preferred basis, we can restrict the choice somewhat by choosing only bases of the form $\mathscr{A}=\{a_1,\ldots,a_g,b_1,\ldots,b_g\}$, satisfying the invariant condition:

$$a_i\cdot a_j=b_i\cdot b_j=0\ ,\qquad a_i\cdot b_j=\delta_{ij}\ ,$$

or

$$\begin{pmatrix}\vec{a}\\ \vec{b}\end{pmatrix}\cdot(\vec{a},\vec{b})=J\quad\text{where}\quad J\equiv\begin{pmatrix}0 & 1\\ -1 & 0\end{pmatrix}\ . \tag{2.1}$$

Any basis $\mathscr{A}$ with property (2.1) will be called "canonical". Any other canonical basis $\tilde{\mathscr{A}}$ will then be related to $\mathscr{A}$ by an integer matrix preserving J:

$$\tilde{\mathscr{A}}=\mathscr{A}\cdot\Lambda^{-1}\ ,\qquad \Lambda^tJ\Lambda=J\ . \tag{2.2}$$

The group of such Λ is the "symplectic modular group" $Sp(2g,\mathbf{Z})$. Letting

$$\Lambda=\begin{pmatrix}A & -B\\ -C & D\end{pmatrix}\ , \tag{2.3}$$

we get

$$\Lambda^{-1}=\begin{pmatrix}D^t & B^t\\ C^t & A^t\end{pmatrix}\ .$$

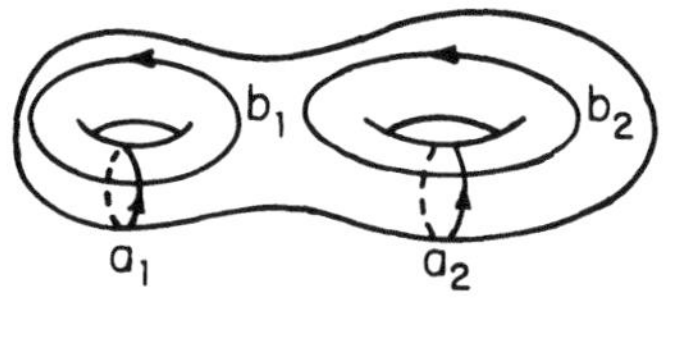

Fig. 1. A Riemann surface of genus two, with a canonical homology basis

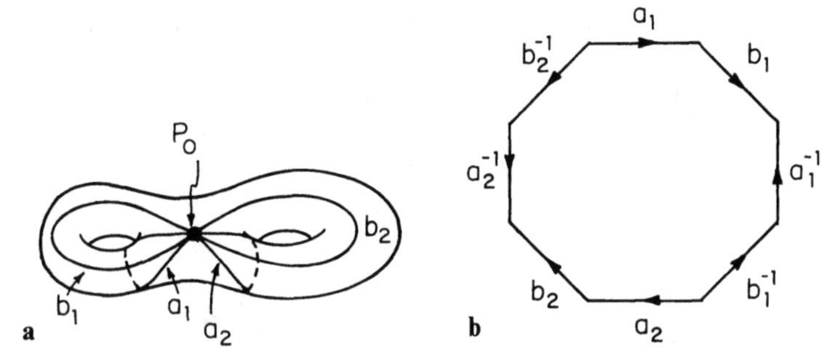

Fig. 2. a Representatives of a basis intersecting at P_0. **b** The cut surface Σ_c

We can also define a dual basis $\mathscr{A}^*$ of $H^1(\Sigma;\mathbf{Z})$ by

$$\mathscr{A}^* = \{\alpha^1,\ldots,\alpha^g,\beta^1,\ldots,\beta^g\}^t , \quad \langle \mathscr{A}^*, \mathscr{A}\rangle = \mathbf{1} , \quad \tilde{\mathscr{A}}^* = \Lambda \mathscr{A}^* .$$

Thus if we expand a cohomology class ψ as $\psi = (n, m)\binom{\alpha}{\beta}$, then with respect to a new basis ψ is described by

$$(\tilde{n}, \tilde{m}) = (n, m)\Lambda^{-1} . \tag{2.4}$$

We will usually append a subscript $\mathscr{A}$ to any object we construct which depends on a choice of basis.

Again, the introduction of a basis $\mathscr{A}$ is a necessary evil needed to parametrize various spaces. It explicitly breaks invariance under diffeomorphisms, since $f_*(a_i)$ is not in general homologous to a_i if f is not connected to the identity. However, since $f_*(a)\cdot f_*(a') = a\cdot a'$, we do know that $f_*\mathscr{A}$ always differs from $\mathscr{A}$ by a transformation in $Sp(2g,\mathbf{Z})$. Thus if we are careful not to make any *further* noninvariant choices beyond that of $\mathscr{A}$, we see that invariance under $Sp(2g,\mathbf{Z})$ suffices to establish invariance under the full group of disconnected coordinate transformations of Σ. The former condition is also called "modular invariance".

Given a homology basis, or "marking"[3], for Σ, we can choose specific curves representing each homology class and all intersecting at one point P. See Fig. 2a. It is then useful to introduce the "cut" surface Σ_c with the topology of a disk (Fig. 2b). Since Σ_c involves more choices than just $\mathscr{A}$, we will have to verify that constructions made with its help are unchanged as we vary the curves in their homology classes.

One useful calculation with Σ_c is the following: let θ, η be closed 1-forms on Σ. Then $\int_\Sigma \theta\wedge\eta = \int_{\Sigma_c} \theta\wedge\eta = \int_{\Sigma_c} d(\varphi\eta)$, where $\theta = d\varphi$ on Σ_c. By Stokes' theorem this is the integral of $\varphi\eta$ around the boundary of Σ_c. Grouping the boundary segments in pairs we get [39]

$$\begin{aligned}\int_\Sigma \theta\wedge\eta &= \sum_i [\varphi(a_i^+) - \varphi(a_i^-)] \oint_{b_i} \eta - \sum_i [\varphi(b_i^+) - \varphi(b_i^-)] \oint_{a_i} \eta \\ &= \sum_i \oint_{a_i} \mathrm{d}\varphi \cdot \oint_{b_i} \eta - (a \leftrightarrow b) ,\end{aligned}$$

[3] This is different from the sense of the word used in [38]

or

$$\int_{\Sigma} \theta \wedge \eta = \sum \left[\oint_{a_i} \theta \oint_{b_i} \eta - \oint_{b_i} \theta \oint_{a_i} \eta \right] = (n, m) J \binom{n'}{m'}, \tag{2.5}$$

where we expanded θ, η in terms of (n,m) and (n',m'). This identity is clearly independent of all choices made.

B. Riemann Surfaces, Sheaves, and Bundles. In order to define a Laplacian, say, we must have a metric. For many purposes, however, the conformal class of a metric is all we need. For example, given a metric class the Hodge theorem gives representative differential forms for the cohomology classes $\vec{\alpha}$, $\vec{\beta}$ dual to $\mathscr{A}$, namely the harmonic forms. Also as is well known a conformal class amounts to a *complex structure* on Σ [16, 31]. From now on Σ will denote a surface endowed with such a structure, which is called a Riemann surface.

The complex coordinate patching functions of Σ are analytic on the overlaps $\mathscr{U}_i \cap \mathscr{U}_j$ of coordinate charts. One can also consider holomorphic line bundles over Σ, complex bundles whose transition functions are always analytic. Such bundles are important because they have a well-defined notion of a holomorphic section. In fact there is a Cauchy-Riemann $\bar{\partial}$ operator on the sections of any holomorphic line bundle ξ, which we call $\bar{\partial}_\xi$; a holomorphic section σ satisfies $\bar{\partial}_\xi \sigma = 0$ (see Sect. 5.C). Clearly, the derivative of the patching functions define such a bundle, the holomorphic tangent K^{-1} of Σ. Its dual K is called the canonical bundle over Σ. We will always consider holomorphic bundles unless otherwise noted.

Thus the bundles K^n correspond to tensors with n lower z indices, where z is a local complex coordinate of Σ. Sections of $K^n \otimes \bar{K}^m$ are called (n,m)-tensors. In order to deal with spin we also define a holomorphic spin bundle[4] L as any bundle such that $L \otimes L \cong K$;, the corresponding tensors then have "half a z index". Before describing these in greater detail, however, we first recall the notion of a sheaf [41].

The language of sheaves is useful for many constructions in geometry. We will only make essential use of it in the last section, however, and the reader may wish to skip the following paragraphs. While we will work on a surface Σ in this section, most of the constructions have analogs in higher dimensions as well.

A sheaf $\mathscr{F}$ of abelian groups on Σ is an assignment of an abelian group $\mathscr{F}(U)$ to every open set $U \subseteq \Sigma$. This is the only kind of sheaf we will consider in this paper. $\mathscr{F}(U)$ is called the group of "sections of $\mathscr{F}$ over U". $\mathscr{F}$ also assigns to pairs of nested sets $V \subseteq U$ a "restriction" map $r_{U,V}: \mathscr{F}(U) \to \mathscr{F}(V)$ in a way which makes sense on overlaps. That is,

a) If $W \subseteq V \subseteq U$, then $r_{U,V} = r_{U,V} \circ r_{V,U}$.

b) If $\sigma_1, \sigma_2, \ldots$ are sections over $U_1, U_2, \ldots$, respectively and each pair σ_i, σ_j have the same restriction to $U_i \cap U_j$, then each σ_i is the restriction of some section ϱ over $U_1 \cup U_2 \cup \ldots$.

c) If ϱ is a section of $\mathscr{F}$ over $U \cup V$ which gives the identity when restricted to both U and V, then ϱ is the identity.

Here are some examples of sheaves which we will use.

[4] See [40, 31] for why this definition coincides with the usual definition of spin structure

a) $\mathcal{O}$. Over $U \subseteq \Sigma$, $\mathcal{O}(U) = \{$analytic functions on $U\}$; the group law is addition.

b) $\mathcal{M}$. Here $\mathcal{M}(U) = \{$meromorphic functions on $U\}$. In between $\mathcal{O}$ and $\mathcal{M}$ we have

c) $\mathcal{O}(P)$. If $P \in U$, then $\mathcal{O}(P)(U) = \{$functions analytic on U except for a possible pole of first order at $P\}$; otherwise $\mathcal{O}(P)(U) = \mathcal{O}(U)$. Note that the constant 1 is a canonical section of $\mathcal{O}(U)$ over any U.

d) $\mathcal{O}(-P)$. If $P \in U$, then $\mathcal{O}(-P)(U) = \{$functions analytic and vanishing to at least first order at $P\}$. Note that $\mathcal{O}(-P)(\Sigma)$ has only the zero section, since Σ is compact.

e) Given any holomorphic bundle ξ on Σ, we can define analogously to (a) $\xi(U)$ = analytic sections of ξ on U. We will not distinguish notationally between a bundle and its sheaf of holomorphic sections.

f) Given a vector space $\mathscr{V}$ and a point $P \in \Sigma$ we can let $\mathscr{F}(U) = \mathscr{V}$, if $P \in U$, and otherwise the zero vector space. The restriction map is either the identity or else the zero map. $\mathscr{F}$ is called the "skyscraper sheaf" and is denoted by $\mathscr{V}|_P$. In particular, given a bundle ξ and a point, we will write $\xi|_P$ to denote both the fiber at P and the corresponding sheaf with support at P.

g) Given any group such as **Z**, **R**, or **C**, we can define $\mathbf{Z}(U) = \mathbf{Z}$ etc. for every connected U. Every restriction is the identity. Sections of **Z** etc. can be thought of as locally constant functions on U.

h) Finally, we can define sheaves where the group law is multiplication, not addition, of functions. These include the constant sheaf $\mathbf{C}^* = \mathbf{C} - \{0\}$ and the sheaf $\mathcal{O}^*$ whose sections are the local analytic functions which never vanish.

Thus roughly speaking the notion of sheaf generalizes that of bundle to include cases where the fiber dimension jumps (example f), as well as cases where only constant local sections are allowed (example g).

In fact $\mathcal{O}(P)$ is the sheaf of analytic sections of a certain bundle ξ, which we construct as follows. Let $U_0 = \Sigma - \{P\}$ and U_1 a small disk neighborhood of P. Trivialize ξ so that a section s is given by functions s_i on U_i with $s_1 = z \cdot s_0$ on $U_0 \cap U_1$, where z is a complex coordinate centered at P. Given a function f in $\mathcal{O}(P)(U)$ we get a section of ξ on $U \cap U_0$, which we then analytically continue to $U \cap U_1$ using the transition function. Then clearly the functions in $\mathcal{O}(P)(U)$ all correspond to *smooth* analytic sections of ξ. In particular, the canonical section 1 *vanishes* once at P. We will write $\mathcal{O}(P)$ to refer either to the sheaf or the bundle, and $\mathbf{1}_{\mathcal{O}(P)}$ for the canonical section. Similarly $\mathcal{O}(-P)$ gives a bundle via $s_1 = z^{-1} \cdot s_0$. It too has a canonical section $\mathbf{1}_{\mathcal{O}(-P)}$ which now blows up at P.

A map between sheaves is a collection of homomorphisms $f_U : \mathscr{F}(U) \to \mathscr{G}(U)$ commuting with restriction. Roughly speaking, a sequence of maps is called exact at $P \in \Sigma$ if

$$\ldots \to \mathscr{F}(U) \to \mathscr{G}(U) \to \ldots$$

is exact for all sufficiently small neighborhood U of P. (See [41] for the precise definition.) An exact sequence of the form

$$0 \to \mathscr{E} \xrightarrow{\alpha} \mathscr{F} \xrightarrow{\beta} \mathscr{G} \to 0 \tag{2.6}$$

is called "short". Recall that this implies that α has no kernel and β is onto, as well as that image $\alpha = \ker \beta$, again on small enough neighborhoods of each point of Σ. We will use three simple exact sequences:

a) $$0 \to 2\pi i \mathbf{Z} \overset{i}{\hookrightarrow} \mathcal{O} \overset{\exp}{\to} \mathcal{O}^* \to 0 \ . \tag{2.7}$$

Here i is inclusion: every constant function is in particular an analytic function. Since $e^{2\pi i x} = 1$ iff x is an integer, this sequence is exact.

b) $$0 \to \xi \otimes \mathcal{O}(-P) \overset{i}{\hookrightarrow} \xi \overset{r}{\to} \xi|_P \to 0 \ , \tag{2.8}$$

where ξ is any holomorphic bundle. Again i includes the sections of ξ vanishing at P into all sections. r restricts a section to its value at P, so the sequence is exact.

c) $$0 \to K \overset{i}{\hookrightarrow} K \otimes \mathcal{O}(P) \overset{\text{res}}{\to} \mathbf{C}|_P \to 0 \ , \tag{2.9}$$

where K is the cotangent bundle. i includes the holomorphic sections into the ones holomorphic except for a possible first-order pole at P. res is the residue map: if f is analytic at P, $\text{res}_P\left(\frac{f}{z}\,dz\right) = f(P)$ is coordinate-invariant.

We can build a cohomology theory for any sheaf $\mathcal{F}$ as follows [41, 37, 42].

We first define the groups of cochains with values in $\mathcal{F}$. Given a cover of Σ by open sets $\{U_\alpha\}$, a 0-cochain $\sigma \in C^0(\mathcal{F})$ is given by associating a section $\sigma_\alpha \in \mathcal{F}(U_\alpha)$ to every U_α. The full cochain group is freely generated by such σ. A 1-cochain $\sigma \in C^1(\Sigma; \mathcal{F})$ is given by associating a section $\sigma_{\alpha\beta} \in \mathcal{F}(U_\alpha \cap U_\beta)$ for every nonempty intersection, and so on. Introduce the coboundary operator

$$\delta : C^p \to C^{p+1} \ ,$$

$$\delta\sigma(U_0, U_1, \ldots, U_{p+1}) = \sum_{k=0}^{p+1} (-1)^k \sigma(U_0, \ldots, \hat{U}_k, \ldots, U_{p+1})|_{U_0 \cap \ldots \cap U_{p+1}} \ .$$

($\hat{U}_k$ means that the $k+1$-entry is deleted.) It is easy to check that $\delta^2 = 0$. If σ is a p-cochain satisfying $\delta\sigma = 0$, we say that σ is a cocyle. If $\sigma_p = \delta\sigma'_{p-1}$ for some $(p-1)$-cochain σ'_{p-1}, then we say that σ_p is a coboundary. The p-th Čech cohomology group associated to the covering $\{U_\alpha\}$ is defined on the group of p-cocyles Z^p modulo p-coboundaries δC^{p-1},

$$H^p(\{U_\alpha\}; \mathcal{F}) = Z^p(\{U_\alpha\}; \mathcal{F}) / \delta C^{p-1}(\{U_\alpha\}; \mathcal{F}) \ .$$

It is possible to define the cohomology groups $H^p(\Sigma; \mathcal{F})$ as the "limit" of these groups as the covering $\{U_\alpha\}$ gets finer and finer [43]. For the constant sheaves of type (g) above these groups are just the usual cohomology groups [43]. For the sheaf associated to a bundle ξ we have that $H^0(\Sigma; \xi)$ is just the space of global holomorphic sections, or in other words that

$$H^0(\Sigma; \xi) = \ker \bar{\partial}_\xi \ . \tag{2.10}$$

The other groups $H^p(\Sigma; \xi)$ are more complicated.

Given a short exact sequence of sheaves (2.6), there is associated a long exact sequence of cohomology groups [43]:

$$\begin{aligned} &0 \to H^0(\Sigma;\mathscr{E}) \to H^0(\Sigma;\mathscr{F}) \to H^0(\Sigma;\mathscr{G}) \to \\ &\to H^1(\Sigma;\mathscr{E}) \to H^1(\Sigma;\mathscr{F}) \to H^1(\Sigma;\mathscr{G}) \to \dots \\ &\dots \to H^p(\Sigma;\mathscr{E}) \to H^p(\Sigma;\mathscr{F}) \to H^p(\Sigma;\mathscr{G}) \to \dots \,. \end{aligned} \tag{2.11}$$

Using that α and β in (2.6) commute with the coboundary operator δ, it is easy to understand how one moves horizontally in the sequence (2.11). The step from $H^p(\Sigma;\mathscr{G})$ to $H^{p+1}(\Sigma;\mathscr{E})$ is more elaborate. Let σ be a p-cocyle in $H^p(\Sigma;\mathscr{G})$. We can represent σ by a cocycle in $C^p(\{U_\alpha\};\mathscr{G})$ for some covering of Σ. By exactness of (2.6) at $\mathscr{G}$, we can find some p-cocycle τ in $C^p(\{U'_\alpha\};\mathscr{F})$ such that $\beta(\tau)=\sigma$, where $\{U'_\alpha\}$ is some covering on Σ finer than $\{U_\alpha\}$. Since $\delta(\sigma)=0$, and β commutes with δ, by exactness of (2.6) at $\mathscr{E}$ and $\mathscr{F}$ there exists a unique $(p+1)$-cocycle $\mu \in C^{p+1}(\{U'_\alpha\};E)$ such that $\alpha(\mu)=\delta(\tau)$. The coboundary operator δ associates the class of μ to the class of σ. It is well-defined and independent of the choices made [43].

As a first application of sheaves we note that the transition functions of a holomorphic line bundle ξ are analytic functions on patch overlaps, i.e., a chain[5] $t_{\alpha\beta}$ in $C^1(\Sigma;\mathcal{O}^*)$. Moreover, the cocyle condition says that $(\delta t)_{\alpha\beta\gamma}\equiv 0$. Also redefining the local trivializations of ξ gives an equivalent set of transition functions $t'_{\alpha\beta}=t_{\alpha\beta}\cdot(\delta v)_{\alpha\beta}$, so that isomorphism classes of bundles are given by the group

$$\mathrm{Pic}\,(\Sigma)\equiv H^1(\Sigma;\mathcal{O}^*)\ . \tag{2.12}$$

Pic Σ is called the Picard group. Multiplication of transition functions $t_{\alpha\beta}\cdot\tilde{t}_{\alpha\beta}$ gives a new line bundle, the tensor product $\xi\otimes\tilde{\xi}$, while inversion $t_{\alpha\beta}^{-1}$ gives the dual bundle ξ^{-1}.

We can learn more about Pic Σ by using the long exact sequence associated to (2.7):

$$\begin{aligned} \dots H^0(\Sigma;\mathcal{O}) &\xrightarrow{\exp} H^0(\Sigma;\mathcal{O}^*) \xrightarrow{0} H^1(\Sigma;2\pi i\mathbf{Z}) \to H^1(\Sigma;\mathcal{O}) \to H^1(\Sigma;\mathcal{O}^*) \\ &\xrightarrow{\deg} H^2(\Sigma;\mathbf{Z}) \to H^2(\Sigma;\mathcal{O}) \dots \,. \end{aligned}$$

Since the first $\exp:\mathbf{C}\to\mathbf{C}^*$ is onto, the next map must be zero. Also $H^2(\Sigma;\mathcal{O})$ is always zero, since by the Dolbeault theorem [41] it is isomorphic to $H^{0,2}_{\bar\partial}(\Sigma)$ and there are no (0,2)-forms: $d\bar z\wedge d\bar z=0$. Hence we get (dropping the normalization $2\pi i$)

$$0\to H^1(\Sigma;\mathcal{O})/H^1(\Sigma;\mathbf{Z})\to \mathrm{Pic}\,(\Sigma)\xrightarrow{\deg} H^2(\Sigma;\mathbf{Z})\to 0\ .$$

Since $H^2(\Sigma;\mathbf{Z})=\mathbf{Z}$, we thus find that Pic (Σ) is disconnected, with identity component a Lie group we will call

$$J(\Sigma)\equiv H^1(\Sigma;\mathcal{O})/H^1(\Sigma;\mathbf{Z})\ .$$

Since $H^1(\Sigma;\mathcal{O})$ is a complex vector space, $J(\Sigma)$ is a complex space called the jacobian of Σ.

---

[5] Here and in the sequel we are somewhat imprecise in our notation: We mean that we have chosen a covering $\{U_i\}$ for Σ and used the transition functions to define a cochain in $C^1(\{U_i\};\mathcal{O}^*)$. It turns out that different choices of covering lead to cohomologous $t_{\alpha\beta}$'s in the limit of fine coverings mentioned above

The image deg ξ of a bundle in $\mathbf{Z}$ is called the degree of ξ. Clearly deg $(\xi_1 \otimes \xi_2) = \deg \xi_1 + \deg \xi_2$, and indeed the degree is just the first Chern class of the $U(1)$ bundle associated to ξ. For example $\mathcal{O}(P)$ has degree 1, as can easily be seen by working through the steps below (2.11). But the corresponding $U(1)$ bundle is the monopole, which has Chern number 1 as well. As another example the degree of the tangent bundle is the Euler number χ, or

$$\deg K = 2(1-g) \ ,$$

an even integer.

We can now return to the study of spin bundles. Suppose that K is described by $g_{\alpha\beta} \in H^1(\Sigma; \mathcal{O}^*)$. We can construct a square root L of K by letting

$$h_{\alpha\beta} = \pm\sqrt{g_{\alpha\beta}} \quad \text{on} \quad U_\alpha \cap U_\beta \ . \tag{2.13}$$

Unfortunately, the cochain h so defined will not in general be closed since

$$(w_2)_{\alpha\beta\gamma} \equiv (\delta h)_{\alpha\beta\gamma} = h_{\beta\gamma} h_{\alpha\gamma}^{-1} h_{\alpha\beta} = h_{\alpha\beta} h_{\beta\gamma} h_{\gamma\alpha}$$

can be ± 1 on $U_\alpha \cap U_\beta \cap U_\gamma$. Hence an arbitrary choice of square roots in (2.13) will not generally define any bundle L.

Given a bad choice of $h_{\alpha\beta} \in C^1(\Sigma; \mathcal{O}^*)$, however, we can try to turn it into a good one by letting

$$h'_{\alpha\beta} = h_{\alpha\beta} \cdot f_{\alpha\beta} \ , \qquad \text{where} \qquad f \in C^1(\Sigma; \mathbf{Z}_2) \ . \tag{2.14}$$

This changes w_2 to $w'_2 = w_2 \cdot (\delta f)$. Hence if w_2 defines a trivial class in $H^2(\Sigma; \mathbf{Z}_2)$, then we can find an appropriate f to shift it away and spin bundles will exist for K. But if we regard deg K as a class in $H^2(\Sigma; \mathbf{Z})$, then working through the definitions shows that

$$w_2 = \exp(i\pi \cdot \deg K) \ .$$

Since the degree of K is always even, w_2 is always trivial. Hence we can always arrange for h to be closed: spin bundles always exist on any Riemann surface.

Now suppose that $h \in H^1(\Sigma; \mathcal{O}^*)$ describes a spin bundle L. Once again we can try modifying it by f as in (2.14), where now f must be closed. This will give a distinct new spin bundle whenever f is not exact in $H^1(\Sigma; \mathbf{Z}_2)$. Thus the *differences* of spin bundles are given by $H^1(\Sigma, \mathbf{Z}_2) \cong (\mathbf{Z}_2)^{2g}$ [37]. Unfortunately there is no canonical, or preferred, spin structure on Σ, so we cannot directly parametrize all L by $H^1(\Sigma, \mathbf{Z}_2)$. Instead we will see that only after the introduction of a homology basis $\mathscr{A}$ will there emerge a special $L_{\mathscr{A}}$; we will then be able to describe other L relative to this one.

Since $\deg K = 2g-2$, we find $\deg L = g-1$ for any L. We will use the term "twisted spin bundle" to refer to *any* ξ of degree $g-1$, not necessarily satisfying $\xi^2 = K$.

Before leaving sheaf theory we will describe without proof one more important theorem. This is Serre duality, which says that for any bundle ξ we have

$$H^1(\Sigma; \xi) \cong [H^0(\Sigma; K \otimes \xi^{-1})]^{-1} \ . \tag{2.15}$$

As usual the inverse refers to the dual vector space. Also a simple argument shows that

$$H^1(\Sigma; \xi) \cong \operatorname{coker} \bar{\partial}_\xi \ . \tag{2.16}$$

(Compare (2.10).) Again one uses the Dolbeault theorem to see that $H^1(\Sigma;\xi) = H^{0,1}_{\bar\partial}(\Sigma;\xi)$. Since all (0,1)-forms are $\bar\partial$-closed, we have all ξ-valued (0,1)-forms modulo the $\bar\partial$-exact ones, which is just coker $\bar\partial_\xi$.

Taking ξ to be trivial, (2.15) says $H^1(\Sigma;\mathcal{O}) \cong [H^0(\Sigma;K)]^{-1}$. The space $H^0(\Sigma;K)$ of holomorphic 1-forms is always g-dimensional. One can see this by noting that if $\eta \in H^0(\Sigma;K)$ then the real and imaginary parts of η are harmonic 1-forms, and so can be expanded in the basis of $2g$ real harmonic 1-forms α^i, β^j mentioned at the beginning of this section. The space of complex linear combinations of α, β with no $\bar z$ piece then has g dimensions. Holomorphic 1-forms are also called "abelian differentials".

Thus the identity component of Pic(Σ) is $J(\Sigma) \cong [H^0(\Sigma;K)]^{-1}/H^1(\Sigma;\mathbf{Z})$, a *torus* of g complex dimensions. Hence it is compact, as promised. We have seen how it parametrizes degree-zero bundles, or *differences* of degree-d bundles for any d.

C. Divisors. In our discussion of sheaves we came across bundles we called $\mathcal{O}(P)$ and $\mathcal{O}(-P)$. It will be useful to generalize these by introducing the notion of a divisor; divisors give a second description of holomorphic line bundles.

A divisor is a formal linear combination of points of Σ with signed multiplicities: $D = \sum_i n_i P_i$. To such a D we associate the line bundle $\mathcal{O}(D) \equiv \otimes_i \mathcal{O}(P_i)^{n_i}$. $\mathcal{O}(D)$ is perfectly regular at the points P_i, but it comes equipped with a section $\mathbf{1}_{\mathcal{O}(D)} \equiv \otimes_i (\mathbf{1}_{\mathcal{O}(P_i)})^{n_i}$ which has zeros (respectively poles) at those P_i with $n_i > 0$ (respectively $n_i < 0$). We can add divisors in the obvious way, whereupon the map $\tilde I: D \mapsto \mathcal{O}(D)$ becomes a homomorphism. Under $\tilde I$ the degree of $\mathcal{O}(D)$ equals $\sum n_i$, as we have noted.

Conversely, given any bundle ξ we can find a divisor as follows. Every ξ has many meromorphic sections [44]. Choose any one section s and let div $(s) = \sum n_i P_i - \sum m_j Q_j$, where $\{P_i\}$ are the zeros of s of order n_i and $\{Q_i\}$ are poles of order m_j. This map inverts I, but it is ambiguous: letting $s' = f \cdot s$ changes div (s) by the divisor of any meromorphic function $f \in \mathcal{M}(\Sigma)$. Thus we define the group of divisor *classes* as all divisors modulo the divisors of meromorphic functions, to get

$$I: \{\text{divisor classes}\} \to \mathrm{Pic}(\Sigma)\ . \tag{2.17}$$

To show that I is an isomorphism we further remark that a nontrivial bundle is never represented by the zero divisor. If it were then the meromorphic section s with no zeros or poles would trivialize ξ.

Note that while two divisors $D \neq D'$ may give rise to isomorphic bundles, still the canonical sections $\mathbf{1}_{\mathcal{O}(D)}$, $\mathbf{1}_{\mathcal{O}(D')}$ are totally different: they vanish and blow up at different places and so cannot correspond to each other under the isomorphism $\mathcal{O}(D) \cong \mathcal{O}(D')$. Conversely, a given bundle has no canonical meromorphic section; only after a specific divisor has been chosen in its class is such a section available.

From now on we will not distinguish notationally between divisor classes and bundles. That is, we will sometimes drop I (and $\tilde I$) from formulas.

Since (2.17) is an isomorphism we can represent any bundle as any other one times some $\mathcal{O}(D)$. We can use this fact to extend a simple result about the cotangent K to arbitrary bundles. Note that $H^0(\Sigma;\mathcal{O}) \cong \mathbf{C}$, since the only analytic functions are the constants, while $\dim H^0(\Sigma;K) = g$ as mentioned earlier. Hence

$\dim H^0(\Sigma;\mathcal{O}) - \dim H^0(\Sigma;K) = \deg \mathcal{O} + 1 - g$, since $\deg \mathcal{O} = 0$. Now consider the corresponding statement for arbitrary ξ:

$$\dim H^0(\Sigma;\xi) - \dim H^0(\Sigma;K\otimes\xi^{-1}) = \deg \xi + 1 - g \ . \tag{2.18}$$

To derive (2.18) from the preceding version express ξ as $\xi \cong K \otimes \mathcal{O}(D)$ for some D. The difference between the two equations can be shown to hold by repeated application of the long exact sequence associated to (2.8). Equation (2.18) together with (2.15) gives the classical Riemann-Roch theorem.

In the remainder of this section we will make the isomorphism (2.17) explicit with the help of a homology basis $\mathscr{A}$. That is, we will define a map $I_{\mathscr{A}}$ from divisor classes to a complex torus $J_{\mathscr{A}}$. While the constructions are not intrinsic, they are helpful for making the connection to theta functions.

We can choose a basis of $H^0(\Sigma;K)$, or abelian differentials, by requiring that

$$\int_{a_i} \omega^j_{\mathscr{A}} = \delta_{ij} \ . \tag{2.19}$$

It is then useful to define the "period matrix"

$$[\tau_{\mathscr{A}}]^{ij} \equiv \int_{b_i} \omega^j_{\mathscr{A}} \ . \tag{2.20}$$

$\tau_{\mathscr{A}}$ is useful because it characterizes the surface Σ. Indeed Torelli's theorem implies that if two marked Riemann surfaces Σ and Σ' of the same genus have the same period matrix then they are isomorphic as Riemann surfaces, although the converse is certainly false ($\tau_{\mathscr{A}} \neq \tau_{\tilde{\mathscr{A}}}$ in general for the same surface Σ with two markings). Using (2.5), $\tau_{\mathscr{A}}$ is easily seen to be symmetric with positive definite imaginary part [39, 6, 31]. Thus we can define

$$Y_{\mathscr{A}} = (\tau_{\mathscr{A}} - \bar{\tau}_{\mathscr{A}})^{-1} \ . \tag{2.21}$$

Note that $(Y_{\mathscr{A}})^{-1}_{ij} = \int_{\Sigma} \bar{\omega}^i \wedge \omega^j$ by (2.5), (2.19), and (2.20).

It is useful to know how things change when we change the marking $\mathscr{A}$. If $\tilde{\mathscr{A}} = \mathscr{A}\Lambda^{-1}$ is a new basis as in (2.2), then the definitions give

$$\vec{\omega}_{\tilde{\mathscr{A}}} = \vec{\omega}_{\mathscr{A}} \cdot (C\tau_{\mathscr{A}} + D)^{-1} \ , \tag{2.22}$$

$$\tau_{\tilde{\mathscr{A}}} = (A\tau_{\mathscr{A}} + B)(C\tau_{\mathscr{A}} + D)^{-1} \ , \quad \text{and} \tag{2.23}$$

$$Y_{\tilde{\mathscr{A}}} = (C\tau_{\mathscr{A}} + D) \cdot Y_{\mathscr{A}} \cdot \overline{(C\tau_{\mathscr{A}} + D)}^t \ . \tag{2.24}$$

We will sometimes drop the subscript $\mathscr{A}$ when it is clear which homology basis is meant.

Given a marked surface $\Sigma_{\mathscr{A}}$ we can now build a complex g-torus:

$$J_{\mathscr{A}} = \mathbf{C}^g / \Gamma_{\mathscr{A}} \ ;$$

$$\Gamma_{\mathscr{A}} = \mathbf{Z}^g \oplus \tau_{\mathscr{A}} \cdot \mathbf{Z}^g \ .$$

Changing homology bases as in (2.2), then we get a map from $J_{\mathscr{A}} \to J_{\tilde{\mathscr{A}}}$ which sends $\vec{z} \in \mathbf{Z}^g$ to

$$\tilde{z} = z \cdot (C\tau_{\mathscr{A}} + D)^{-1} \ . \tag{2.25}$$

We can now define a map $I_{\mathscr{A}}$ from divisor classes of degree zero to $J_{\mathscr{A}}$, commuting with the canonical identifications (2.25). Note that a divisor D of degree zero is the boundary ∂c of some 1-cochain obtained by "connecting the dots". Let

$$\tilde{I}^i_{\mathscr{A}}[D] = \int_c \omega^i_{\mathscr{A}} \mod \Gamma_{\mathscr{A}} . \tag{2.26}$$

Using (2.22), it is clear that (2.26) commutes with (2.25). Also, if we change σ to σ', then $\sigma - \sigma'$ is a cycle and so $\tilde{I}_{\mathscr{A}}[D]$ is unchanged modulo $\Gamma_{\mathscr{A}}$, by (2.19) and (2.20).

Finally $\tilde{I}_{\mathscr{A}}$ defines a map $I_{\mathscr{A}}$ on divisor classes. That is $\tilde{I}_{\mathscr{A}}[D]=0$ if and only if $D = \operatorname{div} f$ for some meromorphic function f. This is called Abel's theorem. Thus $I_{\mathscr{A}}$ sets up an isomorphism between the abstract jacobian $J(\Sigma)$ and its concrete version $J_{\mathscr{A}}$.

D. Curvature and Holonomy. There is one final characterization of holomorphic line bundles which we will use, involving holonomy. It is also time to begin introducing metrics and hence geometry on our bundles.

Let ξ be a holomorphic line bundle with a smooth hermitian norm $\|\cdot\|$. We can describe sections of ξ relative to one local trivializing sections s as $\sigma = f \cdot s$, where f is some function. "Trivializing" means that in some open set s is *analytic* and nonvanishing. With respect to the frame s we can now write down a covariant derivative [45]:

$$\bar{D}\sigma = (\bar{\partial} f)s \ , \qquad D\sigma = (\partial f + \Theta f)s \ ,$$

where

$$\Theta = \partial \log \|s\|^2$$

is the connection 1-form. The corresponding curvature is

$$R = \bar{\partial}\Theta = \bar{\partial}\partial \log \|s\|^2 \ , \tag{2.27}$$

and it is independent of the trivializing section s chosen. If s instead vanishes or has a pole somewhere, then (2.27) must be modified to remove the resulting delta-function singularities.

The Chern-Weil construction represents the Chern number in terms of the integral of R. In fact we can again see that this number equals the degree since

$$\int_{\Sigma} R = -2\pi i \deg \xi \ . \tag{2.28}$$

To prove (2.28), let $\xi = \mathcal{O}(P)$, $s = \mathbf{1}_{\mathcal{O}(P)}$, and integrate over Σ minus a small neighborhood U of P. Since $R = d(\partial \log \|\mathbf{1}_{\mathcal{O}(P)}\|^2)$ away from P we get $-\oint_{\partial U} (\partial_z \log \|\mathbf{1}_{\mathcal{O}(P)}\|^2) \cdot dz$. But $\mathbf{1}_{\mathcal{O}(P)}$ is vanishing linearly near p, so in any smooth metric this equals $-\oint \frac{dz}{z} = -2\pi i$. The reader may want to work out (2.28) explicitly on the sphere with $\xi = K$ to recover $\deg K = -2$.

We now consider a closed curve $\gamma: S^1 \to \Sigma$. Given a vector $v \in \xi|_{\gamma(0)}$ we can transport v around γ while maintaining the relation:

$$\langle Dv, \dot{\gamma} \rangle = 0 \ .$$

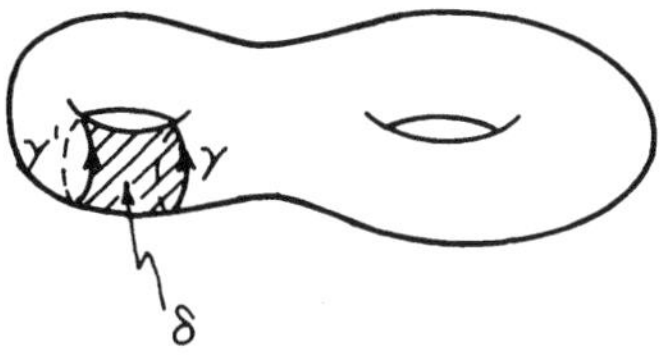

Fig. 3. Deforming a cut on Σ

Then $v(1)$ will again live in $\xi|_{\gamma(0)}$, and we can let

$$v(1)=v(0)e^{2\pi i H(\gamma;\xi)} \tag{2.29}$$

define the holonomy of ξ, with its given metric, around γ. Supposing that γ lies entirely in one trivializing patch, it is easy to show that

$$H(\gamma;\xi)=-\frac{1}{2\pi i}\int_{S^1}\gamma^*\Theta \mod \mathbf{Z}\ . \tag{2.30}$$

Further manipulation shows that $e^{2\pi i H}$ is well-defined under change of trivialization and H is real.

The holonomy changes in a simple way when we deform γ into a nearby γ' (Fig. 3). Suppose δ is the region of Σ lying between γ and γ', with orientation given by $(\dot{\gamma}, \mathbf{n})$, where $\mathbf{n}$ is the outward normal at γ. Then $R=d\Theta$ and Stoke's theorem say that

$$H(\tilde{\gamma},\xi)-H(\gamma,\xi)=-\frac{1}{2\pi i}\int_\delta R\ . \tag{2.31}$$

In particular if $R\equiv 0$, so that $\deg\xi=0$, then H depends only on the homology class of γ. It therefore defines a real cohomology class, modulo an integral class, which we will call $H(\xi)\in H^1(\Sigma;\mathbf{R})/H^1(\Sigma;\mathbf{Z})$.

Actually every degree-zero bundle admits a flat metric, which is unique up to a constant. To see this we choose an arbitrary norm with curvature R, $\int R=0$, and modify the norm by a nonvanishing function $\exp(h)$. Then h should satisfy $\bar{\partial}\partial h=-R$, which can be solved since the right side is orthogonal to the constant function. Thus $H(\xi)$ depends only on ξ.

The flat holonomy $H(\xi)$ vanishes if and only if ξ is trivial. For, if ξ is flat we can let s be a holomorphic, covariantly constant section throughout the cut Riemann surface Σ_c. If $H(\xi)=0$ then s does not jump across the cuts and ξ is trivial, whereas if ξ is trivial we can consider the flat metric as a real function satisfying $\bar{\partial}\partial h=0$, or $h=$constant; then $\Theta=0$ and $H=0$. Furthermore every $H\in H^1(\Sigma;\mathbf{R})/H^1(\Sigma;\mathbf{Z})$ actually arises as $H(\xi)$ for some $\xi:\xi$ is just the bundle with constant transition functions $\exp(-2\pi i\langle H,a_i\rangle)$, $\exp(-2\pi i\langle H,b_i\rangle)$ across the cuts.

We thus have three intrinsic ways of describing degree-zero holomorphic line bundles: as patching data (i.e. $J(\Sigma)$), as divisor classes, and as real 1-forms defined modulo integers. In addition we have a description in terms of a homology basis (namely $J_{\mathscr{A}}$) and maps I, $I_{\mathscr{A}}$, and H making all the viewpoints isomorphic. See Fig. 4, where we have for convenience defined three more maps by requiring that everything commute.

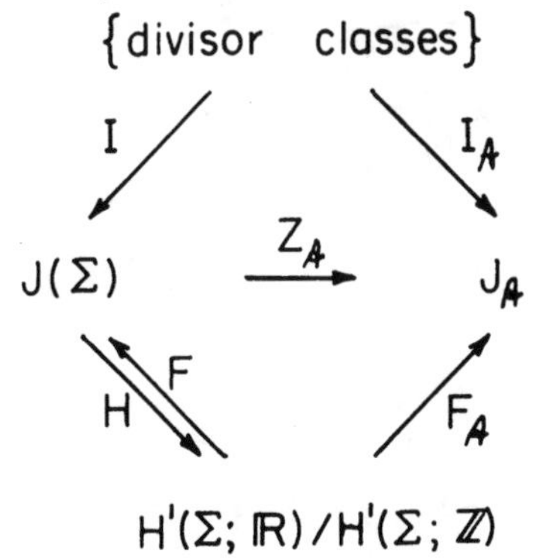

Fig. 4. Three ways to describe flat line bundles on a Riemann surface

Of the new maps in Fig. 4 it will be useful to know $F_{\mathscr{A}}$ explicitly. This takes holonomy data and translates it into a point in $\mathbf{C}^g$ defined modulo $\Gamma_{\mathscr{A}}$:

$$F_{\mathscr{A}}: \psi \mapsto -(\mathbf{1}, \tau_{\mathscr{A}}) J \begin{pmatrix} p \\ q \end{pmatrix} , \tag{2.32}$$

where

$$\psi = (\vec{\alpha}, \vec{\beta}) \begin{pmatrix} \vec{p} \\ \vec{q} \end{pmatrix} .$$

To show that (2.32) is right one must find the holonomy of a flat bundle with a given divisor D, then show that $I_{\mathscr{A}}[D]$ agrees with (2.32). This is done in Appendix A. Note that after a change of marking ψ is represented by $\begin{pmatrix} \tilde{p} \\ \tilde{q} \end{pmatrix} = (\Lambda^{-1})^t \begin{pmatrix} p \\ q \end{pmatrix}$. Using the symplectic property $\Lambda^t J \Lambda = J$, one can show that (2.32) indeed commutes with the identification $J_{\mathscr{A}} \cong J_{\tilde{\mathscr{A}}}$.

Before closing this section we note that the complex torus $J(\Sigma)$ has a natural hermitian norm. In terms of $J_{\mathscr{A}}$ this is

$$B_{\mathscr{A}}(z, z') = 2i\bar{z} \cdot Y_{\mathscr{A}} \cdot z' , \tag{2.33}$$

where $Y_{\mathscr{A}} = (\tau_{\mathscr{A}} - \bar{\tau}_{\mathscr{A}})^{-1}$. One can verify using (2.25), (2.24) that the $B_{\mathscr{A}}$ define an intrinsic norm B on $J(\Sigma)$. Note that one has

$$\operatorname{Im} B(F(\psi), F(\psi')) = \int_{\Sigma} \psi \wedge \psi' = (n, m) J \begin{pmatrix} n' \\ m' \end{pmatrix} . \tag{2.34}$$

E. Theta Functions and Spinors. We will be brief in this subsection; see [39, 6, 31]. The Riemann theta function is defined by

$$\vartheta(z|\tau) = \sum_{\vec{n} \in \mathbf{Z}^g} \exp(i\pi n \cdot \tau \cdot n + 2\pi i n \cdot z) \tag{2.35}$$

and satisfies

$$\vartheta(z + \tau \cdot n + m|\tau) = e^{-i\pi n \cdot \tau \cdot n - 2\pi i n \cdot z} \vartheta(z|\tau) . \tag{2.36}$$

Thus if we insert a period matrix $\tau_{\mathscr{A}}$ of some marked Riemann surface $\Sigma_{\mathscr{A}}$, $\vartheta(z|\tau_{\mathscr{A}})$

defines a section of a holomorphic bundle over the torus $J_{\mathscr{A}}$. This bundle is not trivial since ϑ sometimes vanishes but never blows up. The set $\Theta_{\mathscr{A}} \subseteq J_{\mathscr{A}}$ where it vanishes has complex codimension one and so is the natural generalization of the notion of divisor to g dimensions.

The component Pic_{g-1} of the Picard group consisting of bundles of degree $g-1$ also has a divisor of special points, called simply Θ, the "theta divisor". These are the twisted spin bundles ξ for which $\bar{\partial}_\xi$ has a zero mode, that is, the ones which admit a global holomorphic section. For example, if Σ is the torus, then $J(\Sigma)$ is itself a complex torus of dimension one. Thus we expect Θ to consist of discrete points, and indeed it is exactly one point: the trivial bundle is the only one of degree $g-1=0$ admitting a single-valued holomorphic section.

The Riemann vanishing theorem can be used to characterize the zeros of ϑ. It implies [39, 6, 31] that for any homology basis $\mathscr{A}$ there is a *preferred spin bundle* $L_{\mathscr{A}}$ with the property that

$$\Theta_{\mathscr{A}} = \{ I_{\mathscr{A}}[L_{\mathscr{A}} \otimes L^{-1}], \text{ as } L \text{ runs through } \Theta \} \ . \tag{2.37}$$

(As mentioned, sometimes we will write a bundle for the corresponding divisor class.) That is, for fixed $\tau_{\mathscr{A}}$ $\vartheta(z|\tau_{\mathscr{A}})$ vanishes precisely on a set which is Θ shifted by the preferred spin bundle $L_{\mathscr{A}}$.[6]

We can use (2.37) to parametrize all twisted spin bundles given a marking, as $L = L_{\mathscr{A}} \otimes F_{\mathscr{A}}(\psi)$, where $\psi = (\alpha, \beta)\begin{pmatrix} p \\ q \end{pmatrix}$ is a real cohomology class and $F_{\mathscr{A}}$ is as in Fig. 4. When we change the marking, however, we must be careful to account for the fact that $L_{\mathscr{B}} \not\cong L_{\mathscr{A}}$. Instead, if

$$L = L_{\mathscr{A}} \otimes F_{\mathscr{A}}(\psi) = L_{\mathscr{B}} \otimes F_{\mathscr{B}}(\hat{\psi}) \ , \tag{2.38}$$

$$\psi = (\alpha, \beta) J \begin{pmatrix} \phi \\ \theta \end{pmatrix} \qquad \hat{\psi} = (\alpha, \beta) J \begin{pmatrix} \hat{\phi} \\ \hat{\theta} \end{pmatrix} , \tag{2.39}$$

then (see (2.3))

$$\begin{pmatrix} \hat{\phi} \\ \hat{\theta} \end{pmatrix} = \Lambda \begin{pmatrix} \phi \\ \theta \end{pmatrix} + \delta \ ; \qquad \delta = \begin{bmatrix} \frac{1}{2}(AB^t)_d \\ \frac{1}{2}(CD^t)_d \end{bmatrix} . \tag{2.40}$$

The inhomogeneous term δ added to (2.4) represents the change from $L_{\mathscr{A}}$ to $L_{\mathscr{B}}$. The subscript d means the vector built from the diagonal elements of a matrix. The extra J in the description of ψ is a traditional notation for bundles. From now on $\begin{pmatrix} \phi \\ \theta \end{pmatrix}$ will refer to ψ as in (2.39), while (n, m) still refers to $(n, m)\begin{pmatrix} \alpha \\ \beta \end{pmatrix}$.

Equation (2.40) can be derived from (2.37) and the fact that $\vartheta(\hat{\phi} + \tau_{\mathscr{B}}\hat{\theta}|\tau_{\mathscr{B}})$ is a nonzero factor times $\vartheta(\phi + \tau_{\mathscr{A}}\theta|\tau_{\mathscr{A}})$ [46]. More specifically, one shows that [46]

$$\vartheta\begin{bmatrix} \hat{\theta} \\ \hat{\phi} \end{bmatrix}(0|\tau_{\mathscr{B}}) = \varepsilon \cdot \det(C\tau_{\mathscr{A}} + D)^{1/2} \cdot \vartheta\begin{bmatrix} \theta \\ \phi \end{bmatrix}(0|\tau_{\mathscr{A}}) \ , \tag{2.41}$$

[6] If we choose a point P_0 on Σ then $I_{\mathscr{A}}[L_{\mathscr{A}} \otimes \mathcal{O}((1-g)P_0)]$ is a point in $J_{\mathscr{A}}$ called the "vector of Riemann constants"

where ε is a phase and the "theta function with characteristics" is a convenient modification of (2.35), defined as

$$\vartheta\begin{bmatrix}\theta\\ \phi\end{bmatrix}(z|\tau)\equiv\sum_{\vec{n}\in\mathbf{Z}^g}\exp\left[i\pi(n+\theta)\tau(n+\theta)+2\pi i(n+\theta)(z+\phi)\right]$$
$$=\exp\left[i\pi\theta\cdot(\tau\cdot\theta+2(z+\phi))\right]\vartheta(z+\phi+\tau\theta|\tau)\ . \qquad (2.42)$$

It transforms as

$$\vartheta\begin{bmatrix}\theta\\ \phi\end{bmatrix}(z+m+n\cdot\tau|\tau)=\exp\left[-i\pi n\cdot\tau\cdot n-2\pi i n\cdot(z+\phi)+2\pi i m\cdot\theta\right]\vartheta\begin{bmatrix}\theta\\ \phi\end{bmatrix}(z|\tau)\ .$$

If ϕ, θ have half-integer entries, so that L in (2.38) is a spin bundle, then the phase ε in (2.41) is always an eighth root of unity.

Equation (2.42) is also a useful modification of (2.35) in that its transformation analogous to (2.36) is by a pure phase. Thus its absolute square is an ordinary continuous real function on the torus $J_{\mathscr{A}}$. We will denote this function by

$$\mathscr{N}_{\mathscr{A}}(z)=e^{-4\pi i y\cdot Y\cdot y}|\vartheta(z|\tau)|^2\ , \qquad (2.43)$$

where $y=\operatorname{Im} z$. $\mathscr{N}$ can be regarded as a metric on the bundle over $J_{\mathscr{A}}$ defined by ϑ, or more precisely,

$$\mathscr{N}_{\mathscr{A}}=\left\|\mathbf{1}_{\mathscr{O}(\Theta_{\mathscr{A}})}\right\|^2\ ,$$

a fact we use in Appendix A.

F. Arakelov Metrics. If we are given a metric on the cotangent bundle K, then we get at once metrics on all powers of K, with the property that

$$R_{K^\lambda}=\lambda R_K\ .$$

For fractional λ this is independent of the spin structure chosen, since all spin structures differ by *flat* bundles. In this section we will describe a particularly useful metric first described by Arakelov [20] and used extensively by Faltings [19]. While we will explicitly show that our results remain true for every metric, the use of the Arakelov metric in the intermediate stages will simplify our formulas somewhat in Sect. 5.

Given an arbitrary metric $\|\cdot\|^2$ on K we can define metrics on the bundles $\mathscr{O}(P)$ as follows. Define the (1,1)-form

$$\mu\equiv\frac{1}{4\pi i(1-g)}R_K\ , \qquad (2.44)$$

so that $\int\mu=1$. Define next an electrostatic Green function $\log G$ on Σ by

$$\partial_P\bar{\partial}_P\log G(P,Q)=i\pi(\mu(P)-\delta_Q(P))\ , \qquad (2.45)$$

where δ_Q is the delta function: $\int f\cdot\delta_Q=f(Q)$. $\delta_Q(P)$ is a (1,1)-form at P. The μ is needed in (2.45) so that the right side integrates to zero. Then G equals $|z_P-z_Q|$ times a smooth function as $P\to Q$, and also $G(P,Q)=G(Q,P)$ [20]. We will fix the normalization of G by requiring

$$\int \log G(P,Q)\cdot\mu(P)=0 \ . \tag{2.46}$$

We will need to know how G changes when we rescale the original metric on K. Suppose

$$\tilde{g}^{z\bar{z}}=e^{2\sigma}g^{z\bar{z}} \ . \tag{2.47}$$

Then $\tilde{\mu}=\mu+[2\pi i(1-g)]^{-1}\bar{\partial}\partial\sigma$ and we still have $\int\tilde{\mu}=1$. Working through the definitions one finds that (2.45) implies $\tilde{G}(P,Q)$ is $\exp\left[\frac{1}{2(g-1)}\sigma(P)\right]$ times a function of Q only. Requiring symmetry and the normalization condition (2.46), we get

$$\tilde{G}(P,Q)=\exp\left[\frac{1}{4\pi i(g-1)^2}S_L[\sigma]+\frac{1}{2(g-1)}(\sigma(P)+\sigma(Q))\right]\cdot G(P,Q) \ , \tag{2.48}$$

where the "Liouville action" is

$$S_L\equiv\int_\Sigma(\partial\sigma\wedge\bar{\partial}\sigma+\sigma R) \ , \tag{2.49}$$

and R is the original curvature.

Next, using G we can put a norm on any $\mathcal{O}(P)$ bundle. We declare that for the unit section,

$$\|\mathbf{1}_{\mathcal{O}(P)}\|(Q)=G(P,Q) \ . \tag{2.50}$$

While the right-hand side vanishes at P, so does the section $\mathbf{1}_{\mathcal{O}(P)}$, so the norm defined by (2.50) is nonsingular. We can generalize (2.50) to an arbitrary bundle $\mathcal{O}(D)$ by taking $\mathcal{O}(p+q)\cong\mathcal{O}(p)\otimes\mathcal{O}(q)$ to be an isometry.

We can now define another norm on the cotangent K. Near any point Q, declare that the distance $d'(P,Q)$ between Q and a nearby point P should approach $G(P,Q)$ as $P\to Q$, or in other words that

$$\|dz\|'(Q)=\lim_{P\to Q}\frac{|z_P-z_Q|}{G(P,Q)} \ . \tag{2.51}$$

It is easy to show that (2.51) defines a norm at P which is independent of the choice of coordinate z. Indeed (2.51) just states that the residue map

$$[K\otimes\mathcal{O}(P)]|_P\cong\mathbf{C} \tag{2.52}$$

(see (2.9)) should be an *isometry* in the new norm on K. The new norm $\|\cdot\|'$ is perfectly smooth on Σ. However, it will not in general be simply related to the metric $\|\cdot\|$ we started with.

Arakelov's norm is defined to be the one for which the above procedure reproduces the original metric on K. That is, if we take μ to be related to the Arakelov curvature by (2.44), then we get metrics on the $\mathcal{O}(D)$ bundles such that

(2.51) gives $\|\cdot\|=\|\cdot\|'$. The Arakelov metric is unique, and its curvature is [20]

$$\mu_{\text{Arak}}=\frac{1}{g}\,\bar{\omega}^i_{\mathscr{A}}\,Y^{ij}_{\mathscr{A}}\omega^j_{\mathscr{A}}\ ;\qquad Y_{\mathscr{A}}=(\tau_{\mathscr{A}}-\bar{\tau}_{\mathscr{A}})^{-1}\ . \tag{2.53}$$

The corresponding curvature on the tensors of spin λ is

$$R_{K^\lambda}=4\pi i(1-g)\lambda\mu_{\text{Arak}}\ .$$

More generally, any bundle with metric curvature proportional to μ_{Arak} is said to have an "admissible metric". Since we will be using (2.51) repeatedly, the choice of admissible metrics will simplify many formulas. Specifically, (2.51) expresses the metric on K in terms of the regulated coincident Green function of the bosonic theory.

We emphasize again that the choice of admissible metrics is for convenience only, and that bosonization works in any metric.

3. The Bosonic Action

In this section we will arrive at an action functional for a scalar theory which is to reproduce a general first-order fermionic system. We will begin by reviewing the situation on the sphere. Next we proceed to higher genus, first in spin-1/2 and then for general spin. Throughout, we will consider only free fermions, that is, fermions interacting with background gauge and gravitational fields but without self-interactions such as mass or quartic terms. This is the case of interest for the NSR superstring in flat spacetime.

We begin by reviewing existing results to fix notation. The prototypical fermionic system one might wish to bosonize has action

$$S_f=\int_\Sigma \bar{\psi}i\partial\!\!\!/\psi\ .$$

In euclidean path integrals ψ and $\bar{\psi}$ are independent, and it has become traditional to rename the fields, with $\psi_1\mapsto c$ $\psi_2\mapsto\bar{b}$ $\bar{\psi}_1\mapsto b$ and $\bar{\psi}_2\mapsto\bar{c}$. Using complex notation and rescaling fields we then have

$$S_f=\frac{i}{2\pi}\int_\Sigma(b\bar{\partial}c+\bar{b}\partial\bar{c})\ , \tag{3.1}$$

where b and c are sections of a spin bundle L, and $\bar{b}$ and $\bar{c}$ are sections of $\bar{L}$. More generally we can let c be a section of any holomorphic bundle ξ, and b a section of $K\otimes\xi^{-1}$. In any case the integrand is a (1,1)-form and so can be integrated over Σ without the use of any metric. That is, (3.1) is classically conformally invariant. It is also invariant under the global chiral transformation $b\mapsto e^{i\alpha}b$, $c\mapsto e^{-i\alpha}c$.

Similarly the prototypical scalar action for a single real field φ has the form

$$S_1=\int_\Sigma 2\pi\vec{\nabla}\varphi\cdot\vec{\nabla}\varphi d(\text{vol})=\int_\Sigma 4\pi i\partial\varphi\wedge\bar{\partial}\varphi\ . \tag{3.2}$$

The unusual normalization is for later convenience. The second form makes it clear that S_1 is also classifically conformally invariant. It also has an invariance under shifts $\varphi\mapsto\varphi+\text{constant}$.

The operator analysis of bosonization on $2d$ Minkowski spaces teaches us two important physical lessons. (See for example [1, 47, 48, 49, 25, 26].) First of all, the correspondence should be between the fermionic bilinears and *exponentials* of φ, properly normal-ordered. Secondly, the bosonic field φ is properly to be regarded as a *periodic*, or circle-valued, field. This fact is compatible with the first if the normalization of φ is chosen such that its ambiguity does not affect its exponential. We can see the periodicity of φ either in the periodic sine-Gordon potential of [1] or in the case where spacetime is the Minkowski cylinder [48]. In either case the crucial physical basis of bosonization is that fermions in (3.1) correspond to *solitons*, or states where φ is multiple-valued, in (3.2). In particular, when spacetime has noncontractible loops the partition function of the bosonic system gets important contributions from soliton sectors.

Since we will use a covariant path integral and analyze surfaces with many noncontractible loops, we will sometimes use the term "instanton" to describe any field configuration φ with nontrivial winding numbers in some direction. Clearly we have one independent winding number for every element of a homology basis (Fig. 1). Since these winding numbers are unaffected by a shift of φ by a constant, they amount to specifying the cohomology class of the real 1-form $\mathrm{d}\varphi$.

A. The Sphere. On the sphere, the second observation above is immaterial, since there are no noncontractible loops on Σ and hence no solitons. To make the first observation concrete in our present notation we will assume that we have

$$b\bar{b} \propto e^{q\varphi} \ , \quad c\bar{c} \propto e^{-q\varphi} \ , \tag{3.3}$$

and find q, starting with the case of spin-1/2. First note that (3.2) gives a two-point function with singularity

$$\langle \varphi(z)\varphi(w)\rangle \sim -\frac{1}{16\pi^2}\log|z-w|^2 \ .$$

In proving this we have used the fact that

$$\bar{\partial}_P\partial_P \log|z_P - z_Q|^2 = 2\pi i \delta_Q(P) \ , \tag{3.4}$$

where again $\delta_Q(P)$, the delta function, is a (1,1)-form at P. Equation (3.4) is easily shown by integrating on a small disk and using Stokes' theorem.

The classical stress tensor of (3.2) is

$$T = -8\pi^2 \partial\varphi\partial\varphi \ , \tag{3.5}$$

a (2,0)-tensor. Its quantum version looks the same but with normal ordering to remove self-contractions. T is defined so that in operator products [3]

$$T(z)\psi(w) \sim \frac{h}{(z-w)^2}\psi + \text{less singular terms} \ ,$$

where h is the spin of ψ. One can check the normalization of (3.5) by showing that $\partial\varphi$ has spin one. Then

$$T(z)e^{\pm q\varphi(w)} \sim \left(\frac{1}{z-w}\right)^2 \frac{q^2}{16\pi^2}\cdot\left(-\frac{1}{2}\right) + \dots \ .$$

Thus choosing $q=4\pi i$ gives b, c of spin one half.

It is now simple to show the equivalence of (3.1) and (3.2), still in spin 1/2. First, the zero-point functions agree up to an overall multiplicative constant:

$$Z_b[g]=\text{const.}\cdot Z_f[g] \ . \tag{3.6}$$

Each side of (3.6) is a functional of a metric chosen on Σ to regularize the theories. Equation (3.6) holds because both bosons and nonchiral fermions are free of gravitational anomalies, and both sides have the *same* anomalous variation under Weyl transformations (see Sect. 5.A). Since on the sphere every metric g is related to any reference g_0 by coordinate and Weyl transformations [16], we see that (3.6) really does hold up to a constant. As for the higher correlation functions, they follow from the fact that

$$\langle e^{4\pi i\varphi(z)}e^{-4\pi i\varphi(w)}\rangle=\exp\left[16\pi^2\langle\varphi(z)\varphi(w)\rangle\right]\sim\frac{1}{|z-w|^2}$$

$$\sim\langle b\bar{b}(z)c\bar{c}(w)\rangle \ . \tag{3.7}$$

Next we relax the condition of spin $\lambda=\frac{1}{2}$, to fields b of spin λ and c of spin $1-\lambda$. Clearly (3.6) cannot hold as it stands, since we have not told Z_b about λ. Some modification of (3.2) is needed in order to specify what spin we wish to bosonize. The correct choice is $S=S_1+S_2$, where

$$S_2=(1-2\lambda)\int R_K\varphi \ . \tag{3.8}$$

The easiest way to check the normalization of (3.8) is to note that for $\lambda>1$, say, Z_f is actually zero due to the presence of zero modes of b and c. Using (2.18) with $g=0$ we see that b has $1-2\lambda$ more zero modes than c. Hence we get nonzero answers only if we consider correlation functions with $1-2\lambda$ more insertions of b than of c. To reproduce this behavior in the bosonic system note that the functional integral using (3.8) also vanishes for the zero point function: integrating over the constant mode φ_0 of φ gives $\int_0^{2\pi}\exp\left[(1-2\lambda)\varphi_0\int R_K\right]$, which vanishes by (2.28). To get a nonzero answer we must insert b and $\bar{b}$ fields, to get

$$\int[\mathrm{d}\varphi]e^{-(S_1+S_2)}\prod_{i=1}^{k}e^{4\pi i\varphi(P_i)} \ , \tag{3.9}$$

which with (3.8) is indeed zero unless $k=1-2\lambda$.

Equation (3.8) modifies the stress tensor by adding $-2\pi i(2\lambda-1)\partial^2\varphi$ to it [3]. Computing the operator product one again finds $q=4\pi i$ in (3.3). With the modified action (3.6) continues to hold on the sphere for any spin, since (as we will check later) with S_2 both sides again have the same Weyl transformation and this again suffices on the sphere.

B. The Torus. When we move up in complexity from the sphere to the torus we at once encounter two novel features. First, it is no longer true that every metric is related to every other by coordinate and Weyl symmetries: a residual "moduli space" of inequivalent metrics remains [16]. Secondly, in any given degree there is now a wide variety of inequivalent bundles over Σ in which b and c could take values. Thus we not only need to tell the bosons what spin they are to mimic; we must also tell them about a point in the jacobian $J(\Sigma)$ describing the twists of b, c. Fortunately on the torus we still have a canonical formalism, which we can use to address the problem.

On the torus every spin bundle has degree $2\lambda(g-1)=0$, so we can take a flat metric, $R_K=0$. Then $S_2=0$. A spin bundle L is a flat bundle whose square is trivial; we can parametrize the four possible choices by measuring the difference between L and one particular spin bundle, the trivial one. If we take the torus defined by the unit square in $\mathbf{C}$, we then have

$$b(1)=e^{2\pi i\theta}b(0)\ , \qquad b(i)=e^{-2\pi i\phi}b(0)\ . \tag{3.10}$$

Here θ, ϕ give the holonomy of the flat bundle L as in (2.29) and (2.39). In this introductory section we will restrict to untwisted spin bundles, i.e. θ, $\phi=0$ or $\frac{1}{2}$. The field c then lives in the bundle $K\otimes L^{-1}\cong L$. Also we will not consider $\theta=\phi=0$ since with this choice $Z_f=0$ due to the zero mode; that is, we consider only the three "even" spin structures.

Certainly (3.6) cannot hold as it stands on the torus, since again one side depends on θ, ϕ while the other does not. Instead one expects that the bosonic theory with action S_1 should give the *sum* over all spin structures of the corresponding fermionic theories. Detailed calculation affirms this expectation [6] (see also Sect. 4C). To bosonize just one spin structure one must add to S_1 a new term S_3 depending on θ, ϕ. We will find S_3 by canonically quantizing and applying the physical lesson that fermions correspond to solitons of the field φ.

Since moduli will not play an important role in this subsection we will again take the torus to be the unit square in $\mathbf{C}$, with identifications and (3.10). We will quantize with euclidean time running up the imaginary axis. Then the partition functions $Z_f(\theta,\phi)$ are traces over the Ramond and Neveu-Schwarz Hilbert spaces, for $\theta=0,\frac{1}{2}$ respectively. ϕ on the other hand denotes the boundary conditions in time. We then have

$$Z_f(\tfrac{1}{2},\tfrac{1}{2})=\mathrm{Tr}_{NS}\,e^{-H}\ ,$$

$$Z_f(\tfrac{1}{2},0)=\mathrm{Tr}_{NS}(-)^F e^{-H}\ .$$

Here F is the fermion number operator. Hence $Z_f(\frac{1}{2},\frac{1}{2})\pm Z_f(\frac{1}{2},0)$ is a trace over the even (respectively odd) fermion-number space. It must therefore in the bosonic language receive contributions only from states of even (respectively odd) soliton number.

Recall that the soliton number of a field configuration φ is $2n$, where the cohomology class of $d\varphi$ is $n\alpha+m\beta$, so that a functional integral over φ includes a sum over *all* soliton sectors. Thus our modification to the action S_3 must have the effect of weighting the various winding sectors in such a way as to cancel the odd-soliton contributions to $Z_b(\frac{1}{2},0)+Z_b(\frac{1}{2},\frac{1}{2})$, and so on. A possible set of weighting factors is shown in Fig. 5. In the left column we have shown the spin structures for the fermionic system. On the right the boxes represent the contributions to the bosonic path integral from the winding sectors with $(n,m)=(\frac{1}{2}\cdot\text{even},\ \frac{1}{2}\cdot\text{even})$, $(\frac{1}{2}\cdot\text{even},\ \frac{1}{2}\cdot\text{odd})$, and so on. Each box on the right thus represents a sum $Z_{b,i}^{\text{partial}}$,

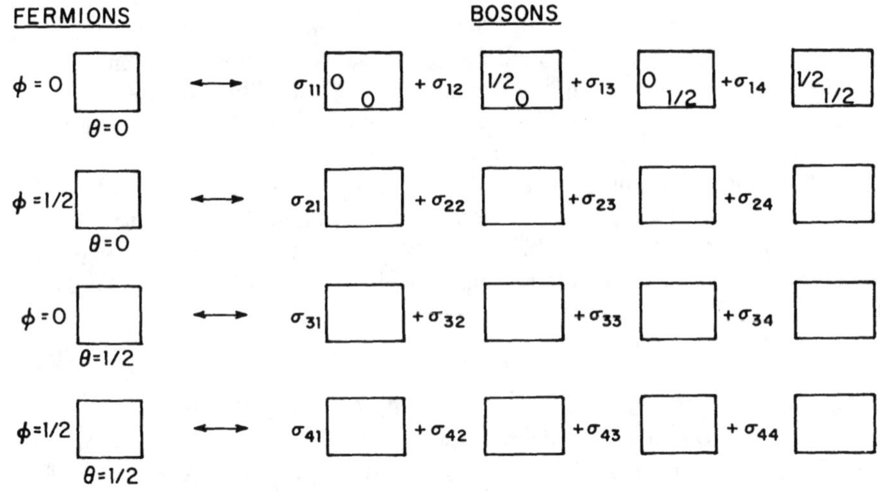

Fig. 5. Weighting the soliton sectors

$i=1, \ldots 4$ over an infinite subclass of field configurations. The effects of S_3 are in the phases σ_{ij}: we have $Z_b(\frac{1}{2},0)=\sum_{i=1}^{4} \sigma_{3i} Z_{b,i}^{\text{partial}}$, etc.

The conditions that $Z_b(\frac{1}{2},0) \pm Z_b(\frac{1}{2},\frac{1}{2})$ have no odd (respectively even) soliton contribution now imply that in these combinations $Z_{b,3}^{\text{partial}}$ and $Z_{b,4}^{\text{partial}}$ must cancel (respectively $Z_{b,1}^{\text{partial}}$ and $Z_{b,2}^{\text{partial}}$), i.e.

$$\sigma_{33} = -\sigma_{43}\ , \qquad \sigma_{34} = -\sigma_{44}$$

$$\sigma_{31} = \sigma_{41}\ , \qquad \sigma_{32} = \sigma_{42}\ .$$

We can get further conditions by letting the torus degenerate with very long time. Then only the ground state contributes to the fermionic partition function. With zeta function regularization the split ground state of the Ramond sector has nonzero energy, so that $Z_f(0,\phi) \to 0$. Take $\phi = \frac{1}{2}$. On the bosonic side, only the zero-soliton sectors contribute, but they do so independently of the time winding $2m$, in the limit. Thus the two contributing partial sums $Z_{b,1}^{\text{partial}}$ and $Z_{b,2}^{\text{partial}}$ must cancel from $Z_b(0,\frac{1}{2})$, so that $\sigma_{21} = -\sigma_{22}$.

We must also impose the condition of modular invariance on the σ_{ij}. Requiring for example that our prescription be unchanged when we exchange the roles of space and time gives relations like $\sigma_{21} = \sigma_{31}$, $\sigma_{22} = \sigma_{33}$, etc. Requiring that the torus with corners $0, 1, i+2, i+1$ give the same answers as the unit square gives $\sigma_{31} = \sigma_{41}$, $\sigma_{33} = \sigma_{44}$, etc. These conditions fix σ_{ij}, $i \neq 1$ up to an overall constant, which we take to be unity:[7]

$$\sigma_{ij} = \begin{cases} -1, & i=j \\ +1, & i \neq j\ . \end{cases} \tag{3.11}$$

[7] We can use the same reasoning to fix the σ_{1j}. However, to fix the relative sign of σ_{1j} relative to the others we must interpolate between the spin structures, as we do in the sequel

We can restate (3.11) in a way which makes its modular invariance obvious. Note again that the four spin structures split into one 'odd" one (the trivial bundle) and three "even" ones. The names indicate that the number of zero modes of $\bar{\partial}_L$ is odd ($=1$) or even ($=0$) in the respective cases [40]. Here $\bar{\partial}_L$ is the Cauchy-Riemann operator coupled to the holomorphic bundle L [16]. Note also that the 1-form $\psi = d\varphi$ corresponds to a flat bundle $F(\psi)$ as in Fig. 4. The prescription (3.11) simply says that we must add to S_1 the topological term

$$S_3 = i\pi\sigma(L \otimes F(\psi)) , \tag{3.12}$$

where $\sigma(L')$ is 0 or 1, depending on whether L' is even or odd. Note that $L \otimes F(\psi)$ really is a spin bundle, since ψ is a half-integral class. Also note that the preferred spin bundle $L_{\mathscr{A}}$ for a marking $\mathscr{A}$ is always even [39].

The proof that $S_1 + S_3$ is the correct bosonic action, as well as the generalization to arbitrary twists θ, ϕ, will come after we generalize everything to genus $g \geqq 1$. We emphasize, however that (3.2) and (3.12) are by now a very plausible prescription on the torus, and that in higher genus essentially no new physics will be needed.

C. Higher Genus. When Σ has more than one handle we can no longer take a flat metric, so we can no longer ignore S_2 (3.8). Also, the trivial bundle no longer serves as a reference spin bundle, since now $g-1 \neq 0$. These two issues will give rise to interlocking subtleties which conspire to make the full action modular-invariant.

The full action should be invariant under constant shifts of φ once we include in it terms for field insertions, as on the sphere (Eq. (3.9)). We will write these terms as

$$S_4 = -4\pi i \left[\sum_{i=1}^{p} \varphi(P_i) - \sum_{i=1}^{q} \varphi(Q_i) \right] . \tag{3.13}$$

Extending the argument of Sect. 3A to any genus, counting fermionic zero modes shows that the numbers of insertions must satisfy

$$p - q = (2\lambda - 1)(g-1) . \tag{3.14}$$

When (3.14) is satisfied, then the full action should depend only in the closed 1-form

$$\psi = d\varphi$$

and not on φ itself.

The only action term besides S_4 which is not shift-invariant is S_2. Thus we are again tempted to take S_2 to be as in (3.8), since again (2.18) gives shift-invariance when (3.14) holds. On a complicated surface like Fig. 1, however, (3.8) is problematic. Given ψ we can recover φ as a function on the cut surface Σ_c by defining

$$\varphi(P) = \int_{\hat{P}}^{P} \psi \tag{3.15}$$

for any point $\hat{P} \in \Sigma$. Shift invariance says that it is immaterial which $\hat{P}$ we choose. However, if we deform slightly the curves used to cut Σ into Σ_c, we run into problems. Since φ in general jumps as we cross the cuts, we get an ambiguity in S_2 proportional to the integral of R over the shaded region of Fig. 6. Thus the action depends not only on the choice of a homology basis, but also on a choice of specific curves representing that basis.

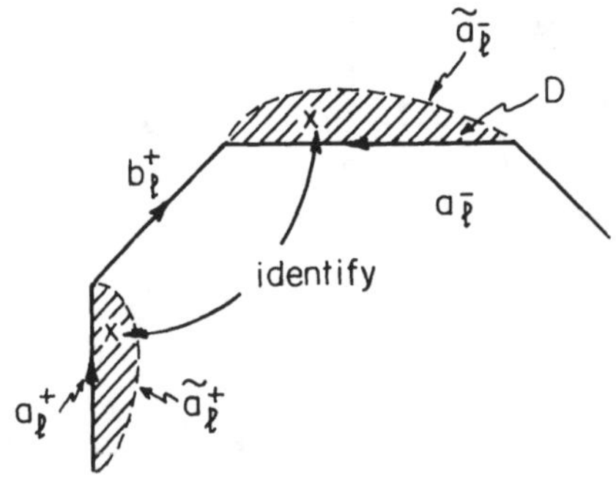

Fig. 6. Ambiguity of S_2 as we change the cut a_l on Σ

We can repair the dependence in the curves by recalling (2.31). We need only to find an appropriate bundle ξ and to add its holonomy around each cut, times the corresponding winding number of φ, to S_2. Specifically consider

$$S_2 \stackrel{?}{=} (1-2\lambda) \int R_K \varphi + 2\pi i [m^i H(a_i;\xi) - n^i H(b_i;\xi)] \ . \tag{3.16}$$

As always R_K is the curvature of the cotangent K and we have expanded the real 1-form $\psi = \mathrm{d}\varphi$ as $\psi = n \cdot \alpha + m \cdot \beta$. Then S_2 will be invariant when we displace the cuts if ξ has spin $1-2\lambda$. A natural choice for ξ exists, namely $\mathscr{L}_c \otimes \mathscr{L}_b^{-1}$, where $\mathscr{L}_b$ is the line bundle where b takes its values and $\mathscr{L}_c = K \otimes \mathscr{L}_b^{-1}$.

Unfortunately (3.16) is not well-defined, even modulo $2\pi i$. The holonomy is defined mod 1, but n^i, m^i are half-intergers. Thus we would prefer to replace the holonomy of ξ by *twice* the holonomy of some other bundle ξ' of spin $1/2-\lambda$. Since there is now no natural choice for ξ', we will let

$$\mathscr{L} \equiv \mathscr{L}_b^{-1} \otimes L_0 \ , \tag{3.17}$$

where L_0 is any even spin structure. Having introduced L_0 we will later have to show that the full action is independent of this choice.

Our candidate for S_2 is thus

$$S_2 = 2 \int R_{\mathscr{L}} \varphi - 4\pi i (\vec{n}, \vec{m}) \cdot J \cdot \begin{pmatrix} H(\vec{a};\mathscr{L}) \\ H(\vec{b};\mathscr{L}) \end{pmatrix} \ . \tag{3.18}$$

Equation (3.18) is invariant when we move the cuts in their homology classes, leaving fixed their common point, P_0. If we move P_0 itself, we can get a new system of cuts by choosing any curve γ from P_0 to $\hat{P}_0$ and attaching it to each end of the existing cuts. Since the new $\hat{\Sigma}_c$ differs from the old by a set of measure zero, the first term of S_2 is unchanged; since each of the new curves traces and retraces γ, the other terms of (3.18) are also unchanged (see Eq. (2.30)). Also (3.18) combined with (3.13) is still shift-invariant, i.e., a functional only of $\psi = \mathrm{d}\varphi$, since clearly this is separately true of the new terms of (3.18).

The only possible nonintrinsic information used in (3.18) is therefore the choice of L_0. We will return shortly to this dependence. First, however, we want to point out another important property of (3.18).

Often in instanton physics it is useful to divide the quantum field into a topologically nontrivial piece satisfying the equations of motion, plus a fluctuation piece which is topologically trivial. By requiring that the former piece stationarize the action we remove cross terms in the total action; for a quadratic system this splits S_1 cleanly into two terms:

$$\begin{gathered}\varphi=\varphi_{nm}+\tilde{\varphi} \\ S_1[\varphi]=S_1[\varphi_{nm}]+S_1[\tilde{\varphi}]\ ,\end{gathered} \tag{3.19}$$

where $\bar{\partial}\partial\varphi_{nm}=0$ and $\tilde{\varphi}$ is single-valued. For our case everything is especially simple, since (3.19) is the Hodge decomposition, and so the harmonic φ_{nm} has no continuous collective coordinates. Indeed we have simply

$$\psi_{nm}\equiv \mathrm{d}\varphi_{nm}=(m,m)\begin{pmatrix}\alpha\\ \beta\end{pmatrix}\ .$$

Of course the linear term S_2 also splits similarly to (3.19).

In the remainder of this section we will focus on $S[\varphi_{nm}]$, leaving $S[\tilde{\varphi}]$ for later.

We can now evaluate $S_2[\varphi_{nm}]$ using any convenient metric. Let $D=\sum \ell_i T_i$ be any divisor representing $\mathscr{L}$ and $\sigma=\mathbf{1}_{\mathcal{O}(D)}$ the corresponding unit section. Then away from the poles and zeros of σ the curvature of $\mathscr{L}$ is (2.27). Let Δ_i be small neighborhoods of the points T_i and $\Sigma_c'=\Sigma_c-\bigcup_i \Delta_i$. Then integrating S_2 by parts, one has for small Δ_i,

$$S_2=-2\sum_i \oint_{\partial\Delta_i} \varphi_{nm}\partial \log \|\sigma\|^2-2\int_{\Sigma} \bar{\partial}\varphi_{nm}\wedge\partial\log\|\sigma\|^2\ .$$

The boundary terms from the cuts of Σ_c have cancelled the explicit holonomy terms of (3.18). Integrating the second term again by parts gives zero, since φ_{nm} is harmonic, $\bar{\partial}\varphi_{nm}$ is single-valued, and the boundary terms near T_i go to zero. Near a zero of σ we have $\partial \log\|\sigma\|^2\sim z^{-1}\mathrm{d}z$, and so adding (3.13) we get

$$\begin{aligned} S_2+S_4&=-4\pi i\left[\sum_i \ell_i\varphi_{nm}(T_i)+\sum_i \varphi_{nm}(P_i)-\sum_i \varphi_{nm}(Q_i)\right] \\ &=-4\pi i\left[\sum_i \ell_i\int^{T_i}\psi_{nm}+\sum_i\int_p^{P_i}\psi_{nm}-\sum_i\int_q^{Q_i}\psi_{nm}\right]\ . \end{aligned}\tag{3.20}$$

The common starting point of these integrals does not matter since we have (3.14). Nor do the paths matter: changing a contour by a homologically trivial circuit gives zero since ψ is closed, while changing it by a generator of $H_1(\Sigma)$ only changes S_2+S_4 by a multiple of $2\pi i$. S_2+S_4 is also manifestly independent of any choice of homology basis. It does however still depend on the chosen even spin bundle L_0 via the divisor class of D. It also potentially depends on the divisor D itself, not just on its class.

To show that S_2+S_4 depends only on the class of D, i.e., on the bundle $\mathscr{L}$, we will cast it in terms of the Jacobi map I_A and use Abel's theorem. Note that any harmonic $\psi=(n,m)\begin{pmatrix}\alpha\\ \beta\end{pmatrix}$ can be expressed in terms of the abelian differentials $\vec{\omega}_{\mathscr{A}}$ as

$$\psi=(m-\bar{\tau}_{\mathscr{A}}n)^t\cdot Y_{\mathscr{A}}\cdot\vec{\omega}_{\mathscr{A}}+\text{c.c.} \tag{3.21}$$

(see (2.21)). To verify (3.21), integrate both sides around a_i, b_i. Inserting into (3.20) and using (2.26) we get

$$S_2+S_4=-4\pi i\{(m-\bar{\tau}_{\mathscr{A}}n)^t\cdot Y_{\mathscr{A}}\cdot I_{\mathscr{A}}[D+D_{\text{ins}}]+\text{c.c.}\}\ , \tag{3.22}$$

where

$$D_{\text{ins}}=\sum_{i=1}^{p} P_i-\sum_{i=1}^{q} Q_i \tag{3.23}$$

is the divisor built from the field insertion points. Thus S_2+S_4 depends only on the divisor class of $D+D_{\text{ins}}$. It is now straightforward to show that

$$S_2+S_4=4\pi i(n\cdot\phi+m\cdot\theta)\ , \tag{3.24}$$

where $I_{\mathscr{A}}[D+D_{\text{ins}}]\equiv-(\phi+\tau_{\mathscr{A}}\cdot\theta)$. One can derive (3.24) directly or using (2.33), (2.34).

We still have not generalized the term S_3 needed on the torus to distinguish the various spin structures. However, the parity $\sigma(L)$ of a spin bundle makes invariant sense in any genus. Thus we can generalize (3.12) to

$$S_3=i\pi\sigma(L_0\otimes F(\psi))\ . \tag{3.25}$$

Equation (3.25) is really the only sensible generalization of (3.12). We can't, for example, use $\mathscr{L}$ in place of L_0, since $\sigma(L)$ is only defined for untwisted spin-1/2 bundles [39]. For the case considered in Sect. 3B, however, $\mathscr{L}_b$ is a bundle of this type. Then we can recover (3.12) from (3.25), (3.22) by taking $g=1$, $\lambda=1/2$, $D_{\text{ins}}=0$, and choosing $L_0=\mathscr{L}_b$ so that (3.22) is zero. ($\mathscr{L}_b$ was called L in (3.12).)

We have now completely defined the bosonic action $S_b=S_1+S_2+S_3+S_4$. One can readily show that this prescription is equivalent to the one given in [5].

D. Consistency and Uniqueness. We have arrived at a modular-invariant action in any genus which reduces to the kinetic term plus (3.8) on the sphere or (3.12) on the torus. The only potential problem with $S_b=S_1+S_2+S_3+S_4$ as a classical action is its dependence on an arbitrary even spin bundle L_0. To see that this dependence is trivial, we express L_0 as $L_0=L_{\mathscr{A}}\otimes F(\hat{\psi})$, where $L_{\mathscr{A}}$ is the preferred spin bundle for some homology basis $\mathscr{A}$. In this basis we expand $\psi=\mathrm{d}\varphi$ as $\psi=(n,m)\begin{pmatrix}\alpha\\ \beta\end{pmatrix}$ and $\hat{\psi}=(\hat{n},\hat{m})\begin{pmatrix}\alpha\\ \beta\end{pmatrix}$. Then the parity equals [39]

$$\begin{aligned}\sigma(L_0\otimes F(\psi))&=\sigma(L_{\mathscr{A}}\otimes F(\psi+\hat{\psi}))=4(n+\hat{n})\cdot(m+\hat{m})\quad(\text{mod }2)\\ &=\sigma(L_{\mathscr{A}}\otimes F(\psi))+4(n\cdot\hat{m}+m\cdot\hat{n})\quad(\text{mod }2)\ ,\end{aligned} \tag{3.26}$$

since L_0 is itself even. At the same time, however, (3.24) changes. Using (2.32) the change in $\begin{pmatrix}\phi\\ \theta\end{pmatrix}$ describing $\mathscr{L}$ is $-J\begin{pmatrix}\hat{n}\\ \hat{m}\end{pmatrix}$, so that the change in S_2+S_3 equals

$$4\pi i(-n\cdot\hat{m}+m\cdot\hat{n})+4\pi i(n\cdot\hat{m}+m\cdot\hat{n})=0\quad(\text{mod }2\pi i)\ .$$

Thus the action S_b is completely intrinsic.

How unique is the bosonic action? Any modification to S_b must have only topological terms, since S_b is locally correct. It must also be intrinsically defined, i.e., modular invariant. Also, to preserve clustering any new global terms in S must *factorize* as a complicated Riemann surface pinches off into many tori. One can see from (3.24), (3.26) that our candidate action already has this property. If we weight each winding sector by an additional phase $E(\vec{n},\vec{m})$, the requirement of factorization says that

$$E(\vec{n},\vec{m})=\prod_{i=1}^{g}\varepsilon(n^i,m^i) \tag{3.27}$$

for some universal function ε. The analysis needed to fix ε has been done in [50]; we will now summarize the relevant case.

We can constrain $\varepsilon(n,m)$ by requiring E to be invariant under the modular transformation given by

$$\Lambda=\begin{bmatrix}1 & 0 & 0 & 0\\ 0 & 1 & 0 & 0\\ -1 & 1 & 1 & 0\\ 1 & -1 & 0 & 1\end{bmatrix}.$$

Using (3.27) in genus 2 this says that

$$\varepsilon(n_1+m_1-m_2,m_1)\varepsilon(n_2-m_1+m_2,m_2)=\varepsilon(n_1,m_1)\varepsilon(n_2,m_2)\ .$$

Without loss of generality we can let $\varepsilon(0,0)=1$. Also, one-loop modular invariance requires

$$\varepsilon(dn+cm,bn+am)=\varepsilon(n,m)\ ,\qquad ad-bc=1\ .$$

Taking $\begin{pmatrix}a & b\\ c & d\end{pmatrix}=\begin{pmatrix}1 & -1\\ 0 & 1\end{pmatrix}$ and $\begin{pmatrix}1 & 0\\ 1 & 1\end{pmatrix}$ we find

$$\varepsilon(n,m-n)=\varepsilon(n,m)\ ,$$

$$\varepsilon(n+m,m)=\varepsilon(n,m)\ ,$$

so

$$\varepsilon(n_1-m_2,m_1)\varepsilon(n_2-m_1,m_2)=\varepsilon(n_1,m_1)\varepsilon(n_2,m_2)\ . \tag{3.28}$$

Taking $n_1=m_1=n_2=0$ in (3.28) gives $\varepsilon(-m_2,0)=1$. Thus

$$\varepsilon(n,n)=\varepsilon(n,0)=\varepsilon(0,n)=\varepsilon(0,0)=1\ . \tag{3.29}$$

Taking $n_1=n_2=0$ in (3.28) gives

$$\varepsilon(n,m)=\varepsilon(-m,\ -n)^{-1}\ .$$

Setting $n_2=0$,

$$\varepsilon(n+n',m)=\varepsilon(n,m)\varepsilon(n',m)\ ,$$

and similarly $\varepsilon(n,m+m')$. Thus $\varepsilon(n,m)=\varepsilon(1,1)^{nm}$. But by (3.29), $\varepsilon(1,1)^{n^2}=1$ for every n. Thus $\varepsilon(1,1)=1$, and the action is unique.

4. Bosonization Formulae

A. Recap. The bosonization of the first-order fermionic system thus proceeds as follows. If b, c are fields of spin $\lambda, 1-\lambda$ with action

$$S_f=\frac{i}{2\pi}\int_\Sigma (b\bar{\partial}c+\bar{b}\partial\bar{c}) \ , \tag{3.1$'$}$$

then we introduce a Bose theory with field φ well-defined up to half-integers and action[8]

$$S_b=S_1+S_2+S_3 \ ,$$

$$S_1=4\pi i\int_\Sigma \partial\varphi\wedge\bar{\partial}\varphi \ , \tag{3.2$'$}$$

$$S_2=2\int R_{\mathscr{L}}\varphi+4\pi i\sum_i [m^i H(a_i;\mathscr{L})-n^i H(b_i;\mathscr{L})] \ , \tag{3.18$'$}$$

$$S_3=i\pi\sigma(L_0\otimes F(\mathrm{d}\varphi)) \ . \tag{3.25$'$}$$

Here $\mathscr{L}$ is the line bundle

$$\mathscr{L}\equiv\mathscr{L}_b^{-1}\otimes L_0 \ , \tag{3.17$'$}$$

$\mathscr{L}_b$ is the bundle of degree $2\lambda(g-1)$ where b takes values, and L_0 is any even spin bundle. $R_{\mathscr{L}}$ and $H(\cdot\,;\mathscr{L})$ are the curvature and holonomy of $\mathscr{L}$, and n_i, m_i are the winding numbers of the field configuration φ about the cycles a_i, b_i. $F(\mathrm{d}\varphi)$ is the flat line bundle with holonomy given by the one-form $\mathrm{d}\varphi$, and $\sigma(L)$ is the parity of a spin bundle L. S_b is well-defined modulo $2\pi i$ once we include appropriate insertions of fields.

The bosonization results we seek to establish say that these two systems have the same correlation functions up to an overall multiplicative constant under the correspondence

$$b(P)\bar{b}(P)\propto e^{4\pi i\varphi(P)} \ , \qquad c(Q)\bar{c}(Q)\propto e^{-4\pi i\varphi(Q)} \ . \tag{3.3$'$}$$

We need p insertions of the first kind and q of the second kind, where

$$p-q=(2\lambda-1)(g-1) \ . \tag{3.14$'$}$$

Sometimes we consider the insertions as a term S_4 in the action (Eq. (3.13)).

Equation (3.3) is not yet completely specified. Here we have some latitude, since neither have we yet specified the normal-ordering prescription to be used in evaluating bosonic correlators. The simplest prescription to use is a coordinate-invariant one, in which all coincident Green functions are replaced by

$$G_r(P,P)\equiv\lim_{Q\to P}[G(P,Q)/d(P,Q)] \ . \tag{4.1}$$

$d(P,Q)$ is the metric distance between two points of Σ. $G(P,Q)$ is the Green function defined using the given metric in (2.44)–(2.46).

[8] O. Alvarez has told us that the terms S_2 and S_3 can also be understood in terms of corrections to the heat kernel on the cut Riemann surface

One reason why (4.1) is so nice is that for the Arakelov metric we have

$$\log G_r(P,P)=0 \ . \tag{4.2}$$

Equation (4.2) follows since the Arakelov metric $\|\cdot\|_{\mathrm{Arak}}$ is by definition the one for which the metric $\|\cdot\|'$ defined by (2.51) again reproduces $\|\cdot\|_{\mathrm{Arak}}$. However we will see that (4.1) is the correct prescription for any metric.

With (4.1) all bosonic correlations will be coordinate scalars. Since $b\bar{b}$ is not a scalar but a (λ,λ)-form, the precise statement of (3.3) is that $e^{4\pi i\varphi(P)}$ should correspond to[9]

$$\|b(P)\|^2 \equiv [g^{z\bar{z}}(P)]^{\lambda} b(P)\bar{b}(P) \ , \quad \|c(Q)\|^2 \equiv [g^{z\bar{z}}(Q)]^{1-\lambda} c(Q)\bar{c}(Q) \ , \tag{4.3}$$

or

$$\langle \|b(P_1)\|^2 \dots \|b(P_p)\|^2 \|c(Q_1)\|^2 \dots \|c(Q_q)\|^2 \rangle_f \stackrel{?}{=} \langle e^{4\pi i\varphi(P_1)} \dots e^{-4\pi i\varphi(Q_q)} \rangle_b \ . \tag{4.4}$$

In the succeeding subsections we write out both sides of (4.4) in some detail to get a set of identities expressing the mathematical content of bosonization.

Note that (4.4) is a nonchiral amplitude. Once we have proven it we can modify (4.1) to eliminate the metric factors in (4.4), then take the holomorphic square root in the variables P_i, Q_i on both sides of the formula. In a sense this gives a bosonic formula for the chiral amplitude on the left-hand side. This is not however the same thing as presenting a bosonic theory which, without any modifications, reproduces chiral amplitudes. We do not know how to write a bosonic path integral whose correlation functions have the appropriate geometrical meaning to do this.

B. Fermion Correlations. We begin with the left side of (4.4). If there are no insertions (this can happen only if the spin $\lambda=\frac{1}{2}$), then the rules of functional Grassmann integration say that the zero-point function is just the functional determinant of $\bar{\partial}^{\dagger}_{\mathscr{L}}\bar{\partial}_{\mathscr{L}}$. We will always use zeta function regulation for determinants, as it fits in best with the methods of Sect. 5. If there are only insertions of b then the p-point function is the antisymmetrized product of the p zero modes of $\bar{\partial}_{\mathscr{L}}$. For example, if the spin λ and the genus g are both greater than 1, then the degree of $\mathscr{L}_c$ is negative and so $\bar{\partial}_{\mathscr{L}_c}$ has no zero modes. Then we can have $q=0$ insertions of c and so, by (2.18), $p=(2\lambda-1)(g-1)$ insertions of b:

$$\langle b(P_1)\bar{b}(P_1)\dots b(P_p)\bar{b}(P_p)\rangle_f = \frac{\det' \bar{\partial}^{\dagger}_{\mathscr{L}_b}\bar{\partial}_{\mathscr{L}_b}}{\det(u_i,u_j)} \cdot \left|\sum_{\pi} (-)^{\pi} u_1(P_{\pi(1)})\dots u_p(P_{\pi(p)})\right|^2 \ .$$

Here $u_1,\dots,u_p$ are the p zero modes of $\bar{\partial}_{\mathscr{L}_b}$ and $(\cdot,\cdot)$ is the inner product on sections of $\mathscr{L}_b$. π runs over permutations, and so the factor inside the absolute square lives in $\overset{\max}{\wedge}\left(\bigotimes_{i=1}^{p} \mathscr{L}_b|_{P_i}\right)$. We will denote this factor by $\det u_i(P_j)$. Using the given norm on $\mathscr{L}_b$ we can now write the left side of (4.4) as

$$\langle \|b(P_1)\|^2 \dots \|b(P_p)\|^2 \rangle_f = \frac{\det' \bar{\partial}^{\dagger}_{\mathscr{L}_b}\bar{\partial}_{\mathscr{L}_b}}{\det(u_i,u_j)} \cdot \|\det u_i(P_j)\|^2 \ . \tag{4.5}$$

[9] We will later comment on alternatives to (4.1)

If $\lambda=1$ then $\bar\partial_{\mathcal{L}_c}$ still has no zero modes unless $\mathcal{L}_b$ is untwisted, i.e. $\mathcal{L}_b=K$. In this case $\mathcal{L}_c$ is trivial and so has the constant zero mode $v_0\equiv 1$. Its norm is $(1,1)=A_\Sigma$, the area of Σ in the given metric, while the factor $\|v_0(Q)\|^2=1$. We thus get

$$\langle\|b(P_1)\|^2\ldots\|b(P_p)\|^2\,\|c(Q)\|^2\rangle_f=\frac{\det'\bar\partial^\dagger\bar\partial}{\det(iY)^{-1}\cdot A_\Sigma}\cdot\|\det\omega^i(P_j)\|^2\ . \tag{4.6}$$

Here we have used the fact that for spin one the u_i are the abelian differentials ω^i and $(\omega^i,\omega^j)=i\int\bar\omega^i\wedge\omega^j=i\oint_{a_k}\bar\omega^i\oint_{b_k}\omega^j-(a\leftrightarrow b)=i(\tau_{ij}-\bar\tau_{ij})\equiv i(Y^{-1})_{ij}$. Also $\det'\bar\partial_K^\dagger\bar\partial_K=\det'\bar\partial^\dagger\bar\partial$, since the nonzero eigenvalues of these operators are the same.

For $\lambda<\frac{1}{2}$ we interchange the roles of b and c.

Finally we can consider the case when more than the minimal number of inserted fields is present. For this we need the fermionic Green function, which is in general more complicated than the spin-1/2 version (see (3.7), [51]) due to the presence of zero modes. Suppose again that $\bar\partial_{\mathcal{L}_c}$ has no zero modes, i.e. $H^0(\Sigma;\mathcal{L}_c)=0$. Let $\mathbf{P}_1$ be the projector to the space orthogonal to the zero modes of $\bar\partial_{\mathcal{L}_b}$. We can unambiguously invert $\bar\partial_{\mathcal{L}_b}$ restricted to this space; call the inverse $\mathcal{G}$. Thus

$$\mathcal{G}\circ\bar\partial_{\mathcal{L}_b}=\mathbf{P}_1\ ,\qquad \bar\partial_{\mathcal{L}_b}\circ\mathcal{G}=\mathbf{1}\ . \tag{4.7}$$

Mathematically $\mathcal{G}$ is a "parametrix" of $\bar\partial_{\mathcal{L}_b}$, an inverse up to a finite-rank term. Its existence is guaranteed by Hodge theory [41]. We can represent $\mathcal{G}$ by an integral kernel $\mathcal{G}(P,Q)$. For fixed Q, $\mathcal{G}(\cdot,Q)$ is a section of $\mathcal{L}_b\otimes\mathcal{O}(Q)\otimes\mathcal{L}_c|_Q$. Its residue at $Q=P$ is therefore a pure number by (2.52), namely $1/2\pi i$.

$\mathcal{G}$ is the basic bc contraction. With it we get

$$\langle\|b(P_1)\|^2\ldots\|b(P_p)\|^2\,\|c(Q_1)\|^2\ldots\|c(Q_q)\|^2\rangle_f$$
$$=\frac{\det'\bar\partial_{\mathcal{L}_b}^\dagger\bar\partial_{\mathcal{L}_b}}{\det(u_i,u_j)}\cdot\left\|\det\begin{pmatrix}u_1(P_1)&\ldots&u_{p-q}(P_1)&\mathcal{G}(P_1,Q_1)&\ldots&\mathcal{G}(P_1,Q_q)\\ \vdots&&\vdots&\vdots&&\vdots\\ u_1(P_p)&\ldots&u_{p-q}(P_p)&\mathcal{G}(P_p,Q_1)&\ldots&\mathcal{G}(P_p,Q_q)\end{pmatrix}\right\|^2\ . \tag{4.8}$$

The large matrix is square, and its determinant is a vector in

$$\left[\bigotimes_{i=1}^{p}\mathcal{L}_b|_{P_i}\right]\otimes\left[\bigotimes_{j=1}^{q}\mathcal{L}_c|_{Q_i}\right]\ .$$

In essence what has happened is that we have manufactured q additional zero modes of $\bar\partial_{\mathcal{L}_b}$. These extra modes $\mathcal{G}(\cdot,Q_i)$ have poles, but this is permitted since unlike (4.5), the left side of (4.8) is supposed to have poles.

Similarly one can generalize (4.6) to the analog of (4.8) when $\lambda=1$. For this one must replace the unit operator in (4.7) by the projector $\mathbf{P}_2$ to the complement of the zero mode space of $\bar\partial_K^\dagger$. One also divides the right-hand side of (4.8) by the area of the Riemann surface and replaces the large determinant by the expression in Eq. (5.5) below.

C. Instanton Sums. We mentioned earlier that the bosonic amplitude splits into the product of a topological part times a fluctuation part when we split φ as in (3.19). In this subsection we work out the former piece.

Recall that in Sect. 3B we argued that an extra topological term in the action was necessary to bosonize a single spin structure. Since the fermionic amplitudes for the different spin structures differ by theta-function factors (e.g. [6, 52]), we want the topological part of the bosonic amplitude to be the absolute square of a theta function for one *single* characteristic, not the sum over spin structures obtained when the topological terms are omitted from the action [6]. The key result of this subsection is that indeed this is what happens.

It will be convenient for explicit computations to choose a homology basis $\mathscr{A}$. We have seen that the action is independent of the choice of $\mathscr{A}$. Since $\mathscr{A}$ will not change, we will sometimes drop it from the notation.

First we substitute (3.21) into (3.2):

$$S_1[\varphi_{nm}]=4\pi i(m-\bar{\tau}n)^t\cdot Y\cdot(m-\tau n)\ .$$

Next we will make a specific choice for the arbitrary even spin bundle in (3.17), namely $L_0=L_{\mathscr{A}}$, the preferred spin bundle for the homology basis $\mathscr{A}$. We have already worked out S_2+S_4 for harmonic φ_{nm} (Eq. (3.22)):

$$S_2+S_4=4\pi i[(m-\bar{\tau}n)^t\cdot Y\cdot z+\text{c.c.}]\ ,$$

where (Eq. (3.17))

$$z\equiv I[\mathscr{L}_b\otimes L_{\mathscr{A}}^{-1}\otimes\mathscr{O}(-D_{\text{ins}})]\ . \tag{4.9}$$

Again D_{ins} is the divisor of insertion points (3.23) and we have written a bundle instead of its divisor class. Parenthetically we note that the topological part of the action can be simply expressed in terms of the natural hermitian form B in (2.33): for harmonic φ we have

$$S_1+S_2+S_4=2\pi[B(F(\mathrm{d}\varphi),F(\mathrm{d}\varphi))+B(F(\mathrm{d}\varphi),z)-B(z,F(\mathrm{d}\varphi))]\ .$$

We will not make explicit use of this form of the action.

Taking L_0 to be $L_{\mathscr{A}}$ has the advantage of making S_3 very simple. By (3.26)

$$S_3=4\pi i n\cdot m\ .$$

Note that e^{-S_3} depends only on the values of n and m modulo 1.

We wish to compute

$$Z_{\text{inst}}\equiv\sum_{d\varphi_{nm}\in H^1(\Sigma;\frac{1}{2}\mathbf{Z})}e^{-S[\varphi_{nm}]}\ ,$$

and in particular to show that it is the absolute square of a theta function.[10] We know however that ϑ is defined by a sum over integer, not half-integer, vectors. Accordingly we will define the 2^{2g} partial sums

$$Z_{\nu,\mu}\equiv\sum_{k,l\in\mathbf{Z}^g}e^{-(S_1+S_2+S_4)[n=k+\nu,m=l+\mu]}\ , \tag{4.10}$$

[10] The derivation below extends easily from the case of a single fermion (the lattice of integers) to many fermions (an arbitrary self-dual lattice)

where every entry of $\vec{\nu}$, $\vec{\mu}$ equals 0 or $\frac{1}{2}$. We omitted e^{-S_3} from (4.10) because as noted it is a function only of ν, μ. Thus

$$Z_{\text{inst}} = \sum_{\nu,\mu\in(\mathbf{Z}_2)^g} e^{4\pi i\nu\cdot\mu} Z_{\nu,\mu} \ . \tag{4.11}$$

$\mathbf{Z}_2$ denotes the group with two elements: $(\frac{1}{2}\mathbf{Z})/\mathbf{Z}$.

In the remainder of this subsection we will prove the following formula for $Z_{\nu,\mu}$:

$$Z_{\nu,\mu} = 2^{-3g/2}(\det 2iY)^{-\frac{1}{2}} \cdot e^{-4\pi i y Y y} \sum_{\varepsilon,\zeta\in(\mathbf{Z}_2)^g} e^{4\pi i\varepsilon\cdot\zeta}\vartheta\begin{bmatrix}\nu+\varepsilon\\ -\mu+\zeta\end{bmatrix}(-z|\tau)\,\overline{\vartheta\begin{bmatrix}\nu+\varepsilon\\ -\mu+\zeta\end{bmatrix}(z|\tau)} \ , \tag{4.12}$$

where $z=x+iy$ defines y. It is not hard to generalize (4.12) to a form useful for nonabelian bosonization and toroidal compactification.

Before proving (4.12), let us pause to see why it is just what we want. Substituting in (4.11), we get

$$Z_{\text{inst}} = (\text{prefactor}) \cdot \sum_{\nu,\mu,\alpha,\beta} e^{4\pi i(\nu\cdot\mu+(\alpha-\nu)(\beta+\mu))}\vartheta\begin{bmatrix}\alpha\\ \beta\end{bmatrix}(-z|\tau)\,\overline{\vartheta\begin{bmatrix}\alpha\\ \beta\end{bmatrix}(z|\tau)} \ .$$

We have changed variables from ε, ζ to $\alpha=\nu+\varepsilon$, $\beta=-\mu+\zeta$. Then

$$Z_{\text{inst}} = (\text{prefactor}) \cdot \sum_{\alpha,\beta} \vartheta\bar{\vartheta}\cdot e^{4\pi i\alpha\cdot\beta}\sum_{\mu} e^{4\pi i\alpha\cdot\mu}\cdot\sum_{\nu} e^{-4\pi i\beta\cdot\nu}$$
$$= (\det iY)^{-\frac{1}{2}} e^{-4\pi i y Y y}|\vartheta(z|\tau)|^2 \ ,$$

as desired. In terms of $\mathcal{N}$ in (2.43), this is[11]

$$Z_{\text{inst}} = (\det iY)^{-\frac{1}{2}}\mathcal{N}(z) \ .$$

Using (2.41) and (2.24) we see that Z_{inst} is independent of the chosen marking, as we have already noted on general grounds.

We can also recover from (4.12) the answer one gets by omitting the topological terms from the action. Setting $z=0$ and dropping the weighting factor from (4.11), one indeed finds that Z_{inst} is then proportional to the sum of the squares of all the even theta functions [6].

The general strategy for proving (4.12) is to diagonalize the action into a sum over two integer vectors $\vec{a}$, $\vec{b}$ of a function of $\vec{a}$ times a function of $\vec{b}$. Roughly speaking we will accomplish this diagonalization by "rotating $\vec{k}$, $\vec{l}$ by 45°". Then $Z_{\nu,\mu}$ becomes essentially the product of two factors, each of which turns out to be a theta function. The tricky part of the procedure lies in the idea of "rotating" a square lattice; this is where ε, ζ will enter.

We start with the observation that

$$\sum_{j\in\mathbf{Z}} e^{-2\pi i j t} = \sum_{\ell\in\mathbf{Z}} \delta(t-\ell) \ .$$

[11] Z_{inst} is essentially the function called $\|\vartheta\|^2$ in [19]

Regarding both sides as functions on S^1 this is just the Fourier transform of the delta function at the origin. Thus for a nice function f,

$$\sum_{\ell \in \mathbf{Z}^g} f(\ell+\mu)=\sum_{j\in \mathbf{Z}^g} e^{2\pi i j\cdot\mu}\int d^g t\, e^{-2\pi i t j} f(t) ,$$

the "Poisson summation formula". We will apply this to the sum over ℓ in (4.10).

Let $\tau=\tau_1+i\tau_2$, $z=x+iy$, and $S_1+S_2+S_4=2\pi(S'+S'')$, where

$$S'(m,n)=m\tau_2^{-1}m-2n\tau_1\tau_2^{-1}m+2im\tau_2^{-1}y,$$

$$S''(n)=n(\tau_2+\tau_1\tau_2^{-1}\tau_1)n+2in(x-\tau_1\tau_2^{-1}y).$$

Then the Poisson fomula applied to S' gives

$$\sum_{\ell} e^{-S'(m=\ell+\mu,n)}=\sum_{j} e^{2\pi i j\mu}\int dt e^{-Q(t)} ;$$

$$Q(t)\equiv 2\pi t\cdot\tau_2^{-1}\cdot t+2\pi(ij-2n\tau_1\tau_2^{-1}+2iy\tau_2^{-1})\cdot t .$$

Performing the gaussian integral we get

$$\begin{aligned}Z_{\nu,\mu}&=\sum_{k,\ell} e^{-S(n=k+\nu,m=l+\mu)}\\&=2^g E\sum_{k,j\in\mathbf{Z}^g}\exp\,[i\pi((k+\nu-\tfrac{1}{2}j)\tau(k+\nu-\tfrac{1}{2}j)-(k+\nu+\tfrac{1}{2}j)\bar\tau(k+\nu+\tfrac{1}{2}j))\\&\quad+4\pi i(\tfrac{1}{2}j\cdot\mu+\tfrac{1}{2}ij\cdot y-(k+\nu)\cdot x)] ;\\E&\equiv 2^{-3g/2}(\det\tau_2)^{\frac{1}{2}}e^{-2\pi y\cdot\tau_2^{-1}\cdot y} .\end{aligned}$$

The "rotation" of the lattice mentioned above is accomplished as follows. We replace the sum over $k\in\mathbf{Z}^g$, $\frac{1}{2}j\in\frac{1}{2}\mathbf{Z}^g$ by a sum over all $\varepsilon\in(\mathbf{Z}_2)^g$ and $a,b\in\mathbf{Z}^g$ such that $a\pm b$ has only even entries. The two lattices so defined are in 1-1 correspondence via

$$k=\tfrac{1}{2}(a+b) , \quad \tfrac{1}{2}j=\tfrac{1}{2}(b-a)+\varepsilon .$$

Also we can enforce the condition that $a\pm b$ be even by performing an unrestricted sum but including the $\mathbf{Z}_2$ delta function:

$$\delta(a-b\ (\mathrm{mod}\ 2))=2^{-g}\sum_{\zeta\in(\mathbf{Z}_2)^g} e^{2\pi i(a-b)\cdot\zeta} .$$

Thus we have

$$\begin{aligned}Z_{\nu,\mu}&=E\cdot\sum_{\substack{a,b\in\mathbf{Z}^g\\ \varepsilon,\zeta\in(\mathbf{Z}_2)^g}}\exp\,[i\pi((a+\nu-\varepsilon)\tau(a+\nu-\varepsilon)-(b+\nu+\varepsilon)\bar\tau(b+\nu+\varepsilon))\\&\quad+2\pi i((b-a+2\varepsilon)(\mu+iy)-(a+b+2\nu)x+(a-b)\zeta)]\\&=E\cdot\sum\exp\,[i\pi(a+\nu-\varepsilon)\tau(a+\nu-\varepsilon)+2\pi i((a+\nu-\varepsilon)(-z-\mu+\zeta)\\&\quad-i\pi(b+\nu+\varepsilon)\bar\tau(b+\nu+\varepsilon)-2\pi i(b+\nu+\varepsilon)(\bar z-\lambda+\zeta)+4\pi i\varepsilon\cdot\zeta] .\end{aligned}$$

Using the definition (2.42) we obtain (4.12). Note that we are permitted to change ε to $-\varepsilon$ since 2ε is an integer vector.

D. Bosonic Correlations. With (4.12) in hand we can now turn to the fluctuation part $\tilde{\varphi}$ of φ, and finally compare the bosonic correlation functions to (4.8).

$\tilde{\varphi}$ has no jumps across the cuts of Σ, and so its action is given by (3.2), (3.8), and (3.13). Hence the right side of (4.4) is given by Z_{inst} times

$$Z_{\text{fluct}} = \int [d\tilde{\varphi}] \exp\left[-\int_\Sigma (4\pi i \partial\tilde{\varphi} \wedge \bar{\partial}\tilde{\varphi} + (1-2\lambda) R_K \tilde{\varphi}) + 4\pi i \left(\sum ip\tilde{\varphi}(P_i) - \sum iq\tilde{\varphi}(Q_i)\right)\right].$$

To do the gaussian integral change variables from $\tilde{\varphi}$ to

$$\hat{\varphi} = \tilde{\varphi} - (2\pi i)^{-1} \left[\sum \log G(P, P_i) - \sum \log G(P, Q_i)\right]$$

and remove the zero mode from the integral over φ. Using (2.44)–(2.46) and integrating by parts several times we obtain

$$Z_{\text{fluct}} = \left(\frac{\det' \bar{\partial}^\dagger \bar{\partial}}{A_\Sigma}\right)^{-\frac{1}{2}} \cdot \frac{\prod_{i,j=1}^{p} G(P_i, P_j) \prod_{i,j=1}^{q} G(Q_i, Q_j)}{\prod G(P_i, Q_j)^2},$$

where A_Σ is the norm squared of the removed zero mode of $\bar{\partial}^\dagger\bar{\partial}$, i.e. the area of Σ in the given metric. The simple form of (4.14) comes from the special normalization (2.46) chosen for the Green function. We have not used the Arakelov condition, however.

Of course as it stands (4.14) equals zero due to the coincident Green functions. We will define the path integral using zeta-function regulation on the determinant and the regulated coincident Green function (4.1). The freedom to make such a choice is the path-integral version of the freedom to choose a normal-ordering convention.

At last we can write out (4.4) in full detail, using (4.8), (4.13), and (4.14). Dropping an overall constant, the statement (4.4) of bosonization says that for spins $\lambda > 1$,

$$\frac{\det' \bar{\partial}^\dagger_{\mathscr{L}_b} \bar{\partial}_{\mathscr{L}_b}}{\det(u_i, u_j)} \cdot \left\| \det \begin{pmatrix} u_1(P_1) & \dots & u_{p-q}(P_1) & \mathscr{G}(P_1, Q_1) & \dots & \mathscr{G}(P_1, Q_q) \\ \vdots & & \vdots & \vdots & & \vdots \\ u_1(P_p) & \dots & u_{p-q}(P_p) & \mathscr{G}(P_p, Q_1) & \dots & \mathscr{G}(P_p, Q_q) \end{pmatrix} \right\|^2$$

$$= \left(\frac{\det' \bar{\partial}^\dagger \bar{\partial}}{\det(iY)^{-1} \cdot A_\Sigma}\right)^{-\frac{1}{2}} \cdot \mathscr{N}(z) \cdot \frac{\prod_{i,j=1}^{p} G(P_i, P_j) \prod_{i,j=1}^{q} G(Q_i, Q_j)}{\prod G(P_i, Q_j)^2}. \tag{4.15}$$

$\mathscr{N}$ is defined in (2.43), and z is defined in (4.9).

For $\mathscr{L}_b = K$ we will give only the formula with no extra insertions, corresponding to (4.6). This case is interesting in that the same determinantal factor appears on both sides of (4.15), so that instead of relating two determinants we get a formula expressing one in terms of special functions on the Riemann surface:

$$\left(\frac{\det' \bar{\partial}^\dagger \bar{\partial}}{\det (iY)^{-1} \cdot A_\Sigma}\right)^{3/2} = \|\det \omega^i(P_j)\|^{-2} \cdot \mathcal{N}(z) \cdot \frac{\prod_{i,j=1}^{g} G(P_i, P_j) \cdot G(Q, Q)}{\prod G(P_i, Q)^2} . \tag{4.16}$$

This formula can in turn be substituted in (4.15) to get formulae for all the determinants.

Finally, for spin $\lambda = \frac{1}{2}$ generically there are no zero modes at all and we can take $p = q = 0$. This gives

$$\det \bar{\partial}^\dagger_{\mathscr{L}_b} \bar{\partial}_{\mathscr{L}_b} = \left(\frac{\det' \bar{\partial}^\dagger \bar{\partial}}{\det (iY)^{-1} \cdot A_\Sigma}\right)^{-\frac{1}{2}} \cdot \mathcal{N}(\mathscr{L}_b \otimes L_{\mathscr{A}}^{-1}) , \tag{4.17}$$

the "spin-$\frac{1}{2}$ bosonization formula". This has already been derived (cf. [6, 7]) and it forms the basis for our proof in the next section of (4.15) and (4.16). At one loop this formula is essentially the Jacobi triple product formula [25].

5. Mathematical Proof of Bosonization

We now present a mathematical proof of the bosonization formulae obtained in the previous sections. We will prove (4.4) by proving its explicit restatements (4.15) and (4.16). Actually, for technical reasons we prove these identities only up to an overall constant depending on the genus, the spin and the number of field insertions. This is adequate for proving the equivalence of two given field theories. For string applications, where one wants to relate different genera, factorization of amplitudes as a surface degenerates will fix the relative normalizations.

A. Weyl Invariance. As a first consistency check on (4.15) we now show that once it holds for any metric, it then holds for any conformally-related metric. This is a simple application of the conformal anomaly formula [53, 54, 55], which says that if $\tilde{g}^{z\bar{z}} = e^{2\sigma} g^{z\bar{z}}$, then the zeta-regulated determinant behaves as

$$\left[\frac{\det' \bar{\partial}^\dagger_{\mathscr{L}} \bar{\partial}_{\mathscr{L}}}{\det (u_i, u_j) \cdot \det (v_i, v_j)}\right]^{\sim} = \exp\left[\frac{6\lambda(\lambda-1)+1}{6\pi i} S_L\right] \cdot \frac{\det' \bar{\partial}^\dagger_{\mathscr{L}} \bar{\partial}_{\mathscr{L}}}{\det (u_i, u_j) \cdot \det (v_i, v_j)} . \tag{5.1}$$

Here $\{u_i\}$, $\{v_i\}$ are zero modes of $\bar{\partial}_{\mathscr{L}}$ and $\bar{\partial}^\dagger_{\mathscr{L}}$, λ is the spin of $\mathscr{L}$, and S_L is defined in (2.49). We also have the result (2.48) on the rescaling of the Green function. Equation (2.48) must however be modified for coincident points because of the regulator (4.1), which is not Weyl-invariant:

$$\tilde{G}_r(P, P) = \exp\left[\frac{1}{4\pi i(g-1)^2} S_L[\sigma] + \left(\frac{1}{g-1} + 1\right)\sigma(P)\right] \cdot G_r(P, P) .$$

The net number of Green functions in (4.14) is $p^2+q^2-2pq=(2\lambda-1)^2(g-1)^2$ by (3.14). Collecting factors of S_L and $\sigma(P_i)$ we therefore see that

$$\tilde{Z}_{\text{fluct}}=\exp\left[-\frac{1}{12\pi i}S_L+(2\lambda-1)^2(g-1)^2\frac{1}{4\pi i(g-1)^2}S_L\right]$$

$$\cdot\exp\left[\frac{1}{2(g-1)}(2p\sum\sigma(P_i)+2q\sum\sigma(Q_i)-2q\sum\sigma(P_i)-2p\sum\sigma(Q_i))\right.$$

$$\left.+\sum\sigma(P_i)+\sum\sigma(Q_i)\right]\cdot Z_{\text{fluct}}$$

$$=\exp\left[\frac{1}{6\pi i}(6\lambda^2-6\lambda+1)S_L\right]$$

$$\cdot\exp\left[2\lambda\sum\sigma(P_i)+(2-2\lambda)\sum\sigma(Q_i)\right]\cdot Z_{\text{fluct}}\ .$$

The Liouville part matches (5.1), while the remaining factor gives the correct rescaling properties of (4.3).

Thus we can require our metric to be in any convenient conformal slice. Only now will we use this freedom to choose the Arakelov metric, so that coincident Green functions equal one [Eq. (4.2)].

Had we used an alternate normal-ordering prescription to (4.1), we would have gotten coordinate-dependent factors and a different metric dependence at the P_i, Q_i. For example, in [5] we used a modification of (4.1) to get expressions for $b(P)\bar{b}(P)$, not (4.3). The two prescriptions are completely equivalent.

B. Outline of Proof. The main ingredient in the proof is Quillen's treatment of a holomorphic family of Cauchy-Riemann operators on compact Riemann surfaces: the zeta-function regulated determinant det′ $\bar{\partial}_\xi^\dagger\bar{\partial}_\xi$ is used to define a metric, the "Quillen metric" on the determinant line bundle of ξ.[12] The bosonization formulae can be seen as asserting that some natural isomorphisms of determinant line bundles are isometries when one uses the Quillen metrics. We will establish these isometries from two basic results:

1. *The spin-1/2 bosonization formula* (4.17). Recall that in this formula $\mathscr{L}_b$ is any twisted spin bundle, i.e. any line bundle on Σ of degree $g-1$. Σ is equipped with an *arbitrary* metric. This metric in turn defines a metric on the preferred spin bundle $L_{\mathscr{A}}$ (the one which makes the isomorphism $L_{\mathscr{A}}^2\cong K$ an isometry), and hence a metric on $\mathscr{L}_b$. The left hand side of (4.17) is computed using this metric. On the right hand side $(iY)^{-1}$ is the period matrix of Σ, A_Σ is the metric area of Σ, and $\mathscr{N}$ is defined in (2.43).

We note that all of the bosonization formulae involve the same function $(\ldots)^{-\frac{1}{2}}\mathscr{N}$, which is essentially the spin-1/2 determinant [6, 52, 7]. Thus it seems that one could prove all the results we need by referring each spin to the known case of spin 1/2. This is accomplished by

---

[12] This metric is closely tied to the analytic torsion of [56]

2. *The insertion theorem*, given below in subsection D, which relates $\det' \bar{\partial}_\xi^\dagger \bar{\partial}_\xi$ and $\det' \bar{\partial}_{\xi\otimes\mathcal{O}(P)}^\dagger \bar{\partial}_{\xi\otimes\mathcal{O}(P)}$ for any line bundle ξ and any point P on Σ. It is in this theorem that the Arakelov metric plays a simplifying role.

The insertion theorem is the mathematical counterpart of the insertions of $b(P)$ or $c(Q)$ in the functional determinants of Sect. four. It is also closely related to the third axiom defining metrics on direct image bundles in Falting's work on arithmetic geometry [19]. In fact, this theorem allows one to prove that the norm used by Faltings differs from the Quillen norm only by a multiplicative factor depending only on the surface Σ [55].

We will prove formulae slightly more general than the bosonization identities of Sect. four. Indeed, we consider not only the line bundles $\mathscr{L}_b$ with degree a multiple of $g-1$, but arbitrary line bundles ξ on Σ of degree $d \geqq g-1$. In general ξ does not have an Arakelov metric. Instead we will demand that the metric on ξ be "admissible", which means that its curvature is proportional to the Arakelov curvature form μ_{Arak} in (2.53). Thus the Arakelov metric itself is admissible. Admissible metrics always exist; they are unique up to a constant since Σ is compact.

To state our more general formula le us begin by supposing that

$$H^1(\Sigma;\xi)=0 \ . \tag{5.2}$$

Recall from (2.16) that this condition means that the adjoint $\bar{\partial}_\xi^\dagger$ has no zero mode. This statement is independent of the metrics chosen on Σ and ξ. It is satisfied if $d>2g-2$ or if $d=2g-2$ but $\xi \not\cong K$, or if $d=g-1$ and ξ does not belong to the theta divisor. We will sketch later the necessary modification to the proof when (5.2) is not satisfied – for example in the proof of (4.16).

We also have from (2.10) that $\dim H^0(\Sigma;\xi)=\dim\ker \bar{\partial}_\xi \equiv k$. Using the simplifying condition (5.2) and the Riemann-Roch theorem (2.18) we then get that $k=d+1-g$. Let $u_1,\dots,u_k$ be a basis of $H^0(\Sigma;\xi)$.

For any integer $q\geqq 0$, let $p=q+k$. Suppose we are given $p+q$ pairwise distinct points on Σ, $P_1,\dots,P_p$, $Q_1,\dots,Q_q$. From these points we can build a determinant generalizing the one in (4.8):

$$\operatorname{Det}(u_i,P_j,Q_l)\equiv\det\begin{pmatrix} u_1(P_1) & \dots & u_k(P_1) & \mathscr{G}(P_1,Q_1) & \dots & \mathscr{G}(P_1,Q_q)\\ \vdots & & \vdots & \vdots & & \vdots\\ u_1(P_p) & \dots & u_k(P_p) & \mathscr{G}(P_p,Q_1) & \dots & \mathscr{G}(P_p,Q_q)\end{pmatrix}. \tag{5.3}$$

The parametrix $\mathscr{G}$ is again defined by (4.7), where now $\mathscr{L}_b$ is replaced by the arbitrary line bundle ξ.[13] Again the residue of $\mathscr{G}$ as $P\to Q$ is $1/2\pi i$. $\operatorname{Det}(u_i,P_j,Q_l)$ is an element of $\left[\bigotimes_{j=1}^{p}\xi|_{P_i}\right]\otimes\left[\bigotimes_{j=1}^{q}(\xi^{-1}\otimes K)|_{Q_j}\right]$.

We also suppose that Σ is equipped with the Arakelov metric, and that ξ is equipped with an admissible metric.

[13] Again when (5.2) is not satisfied we replace the unit operator in (4.7) by the projector off the zero modes of $\bar{\partial}_\xi^\dagger$

The generalization of (4.15) which we will prove then says that with the above choices, when (5.2) is satisfied we have

$$\|\mathrm{Det}\,(u_i, P_j, Q_\ell)\|^2 \cdot \frac{\det' \bar\partial_\xi^\dagger \bar\partial_\xi}{\det\,(u_i, u_j)} = A(g,d,q)\left(\frac{\det' \bar\partial^\dagger \bar\partial}{\det\,(iY)^{-1}\cdot A_\Sigma}\right)^{-\frac{1}{2}}$$

$$\cdot \mathcal{N}(\xi \otimes \mathcal{O}(-D_{\mathrm{ins}}) \otimes L_{\mathcal{A}}^{-1}) \frac{\prod_{i<j} G(P_i,P_j)^2 \prod_{i<j} G(Q_i,Q_j)^2}{\prod G(P_i,Q_j)^2}, \tag{5.4}$$

where $A(g,d,q)$ is a constant which depends only on g, d, and q, and

$$D_{\mathrm{ins}} = \sum_{i=1}^{p} P_i - \sum_{i=1}^{q} Q_i \ . \tag{3.23$'$}$$

Taking $\xi = \mathcal{L}_b$ gives (4.15). This follows because with the Arakelov norm coincident Green functions vanish, while each noncoincident function in the numerator of (4.15), (4.16) appears twice.

The proof of (5.4) will go roughly as follows. Thanks to the insertion theorem applied $p+q$ times, we can relate $\det' \bar\partial_\xi^\dagger \bar\partial_\xi$ and $\det' \bar\partial_{\xi'}^\dagger \bar\partial_{\xi'}$ where $\xi' = \xi \otimes \mathcal{O}(-D_{\mathrm{ins}})$. Next, as the degree of ξ' is $d+q+p = d-k = g-1$, we can relate $\det' \bar\partial_{\xi'}^\dagger \bar\partial_{\xi'}$ to $\det' \bar\partial^\dagger \bar\partial$ by the spin-1/2 bosonization formula. The finite dimensional determinants occur in (5.4) because the precise definitions of the determinant line bundle and of the Quillen metric on it involve the finite dimensional spaces of zero modes of $\bar\partial$ operators and their adjoints, so that we have to take care of them when we apply the insertion theorem. The bosonic Green functions occur because this theorem makes essential use of the admissible metrics on the $\mathcal{O}(P)$'s defined by (2.50).

We can also derive formulae analogous to (5.4) when the condition (5.2) is not satisfied. We will only consider the case $\xi = K$, equipped with the Arakelov metric. Again let $q>0$ be an integer, let $p = g-1+q$, and let $P_1, \ldots, P_p, Q_1, \ldots, Q_q$ be pairwise distinct points of Σ. We can then build the determinant

$$\mathrm{Det}\,(\omega_i, P_j, Q_l) \equiv \det \begin{pmatrix} \omega_1(P_1) & \ldots & \omega_g(P_1) & \mathcal{G}(P_1, Q_1) & \ldots & \mathcal{G}(P_1, Q_q) \\ \vdots & & \vdots & \vdots & & \vdots \\ \omega_1(P_p) & \ldots & \omega_g(P_p) & \mathcal{G}(P_p, Q_1) & \ldots & \mathcal{G}(P_p, Q_q) \\ 0 & \ldots & 0 & 1 & \ldots & 1 \end{pmatrix}. \tag{5.5}$$

It belongs to $\bigotimes_{i=1}^{g-1+q} K|_{P_i}$.

Then we have the equality

$$\|\mathrm{Det}\,(\omega_i, P_j, Q_l)\|^2 \cdot \frac{\det' \bar\partial_K^\dagger \bar\partial_K}{(1,1)\cdot \det\,(\omega_i, \omega_j)} = B(g,q)\left(\frac{\det' \bar\partial^\dagger \bar\partial}{\det\,(iY)^{-1}\cdot A_\Sigma}\right)^{-\frac{1}{2}}$$

$$\cdot \mathcal{N}(\mathcal{O}(-D_{\mathrm{ins}}) \otimes L_{\mathcal{A}}) \frac{\prod_{i<j} G(P_i,P_j)^2 \prod_{i<j} G(Q_i,Q_j)^2}{\prod G(P_i,Q_j)^2}, \tag{5.6}$$

where $B(g,q)$ is a constant depending only on g,q. When $q=1$ we then recover (4.16). [See the comments surrounding (4.6).]

Strictly speaking in this paper we will prove (5.4) and (5.6) only for $g>2$.[14] To get complete proofs when $g \leqq 2$, one needs estimates on the growth of regularized determinants and Green functions when Σ degenerates into a Riemann surface with one node, which we will not discuss here.

Note that (5.4) and (5.6) are closely related to some classical identities in the theory of abelian functions on Riemann surface [57], in particular the trisecant identity [23, 22].

C. The Local Riemann-Roch Theorem. In this subsection we review some basic facts about determinant line bundles, Quillen metrics, holomorphic families of $\bar{\partial}$ operators on compact Riemann surface, and the Riemann-Roch theorem for families (cf. [13, 58, 17, 63, 18]).

If $\mathscr{D}$ is an elliptic differential operator on a compact manifold, one defines the one dimensional vector space

$$\mathrm{DET}\,\mathscr{D} = \left(\bigwedge^{\max} \ker \mathscr{D}\right)^{-1} \otimes \left(\bigwedge^{\max} \operatorname{coker} \mathscr{D}\right). \tag{5.7}$$

Formally DET $\mathscr{D}$ is the dual of the "top exterior power" of the family index of $\mathscr{D}$.

It is important for our purposes to consider not only one particular $\mathscr{D}$ but a parametrized *family* of operators. We therefore need a notion of a family $\{\Sigma_s\}$ of Riemann surfaces, with a family of line bundles $\{\xi_s\}$ on them. We can glue together all the Riemann surfaces into a large space X, and glue the ξ_s, into a single bundle E over the total space X. In this paper we will actually consider *holomorphic* families of Riemann surface and bundles. Thus we let $\pi: X \to S$ be a proper holomorphic submersion, the fibers of which are compact and of complex dimension one. We also take E to be a holomorphic vector bundle on X and $F = E \otimes \bar{K}_{X|S}$, where $K_{X|S}$ is the line bundle of vertical (1,0)-forms on X. For each $s \in S$ we then get an elliptic operator $\bar{\partial}_s : C^\infty(\Sigma_s;\xi_s) \to C^\infty(\Sigma_s;\xi_s \otimes \bar{K})$ on the Riemann surface $\Sigma_s = \pi^{-1}(s)$. $\bar{\partial}_s$ is called the Cauchy-Riemann operator "coupled to ξ_s" and the family so defined is denoted by $\bar{\partial}_E$. The determinant line bundle DET $\bar{\partial}_E$ has a canonical holomorphic structure [59, 17]. This construction was introduced first within the framework of algebraic geometry [60, 59]; the construction for smooth families appears e.g. in [61, 58].

One can define a norm on the determinant line bundle as follows. Suppose we are given a smooth family of riemannian metrics on the fibers Σ_s (i.e. a smooth metric on the vertical tangent bundle $K_{X|S}^{-1}$), and a smooth hermitian metric on E. Then for any $s \in S$, $\bar{\partial}_s^\dagger$ is defined, and ker $\bar{\partial}_s$ and coker $\bar{\partial}_s$ have natural L^2 metrics, which define a metric $\|\cdot\|_{L^2}$ on DET $\bar{\partial}_s$. Generally, because of the jumps of the dimension of ker $\bar{\partial}_s$, this norm is not smooth on all of S. However, the "Quillen norm"

$$\|\cdot\|_Q^2 \equiv \det{}' \bar{\partial}_s^\dagger \bar{\partial}_s \cdot \|\cdot\|_{L^2}^2 \tag{5.8}$$

is always a smooth metric on DET $\bar{\partial}_E$ [13, 58].

---

[14] See Lemma 2 below

We can now state a local Riemann-Roch theorem for families of curves. Recall that for any hermitian metric on a holomorphic vector bundle like E there is a unique unitary connection on E compatible with its holomorphic structure. Using this connection and the Chern-Weil formulae for characteristic classes, we can associate to a line bundle E and its metric the Chern forms $c_1(E, \|\cdot\|_E)$, the Chern character form $\mathrm{Ch}(E, \|\cdot\|_E)$, and the Todd form $\mathrm{Td}(E, \|\cdot\|_E)$ using the polynomials $\mathrm{Ch}(x)=e^x$, $\mathrm{Td}(x)=1+\frac{1}{2}x+\frac{1}{12}x^2+\dots$. It is important to note that in the family setting these forms are constructed from the full curvature of the bundle E over X, not just its vertical parts. If σ is any differential form, we denote by $\sigma^{(k)}$ its component of degree k.

Theorem. *Let $\pi: X\to S$ be a holomorphic family of Riemann surfaces and E a holomorphic vector bundle over X. Let $\|\cdot\|$ be any smooth hermitian metric on the tangent bundle $K_{X|S}^{-1}$ and $\|\cdot\|_E$ a smooth hermitian metric on E. Let $\|\cdot\|_Q$ be the Quillen metric they define on* DET $\bar\partial_E$. *Then one has the formula*

$$c_1(\mathrm{DET}\,\bar\partial_E, \|\cdot\|_Q) = -\int_{X|S} \{\mathrm{Ch}(E, \|\cdot\|_E)\wedge \mathrm{Td}(K_{X|S}^{-1}, \|\cdot\|)\}^{(4)} . \tag{5.9}$$

Here $\int_{X|S}$ denotes integration of a form along the fibers Σ_s of π. Note that (5.9) makes no use of the admissibility condition; it works for any metrics.

The cohomological form of (5.9) is a direct consequence of the Atiyah-Singer index theorem for families, or of the Riemann-Roch-Grothendieck theorem (which gives a more precise formula, true in the rational Chow group of S). The formula (5.9) was proved by Quillen [13] when $X=X_0\times S$ and the metrics are fixed, and by Belavin and Knizhnik [14] when $E=K^n$. Bismut and Freed have proven an analogous statement for families of Dirac operators [58], from which one can deduce the general formula [15, 17, 63, 18].

D. The Insertion Theorem. Let Σ be a compact connected Riemann surface of genus $g>0$, equipped with its Arakelov metric, and let ξ be an arbitrary line bundle on Σ, equipped with an admissible metric. For any point P of Σ, we get an admissible metric on $\xi\otimes\mathcal{O}(-P)$ by multiplying the given metric on ξ by the canonical metric on $\mathcal{O}(-P)$, i.e. the metric dual to the metric on $\mathcal{O}(P)$ given by (2.50). From these data, we obtain Quillen metrics on the one-dimensional spaces DET $\bar\partial_\xi$ and DET $\bar\partial_{\xi\otimes\mathcal{O}(-P)}$.

On the other hand, the long exact sequence

$$0\to H^0(\Sigma;\xi\otimes\mathcal{O}(-P))\to H^0(\Sigma;\xi)\to\xi|_P\to H^1(\Sigma;\xi\otimes\mathcal{O}(-P))\to H^1(\Sigma;\xi)\to 0$$

associated to (2.8) gives rise to a canonical isomorphism of one dimensional vector spaces, by taking the top exterior power:

$$I: \mathrm{DET}\,\bar\partial_{\xi\otimes\mathcal{O}(-P)}\cong(\mathrm{DET}\,\bar\partial_\xi)\otimes\xi|_P . \tag{5.10}$$

We have used (2.16) to replace the cokernel in (5.7) by H^1.

Insertion Theorem. *The isomorphism* (5.10) *is an isometry when ξ_P has the given metric and the determinant spaces are given the Quillen metrics (up to a multiplicative constant depending only on the genus of Σ and on the degree of ξ).*

For $g>2$, the theorem is a direct consequence of the following lemmas:

Lemma 1. *Let $\pi: X \to S$ be a holomorpic family of compact connected Riemann surfaces of genus $g>0$, E a holomorpic line bundle on X, and $\sigma: S \to X$ a holomorpic section of π. Let $\Sigma_s = \pi^{-1}(s)$ and $\xi_s = E|_{\Sigma_s}$.*

i) *The family consisting of the Arakelov metrics on the Riemann surfaces Σ_s defines a smooth metric on the vertical tangent bundle $K_{X|S}^{-1}$.*

ii) *The family of canonical metrics* (2.50) *on the line bundles $\mathcal{O}(-\sigma(s))$ over Σ_s, $s \in S$, defines a smooth metric on $\mathcal{O}(-\sigma(S))$ (note that $\mathcal{O}(-\sigma(S))|_{\Sigma_s} \cong \mathcal{O}(-\sigma(s))$).*

iii) *The family of isomorphisms*

$$I_s: \mathrm{DET}\ \bar{\partial}_{\xi_s \otimes \mathcal{O}(-\sigma(s))} \otimes (\mathrm{DET}\ \bar{\partial}_{\xi_s}) \otimes \xi_s|_{\sigma(s)}$$

(cf. (5.10)*) defines an isomorphism of holomorphic line bundles on S*

$$I: \mathrm{DET}\ \bar{\partial}_{E'} \otimes \mathrm{DET}\ \bar{\partial}_E \otimes \sigma^*(E)\ , \tag{5.11}$$

where $E' = E \otimes \mathcal{O}(-\sigma(S))$.

iv) *Suppose that E is equipped with a smooth metric $\|\cdot\|_E$ whose restriction to any Σ_s is an admissible metric on ξ_s, and that $K_{X|S}$ and $\mathcal{O}(-\sigma(S))$ are equipped with the Arakelov metric and the canonical metric defined in (i) and (ii). Using these metrics, we obtain Quillen metrics $\|\cdot\|_Q$ and $\|\cdot\|'_Q$ on the line bundles* DET $\bar{\partial}_E$ *and* DET $\bar{\partial}_{E'}$ *on S. Then we have the equality of differential forms*

$$c_1(\mathrm{DET}\ \bar{\partial}_{E'}, \|\cdot\|'_Q) = c_1(\mathrm{DET}\ \bar{\partial}_E, \|\cdot\|_Q) + \sigma^* c_1(E, \|\cdot\|_E)\ . \tag{5.12}$$

Hence when S is compact and connected, the isomorphism I in (5.11) is an isometry up to an overall constant. The next lemma says that for $g>2$ we can always take S to be compact and connected.

Lemma 2. *Let Σ_0, Σ_1 be two compact connected Riemann surfaces with the same genus $g>2$ and let ξ_0, ξ_1 be holomorphic line bundles on them with the same degree d. Suppose each ξ_i is equipped with an admissible metric $\|\cdot\|_i$ and a point P_i. There exists a compact and connected complex manifold S, a holomorphic family $\pi: X \to S$ of compact connected Riemann surfaces, and a holomorphic line bundle E on X equipped with a smooth metric $\|\cdot\|_E$ which, restricted to any $\Sigma_s = \pi^{-1}(s)$, is admissible, and two points $s_i \in S$, $i=1,2$ and two isomorphisms $\varphi_i: \Sigma_i \to \Sigma_{s_i}$ such that*

$$(\xi_i, \|\cdot\|_i) \cong \varphi_i^*(E, \|\cdot\|_E)\ .$$

Lemma 2 is an easy consequence of the hard fact that, for $g>2$, for any two points in the moduli space of smooth irreducible curves $\mathcal{M}_g$ there exists a complete curve which contains those two points [62].

The first two assertions of Lemma 1 are consequences of the definitions and of the theory of families of elliptic operators. The third assertion is a consequence of the definition of the holomorphic structure on the determinant line bundle. We will not enter here into the details of the proof of these assertions. The fourth assertion is a consequence of the local Riemann-Roch theorem of subsection C and the choice of Arakelov (respectively admissible) metrics on K, ξ. We now give the proof of the equality (5.12).

Let D denote the hypersurface $\sigma(S)$, and let $[D]$ be the current, or form-valued distribution, associated to D. That is, for any form ω of degree the real dimension of D one has $\int_X [D]\omega = \int_D \omega$.

Let $\tilde{G}_D(x) = \|\mathbf{1}_{\mathcal{O}(D)}\|(x)$, where $\mathcal{O}(D)$ is equipped with the canonical metric. The function $\tilde{G}_D$ is smooth on $X-D$, and it vanishes on D. If G is the Arakelov Green function on Σ_s, we have

$$\tilde{G}_D(x) = G(x, \sigma \circ \pi(x)) \ . \tag{5.13}$$

Furthermore, the following statements are easily proved:

a) The first Chern form of the line bundle $\mathcal{O}(-D)$ equipped with the canonical metric satisfies the following equality:

$$c_1(\mathcal{O}(-D), \|\cdot\|) = -\frac{1}{\pi i}\,\partial\bar{\partial}\log\tilde{G}_D - [D] \ . \tag{5.14}$$

[See the remark after (2.27).] The right-hand side is the sum of two non-smooth currents.

b) For any smooth differential form ω on X

$$\int_{X|S} [D]\omega = \sigma^*\omega \ . \tag{5.15}$$

c) For any closed differential form ω of type $(1,1)$ on X, and for any distribution ϕ on X,

$$\int_{X|S} (\partial\bar{\partial}\phi)\omega = \partial\bar{\partial}\int_{X|S}\phi\omega \ . \tag{5.16}$$

The equality (5.12) now follows from the following computation of first Chern forms, where we do not write explicitly the metrics on the various line bundles (they are the metrics specified in the statement of Lemma 1):

$$\begin{aligned} c_1(\mathrm{DET}\,\bar{\partial}_E) - c_1(\mathrm{DET}\,\bar{\partial}_{E'}) &= -\int_{X|S}\{[\mathrm{Ch}\,E - \mathrm{Ch}\,E']\mathrm{Td}\,K_{X|S}^{-1}\}^{(4)} \\ &= -\int_{X|S}\{[1-\mathrm{Ch}(\mathcal{O}(-D))]\mathrm{Ch}\,E\ \mathrm{Td}\,K_{X|S}^{-1}\}^{(4)} \\ &= \int_{X|S} c_1(\mathcal{O}(-D))\,[c_1(E) - \tfrac{1}{2}c_1(K_{X|S}) + \tfrac{1}{2}c_1(\mathcal{O}(-D))] \ . \end{aligned}$$

The first equality follows from the local Riemann-Roch theorem, the second from the multiplicativity of the Chern character, and the last from the expressions for Ch and Td. Using (5.14), we can rewrite the integral (5.15) as the sum of

$$\omega_1 = -\int_{X|S}\frac{1}{\pi i}\,\partial\bar{\partial}\log\tilde{G}_D\cdot\left[c_1(E) - \frac{1}{2}c_1(K_{X|S}) + \frac{1}{2}c_1(\mathcal{O}(-D))\right]$$

and

$$\omega_2 = -\int_{X|S}[D]\cdot\left[c_1(E) - \frac{1}{2}c_1(K_{X|S}) + \frac{1}{2}c_1(\mathcal{O}(-D))\right] \ .$$

The identity (5.16) shows that $\omega_1 = \partial\bar{\partial}F$, where

$$f = -\frac{1}{\pi i} \int_{X|S} \log \tilde{G}_D \cdot [c_1(E) - \tfrac{1}{2} c_1(K_{X|S}) + \tfrac{1}{2} c_1(\mathcal{O}(-D))] \ .$$

The restriction of the quantity in brackets to any Σ_s is a multple of μ_{Arak}, thanks to the admissibility hypothesis on the metrics on E, $K_{X|S}$, and $\mathcal{O}(-D)$. Next the formula (5.13) and the normalization condition (2.46) show that $F \equiv 0$. Hence $\omega_1 = 0$. On the other hand, the identity (5.15) shows that

$$\omega_2 = -\sigma^* c_1(E) + \tfrac{1}{2} c_1(\sigma^*(K_{X|S} \otimes \mathcal{O}(D))) \ .$$

Recall that the Arakelov metric and the canonical metric are such that the residue map (2.52) is an isometry. This implies that the line bundle with metric $\sigma^*(K_{X|S} \otimes \mathcal{O}(D))$ is canonically isomorphic to the trivial bundle on S, with the trivial metric. Hence its first Chern form is zero, and $\omega_2 = -\sigma^* c_1(E)$.

Finally we get

$$c_1(\text{DET}\ \bar{\partial}_E) - c_1(\text{DET}\ \bar{\partial}_{E'}) = \omega_1 + \omega_2 = -\sigma^* c_1(E) \ ,$$

as was to be proved.

Having established (5.12) we now invoke Lemma 2 to say that metrics with the same curvature on S must be equal up to a constant. This establishes the insertion theorem.

E. Proof of (5.4). Using the spin-1/2 bosonization formula (4.17) we see that the desired formula (5.4) follows from the following equality:

$$\|\text{Det}(u_i, P_j, Q_\ell)\|^2 \cdot \frac{\det' \bar{\partial}_\xi^\dagger \bar{\partial}_\xi}{\det(u_i, u_j)}$$

$$= C(g, d, q) \frac{\prod_{i<j} G(P_i, P_j)^2 \prod_{i<j} G(Q_i, Q_j)^2}{\prod G(P_i, Q_j)^2} \cdot \det' \bar{\partial}_{\xi'}^\dagger \bar{\partial}_{\xi'} \ , \tag{5.17}$$

when $\xi' \equiv \xi \otimes \mathcal{O}(-D_{\text{ins}})$ is equipped with the product of the given admissible metric on ξ and the canonical metric on $\mathcal{O}(-D_{\text{ins}})$. Again D_{ins} is the divisor of insertion points (3.23). To prove (5.17) we use the insertion theorem.

Consider the following short exact sequence of sheaves [cf. (2.8)]:

$$0 \to \zeta \otimes \mathcal{O}(-\textstyle\sum P_i) \to \zeta \to \zeta|_{\Sigma P_i} \to 0 \ . \tag{5.18}$$

Taking $\zeta = \xi \otimes \mathcal{O}(\sum Q_i)$ we get

$$0 \to \xi \otimes \mathcal{O}(-D_{\text{ins}}) \to \xi \otimes \mathcal{O}(\textstyle\sum Q_i) \to [\xi \otimes \mathcal{O}(\sum Q_i)]|_{\Sigma P_i} \to 0 \ . \tag{5.19}$$

Setting all the $Q_i = P_i$ we get

$$0 \to \xi \to \xi \otimes \mathcal{O}(\textstyle\sum Q_i) \to [\xi \otimes \mathcal{O}(\sum Q_i)]|_{\Sigma Q_i} \to 0 \ . \tag{5.20}$$

From the cohomology long exact sequences associated to these short exact sequences we deduce canonical isomorphisms of the one-dimensional vector spaces [cf. (5.10)]:

$$I_1 : \mathrm{DET}\ \bar\partial_{\xi\otimes\mathcal{O}(-D_{\mathrm{ins}})} \xrightarrow{\sim} \mathrm{DET}\ \bar\partial_{\xi\otimes\mathcal{O}(\Sigma Q_j)} \otimes \bigwedge^{p}\left(\bigoplus_{j=1}^{p}\left(\xi\otimes\mathcal{O}\left(\sum_{i=1}^{q} Q_i\right)\right)\bigg|_{P_j}\right)$$

$$\cong \mathrm{DET}\ \bar\partial_{\xi\otimes\mathcal{O}(\Sigma Q_j)} \otimes \left[\bigotimes_{i=1}^{p} \xi|_{P_i}\right], \tag{5.21}$$

$$I_2 : \mathrm{DET}\ \bar\partial_{\xi} \xrightarrow{\sim} \mathrm{DET}\ \bar\partial_{\xi\otimes\mathcal{O}(\Sigma Q_j)} \otimes \bigwedge^{q}\left(\bigoplus_{j=1}^{q}\left(\xi\otimes\mathcal{O}\left(\sum_{i=1}^{q} Q_i\right)\right)\bigg|_{Q_j}\right)$$

$$\cong \mathrm{DET}\ \bar\partial_{\xi\otimes\mathcal{O}(\Sigma Q_j)} \otimes \left[\bigotimes_{i=1}^{q} (\xi\otimes K^{-1})|_{Q_i}\right]. \tag{5.22}$$

In the second lines of (5.21), (5.22) we have used the unit section to trivialize $\mathcal{O}(\sum Q_i)$ away from the Q_i. In the second line of (5.22) we have used the canonical isomorphism (2.52).

From I_1 and I_2 one can build an isomorphism

$$I : \mathrm{DET}\ \bar\partial_{\xi} \xrightarrow{\sim} \mathrm{DET}\ \bar\partial_{\xi'} \otimes \left[\bigotimes_{i}^{p} \xi \bigotimes_{i}^{p} \xi^{-1}|_{P_i}\right] \otimes \left[\bigotimes_{1}^{q} (\xi\otimes K^{-1})|_{Q_j}\right].$$

Then we have the *general insertion formula*, when the determinant bundles are equipped with the Quillen metrics, ξ with the given admissible metric, and K with the Arakelov metric:

$$\|I(v)\| = D(g,d,p,q)\cdot\frac{\prod\limits_{i<j} G(P_i,P_j)\prod\limits_{i<j} G(Q_i,Q_j)}{\prod G(P_i,Q_j)}\cdot\|v\|\ . \tag{5.23}$$

This formula is true for any ξ and any collection $\{P_1,\ldots,P_p,\ldots,Q_q\}$ of pairwise distinct points on Σ; the conditions (5.2) and $p=q+d+1-g$ do not matter here.

The general insertion formula (5.23) follows from the insertion theorem by induction on p,q. Hence it too relies on the Arakelov condition. We now give the details for the case $p=2$, $q=0$ to show how the bosonic Green function enters.

The isomorphism (5.21) in this case reads

$$I_1 : \mathrm{DET}\ \bar\partial_{\xi\otimes\mathcal{O}(-P_1-P_2)} \xrightarrow{\sim} \mathrm{DET}\ \bar\partial_{\xi}\otimes\xi|_{P_1}\otimes\xi|_{P_2}\ . \tag{5.21$'$}$$

It is obtained by composition and tensor product from the "insertion isomorphisms"

$$\mathrm{DET}\ \bar\partial_{\xi\otimes\mathcal{O}(-P_1)} \cong \mathrm{DET}\ \bar\partial_{\xi}\otimes\xi|_{P_1}\ , \tag{5.24a}$$

$$\mathrm{DET}\ \bar\partial_{\xi\otimes\mathcal{O}(-P_1-P_2)} \cong \mathrm{DET}\ \bar\partial_{\xi\otimes\mathcal{O}(-P_1)}\otimes(\xi\otimes\mathcal{O}(-P_1))|_{P_2}\ , \tag{5.24b}$$

and from the canonical isomorphism

$$(\xi\otimes\mathcal{O}(-P_1))|_{P_2}\cong\xi|_{P_2}\ ,\qquad v\otimes\mathbf{1}_{\mathcal{O}(-P_1)}(P_2)\mapsto v\ . \tag{5.24c}$$

By the insertion theorem, (5.24a) and (5.24b) are isometries up to constants. On the other hand, (5.24c) multiplies the norms by $\|\mathbf{1}_{\mathcal{O}(-P_1)}(P_2)\|^{-1}=G(P_1,P_2)$. So the isomorphism I multiplies the norms by (a constant times) $G(P_1,P_2)$.

Proceeding in this way, and using (5.22) we obtain (5.23).

Now we can complete the proof of (5.17). It is enough to prove this formula when

$$H^0(\Sigma;\xi\otimes\mathcal{O}(-D_{\text{ins}}))=0\ , \tag{5.25}$$

i.e. when $\xi'=\xi\otimes\mathcal{O}(-D_{\text{ins}})$ does not belong to the theta divisor. Indeed this condition is satisfied for a generic choice of insertion points since $H^1(\Sigma;\xi)=0$. Moreover, one can see directly that when (5.25) is not satisfied, the two sides of (5.4) are both zero.

The condition (5.25) implies that DET $\bar\partial_{\xi'}$ is canonically trivial. The Quillen norm on this space is thus given by

$$\|\mathbf{1}_{\mathcal{O}(\Theta)}\|_Q^2=\det\bar\partial_{\xi'}^\dagger\bar\partial_{\xi'}\ . \tag{5.26}$$

The condition (5.2) gives that

$$\text{DET}\ \bar\partial_\xi\cong\left(\bigwedge^k H^0(\Sigma;\xi)\right)^{-1}\ , \tag{5.27}$$

so $(u_1\wedge\ldots\wedge u_k)^{-1}$ is a basis of DET $\bar\partial_\xi$. Its Quillen norm is

$$\|(u_1\wedge\ldots\wedge u_k)^{-1}\|_Q^2=\det{}'\,\bar\partial_\xi^\dagger\bar\partial_\xi\cdot(\det(u_i,u_j))^{-1}\ . \tag{5.28}$$

Finally we see that the formula (5.17) is a consequence of the generalized insertion formula (5.23), of (5.26) and (5.28) and of the following lemma:

Lemma 3.

$$I((u_1\wedge\ldots\wedge u_k)^{-1})=(2\pi i)^{-q}\,\text{Det}(u_i,P_j,Q_l)^{-1}\ .$$

This lemma is a consequence of the following observations:

i) $H^1(\Sigma;\xi\otimes\mathcal{O}(\sum Q_i))=0$ because of (5.25). Thus

$$\text{DET}\ \bar\partial_{\xi\otimes\mathcal{O}(\sum Q_i)}\cong\left(\bigwedge^p H^0\left(\Sigma;\xi\otimes\mathcal{O}\left(\sum_{i=1}^q Q_i\right)\right)\right)^{-1}\ . \tag{5.29}$$

ii) The map

$$I_2:[\text{DET}\ \bar\partial_{\xi\otimes\mathcal{O}(\sum Q_i)}]^{-1}\to\bigotimes_{i=1}^p\xi|_{P_i}$$

is the p-th exterior power of the restriction map

$$\mathrm{r}:H^0(\Sigma;\xi\otimes\mathcal{O}(\textstyle\sum Q_i))\to[\xi\otimes\mathcal{O}(\sum Q_i)]|_{\sum P_j}\cong\bigoplus_{i=1}^p\xi|_{P_i}\ .$$

I_2' is an isomorphism when the insertion points satisfy (5.25), and in that case it is the isomorphism (5.21).

iii) The map $\mathrm{I}_2:\text{DET}\ \bar\partial_\xi\xrightarrow{\sim}\text{DET}\ \bar\partial_{\xi\otimes\mathcal{O}(\sum Q_i)}\otimes\left[\bigotimes_{i=1}^q(\xi^{-1}\otimes K)|_{Q_i}\right]$ can be written, thanks to the identifications (5.27) and (5.29):

$$\begin{aligned}I_1((u_1\wedge\ldots\wedge u_k)^{-1})=&(u_1\wedge\ldots\wedge u_k\wedge 2\pi i\mathcal{G}(\cdot,Q_1)v_1\wedge\ldots\wedge 2\pi i\mathcal{G}(\cdot,Q_q)v_q)^{-1}\\&\otimes(v_1\otimes\ldots\otimes v_q)^{-1}\end{aligned}$$

for any choice of nonzero $v_i\in(\xi\otimes K^{-1})|_{Q_i}$.

The last assertion is a consequence of the fact that $2\pi i\mathscr{G}(\cdot, Q_j)v_j$ is an element of $H^0(\Sigma; \xi\otimes\mathcal{O}(\sum Q_i))$, which has $[\delta_{ij}v_i]_{i=1,\ldots,q}\in\bigoplus_{i=1}^{q}(\xi\otimes K^{-1})|_{Q_i}$ as its image by the residue map at the points $Q_1,\ldots,Q_q$. See the remark following (4.7).

This completes the proof of (5.17), hence in particular of (4.15) and so the bosonization identity (4.4), when the condition (5.2) is satisfied.

When (5.2) is not satisfied, e.g. for $\xi=K$, the proof of the bosonization formula (5.6) follows the same lines as the proof of (5.17). The only real difference occurs in the construction of the isomorphism I_1:

$$I_1 : \mathrm{DET}\,\bar{\partial}_K \xrightarrow{\sim} \mathrm{DET}\,\bar{\partial}_{K\otimes\mathcal{O}(\sum Q_j)} \,.$$

We need the identifications

$$\mathrm{DET}\,\bar{\partial}_K \cong \left(\bigwedge^{g} H^0(\Sigma; K)\right)^{-1},$$

$$\mathrm{DET}\,\bar{\partial}_{K\otimes\mathcal{O}(\sum Q_j)} \cong \left(\bigwedge^{g-1+q} H^0(\Sigma; K\otimes\mathcal{O}(\textstyle\sum Q_j))\right)^{-1},$$

$$I_1(\omega_1\wedge\ldots\wedge\omega_g)^{-1}) = (2\pi i)^{1-q}[\omega_1\wedge\ldots\wedge\omega_g\wedge(\mathscr{G}(\cdot,Q_1)-\mathscr{G}(\cdot,Q_q))\wedge\ldots$$
$$\wedge(\mathscr{G}(\cdot,Q_{q-1})-\mathscr{G}(\cdot,Q_q))]^{-1} \,.$$

Note that the norm of the constant function 1 appears in the denominator of the left-hand side of (5.6), not in the numerator as (5.7), (5.8) might seem to imply. This is because we have represented a basis of coker $\bar{\partial}_K$ by a basis of the *dual* space ker $\bar{\partial}$ using Serre duality (2.15).

Appendix A. Proof of (2.32)

We are to establish that Fig. 4 commutes when $F_{\mathscr{A}}$, H, and $I_{\mathscr{A}}$ are defined as in (2.32), (2.29), and (2.26) respectively. Consider the following complex function on the cut surface Σ_c:

$$f(P) = \vartheta(z + I_{\mathscr{A}}[P-P_0]|\tau_{\mathscr{A}})/\vartheta(I_{\mathscr{A}}[P-P_0]|\tau_{\mathscr{A}}) \,.$$

By the Riemann vanishing theorem f has exactly g zeros and poles for generic z. The zeros are at P_i where

$$z + I_{\mathscr{A}}\left[\sum_i P_i - P_0 - D_{\mathscr{A}}\right] = 0 \,,$$

and similarly the poles, with z replaced by 0. $D_{\mathscr{A}}$ is a divisor representing the preferred spin bundle $L_{\mathscr{A}}$. Hence

$$I_{\mathscr{A}}[\mathrm{div} f] = -z \,. \tag{5.30}$$

As P goes around the cycle $n\cdot a + m\cdot b$, f jumps by $e^{-2\pi i m\cdot z}$. Build a bundle ξ with transition functions such that f defines a meromorphic section σ on all of Σ.

Next put a norm on ξ: let

$$\|\sigma\|^2 = \exp\left\{2\pi i(z-\bar{z})\cdot Y_{\mathscr{A}}\cdot\left[\int_{P_0}^{P}\omega_{\mathscr{A}} - \int_{P_0}^{P}\bar{\omega}_{\mathscr{A}}\right]\right\}|f|^2 ,$$

where $Y_{\mathscr{A}} = (\tau_{\mathscr{A}} - \bar{\tau}_{\mathscr{A}})^{-1}$. Note that $\|\sigma\|^2$ is the quotient of two of the $\mathscr{N}$ functions defined in (2.43). We then get the connection

$$\Theta = \partial \log \|\sigma\|^2 = 2\pi i(z-\bar{z})\, Y_{\mathscr{A}}\omega_{\mathscr{A}} + \partial \log f .$$

Integrating along contours which avoid the poles and zeros of σ, we find the holonomy

$$H(\xi) = (\alpha, \beta)\begin{pmatrix} -Y_{\mathscr{A}}(z-\bar{z}) \\ -\tau_{\mathscr{A}} Y_{\mathscr{A}}(z-\bar{z}) + z \end{pmatrix} .$$

Applying (2.32) to $H(\xi)$ we recover $-z \in J_{\mathscr{A}}$, which indeed agrees with (5.30).

Acknowledgements. We would like to thank S. Coleman, S. Dellapietra, V. Dellapietra, D. Freed, J. Harris, D. Kazhdan, E. Martinec, I. Singer, and C. Soulé for many valuable discussions.

References

1. Coleman, S.: Phys. Rev. D **11**, 2088 (1975)
2. Marnelius, R.: Nucl. Phys. B **211**, 14 (1983)
3. Friedan, D., Martinec, E., Shenker, S.: Conformal invariance, supersymmetry, and string theory. Nucl. Phys. B **271**, 93 (1986)
4. Siegel, W., Zwiebach, B.: Nucl. Phys. B **263**, 105 (1986)
5. Alvarez-Gaumé, L. Bost, J.B., Moore, G., Nelson, P., Vafa, C.: Bosonization in arbitrary genus. Phys. Lett. B **178**, 41 (1986)
6. Alvarez-Gaumé, L., Moore, G., Vafa, C.: Theta functions, modular invariance, and strings. Commun. Math. Phys. **106**, 1 (1986)
7. Bost, J., Nelson, P.: Spin-1/2 bosonization on compact surfaces. Phys. Rev. Lett. **57**, 795 (1986)
8. Shankar, R.: Phys. Lett. **92** B, 333 (1980); Witten, E.; In: Fourth workshop on grand unification. Weldon, H., Langacker, P., Steinhardt, P. (eds.). Boston: Birkhäuser 1983)
9. Nepomechie, R.: Nonabelian bosonization, triality, and superstring theory. Phys. Lett. **178** B, 207 (1986); **180** B, 423 (1986); L. Brown, R. Nepochemie, Non-abelian bosonization: Current correlation functions. Phys. Rev. D **35**, 3239 (1987)
10. Gross, D., Harvey, J., Martinec, E., Rohm, R.: The heterotic string. Phys. Rev. Lett. **54**, 502 (1985); Heterotic string theory I, II. Nucl. Phys. B **256**, 253 (1985); B **267**, 75 (1986)
11. Friedan, D., Martinec, E., Shenker, S.: Phys. Lett. B **160**, 55 (1985)
12. Knizhnik, V.: Covariant superstring amplitudes from the sum over fermionic surfaces. Phys. Lett. **178** B, 21 (1986)
13. Quillen, D.: Determinants of Cauchy-Riemann operators on Riemann surfaces. Funk. Anal. i Prilozen **19**, 37 (1985) [=Funct. Anal. Appl. **19**, 31 (1986)]
14. Belavin, A., Knizhnik, V.: Phys. Lett. B **168**, 201 (1986); Complex geometry and theory of quantum strings. Landau Inst. preprint submitted to ZETF
15. Bost, J., Jolicœur, J.: Phys. Lett. B **174**, 273 (1986)
16. Nelson, P.: Lectures on strings and moduli space. Phys. Reports **149**, 337 (1987)
17. Bost, J.B.: Fibrés déterminants régularisés et mesures sur les espaces de modules des courbes complexes, sém. Bourbaki, 1986-7, n° 676

18. Freed, D.: On determinant line bundles, preprint to appear In: Mathematical aspects of string theory. Yau, S.-T. (ed.). (New York: World 1987)
19. Faltings, G.: Calculus on arithmetic surfaces. Ann. Math. **119**, 387 (1984)
20. Arakelov, S.: Izv. Akad. Nauk. SSSR Ser. Mat. **38** (1974) [= Math. USSR Izv. **8**, 1167 (1974)
21. Martinec, E.: Conformal field theory on a (super-)Riemann surface. Nucl. Phys. B **281**, 157 (1987)
22. Eguchi, T., Ooguri, H.: Chiral bosonization on a Riemann surface. Phys. Lett. **187** B, 127 (1987)
23. Verlinde, E., Verlinde, H.: Chiral bosonization, determinants, and the string partition function, Nucl. Phys. B **288**, 357 (1987)
24. Alvarez-Gaumé, L., Gomez, C., Reina, C.: Loops groups, grassmannians, and string theory. Phys. Lett. **190** B, 55 (1987)
25. Frenkel, I.: J. Funct. Anal. **14**, 259 (1981)
26. Goddard, P., Olive, D.: In: Vertex operators in mathematics and physics. Lepowsky, J., Mandelstam, S., Singer, I.M. (eds.). Berlin, Heidelberg, New York: Springer 1985
27. Ishibashi, N., Matsuo, Y., Ooguri, H.: Soliton equations and free fermions on Riemann surfaces. Tokyo UT-499
28. Vafa, C.: Operator formulation on Riemann surfaces. Phys. Lett. **190** B, 47 (1987)
29. Manin, Yu.: Critical dimensions of string theories. Funk. Anal. **20**, 88 (1986)
30. Beilinson, A., Manin, Yu., Shechtman, V.: Localization of the Virasoro and Neveu-Schwraz algebras, preprint
31. Alvarez-Gaumé, L., Nelson, P.: Riemann surfaces and string theories. In: Grisaru, M., de Wit, B. (eds.) Supersymmetry, supergravity, and superstrings '86 Singapore: World Scientific 1986
32. Knizhnik, V.: Analytic fields on Riemann surfaces. Phys. Lett. **180** B, 247 (1986)
33. Redlich, A., Schnitzer, H., Tsokos, K.: Bose-fermi equivalence on the two-dimensional torus for simply-laced groups. Nucl. Phys. B **289**, 397 (1987); Schnitzer, H. Tsokos, K.: Partition functions and fermi-bose equivalence for simply-laced groups on compact Riemann surfaces. Brandeis BRX TH-215
34. Kostelecky, V., Lechtenfeld, O., Lerche, W., Samuel, S., Watamura, S.: Conformal techniques, bosonization, and three-level string amplitudes. Nucl. Phys. B **288**, 173 (1987)
35. Bagger, J., Nemeschansky, D., Seiberg, N., Yankielowicz, S.: Bosons, fermions, and Thirring strings. Nucl. Phys. B **289**, 53 (1987)
36. Dugan, M., Sonoda, H.: Functional determinants on Riemann surfaces. Nucl. Phys. B **289**, 227 (1987)
37. Bott, R., Tu, L.: Differential forms in algebraic topology. Berlin, Heidelberg, New York: Springer 1982
38. Bers, L.: Bull. Lond. Math. Soc. **4**, 257 (1972)
39. Mumford, D.: Tata lectures on theta. Boston, MA: Birkhäuser 1983
40. Atiyah, M.: Riemann surfaces and spin structures. Ann. Sci. Ec. Norm. Sup. 4e série **4**, 47 (1971)
41. Wells, R.O.: Differential analysis on complex manifolds. Berlin, Heidelberg, New York: Springer 1980
42. Alvarez, O.: Topological quantization and cohomology. Commun. Math. Phys. **100**, 279 (1985)
43. Gunning, R.: Lectures on Riemann surfaces. Princeton, NS: Princeton University Press 1966
44. Farkas, H., Kra, I.: Riemann surfaces. Berlin, Heidelberg, New York: Springer 1980
45. Griffiths, P., Harris, J.: Principles of algebraic geometry. New York: Wiley 1978
46. Igusa, J.: Theta functions. Berlin, Heidelberg, New York: Springer 1972
47. Mandelstam, S.: Phys. Rev. D **11**, 3026 (1975)
48. Wolf, D., Zittartz, J.: Zeit. Phys. B **51**, 65 (1983)
49. Banks, T. et al.: Nucl. Phys. B **108**, 119 (1976)
50. Vafa, C.: Modular invariance and discrete torsion on orbifolds. Nucl. Phys. B **273**, 592 (1986)
51. Sonoda, H.: Calculation of a propagator on a Riemann Surface. Phys. Lett. 178 **B**, 390 (1986)

52. Namazie, M., Narain, K., Sarmadi, N.: Phys. Lett. 177**B**, 329 (1986)
53. Polyakov, A.: Phys. Lett. **103**B, 207 (1981)
54. Alvarez, O.: Theory of strings with boundary. Nucl. Phys. B**216**, 125 (1983)
55. Bost, J.-B.: Talk presented at the 8th International Congress on Mathematical Physics. Marseille 1986
56. Ray, D., Singer, I.: Analytic torsion. Ann. Math. **98**, 154 (1973)
57. Fay, J.: Theta functions on Riemann surfaces. Berlin, Heidelberg, New York: Springer 1973
58. Bismut, J.-M., Freed, D.: Geometry of elliptic families I, II. Commun. Math. Phys. **106**, 159 (1986); **107**, 103 (1986)
59. Knudsen, F., Mumford, D.: The projectivity of the moduli space of stable curves, I. Math. Scand. **39**, 19 (1976)
60. Séminaire de Géométrie Algèbrique 6. Lecture Notes in Mathematics, Vol. 225. Berlin, Heidelberg, New York: Springer 1971
61. Atiyah, M., Singer, I.: Dirac operators coupled to vector potentials. Proc. Nat. Acad. Sci. USA, **81**, 2597 (1984)
62. Harris, J.: In: Proc. Int. Cong. of Mathematicians 1983, Warszawa. Olech, C., Ciesielski, Z. (eds.). London: Elsevier 1984
63. Bismut, J., Gillet, H., Soulé, C.: Torsion analytique et fibrés déterminants holomorphes. CR Acad. Sci. Paris **305** Sér. I, 81 (1987); Analytic torsion and determinant line bundles. Orsay preprint 87T8
64. Ohrndorf, T.: On the compactification of the bosonic string at higher loops. Nucl. Phys. B**281**, 470 (1987)
65. Craigie, N., Nahm, W., Narain, K.: Ann. Phys. **147**, 78 (1987)

Communicated by A. Jaffe

Received March 20, 1987; in revised form March 30, 1987

Notes added in proof

1) Additional terms in the bosonic action similar to ours have also been considered in [64] and [65]. Also Gawedzki has recently reformulated the extra action terms (to appear).

2) The insertion isomorphism following (2.22) is in a sense the square root of the nonchiral bosonization formula. In fact it can be seen that the chiral bosonization formula of [23] [Eq. (6.21)] follows when this isomorphism is written explicitly.

3) J. Fröhlich has produced chiral fermion correlation functions from a bosonic path integral by the introduction of disorder operators (unpublished). His prescription can be formulated in terms of additional terms in the bosonic action similar to the ones considered here.

BOSONIZATION IN ARBITRARY GENUS

Luis ALVAREZ-GAUMÉ, Gregory MOORE, Philip NELSON, Cumrun VAFA
Lyman Laboratory of Physics, Harvard University, Cambridge, MA 02138, USA

and

J.B. BOST
Ecole Normale Supérieure, 45 Rue d'Ulm, F-75230 Paris Cedex 05, France

Received 17 June 1986

The equivalence is proved between fermionic and scalar field theories on Riemann surfaces of arbitrary topology. The effects of global topology include a modification of the bosonic action.

Fermi–Bose equivalence has recently played an important role in several aspects of string theory. For example, bosonization figures prominently in the formulation of current algebras on the world sheet and in the construction of spin operators [1–5].

In the string path integral one integrates the partition and correlation functions ot two-dimensional quantum field theories over the moduli space $\mathcal{M}$ of conformally inequivalent Riemann surfaces. Recently it has become clear that physicists can use the rich and beautiful structure of $\mathcal{M}$ to gain insight into problems of 2D field theory such as bosonization. Most importantly, $\mathcal{M}$ is a complex space, so that methods from algebraic geometry become applicable. The exact link between field theory and complex geometry comes from a remarkable theorem of Quillen [6,7], which we use. Quillen's theorem describes the determinant of a family of wave operators in complex-analytic terms. His result was later generalized by Belavin and Knizhnik to include families parametrized by $\mathcal{M}$, the case of interest in string theory [8–12]. On the other hand, it has been known for some time that the work of Quillen is closely related to Faltings' work on Arakelov geometry [13]. Faltings describes the bundles in which the determinants live in an inductive way, building them up from simpler ones. It has been suggested that some combination of Quillen's and Faltings' approaches would be of use in string theory [9].

In this letter we will derive the bosonization laws on arbitrary genus Riemann surfaces for fermionic fields of arbitrary spin, using a combination of Quillen's and Faltings' work to justify the procedure rigorously [14]. One can see that such a combination could well lead to bosonization by examining the recent formulation of the string integrand given by Manin [15]. Manin's formula is strongly reminiscent of bosonization. For example it contains the exponential of the Green function which we would expect from correlation functions of the form $\langle e^{\varphi} e^{\varphi} \rangle$; these look like the insertions of fields needed to soak up zero modes in a fermionic system with an index.

In the following we will first describe what bosonization says about the partition function of a generalized ghost system. The method extends to give similar formulae for the correlation functions [16]. The answers we will work out necessarily involve some new global terms in the bosonic action. We will then sketch a mathematical proof that the bosonization formulae are rigorously correct, deferring the details to a later publication [16]. The insertion of fields mentioned above corresponds precisely to the insertion of points used by Faltings to build up arbitrary-spin determinant bundles. We think this is a very pleasant interaction of mathematics and physics, one which is likely to yield further results about two-dimensional field theory in the future.

0370-2693/86/$ 03.50 © Elsevier Science Publishers B.V.
(North-Holland Physics Publishing Division)

We will now outline the bosonization procedure on a Riemann surface Σ for a first-order Fermi system of weight λ [4]. Thus we consider two anticommuting fields b, c on Σ, where b is a λ-form, and c is a $(1-\lambda)$-form with action $S_{bc} = \int_\Sigma b\bar{\partial}c$, where $\bar{\partial}$ is the Cauchy–Riemann operator coupled to $\mathcal{L}_c$, a power of a twisted spin bundle. The ghost number current has an anomaly leading to a net violation of ghost charge $\Delta Q = k \equiv (2\lambda-1)(g-1)$ for a surface of genus g. Therefore we must insert the appropriate number of b and c fields into the partition function to obtain a nonvanishing path integral. On a higher genus Riemann surface we require no insertions for $\lambda = \frac{1}{2}$; g insertions of b and one insertion of c for $\lambda = 1$; and k insertions of b for $\lambda > 1$. Thus, denoting by ω_i and by ψ_i a basis of holomorphic one-forms and λ-forms respectively we obtain for the partition functions with fields inserted at points P_i:

$$Z^{bc}_{\lambda=1/2} = \det \bar{\partial}^\dagger_{\mathrm{L}}\bar{\partial}_{\mathrm{L}},$$

$$Z^{bc}_{\lambda=1} \equiv \left| \int [db]\,[dc] \prod_{i=1}^{g} b(P_i)c(Q)\exp(-S_{bc}) \right|^2 = \frac{\det\omega_i(P_j) \wedge \det\bar{\omega}_i(P_j)}{\det\langle\omega_i|\omega_j\rangle} \left(\frac{\det' - \nabla^2}{\int\sqrt{g}}\right),$$

$$Z^{bc}_{\lambda} \equiv \left| \int [db]\,[dc] \prod_{i=1}^{k} b(P_i)\exp(-S_{bc}) \right|^2 = \frac{\det\psi_i(P_j) \wedge \det\bar{\psi}_i(P_j)}{\det\langle\psi_i|\psi_j\rangle} \det \bar{\partial}^\dagger_{\mathcal{L}_c}\bar{\partial}_{\mathcal{L}_c}, \tag{1}$$

Since we have inserted fields b which are differential forms, the partition functions should be regarded as $(k\lambda, k\lambda)$-forms on Σ^k (for $\lambda > 1$).

Bosonization is the statement that the above first-order field theories can be replaced by equivalent scalar field theories. It was shown in ref. [4] that the local properties of the weight λ system are reproduced by a scalar action which is a sum of the usual action $S_1 = 4\pi\mathrm{i}\int\partial\phi\,\bar{\partial}\phi$ and an anomaly term S_2

$$S[\phi] = S_1 + S_2 \equiv S_1 + (2\lambda - 1)\int R_{K^{-1}}\phi\,. \tag{2}$$

The second term accounts for the local anomaly in the ghost number current. $R_{K^{-1}}$ is the curvature of the holomorphic tangent bundle K^{-1}, normalized so that $\int R_{K^{-1}} = -2\pi\mathrm{i}\deg K^{-1} = 4\pi\mathrm{i}(g-1)$. We will see below that the scalar field ϕ is only defined up to an integer or half-integer so that we must include "instanton" configurations ϕ_{nm} in which ϕ_{nm} shifts by n_i around the a_i-cycle and m_j around the b_j-cycle (fig. 1), where $n_i, m_i \in \frac{1}{2}\mathbf{Z}$. To evaluate the $R\phi$ term in an instanton sector we must define the multiple-valued field ϕ by choosing a system of curves intersecting in a single point A (fig. 1). Such a choice lets us cut open the surface to obtain a polygon Σ_c. We must also choose a basepoint P_0; then $\phi = \int_{P_0}^{P} d\phi$ is well-defined on Σ_c and we may evaluate the action (2).

Let us now investigate the dependence of this action on the various choices we have made. First, a change of basepoint shifts ϕ, and therefore S, by a constant. This is simply a reflection of the (integrated) U(1) anomaly and is compensated by the bosonized insertions of the fermionic fields that soak up zero modes. Next, let us consider the dependence on the curves a_i, b_i chosen to represent the homology basis. If we view ϕ_{nm} as a discontinuous function on Σ, then deforming a cycle through the discontinuity produces a change in the action. Alternatively,

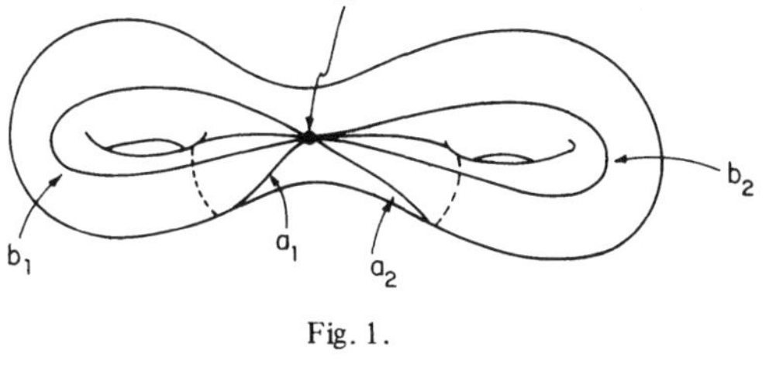

Fig. 1.

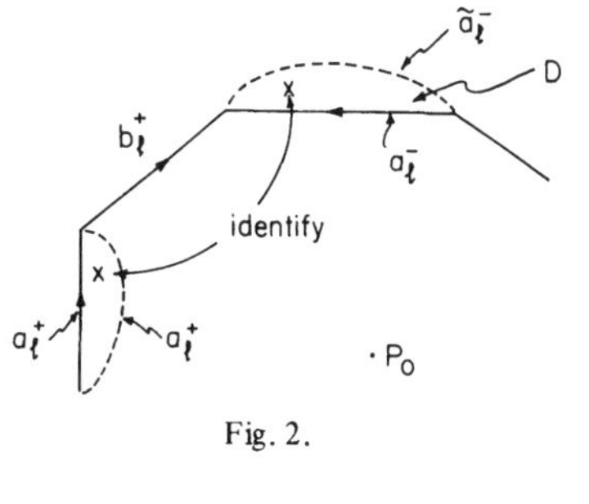

Fig. 2.

consider the two choices of representatives for the a_l cycle in fig. 2. If ϕ has a winding number around b_l then there is a discrepancy in the actions because

$$\int_{\tilde{\Sigma}_c} R\phi - \int_{\Sigma_c} R\phi = \int_{b_l} d\phi \int_D R, \tag{3}$$

where D is the region bounded by $\tilde{a}_l$ and a_l. Thus, in the instanton sectors the anomaly term depends on the unphysical choice of curves used to cut open Σ. We may compensate for this by modifying the action so that

$$S_2 = (2\lambda - 1) \int_{\Sigma_c} R_{K^{-1}} \phi - \int_{b_k} d\phi f[a_k] + \int_{a_k} d\phi f[b_k], \tag{4}$$

where f is any functional of the curves such that if $\tilde{\gamma}$ is homologous to γ then

$$f[\tilde{\gamma}] - f[\gamma] = (2\lambda - 1) \int_D R_{K^{-1}} \mod 4\pi i \mathbf{Z}, \tag{5}$$

where D is the region enclosed by the two curves. Since only $\exp(-S)$ is of physical relevance and ϕ has half-integral winding numbers, f need only be defined modulo $4\pi i\mathbf{Z}$. One can also show that if f satisfies (5) then the modified action (4) is also independent of the intersection point A.

A natural choice of f may be described in terms of the metric properties of holomorphic line bundles. Recall that the metric on any holomorphic line bundle $\mathcal{L}$ is specified up to a constant by its curvature $R_{\mathcal{L}}$, which satisfies $\int_\Sigma R_{\mathcal{L}} = -2\pi i \deg \mathcal{L}$. If we choose the unique holomorphic connection compatible with the metric then the holonomy of $\mathcal{L}$ around a curve γ, denoted by $h[\gamma;\mathcal{L}]$, satisfies $h[\tilde{\gamma};\mathcal{L}] - h[\gamma;\mathcal{L}] = \int_D R_{\mathcal{L}} \mod 2\pi i\mathbf{Z}$. Thus for example we can take $\mathcal{L}$ to be $\mathcal{L}_b$ or $\mathcal{L}_c$, the bundles of λ- and $(1-\lambda)$-forms, and give $\mathcal{L}$ a metric and connection so that $R_{\mathcal{L}}$ is proportional to $R_{K^{-1}}$. Then the natural choice $f[\gamma] = h[\gamma;\mathcal{L}_c] - h[\gamma;\mathcal{L}_b]$ satisfies (5), but is only well-defined modulo $2\pi i\mathbf{Z}$ so that with this choice the sign of $\exp(-S_2)$ is ill-defined. We can fix this sign ambiguity as follows. Note that for any $\mathcal{L}$, $2h[\gamma;\mathcal{L}]$ is defined modulo $4\pi i\mathbf{Z}$, so that if we can find a preferred bundle of spin $\frac{1}{2} - \lambda$, we will get a candidate for f. Since we have chosen a marking of Σ there is indeed a preferred choice of spin structure $\mathcal{L}_\Delta$ associated to the vector of Riemann constants [17,10]. Thus we can write

$$f[\gamma] = 2h[\gamma;\mathcal{L}_c \otimes \mathcal{L}_\Delta^{-1}] + 4\pi i v[\gamma] \mod 4\pi i\mathbf{Z}, \tag{6}$$

where for generality we have allowed a function $v[\gamma]$ of the homology class of γ. Had we chosen a spin bundle other than $\mathcal{L}_\Delta$ the difference would be accounted for by a change of $v[\gamma]$.

Finally, we must check that the candidate action describes a modular invariant theory: the same instanton configuration described in two different markings should contribute the *same* action to the partition function. The change of $S_2[\phi_{nm}]$ under a change of marking is not obvious and requires a computation. Since $S_2[\phi]$ is Weyl-invariant for on-shell fields ϕ we may do the computation in any convenient metric. A natural metric from the standpoint of both physics and mathematics is the Arakelov metric [18,13] which is defined up to a constant by specifying the curvature of the holomorphic line bundle $\mathcal{L}_c$ by $R = 4\pi i(\lambda - 1)(g - 1)\mu$ where μ is the $(1,1)$-form $\mu \equiv (i/2g)\omega^i(\mathrm{Im}\,\Omega)^{-1}_{ij}\bar{\omega}^j$, normalized so that $\int\mu = 1$. Every metric is gauge-equivalent to a single Arakelov metric.

We also need the preferred spin bundle $\mathcal{L}_\Delta$ in order to parametrize the possible $\mathcal{L}_c$'s. Write $\mathcal{L}_c$ as $\mathcal{L}_\Delta^{2-2\lambda} \otimes F_{\theta_1,\theta_2}$, where F_{θ_1,θ_2} is the holomorphic flat bundle with holonomy $2\pi i\theta_1$, $-2\pi i\theta_2$ around the a, b cycles [19, 10]. A short computation in the Arakelov metric then shows that

$$S_2[\phi_{nm}] = 4\pi i\{(m - \bar{\Omega}n)(\Omega - \bar{\Omega})^{-1}[(2\lambda - 1)\Delta - (\theta_1 + \Omega\theta_2)] + \text{c.c.}\}$$

$$+ 4\pi i[(m - \bar{\Omega}n)(\Omega - \bar{\Omega})^{-1}(v[b] - \Omega v[a]) + \text{c.c.}] \mod 2\pi i\mathbf{Z}, \tag{7}$$

where Δ is the vector of Riemann constants, considered as a point in the jacobian. Using the transformation law of Δ under a change of marking one can show that the choice $v[b_i] = \frac{1}{2}m_i$, $v[a_j] = -\frac{1}{2}n_j$ guarantees modular covariance. This solution is equivalent to taking $v = 0$ and adding a third term to the action $S_3 = 4\pi i \int_{a_k} d\phi \int_{b_k} d\phi = 4\pi i nm$. The new term $\exp(-S_3)$ weights "even" and "odd" instantons with opposite sign. Thus we have finally arrived at an action with the property that $\exp(-S)$ is independent of *all* choices. We must now determine if there are other solutions for v to the modular covariance equations. Other solutions amount to changes in the weighting of the instanton sectors by phases $\epsilon(n,m)$. If we combine the requirement of modular covariance with that of factorization of the instanton contributions on the boundary of moduli space then the analysis of ref. [20] can be used to show that there are no other solutions, and $\epsilon(n,m) = 1$.

To complete the Fermi–Bose correspondence we must express the Fermi fields b, c in terms of the Bose field ϕ. For the nonchiral bosonization discussed in this letter it suffices to know the bosonization rules for the generalized fermion mass terms

$$b\bar{b} = (dz)^{\lambda}(d\bar{z})^{\lambda} :\exp(4\pi i\phi):, \qquad c\bar{c} = (dz)^{1-\lambda}(d\bar{z})^{1-\lambda} :\exp(-4\pi i\phi): . \tag{8}$$

The factor of $4\pi i$ is determined by demanding that the expressions on the right-hand side have the correct conformal weight, and the normal ordering prescription : : cancels the coordinate-dependence so that b, c are well-defined differential forms. In particular we see from (8) that the winding numbers n_i, m_i must be integer or half-integer, as mentioned above.

We may now evaluate the gaussian path integral with insertions by introducing Arakelov's Green function [13] which satisfies $\partial\bar{\partial} \log G(P,Q) = i\pi\mu(P) - i\pi\delta(P,Q)$ and $\int \mu \log G(\cdot, Q) = 0$. The normal-ordering of the Green functions at coincident points is fixed by the requirement that the expression be coordinate-independent and finite:

$$:\log G(P,P): = \lim_{Q\to P} [\log G(Q,P) - 2\lambda \log|z(P) - z(Q)| - (1-2\lambda)\log d(P,Q)], \tag{9}$$

where $d(P,Q)$ is the invariant distance. This prescription can also be derived by demanding that the bosonic partition functions in eq. (10) below have the correct conformal anomaly [‡1]. The Arakelov metric has the important property that the normal ordering of the scalar ($\lambda = 0$) field gives zero at the coincident points. Thus the Bose partition functions are

$$Z^{\text{Bose}}_{\lambda=1/2} = \left(\frac{\det' -\nabla^2}{\int\sqrt{g}}\right)^{-1/2} Z_{\text{inst}},$$

$$Z^{\text{Bose}}_{\lambda=1} = \prod_{i=1}^{g} \frac{\sqrt{-1}}{2}\, dz_i(P_i) \wedge d\bar{z}_i(P_i) \exp[:\log G(P_i,P_i):] \left(\frac{\det' -\nabla^2}{\int\sqrt{g}}\right)^{-1/2} \frac{\Pi_{i<j} G(P_i,P_j)^2}{\Pi_i G(P_i,Q)^2} Z_{\text{inst}},$$

$$Z^{\text{Bose}}_{\lambda} = \prod_{i=1}^{k} \left(\frac{\sqrt{-1}}{2}\right)^{\lambda} (dz_i(P_i))^{\lambda} \wedge (d\bar{z}_i(P_i))^{\lambda} \exp[:\log G(P_i,P_i):] \left(\frac{\det' -\nabla^2}{\int\sqrt{g}}\right)^{-1/2} \prod_{i<j} G(P_i,P_j)^2 Z_{\text{inst}}, \tag{10}$$

where Z_{inst} denotes the instanton sum. This may be expressed as

$$Z_{\text{inst}} = \sum_{n,m\in(\frac{1}{2}\mathbf{Z})^g} \exp\{-S_1[\phi_{nm}] + 4\pi i[(m - \bar{\Omega}n)(\Omega - \bar{\Omega})^{-1}z + \text{c.c.}] + 4\pi i mn\}, \tag{11}$$

where z may be expressed in terms of the insertion points P_i, Q and the jacobian map $I[\cdot]$ as

‡1 Using this normal-ordering prescription one can show that $:\exp(4\pi i q\phi):$ ia a $(\frac{1}{2}q(q+2\lambda-1), \frac{1}{2}q(q+2\lambda-1))$-form, in agreement with ref. [4].

$$z_{\lambda=1/2} = \theta_1 + \Omega\theta_2, \quad z_{\lambda=1} = I\left(\sum_{i=1}^{g} P_i - Q\right) + \theta_1 + \Omega\theta_2 - \Delta, \quad z_\lambda = I\left(\sum_{i=1}^{k} P_i\right) + \theta_1 + \Omega\theta_2 - (2\lambda - 1)\Delta. \tag{12}$$

After an application of the Poisson summation formula this sum may be expressed in terms of a function $\mathcal{N}[z]$:

$$Z_{\text{inst}} = (\det \operatorname{Im} \Omega)^{1/2} \mathcal{N}[z] \equiv (\det \operatorname{Im} \Omega)^{1/2} \exp[-2\pi(\operatorname{Im} z)(\operatorname{Im} \Omega)^{-1}(\operatorname{Im} z)]\, |\vartheta(z|\Omega)|^2, \tag{13}$$

where ϑ is the Riemann theta function. It is straightforward to carry out the derivation of (10) for any metric by expressing it as a Weyl transformation of the Arakelov metric. One finds the above expressions multiplied by the Liouville action for the conformal factor.

Bosonization states that the Fermi and Bose partition functions are equal. Equating these we obtain the following formulae for the determinants of the laplacians for any spin [‡2]

$$\det \bar\partial_L^\dagger \bar\partial_L = \left(\frac{\det' - \nabla^2}{\int\sqrt{g}\, \det \operatorname{Im} \Omega}\right)^{-1/2} \mathcal{N}[\theta_1 + \Omega\theta_2], \tag{14a}$$

$$\left(\frac{\det' - \nabla^2}{\int\sqrt{g}\, \det \operatorname{Im} \Omega}\right)^{3/2} = \frac{\Pi_{i<j} G(P_i, P_j)^2}{\|\det \omega_i(P_j)\|^2}\, \frac{\mathcal{N}[I(\Sigma_1^g P_i - Q) - \Delta]}{\Pi G(P_i, Q)^2}, \tag{14b}$$

$$\frac{\det \bar\partial_{\mathcal{L}_c}^\dagger \bar\partial_{\mathcal{L}_c}}{\det\langle\psi_i|\psi_j\rangle} = \left(\frac{\det' - \nabla^2}{\int\sqrt{g}\, \det \operatorname{Im} \Omega}\right)^{-1/2} \frac{\Pi_{i<j} G(P_i, P_j)^2}{\|\det \psi_i(P_j)\|^2}\, \mathcal{N}\left[I\left(\sum_1^k P_i\right) - (2\lambda - 1)\Delta\right], \tag{14c}$$

where for simplicity we have set the twists to zero for $\lambda \neq \frac{1}{2}$. The quantity $\|\det \psi_i(P_j)\|^2$ is the ratio of differential forms

$$\frac{\det \psi_i(P_j) \wedge \det \bar\psi_i(P_j)}{\Pi_{i=1}^k (dz_i(P_i))^\lambda \wedge (d\bar z_i(P_i))^\lambda \exp[:\log G(P_i, P_i):]}.$$

Note that (14b) expresses the determinant of the laplacian purely in terms of Green's functions, theta-functions, and zero-mode determinants. From (14a), (14b) we obtain similar formulae for determinants of operators of any spin. Note too that the expressions on the right hand side of (14) are independent of the choice of insertion points P_i, Q.

Actually, bosonization asserts the equality of *all* the correlation functions of the two theories. These may be easily computed using the above rules. We will now justify our bosonization procedure by describing how the identities (14) and the equalities of correlation functions can be proved rigorously à la Faltings.

The spin-$\frac{1}{2}$ partition function, eq. (14a), has already been derived in refs. [10,21]. Thus we would like to find a mathematical operation corresponding to the insertion of fields at points P_i. We can then use such an operation to build up arbitrary spin bundles $\mathcal{L}_c$ from the known case.

U(1) line bundles on a 2D surface are familiar from the theory of magnetic monopoles, where Σ is the sphere. The total magnetic charge inside Σ can be found by counting the net number of string singularities, which are points P_i of Σ. Turning this around, we can specify the bundle by naming k points $P_1, ..., P_k$ and putting transition functions $\exp(i \arg z_{(i)})$ near each. $z_{(i)}$ is a local coordinate vanishing at P_i. Similarly in the analytic case we can choose $P_1, ..., P_k$ with transitions simply given by $z_{(i)}$. Let $\mathcal{O}(P_1 + ... + P_k)$ denote the resulting line bundle. Its smooth sections can also be viewed as ordinary functions, possibly with simple poles at $\{P_i\}$. In particular the section $\sigma^{(P)}$, which equals one away from P, vanishes at P. We can then put a nonsingular metric on $\mathcal{O}(P)$ by setting [13,18]

‡2 These equalities hold up to a constant which depends only on g and λ.

$$\|\sigma^{(P)}\|(Q) = G(P, Q), \tag{15}$$

and similarly for $\mathcal{O}(P_1 + ... + P_k)$.

We can now ask what happens to the fermion determinant when we replace $\mathcal{L}_c$ by $\mathcal{L}_c \otimes \mathcal{O}(P)$. Once we know that, we can apply the operation k times to raise $\mathcal{L}_c$ up to degree $g-1$, i.e. to a twisted spin bundle, then apply the known formula. For conciseness we will only give the answer for spin $\lambda \geqslant \frac{3}{2}$, but the general formula needed to prove (14) is not much harder.

If $\lambda \geqslant \frac{3}{2}$, then $\bar\partial' \equiv \bar\partial_{\mathcal{L}_0 \otimes \mathcal{O}(P)}$ and $\bar\partial \equiv \bar\partial_{\mathcal{L}_0}$ have no zero modes, while $\bar\partial'^{\dagger}$ has exactly one fewer zero mode than $\bar\partial^{\dagger}$. We will denote by χ_i $i = 1, ..., k-1$ the zero modes of $\bar\partial'^{\dagger}$, by ψ_i, $i = 1, ..., k-1$ the corresponding zero modes of $\bar\partial^{\dagger}$, and by ψ_k the extra zero mode. There is an Arakelov norm on $\{\psi_i\}$, and the same norm times (15) on $\{\chi_i\}$. With this notation, we get

$$\det \bar\partial^{\dagger}\bar\partial/\det\langle\psi_i|\psi_j\rangle = C_{g,k}(\det \bar\partial'^{\dagger}\bar\partial'/\det\langle\chi_i|\chi_j\rangle)\,\|\psi_k(P)\|^{-2}. \tag{16}$$

The last factor is defined below (14). The proof of (16) follows lines similar to ref. [21]. In particular the key step equates the curvatures of two Quillen norms. Since curvature only determines a norm up to a constant, we have an undetermined $C_{g,k}$ depending on the genus and the index. The form of (16) would have been more complicated had we not used the Arakelov metric slice.

Eq. (16) remains valid when we replace $\mathcal{L}_c$ by a more general bundle. Applying it a second time we get the determinant for $\mathcal{L}_c \otimes \mathcal{O}(P_1 + P_2)$ expressed in terms of the norm $\|\psi_{k-1}\|^2$ in $\mathcal{L}_c \otimes \mathcal{O}(-P_1)$. We can rewrite this in terms of the original norm times $G(P_1, P_2)^{-1}$ using (15). Continuing in this way we arrive at formulae (14b), (14c) when proper care is taken with the last few steps, when $\lambda \leqslant 1$. Thus we have put the bosonization procedure on a completely rigorous footing.

Bosonization should prove useful in investigating properties of multiloop string amplitudes. For example, using either Faltings' approach or the present one it is possible to write various expressions for the string integrand similar to those in ref. [15]. One can also use (14) to investigate the behavior of the string integrand on the boundary of moduli space. Moreover (14c) with $\lambda = \frac{3}{2}$ and similar identities for correlation functions should be useful for investigations of the modular invariance of multiloop superstring amplitudes. Finally, these formulae should help further our understanding of the ultraviolet structure of string perturbation theory and superstring finiteness.

In conclusion, we have shown that the algebraic geometry of determinant line bundles and the physics of bosonization are intimately related. The mathematics allows us to prove bosonization rigorously while the physics suggests both the existence of new identities for determinants, zero-modes, and Green functions, as well as an as yet unexplored connection with the representation theory of Kac–Moody algebras. We believe that bosonization will be a useful tool for further exploration of the deep connection between algebraic geometry and string theory.

We would like to thank S. Coleman, J. Harris, D. Kazhdan, E. Martinec, S. Shenker, I. Singer, and C. Soulé for discussions. G.M., P.N., and C.V. were supported in part by the Harvard Society of Fellows. L.A., G.M., P.N., and C.V. were supported by NSF grant PHY-85-15249. J.B. acknowledges the support of DOE contract DE-FG02-84-ER-40164-A001 and the hospitality of the MIT mathematics department.

References

[1] R. Shankar, Phys. Lett. B 92 (1980) 333;
E. Witten, in: Fourth Workshop on Grand unification, eds. H. Weldon, P. Langacker and P. Steinhardt (Birkhauser, Basel, 1983).

[2] D. Gross, J. Harvey, E. Martinec and R. Rohm, Phys. Rev. Lett. 54 (1985) 502; Nucl. Phys. B 256 (1985) 253; B 267 (1986) 75.

[3] R. Marnelius, Nucl. Phys. B 211 (1983) 14.

[4] D. Friedan, E. Martinec and S. Shenker, Phys. Lett. B 160 (1985) 55; Nucl. Phys. B 271 (1986) 93.

[5] W. Siegel and B. Zwiebach, Nucl. Phys. B 263 (1986) 105.

[6] D. Quillen, Funk. Anal. Prilozen 19 (1985) 37 [Funct. Anal. Appl. 19 (1986) 31].
[7] J.-M. Bismut and D. Freed, Geometry of elliptic families I, II, Orsay preprint 85T47.
[8] A. Belavin and V. Knizhnik, Phys. Lett. B 168 (1986) 201.
[9] J. Bost and T. Jolicoeur, Phys. Lett. B 174 (1986) 273.
[10] L. Alvarez-Gaumé, G. Moore and C. Vafa, Harvard preprint HUTP-86/A017.
[11] R. Catenacci, M. Cornalba, M. Martinelli and C. Reina, Pavia preprint.
[12] D. Friedan and S. Shenker, Chicago preprint EFI-86-18B.
[13] G. Faltings, Ann. Math. 119 (1984) 387.
[14] E. Martinec, in: Conformal field theory on a (super-) Riemann surface, Princeton preprint.
[15] Yu. Manin, Phys. Lett. B 172 (1986) 184.
[16] L. Alvarez-Gaumé, J.-B. Bost, G. Moore, P. Nelson and C. Vafa, to be published.
[17] D. Mumford, Tata Lectures on Theta (Birkhäuser, Basel, 1983).
[18] S. Arakelov, Izv. Akad. Nauk. SSSR Ser. Mat. 38 (1974) 1 [Math. USSR Izv. 8 (1974) 1167].
[19] R. Gunning, Introduction to Riemann surfaces, Princeton Mathematics (1966).
[20] C. Vafa, Modular invariance and discrete torsion on orbifolds, Harvard preprint, HUTP-86/A011.
[21] J. Bost and P. Nelson, Harvard preprint HUTP-86/A044.

OPERATOR FORMULATION ON RIEMANN SURFACES

Cumrun VAFA
Lyman Laboratory of Physics, Harvard University, Cambridge, MA 02138, USA

Received 27 February 1987

We use symmetries of the path-integral on Riemann surfaces with boundaries to develop an operator formulation for higher loop Riemann surfaces.

Over the last few years, a relatively clear picture of string perturbation theory has been developed starting from the path-integral approach of Polyakov [1]. This seems to be a very different approach from the early days of string theory, where the string theory was defined through operators acting on the string Hilbert space (for the tree and one-loop diagrams). Therefore it is natural to ask whether there exists an operator formulation of string theory at higher loops which extends the treatment at tree and one-loop levels.

Ideas to put such a setup in string theory had previously been suggested in ref. [2]. Recently there has been some major progress in this direction in some interesting works [3–5]. In these papers use has been made of the following fact (for this fact and its relation to KP hierarchy see ref. [6]): To any Riemann surface with a spin bundle together with a point and a choice of analytic coordinates at that point corresponds in a one-to-one way different choices for the vacuum space of a single fermion (which can be obtained by Bogoliubov transformations of the ordinary fermionic vacuum). This space is called UGM, the universal grassmannian manifold. This map is not onto and there are points on the UGM which do not come from any Rieman surface. However the image of this map is dense in UGM, and one may consider redefining string theory to include the full UGM (or more precisely all possible choices for the relevant vacuum). This may then provide a way to formulate new and possibly non-perturbative questions in string theory [‡1]. Using this relation with the UGM a prescription was given in refs. [4,5] to compute the correlation functions for arbitrary number of fermions on a Riemann surface as expectation values of the corresponding operators in a Hilbert space. One can use the standard bosonization on a circle to relate a point in the UGM to a bosonic state, and could transfer the computation of the fermionic correlation functions to the correlation function between bosonic vertex operators. This was in fact the approach followed in these two references.

It is important to understand the connection of these two states, one in a bosonic Hilbert space and the other in the fermionic Hilbert space, with the quantum field theory of bosons and fermions on a Riemann surface.

The connection in the fermionic case has been essentially given by the work of mathematicians [6] which was interpreted physically in refs. [4,5]. The point of UGM was interpreted as a kind of Dirac vacuum defined by the Riemann surface. However, the bosonic side of the story has not been worked out. In this paper using a path-integral approach, we will rederive the results of refs. [4,5] in the fermionic case and justify their prescription in the bosonic case. This gives a simple unified treatment of both the bosonic and fermionic theories and may be viewed as an alternative proof of bosonization on higher genus Riemann surfaces [8,12]. Even though our work is phrased in the physics language these concepts are easily put on a mathematically rigorous footing

[‡1] UGM in some ways resembles the "universal moduli space" proposed in ref. [7] which includes Riemann surfaces with different genus (including the infinite genus case).

0370-2693/87/$ 03.50 © Elsevier Science Publishers B.V.
(North-Holland Physics Publishing Division)

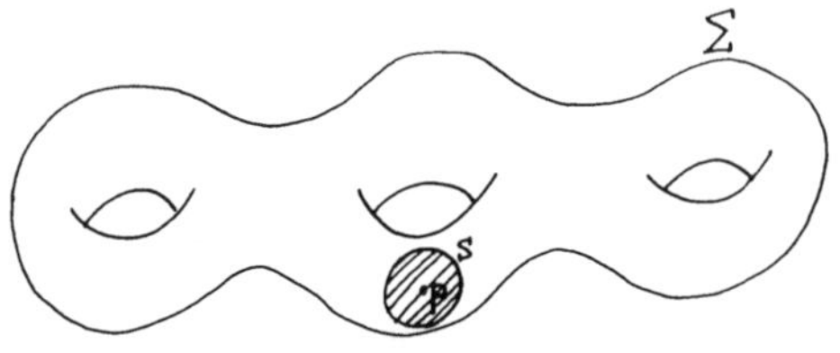

Fig. 1. Path-integral on the Riemann surface Σ minus the disc centered at p (the shaded region) gives a state on the circle S.

and may be viewed as a generalization of the works in ref. [6] to the bosonic case.

The basic idea is that a path integral of a quantum field theory on a space with boundaries results in a state in the Hilbert space based on the boundary, i.e., the space of functionals of fields on the boundary. In other words if we specify the values of fields on the boundary, the result of the path-integral with that boundary condition, is a number. This then defines a functional on the space of field values at the boundary, and thus specifies a state of the Hilbert space.

We will apply this idea to Riemann surfaces with boundaries. For the most part we consider the case discussed in refs. [4,5] and comment on the generalizations at the end of the paper. Consider a Riemann surface Σ without boundaries, together with a point p and an analytic coordinate t vanishing at p. Consider the Riemann surface minus the disc D, defined by $|t|<1$ (see fig. 1). Any quantum field theory on $\Sigma - \mathrm{D}$ through path-integral gives rise to a state on the Hilbert space based on the circle S defined by $|t|=1$. We shall find this state (up to an overall normalization) by finding an infinite number of symmetries of the path-integral on $\Sigma - \mathrm{D}$. In particular we shall find that for conformal field theories on the surface, the normalization of the state depends on the choice of metric, but the direction of the state depends only on the conformal structure of the surface. So the conformal structure is sufficient to define a ray in the Hilbert space. The overall normalization will cancel out of the correlation functions, and so to compute the correlation functions we will only need the conformal structure of the surface, as expected.

In this paper we shall focus on conformal field theory of spin 1/2 fermions and spin-0 bosons. The fermionic action is

$$S = \int b\bar{\partial}c + c\bar{\partial}b + \text{c.c.}\,, \tag{1}$$

where b and c are spin 1/2 fields which are sections of some (twisted) spin bundle. For simplicity we shall assume that the Dirac operator has no zero modes, but we shall allow the possibility of twisting the boundary conditions. c is a section of a twisted spin bundle which will be denoted by the theta characters (θ, ϕ) (see for example refs. [10,11]), with respect to a choice of canonical homology basis (we have used the notation b and c for spin 1/2 fermionic fields instead of the standard one in order to emphasize that b is not the complex conjugate of c).

By Riemann–Roch (and our assumption about the absence of zero modes of the Dirac operator) it follows that for each integer n there exists a meromorphic section ψ_n of the spin bundle which has a pole only at $t=0$ of order n. An explicit basis for this is given by the derivatives of the Szego kernel

$$\psi_n = \frac{1}{(n-1)!}\,\partial_y^{n-1}\left.\frac{\vartheta[^{\theta}_{\phi}](\int_y^t \omega\,|\,\tau)}{E(t,y)\,\vartheta[^{\theta}_{\phi}](0\,|\,\tau)}\right|_{y=0}, \tag{2}$$

where $E(t, y)$ is the prime form [10] and ω are the holomorphic one forms adapted to this basis, and ϑ is the theta function (with characteristics) and τ specifies the period matrix. That ψ_n is a section of the relevant spin bundle is a consequence of Riemann's vanishing theorem. We can expand ψ_n in powers of t, near p to get

$$\psi_n = t^{-n} + \sum_{m=1}^{\infty} B_{nm} t^{m-1}\,, \tag{3}$$

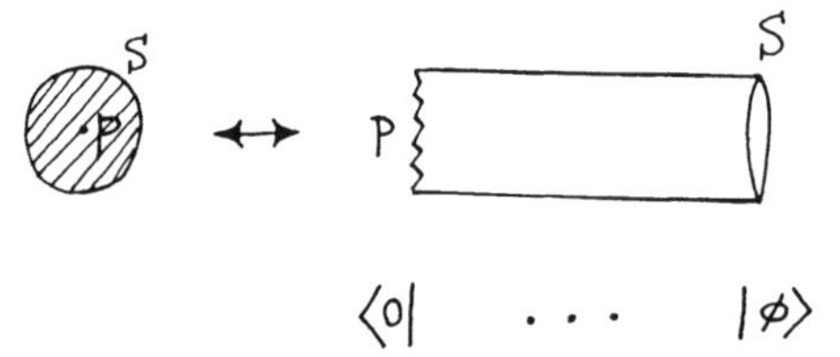

Fig. 2. Path-integral on the disc is equivalent to the operator formulation on the semi-infinite cylinder, one end of which is identified with the circle S and the other at infinity is identified with the point p.

where

$$B_{nm}=\frac{1}{(n-1)!(m-1)!}\,\partial_t^{m-1}\partial_y^{n-1}\left(\frac{\vartheta[^{\theta}_{\phi}](\int_y^t\omega\,|\,\tau)}{E(t,y)\vartheta[^{\theta}_{\phi}](0\,|\,\tau)}-\frac{1}{(t-y)}\right)\Bigg|_{t=y=0} \tag{4}$$

(we have suppressed $\sqrt{dt}$ from the expression for ψ_n and continue to suppress similar differentials in the following). Since ψ_n is holomorphic on $\Sigma-\mathrm{D}$ it solves the wave equation

$$\bar\partial\psi_n=0$$

in that region. This implies that the path-integral on $\Sigma-\mathrm{D}$ has the symmetry

$$c\to c+\epsilon_n\psi_n\,. \tag{5}$$

This gives rise to an infinite number of conserved currents

$$j_n=b\psi_n\,. \tag{6}$$

Let us denote by $|\phi\rangle$ the state that we get on S as a result of the path integral on $\Sigma-\mathrm{D}$. Then using the infinite number of conserved currents we are going to get an infinite number of charges on the circle S, which should annihilate the state $|\phi\rangle$ (of course this can be seen directly from the path integral by shifting the fields). This means that we have

$$Q_n|\phi\rangle=0 \tag{7}$$

with

$$Q_n=\frac{1}{2\pi\mathrm{i}}\int_{\mathrm{S}} b\psi_n\,. \tag{8}$$

We can now continue doing the path-integral for the quantum field theory on the disc. However, instead of doing the path-integral on the disc, we can equivalently use the standard operator formulation which is worked out for the disc. We just view the disc as a semi-infinite cylinder by a conformal transformation, and view $t=0$ as the outgoing state, and the state $|\phi\rangle$ at $|t|=1$ as the incoming state (see fig. 2). We can expand the field variables for $|t|\leqslant 1$ in terms of the creation and annihilation operators $b(t)=\sum_{n\in\mathrm{Z}}b_{n+1/2}t^n$ and $c(t)=\sum_{n\in\mathrm{Z}}c_{n+1/2}t^n$, with the anti-commutation relations

$$[b_{n+1/2},c_{m-1/2}]_+=\delta_{n+m,0},\quad [b_{n+1/2},b_{m+1/2}]_+=[c_{n+1/2},c_{m+1/2}]_+=0\,. \tag{9}$$

Using (8) we can now express the conserved charges in terms of the creation and annihilation operators

$$Q_n=b_{n-1/2}+B_{nm}b_{-m+1/2}\,. \tag{10}$$

Similarly by shifting b with holomorphic section on $\Sigma-\mathrm{D}$ we get infinitely many conserved charges

$$\tilde Q_n=c_{n-1/2}-c_{-m+1/2}B_{mn}\,. \tag{11}$$

The Q's and $\tilde{Q}$'s anti-commute among themselves and with each other. Eqs. (7) and the corresponding ones for $\tilde{Q}_n$ can be solved (up to an overall normalization) for $|\phi\rangle$. If we define the standard vacuum by $b_{n-1/2}|0\rangle = c_{n-1/2}|0\rangle = 0$ for all $n \geq 1$, then $|\phi\rangle$ is given by a Bogoliubov transformation of the standard vacuum,

$$|\phi\rangle = C \exp(-c_{-n+1/2} B_{nm} b_{-m+1/2})|0\rangle . \tag{12}$$

The overall constant C depends not just on the conformal data but also on the actual form chosen for the metric. In addition the phase of C would be ill-defined due to local gravitational anomalies, unless we put the complex conjugate fields into the theory, in which case $|\phi\rangle$ gets multiplied by the corresponding state in the Fock space of anti-holomorphic degrees of freedom.

Now suppose we are interested in computing correlation functions of fermions on the disc D. The path-integral on $\Sigma - \mathrm{D}$ has had the effect of preparing the state $|\phi\rangle$ on S, so the correlation functions can be computed in an operator formulation by

$$\langle b(t_1)c(t_2)b(t_3)c(t_4)\ldots\rangle = \frac{\langle 0|b(t_1)c(t_2)b(t_3)c(t_4)\ldots|\phi\rangle}{\langle 0|\phi\rangle} . \tag{13}$$

This is the prescription suggested in refs. [4,5], and leads to the standard results. One can extend this correlation function outside the disc by using analytic continuation of the variables in the correlation function.

Now let us move on to the bosonic theory. We shall first consider a theory of non-chiral real scalars X (with no windings). The action is

$$S = \frac{\mathrm{i}}{4\pi} \int X \partial \bar{\partial} X . \tag{14}$$

The wave equation is now

$$\partial \bar{\partial} X = 0 . \tag{15}$$

We can solve (15) on $\Sigma - \mathrm{D}$ by finding a meromorphic function on Σ which is allowed to have poles only at $t=0$. This again gives us an infinite number of conserved charges which apparently can be used to find the corresponding ray on the circle. However, we run into the following problem: In this way we get $\infty - g$ meromorphic solutions to (15), which means that if we are interested in a meromorphic function which has a pole only at $t=0$ of order n, by the Weierstrass gap theorem [12], for g values values of n between 1 and $2g$ there are no solutions. This means that we will not be able to solve for the ray in this way. However we note that the solutions to eq. (15) allow a holomorphic plus an anti-holomorphic function, and we shall use this freedom to avoid this problem. This will imply that the ray that we get is not going to be a simple tensor product of a state from the holomorphic degrees of freedom times a state from the anti-holomorphic degrees of freedom, but rather a sum of such products. Let

$$X_n(t) = \int^{t} [\eta_n - \pi A_n (\mathrm{Im}\, \tau)^{-1} (\omega - \bar{\omega})] , \tag{16}$$

where η_n is a meromorphic differential with only a single pole of order $n+1$ at $t=0$ and residue $(-n)$ and zero a-periods, and A_n is defined by

$$\omega = \sum_{n=1}^{\infty} A_n t^{n-1} \mathrm{d}t , \tag{17}$$

where we have suppressed an index running from 1 to g for ω and A_n. We can write η_n explicitly using the prime form

$$\eta_n = \frac{-1}{(n-1)!} \partial_t \partial_y^n \log E(t,y) \bigg|_{y=0} . \tag{18}$$

It is then easy to check that (16) defines a single-valued function on the Riemann surface by using the transformation properties of the prime form as we go around the b cycles. This defines, similar to the fermionic case, an infinite number of conserved currents by considering the transformation

$$X \to X + \epsilon_n X_n \tag{19}$$

(if one is bothered by the fact that this is not a real shift in X one can consider addition (or subtraction) of its complex conjugate (multiplied by 1/2i)). If we denote by X_n^{h} and X_n^{a} the holomorphic and anti-holomorphic parts of X_n, the corresponding conserved currents are given by

$$j_n = (\partial X) X_n^{\mathrm{h}} + (\bar{\partial} X) X_n^{\mathrm{a}} , \tag{20}$$

and we have

$$X_n^{\mathrm{h}} = t^{-n} - \sum_{m=1}^{\infty} [2Q_{nm} + \pi A_n (\mathrm{Im}\,\tau)^{-1} A_m] t^m / m, \quad X_n^{\mathrm{a}} = \pi \sum_{m=1}^{\infty} A_n (\mathrm{Im}\,\tau)^{-1} \bar{A}_m (\bar{t})^m / m ,$$

with

$$Q_{nm} = \frac{1}{2(m-1)!(n-1)!} \partial_t^m \partial_y^n \log \frac{E(t,y)}{(t-y)} \bigg|_{t=y=0} .$$

Once again we find infinitely many charges Q_n which annihilate the state $|\phi\rangle$ which comes from the path-integral. To be more explicit we will write X in the operator language inside the disc by

$$X(t) = x + \mathrm{i} p \log t + \sum_{n=1}^{\infty} \left(t^{-n} \frac{a_n^\dagger}{\sqrt{n}} + t^n \frac{a_n}{\sqrt{n}} + (\bar{t})^{-n} \frac{\bar{a}_n^\dagger}{\sqrt{n}} + (\bar{t})^n \frac{\bar{a}_n}{\sqrt{n}} \right),$$

with the only non-vanishing commutation relations

$$-\mathrm{i}[x,p] = [a_n, a_n^\dagger] = [\bar{a}_n, \bar{a}_n^\dagger] = 1$$

we have

$$Q_n = \sqrt{n} a_n + [2Q_{nm} + \pi A_n (\mathrm{Im}\,\tau)^{-1} A_m] a_m^\dagger / \sqrt{m} - \pi A_n (\mathrm{Im}\,\tau)^{-1} \bar{A}_m \bar{a}_m^\dagger / \sqrt{m} . \tag{21}$$

Similarly we can shift X by the complex conjugate solution $\bar{X}_n$ to obtain the charge $\bar{Q}_n$. These charges all commute with one another (by using $Q_{nm} = Q_{mn}$) and can be used to solve for $|\phi\rangle$ up to an overall constant. First we note that the state $|\phi\rangle$ has zero momentum, which follows from another symmetry which is simply a constant shift $X \to X + a$ giving rise to the conserved charge which is simply the momentum p. We will solve for $|\phi\rangle$ by making use of the isomorphism of the algebra of creation and annihilation operators with

$$a_n \leftrightarrow (1/\mathrm{i}\sqrt{n}) \partial / \partial x_n, \quad a_n^\dagger \leftrightarrow \mathrm{i}\sqrt{n} x_n ,$$

and similarly for $\bar{a}$ and $\bar{a}^\dagger$. In this language any state in the Hilbert space will be represented by a function of the x_n and $\bar{x}_n$. The standard vacuum is described by the function (1). We solve for $|\phi\rangle$ and obtain

$$|\phi\rangle = C \exp\left([x, \bar{x}] M \begin{bmatrix} x \\ \bar{x} \end{bmatrix} \right), \tag{22}$$

where

$$M = \begin{pmatrix} Q_{nm} + \frac{1}{2}\pi A_n (\mathrm{Im}\,\tau)^{-1} A_m & -\frac{1}{2}\pi A_n (\mathrm{Im}\,\tau)^{-1} \bar{A}_m \\ -\frac{1}{2}\pi \bar{A}_n (\mathrm{Im}\,\tau)^{-1} A_m & \bar{Q}_{nm} + \frac{1}{2}\pi \bar{A}_n (\mathrm{Im}\,\tau)^{-1} \bar{A}_m \end{pmatrix} .$$

Again the overall normalization of the state is undetermined. Using this expression we can again compute the correlation functions. We will find, in particular, that

$$\langle \exp[-\mathrm{i}X(t_1)]\,\exp[\mathrm{i}X(t_2)]\rangle = \left| \frac{\exp(-\pi \mathrm{Im}(\int_{t_1}^{t_2}\omega)(\mathrm{Im}\,\tau)^{-1}\mathrm{Im}(\int_{t_1}^{t_2}\omega))}{E(t_2,t_1)} \right|^2 ,$$

which is the correct result [13].

Now we come to the case where the bosonic field is allowed to have non-trivial windings in order to reproduce the fermionic theory. In that case, we have instantons to sum over. We will have to shift the periodic function X by a classical solution which corresponds to windings about the a- and b-cycles. We label the instantons by their windings n, m. The path-integral in this case gives $\sum_{n,m} |\phi_{nm}\rangle$ where $|\phi_{nm}\rangle$ is the result of the path-integral with the corresponding background instanton. Now we will have to find the relation between $|\phi_{nm}\rangle$ and $|\phi\rangle$. To find it, we note that the effect of the instantons is to put a weighting factor (with the proper phases) which has been worked out in ref. [8]. There is only one additional modification, and that is because we have shifted our variables so that X becomes periodic. Therefore in the presence of instantons we will have to modify the vacuum so that $\langle 0|X|\phi\rangle = X_{\mathrm{classical}} = \int (n\alpha + m\beta)$ where α and β are a basis for the first cohomology dual to the a- and b-cycles. This is achieved by using the coherent states of the form $\exp(\lambda_n a_n^\dagger + \bar{\lambda}_n \bar{a}_n^\dagger)|\phi\rangle$ for appropriate λ_n ($\bar{\lambda}_n$). Putting all this together and doing the sum we find that the terms which mixed left and right Hilbert spaces now disappear, and we are left with

$$\sum_{nm} |\phi_{nm}\rangle = C\tau(x_n)\tau(\bar{x}_n)\,, \tag{23}$$

where

$$\tau(x_n) = \exp(x_n Q_{nm} x_m)\vartheta[^{\theta}_{\phi}](A_n x_n|\tau) \tag{24}$$

is the famous tau function. That this bosonic state corresponds to the fermionic $|\phi\rangle$ computed earlier under standard bosonization on circle is a non-trivial fact, and may be viewed as one of the major results in ref. [6]. In fact the proof of this relies on the fact that both of them provide solutions to the KP hierarchy, which is equivalent to Hirota's bilinear conditions [14]. In particular the equivalence of these two states involves theta function identities (Fay's trisecant identity and its infinitesimal versions) which hold only for theta functions coming from Riemann surfaces. Alternatively, the results in refs. [8,9] establishing the equivalence of bosonic and fermionic correlation functions together with our derivation here provides another proof of the equivalence of the tau function with the corresponding fermionic state, without using KP hierarchy.

We also note that Hirota's bilinear condition admits a very simple physical interpretation in the fermionic language: Consider doubling the number of fermions. Then the state that we get on the circle by path-integral on $\Sigma - \mathrm{D}$ belongs to the tensor product of the Fock space of one fermion times the other, and is simply the tensor product of the states $|\phi\rangle_1$ we got for one fermion, times the corresponding state $|\phi\rangle_2$ for the other fermion. Now we have an additional global symmetry which rotates one fermion to another, and this gives rise to an additional conserved current. The condition that the corresponding charge Q annihilate the state

$$Q|\phi\rangle_1|\phi\rangle_2 = 0$$

is Hirota's condition.

Here we have dealt with non-chiral bosons because there is no path-integral for a chiral boson. However if we wished to give a prescription for a chiral boson in this setup, we will have to consider only holomorphic instantons, and this implies that we should allow X to have arbitrary period in the b direction. This together with the right weighting factor for holomorphic instantons would give us $\tau(x_n)$.

We also note that it is simple to compute the corresponding $|\phi\rangle$ when X corresponds to the coordinates of a torus (even in the presence of flat background fields). We will leave it to the reader to write it explicitly. It

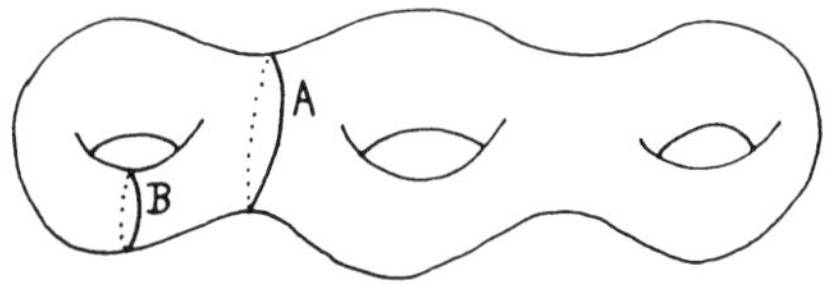

Fig. 3. We can consider other cycles, such as A, which is a homologically trivial but homotopically nontrivial curve, and B which is homologically non-trivial.

would also be interesting to apply these ideas to arbitrary non-linear sigma models. This may allow a reformulation of non-linear sigma models.

We conclude by noting that the construction we have done here can be generalized. For example we can consider instead of the circle $|t|=1$, some other homologically trivial (but homotopically nontrivial) circle, for example the one showed in fig. 3 (curve A), and apply the same ideas again [‡2]. Again we can use the same idea of infinitely many conserved charges (on each side of the circle) and get two rays in the Hilbert space. Or we can consider a narrow band near the curve A and compute correlation functions on the band by sandwiching vertex operators between these two states. However in this case it may not be so simple to choose a convenient basis for the solutions to the wave equation on each side of the Riemann surface. Also we can consider taking homologically non-trivial cycles (see fig. 3 curve B). This will be qualitatively different from the case we have considered in this paper, and the resulting path-integral on the Riemann surface will define an operator on the Hilbert space based on this circle. It would be interesting to work out this case in detail, because this is what corresponds to (diagonal elements of) the operator q^{H} in the case of the torus, and so one can identify this operator with the "time evolution operator" on the handles for homologically non-trivial cycles on higher genus Riemann surfaces. Also one can play other games like cutting n disks out of the surface. The result of the path-integral in this case is a state in the tensor product of the n Hilbert spaces based on each of the circles.

In this paper we have focused on spin-1/2 fermions and spin-0 bosons. It is possible to generalize these to fermions of higher spin [16] and derive the bosonic prescription suggested in ref. [5]. Also ideas to fix the overall normalization of the state $|\phi\rangle$ as well as applications to string theories will be discussed in ref. [16].

We have greatly benefited from discussions with L. Alvarez-Gaumé, S. Davis, D. Kazhdan, G. Moore and P. Nelson. This research was supported in part by NSF contract PHY-82-15249, and by a fellowship from the Harvard Society of Fellows.

‡2 This has also been considered by Kazhdan [15].

References

[1] A.M. Polyakov, Phys. Lett. B 103 (1981) 207, 211.
[2] Y.I. Manin, Critical dimensions of string theories and the dualizing sheaf on the moduli space of curves, preprint (1985).
[3] A.A. Beilinson, Y.I. Manin and Y.A. Shechtman, in preparation.
[4] N. Ishibashi, Y. Matsuo and H. Ooguri, University of Tokyo preprint UT-499 (1986).
[5] L. Alvarez-Gaumé, C. Gomez and C. Reina, preprint CERN TH-4641/87.
[6] G. Segal and G. Wilson, Publ. IHES 61 (1985);
T. Shiota, Inv. Math. 83 (1986) 333;
M. Mulase, J. Diff. Geom. 19 (1984) 403;
I.M. Krichever, Russ. Math. Surv. 32 (1977) 185;
B.A. Dubrovin, I.M. Krichever and S.P. Novikov, Sov. Sci. Rev. 3 (1982);
E. Date, M. Jimbo, M. Kashiwara and T. Miwa, J. Phys. Soc. Japan 50 (1981) 3806.
[7] D. Friedan and S. Shenker, Phys. Lett. B 175 (1986) 287; Nucl. Phys. B 281 (1987) 509.
[8] L. Alvarez-Gaumé, J.B. Bost, G. Moore, P. Nelson and C. Vafa, Phys. Lett. B 178 (1986) 41; preprint HUTP-86/A062.
[9] E. Verlinde and H. Verlinde, Utrecht preprint (1986);
T. Eguchi and H. Ooguri, preprint LPTENS 86.39. (December 1986);
M. Dugan and H. Sonoda, LBL preprint (1987).

[10] D. Mumford, in: Tata Lectures on Theta, Vols. I, II (Birkhauser, Basel, 1984);
J. Fay, in: Theta functions on Riemann surfaces, Springer Notes in Mathematics, Vol. 352 (Springer, Berlin, 1973).
[11] L. Alvarez-Gaumé, G. Moore and C. Vafa, Commun. Math. Phys. 106 (1986) 1.
[12] H.M. Farkas and I. Kra, in: Riemann surfaces (Springer, Berlin, 1980).
[13] S. Hamidi and C. Vafa, Nucl. Phys. B 279 (1987) 465;
M.A. Namazie, K.S. Narain and M.H. Sarmadi, Phys. Lett. B 177 (1986) 329;
H. Sonoda, Phys. Lett. B 178 (1986) 390.
[14] R. Hirota, in: Lecture Notes in Mathematics, Vol. 515 (Springer, Berlin, 1976).
[15] D. Kazhdan, private communication.
[16] L. Alvarez-Gaumé, C. Gomez, C. Reina and C. Vafa, in preparation.

LOOP GROUPS, GRASSMANIANS AND STRING THEORY

L. ALVAREZ-GAUMÉ [1], C. GOMEZ
CERN, CH-1211 Geneva 23, Switzerland

and

C. REINA
Department of Physics, University of Milan, Via Celoria 16, I-20133 Milan, Italy

Received 8 February 1987

The theory of representations of loop groups provides a framework where one can consider Riemann surfaces with arbitrary numbers of handles and nodes on the same footing. Using infinite grassmanians we present a general formulation of some conformal field theories on arbitrary surfaces in terms of an operator formalism. As a by-product, one can obtain some general results for the chiral bosonization of fermions using the vertex operator representation of infinite dimensional groups. We believe that this set-up provides the natural arena where the recent proposal of Friedan and Shenker of formulating string theory in the universal moduli space can be discussed.

1. One of the open problems in the theory of strings is how to extend the first quantized formulation [1][‡1] to obtain semiclassical or non-perturbative information. In the first quantized approach one computes the partition and correlation functions for a conformally invariant two-dimensional field theory on a genus g Riemann surface, and then sums over arbitrary genus. Since one would expect very important non-perturbative effects as a consequence of string dynamics, the perturbative series is unlikely to make much sense. In a recent proposal, Friedan and Shenker [3] try to formulate string theory in some kind of universal moduli space, which would include surfaces of arbitrary genus, with and without nodes, and including infinite genus surfaces as well. In this approach, the generating functional for the string S-matrix is written formally as

$$S=\int_{\mathrm{R}} h(\psi, \psi^*)\ , \qquad (1)$$

where R is the "universal moduli" space for Riemann surfaces, and ψ is a section of a holomorphic vector bundle over R. The bundle group would be the modular group of R. In this letter we present a framework where this appealing formulation can be realized more explicitly. Since one of the main motivations for this formulation is provided by the treatment of string theory as a conformal field theory, there are a number of basic issues that should be addressed. In refs. [4,5] one finds a general treatment of conformally invariant field theories on the twice punctured sphere (conformally equivalent to a cylinder). The extension of this analysis to higher genus surfaces encounters a number of problems, for example: (i) The treatment of the zero modes of the conformal fields, responsible for the inclusion of moduli and supermoduli parameters in string path integrals [6]. (ii) The lack of a natural operator formalism. (iii) The meaning of the Virasoro algebra still remains quite obscure... . There has been some progress in these problems [7,8], but we are still far from a complete understanding. One would like to have a general framework which treats Riemann surfaces with and without nodes on the same footing, and which incorporates the main features of the

[1] Alfred P. Sloan Foundation Fellow.
[‡1] For a review of string theories, see ref. [2].

operator methods that can be developed in a local coordinate patch, but without losing track of any of the global information implicit in the fact that we want to deal with surfaces of complicated topology. Mathematical objects with these properties appear naturally in the theory of representations of infinite dimensional groups [8]. These objects are the infinite grassmanians.

Some of the features which make these objects useful for the study of conformal field theories are: (i) The collection of all Riemann surfaces, with and without nodes, together with the local data provided by a conformal field theory (for example the line bundle on which the primary fields take their values), generates a dense set. (ii) In analogy with finite dimensional moduli spaces, the infinite grassmanian is quasi-compact i.e., the only globally defined holomorphic functions are constants. (iii) It provides a topology where one may start asking questions concerning the role of infinite genus surfaces, (for example, it is possible to study the limits of sequences of surfaces with growing numbers of handles). (iv) Using representation theory for infinite dimensional groups it is possible to reconstruct the correlation functions for free conformal field theories and ghost systems on arbitrary surfaces, thus providing the analog of an operator formalism for higher genus. The formulae obtained using these methods coincide with the results derived using bosonization techniques [9–11], thus lending support to the general framework advocated in this letter.

2. In the formulation of conformal field theory on the twice punctured sphere, one can associate an "in" vacuum to the north pole, and an out vacuum to the south pole. Since the north and south poles are related by an SL_2 transformation, if we use an SL_2 invariant vacuum, we can relate what happens in the north and south poles using standard operator methods. On a higher genus surface Σ, we will try to follow as closely as possible our intuition from the operator approach. In order to make the discussion concrete, we will concentrate on the conformal properties of first-order anticommuting b–c systems [5] of some given conformal spin j and $1-j$, respectively. Geometrically b and c are sections of holomorphic line bundles L_j, L_{1-j} over Σ. We also pick a point p (the analog of the north pole on the sphere), and a local coordinate t vanishing at p. Choosing the local parameter conveniently, we can assume that it extends to a local neighborhood U_0 of p homeomorphic to a disk in the complex plane and containing the unit circle $S^1=\{t;\ |t|=1\}$. In the functional integral approach, the partition function of the b–c system is given in terms of the determinant of the Cauchy–Riemann operator $\bar\partial_j$. A convenient way of thinking about DET $\bar\partial_j$ is as a line bundle over moduli space: if Ker $\bar\partial_j$ and Cok $\bar\partial_j$ are the kernel and cokernel of $\bar\partial_j$ respectively, DET $\bar\partial_j$ is defined as a section of $(\lambda\,\mathrm{Ker}\,\bar\partial_j)^{-1}\otimes(\lambda\,\mathrm{Cok}\,\bar\partial_j)$ (λV is the highest exterior power of the vector space V, and V^{-1} is the vector space dual to V). The first thing we would like to do is to capture the global information contained in Ker $\bar\partial_j$ and Cok $\bar\partial_j$ in terms of local data. To achieve this, let $U_1=\Sigma-p$; then U_0, U_1 provide an open cover of Σ, intersecting in a disc minus a point. Let $\Gamma(U_i, L)$ and $\Gamma(U_0\cap U_1, L)$ be the spaces of holomorphic sections of L over U_0, U_1 and $U_0\cap U_1$, respectively. Given a nowhere vanishing local holomorphic section σ_0 of L_j over U_0, we can identify sections of L with complex valued analytic functions f via $\sigma=f\sigma_0$. Thus we can identify $\Gamma(U_0, L)$ with convergent power series in U_0, $f=\sum f_n t^n$, $n=0, 1, \ldots$, and $\Gamma(U_0\cap U_1, L)$ with convergent series in $U_0\cap U_1$, $f=\sum f_n t^n$, $n\in\mathbb{Z}$. If pr denotes the projection from $\Gamma(U_1, L)$ into $\Gamma(U_0\cap U_1, L)/\Gamma(U_0, L)$, we have the exact sequence

$$0\to\mathrm{Ker}\,\bar\partial_j\to\Gamma(U_1, L)$$

$$\xrightarrow{\mathrm{pr}}\Gamma(U_0\cap U_1, L)/\Gamma(U_0, L)\to\mathrm{Cok}\,\bar\partial_j\to 0\,, \qquad (2)$$

and we conclude that Ker pr = Ker $\bar\partial_j$, Cok pr = Cok $\bar\partial_j$, and DET pr = DET $\bar\partial_j$. The construction of the infinite grassmanian of interest proceeds as follows [‡2]. Let $H=L^2(S^1, \mathbb{C})$ be the space of square integrable functions on the circle. A basis for this space is given by $\{t^k, k\in\mathbb{Z}\}$. Set $H=H_+\otimes H_-$, with H_+ (respectively H_-) spanned by t^k, $k=0, 1, 2,\ldots$ (respectively t^k, $k=-1, -2, \ldots$). A linear subspace w of H belongs to the grassmanian Gr if two conditions are met: (i) The orthogonal projection pr_-

‡2 For this and many more details about infinite grassmanians see ref. [12].

from w to H_- is a Fredholm operator, i.e., it has finite dimensional kernel and cokernel. (ii) The orthogonal projection pr_+ from w into H_+ is compact. The index of pr_- is called the virtual dimension of w. Thus Gr splits into an infinite number of components labelled by ind pr. Since pr_- must be Fredholm, the full group of invertible operators acting on H does not act on Gr, but there is still a rather large group acting on Gr: GL_{res}. Writing its elements in a decomposition adapted to the splitting $H_+ \oplus H_-$ as $\begin{pmatrix} a & b \\ c & d \end{pmatrix}$, then a and d must be Fredholm and b, c must be compact. There are several subgroups of GL_{res} which will be important in our discussion: Γ_+, Γ_-, $GL(\infty)$ and their central extensions. The group $\Gamma_+(\Gamma_-)$ is defined as multiplication by nowhere vanishing analytic functions on the circle which extend holomorphically into the disc $|t| \leqslant 1$ ($|t| \geqslant 1$) and giving the identity at $t=0$ ($t=\infty$). The group Γ_+ acts on Gr without fixed points. $GL(\infty)$ is a Kac–Moody group of infinite rank whose elements are infinite dimensional invertible matrices g such that $g-1$ contains only a finite number of non-vanishing entries. With respect to the basis of H we have introduced, the generators of $GL(\infty)$ are the operators $E_{ij}(t^j)=t^i$; or in matrix form $(E_{ij})_{kl}=\delta_{ik}\delta_{jl}$, for any integers i, j, k, l. We can think of the space H as the space of wave functions for a free field moving on S^1, where H_- (H_+) represents the space of positive (negative) frequency solutions of the wave equation. Thus we can construct a Fock space F out of H, which admits several useful and equivalent representations. We can represent the elements in the Fock space F as semi-infinite forms $v_{i_1} \wedge v_{i_2} \wedge v_{i_3} \wedge \ldots$ for a basis v_i of H. If v_i corresponds to the basis introduced above for H (the element t^i is a solution to the wave equation with energy $-i$. We will identify v_i with the wave function of energy i), the state $v_0 \wedge v_{-1} \wedge v_{-2} \wedge \ldots$ can be identified with the filled Dirac sea. The group $GL(\infty)$ acts naturally on F because by definition the elements of this group have well-defined determinants [3]. Similarly, as in field theory we can introduce a set of fermionic creation and annihilation operators ψ_n, ψ_n^* and field operators $\psi(z)=\sum \psi_n z^{-n-1/2}$, $\psi^*(z)=\sum \psi_n^* z^{-n-1/2}$, $n \in \mathbb{Z}+\frac{1}{2}$ so that the semi-infinite forms are represented by acting with creation and annihilation operators on the Dirac vacuum (the vacuum conditions are $\psi_n|0\rangle=\psi_n^*|0\rangle=0$, $n>0$). In this way, the generators of $gl(\infty)$ are represented by fermion bilinears $E_{ij}=\psi_i\psi_j^*$. Finally a third useful way of thinking about the same states is in terms of the determinant line bundle of Gr: DET. Its fiber at w in Gr is a one dimensional space spanned by elements of the form $\Lambda f=f_1 \wedge f_2 \wedge f_3 \ldots$ for a basis of w. Again the group $GL(\infty)$ acts naturally on the space of sections of DET, $\Gamma(\text{DET})$, and we can certainly think of semi-infinite forms as sections of DET. The importance of DET is that it can easily be shown that DET pr is a section of DET^{-1}. The group GL_{res} does not act naturally on DET, because its elements will in general induce changes of basis by means of operators without well defined determinants; nevertheless, for a certain subgroup of GL_{res}, (generated by elements such that b, c are trace class) it is possible to define a projective action on DET [12].

Contact with conformal field theory follows by noting that a local conformal field theory gives a set of geometrical data (Σ, p, t, L, σ_0). To this data we can associate a point w in Gr, given by the set of analytic functions f on S^1 such that $f\sigma_0$ extends to a section in $\Gamma(U_1, L)$. The converse is not true in general. However, Segal and Wilson show [12] that any point on $Gr_* = \{w \in Gr;\ t^{-q}H_- \subset w \subset t^q H_-$ for some $q>0\}$ comes from geometrical data. This result is useful because Gr_* is dense on Gr, and it is also the union of a sequence of finite dimensional grassmanians. These two facts together imply that any globally defined holomorphic function on Gr is a constant, one of the distinctive features of Gr alluded to in the introduction. As a consequence of these remarks, Gr contains the moduli space of all Riemann surfaces, including singular ones.

Let us for the time being restrict our attention to the zero component of Gr, $Gr^{(0)}$ (points of Gr of virtual dimension zero). From the exact sequence (2) we know that we are dealing with line bundles of spin 1/2 over Σ on $Gr_*^{(0)}$. We will later extend the construction to higher spin fields. Using the realization of the space of sections of DET in terms of semi-infinite forms, we can set up an abstract fermionic free field theory on $Gr^{(0)}$. Since we want to have a fermion number current and an energy-momentum tensor with a well-defined action on the vacuum, we

[3] For more details on the infinite wedge representation and the K–P hierarchy see ref. [13].

introduce the Lie algebra of $A(\infty)$, (a central extension of $gl(\infty)$ [15]), generated by $E_{ij}=:\psi_i\psi_j^*:$ (normal ordering with respect to the Dirac vacuum). The fermionic states can also be represented in terms of sections of DET^{-1}. If w is a point in Gr, we can associate with it the section

$$\psi_w(w')=\det\langle w, w'\rangle\ , \tag{3}$$

with the determinant defined with respect to a choice of admissible basis. Two bases are admissible if the operator relating them has a determinant. Thus the vacuum section is defined for $w_0=H_-$. Introducing creation and annihilation operators on $\Gamma(DET^{-1})$, we can construct a mapping from the set of semi-infinite forms into the fermionic Fock space with respect to the vacuum ψ_{w_0}. If j_n represents the Fourier components of the fermion number current $j(z)=:\psi(z)\psi^*(z)$: the action of Γ_+ and Γ_- in the Fock space can be written in terms of the j_n. A generic element of Γ_+ (Γ_-) takes the form $\exp(\sum_1^\infty x_n t^n)$ ($\exp(\sum_1^\infty y_n t^{-n})$) where the x's and y's are the parameters of $\Gamma_\pm$. The corresponding representation in F is respectively given by $H(x)=\sum_{n=1}^\infty x_n j_n$, $H(y)=\sum_{n=1}^\infty y_n j_{-n}$. In this abstract field theory defined on $Gr^{(0)}$, the proof of bosonization and the computation of fermionic amplitudes can be done in terms of the vertex operator representation of $A(\infty)$, (see refs. [13,14] for details and references). The first step consists of identifying the space of semi-infinite forms with the space of formal power series $C[x_1, x_2, ...]$. If $S=\{i_0, i_1, i_2, ...\}$ represents a Dirac partition, i.e. an infinite sequence of integers such that $i_0>i_{-1}>i_{-2}...$ and satisfying $i_n=n$ for $|n|$ large enough, we can construct the state $v_{i_0}\wedge v_{i_{-1}}\wedge ...$ (recall that in our notation v_n is a solution to the free wave equation with energy n). The associated element in $C[x_1, x_2,...]$ is

$$F_S(x_1,...)=S_{i_0,i_{-1}+1,i_{-2}+2\,...}(x)\ , \tag{4}$$

where $S_{\lambda_1,\lambda_2,...}(x)$ is a Schur function defined for any finite sequence of non-negative integers $\lambda_1\geqslant\lambda_2\geqslant ...$ as

$$S_{\lambda_0,\lambda_1,...}(x)=\det_{i,j}(P_{\lambda_i-i+j}(x))\ ,\quad 0\leqslant i,j\leqslant r-1\ ,$$

$$\exp\left(\sum_{k=1}^{\infty} z^k x_k\right)=\sum_{n=0}^{\infty} z^n P_n(x),\quad P_0(x)=1\ , \tag{5}$$

and r is chosen large enough so that $\lambda_i=0$ for $i>r-1$.

The Schur function for a given set of λ's is the same as the character formulae for $GL_N(\mathbb{C})$ for N large. On $C[x]$ the generators of Γ are represented by an infinite Heisenberg algebra. For n positive, j_n becomes multiplication by nx_n and j_{-n} becomes $\partial/\partial x_n$, and the rest of the algebra of $gl(\infty)$ is obtained in terms of vertex operators

$$\sum_{i,j} u^{-i}v^{-j}E_{ij}=-\frac{(uv)^{-1}}{1-u/v}\Gamma_0(u,v)$$
$$=\Gamma_0(u)\Gamma_0(v)^*\ , \tag{6a}$$

$$\Gamma_0(u,v)=\exp\left(\sum_{j\geqslant 1}(u^{-j}-v^{-j})x_j\right)$$
$$\times\exp\left(-\sum_{j\geqslant 1}\frac{1}{j}(u^j-v^j)\frac{\partial}{\partial x_j}\right), \tag{6b}$$

$$\Gamma_0(u)=\frac{i}{u}\exp\left(\sum_{j\geqslant 1}u^{-j}x_j\right)\exp\left(\sum_{j\geqslant 1}\frac{1}{j}u^j\frac{\partial}{\partial x_j}\right), \tag{6c}$$

$$\Gamma_0(u)^*=\Gamma(u,-x)\ .$$

For the normal ordered generators of $A(\infty)$, one obtains instead

$$\sum_{i,j} u^{-i}v^{-j}\hat{E}_{ij}=-\frac{(uv)^{-1}}{1-u/v}[\Gamma_0(u,v)-1]\ . \tag{7}$$

The hermitian form in Fock space satisfying $j_n^+ = j_{-n}$ induces a hermitian form in $C[x]$:

$$\langle f(x_1,...), g(x_1,...)\rangle$$
$$=f\left(\frac{\partial}{\partial x_1},\frac{1}{2}\frac{\partial}{\partial x_2},...\frac{1}{n}\frac{\partial}{\partial x_n},...\right)g(x_1,...)\Big|_{x_i=0}\ , \tag{8}$$

thus, if ψ_w is any section of DET^{-1}, the correspondence just described implies that

$$\langle\psi_0,\psi_w\rangle=\langle 1, F_{S(w)}(x)\rangle=F_{S(w)}(0)\ , \tag{9}$$

and $S(w)$ is the Dirac partition associated to w. Notice that (6a) is nothing but the bosonic representation of $\psi(u)\psi^*(v)$. Therefore matrix elements of fermionic operators between any state and the vacuum can be represented as

$$\langle\psi_0,\psi(u)\psi^*(v)\psi_w\rangle$$
$$=\Gamma_0(u)\Gamma_0(v)^*F_{S(w)}(x)|_{x=0}\ . \tag{10}$$

In particular we expect all of these formulae to

reproduce the field theoretic results obtained on the twice punctured sphere. Here we find that (6)–(10) do not completely yield the answer that one would expect. The reason is very simple. If one computes for example $\langle \psi_0, \psi(u)\psi(v)^*\psi_0 \rangle$ one gets

$$\Gamma_0(u)\Gamma_0^*(v)\mathbb{1}|_{x_i=0, i>0} = \frac{-1}{1-u/v}\frac{1}{uv}, \quad (11)$$

therefore, even though one gets the correct singularity, the residue at the pole is incorrect. This is easily fixed by including one more variable on the vertex operators but not on the Schur functions. The new variable x_0 corresponds to enlarging Γ_+ so that $g(t)\in\Gamma_+$ can now take any non-zero value at $t=0$. From the fermionic Fock space point of view this means that we include in $H(x)$ a term $x_0 j_0$ which counts the fermionic charge of the state where it acts. If we now define

$$\Gamma(u, x) = \exp\left(\sum_{n=0}^{\infty} x_n u^{-n}\right)$$

$$\times \exp\left(-\ln u \frac{\partial}{\partial x_0} - \sum_{n=1}^{\infty} \frac{1}{n} u^n \frac{\partial}{\partial x_n}\right), \quad (12)$$

and in the hermitian form (8) one requires at the end that x_0 vanishes along with the other x_i. Using (12) one now gets the desired result

$$\Gamma(u)\Gamma^*(v)\cdot\mathbb{1}|_{x_i=0, i\geqslant 0} = \frac{1}{u-v}, \quad (13)$$

and the rest of the correlation functions at tree level then follow from standard vertex operator manipulations.

The reason why this abstract fermionic system is interesting is because to the data $(\Sigma, p, t, L_{1/2}, \sigma_0)$ we can associate a point in $\mathrm{Gr}_*^{(0)}$. What is more interesting yet, is that using Krichever's construction of solutions to the K–P hierarchy [16], the Schur function $F_{S(w)}(x)$ for a point $w\in\mathrm{Gr}_*^{(0)}$ is given in by the Riemann theta function of the surface Σ it represents with characteristics determined by the spinor bundle $L_{1/2}$. In this case the Schur function is known as a τ-function for the K–P hierarchy, and it is explicitly given by [‡4]

$$F_S(x) = \exp\left(\sum_{n,m=1}^{\infty} Q_{nm} x_n x_m\right) \vartheta[^\alpha_\beta]$$

$$\times \left(\sum_{n=1}^{\infty} A_n x_n | \Omega\right). \quad (14)$$

The various factors in (14) are defined as follows: $\omega_i(t)\mathrm{d}t$ denotes the abelian differentials on Σ normalized as

$$\int_{a_i} \omega_j = \delta_{ij}, \quad \int_{b_i} \omega_j = \Omega_{ij}, \quad (15)$$

with respect to a canonical homology basis; the coefficients A appearing in (14) are given by the coefficients of the Taylor expansion of the ω's around p

$$\omega_i = (A_{i1} + A_{i2} t + \ldots)\mathrm{d}t, \quad (16)$$

and the quadratic form Q_{nm} is constructed using the normalized differentials of the second kind. η_n is the differential of the second kind with zero a-periods and with a single pole at p of order $n+1$. η_n is normalized so that close to p its integral takes the form

$$\int^q \eta_n = t^{-n} - 2\sum_{m=1}^{\infty} Q_{nm} t^m/m, \quad (17)$$

recalling that the differentials of the second kind are obtained by taking logarithmic derivatives of the prime form of the surface (see for example refs. [18,19]), it is easy to check that the fermion propagator is given by

$$\frac{\langle \psi_{\omega 0}, \psi(z)\psi(w)^*\psi_\omega \rangle}{\langle \psi_{\omega 0}, \psi_\omega \rangle} = \frac{\vartheta[^\alpha_\beta](\int_w^z \omega | \Omega)}{\vartheta[^\alpha_\beta](0|\Omega)E(z, w)} \quad (18)$$

and similarly, if we consider multiple point functions we obtain an answer involving prime forms and theta functions

$$\frac{\langle \psi_{\omega 0}, \psi(z_1)\psi^*(w_1)\ldots\psi(z_n)\psi^*(w_n)\psi_\omega \rangle}{\langle \psi_{\omega 0}, \psi_\omega \rangle}$$

$$= \frac{\prod_{i<j} E(z_i, z_j) \prod_{k<l} E(w_k, w_l)}{\vartheta[^\alpha_\beta](0|\Omega) \prod_{i,k} E(z_i, w_k)}$$

$$\times \vartheta[^\alpha_\beta]\left(\sum_i z_i - \sum_k w_k\right), \quad (19)$$

[‡4] For details and references see ref. [17].

as expected from the general results of bosonization in higher genus surfaces [9–11]. Note however that (19) follows from straightforward vertex operator algebra, and that we can deal with chiral fermions from the beginning. Thus by working with the abstract field theory provided by the grassmanian, we can use operator methods, to obtain the desired correlation functions on a Riemann surface, by computing matrix elements between the vacuum $\psi_{\omega 0}$ and the state ψ_{ω} representing in Gr the geometrical data. It is in this sense that one can develop an operator formalism for higher genus surfaces.

3. Before extending the previous results to higher spin fields on Σ, we would like to give a geometrical picture of what is going on, and understand where the Virasoro algebra appears. Since in our construction we always need a triple (Σ, p, t), let $\hat{\mathrm{M}}_g$ be the moduli space of such triples for genus g surfaces. If M_g denotes the moduli space for curves, there is a natural projection map ρ: $\hat{\mathrm{M}}_g \to \mathrm{M}_g$. The universal curve $\hat{\mathscr{C}}$ over $\hat{\mathrm{M}}_g$ is a complex manifold fibered over $\hat{\mathrm{M}}_g$, whose fiber at $\hat{m}$ is a copy of the triple (Σ, p, t) parametrized by $\hat{m}$. If $\mathscr{C}$ is the universal curve over M_g, then $\rho^*(\mathscr{C}) = \hat{\mathscr{C}}$. Along the fibers of $\hat{\mathscr{C}}$ we can consider the line bundle of fields of spin j. These fit together to give a line bundle over $\hat{\mathscr{C}}$: $\hat{\omega}^j$ ($\hat{\omega}$ stands for the line bundle of differentials). At $\hat{m} \in \hat{\mathrm{M}}_g$, we have the space of sections $\Gamma(\hat{\mathscr{C}}_{\hat{m}} - p, \hat{\omega}^j|_{\hat{m}})$. The local coordinate t gives a trivializing section $\sigma_0 = \mathrm{d}t^j$, and thus we can get a point in Gr as explained before. The holomorphic map k_j: $\hat{\mathrm{M}}_g \to \mathrm{Gr}^{(i)}$ ($i = (2j-1)(g-1) =$ index of $\bar{\partial}_j$) obtained this way will be called the Krichever map and the set of points obtained on Gr the Krichever locus. The family version of the exact sequence (2) implies that the pullback $k_j^*(\mathrm{DET})$ is a line bundle over $\hat{\mathrm{M}}_g$ whose fiber at $\hat{m}$ is $(\mathrm{DET}\ \bar{\partial}_j)^{-1}$. Since DET $\bar{\partial}_j$ only involves coordinate independent information on Σ, it follows that $k_j^*(\mathrm{DET})$ is trivial along the fibers of ρ, and we have $k_j^*(\mathrm{DET}) \simeq \rho^*(\lambda_j)$ where $\lambda_j = (\mathrm{DET}\ \bar{\partial}_j)^{-1}$ is the determinant line bundle for spin j fields over M_g. Using the Grothendieck-Riemann-Roch theorem [20], or using the Quillen metric on λ_j [21], we can compute the first Chern class of $k_j^*(\mathrm{DET})$. The result is

$$c_1(k_j^*\ \mathrm{DET}) = c_1(\rho^*\lambda_j)$$

$$= (6j^2 - 6j + 1)\rho^*(c_1(\lambda_1)). \tag{20}$$

Again by (2) we have the identification $k_j^*(\mathrm{DET\ pr}) \simeq \rho^*(\mathrm{DET}\ \bar{\partial}_j)$. The disadvantage of the Krichever map k_j introduced is that it maps M to different components of Gr according to the value of j. We know from bosonization in higher genus surfaces [9–11], that the partition and correlation functions for a general b–c system can be expressed in terms of either spin 1/2 or spin 0 correlation functions. This is achieved in this context by introducing a modified Krichever map [22]. The universal curve $\hat{\mathscr{C}}$ has a natural section s: $\hat{\mathrm{M}}_g \to \hat{\mathscr{C}}$ given by $s(\Sigma, p, t) = p \in \Sigma$, the image $\Delta = s(\hat{\mathrm{M}}_g)$ is a divisor on $\hat{\mathscr{C}}$. The line bundle $\hat{\omega}^j(-np)$ (the family of j-differentials vanishing to at least order n at p), has degree $2j(g-1) - n$. Thus if we choose $n = n_j = (2j-1)(g-1)$, we have that $\mathscr{L}_j = \hat{\omega}^j(-n_j\Delta)$ has degree $g-1$ and zero index, and the Krichever map $k(\mathscr{L}_j)$: $\hat{\mathrm{M}}_g \to \mathrm{Gr}^{(0)}$ given by $\mathscr{L}_j$ with $\sigma_0 = t^{n_j}\mathrm{d}t^j$ takes values on $\mathrm{Gr}^{(0)}$. It is possible to show that [22]

$$k(\mathscr{L}_j)^*(\mathrm{DET}) \simeq \rho^*(\lambda_j) \otimes s^*\hat{\omega}^{-\alpha},$$

$$\alpha = (2j-1)^2 g(g-1)/2. \tag{21}$$

This result agrees with the general bosonization rules of refs. [9–11] in a particular limit as will be discussed later. A geometric realization of the Virasoro algebra for any conformal field theory on Σ can now be described. At $\hat{m} \in \hat{\mathrm{M}}$ we have the Lie algebra of meromorphic vector fields on $\mathrm{U}_0 \cap \mathrm{U}_1$: $\mathscr{D} = \Gamma(\mathrm{U}_0 \cap \mathrm{U}_1, \hat{\omega}_{\hat{m}}^{-1})$, ($\hat{\omega}_{\hat{m}}^{-1}$ are holomorphic vectors along the curve labelled by $\hat{m}$). A basis for this algebra is given by $X_n = -t^{n+1}\mathrm{d}/\mathrm{d}t$ satisfying $[X_n, X_m] = (n-m)X_{n+m}$. For surfaces of genus $g > 1$ we can single out the following subspaces [23]:

(i) $\mathscr{D}_1 = \Gamma(\mathrm{U}_1, \hat{\omega}_{\hat{m}}^{-1})$. The subalgebra of vector fields which induce holomorphic automorphisms of $\mathrm{U}_1 = \hat{\mathscr{C}}_{\hat{m}} - p$. Its action does not change any of the data registered in M.

(ii) $\mathscr{D}_0 = \Gamma(\mathrm{U}_0, \hat{\omega}_{\hat{m}}^{-1})$. The subalgebra of vector fields extending holomorphically to U_0. Those vector fields vanishing at p induce infinitesimal variations in the local parameter t, while the constant vector X_{-1} induces infinitesimal translations of p. Hence $\mathscr{D}_0$ can be identified with the vertical tangent space $T^{v}_{\hat{m}}\hat{M}_g$ to the fiber of p at $\hat{m}$.

(iii) The remaining linear space $\mathscr{D}/\mathscr{D}_0 \oplus \mathscr{D}_1$ is isomorphic to $H^1(\mathscr{C}_{\hat{m}}, \hat{\omega}_{\hat{m}}^{-1})$ the tangent space to moduli space at $m=p(\hat{m})$.

Summarizing, we see that at any m there is an isomorphism $T_{\hat{m}}\hat{M}_g \simeq \mathscr{D}/\mathscr{D}_0$. Letting $\hat{m}$ vary in $\hat{M}_g$, the collection $X_n(\hat{m}) = -t^{n+1}\mathrm{d}/\mathrm{d}t|_{\hat{m}}$ gives rise to a vector field on $\hat{M}_g$. Its action on $\hat{M}_g$ and on the determinant line bundle can be studied directly on the Krichever locus of $\mathrm{Gr}^{(0)}$ by pushing $X_n(\hat{m})$ forward using $k(\mathscr{L}_j)$. To construct this vector field $k(\mathscr{L}_j)_*(X_n)$, notice that at any $\hat{m}$ the algebra $\mathscr{D}$ acts naturally on the space $\Gamma(U_0 \cap U_1, \hat{\omega}_{\hat{m}}(-n_j p))$ by the Lie derivative along the fibers of $\mathscr{C}$. If $\sigma = t^k \sigma_0$ ($\sigma_0 = t^{n_j}\,\mathrm{d}t^j$ is a local section of $\hat{\omega}_{\hat{m}}{}^j(-n_j p)$ we have

$$\mathscr{L}_{X_n}\sigma = (k+n_j+(n+1)j)t^{k+n}(t^{n_j}\mathrm{d}t^j)\ , \qquad (22)$$

inducing an infinitesimal action on the basis of H

$$\delta_n t^k = (k+n_j+(n+1)j)t^{n+k}\ .$$

This gives a representation of $\mathscr{D}$ in $\mathrm{GL}_{\mathrm{res}}$. If we compute the action of this algebra DET, one obtains a central extension for the image of the Virasoro algebra proportional to $6j^2-6j+1$ as should be expected from (20). Making contact with the bosonization formulae [9–11] is slightly more difficult in this case; and it is easy to understand why. In the approach of ref. [9] one starts with a line bundle ξ over Σ of degree $>2g-2$ and a set of points $P_1, ..., P_k$, $k=\mathrm{ind}\ \bar{\partial}_j$ in general position. This means that $\xi(-\Sigma P_i)$ is a line bundle of degree $g-1$ without a holomorphic section. In the construction of ref. [22], all of these points are identified with p. Since the P's are used in the field theoretic approach to insert b-fields and remove the zero modes, it is clear that in order to compare the formulae obtained from $k(\mathscr{L}_j)$ on $\hat{M}_g$ and those obtained by bosonization, we have to take the limiting case where all the P's coincide. The section of $\hat{\omega}^{\alpha}$, $\alpha=(2j-1)^2(g-1)g/2$ that one has to pull-back to the surface in (21) [22], is the one whose locus of zero modes coincides with the Weierstrass points of ω^j. This gives the clue for the construction of the Schur function in this case. Even though there is yet no complete derivation from the grassmanian point of view of the Schur function that we will write down, the proof should be a consequence of extending the construction of ref. [22] so that $n_j p$ is replaced by a divisor of degree n_j in general position. Combining the arguments presented for spin 1/2 together with the results of refs. [9–11] we obtain a prescription for the computation of correlation functions for a b–c system of spin j

$$\left\langle \prod_{i=1}^{n} b(z_i) \prod_{j=1}^{m} c(w_j) \right\rangle \quad n-m=n_j \qquad (23)$$

on a general surface in terms of the vertex operator representation of $A(\infty)$. Let ψ_ω be the section of DET^{-1} associated to the geometrical data embodied in the b–c system. Since ψ_ω has fermion number charge $-n_j$, choose n_j points on the surface (or add n_j formal parameters to $C[x]$) $u_1, u_2, ..., u_n$. Define the Schur function as

$$\Gamma^*(u_1)...\Gamma^*(u_{n_j})F(x)\ , \qquad (24)$$

where $F(x)$ is the right-hand side of (14), and the characteristics depend on how we chose to parametrize holomorphic line bundles on Σ (see for instance ref. [9] for some useful choices). Now (23) becomes

$$\prod_{i=1}^{m}\Gamma(z_i)\prod_{l=1}^{n}\Gamma^*(w_l)\prod_{k=1}^{n_j}\Gamma^*(u_k)F(x)\Big|_{x=0}\ . \qquad (25)$$

Comparing (25) with the chiral bosonization formulae results in refs. [9–11] (in ref. [9] only non-chiral bosonization formulae are derived. It is, however, straightforward to extract their "meromorphic" square root, and one obtains the results of refs. [10,11]), we find that after some theta function gymnastics they agree. In particular if $\varphi_i(z)$, $i=1, ..., n_j$ are the holomorphic j-differentials, then (25) coincides with the holomorphic square root of

$$|\det \varphi_i(z_j)|^2 \frac{\det' \bar{\partial}_j^+ \bar{\partial}_j}{\det\langle \varphi_i | \varphi_j \rangle}$$

$$\times \left(\frac{\det' \bar{\partial}_0^+ \bar{\partial}_0}{\int\sqrt{g}\,\det \mathrm{Im}\,\Omega} \right)^{1/2} \qquad (26)$$

If we take the limit as the u_i become equal to p, both (25), (26) vanish. The leading non-vanishing terms are proportional to the wronskian of the holomorphic j-differentials at p. When p moves over Σ, the divisor of zeros of the wronskian is given by the Weierstrass points on the surface as implied by (21). As in the case of spin 1/2, the rather simple prescription embodied in (25) allows us to compute arbitrary correlation functions of any free b–c system on arbitrary surfaces using the vertex operator representation of A(∞). The only non-trivial step in this operator construction of the correlation function (23) is the identification of the state ψ_ω and the corresponding Schur function associated to the geometrical information conveyed by the b–c system.

We believe that the approach advocated in this letter provides the natural framework to formulate conformal field theory and string theory on higher genus surfaces using operator methods. Further details will be presented elsewhere [24].

While this work was being completed, we received the work of Ishibashi, Matsuo and Ooguri [25], which also analyzes some conformal theories on higher genus surfaces using similar techniques. We have also received a recent paper by Dugan and Sonoda [26] which also presents a derivation of the chiral bosonization formulas for b–c systems using field theoretic methods. We would like to thank P. Ginsparg for bringing this reference to our attention.

We would like to thank Professor E. Arbarello and Professor Yu. Manin for sending us their work prior to publication.

References

[1] A.M. Polyakov, Phys. Lett. B 103 (1981) 207, 211.
[2] E.g. J. Schwarz, ed., Superstrings, the first fifteen years, Vols. I, II (World Scientific, Singapore, 1985).
[3] D. Friedan and S. Shenker, Phys. Lett. B 175 (1986) 287; Nucl. Phys. B 281 (1987) 509.
[4] A.A. Belavin, A.M. Polyakov and A.B. Zamolodchikov, Nucl. Phys. B 241 (1984) 333.
[5] D. Friedan, E. Martinec and S. Shenker, Nucl. Phys. B 271 (1986) 93.
[6] D. Friedan, Lectures on Riemann surfaces, EFI preprint (1987).
[7] E. Martinec, Conformal field theory on a (super)Riemann surface, Princeton preprint (1986).
[8] T. Eguchi and H. Ooguri, University of Tokyo preprint UT-491;
H. Sonoda, Nucl. Phys. B 281 (1987) 546.
[9] L. Alvarez-Gaumé, J.B. Bost, G. Moore, P. Nelson and C. Vafa, preprint HUTP86/A039, and in preparation.
[10] E. Verlinde and H. Verlinde, Utrecht preprint (October 1986).
[11] T. Eguchi and H. Ooguri, preprint LPTENS 86.39. (December 1986).
[12] G. Segal and G. Wilson, Publ. IHES 61 (1985) 1.
[13] V.G. Kac and D.H. Peterson, preprint IHES/M/85/63.
[14] E. Date, M. Jimbo, M. Kashiwara and T. Miwa, J. Phys. Soc. Japan 50 (1981) 3806.
[15] V.G. Kac, Infinite dimensional algebras (Birkhauser, Basel, 1983).
[16] I.M. Krichever, Russ. Math. Surveys 32 (1977) 185;
B.A. Dubrovin, I.M. Krichever and S.P. Novikov, Topological and algebraic geometric methods in contemporary mathematical physics II, Sov. Sci. Rev. 3 (1982) 1.
[17] T. Shiota, Inv. Math. 83 (1986) 333.
[18] D. Mumford, Tata lectures on theta, Vol. II (Birkhauser, Basel, 1984).
[19] J. Fay, Theta functions on Riemann surfaces (Springer, Berlin, 1973).
[20] F. Hirzebruch, Topological methods in algebraic geometry (Springer, Berlin, 1966).
[21] D. Quillen, Funct. Anal. Appl. 19 (1986) 31;
J. Bismut and D. Freed, Commun. Math. Phys. 106 (1986) 159.
[22] E. Arbarello, C. DeConcini, V.G. Kac and C. Procesi, in preparation; and private communication.
[23] A.A. Beilinson, Yu.I. Manin and Y.A. Shechtman, Localization of the Virasoro and Neveu–Schwarz algebras, in preparation.
[24] L. Alvarez-Gaumé, C. Gomez and C. Reina, in preparation.
[25] N. Ishibashi, Y. Matsuo and H. Ooguri, University of Tokio preprint UT-499 (December 1986).
[26] M. Dugan and H. Sonoda, LBL preprint.

PHYSICAL REVIEW C VOLUME 24, NUMBER 4 OCTOBER 1981

Path integrals for the nuclear many-body problem

J. P. Blaizot*
Department of Physics, University of Illinois at Urbana-Champaign, Urbana, Illinois 61801

H. Orland
Service de Physique Theorique, Saclay, 02, 91190 Gif-sur-Yvette, France
(Received 22 December 1980)

We present a general method for constructing path intergrals for the nuclear many-body problem. This method uses continuous and overcomplete sets of vectors in the Hilbert space. The state labels play the role of classical coordinates which are quantized as bosons. The equations of motion for the classical coordinates are obtained by calculating the functional integral in the saddle point approximation. In the particular case where the overcomplete set considered is the set of all Slater determinants, the classical equations of motion are the time-dependent Hartree-Fock equations. The functional integral provides a way of requantizing these classical equations. This quantization involves boson degrees of freedom and is in some cases very similar to the method of boson expansion. It is shown that the functional integral formalism provides a unifying framework to describe various approaches to the nuclear many-body problem.

[NUCLEAR STRUCTURE Functional integrals on continuous overcomplete sets. Time-dependent Hartree and Hartree-Fock theories. Boson representations for fermion systems.]

I. INTRODUCTION

The present work examines the application of path integrals to the nuclear many-body problem. It has been motivated partly by the recent developments in the time-dependent mean field theories which have been applied to the description of large amplitude collective motion or heavy ions reactions.[1-4] One of such theories is the time-dependent Hartree-Fock theory hereafter referred to as TDHF. As is well known, the mean field approximation to the many-body problem leaves out definite effects which are usually interpreted in terms of quantum mechanics. For example the vibrational and rotational modes are not quantized in TDHF. To make connection with quantum spectra, a "requantization" is obviously required. The procedure followed for this requantization is often empirical and mostly unjustified. This originates from the fact that most of the derivations of the time-dependent mean field equations do not allow for a systematic expansion beyond the mean field level. An exception to this criticism is the boson expansion method.[5-7] In this method, the time-dependent mean field equations arise from the replacement of the boson operators by c numbers. Moreover, and this is a major point, it can be shown that the boson expansion retrieves exactly the original many-body problem of interacting fermions. In other words boson expansions provide an exact quantization scheme for the time-dependent mean field equations.

Path integrals provide other possible quantization schemes. The standard procedure is to calculate first the functional integral using the saddle-point approximation. This provides the "classical" approximation of the theory. Knowing the classical solution, one can get a semiclassical expression for the transition amplitudes and apply a generalization of the Wentzel-Kramers-Brillouin (WKB) method to quantize periodic motions. Further quantum effects are recovered by calculating the successive corrections to the saddle-point approximation. In the calculation of these corrections, boson degrees of freedom appear naturally. Actually, as we shall see, the boson expansion method is closely related to the quantization through path integrals.

Path integrals have been extensively used in many different areas of physics, in particular in quantum field theory and statistical mechanics. As is well known they are mathematically ill-defined objects and some of the manipulations one usually performs on ordinary integrals are not necessarily allowed, since they may lead to completely erroneous results. This is an important point which must always be kept in mind. To circumvent part of the mathematical difficulties associated with the definition of the functional integral, one usually identifies the functional integral with the formal perturbation expansion. All the manipulations on the integral which can be interpreted as manipulations on the perturbation expansion are then allowed. Other manipulations should be examined with great care. This does not imply that the use of the functional integral is restricted to perturbative approximations. It only guarantees that the properties of the integral are identical to those of the perturbation expansion.

In this paper, we discuss a general method for constructing path integrals for the nuclear many-body problem. This method, due to Klauder,[8,9] makes use of continuous and overcomplete sets of vectors of the Hilbert space. Among those, coherent states or generalized coherent states are particularly important sets. Thus the vectors of the Hilbert space are parametrized by a set of complex numbers which play the role of classical coordinates in a generalized phase space. According to the choice of the overcomplete set, different "classical approximations" are generated from the functional integral. In the present context one should remember that the word classical does not imply that something is small compared to $\hbar$. For example, choosing the set of all the vectors in the Hilbert space as the overcomplete set, one gets as the classical approximation to the Schrödinger equation, the Schrödinger equation itself. This is certainly an extreme case and most of the interesting approximations leave out genuine quantum effects. One of the purposes of the present work is to analyze these effects in the case of the time-dependent mean field approximations.

Functional integrals have been used recently in nuclear physics by several authors.[10–15] It will be seen that all the methods used by these authors are actually particular cases of the general method presented here which has much more flexibility.

This work is organized as follows. In Sec. II of this paper we discuss the properties of some overcomplete sets which are relevant to the discussion of the nuclear many-body problem. In Sec. III we construct the functional integral. Several specific forms of the functional integral are explicitly given. In Sec. IV we discuss the link between the path integral and the formal perturbation expansion. We analyze the difficulties associated with the quantization of the time-dependent Hartree-Fock theory.

In Sec. V, we analyze the successive corrections to the mean field approximation and discuss the physical nature of the quantum effects which are left out in this approximation. We also briefly discuss the connection between path integrals and the boson expansion methods. Section VI summarizes the conclusions. Let us finally mention that a partial account of this work can be found in Refs. 16–19.

II. OVERCOMPLETE SETS

Let $\{ \, |\psi(z)\rangle \, \}$ be an overcomplete set of vectors in the Hilbert space $\mathscr{H}$, depending upon a family of parameters which we denote collectively by z. We shall call the parameters z classical coordinates, and the space of variations of z the generalized phase space. The justification for this will appear in Sec. III. The overcompleteness means that any vector of $\mathscr{H}$ can be expanded on the states $|\psi(z)\rangle$ and that the states $|\psi(z)\rangle$ are linearly dependent. We assume that the parameters z vary continuously and that there exists a measure $\mu(z)$ on the space where z is defined, such that

$$\int d\mu(z)\,|\,z\,\rangle\langle\,z\,| = 1 \ , \tag{2.1}$$

where 1 denotes the unit operator in $\mathscr{H}$. In (2.1) as well as in the following, we use the abridged notation $|\,z\,\rangle$ for the state $|\psi(z)\rangle$. We give below examples of overcomplete sets which are useful in our discussion.

A. Coherent states of the harmonic oscillator

This is a well known and typical example of an overcomplete set. We recall briefly its properties. The coherent states are thus defined:

$$|\,z\,\rangle = e^{zc^\dagger}|\,0\rangle = \sum_n \frac{z^n}{\sqrt{n\,!}}\,|\,n\,\rangle \ , \tag{2.2}$$

where $|\,0\rangle$ is the oscillator ground state and $|\,n\,\rangle$ the state with n quanta. $c^\dagger$ is the raising operator. The closure relation is written in terms of the states

$|z\rangle$ using Bargman's measure[20]:

$$\int \frac{dz\,dz^*}{2\pi i} e^{-z^*z} |z\rangle\langle z| = 1 \ , \tag{2.3}$$

where

$$\frac{dz\,dz^*}{2\pi i} = \frac{d\mathrm{Re}z\, d\mathrm{Im}z}{\pi}$$

and the integration is carried over the whole complex plane. The overlap of two coherent states is given by

$$\langle z | z' \rangle = e^{z^* z'} \ . \tag{2.4}$$

The coherent state $|z\rangle$ is an eigenstate of the lowering operator c with eigenvalue z,

$$c\,|z\rangle = z\,|z\rangle \ . \tag{2.5a}$$

The matrix element of an operator $A(c^\dagger, c)$, in which the operators $c^\dagger$ and c are written in normal order (the $c^\dagger$ on the left of the c's) is therefore given by

$$\langle z | A(c^\dagger, c) | z' \rangle = e^{z^* z'} A(z^*, z') \ . \tag{2.5b}$$

B. Bosons coherent states

Let us consider the boson Fock space generated by the repeated action of the creation operators $c_\alpha^\dagger$ on the vacuum $|0\rangle$, the index α running over a complete set of single particle states. The operators $c_\alpha^\dagger$ and their Hermitian conjugates c_α obey boson commutation rules

$$[c_\alpha, c_\beta] = 0,\ [c_\alpha^\dagger, c_\beta^\dagger] = 0,\ [c_\alpha, c_\beta^\dagger] = \delta_{\alpha\beta} \ . \tag{2.6}$$

Boson coherent states are defined by

$$|Z\rangle = \exp\left[\sum_\alpha z_\alpha c_\alpha^\dagger\right] |0\rangle \ . \tag{2.7}$$

The properties of these coherent states generalize those of the preceding section. The closure relation in Fock space can be written

$$\int \prod_\alpha \frac{dz_\alpha^*\, dz_\alpha}{2\pi i} \exp\left[-\sum_\alpha z_\alpha^* z_\alpha\right] |Z\rangle\langle Z| = 1 \ . \tag{2.8}$$

The state (2.7) is an eigenstate of the destruction operator c_α with the eigenvalue z_α

$$c_\alpha |Z\rangle = z_\alpha |Z\rangle \ . \tag{2.9}$$

The overlap of two coherent states (2.7) is,

$$\langle Z | Z' \rangle = \exp\left[\sum_\alpha z_\alpha^* z_\alpha'\right] \ . \tag{2.10}$$

Therefore the matrix element of a normal ordered operator $A(c^\dagger, c)$ is

$$\langle Z | A(c^\dagger, c) | Z' \rangle = A(Z^*, Z') \exp\left[\sum_\alpha z_\alpha^* z_\alpha'\right] \ . \tag{2.11}$$

C. Fermions coherent states

The coherent states of fermions are defined by analogy with the coherent states of boson.[21] Let $a_\alpha^\dagger$, a_α be the fermion creation and destruction operators. They satisfy the anticommutation relations

$$[a_\alpha, a_\beta]_+ = 0,\ [a_\alpha^\dagger, a_\beta^\dagger]_+ = 0,\ [a_\alpha, a_\beta^\dagger]_+ = \delta_{\alpha\beta} \ . \tag{2.12}$$

Let us consider the state

$$|Z\rangle = \exp\left[\sum_\alpha z_\alpha a_\alpha^\dagger\right] |0\rangle \ , \tag{2.13}$$

where $|0\rangle$ is the vacuum of the fermion Fock space. Since $a_\alpha^2 = 0$, $|z\rangle$ can be an eigenstate of a_α only if $z_\alpha^2 = 0$. This can be realized using anticommuting Grassman variables. The rules for calculating with these objects have been widely discussed in the literature. All the formulas of Sec. (II B) hold for the fermion coherent states, provided z is understood as a Grassman variable. Note that fermion coherent states do not belong to the Fock space. However, they allow for a decomposition of the identity in Fock space

$$\int \prod_\alpha dz_\alpha^*\, dz_\alpha \exp\left[-\sum_\alpha z_\alpha^* z_\alpha\right] |Z\rangle\langle Z| = 1 \ . \tag{2.14}$$

D. Boson representation for fermions

Fermion states are usually represented by vectors of a fermion Fock space, constructed from a complete set of single particle states $\{\, |\alpha\rangle\, \}$. It is also possible to represent fermion states as vectors belonging to a subspace (called the physical subspace) of a large boson Fock space. We consider in this section the representation which has been described in Ref. 17, and which is a generalization of that introduced in Ref. 7.

To construct the boson image of an N-fermion state, we consider a large space G, product of N boson Fock spaces B_i, associated with each of the particles i

$$G = \mathscr{B}_1 \otimes \mathscr{B}_2 \otimes \cdots \otimes \mathscr{B}_N \ . \tag{2.15}$$

We call $C_i^\dagger(\alpha)$, $C_i(\alpha)$ the creation and annihilation operators acting in $\mathscr{B}_i$. These operators satisfy the commutation relations

$$\begin{aligned}[][C_i(\alpha),C_j^\dagger(\beta)] &= \delta_{ij}\delta_{\alpha\beta} \ , \\ [C_i(\alpha),C_j(\beta)] &= [C_i^\dagger(\alpha),C_j^\dagger(\beta)] = 0 \ .\end{aligned} \tag{2.16}$$

The following states

$$|\psi\rangle = \sum_P (-)^P C_1^\dagger(\alpha_{P_1})\ldots C_N^\dagger(\alpha_{P_N})\,|0\rangle_B \ , \tag{2.17}$$

where $\sum_P$ is a sum over all the possible permutations of the indices $\alpha_1\ldots\alpha_N$, and $|0\rangle_B$ is the boson vacuum, are in one-to-one correspondence with the N-fermion states of the fermion Fock space. They span the physical subspace. They are characterized by two properties. There is one and only one particle per subspace $\mathscr{B}_i$,

$$\sum_\alpha C_i^\dagger(\alpha)C_i(\alpha)\,|\psi\rangle = |\psi\rangle \quad (i=1,\ldots,N) \ . \tag{2.18}$$

The state changes sign in any transposition of the particle indices. The operator which realizes such a transposition is

$$\begin{aligned} P_{ij} &= \sum_{\alpha\beta} C_i^\dagger(\alpha)C_j^\dagger(\beta)C_j(\alpha)C_i(\beta) \\ &= -\sum_\alpha C_i^\dagger(\alpha)C_i(\alpha) \\ &\quad + \sum_\alpha C_i^\dagger(\alpha)C_j(\alpha)\sum_\beta C_j^\dagger(\beta)C_i(\beta) \ . \end{aligned} \tag{2.19}$$

In view of Eqs. (2.18) and (2.19), the condition

$$P_{ij}\,|\psi\rangle = -\,|\psi\rangle \tag{2.20}$$

is equivalent to the condition

$$\sum_\alpha C_i^\dagger(\alpha)C_j(\alpha)\,|\psi\rangle = 0 \quad (i\neq j) \ . \tag{2.21}$$

Thus the states of the physical subspace are characterized by the following set of equations

$$\sum_\alpha C_i^\dagger(\alpha)C_j(\alpha)\,|\psi\rangle = \delta_{ij}\,|\psi\rangle \begin{bmatrix} i=1,\ldots,N \\ j=1,\ldots,N \end{bmatrix} \ . \tag{2.22}$$

The operators $d_{ij} = \sum_\alpha C_i^\dagger(\alpha)C_j(\alpha)$ satisfy the U(N) algebra,

$$[d_{ij},d_{kl}] = d_{il}\delta_{jk} - d_{kj}\delta_{li} \ . \tag{2.23}$$

They are the generators of the transformation which mixes the various components of a state vector in G. These operators can be used to construct explicitly a projector onto the physical subspace

$$P = \int \prod_{i,j} dA_{ij}\, e^{-i\,\mathrm{Tr}A} \exp\left[i \sum_{ij} A_{ij} d_{ij} \right] \ , \tag{2.24}$$

where the matrix A may be chosen to be a real matrix and the integration carried from $-\pi$ to π. Other forms are of course possible for P. The Hamiltonian in G takes the following form

$$\begin{aligned} H_B &= \sum_{i=1}^N \sum_{\alpha\beta} T_{\alpha\beta} C_i^\dagger(\alpha)C_i(\beta) \\ &\quad + \tfrac{1}{2}\sum_{i,j}\sum_{\alpha\beta\gamma\delta} (\alpha\beta\,|\,V\,|\,\gamma\delta) C_i^\dagger(\alpha)C_j^\dagger(\beta)C_j(\delta)C_i(\gamma) \ , \end{aligned} \tag{2.25}$$

where $(\alpha\beta\,|\,V\,|\,\gamma\delta)$ denotes the nonantisymmetrized matrix element of the two-body interaction V. It is easily verified that H_B has the same matrix element within the physical subspace as the Hamiltonian

$$H = \sum_{\alpha\beta} T_{\alpha\beta} a_\alpha^\dagger a_\beta + \tfrac{1}{2}\sum_{\alpha\beta\gamma\delta} (\alpha\beta\,|\,V\,|\,\gamma\delta) a_\alpha^\dagger a_\beta^\dagger a_\delta a_\gamma \tag{2.26}$$

has in the Fermion Fock space. It is also easily checked that H_B commutes with the projector P given by (2.24), that is, H_B has no matrix elements between physical and unphysical states. This follows from the fact that physical and unphysical states belong to different representations of the unitary group, and H_B commutes with the generators d_{ij}.

The closure relation in G is conveniently written

with the help of the coherent states:

$$|Z\rangle = \exp\left[\sum_{k=1}^{N}\sum_{\alpha} Z_k(\alpha)C_k^{\dagger}(\alpha)\right]|0\rangle_B \ , \tag{2.27a}$$

where the index α runs over all the single particle states and $|0\rangle_B$ is the boson vacuum. We shall also use continuous representation with the notation

$$|\varphi\rangle = \exp\left[\sum_{k=1}^{N}\int dx\, \varphi_k(x)\psi_k^{\dagger}(x)\right]|0\rangle_B \ . \tag{2.27b}$$

The closure relation in G reads [see Eq. (2.8)]

$$\begin{aligned} 1_G &= \int \prod_{k=1}^{N}\prod_{\alpha}\frac{dZ_k^*(\alpha)dZ_k(\alpha)}{2\pi i} \times \exp\left[-\sum_k (Z_k\,|\,Z_k)\right]|Z\rangle\langle Z| \\ &= \int \prod_{k=1}^{N}\prod_{x}\frac{d\varphi_k^*(x)d\varphi_k(x)}{2\pi i} \times \exp\left[-\sum_k (\varphi_k\,|\,\varphi_k)\right]|\varphi\rangle\langle\varphi| \ , \end{aligned} \tag{2.28}$$

where we have used the abridged notations

$$(Z_k\,|\,Z_k) = \sum_{\alpha} Z_k^*(\alpha)Z_k(\alpha) \ , \tag{2.29}$$

$$(\varphi_k\,|\,\varphi_k) = \int dx\, \varphi_k^*(x)\varphi_k(x) \ .$$

The closure relation (2.28) induces a closure relation in the physical subspace of G, obtained by applying the projector P onto the physical subspace on both sides of (2.28),

$$P = \int \prod_{k=1}^{N}\prod_{\alpha}\frac{dZ_k^*(\alpha)dZ_k(\alpha)}{2\pi i} \times \exp\left[-\sum_k (Z_k\,|\,Z_k)\right]P|Z\rangle\langle Z|P \ . \tag{2.30}$$

Now we note that

$$P|Z\rangle = \det[Z_k(\alpha_i)]C_1^{\dagger}(\alpha_1)\cdots C_N^{\dagger}(\alpha_N)|0\rangle \ . \tag{2.31}$$

Thus, in a scale transformation

$$Z(\alpha) = \Lambda Z'(\alpha), \tag{2.32}$$

$P|Z\rangle$ scales as $\det\Lambda$. This property can be verified using the explicit form of P given by Eq. (2.26). One has indeed

$$P|\Lambda Z\rangle = \int d\mu(Z')|Z'\rangle\langle Z'|P|\Lambda Z\rangle \tag{2.33}$$

and

$$\langle Z'|P|\Lambda Z\rangle = \int dA e^{-i\,\mathrm{Tr}A}\prod_{\alpha} e^{Z'^{\dagger}(e^{iA}\Lambda)Z} \ , \tag{2.34}$$

where we have used the property

$$e^{C_k^{\dagger}A_{kl}C_l} = :e^{(e^A-1)_{kl}C_k^{\dagger}C_l}: \ . \tag{2.35}$$

Changing the integration variable A into A-$i\ln\Lambda$ and using the property $\det\Lambda = \exp\mathrm{tr}\ln\Lambda$, one obtains the desired equation

$$P|\Lambda Z\rangle = (\det\Lambda)P|Z\rangle \ . \tag{2.36}$$

Equation (2.30) may then be written as follows

$$P = \int \prod_{k=1}^{N}\prod_{\alpha}\frac{dZ_k^*(\alpha)dZ_k(\alpha)}{2\pi i} \times \int d\Lambda \prod_{k,l}\delta[\Lambda_{kl}-(Z_k^*\,|\,Z_l)] \times e^{-\mathrm{Tr}\Lambda}P|Z\rangle\langle Z|P \ , \tag{2.37}$$

where the integration over Λ runs over all the positive definite Hermitian matrices. Making the change of variable

$$Z(\alpha) = \Lambda^{1/2}Z'(\alpha), \quad Z^*(\alpha) = Z'^*(\alpha)\tilde{\Lambda}^{1/2} \ , \tag{2.38}$$

where $\tilde{\Lambda}$ denotes the transpose of the matrix Λ one gets

$$P = \left[\int d\Lambda(\det\Lambda)^n e^{-\mathrm{Tr}\Lambda}\right] \times \int \prod_{k=1}^{N}\prod_{\alpha}\frac{dZ_k'^*(\alpha)dZ_k'(\alpha)}{2\pi i} \times \prod_{l=1}^{N}\delta[(Z_k'\,|\,Z_l')-\delta_{kl}] \times P|Z'\rangle\langle Z'|P \ , \tag{2.39}$$

where n is the total number of single particle states. The integral over Λ is just a normalization constant

We thus arrive at the result

$$P=\mathcal{N}\int\prod_{k=1}^{N}\prod_{\alpha}\frac{dZ_k^*(\alpha)dZ_k(\alpha)}{2\pi i}\prod_{l=1}^{N}\delta[(Z_k\,|\,Z_l)-\delta_{kl}]P\,|\,Z\rangle\langle Z\,|\,P\ . \tag{2.40}$$

This result will be rederived in a different way in the next section.

E. Independent particle states

In this section we consider the overcomplete set formed by all the Slater determinants describing systems with a fixed number of particles N. This set can be parametrized in many ways. We give below some parametrizations which are useful in practice, together with the corresponding closure relations. The derivations are reported in the Appendix.

Let $|\phi_0\rangle$ be a particular Slater determinant. $|\phi_0\rangle$ is composed of N orthonormalized single particle orbitals, which we call "hole" states,

$$|\phi_0\rangle=\prod_h a_h^\dagger\,|0\rangle\ . \tag{2.41}$$

We call "particle" states the states such that

$$a_p\,|\phi_0\rangle=0\ . \tag{2.42}$$

We assume that the number of single particle states is finite. We call n_h the number of hole states and n_p the number of particle states. It is known that any Slater determinant nonorthogonal to $|\phi_0\rangle$ can be written[22]

$$|Z\rangle=\exp\sum_{ph}(Z_{ph}a_p^\dagger a_h)\,|\phi_0\rangle\ . \tag{2.43}$$

The states (2.43) are not normalized. The overlap between two of them is

$$\langle Z\,|\,Z'\rangle=\det(1+Z^\dagger Z')\ , \tag{2.44}$$

where Z denotes the complex $n_p\times n_h$ matrix made out of the Z_{ph} amplitudes. In terms of the states (2.43) the closure relation in the Hilbert space of N-fermions states takes the following form (see the Appendix),

$$\int\prod_{ph}\frac{dZ_{ph}^*\,dZ_{ph}}{2\pi i}[\det(1+Z^\dagger Z)]^{-(n_p+n_h+1)}\times|\,Z\rangle\langle Z\,|=1\ . \tag{2.45}$$

Using Wick's theorem one can express the matrix elements of any operator between two states Z and Z′ in terms of the one-body density matrix defined thus

$$\rho_{\beta\alpha}(Z^\dagger,Z')=\frac{\langle Z\,|\,a_\alpha^\dagger a_\beta\,|\,Z'\rangle}{\langle Z\,|\,Z'\rangle}\ . \tag{2.46}$$

Thus, for example, one has

$$\frac{\langle Z\,|\,a_\alpha^\dagger a_\beta^\dagger a_\gamma a_\delta\,|\,Z'\rangle}{\langle Z\,|\,Z'\rangle}=\rho_{\delta\alpha}(Z^\dagger,Z')\rho_{\gamma\beta}(Z^\dagger,Z')-\rho_{\delta\beta}(Z^\dagger,Z')\rho_{\gamma\alpha}(Z^\dagger,Z')\ . \tag{2.47}$$

The matrix elements $\rho_{\alpha\beta}(Z^\dagger,Z')$ have the following expressions[6]

$$\begin{aligned}\rho_{ph}&=[Z'(1+Z^\dagger Z')^{-1}]_{ph}\ ,\\ \rho_{hp}&=[(1+Z^\dagger Z')^{-1}Z^\dagger]_{hp}\ ,\\ \rho_{pp'}&=[Z'(1+Z^\dagger Z')^{-1}Z^\dagger]_{pp'}\ ,\\ \rho_{hh'}&=[(1+Z^\dagger Z')^{-1}]_{hh'}\ .\end{aligned} \tag{2.48}$$

Performing the change of variable,

$$\beta_{ph}=Z_{ph'}[(1+Z^\dagger Z)^{-1/2}]_{h'h}\ , \tag{2.49}$$

one can simplify (2.45). In the variables β, the closure relation takes the form

$$\int\prod_{ph}\frac{d\beta_{ph}^*\,d\beta_{ph}}{2\pi i}\,|\,\widetilde{\beta}\rangle\langle\widetilde{\beta}\,|=1\ , \tag{2.50}$$

where the states $|\widetilde{\beta}\rangle$ are obtained from (2.43) by expressing Z in terms of β and normalizing. This parametrization has been used in works on boson expansions.[5] The density matrix elements have the following expressions in terms of the β's

$$\begin{aligned}\rho_{\alpha\beta}&=\langle\widetilde{\beta}\,|\,a_\beta^\dagger a_\alpha\,|\,\widetilde{\beta}\rangle\ ,\\ \rho_{ph}&=[\beta(1-\beta^\dagger\beta)^{1/2}]_{ph},\ \ \rho_{hp}=\rho_{ph}^*\ ,\\ \rho_{pp'}&=(\beta\beta^\dagger)_{pp'}\ ,\\ \rho_{hh'}&=\delta_{hh'}-(\beta^\dagger\beta)_{hh'}\ .\end{aligned} \tag{2.51}$$

Note that the measure in (2.50) is extremely simple. This results from the fact that the β_{ph} are the coefficients of the unitary transformation which carries $|\phi_0\rangle$ into the state $|\widetilde{\beta}\rangle$. The domain of integration is complicated, however, since the matrix β must satisfy

$$1-\beta^{\dagger}\beta\rangle 0 \ , \tag{2.52}$$

while in the parametrization (2.43) the parameters Z_{ph} vary over the whole complex plane. Note also that the expression of the density matrix (2.48) is formally the same, whether Z' differs from Z or not. The density matrix (2.51) has a simple expression only if the bra and the ket in (2.51) are Hermitian conjugates of one another.

Using a further change of variable, one arrives at a parametrization in which the coordinates are the single particle wave functions which build up the determinant. The closure relation can be written (see Appendix), with respect to a normalization constant,

$$\int \prod_{k=1}^{N}\prod_{x}\frac{d\varphi_k^*(x)d\varphi_k(x)}{2\pi i}\prod_{l=1}^{N}\delta[(\varphi_k\,|\,\varphi_l)-\delta_{kl}] \times |\varphi\rangle\langle\varphi| \sim 1 \ , \tag{2.53}$$

which is identical to the relation (2.40).

III. PATH INTEGRALS

In this section we give functional integral representations for the matrix elements of the evolution operator e^{-iHt} between some initial state $|Z_i\rangle$ and some final state $\langle Z_f|$,

$$\langle Z_f|e^{-iH(t_f-t_i)}|Z_i\rangle \ , \tag{3.1}$$

where $|Z_f\rangle$ and $|Z_i\rangle$ belong to the class of states described in the previous section. The general procedure for constructing path integrals is quite standard. First one factorizes the operator $e^{-iH(t_f-t_i)}$ into N terms $e^{-i\epsilon H}$, where $\epsilon=(t_f-t_i)/N$. Then one inserts the closure relation (2.1) between each of the factors and gets

$$\langle Z_f|e^{-iH(t_f-t_i)}|Z_i\rangle=\int\prod_{k=1}^{N}d\mu(Z_k)\langle Z_f|Z_N\rangle\langle Z_N|e^{-i\epsilon H}|Z_{n-1}\rangle\cdots\langle Z_{k+1}|e^{-i\epsilon H}|Z_k\rangle\cdots\langle Z_1|e^{-i\epsilon H}|Z_i\rangle . \tag{3.2}$$

In the limit $N\to\infty$, $\epsilon\to 0$ and

$$\frac{\langle Z_{k+1}|e^{-i\epsilon H}|Z_k\rangle}{\langle Z_{k+1}|Z_k\rangle}=\exp\left[-i\epsilon\frac{\langle Z_{k+1}|H|Z_k\rangle}{\langle Z_{k+1}|Z_k\rangle}\right]+0(\epsilon^2) \ . \tag{3.3}$$

One then arrives at

$$\langle Z_f|e^{-iH(t_f-t_i)}|Z_i\rangle=\lim_{N\to\infty}\int\prod_{k=1}^{N}d\mu(Z_k)\prod_{k=0}^{N}\langle Z_{k+1}|Z_k\rangle\exp\left[-i\epsilon\sum_{k=0}^{N-1}\frac{\langle Z_{k+1}|H|Z_k\rangle}{\langle Z_{k+1}|Z_k\rangle}\right] , \tag{3.4}$$

where $|Z_0\rangle\equiv|Z_i\rangle$ and $\langle Z_{N+1}|\equiv\langle Z_f|$. One defines

$$|\delta Z_{k+1}\rangle=|Z_{k+1}\rangle-|Z_k\rangle \ , \tag{3.5}$$

so that (3.3) may be rewritten as follows:

$$\langle Z_f|e^{-iH(t_f-t_i)}|Z_i\rangle=\lim_{N\to\infty}\int\prod_{k=1}^{N}[d\mu(Z_k)\langle Z_k|Z_k\rangle]\langle Z_f|Z_N\rangle \times\exp\left\{\sum_{k=1}^{N}\left[\ln\left[1-\frac{\langle Z_k|\delta Z_k\rangle}{\langle Z_k|Z_k\rangle}\right]-i\epsilon\frac{\langle Z_k|H|Z_{k-1}\rangle}{\langle Z_k|Z_{k-1}\rangle}\right]\right\} . \tag{3.6}$$

A further simplification is achieved if one admits that the major contribution to the integral comes from those "paths" for which $|\delta Z_k\rangle$ is of order ϵ for almost all k, that is, assuming that only piecewise continuous paths contribute in (3.6). Setting $d|Z\rangle/dt=|\delta Z\rangle/\epsilon$ and keeping only lowest order terms in ϵ, one finally ends up with the continuous expression

$$\langle Z_f | e^{-iH(t_f-t_i)} | Z_i \rangle = \int_{|Z(t_i)\rangle = |Z_i\rangle}^{\langle Z(t_f)| = \langle Z_f|} \mathscr{D}(Z^*(t),Z(t)) e^{iS[Z^*,Z]} \;, \tag{3.7}$$

where the action S is given by

$$S[Z^*,Z] = \int_{t_i}^{t_f} dt \frac{\langle Z(t) | i\partial_t - H | Z(t)\rangle}{\langle Z(t) | Z(t)\rangle} - i \ln\langle Z_f | Z(t_f)\rangle \;, \tag{3.8}$$

and the integration measure is

$$\mathscr{D}(Z^*,Z) = \prod_{t_i<t<t_f} d\mu[Z^*(t),Z(t)] \langle Z(t) | Z(t)\rangle \;. \tag{3.9}$$

The integration in (3.7) is carried over all the paths $\langle Z(t)|$ and $|Z(t)\rangle$ in the overcomplete set, subject to the boundary conditions

$$|Z(t_i)\rangle = |Z_i\rangle, \quad \langle Z(t_f)| = \langle Z_f| \;. \tag{3.10}$$

Note that in this formulation $\langle Z(t)|$ and $|Z(t)\rangle$ have to be considered as independent variables, e.g., there are no constraints on $|Z(t_f)\rangle$ and $\langle Z(t_i)|$. It is important to keep this point in mind when applying the saddle-point approximation. (See Sec. V and Refs. 9 and 23.)

The action (3.8) may be given a more symmetrical form with respect to the boundary conditions by an integration by parts

$$S[Z^*,Z] = \int_{t_i}^{t_f} dt \frac{\langle Z(t) | i\overleftrightarrow{\partial}_t - H | Z(t)\rangle}{\langle Z(t) | Z(t)\rangle} - \frac{i}{2} \ln\langle Z(t_i) | Z_i\rangle\langle Z_f | Z(t_f)\rangle \;, \tag{3.8$'$}$$

where we have used the notation

$$\langle Z(t) | \overleftrightarrow{\partial}_t | Z(t)\rangle = \tfrac{1}{2}\left[\left\langle Z(t) \middle| \frac{dZ}{dt}\right\rangle - \left\langle \frac{dZ}{dt} \middle| Z\right\rangle\right] \;.$$

It is worth emphasizing that the expression (3.7) has gotten no rigorous mathematical meaning from its derivation. This is known to lead to difficulties when some "unallowed" manipulations are performed on the functional integral. An example of such difficulties will be encountered in Sec. IV.

We examine now various explicit forms of the functional integral (3.8) obtained with some of the overcomplete sets described in Sec. II.

Let us first consider the form of the functional integral obtained when one uses coherent states as an overcomplete set. Due to the special form of the overlap (2.10), the integration measure simplifies into

$$\mathscr{D}(Z^*,Z) = \prod_t \frac{dZ^*(t)dZ(t)}{2\pi i} \;. \tag{3.11}$$

The action reads

$$S[Z^*,Z] = \int_{t_i}^{t_f} dt \left\{ \frac{i}{2}(Z^*\dot{Z} - \dot{Z}^*Z) - H(Z^*,Z) - \frac{i}{2}[Z_f^*Z(t_f) + Z^*(t_i)Z_i] \right\} , \tag{3.12}$$

where $H(Z^*,Z)$ is the normal form of the second quantized Hamiltonian, with the creation and annihilation operators $a^\dagger$ and a replaced by Z^* and Z, respectively. The formulas above hold for bosons and fermions. In the latter case, the variable Z has to be understood as a Grassman variable. This formulation has been used in Refs. 11 and 13. The formulas (3.11) and (3.12) hold also for the coherent states (2.27) described in Sec. II D. However, in this latter case, special attention must be given to the boundary conditions. Indeed, one is not interested in the matrix element of the evolution operator between two coherent states (2.27), but rather in this matrix element between two physical states. Let $|\phi\rangle$ and $\langle\psi|$ be two coherent states (2.27) and $|\Phi\rangle$ and $\langle\Psi|$ the Slater determinants built from the same single particle orbitals; that is,

$$|\Phi\rangle = \sum_P (-)^r |\phi_{P_1}\phi_{P_2}\cdots\phi_{P_N}\rangle = P|\phi\rangle \;,$$
$$\langle\Psi| = \sum_P (-)^r \langle\psi_{P_1}\psi_{P_2}\cdots\psi_{P_N}| = (\psi|P \;, \tag{3.13}$$

where P denotes the projector on the physical subspace (see Sec. II D). We are interested in the matrix elements

$$\langle\Psi | e^{-iH(t_f-t_i)} | \Phi\rangle \;, \tag{3.14}$$

which can be written

$$\int d\mu(\phi^*,\phi) d\mu(\psi^*,\psi) \times \langle\Psi|\psi)(\psi| e^{-iH_B(t_f-t_i)} |\phi)(\phi|\Phi\rangle \;, \tag{3.15}$$

where H_B is the boson image of H, given by Eq. (2.25). $(\psi| e^{-iH_B(t_f-t_i)} |\phi)$ may be represented by the functional integral

$$\int \mathscr{D}(\varphi^*,\varphi)e^{iS[\varphi^*,\varphi]} ,$$

where the measure and the action are given, respectively, by (3.11) and (3.12), except for an obvious change of notation. The overlaps $\langle \Psi \mid \psi)$ and $(\phi \mid \Phi\rangle$ determine the boundary condition

$$\varphi_k(t_i) = \phi_{P(k)},\ \varphi_k^*(t_f) = \psi_{P'(k)}^* ,$$

where $p(k)$ and $p'(k)$ denote two permutations of the particle indices $1,2,\ldots,N$. The expression (3.15) thus contains an obvious summation over all such permutations. An alternative way of calculating (3.14) is to use the explicit form (2.24) for the projector P onto the physical subspace. Furthermore, since P commutes with H, it needs to be inserted only once. One then arrives at the expression

$$\begin{aligned}\langle \Psi \mid e^{-iH(t_f-t_i)} \mid \Phi\rangle &= (\psi \mid P e^{-iH_B(t_f-t_i)} \mid \phi) \\ &= \int dA e^{-i\,\mathrm{Tr}A}(\psi \mid e^{-i(t_f-t_i)\{H_B-[A/(t_f-t_i)]\}} \mid \phi) ,\end{aligned} \quad (3.16)$$

where A is the following operator

$$A = \sum_{kl}\sum_{\alpha} C_k^\dagger(\alpha)A_{kl}C_l(\alpha) . \quad (3.17)$$

The matrix element $(\psi \mid e^{-it[H_B-(A/t)]} \mid \varphi)$ has the following functional integral representation

$$(\psi \mid e^{-it[H_B-(A/t)]} \mid \phi) = \int \mathscr{D}(\varphi^*,\varphi)e^{iS[\varphi^*,\varphi;A]} \quad (3.18)$$

with

$$S[\varphi^*,\varphi;A] = S[\varphi^*,\varphi] - \frac{i}{t}\sum_{kl} A_{kl}(\varphi_k \mid \varphi_l) , \quad (3.19)$$

where $S[\varphi^*,\varphi]$ is the action (3.12), except for an obvious change of notations. The expression (3.18) describes the evolution of a system of bosons subject to special constraints represented by the "external" field A. When Fourier transformed [see Eq. (3.16)] with respect to A this expression retrieves the original fermion dynamics.

In the two formulations above [cf. Eqs. (3.15) and (3.18)], the paths are allowed to lie outside the physical subspace; the projection onto the physical subspace is done by the overall integral over A in the case of (3.18), or by the summation over specific boundary conditions in the case of (3.15). Now it is possible to constrain the path at each time t so that it lies entirely within the physical subspace. This is achieved by inserting the projector P at each time step in the construction of the path integral. One then arrives at the following expression

$$\langle \Psi \mid e^{-iH(t_f-t_i)} \mid \Phi\rangle = \int \mathscr{D}(\varphi^*,\varphi)\prod_t \langle \varphi \mid P \mid \varphi\rangle e^{iS[\varphi^*,\varphi]} \quad (3.20)$$

with

$$S[\varphi^*,\varphi] = \int_{t_i}^{t_f} \frac{\langle \varphi \mid (i\overleftrightarrow{\partial}_t - H_B)P \mid \varphi\rangle}{\langle \varphi \mid P \mid \varphi\rangle} - \frac{i}{2}\mathrm{Tr}\ln\langle \Psi \mid \varphi(t_f)\rangle\langle \varphi(t_i) \mid \Phi\rangle . \quad (3.21)$$

Now let us perform the same change of variable as in Sec. II D, namely, $\varphi = \Lambda^{1/2}\varphi'$ [see Eq. (2.38)]. In this change of variable, $\langle \varphi \mid P \mid \varphi\rangle$ scales as $\det\Lambda$, as $\langle \varphi \mid HP \mid \varphi\rangle$ and $\langle \varphi \mid \overleftrightarrow{\partial}_t P \mid \varphi\rangle$ do. In this later case, it is easily verified that the possible time derivative of Λ cancel. Thus Eq. (3.21) can be rewritten as follows (with respect to an overall constant, namely, the integral over Λ; see Sec. II D)

$$\langle \Psi \mid e^{-iH(t_f-t_i)} \mid \Phi\rangle = \int \mathscr{D}(\varphi^*,\varphi)\delta[(\varphi_k \mid \varphi_l) - \delta_{kl}]e^{iS[\varphi^*,\varphi]} , \quad (3.22)$$

where, ignoring the boundary term

$$S[\varphi^*,\varphi] = \sum_k (\varphi_k \mid i\overleftrightarrow{\partial}_t \mid \varphi_k) - \sum_k (\varphi_k \mid T \mid \varphi_k) - \frac{1}{2}\sum_{kl} \langle \varphi_k\varphi_l \mid V \mid \varphi_k\varphi_l\rangle , \quad (3.23)$$

where now the *antisymmetrized* matrix element of the two-body interaction occurs. In contrast, the action (3.19) involves only the direct matrix elements of the two-body interaction. One recognizes in the expression (3.22) the functional integral one would have obtained working directly with Slater determinants and the measure (2.53).

The functional integrals (3.18) and (3.22) are *a priori* equivalent, i.e., they correspond to the same Schrödinger equation. However, we shall see in the next section that they have actually very different structures. Let us remark here that they differ essentially by the way the constraints are handled. In (3.18) the constraints are imposed in a global way while in (3.22) they are imposed locally (in time). One may also notice that the constraints

$(\varphi_k \mid \varphi_l)$ are constants of motion for the classical equations of motion. This situation is very much reminiscent of what happens in gauge theory; here the gauge group is the group U(N) which mixes the single particle orbitals. We shall not further develop this point of view here. There is still another way to take care of the constraints, namely, choose a system of coordinates in which the constraints are automatically satisfied. This is realized by the parametrizations (2.43) and (2.49) of Slater determinants. We give below the explicit form of the functional integrals in these two representations.

For the representation (2.43) the integration measure reads

$$\mathscr{D}(Z^\dagger,Z) = \prod_t \prod_{ph} \frac{dZ^*_{ph}(t)\,dZ_{ph}(t)}{2\pi i} \times \det[1+Z^\dagger(t)Z(t)]^{-(n_p+n_h)} , \quad (3.24)$$

and the action is

$$S[Z^\dagger,Z] = \int_{t_i}^{t_f} dt \frac{i}{2}\mathrm{Tr}(1+Z^\dagger Z)^{-1}(Z^\dagger\dot{Z} - \dot{Z}^\dagger Z) - E[\rho(Z^\dagger,Z)] - \frac{1}{2}\mathrm{Tr}\ln[1+Z_f^\dagger Z(t_f)][1+Z^\dagger(t_i)Z_i] , \quad (3.25)$$

where ρ is the density matrix (2.48) and $E[\rho]$ is the HF energy calculated with this density matrix

$$E[\rho] = \sum_{\alpha\beta} T_{\alpha\beta}\rho_{\beta\alpha} + \tfrac{1}{2}\sum_{\alpha\beta\gamma\delta} \langle\alpha\beta \mid V \mid \gamma\delta\rangle \rho_{\gamma\alpha}\rho_{\delta\beta} \quad (3.26)$$

and $\langle\alpha\beta \mid V \mid \gamma\delta\rangle$ is the antisymmetrized matrix element of the two-body interaction V. For later purposes, we write $E[\rho]$ using the following matrix notation

$$E[\rho] = T\cdot\rho + \tfrac{1}{2}\rho\cdot V\cdot\rho , \quad (3.27)$$

where V is the (symmetrical) matrix,

$$V_{\alpha\gamma,\beta\delta} = \langle\alpha\beta \mid V \mid \gamma\delta\rangle = V_{\beta\delta,\alpha\gamma} . \quad (3.28)$$

The action (3.25) is the one used in Ref. 14, except for the boundary term.

For the representation (2.49) the measure is simply

$$\mathscr{D}(\beta^\dagger,\beta) = \prod_t \prod_{ph} \frac{d\beta^*_{ph}(t)\,d\beta_{ph}(t)}{2\pi i} \quad (3.29)$$

and the action reads

$$S[\beta^\dagger,\beta] = \int_{t_i}^{t_f} dt \left\{ \frac{i}{2}(\beta^\dagger\dot{\beta} - \dot{\beta}^\dagger\beta) - E[\rho(\beta^\dagger,\beta)] \right\} - \tfrac{1}{2}\mathrm{tr}\ln(1-\beta_f^\dagger\beta(t_f))(1-\beta^\dagger(t_i)\beta_i) \quad (3.30)$$

Almost all the functional integrals described in this section describe boson theories with particular constraints. Indeed, the elementary fields, or coordinates, are represented by complex numbers which are quantized as boson. This boson structure has been explicitly analyzed in Sec. (II D) for the representation (3.18). The representations underlying (3.25) and (3.30) are familiar in nuclear physics for their intimate connection with perturbative boson expansions.[24,7] The method we have used to generate path integrals clearly generate at the same time boson expansions, or more precisely boson representations. In these representations, the bosons are just the quantum version of the classical parameters which label the quantum states of the overcomplete set used in the functional integral. The role of the bosons in the functional integrals, and in particular of coherent state of bosons, will be seen in the next sections.

IV. PERTURBATION EXPANSION

In this section we compare the structure of the path integrals described in the preceding section with that of the formal perturbation expansion. Let us consider the expression

$$\langle Z_f | e^{-iH(t_f-t_i)} | Z_i\rangle = \int_{\substack{Z^*(t_f)=Z_f^* \\ Z(t_i)=Z_i}} \mathscr{D}(Z^*,Z)\exp\left\{ i\int_{t_i}^{t_f} L[Z^*,Z]dt + \ln\langle Z_f \mid Z(t_f)\rangle \right\} , \quad (4.1)$$

where

$$L[Z^*,Z]=\frac{\langle Z\,|\,i\partial_t-H\,|\,Z\rangle}{\langle Z\,|\,Z\rangle} \tag{4.2}$$

We can rewrite L as follows

$$\begin{aligned}L[Z^*,Z]&=\frac{\langle Z\,|\,i\partial_t-H_0|\,Z\rangle}{\langle Z\,|\,Z\rangle}-\frac{\langle Z\,|\,V\,|\,Z\rangle}{\langle Z\,|\,Z\rangle}\\ &\equiv L_0[Z^*,Z]-\frac{\langle Z\,|\,V\,|\,Z\rangle}{\langle Z\,|\,Z\rangle}\end{aligned}$$

and expand $\exp[-i\int\langle Z\,|\,V\,|\,Z\rangle/\langle Z\,|\,Z\rangle]$ in (4.1) in powers of V. One gets

$$\begin{aligned}\langle Z_f\,|\,e^{-iH(t_f-t_i)}\,|\,Z_i\rangle&=\sum_n\frac{(-i)^n}{n!}\int_{t_i}^{t_f}dt_1\ldots dt_n\int\mathscr{D}(Z^*,Z)\exp\left\{i\int_{t_i}^{t_f}L_0[Z^*,Z]+\ln\langle Z_f\,|\,Z(t_f)\rangle\right\}\\ &\quad\times\frac{\langle Z(t_n)\,|\,V\,|\,Z(t_n)\rangle}{\langle Z(t_n)\,|\,Z(t_n)\rangle}\cdots\frac{\langle Z(t_1)\,|\,V\,|\,Z(t_1)\rangle}{\langle Z(t_1)\,|\,Z(t_1)\rangle}\\ &=\sum_n\frac{(-i)^n}{n!}\int_{t_i}^{t_f}dt_1\ldots dt_n\int\mathscr{D}(Z^*,Z)\langle Z_f\,|\,Z(t_f)\rangle\\ &\quad\times e^{i\int_{t_n}^{t_f}L_0}\frac{\langle Z(t_n)\,|\,V\,|\,Z(t_n)\rangle}{\langle Z(t_n)\,|\,Z(t_n)\rangle}e^{i\int_{t_{n-1}}^{t_n}L_0}\frac{\langle Z(t_{n-1}\,|\,V\,|\,Z(t_{n-1})\rangle}{\langle Z(t_{n-1})\,|\,Z(t_{n-1})\rangle}\cdots\\ &\quad\times e^{i\int_{t_1}^{t_2}L_0}\frac{\langle Z(t_1)\,|\,V\,|\,Z(t_1)\rangle}{\langle Z(t_1)\,|\,Z(t_1)\rangle}e^{i\int_{t_i}^{t_1}L_0}\ .\end{aligned} \tag{4.3}$$

By going back to the discretized form of the functional integral (Sec. III), one easily shows that (4.3) can be rewritten as follows

$$\begin{aligned}\langle Z_f\,|\,e^{-iH(t_f-t_i)}\,|\,Z_i\rangle=\sum_n\frac{(-i)^n}{n!}\int_{t_i}^{t_f}dt_1\ldots dt_n\int d\mu(Z_n)\ldots d\mu(Z_1)d\mu(Z_n')\ldots d\mu(Z_1')\\ \times\langle Z_f\,|\,e^{-iH_0(t_f-t_n)}\,|\,Z_n\rangle\langle Z_n\,|\,V\,|\,Z_n'\rangle\\ \times\langle Z_n'\,|\,e^{-iH_0(t_n-t_{n-1})}\,|\,Z_{n-1}\rangle\ldots\\ \times\langle Z_1'\,|\,e^{-iH_0(t_1-t_i)}\,|\,Z_i\rangle\ ,\end{aligned} \tag{4.4}$$

where we have used the expression

$$\langle Z_n\,|\,e^{-iH_0(t_n-t_{n-1})}\,|\,Z_{n-1}\rangle=\int_{\substack{Z^*(t_n)=Z_n^*\\ Z(t_{n-1})=Z_{n-1}}}\mathscr{D}(Z^*,Z)\exp\left[i\int_{t_{n-1}}^{t_n}L_0+\ln\langle Z_n\,|\,Z(t_n)\rangle\right]. \tag{4.5}$$

The closure relations over $|\,Z_n\rangle\langle Z_n\,|$ can now be removed. One then ends up with

$$\begin{aligned}\langle Z_f\,|\,e^{-iH(t_f-t_i)}\,|\,Z_i\rangle&=\sum_n\frac{(-i)^n}{n!}\int_{t_i}^{t_f}dt_1\ldots dt_n\langle Z_f\,|\,e^{-iH_0t_f}T[V(t_n)\ldots V(t_1)]e^{iH_0t_i}\,|\,Z_i\rangle\\ &=\langle Z_f\,|\,e^{-iH_0t_f}T\exp\left[-\int_{t_i}^{t_f}V(t)dt\right]e^{iH_0t_i}\,|\,Z_i\rangle\ ,\end{aligned} \tag{4.6}$$

where $V(t)$ is the interaction representation of V,

$$V(t)=e^{iH_0t}Ve^{-iH_0t} \ . \tag{4.7}$$

One recognizes in the expression (4.6) the standard perturbation expansion in powers of V. This shows that the functional integral preserves the structure of the formal perturbation expansion. This follows from the fact that the functional integral, by construction, preserves the structure of the T product, and that we have the following identity[25]:

$$e^{-iHt}=\lim_{N\to\infty}(e^{-i(t/N)H_0}e^{-i(t/N)V})^N \ , \tag{4.8}$$

that is, in the continuous limit, one can neglect the noncommutation of the operators V and H_0. Had one started from the expression (4.8) instead of using

$$e^{-iHt}=\lim_{N\to\infty}(e^{-iHt/N})^N$$

for constructing the path integral, one would have obtained directly (4.6).

It should be stressed that the identification of the perturbation series obtained with the functional integral and operator methods has made explicit reference to the discretized form, which was needed to disentangle the integration over Z and Z' at different times. This is therefore not a check of the continuous limit.

In the remaining part of this section, we are going to rearrange the perturbation expansion using operator identities. The rearrangement which will be performed can be interpreted as a change of variable in the functional integral. We shall see that this change of variable is not always allowed.

Let us first notice that the T exponential may be written

$$\begin{aligned} T\exp -i\int_{t_i}^{t_f} V(t)dt &= \lim_{N\to\infty} T\prod_{k=1}^{N}[1-i\epsilon V(t_k)] \\ &= \lim_{N\to\infty} T\prod_{k=1}^{N} :e^{-i\epsilon V(t_k)}: \left[\epsilon=\frac{t_i-t_f}{N}\right] . \end{aligned} \tag{4.9}$$

The second line differs from the first one by terms which are negligible in the limit $\epsilon\to 0$. We shall keep them, however, for reasons which will become clear soon. Note that the second line defines the T product of two operators at equal times as their normal product:

$$T[A(t)B(t)]=:A(t)B(t): \ . \tag{4.10}$$

This refinement clearly does not affect the preceding discussion. But it is going to be of crucial importance in the following.

We now consider an alternative form of the perturbation expansion which relies on the following identity

$$T\exp -\frac{i}{2}\int_{t_i}^{t_f} V(t)dt = N\int \mathscr{D}W \exp\left[\frac{i}{2}\int_{t_i}^{t_f} W(t)\cdot V^{-1}\cdot W(t)dt\right] T\exp\left[-i\int_{t_i}^{t_f} W_{\alpha\beta}(t)a_\alpha^\dagger(t)a_\beta(t)dt\right] , \tag{4.11}$$

where the normalization constant N is given by

$$N^{-1}=\int \mathscr{D}W \exp\left[\frac{i}{2}\int_{t_i}^{t_f} dt\, W(t)\cdot V^{-1}\cdot W(t)\right] \tag{4.12}$$

and V^{-1} is the inverse of the matrix

$$V_{\alpha\gamma,\beta\delta}=(\alpha\beta\,|\,V\,|\,\gamma\delta) \ . \tag{4.13}$$

$(\alpha\beta\,|\,V\,|\,\gamma\delta)$ is the nonantisymmetrized matrix element of V and in (4.11) $\frac{1}{2}V(t)$ stands for $\frac{1}{2}\sum_{\alpha\beta\gamma\delta}(\alpha\beta\,|\,V\,|\,\gamma\delta)a_\alpha^\dagger(t)a_\beta^\dagger(t)a_\delta(t)a_\gamma(t)$. The identity (4.11) is easily proved. It is very similar to the identity used in Ref. 26. Let us simply remark here that the Gaussian integration over W operates like a Wick's theorem, the elementary contraction being

$$\begin{aligned} \langle W_{\alpha\beta}(t_1)W_{\gamma\delta}(t_2)\rangle &= N\int \mathscr{D}W \exp\left[\frac{i}{2}\int_{t_i}^{t_f} W(t)\cdot V^{-1}\cdot W(t)dt\right] W_{\alpha\beta}(t_1)W_{\gamma\delta}(t_2) \\ &= -\delta(t_1-t_2)V_{\alpha\beta,\gamma\delta} \ . \end{aligned} \tag{4.14}$$

Thus the integration over W reconstructs the original two-body potential. Now it is important to realize that the integration over W involves two W at the same time, and therefore, depends crucially upon the way the T product at equal time has been defined. The formula (4.4) follows then from the application of the two identities

$$T\exp -i\int_{t_i}^{t_f} V(t)dt = \lim_{N\to\infty} T\prod_{k=1}^{N} :e^{-i\epsilon V(t_k)}: , \tag{4.15}$$

$$:e^{-i\epsilon V}: = N\int \mathcal{D}w e^{i/2\epsilon W V^{-1} W} :e^{-i\epsilon W a^\dagger a}: . \tag{4.16}$$

Let us now consider the functional integral representation of $\langle Z_f | e^{-iH(t_f-t_i)} | Z_i\rangle$ in the overcomplete set of Slater determinants. To avoid complications with the constraints, let us use for example the parametrization (3.25) or (3.30). Using the identity

$$\exp\left[-\frac{i}{2}\int \rho\cdot V\cdot\rho\, dt\right] = N\int \mathcal{D}W \exp\left[\frac{i}{2}\int W\cdot V^{-1}\cdot W dt - i\int W\cdot\rho\, dt\right]$$

one obtains the following expression

$$\langle Z_f | e^{-iH(t_f-t_i)} | Z_i\rangle = \int_{\substack{Z^*(t_f)=Z_f^* \\ Z(t_i)=Z_i}} \int \mathcal{D}W \exp\left[\frac{i}{2}\int W(t)\cdot V^{-1}\cdot W(t)dt\right]$$
$$\times \exp\left[i\int_{t_i}^{t_f} \frac{\langle Z | i\partial_t - H_0 - W | Z\rangle}{\langle Z | Z\rangle} + \ln\langle Z_f | Z(t_f)\rangle\right] . \tag{4.17}$$

Note that in the above formula, the matrix V is constructed with antisymmetrized matrix elements of the two-body interaction. It is extremely tempting at this stage to interchange the orders of the integrations over W and Z. Writing

$$\langle Z_f | e^{-iH(t_f-t_i)} | Z_i\rangle = \int \mathcal{D}W \exp\left[\frac{i}{2}\int W(t)\cdot V^{-1}\cdot W(t)dt\right]$$
$$\times \int \mathcal{D}(Z^*,Z) \exp\left[i\int_{t_i}^{t_f} \frac{\langle Z | i\partial_t - H_0 - W | Z\rangle}{\langle Z | Z\rangle} + \ln\langle Z_f | Z(t_f)\rangle\right] . \tag{4.18}$$

Now the integral over Z is the matrix element between $|Z_i\rangle$ and $\langle Z_f|$ of the evolution operator for noninteracting particles in the fluctuating field $W(t)$. It is, therefore, equal to

$$\langle Z_f | e^{-iH_0 t_f}\left[T\exp -i\int_{t_i}^{t_f} W(t)\right] e^{iH_0 t_i} | Z_i\rangle . \tag{4.19}$$

However, a careful analysis of the first terms of the perturbation expansion reveals overcounting, a signal that nonallowed manipulations have been performed. The origin of the trouble can be traced back to the fact that the contributions to the integral over W in (4.8) comes from terms which are of order $(dt)^2$, or ϵ^2 in the discretized version. Thus the integral over Z in (4.18) contains terms like

$$e^{-\epsilon W\langle a^\dagger a\rangle} \sim 1 - \epsilon W\langle a^\dagger a\rangle + \frac{\epsilon^2}{2} W\langle a^\dagger a\rangle W\langle a^\dagger a\rangle . \tag{4.20}$$

If one replaces this integral by the expression (4.19) one gets instead terms of the form

$$\langle e^{-\epsilon W a^\dagger a}\rangle \sim 1-\epsilon W\langle a^\dagger a\rangle + \frac{\epsilon^2}{2}W_{\alpha\beta}W_{\gamma\delta}\langle a_\alpha^\dagger a_\beta\rangle\langle a_\gamma^\dagger a_\delta\rangle + \frac{\epsilon^2}{2}W_{\alpha\beta}W_{\gamma\delta}\langle a_\alpha^\dagger a_\delta\rangle\langle a_\beta a_\gamma^\dagger\rangle \ , \tag{4.21}$$

that is, one obtains two terms corresponding to the two possible contractions, and this is the origin of the overcounting. Thus one cannot replace the integral over Z by the expression (4.19) which is really troublesome, since the integral over Z in (4.18) is really the one which in our formalism represents (4.19). Another way of stating the difficulty is to consider that the change of variable involved in (4.17) is not allowed for the integral over Slater determinants, or more precisely that one is not allowed to interchange the order of integration over Z and W in (4.17).

It is easily seen that all these difficulties disappear when one is working with a path integral constructed with coherent states. Indeed taking the matrix element of Eq. (4.16) between two coherent states yields

$$\langle :e^{-i\epsilon V}:\rangle = e^{-i\epsilon\langle V\rangle} = N\int \mathscr{D}W e^{i/2\epsilon W V^{-1}W} e^{-i\epsilon W\langle a^\dagger a\rangle} \ , \tag{4.22}$$

that is, the functional integral preserves exactly the operator identities. In this particular case, the change of variable involved in (4.17) is therefore perfectly allowed. Note that this holds for any kind of coherent states, of boson or fermions. It holds in particular for the coherent states (2.27).

V. MEAN FIELD THEORIES AND BEYOND

In the preceding section, we made explicit the similarities in the structures of the functional integral and the formal perturbation expansion. However, the most interesting feature of the functional integral is to suggest approximation schemes which are different from those of conventional perturbation theory. In this section, we examine in particular the saddle-point approximation and its successive corrections. As well known, this approximation, when performed on the standard Feynman path integral, retrieves classical mechanics. In the many-body problem, the classical approximation obtained depends on the choice of the overcomplete set of states which have been chosen to construct the functional integral. If independent particle wave functions are used, the classical equations are the time-dependent mean field equations. It turns out that these nonlinear equations have definite classical features which we analyze. We also discuss in this section the connection between path integrals and perturbative boson expansions.

Let us then apply the saddle-point approximation and its successive corrections to the calculation of the functional integral

$$\int_{\substack{Z^*(t_f)=Z_f^*\\ Z(t_i)=Z_i}} \mathscr{D}(Z^*,Z)e^{iS[Z^*,Z]} = \langle Z_f|e^{-iH(t_f-t_i)}|Z_i\rangle \ . \tag{5.1}$$

The saddle points are given by the following equations, with their boundary condition

$$\frac{\delta S}{\delta Z}=0,\ \ Z^*(t_f)=Z_f^* \ , \tag{5.2a}$$

$$\frac{\delta S}{\delta Z^*}=0,\ \ Z(t_i)=Z_i \ . \tag{5.2b}$$

We call $Z_C^{(+)*}$ and $Z_C^{(-)}$ the solutions of Eqs. (5.2a) and (5.2b), respectively. Note that $Z_C^{(+)*}(t)$ and $Z_C^{(-)}(t)$ are not, in general, complex conjugates of each other. We then expand the action $S[Z^*,Z]$ around the classical solution $(Z_C^{(+)*}, Z_C^{(-)})$:

$$S=S_C+\sum_{n\geq 2} S_n[Z^*,Z] \ , \tag{5.3}$$

where we have set

$$S_C \equiv S[Z_C^{(+)*},Z_C^{(-)}] \ ,$$

$$S_n[Z^*,Z]=\frac{1}{n!}\sum_{p=1}^{N} C_N^P (Z^*)^P \frac{\delta^n S}{\delta Z^{*P}\delta Z^{n-p}}\bigg|_C Z^{n-p} \ , \tag{5.4}$$

and the functional derivatives are evaluated for $Z^*=Z_C^{(+)*}$, $Z=Z_C^{(-)}$. The functional integral (5.1) then takes the form

$$\langle Z_f|e^{-iH(t_f-t_i)}|Z_i\rangle = e^{iS_C}\int_{\substack{Z^*(t_f)=0\\ Z(t_i)=0}} \mathscr{D}(Z^*,Z)\exp\left[i\sum_{n\geq 2}S_n(Z^*,Z)\right] \ . \tag{5.5}$$

Note that the boundary conditions are now independent of Z_i and Z_f^*. That is, all the dependence of the expression (5.5) on Z_i and Z_f^* is contained in S_C and the possible boundary terms which subsist in S_n.

It is interesting to notice that Eqs. (5.2a) and (5.2b) correspond to the time-dependent variational principle (very similar to the ones developed in Ref. 27):

$$\delta S[Z^*,Z]=0 \ , \tag{5.6}$$

where $S[Z^*,Z]$ reads explicitly:

$$S[Z^*,Z]=\int_{t_i}^{t_f} \frac{\langle Z \mid i\partial_t - H \mid Z\rangle}{\langle Z \mid Z\rangle} - i\ln\langle Z_f \mid Z(t_f)\rangle \ . \tag{5.7}$$

It is easily verified that Eq. (5.6) leads back to the time-dependent Schrödinger equation if $|Z\rangle$ is assumed to represent any state of the Hilbert space, i.e., in the case of unrestricted variations. When $|Z\rangle$ is chosen in a given class of states, the solution of the Eq. (5.6) provides an approximation for the transition amplitude $\langle Z_f | e^{-iH(t_f-t_i)} | Z_i\rangle$. This is given by e^{iS_C}. The usefulness of this expression lies in the fact that it is a stationary quantity. It appears then clearly that the choice of an overcomplete set for the construction of the functional integral is equivalent to the choice of a class of trial states in the use of the time-dependent variational principle. Therefore the separation into a classical motion and quantum corrections, implied by Eq. (5.3) does not require that some quantity is small compared to $\hbar$. The nature of the classical approximation discussed here, or the type of quantum effects which are left out in this approximation, are entirely determined by the specific choice of an overcomplete set in the Hilbert space. In particular, if the overcomplete set is the Hilbert space itself, the classical equations of motion are identical with the Schrödinger equation.

The limitation of the time-dependent variational principle (5.6) is that it does not provide a way of estimating the error associated with a given choice of trial states. This is precisely what the functional integral (5.5) does. Although the corrections to the classical approximation would be in most cases hard to evaluate, the functional integral provides the possibility of analyzing them, and therefore, allows for a better understanding of the classical approximation itself. We shall illustrate these considerations in the case of the mean field approximations to the many-body problem.

Let us then consider that the coordinates $\{Z\}$ represent a Slater determinant, that is, $\{Z\}$ denotes any of the sets of coordinates discussed in Sec. II E. [Actually the equations of motion given below only hold if the action (3.30) or (3.23) are used. If the action (3.25) is used, extra kinematical terms appears in front of the time derivatives.] The classical equations of motion are the time-dependent Hartree-Fock equations

$$i\dot{Z} - \frac{\delta H(Z^*,Z)}{\delta Z^*}=0 \ , \tag{5.8}$$

$$i\dot{Z}^* + \frac{\delta H(Z^*,Z)}{\delta Z}=0 \ , \tag{5.9}$$

where $H(Z^*,Z)=\langle Z|H|Z\rangle/\langle Z|Z\rangle$ is given explicitly in Sec. II E. The state vectors $|Z(t)\rangle$ which make the action (5.7) stationary are of the form

$$|Z(t)\rangle = |Z_0(t)\rangle \exp\left[-i\int_{t_i}^{t} f(t')dt'\right] \ , \tag{5.10}$$

where $f(t)$ is an arbitrary function of time and $Z_o(t)$ is a solution of the Eq. (5.8). This arbitrariness in the phase of $|Z(t)\rangle$ reflects the invariance of the action (5.7) with respect to the choice of phase of the state vectors. Equations (5.8) and (5.9) can be easily transformed into an equation for the one-body density matrix

$$i\dot{\rho}=[h,\rho] \tag{5.11}$$

with

$$\rho_{\alpha\beta}(t)=\frac{\langle Z_C^{(+)}(t)\,|\,a_\alpha^\dagger a_\beta\,|\,Z_C^{(-)}(t)\rangle}{\langle Z_C^{(+)}(t)\,|\,Z_C^{(-)}(t)\rangle} \ , \tag{5.12}$$

and $h=\delta E/\delta\rho$ is the usual Hartree-Fock Hamiltonian calculated with the density matrix (5.12). Note that the density matrix is not Hermitian, so that the Hartree-Fock Hamiltonian is in general not real. Equation (5.11) is a generalization of the ordinary time-dependent Hartree-Fock equation, appropriate to the calculation of scattering amplitudes. This equation has already been considered in Ref. 12. Note that the standard TDHF equations are recovered if one chooses the boundary conditions such that $Z(t_f)=Z_f$. Then $Z_C^{(-)}(t)$ and $Z_C^{(+)*}(t)$ are complex conjugates of each other and the density matrix, as well as the Hartree-Fock Hamiltonian are Hermitian. The classical solutions $[Z_C^{(+)}(t), Z_C^{(-)}(t)]$ can be used to get a semiclassical approximation to the transition amplitude $\langle Z_f t_f | Z_i t_i \rangle$. It can also be used to obtain semiclassical approximations to the bound state energies of the system, applying a generalization of the WKB method developed in Ref. 28. Typically one arrives at semiclassical quantization rules for the periodic trajectories of the time-dependent mean-field equations. This method has already been applied in different ways to several simple cases.[13,12,29–31]

Let us now calculate the corrections to the mean field theory. This is obtained by expanding the action S around the classical solution $[Z_C^{(+)*}(t), Z_C^{(-)}(t)]$, as indicated by Eq. (5.3). We shall limit ourselves first to the quadratic corrections, and to simplify the discussion, we shall consider the fluctuations around a static solution of Eq. (5.8). We call $|\phi_0\rangle$ the corresponding state and we calculate

$$\langle \phi_0 | e^{-\beta H} | \phi_0 \rangle \approx e^{-\beta E_{\rm HF}} \int_{\substack{Z^*(\beta)=0 \\ Z(0)=0}} \mathscr{D}(Z^*,Z) \exp\left\{ -\int_0^\beta [Z^*\dot{Z} + H_2(Z^*,Z)]dt \right\} , \tag{5.13}$$

where $E_{\rm HF}$ is the Hartree-Fock energy of the state $|\phi_0\rangle$ and we have assumed $\langle \phi_0 | \phi_0 \rangle = 1$. $H_2(Z^*,Z)$ is the quadratic form obtained by expanding $H(Z^*,Z)$ around $Z=0$ ($\equiv |\phi_0\rangle$). In terms of the amplitudes Z_{ph}, $H_2(Z^*,Z)$ have the explicit form

$$H_2(Z^*,Z) = \tfrac{1}{2}(Z_{ph}^*, Z_{ph}) \begin{bmatrix} A & B \\ B^* & A^* \end{bmatrix} \begin{bmatrix} Z_{ph} \\ Z_{ph}^* \end{bmatrix} , \tag{5.14}$$

where the matrices A and B are the usual matrices of the random phase approximation. Note that at this level of approximation, all the parametrizations considered in Sec. II E, with proper inclusion of the constraints when necessary, yield the same result, Eq. (5.14). Now the functional integral (5.13) is identical to that of a system of coupled harmonic oscillators; more precisely it can be written

$$\int_{\substack{Z^*(\beta)=0 \\ Z(0)=0}} \mathscr{D}(Z^*,Z) \exp\left\{ -\int_0^\beta [Z^*\dot{Z} + H_2(Z^*,Z)]dt \right\}$$

$$= {}_B\langle 0 | \exp -\beta(C^\dagger \cdot A \cdot C + \tfrac{1}{2} C^\dagger \cdot B \cdot C^\dagger + \tfrac{1}{2} C \cdot B \cdot C) | 0 \rangle_B , \tag{5.15}$$

where we have used the matrix notation

$$C^\dagger \cdot A \cdot C = \sum_{\substack{ph \\ p'h'}} C_{ph}^\dagger A_{ph,p'h'} C_{p'h'} .$$

$C_{ph}^\dagger$ and C_{ph} denote boson creation and annihilation operators and $|0\rangle_B$ is the boson vacuum. By using the canonical form which diagonalizes the quadratic form in (5.15) one easily obtains:

$${}_B\langle 0 | \exp(C^\dagger A C + \tfrac{1}{2} C^\dagger B C^\dagger + \tfrac{1}{2} C B^\dagger C) | 0 \rangle_B = e^{\Delta E_0} , \tag{5.16}$$

where ΔE_0 is the correlation energy associated with the random phase approximation (RPA) vibrations, that is,

$$\Delta E_0 = \tfrac{1}{2} \sum_N \omega_N - \tfrac{1}{2} {\rm Tr} A . \tag{5.17}$$

This expression is easily shown to be equal to the sum of all the ring diagrams calculated with antisymmetrized matrix element and including the well known double counting of the second order term.

The boson degrees of freedom which appear naturally in the calculation of the integral (5.15) are the usual RPA phonons. The successive corrections to the expression (5.15) represent the various couplings between these RPA phonons. A systematic expansion can be derived in the following way. We first expand $H(Z^*,Z)$ to all order in Z^* and Z. Since we have treated explicitly the terms of order 2, this expansion starts at third order. These higher order terms can be treated in perturbation, which leads to the expression

$$\langle\phi_0|e^{-\beta H}|\phi_0\rangle = e^{-\beta E_{HF}} \exp\left[-\int_0^\beta dt \sum_{n>3} H_n\left[\frac{\delta}{\delta j},\frac{\delta}{\delta j^*}\right]\right] \times \int_{\substack{Z^*(\beta)=0\\ Z(0)=0}} \mathscr{D}(Z^*,Z)\exp\left\{-\int_0^\beta [Z^*\dot{Z}+H_2(Z^*,Z)+j^*\cdot Z+Z^*\cdot j]\right\}\Bigg|_{\substack{j=0\\ j^*=0}} . \quad (5.18)$$

The expression (5.18) is very reminiscent of the familiar perturbative boson expansion. The unperturbed propagator for the bosons is the RPA propagator and the term H_n describes a coupling between n RPA bosons. The occurrence of n-body interactions between the RPA bosons arises from the Pauli principle, or in other words from the constraints necessary to project onto the physical subspace. It must be kept in mind that we are not making here an exact connection between our formalism and a perturbative boson expansion. Indeed, when going beyond the quadratic approximation, technical problems arise with the treatment of the constraints, the integration measure or the domain of integration, depending upon whether one chooses, respectively, the parametrization (2.53), (2.45), or (2.50) for the Slater determinant. In the absence of a careful treatment of these points, we consider the expression (5.18) as approximate. It is clear, however, that the physical content of (5.18) will not be very much altered by a more rigorous derivation. This physical content is indeed quite transparent. The functional integral "quantizes" as bosons the coordinates which were introduced to parametrize the states of the overcomplete set. Inversely, the classical limit obtained in the saddle-point approximation is achieved by replacing the boson operators by c numbers (see Ref. 18).

The technical difficulties mentioned above do not show up when one considers the expansion around a solution of the Hartree equation. In this case, the expansion can then be given easily a diagrammatic interpretation, using the standard technics of perturbation theory. We shall again restrict ourselves to a time-independent problem and consider the expression

$$\langle\phi_0|e^{\beta H_0}e^{-\beta H}|\phi_0\rangle = N\int \mathscr{D}W \exp\left[\frac{1}{2}\int_0^\beta W(t)\cdot V^{-1}\cdot W(t)dt\right] e^{\mathrm{Tr}\ln(1-WG_0)} , \quad (5.19)$$

which follows trivially from (4.11) and the identity

$$\langle\phi_0|T\exp\int_0^\beta W(u)a^\dagger(u)a(u)du\,|\phi_0\rangle = \exp\mathrm{Tr}\ln(1-WG_0) , \quad (5.20)$$

where G_0 is the single particle Green's function:

$$G^0_{\alpha\beta}(u_1-u_2) = \langle\phi_0|Ta_\alpha(u_1)a^\dagger_\beta(u_2)|\phi_0\rangle . \quad (5.21)$$

Equation (5.19) can also be derived from (3.16) (see Ref. 18). Application of the saddle-point approximation on the integral over W leads to the equation

$$W\cdot V^{-1} = V^{-1}\cdot W = G_0(1-WG_0)^{-1} = G[W] , \quad (5.22)$$

where $G[W]$ is the single particle Green's function in presence of the external field W:

$$G^{-1}[W] = G_0{}^{-1} - W . \quad (5.23)$$

The density matrix is related to G by

$$\rho(t) = \lim_{\tau\to 0_+} G\left[t-\frac{\tau}{2},t+\frac{\tau}{2}\right] . \quad (5.24)$$

It satisfies the equation of motion

$$\partial_t\rho + [H_0 - W,\rho] = 0 , \quad (5.25)$$

which is the time-dependent Hartree equation written in imaginary time. The expansion around a static solution is obtained easily. Let W_0 a static field, solution of

$$[H_0 - W_0,\rho_0] = 0 . \quad (5.26)$$

The expansion of (5.19) in powers of $W' = W - W_0$ reads

$$\langle\phi_0|e^{\beta H_0}e^{-\beta H}|\phi_0\rangle = Ne^{(\beta/2)\rho_0\cdot V\cdot\rho_0}\int\mathscr{D}W\exp\left[\int_0^\beta \frac{1}{2}W'\cdot V^{-1}\cdot W' - \frac{1}{2}\int_0^\beta \mathrm{Tr}W'\cdot G(W_0)\cdot W'\cdot G(W_0)\right] \times\exp\left\{\int_0^\beta \mathrm{Tr}\sum_{n>2} -\frac{1}{n}[G(W_0)\cdot W']^n\right\} . \quad (5.27)$$

In order to calculate the remaining integral, we first regroup the two quadratic terms defining

$$\Gamma^{-1} = V^{-1} - Q \ , \tag{5.28}$$

where

$$Q_{\alpha\beta,\gamma\delta}(u_1 - u_2) = G_{\beta\gamma}[W_0;u_1 - u_2]G_{\gamma\alpha}[W_0;u_2 - u_1] \ . \tag{5.29}$$

Using a standard procedure, one introduces a source term for the field W' and treats in perturbation the terms of order higher than 2 in W', in the exponent of (5.27). One then gets

$$\langle \phi_0 | e^{\beta H_0} e^{-\beta H} | \phi_0 \rangle = e^{-(\beta/2)\rho_0 \cdot V \cdot \rho_0} e^{-1/2 \, \mathrm{Tr} \ln(1 - VQ_0)}$$
$$\times \exp\left\{ -\sum_{n>2} \int_0^\beta \mathrm{tr} \frac{1}{n} \left[G(W_0) \cdot \frac{\delta}{\delta j} \right]^n \right\} e^{-(1/2) j \cdot \underline{\Gamma} \cdot j} \Bigg|_{j=0} \ , \tag{5.30}$$

where the factor $e^{-1/2 j \Gamma j}$ comes from the Gaussian integral over W'

$$e^{-(1/2) j \cdot \Gamma \cdot j} = \frac{\int \mathscr{D} W e^{(1/2) W \cdot \Gamma^{-1} \cdot W + j \cdot W}}{\int \mathscr{D} W e^{(1/2) W \cdot \Gamma^{-1} \cdot W}} \ . \tag{5.31}$$

The diagrammatic interpretation of the formula (5.30) is very simple (we consider vacuum-vacuum diagrams corresponding to the ground state energy). The first term is the Hartree energy

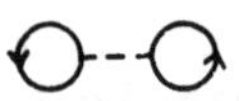 (5.32)

The second term is the sum of all ring diagrams (calculated here with direct matrix elements), plus actually the exchange counterpart of (5.32)

$\mathrm{tr} \ln (1 - VQ_0)$ = 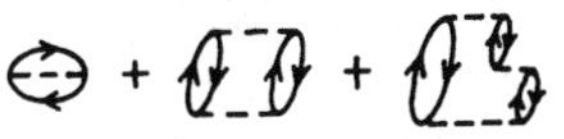 . . (5.33)

To pursue the analysis we give the following representation of Γ

$\Gamma \equiv$ 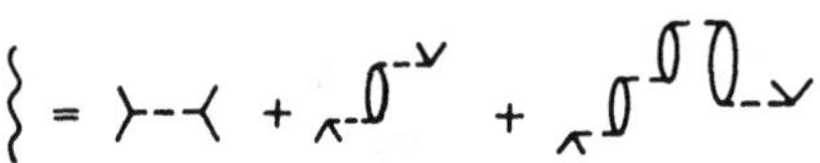 . (5.34)

Thus (5.33) can be represented by

 (5.35)

and

$$\sum_{n>2} \frac{1}{n} \mathrm{Tr} \left[G(W_0) \cdot \frac{\delta}{\delta j} \right]^n e^{-(1/2) j \cdot \Gamma \cdot j} \Bigg|_{j=0} \tag{5.36}$$

is the sum of all diagrams with one closed fermion loop and an arbitrary number of Γ lines:

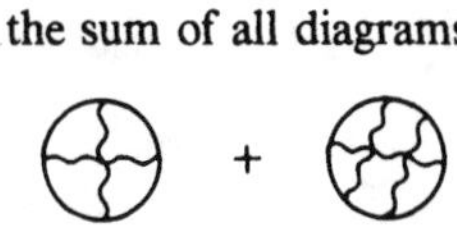 + ⋯ . (5.37)

The single particle Green's function can be written as follows:

$$G_{\alpha\beta}(u_1 - u_2) = \frac{\int \mathscr{D} W \exp\left[\frac{1}{2} \int_0^\beta W \cdot V^{-1} \cdot W + \mathrm{Tr} \ln(1 - WG_0) \right] G_{\alpha\beta}[W; u_1 - u_2]}{\int \mathscr{D} W \exp\left[\frac{1}{2} \int_0^\beta W \cdot V^{-1} \cdot W + \mathrm{Tr} \ln(1 - WG_0) \right]} \ . \tag{5.38}$$

Following a derivation similar to the one which leads to (5.30), one obtains the following expression:

$$G_{\alpha\beta}(u_1-u_2)=\left\{\exp\left\{-\int_0^{\beta}\mathrm{tr}\sum_{n>2}\frac{1}{n}\left[G(W_0)\cdot\frac{\delta}{\delta j}\right]^n\right\}G_{\alpha\beta}\left[W_0+\frac{\delta}{\delta j}\right]e^{-(1/2)j\cdot\Gamma\cdot j}\bigg|_{j=0}\right\}_L\ , \tag{5.39}$$

where the symbol $\{\ \}_L$ means that we have to consider only the linked diagrams. G has the following diagrammatic representation:

G =  + ⋯ .

The first term may be veiwed as the classical propagator. It describes the motion of a particle in the field W_0. The other terms which describes the coupling of a particle to a vibration, *with propagation of the vibration,* are the quantum effects which are left out in the classical approximation.

VI. CONCLUSIONS

The functional integrals built on overcomplete sets of the Hilbert space provide a unifying understanding of different approaches to the nuclear many body problem. The role and the significance of the overcomplete set are best understood when calculating the functional integral using the saddle-point approximation, and its successive corrections. Then, it can be seen that the parameters which are used to label the states of the overcomplete set obey classical equations of motion. The state labels may then be viewed as classical coordinates in a generalized phase space. The classical equations of motion are identical to those obtained applying a time-dependent variational principle, using as trial states the states of the overcomplete set. But in contrast to the variational principle, the functional integral does provide a way of calculating corrections to the variational solution. A proper treatment of the fluctuations around the classical path introduces a quantization of the classical coordinates in terms of boson degrees of freedom.

As we have seen throughout this paper bosons play an important role in the functional integral formalism. In particular boson coherent states appear to be very useful because they are eigenstates of the destruction operators. This greatly facilitates the calculation of matrix elements. But more than that, it makes the structure of the functional integral simpler. Also we have seen that some changes of variables are allowed only if the overcomplete set is a set of coherent states. We have also shown that the functional integral transforms a fermion theory into a boson theory in very much the same way as the usual boson expansions do.

We have also obtained a clear physical interpretation of the classical features of the mean field approximations. In the language of boson representations, this approximation is obtained by replacing the boson propagators, e.g., the propagators corresponding to the RPA vibrations, by their classical approximation. This implies that only the static part of the particle-vibration interactions are taken into account in the mean field approximation. This point is further illustrated by diagrammatic expansion around the mean field. The processes involving a real propagation of a phonon between the time when it is emitted and the time when it is absorbed appear as quantum corrections to the mean field. Another equivalent statement, also suggested by the functional integral formalism, is that the mean field has at each time a given classical value. The functional integral allows for possible approximate schemes for calculating the "quantum" fluctuations around this value.

ACKNOWLEDGMENTS

One of the authors (J.P.B.) gratefully acknowledges the warm hospitality of the Physics Department of the University of Illinois at Urbana-Champaign where this work was completed.

APPENDIX

We construct explicitly the measures which have been used in Sec. II E to construct closure relations. The general idea underlying the method is to associate the parameters Z with some group operation and to construct the invariant measure over the group. In the case of the Slater determinants the group to be considered is the group of unitary transformations in the space of single particle states. A general element of the group is represented by the matrix

$$U=\begin{bmatrix}A & B\\ C & D\end{bmatrix},\quad UU^{\dagger}=U^{\dagger}U=1\ , \tag{A1}$$

where A, B, C, and D are $n_h \times n_h$, $n_h \times n_p$, $n_p \times n_h$, and $n_p \times n_p$ matrices, respectively. These matrices satisfy:

$$\begin{aligned} AA^\dagger + BB^\dagger = 1,\ CA^\dagger + DB^\dagger = 0\ ,\\ AC^\dagger + BD^\dagger = 0,\ CC^\dagger + DD^\dagger = 1\ ,\\ A^\dagger A + C^\dagger C = 1,\ A^\dagger B + C^\dagger D = 0\ ,\\ B^\dagger A + D^\dagger C = 0,\ B^\dagger B + D^\dagger D = 1\ . \end{aligned} \tag{A2}$$

Let us now consider the states (2.43), normalized:

$$|\widetilde{Z}\rangle = N e^{Z_{ph} a_p^\dagger a_h} |\phi_0\rangle\ , \tag{A3}$$

where N is a normalization constant. Let S be the unitary transformation which carries $|\widetilde{Z}\rangle$ into $|\widetilde{Z}'\rangle$:

$$|\widetilde{Z}'\rangle = S|\widetilde{Z}\rangle\ . \tag{A4}$$

We look for an invariant measure $\mu(Z)$ such that

$$\begin{aligned}\mu(Z) &= \mu(Z')\,|J(Z',Z)|\\ &= \mu(0)\,|J(0,Z)|\ ,\end{aligned} \tag{A5}$$

where $J(Z',Z)$ is the Jacobian of the transformation which transforms Z into Z'.

The law of transformation of the coordinates Z the transformation (A4) is easily derived. Indeed $|\widetilde{Z}\rangle$ can be written

$$|\widetilde{Z}\rangle = N\prod_h (a_h^\dagger + Z_{ph} a_p^\dagger)|0\rangle\ . \tag{A6}$$

Under the unitary transformation (A4), this becomes

$$|\widetilde{Z}\rangle = N\prod_h (a_h^\dagger + Z_{ph} a_p^\dagger)|0\rangle\ . \tag{A7}$$

where

$$b_h^\dagger = S a_h^\dagger S^\dagger,\ b_p^\dagger = S a_p^\dagger S^\dagger,\ S|0\rangle = |0\rangle\ . \tag{A8}$$

To the operator S is associated a matrix U of the form (A1) which realizes the linear transformation of the creation operators

$$(b_h^\dagger b_p^\dagger) = (a_h^\dagger a_p^\dagger)\begin{pmatrix} A & B\\ C & D\end{pmatrix}\ . \tag{A9}$$

Replacing $b_h^\dagger$ and $b_p^\dagger$ in the equation (A7) by their expression in terms of $a_h^\dagger$ and $a_p^\dagger$ given above, one gets:

$$\begin{aligned}|\widetilde{Z}'\rangle &= N\prod_h [a_{h'}^\dagger (A + BZ)_{h'h}\\ &\qquad + b_{p'}^\dagger (C + DZ)_{p'h}]|0\rangle\\ &= N'\prod_h (a_h^\dagger + Z'_{ph} a_p^\dagger)|0\rangle\ ,\end{aligned} \tag{A10}$$

where

$$Z' = (C + DZ)(A + BZ)^{-1}\ . \tag{A11}$$

This is the desired transformation law. From this it is easy to evaluate the Jacobian which appears in (A5). First we write

$$Z'(A + BZ) = C + DZ$$

then differentiate,

$$dZ'(A + BZ) + Z'BdZ = DdZ\ .$$

We replace Z by its expression in terms of Z' by inverting the equation (A11) and finally put $Z' = 0$. We then get

$$|J(0,Z)| = |(\det D)^{2n_h} \det(A - BD^{-1}C)^{-2n_p}|\ . \tag{A12}$$

Using the relations (A2) one easily shows that

$$|\det(A - BD^{-1}C)| = |\det A|^{-1} = |\det D|^{-1}\ , \tag{A13}$$

so that the Jacobian takes the form

$$|J(0,Z)| = |\det A|^{2(n_p + n_h)}\ . \tag{A14}$$

It remains to relate the matrix A to the matrix Z. For that purpose one can use the following coset decomposition:

$$\begin{pmatrix} A & B\\ C & D\end{pmatrix} = \begin{pmatrix} 1 & -Z^\dagger\\ Z & 1\end{pmatrix}\begin{pmatrix} U & 0\\ 0 & U'\end{pmatrix}\begin{pmatrix} A_1 & 0\\ 0 & D_1\end{pmatrix}\ , \tag{A15}$$

where U and U' are, respectively, $n_h \times n_h$ and $n_p \times n_p$ arbitrary unitary matrices. A_1 and D_1 are, respectively, $n_h \times n_h$ and $n_p \times n_p$ matrices to be determined so that the matrix

$$\begin{pmatrix} A & B\\ C & D\end{pmatrix}$$

satisfies the conditions (A2). One solution is

$$A_1 = (1 + Z^\dagger Z)^{-1/2},\ D_1 = (1 + Z^\dagger Z)^{-1/2}\ . \tag{A16}$$

These equations define the matrices A and D, with respect to an arbitrary transformation of the form

$$\begin{bmatrix} U & 0 \\ 0 & U' \end{bmatrix}$$

which does not change the state of the system and which can be ignored. It is easily checked that the transformation thus defined carries the state $|Z\rangle$ into $|\phi_0\rangle$. Therefore, the Jacobian (A14) can be written

$$|J(0,Z)| = [\det(1+Z^\dagger Z)]^{-(n_p+n_h)} = [\det(1+ZZ^\dagger)]^{-(n_p+n_h)} . \quad \text{(A17)}$$

The expression of the measure used in (2.24) follows trivially. This measure can also be obtained by identifying the set of Slater determinants with a complex Grassman manifold.[32] This method was used in Ref. 14. The method presented here is more elementary and similar to the methods used in Ref. 33 and Ref. 34. (See also Ref. 35.) We consider now the change of variables (2.27)

$$\beta_{ph} = \sum_{h'} Z_{ph'}[(1+Z^\dagger Z)^{-1/2}]_{h'h} . \quad \text{(A18)}$$

The expression of Z_{ph} in terms of β_{ph} is

$$Z_{ph} = \sum_{h'} \beta_{ph'}[(1-\beta^\dagger\beta)^{-1/2}]_{h'h} . \quad \text{(A19)}$$

In terms of these new variables, the matrix

$$\begin{bmatrix} A & B \\ C & D \end{bmatrix}$$

of Eq. (A15) takes the form

$$\begin{bmatrix} (1-\beta^\dagger\beta)^{1/2} & -\beta^\dagger \\ \beta & (1-\beta\beta^\dagger)^{1/2} \end{bmatrix} . \quad \text{(A20)}$$

It is easily seen that the Jacobian of the transformation (A18) is precisely given by (A17). When the β are chosen as coordinates, the measure is, therefore, extremely simple. The domain of integration is complicated, however. The volume θ of this domain can be calculated. This fixes the arbitrary constant in the measure. One has[34]

$$\theta = \frac{1!2!\ldots(n_h-1)!1!2!\ldots(n_p-1)!}{1!2!\ldots(n_p+n_h-1)!}\pi^{n_p n_h} . \quad \text{(A21)}$$

Finally it is convenient to introduce new variables $\tilde{\alpha}$ and $\tilde{\beta}$ defined as follows:

$$\tilde{\alpha} = (1-\beta^\dagger\beta)^{1/2}U, \quad \tilde{\beta} = \beta U , \quad \text{(A22)}$$

where U is a $n_h \times n_h$ unitary matrix. It is easily seen that the integral over β transforms into

$$\int d\beta\, d\beta^* = \int d\tilde{\beta}\, d\tilde{\beta}^* d\tilde{\alpha}\, d\tilde{\alpha}^* \delta(\tilde{\alpha}^\dagger\tilde{\alpha} + \tilde{\beta}^\dagger\tilde{\beta} - 1) . \quad \text{(A.23)}$$

But $\tilde{\alpha}$ and $\tilde{\beta}$ are the expansion coefficients of a set of N single particle states on a fixed basis. Any basis may be used to write (A23). In particular, we can choose a wave function representation, in which case we shall write the integration measure (A23) as follows:

$$\int \prod_{k=1}^{N} \prod_x d\varphi_k^*(x) d\varphi_k(x) \prod_{l=1}^{N} \delta(\langle \varphi_k | \varphi_l \rangle - \delta_{kl}) . \quad \text{(A24)}$$

---

*On leave from Service de Physique Theorique, CEN Saclay, France.

[1] A. K. Kerman and S. E. Koonin, Ann. Phys. (N.Y.) 100, 332 (1976).
[2] F. Villars, Nucl. Phys. A285, 269 (1977).
[3] K. Goeke and P. G. Reinhard, Ann. Phys. (N.Y.) 112, 328 (1978).
[4] M. Baranger and M. Veneroni, Ann. Phys. (N.Y.) 114, 123 (1978).
[5] E. R. Marshalek and J. Weneser, Phys. Rev. C 2, 1682 (1970).
[6] D. Janssen, F. Dönau, S. Frauendorf, and R. V. Jolos, Nucl. Phys. A172, 145 (1971).
[7] J. P. Blaizot and E. R. Marshalek, Nucl. Phys. A309, 422 (1978); A309, 453 (1978).
[8] J. R. Klauder, Ann. Phys. (N.Y.) 11, 123 (1960); J. Math. Phys. 4, 1055 (1963); 4, 1058 (1963).
[9] J. R. Klauder, in *Path Integrals and their Application in Quantum, Statistical and Solid State Physics*, edited by G. Papadopoulos and J. Devrese (Plenum, New York, 1977).
[10] H. Kleinert, Phys. Lett. 69B, 9 (1977).
[11] H. Reinhardt, Nucl. Phys. A251, 317 (1975).
[12] S Levit, Phys. Rev. C 21, 1594 (1980); S. Levit, J. W. Negele, and Z. Paltiel, *ibid.* 21, 1603 (1980).
[13] H. Kleinert and H. Reinhardt, Nucl. Phys. A332, 331 (1979).
[14] H. Kuratsuji and T. Suzuki, Phys. Lett. 92B, 19 (1980).
[15] A. Kerman and T. Troudet (unpublished).

[16]J. P. Blaizot and H. Orland, J. Phys. Lett. 41, 53 (1980); 41, 401 (1980).
[17]J. P. Blaizot and H. Orland, J. Phys. Lett. 41, 523 (1980).
[18]J. P. Blaizot and H. Orland, Phys. Lett. 100B, 195 (1981).
[19]H. Orland, in *Méthodes Mathématiques de la Physique Nucléaire,* edited by B. Giraud and P. Quentin (Collége de France, Paris, 1980).
[20]V. Bargman, Commun. Pure Appl. Math. 14, 187 (1961).
[21]Y. Ohnuki and T. Kashiwa, Prog. Theor. Phys. 60, 548 (1978).
[22]D. J. Thouless, Nucl. Phys. 21, 225 (1960).
[23]L. D. Faddeev, in *Les Houches 1975,* Proceedings of the Methods in Field Theory, edited by R. Balian and J. Zinn-Justin (North-Holland, Amsterdam, 1970), p. 1.
[24]E. R. Marshalek and G. Holzwarth, Nucl. Phys. A191, 438 (1972).
[25]H. Trotter, Proc. Amer. Math. Soc. 10, 541 (1959).
[26]J. Hubbard, Phys. Rev. Lett. 3, 77 (1959).
[27]B. A. Lippman and J. Schwinger, Phys. Rev. 79, 469 (1950).
[28]R. F. Dashen, B. Hasslacher, and A. Neveu, Phys. Rev. 12D, 2443 (1975).
[29]R. Shankar, Phys. Rev. Lett. 45, 1088 (1980).
[30]K. K. Kan, J. J. Griffin, P. C. Lichtner, and M. Dworzecka, Nucl. Phys. A332, 109 (1979).
[31]H. Reinhardt, Nucl. Phys. A346, 1 (1980).
[32]S. Kobayashi and K. Nomizu, *Foundations of Differential Geometry* (Interscience, New York, 1969), Vol. II.
[33]M. I. Monastyrsky and A. M. Perelomov, Rep. Math. Phys. 6, 1 (1974).
[34]L. K. Hua, *Harmonic Analysis of Functions of Several Complex Variables in Classical Domains,* translation of Mathematical Monographs (American Mathematical Society, Providence, 1963), Vol. 6, p.46.
[35]The parametrization (A3) of Slater determinants and the derivation of the corresponding invariant measure has been extensively studied by H. Kuratsuji and T. Suzuki. These authors have also derived the invariant measure following arguments similar to those used in the first part of this appendix (private communication of T. Suzuki to one of us).

TRANSFORMATION GROUPS FOR SOLITON EQUATIONS

Etsuro Date
Department of Mathematics, College of Liberal Arts
Kyoto University, Kyoto 606, Japan

Masaki Kashiwara, Michio Jimbo and Tetsuji Miwa
Research Institute for Mathematical Sciences
Kyoto University, Kyoto 606, Japan

in

Proceedings of RIMS Symposium on Non-Linear Integrable Systems-Classical Theory and Quantum Theory, Kyoto, Japan May 13 - May 16, 1981, ed. by M. Jimbo and T. Miwa (World Science Publishing Co., Singapore, 1983).

Contents

Transformation Groups for Soliton Equations

By

Etsuro Date
Michio Jimbo
Masaki Kashiwara
Tetsuji Miwa

41

§0. Introduction.

Through the latest developments in soliton theory, much attention is called to its link with infinite dimensional Lie algebras [1]-[8]. It has been shown by several authors [1]-[4] that these Lie algebras appear as infinitesimal transformations of solutions for soliton equations. The purpose of this paper is to give a survey of one such approach developed in [1], with emphasis on τ functions, method of field theory and vertex operators.

To simplify the exposition, we shall mainly deal with the Kadomtsev-Petviashvili (KP) equation

$$\frac{3}{4}\frac{\partial^2 u}{\partial y^2} = \frac{\partial}{\partial x}\left(\frac{\partial u}{\partial t} - \frac{3}{2}u\frac{\partial u}{\partial x} - \frac{1}{4}\frac{\partial^3 u}{\partial x^3}\right) \qquad (0.0.1)$$

and its relatives, such as the Korteweg-de Vries (KdV) equation

$$\frac{\partial u}{\partial t} = \frac{3}{2}u\frac{\partial u}{\partial x} + \frac{1}{4}\frac{\partial^3 u}{\partial x^3} \,. \qquad (0.0.2)$$

The reason why we pick up this particular equation is because of the basic role it plays, as will become clearer in the sequel.

§0.1. From the viewpoint of soliton theory, we aim at revealing the hidden symmetry (=the group of transformations of solutions) for various completely integrable nonlinear equations. For the KP equation (0.0.1), this is achieved by following the three steps.

(i) Introducing the hierarchy.

The KP equation (0.0.1) is the integrability condition for the linear system

$$\begin{aligned}\frac{\partial w}{\partial y} &= \left(\frac{\partial^2}{\partial x^2} + u\right)w\\ \frac{\partial w}{\partial t} &= \left(\frac{\partial^3}{\partial x^3} + \frac{3}{2}u\frac{\partial}{\partial x} + v\right)w.\end{aligned} \tag{0.1.1}$$

Calling x_1, x_2, x_3 the variables x, y, t, respectively, we extend (0.1.1) into a series of linear equations with respect to "higher time variables" $x_4, x_5, \cdots$

$$\frac{\partial w}{\partial x_n} = \left(\frac{\partial^n}{\partial x_1^n} + \frac{n}{2}u\frac{\partial^{n-2}}{\partial x_1^{n-2}} + \cdots\right)w, \qquad n=1,2,3,\cdots, \tag{0.1.2}$$

where now w and the coefficients $u, v, \cdots$ are functions in $x=(x_1, x_2, x_3, \cdots)$. The integrability condition gives rise to a set of nonlinear differential equations for the unknown functions $u, v, \cdots$, which we call the KP hierarchy.

(ii) τ function.

We rewrite the KP hierarchy by introducing a new dependent variable $\tau(x)$ (the τ function)

$$u = 2\frac{\partial^2}{\partial x_1^2}\log\tau, \quad v - \frac{3}{4}\frac{\partial u}{\partial x_1} = \frac{3}{2}\frac{\partial^2}{\partial x_1 \partial x_2}\log\tau, \cdots.$$

In terms of $\tau(x)$, the KP hierarchy takes the bilinear form of Hirota

$$(D_1^4 + 3D_2^2 - 4D_1D_3)\tau\cdot\tau = 0, \quad (D_1^3D_2 + 2D_2D_3 - 3D_1D_4)\tau\cdot\tau = 0, \cdots. \tag{0.1.3}$$

Here Hirota's bilinear operator ([9]) is defined by

$$P(D_1, D_2, \cdots)f\cdot g = P(\partial_{y_1}, \partial_{y_2}, \cdots)(f(x+y)g(x-y))\big|_{y_1=y_2=\cdots=0}.$$

(iii) Vertex operator.

The following linear differential operator of infinite order

$$Z(p,q) = \frac{q/p}{1-q/p}\left[\exp\left(\sum_{n=1}^{\infty}(p^n-q^n)x_n\right)\exp\left(-\sum_{n=1}^{\infty}\frac{1}{n}(p^{-n}-q^{-n})\frac{\partial}{\partial x_n}\right) - 1\right] \tag{0.1.4}$$

p,q: arbitrary parameters

is called the vertex operator. We show that the bilinear KP hierarchy allows the transformation

$$\tau(x) \mapsto e^{aZ(p,q)}\tau(x)$$

which sends a solution of (0.1.3) to another solution. The vertex operator (0.1.4) generates an infinitesimal Bäcklund transformation. The totality of $Z(p,q)$'s and 1 (multiplication by a constant) constitutes a non-trivial Lie algebra isomorphic to $\mathfrak{gl}(\infty)$.

To sum up, $\mathfrak{gl}(\infty)$ is the hidden symmetry of the KP hierarchy.

§0.2. One can give another description of the KP τ function from a group theoretical viewpoint. Here, field operators provide the most appropriate language.

(i) Free fermion.

We consider the infinite dimensional Clifford algebra $\mathbb{A}$, having generators ψ_n, ψ_n^* $(n \in \mathbb{Z})$ with the defining relation

$$[\psi_m, \psi_n]_+ = 0, \; [\psi_m, \psi_n^*]_+ = \delta_{mn}, \; [\psi_m^*, \psi_n^*]_+ = 0.$$

To $\mathbb{A}$ is associated the standard Fock representation: an irreducible left (resp. right) $\mathbb{A}$-module $\mathcal{F}$ (resp. $\mathcal{F}^*$) with the cyclic vector $|vac\rangle$ (resp. $\langle vac|$) satisfying

$$\psi_n|vac\rangle = 0 \quad (n<0), \quad \psi_n^*|vac\rangle = 0 \quad (n\geq 0)$$

$$\langle vac|\psi_n = 0 \quad (n\geq 0), \quad \langle vac|\psi_n^* = 0 \quad (n<0).$$

$\mathcal{F}$ and $\mathcal{F}^*$ are equipped with a nondegenerate pairing $\mathcal{F}\times\mathcal{F}^* \to \mathbb{C}$ such that $(\langle vac|, |vac\rangle) \mapsto 1$.

(ii) Realization of the Fock representation.

The set of quadratic elements $\sum a_{ij}\psi_i\psi_j^*$ and 1 spans a Lie algebra $\mathfrak{g}(V,V^*)$ isomorphic to $\mathfrak{gl}(\infty)$. The elements $\Lambda_n = \sum_{j\in\mathbb{Z}}\psi_j\psi_{j+n}^*$ $(n=1,2,\cdots)$ span an Abelian subalgebra thereof. We set

$$H(x) = \sum_{n=1}^{\infty} x_n\Lambda_n.$$

Then the Fock representation is realized on the vector space of polynomials $\mathbb{C}[x_1,x_2,\cdots,u,u^{-1}]$ as

$$\begin{aligned} \mathcal{F} &\xrightarrow{\sim} \mathbb{C}[x_1,x_2,\cdots,u,u^{-1}] \\ |a\rangle &\longmapsto f_a(x,u) = \sum_{m\in\mathbb{Z}} \langle m|e^{H(x)}|a\rangle u^m \end{aligned} \qquad (0.2.1)$$

where $\langle m| = \langle vac|\psi_{-1}^*\cdots\psi_m^*$ $(m<0)$, $= \langle vac|$ $(m=0)$, $= \langle vac|\psi_0\cdots\psi_{m-1}$ $(m\geq 1)$, and $\langle m|e^{H(x)}|a\rangle$ signifies the pairing of vectors $\langle m|$ and $e^{H(x)}|a\rangle$. In the right hand side of (0.2.1), "free fermion operators" $\psi(k) = \sum_{n\in\mathbf{Z}}\psi_n k^n$, $\psi^*(k) = \sum_{n\in\mathbf{Z}}\psi_n^* k^{-n}$ have the following realization.

$$\psi(k) : f(x,u) \mapsto f(x_1 - \tfrac{1}{k}, x_2 - \tfrac{1}{2k^2}, \cdots, ku)u \exp(\sum_{n=1}^{\infty} k^n x_n)$$

$$\psi^*(k) : f(x,u) \mapsto f(x_1 + \tfrac{1}{k}, x_2 + \tfrac{1}{2k^2}, \cdots, k^{-1}u)ku^{-1} \exp(-\sum_{n=1}^{\infty} k^n x_n).$$

(iii) The space of τ functions.

The Lie group corresponding to $\mathfrak{g}(V,V^*)$ is the Clifford group given by

$$G(V,V^*) = \{g \in \mathbb{A} \mid \exists\, a_{mn} \in \mathbb{C} \text{ such that}$$

$$g\psi_n = \sum_{m\in\mathbb{Z}} \psi_m g a_{mn},\quad \psi_n^* g = \sum_{m\in\mathbb{Z}} g\psi_m^* a_{nm}\}.$$

For any $g \in G(V,V^*)$, the realization of the vector $g|vac\rangle$ by (0.2.1)

$$\tau_g(x) = \sum_{m\in\mathbb{Z}} \langle m|e^{H(x)}g|vac\rangle u^m = \langle vac|e^{H(x)}g|vac\rangle$$

satisfies the bilinear KP hierarchy (0.1.3), and vice versa. In other words, the space of KP τ functions is the $G(V,V^*)$-orbit of the vacuum vector $|vac\rangle$

$$G(V,V^*)|vac\rangle \simeq \{\tau_g(x)\}_{g\in G(V,V^*)},$$

which is naturally equipped with the structure of the Grassmann manifold (Sato [2]).

§0.3. When we require the condition $\frac{\partial u}{\partial y} = 0$ on the KP equation (0.0.1), it is reduced to the KdV equation (0.0.2). The general solution involves functional parameters in two variables in the KP case, and in one variable in the KdV case. The former is called sub-sub-holonomic, while the latter is called sub-holonomic, according to this difference in nature. On the level of transformation groups, the reduction procedure above amounts to considering a subfamily

46

of τ functions obtained as the orbit of $|vac\rangle$ by a subalgebra $\mathfrak{g}$ of $\mathfrak{gl}(\infty)$. A number of sub-holonomic soliton equations are systematically generated in this way.

Let ℓ be a positive integer. The ℓ-reduced KP hierarchy ($(KP)_\ell$) is defined by one of the following equivalent conditions,

(i) Impose $\frac{\partial w}{\partial x_n} = k^n w$ $(n \equiv 0 \mod \ell)$ on the linear problem (0.1.2).

(ii) Impose $\frac{\partial}{\partial x_n}\tau(x) = 0$ $(n \equiv 0 \mod \ell)$.

(iii) Restrict the vertex operators (0.1.4) to

$$Z(p,\omega p), \quad \omega^\ell = 1.$$

(iv) Infinitesimal transformations in $\mathfrak{g} \subset \mathfrak{g}(V,V^*)$ are generated by elements invariant under the change $\psi_i \mapsto \psi_{i+\ell}$, $\psi_j^* \mapsto \psi_{j+\ell}^*$.

(v) $\mathfrak{g}$ is isomorphic to the Euclidean Lie algebra $A_{\ell-1}^{(1)}$.

Among others,

$(KP)_2$ = the KdV hierarchy

$(KP)_3$ = the Boussinesq hierarchy

are well known.

Viewed from the other way, a concrete representation of the Lie algebra $A_{\ell-1}^{(1)}$ is obtained by specializing the realization (0.2.1) (this, in fact, coincides with the basic representation constructed in [10]). For instance, the coefficients of

$$Z(p,p) = \sum_{n=1}^{\infty} p^n x_n + \sum_{n=1}^{\infty} p^{-n}\frac{\partial}{\partial x_n}$$

$$Z(p,-p) = \frac{1}{2}\,[\exp(2\sum_{n\ \mathrm{odd}} p^n x_n)\exp(-2\sum_{n\ \mathrm{odd}}\frac{1}{n}p^{-n}\frac{\partial}{\partial x_n}) - 1$$

47

give the basic representation of $A_1^{(1)}$ on $\mathbb{C}[x_1,x_3,x_5,\cdots]$ first constructed by Lepowsky-Wilson [11].

§0.4. Such is the outline of our story. Naturally one may think of several variants:

(1) to find sub-sub-holonomic hierarchies corresponding to groups other than $\mathfrak{gl}(\infty)$,

(2) to extend this framework to the multicomponent case (the matrix system),

(3) to find other types of sub-holonomic reductions.

The problems (1), (2) are treated in [1]. In general, we must introduce several kinds of τ functions and consider Hirota's equations of the form $\sum P_{ij}(D)\tau_i\cdot\tau_j=0$. If we restrict ourselves to hierarchies described by a single τ function, three sub-sub-holonomic ones are known: the KP, the BKP and the 2-component BKP hierarchies. The corresponding Lie algebras are $\mathfrak{gl}(\infty)$, $\mathfrak{o}(\infty)$ and $\mathfrak{o}(2\infty)$, respectively. As for (3), an example of reduction related to an elliptic curve (the Landau-Lifshitz equation) is discussed in [12].

We postpone detailed discussions of these topics to another occasion.

§0.5. This paper is organized as follows.

In §1, we give a systematic study of the KP hierarchy. Here it is formulated to be the integrability of linear systems ([2])

$$L(x,\partial)w = kw,$$

$$\frac{\partial w}{\partial x_n} = B_n(x,\partial)w, \qquad \partial = \partial/\partial x_1 \tag{0.5.1}$$

where $L(x,\partial)=\partial+u_2(x)\partial^{-1}+u_3(x)\partial^{-2}+\cdots$ is a pseudo-differential operator, and

$B_n(x,\partial)$ = the differential operator part of $L(x,\partial)^n$.

We prove the existence of τ function, and the expression of the wave function (solution to (0.5.1))

$$w(x,k) = \frac{\tau(x_1-\frac{1}{k},x_2-\frac{1}{2k^2},x_3-\frac{1}{3k^3},\cdots)}{\tau(x_1,x_2,x_3,\cdots)} \exp(\sum_{n=1}^{\infty} k^n x_n).$$

The derivation is based on the basic identity

$$\mathrm{Res}_{k=\infty} w(x,k)w^*(x',k)dk = 0 \qquad \text{for any } x,x', \tag{0.5.2}$$

where the adjoint wave function $w^*(x,k)$ is defined by

$$w^*(x,k) = \frac{\tau(x_1+\frac{1}{k},x_2+\frac{1}{2k^2},x_3+\frac{1}{3k^3},\cdots)}{\tau(x_1,x_2,x_3,\cdots)} \exp(-\sum_{n=1}^{\infty} k^n x_n).$$

Rewriting (0.5.2), we obtain the generating function for the bilinear KP equations for $\tau(x)$.

We also discuss briefly the BKP hierarchy, and state the relation between the KP and the BKP τ functions. Finally, we describe construction of quasi-periodic solutions for both of them.

§2 is devoted to the formulation in terms of free fermions operators. As mentioned in §0.2, the KP hierarchy is described by using charged free fermions ψ_n, ψ_n^*, while the BKP hierarchy is done in terms of neutral free fermions

ϕ_n. Construction of soliton or polynomial solutions is particularly simple in this language. We show how the character polynomials for the general linear group appear as KP τ functions (Sato [2]).

In §3, we discuss the reduction problem. As the subalgebras corresponding to the ℓ-reduced KP or BKP hierarchies, we find the Euclidean Lie algebras of the type $A^{(1)}_{\ell-1}$, $A^{(2)}_{2\ell}$ or $D^{(2)}_{\ell}$.†) In the latter half of this §3, we give one application of this connection with Euclidean Lie algebra theory: to count the number of linearly independent Hirota equations of a given homogeneous degree

$$h_S(m) = \dim_{\mathbb{C}}\{P(D) \in \mathbb{C}[D] \mid \deg P = m,\ P(D)\tau\cdot\tau = 0 \text{ for any } \tau \text{ function of the hierarchy } S\}.$$

Here we count the degree of D_j to be j. This problem, originally proposed by Sato [13], is shown to reduce to the Weyl-Kac character formula for Euclidean Lie algebras.

---

†) The Lie algebras of the type $A^{(2)}_{2n-1}$ and $D^{(1)}_n$ are obtained as reductions of the 2-component BKP hierarchy [1].

§1. The Kadomtsev-Petviashivili equation.

1.1 Introduction. In this section we give a description of the hierarchy of the Kadomtsev-Petviashivili (KP) equation, or the two-dimensional KdV equation.

Following Sato [2], we start with a spectral problem of a first order pseudo-differential operator and its deformation equations. These constitute a linear system whose compatibility condition gives the KP hierarchy.

We consider a solution $w(x,k)$ (where $x = (x_1, x_2, \ldots)$ is the deformation variables and k is the spectral parameter) to the linear system which has a certain exponential behavior at $k = \infty$. This solution is called the wave function. The key idea is to formulate a bilinear identity for the wave function and its adjoint.

The bilinear identity is equivalent not only to the linear system but also to the whole family of Hirota's bilinear differential equations. This fact is established by introducing the τ function (Hirota's dependent variable) and by relating it with the wave function through a shift of the deformation variables.

The bilinear identity is also useful in order to show that a kind of infinite order linear differential operators act infinitesimally on the space of τ functions of the KP hierarchy. These operators are called vertex operators. In §2 we discuss an infinite dimensional Lie algebra of vertex operators in detail.

The last two paragraphs of §1 deal with the KP hierarchy

of B type and the quasi-periodic solutions.

1.2. KdV equation. Among many ways of introducing the KdV equation, this is obtained as the compatibility condition of the system of linear differential equations:

$$\left(\frac{\partial^2}{\partial x^2} + a(t,x)\right)w(t,x,\lambda) = \lambda w(t,x,\lambda), \qquad (1.2.1)$$

$$\frac{\partial}{\partial t}w(t,x,\lambda) = \left(\frac{\partial^3}{\partial x^3} + b(t,x)\frac{\partial}{\partial x} + c(t,x)\right)w(t,x,\lambda).$$

If we replace the third order differential operator with a higher order differential operator of odd order, we obtain the so called higher order KdV equation. In order to treat all the higher order KdV equations at once, we introduce infinitely many independent variables $x = (x_1,x_3,x_5,\dots)$ and consider the compatibility conditions of the following system of linear differential equations:

$$\left(\frac{\partial^2}{\partial x_1^2} + a(x)\right)w(x,\lambda) = \lambda w(x,\lambda) \qquad (1.2.2)$$

$$\frac{\partial}{\partial x_n}w(x,\lambda) = B_n(x,\partial)w(x,\lambda) \qquad (n = 3,5,\dots)$$

where $B_n(x,\partial) = \partial^n + b_{n,n-2}(x)\partial^{n-2} + \dots + b_{n,0}(x)$ and $\partial = \frac{\partial}{\partial x_1}$. The system of non-linear differential equations thus obtained with $b_{n,j}(x)$'s as dependent variables is called the KdV hierarchy.

1.3. KP equation. The KP equation is obtained through a

generalization of the above formalism. The KdV equation is an equation which describes the eigenvalue preserving deformation of a second order differential operator, while the KP equation describes that of a first order pseudo-differential operator.

We consider a system of linear equations for $w(x,k)$

$$L(x,\partial)w(x,k) = kw(x,k)$$
$$\frac{\partial}{\partial x_n}w(x,k) = B_n(x,\partial)w(x,k) \qquad (1.3.1)$$

where $x = (x_1,x_2,x_3,\ldots)$, $\partial = \frac{\partial}{\partial x_1}$, and

$$L(x,\partial) = \partial+u_1(x)+u_2(x)\partial^{-1}+u_3(x)\partial^{-2}+\ldots,$$

$$B_n(x,\partial) = \partial^n+\sum_{j=0}^{n-1} b_{nj}(x)\partial^j.$$

The KP hierarchy is a system of non-linear differential equations for $u_j(x)$ and $b_{nj}(x)$, resulting from the compatibility conditions of (1.3.1):

$$\frac{\partial L}{\partial x_n} = [B_n,L], \qquad (1.3.2)$$

$$\frac{\partial B_m}{\partial x_n} - \frac{\partial B_n}{\partial x_m} = [B_n,B_m]. \qquad (1.3.3)$$

If we assume that $B_2 = L^2$ is a differential operator and set $\lambda = k^2$, the KP hierarchy reduces to the KdV hierarchy.

For a pseudo-differential operator $P(x,\partial) = \sum_{j\in\mathbb{Z}} a_j(x)\partial^j$, we denote by $(P)_+$ the differential operator part and by $(P)_-$ the residual part;

$$(P)_+ = \sum_{j \geq 0} a_j(x)\partial^j, \tag{1.3.4}$$

$$(P)_- = P-(P)_+ = \sum_{j<0} a_j(x)\partial^j.$$

Since $e^{\varphi}Le^{-\varphi} = \partial+(u_1(x)-\frac{\partial\varphi}{\partial x_1})+\ldots$, we may assume that the coefficient $u_1(x)$ of ∂^0 in L equals to zero by multiplying a function in x to $w(x,k)$. Then $[B_n-L^n,L]$ becomes of order < 0 by (1.3.2). If we put $B_n-L^n = a(x)\partial^r$ +(lower order terms) $(r < n)$, we have

$$[B_n-L^n,L] = -\frac{\partial a(x)}{\partial x_1}\partial^r+(\text{lower order terms}).$$

This shows that $a(x)$ depends trivially on x_1 if $r \geq 0$. Therefore, it is natural to impose $a(x) = 0$. By repeating this procedure, we may assume $r < 0$, or

$$B_n(x,\partial) = (L(x,\partial)^n)_+. \tag{1.3.5}$$

Under this assumption, (1.3.3) is a consequence of (1.3.2). In fact, (1.3.2) implies $\frac{\partial}{\partial x_n}L^m = [B_n,L^m]$ for any m. Hence we have

$$\frac{\partial B_m}{\partial x_n} - \frac{\partial B_n}{\partial x_m} - [B_n,B_m]$$

$$= (\partial_n L^m - \partial_m L^n - [B_n,B_m])_+$$

$$= ([B_n, L^m] - [B_m, L^n] - [B_n, B_m])_+$$

$$= ([B_n - L^n,\ B_m - L^m])_+$$

$$= 0.$$

Thus the KP hierarchy is a deformation equation of a pseudo-differential operator

$$L(x,\partial) = \partial + u_2(x)\partial^{-1} + u_3(x)\partial^{-2} + \dots$$

under the condition (1.3.2) together with (1.3.5).

1.4. <u>The wave function and the adjoint wave function</u>. The system of linear euqaitons (1.3.1) has a solution $w(x,k)$ of the form

$$w(x,k) = \hat{w}(x,k)e^{\xi(x,k)}, \qquad (1.4.1)$$

$$\hat{w}(x,k) = 1 + w_1(x)k^{-1} + w_2(x)k^{-2} + \dots,$$

where

$$\xi(x,k) = x_1 k + x_2 k^2 + x_3 k^3 + \dots .$$

In fact, (1.3.2), (1.3.3) and (1.3.5) show that the system of the equations for P

$$LP = P\partial \qquad (1.4.2)$$

$$\partial_n P = -(L^n)_- P \qquad (n = 2,3,4,\dots)$$

is compatible and has a pseudo-differential operator

$$P = 1+w_1(x)\partial^{-1}+w_2(x)\partial^{-2}+\dots \tag{1.4.3}$$

as a solution. By setting $w(x,k) = Pe^{\xi(x,k)} = (1+w_1(x)k^{-1}+w_2(x)k^{-2}+\dots)e^{\xi(x,k)}$, we obtain the solution to (1.4.1). In fact, we have

$$Lw(x,k) = LPe^{\xi(x,k)} = P\partial e^{\xi(x,k)} = kw(x,k)$$

and

$$\frac{\partial}{\partial x_n}w(x,k) = \frac{\partial P}{\partial x_n}e^{\xi(x,k)}+Pk^n e^{\xi(x,k)}$$

$$= -(L^n)_- Pe^{\xi(x,k)}+P\partial^n e^{\xi(x,k)}$$

$$= -(L^n)_- Pe^{\xi(x,k)}+L^n Pe^{\xi(x,k)}$$

$$= B_n w(x,k).$$

We shall call $w(x,k)$ the wave function. The wave function is uniquely determined up to multiplication by a function in k.

Denoting by $P^*(x,\partial) = 1+(-\partial)^{-1}w_1(x)+(-\partial)^{-2}w_2(x)+\dots$ the formal adjoint of $P(x,\partial)$, we set

$$w^*(x,k) = P^*(x,\partial)^{-1}e^{-\xi(x,k)},$$

and call it the adjoint wave function. The adjoint wave function has the form

$$w^*(x,k) = \hat{w}^*(x,k)e^{-\xi(x,k)} \tag{1.4.4}$$

$$\hat{w}^*(x,k) = 1+w_1^*(x)k^{-1}+w_2^*(x)k^{-2}+\ldots$$

and satisfies the equations

$$L^*w^*(x,k) = kw^*(x,k), \tag{1.4.5}$$

$$\frac{\partial}{\partial x_n}w^*(x,k) = -B_n^*(x,\partial)w^*(x,k).$$

1.5. Bilinear identity. The wave function and the adjoint wave function for the KP hierarchy satisfy the "bilinear identity"

$$\int w(x,k)w^*(x',k)dk = 0 \tag{1.5.1}$$

for any x and x'. Here the integral is taken along a contour around $k = \infty$.

In order to prove (1.5.1), we note the following

Lemma 1.1. For pseudo-differential operators $P(t,\partial_t)$ and $Q(t,\partial_t)$, we have

$$\int (P(t,\partial_t)e^{tk}(Q(t',\partial_{t'})e^{-t'k})\frac{dk}{2\pi i} \tag{1.5.2}$$

$$= f(t,t')$$

57

where $f(t,t')$ is a function given by

$$f(t,t')Y(t-t') = (P(t,\partial_t)Q^*(t,\partial_t))_-\delta(t-t')$$

with $Y(t) = \partial_t^{-1}\delta(t)$ denoting the Heaviside function.

Proof. Set $P(t,\partial_t) = \sum a_j(t)\partial_t^j$ and $Q(t,\partial_t) = \sum b_j(t)(-\partial_t)^j$. Then we have $P(t,\partial_t)e^{tk} = (\sum a_j(t)k^j)e^{tk}$ and $Q(t,\partial_t)e^{-tk} = \sum b_j(t)k^j e^{-tk}$. Hence the integral (1.5.2) is equal to

$$\sum a_j(t)b_i(t')\int k^{j+i}e^{(t-t')k}\frac{dk}{2\pi i}$$

$$= \sum_{j+i\leq -1} a_j(t)b_i(t')\frac{(t-t')^{-1-j-i}}{(-1-j-i)!}.$$

On the other hand, we have

$$P(t,\partial_t)Q^*(t,\partial_t)\delta(t-t')$$

$$= P(t,\partial_t)Q(t',\partial_{t'})\delta(t-t')$$

$$= \sum a_j(t)b_i(t')\partial_t^{j+i}\delta(t-t')$$

$$= \sum_{j+i\geq 0} a_j(t)b_i(t')\partial_t^{j+i}\delta(t-t')$$

$$+ \sum_{j+i\leq -1} a_j(t)b_i(t')\frac{(t-t')^{-1-j-i}}{(-1-j-i)!}Y(t-t').$$

This implies $f(t,t') = \sum_{j+i\leq -1} a_j(t)b_i(t')\frac{(t-t')^{-1-j-i}}{(-1-j-i)!}$.

Lemma 1.1 shows that (1.5.1) is true when $x_j = x_j'$ for $j \geq 2$. Moreover we have

$$\int (\frac{\partial}{\partial x_1})^\alpha w(x,k) \cdot w^*(x,k) dk = 0 \qquad (1.5.3)$$

for any $\alpha \geq 0$. In order to prove (1.5.1) for any x and x', it is sufficient to show

$$\int (\frac{\partial}{\partial x_1})^{\alpha_1} \dots (\frac{\partial}{\partial x_n})^{\alpha_n} w(x,k) \cdot w^*(x,k) dk = 0 \qquad (1.5.4)$$

for any n and $\alpha_1, \dots, \alpha_n \geq 0$. Since $w(x,k)$ satisfies (1.3.1), $(\frac{\partial}{\partial x_1})^{\alpha_1} \dots (\frac{\partial}{\partial x_n})^{\alpha_n} w(x,k)$ is written in the form $\sum c_j(x)(\frac{\partial}{\partial x_1})^j w(x,k)$. Therefore (1.5.4) follows from (1.5.3).

Conversely, we have

<u>Proposition 1.2</u>. Let $w(x,k)$ and $w^*(x,k)$ be functions of the form (1.4.1) and (1.4.4). If the bilinear identity (1.5.1) is satisfied, then $w(x,k)$ is a wave function for the KP hierarchy, and $w^*(x,k)$ is its adjoint wave function.

<u>Proof</u>. Define $P(x,\partial)$ and $Q(x,\partial)$ by $w(x,k) = P(x,\partial)e^{\xi(x,k)}$ and $w^*(x,k) = Q(x,\partial)e^{-\xi(x,k)}$. Then by applying Lemma 1.1, we have $(PQ^*)_- = 0$, i.e. $PQ^* = 1$. Set $L = P\partial P^{-1}$ and $B_n = (L^n)_+$. Then (1.5.1) implies

$$\int (\frac{\partial}{\partial x_n} - B_n) w(x,k) \cdot w^*(x',k) dk = 0.$$

We have

$$(\frac{\partial}{\partial x_n} - B_n)w(x,k)$$

$$= (\frac{\partial}{\partial x_n}P+k^nP-B_nP)e^{\xi(x,k)}$$

$$= (\frac{\partial}{\partial x_n}P+P\partial^n-B_nP)e^{\xi(x,k)}$$

$$= (\frac{\partial}{\partial x_n}P+L^nP-B_nP)e^{\xi(x,k)}$$

$$= (\frac{\partial}{\partial x_n}P+(L^n)_-P)e^{\xi(x,k)}.$$

Since $\frac{\partial}{\partial x_n}P+(L^n)_-P$ is of order < 0, we have $\frac{\partial}{\partial x_n}P+(L^n)_-P = 0$, again by using Lemma 1.1. This shows that $w(x,k)$ is a wave function of the KP hierarchy.

Thus, the KP hierarchy, which is a system of infinitely many linear equations and its compatibility condiiton, is rephrazed by the bilinear identity (1.5.1).

1.6. <u>The τ function for the KP hierarchy</u>. We shall introduce the τ function for the KP hierarchy. We define a one form

$$\omega(x,dx)$$

$$= \sum_{i\geq 1} dx_i \mathrm{Res}_{k=\infty} k^i (\sum_{j\geq 1} k^{-j-1}\frac{\partial}{\partial x_j} - \frac{\partial}{\partial k})\log \hat{w}(x,k)dk. \tag{1.6.1}$$

Then, as shown later, $\omega(x,dx)$ is a closed one form, and the τ function is then defined by

$$d \log \tau(x) = \omega(x,dx) \tag{1.6.2}$$

or

$$\frac{\partial}{\partial x_n} \log \tau(x) = \mathrm{Res}_{k=\infty} k^n \Big(\sum_{j\geq 1} k^{-j-1} \frac{\partial}{\partial x_j} - \frac{\partial}{\partial k} \Big) \log \hat{w}(x,k)dk.$$

Hence, for a given $w(x,k)$, $\tau(x)$ is determined up to a constant multiple.

Conversely, the wave function and the adjoint wave function are neatly expressed in terms of the τ function:

$$w(x,k) = \frac{\tau(x_1 - \frac{1}{k}, x_2 - \frac{1}{2k^2}, x_3 - \frac{1}{3k^3}, \ldots)}{\tau(x)} e^{\xi(x,k)},$$

$$w^*(x,k) = \frac{\tau(x_1 + \frac{1}{k}, x_2 + \frac{1}{2k^2}, x_3 + \frac{1}{3k^3}, \ldots)}{\tau(x)} e^{-\xi(x,k)}. \tag{1.6.3}$$

Note that (1.6.3) is nothing but a rephrase of (1.6.2). In the formula (1.6.3), introduction of infinitely many variables $x_1, x_2, x_3, \ldots$ is indispensable. Nevertheless, the closedness of ω, and hence the definition (1.6.2) of $\tau(x)$, is valid and meaningful even if we introduce only finitely many independent variables.

Now, we shall prove these facts using the bilinear identity (1.5.1). First we prepare the δ-function with $t = 1$ as its support:

$$\hat{\delta}(t) = \sum_{n\in\mathbb{Z}} t^n = \frac{1}{1-t} + \frac{t^{-1}}{1-t^{-1}}.$$

Here $\frac{1}{1-t}$ and $\frac{t^{-1}}{1-t^{-1}}$ are regarded as their Taylor expansions with respect to t and t^{-1}, respectively. This $\hat{\delta}(t)$

satisfies the following property.

$$f(t,k)\hat{\delta}(t/k) = f(t,t)\hat{\delta}(t/k) \qquad (1.6.4)$$

$$\int f(t,k)\hat{\delta}(t/k)\frac{dk}{2\pi ik} = f(t,t). \qquad (1.6.5)$$

We shall apply (1.5.1) by setting $x_j' = x_j - \frac{1}{j}t_1^{-j} - \frac{1}{j}t_2^{-j}$. Then we have

$$\int \hat{w}(x,k)G(t_1)G(t_2)\hat{w}^*(x,k)\frac{dk}{(1-k/t_1)(1-k/t_2)} = 0,$$

where $G(t)$ is defined by

$$G(t)f(x) = f(x_1 - \frac{1}{t}, x_2 - \frac{1}{2t^2}, x_3 - \frac{1}{3t^3}, \ldots). \qquad (1.6.6)$$

We have

$$\frac{1}{(1-k/t_1)(1-k/t_2)} = \frac{t_1/k}{1-t_1/t_2}(\hat{\delta}(k/t_1) - \hat{\delta}(k/t_2)) + \frac{t_1t_2/k^2}{(1-t_1/k)(1-t_2/k)}.$$

Since the last term gives no effect to the integration, we obtain

$$\hat{w}(x,t_1)G(t_1)G(t_2)\hat{w}^*(x,t_1) = \hat{w}(x,t_2)G(t_1)G(t_2)\hat{w}^*(x,t_2). \qquad (1.6.7)$$

By setting $t_1 = k$ and $t_2^{-1} = 0$ we have

$$\hat{w}(x,k)^{-1} = G(k)\hat{w}^*(x,k). \tag{1.6.8}$$

If we eliminate $\hat{w}^*$ in (1.6.7) by using (1.6.8), we obtain

$$\frac{\hat{w}(x,t_1)}{G(t_2)\hat{w}(x,t_1)} = \frac{\hat{w}(x,t_2)}{G(t_1)\hat{w}(x,t_2)}. \tag{1.6.9}$$

We set $f(x,k) = \log \hat{w}(x,k)$. Then (1.6.9) is equivalent to

$$f(x,k) = G(t)f(x,k)+f(x,t)-G(k)f(x,t). \tag{1.6.10}$$

Therefore we have

$$(\sum k^{-\nu-1}\frac{\partial}{\partial x_\nu} - \frac{\partial}{\partial k})f(x,k)$$

$$= G(t)(\sum k^{-\nu-1}\frac{\partial}{\partial x_\nu} - \frac{\partial}{\partial k})f(x,k) + \sum k^{-\nu-1}\frac{\partial}{\partial x_\nu}f(x,t),$$

which implies

$$\omega(x,dx) = G(t)\omega(x,dx) - df(x,t). \tag{1.6.11}$$

Here d denotes the exterior differentiation with respect to x.

By differentiating again, we obtain

$$d\omega(x,dx) = G(t)d\omega(x,dx),$$

which implies that the coefficients of $d\omega(x,dx)$ are constant functions:

$$d\omega(x,dx) = \sum a_{ij} dx_i \wedge dx_j$$

with $a_{ij} = -a_{ji} \in \mathbf{C}$. Hence

$$\omega(x,dx) = \sum a_{ij} x_i dx_j + dh(x)$$

for some function $h(x)$. Substituting this into (1.6.11) we have

$$df(x,t) = d(G(t)h(x)-h(x)-\sum \frac{1}{it^i} a_{ij} x_j).$$

Hence we obtain

$$f(x,t) = G(t)h(x)-h(x)-\sum \frac{1}{it^i} a_{ij} x_j + \varphi(t)$$

for some function $\varphi(t)$. Substituting this into (1.6.10) we obtain $a_{ij} = 0$. Thus we have shown $d\omega(x,dx) = 0$.

1.7. Bilinear equations for the τ function. In 1.3 we showed that the KP hierarchy is a system of non linear equations for the coefficients $u_2(x)$, $u_3(x)$,... of $L(x,\partial)$. Since $L = P\partial P^{-1}$, (1.4.3) and (1.6.3) imply that the KP hierarchy is described by only one dependent variable $\tau(x)$. More explicitly, $u_n(x)$ are recovered from $\tau(x)$ by the formula

$$\frac{\partial^2}{\partial x_1 \partial x_n}\log \tau(x) = \text{the coefficient of } \partial^{-1} \text{ in } L(x,\partial)^n$$

$$= nu_{n+1}(x) + \binom{n}{2}\frac{\partial}{\partial x_1}u_n(x)$$

$$+ (\text{terms involving } u_2(x),\ldots,u_{n-1}(x)).$$

In general, $\frac{\partial^2}{\partial x_j \partial x_k}\log \tau(x)$ is expressible as a differential polynomial of $u_n(x)$.

Now we shall derive the equations for the τ function itself. The remarkable fact is that they are written in Hirota's bilinear form as we shall see shortly.

By (1.5.1) and (1.6.3) we obtain

$$\int \tau(x_1-\tfrac{1}{k}, x_2-\tfrac{1}{2k^2},\ldots)\tau(x_1'+\tfrac{1}{k}, x_2'+\tfrac{1}{2k^2},\ldots)e^{\xi(x-x',k)}dk = 0.$$

By changing the variables, we have

$$0 = \int \tau(x_1-y_1-\tfrac{1}{k}, x_2-y_2-\tfrac{1}{2k^2},\ldots)$$

$$\times\tau(x_1+y_1+\tfrac{1}{k}, x_2+y_2+\tfrac{1}{2k^2},\ldots)e^{-2\xi(y,k)}dk$$

$$= \int e^{\xi(\tilde{\partial}_y,k^{-1})}(\tau(x+y)\tau(x-y))e^{-2\xi(y,k)}dk$$

$$= \int (\textstyle\sum k^{-j}p_j(\tilde{\partial}_y))(\tau(x+y)\tau(x-y))(\sum k^{\ell}p_{\ell}(-2y))dk.$$

Here $p_j(x)$ is defined by

$$e^{\xi(x,k)} = \sum k^j p_j(x),$$

and $\tilde{\partial}_y = (\partial_{y_1}, \frac{1}{2}\partial_{y_2}, \frac{1}{3}\partial_{y_3}, \ldots)$. Thus we obtain

$$\sum_{j=0}^{\infty} p_j(-2y)p_{j+1}(\tilde{\partial}_y)\tau(x+y)\tau(x-y) = 0$$

or

$$\sum_{j=0}^{\infty} p_j(-2y)p_{j+1}(\tilde{\partial}_z)e^{\sum_{\ell=1}^{\infty} y_\ell \frac{\partial}{\partial z_\ell}}\tau(x+z)\tau(x-z)\Big|_{z=0} = 0$$

If we employ Hirota's bilinear operator

$$P(D)f(x)\cdot g(x) = P(\partial_y)(f(x+y)g(x-y))\Big|_{y=0}$$

we obtain

$$\sum_{j=0}^{\infty} p_j(-2y)p_{j+1}(\tilde{D}_x)e^{\sum_{\ell=1}^{\infty} y_\ell D_x}\tau(x)\cdot\tau(x) = 0, \qquad (1.7.1)$$

where $\tilde{D}_x = (D_1, \frac{1}{2}D_2, \frac{1}{3}D_3, \ldots)$ signify Hirota's operators with respect to x. The first two non trivial equations are

$$(D_1^4+3D_2^2-4D_1D_3)\tau(x)\cdot\tau(x) = 0,$$

$$(D_1^3D_2+2D_2D_3-3D_1D_4)\tau(x)\cdot\tau(x) = 0.$$

1.8. Vertex operators and soliton solutions. The key in the analysis of 1.6 and 1.7 was that the τ function $\tau(x)$ and the wave function $w(x,k)$ are related with each other by an infinite order linear differential operator:

$$w(x,k) = X(k)\tau(x)/\tau(x)$$

66

where

$$X(k) = e^{\xi(x,k)}e^{-\xi(\tilde{\partial},k^{-1})},$$

$$\tilde{\partial} = (\frac{\partial}{\partial x_1},\frac{1}{2}\frac{\partial}{\partial x_2},\frac{1}{3}\frac{\partial}{\partial x_3},\ldots).$$

Such an operator is called a vertex operator.

Now we introduce another vertex operator:

$$X(p,q) = e^{\xi(x,p)-\xi(x,q)}e^{-\xi(\tilde{\partial},p^{-1})+\xi(\tilde{\partial},q^{-1})}. \quad (1.8.1)$$

The fact is that this vertex operator acts infinitesimally on the space of τ functions of the KP hierarchy. Namely, we claim the following.

$$\int G(k)(X(p.q)\tau(x))G^*(k)\tau(x')e^{\xi(x-x',k)}dk$$

$$+\int G(k)\tau(x)G^*(k)(X'(p,q)\tau(x'))e^{\xi(x-x',k)}dk = 0. \quad (1.8.2)$$

Here we set

$$G(k)f(x) = f(x_1-\frac{1}{k},x_2-\frac{1}{2k^2},x_3-\frac{1}{3k^3},\ldots),$$

$$G^*(k)f(x) = f(x_1+\frac{1}{k},x_2+\frac{1}{2k^2},x_3+\frac{1}{3k^3},\ldots), \quad (1.8.3)$$

and we mean by $X'(p.q)$ the vertex operator (1.8.1) with x replaced by x'.

Let us check (1.8.2). The first integral reads as

67

$$\int G(k)G(p)G^*(q)\tau(x)\cdot G^*(k)\tau(x')$$

$$\times e^{\xi(x-x',k)+\xi(x,p)-\xi(x,q)}\frac{1-p/k}{1-q/k}dk. \qquad (1.8.4)$$

On the other hand, by operating $G(p)G^*(q)$ to the bilinear identity (1.5.1), we have

$$\int G(k)G(p)G^*(q)\tau(x)\cdot G^*(k)\tau(x')$$

$$\times e^{\xi(x-x',k)}\frac{1-k/p}{1-k/q}dk = 0.$$

Since $\frac{-p/k}{1-q/k} = \frac{q-p}{k}\hat{\delta}(q/k)+\frac{p(1-k/p)}{q(1-k/q)}$, the integral (1.8.4) reduces to

$$2\pi i(q-p)G(p)\tau(x)\cdot G^*(q)\tau(x')\cdot e^{\xi(x,p)-\xi(x',q)}.$$

The second term in (1.8.2) can be similarly calculated and cancels the first term.

The constant function 1 is a τ function of the KP hierarchy. Since the vertex operator $X(p,q)$ acts on the space of τ functions of the KP hierarchy,

$$\tau(x;a_1,p_1,q_1;\ldots;a_N,p_N,q_N) = e^{\sum_{j=1}^{N} a_j X(p_j,q_j)}1 \qquad (1.8.5)$$

gives another τ function of the KP hierarchy.

A direct calculation yields

$$X(p,q)e^{\sum_j(\xi(x,p_j)-\xi(x,q_j))} = \prod_j\frac{(p-p_j)(q-q_j)}{(p-q_j)(q-p_j)}$$

$$\times e^{\xi(x,p)-\xi(x,q)+\sum_j(\xi(x,p_j)-\xi(x,q_j))},$$

and hence $X(p,q)^2 e^{\sum(\xi(x,p_j)-\xi(x,q_j))} = 0$. Therefore we have

$$\tau(x;a_1,p_1,q_1;\ldots;a_N,p_N,q_N)$$

$$= 1+\sum_{j=1}^{N} a_j e^{\xi_j} + \sum_{1\le i<j\le N} c_{ij} a_i a_j e^{\xi_i+\xi_j} + \ldots$$

$$= \sum_{\substack{0\le r\le N \\ 1\le j_1<\ldots<j_r\le N}} \prod_{\nu=1}^{r} a_{j_\nu} \prod_{1\le\nu<\mu\le r} c_{j_\nu j_\mu} e^{\sum_{\nu=1}^{r}\xi_{j_\nu}}, \qquad (1.8.6)$$

where $\xi_j = \xi(x,p_j)-\xi(x,q_j)$ and

$$c_{jk} = \frac{(p_j-p_k)(q_j-q_k)}{(p_j-q_k)(q_j-p_k)}.$$

The solution (1.8.6) is called the N-soliton solution of the KP hierarchy ([14]). It is easy to see that

$$e^{aX(p,q)}\tau(x;a_1,p_1,q_1;\ldots;a_N,p_N,q_N)$$

$$= \tau(x;a_1,p_1,q_1;\ldots;a_N,p_N,q_N;a,p,q).$$

Thus the vertex operator represents the infinitesimal transformation of adding 1 soliton, which is knwon as the Bäcklund transformation.

In usual textbooks it is said that two Bäcklund transformations commute. This is true in the sense

$$[X(p_1,q_1),X(p_2,q_2)] = 0$$

if $p_1 \neq q_2$ and $p_2 \neq q_1$. But a careful calculation of the bracket $[X(p_1,q_1),X(p_2,q_2)]$ shows that there remain δ function terms:

$$[X(p_1,q_1),X(p_2,q_2)]$$

$$= \frac{(1-q_1/p_1)(1-q_2/p_2)}{1-q_2/p_1} X(p_1,q_2)\delta(p_2/q_1)$$

$$- \frac{(1-q_1/p_1)(1-q_2/p_2)}{(1-q_1/p_2)} X(p_2,q_1)\delta(p_1/q_2)$$

Thus, the infinitesimal Bäcklund transformations form a non-commutative Lie algebra. In fact, we shall see in §2 that it is isomorphic to $\mathfrak{gl}(\infty)$. Thus we can state: The space of τ functions of the KP hierarchy is the orbit space of the constant function 1 by the infinitesimal action of $\mathfrak{gl}(\infty)$.

1.9. BKP hierarchy. So far we have constructed the KP hierarchy. A similar construction leads us to another hierarchy, which we call the BKP hierarchy. Here B signifies B for the odd dimensional orthogonal group. In contrast, we call sometimes the KP hierarchy the AKP hierarchy.

The BKP hierarchy is the system of non-linear differential equations obtained as the compatibility conditions of

$$Lw(x,k) = kw(x,k) \tag{1.9.1}$$

$$\frac{\partial}{\partial x_n} w(x,k) = B_n(x,\partial)w(x,k) \qquad (n = 1,3,5,\ldots)$$

where $x = (x_1, x_3, x_5, \ldots)$, $L = \partial + u_2(x)\partial^{-1} + \ldots$ and $B_n(x,\partial) = [L(x,\partial)^n]_+$, under the condition

$$B_n(x,\partial)1 = 0. \qquad (n = 1,3,5,\ldots) \tag{1.9.2}$$

Namely, we impose the condition that the constant term $b_{n0}(x)$ of $B_n(x,\partial) = \partial^n + \sum_{j=0}^{n-2} b_{nj}(x)\partial^j$ should vanish. In order to compensate this additional condition, we have to suppress the even time evolution.

<u>Lemma 1.3</u>. For a pseudo-differential operator $L = \partial + u_1(x) + u_2(x)\partial^{-1} + \ldots$,

$$(L^n)_+ 1 = 0$$

for any $n = 1,3,5,\ldots$ if and only if

$$L^* = -\partial L \partial^{-1}.$$

<u>Proof</u>. If $L^* = -\partial L \partial^{-1}$ is satisfied, then $L^n\partial^{-1} = (-)^n \partial^{-1} L^{*n}$. Hence, for an odd n, $Q = (L^n\partial^{-1})_- = b_{n,0}(x)\partial^{-1}$ +(lower order terms) satisfies $Q^* = -Q$. This implies $b_{n,0}(x) = 0$.

Conversely, assuming $b_{n,0}(x) = 0$ for all odd n, we shall prove the vanishing of $R = \partial^{-1}L^* + L\partial^{-1}$. If $R = a(x)\partial^{-n}$ +(lower order terms) with $a(x) \not\equiv 0$, then $R = -R^*$ would imply that n is odd. Then

71

$$L^n\partial^{-1} = (R\partial-\partial^{-1}L^*\partial)^n\partial^{-1}$$

$$= -(\partial^{-1}L^*\partial)^n\partial^{-1}+nR(\partial^{-1}L^*\partial)^{n-1}$$

$$+ \text{ an operator of order } < -n-1$$

$$= (L^n\partial^{-1})^* + na(x)\partial^{-1} + \text{lower order terms},$$

which contradicts the assumption $a(x) \not\equiv 0$. Thus we have $R = 0$.

Lemma 1.3 tells us that the BKP hierarchy is a system describing the spectrum preserving deformation of a pseudo-differential operator L with $L^* = -\partial L\partial^{-1}$.

Bilinear identity and Hirota's equations.

The linear equation (1.9.1) has a solution $w(x,k)$ of the form

$$w(x,k) = \hat{w}(x,k)e^{\tilde{\xi}(x,k)} \qquad (1.9.3)$$

$$\hat{w}(x,k) = 1+w_1(x)k^{-1}+w_2(x)k^{-2}+\dots$$

and

$$\tilde{\xi}(x,k) = x_1k+x_3k^3+x_5k^5+\dots \ .$$

Similarly to the KP case, we have the bilinear identity

$$\int w(x,k)w(x',-k)\frac{dk}{2\pi ik} = 1. \qquad (1.9.4)$$

72

The one form

$$\omega_{BKP}(x,dx)$$

$$= \sum_{i\geq 1,odd} dx_i \operatorname{Res}_{k=\infty} k^i \Big(\sum_{j=1,odd} k^{-j-1}\frac{\partial}{\partial x_j} - \frac{1}{2}\frac{\partial}{\partial k}\Big)\log \hat{w}(x,k)dk$$

is closed. If we define the τ function $\tau_{BKP}(x)$ by

$$d \log \tau_{BKP}(x) = \omega_{BKP}(x,dx), \qquad (1.9.5)$$

then the wave function is written as

$$w(x,k) = \frac{\tau_{BKP}(x_1-\frac{2}{k}, x_3-\frac{2}{3k^3}, x_5-\frac{2}{5k^5},\dots)}{\tau_{BKP}(x)} e^{\tilde{\xi}(x,k)}. \qquad (1.9.6)$$

Thus, the BKP hierarchy is described by only one dependent variable $\tau_{BKP}(x)$.

By using the expression in terms of the τ function (1.9.6), we can rewrite the bilinear identity (1.9.4) in Hirota's bilinear form:

$$\sum_{j\geq 1} \tilde{p}_j(2y)\tilde{p}_j(-2\tilde{D}_x)e^{\sum_{\ell:odd} y_\ell D_{x_\ell}}\tau(x)\cdot\tau(x) = 0, \qquad (1.9.7)$$

where $\tilde{p}_j(y)$ is given by $e^{\tilde{\xi}(y,k)} = \sum \tilde{p}_j(y)k^j$ and $\tilde{D}_x = (D_{x_1}, \frac{1}{3}D_{x_3}, \frac{1}{5}D_{x_5}, \dots)$. The first two non-trivial equations are

$$(D_1^6-5D_1^3D_3-5D_3^2+9D_1D_5)\tau(x)\cdot\tau(x) = 0,$$

$$(D_1^8+7D_1^5D_3-35D_1^2D_3^2-21D_1^3D_5-42D_3D_5+90D_1D_7)\tau(x)\cdot\tau(x) = 0.$$

The relation between the τ functions of two hierarchies.

The τ function of the KP hierarchy and that of the BKP hierarchy are related by

$$\tau_{KP}(x_1,0,x_3,0,x_5,\dots)$$

$$= \text{const.}\ \tau_{BKP}(x_1,x_3,x_5,\dots)^2, \qquad (1.9.8)$$

if the wave function for the KP hierarchy satisfies the BKP condition (1.9.2) when we freeze x_{even} to be zero.

In fact, we have, by (1.6.3) and (1.9.6),

$$\frac{\tau_{BKP}(x_1-\frac{2}{k},x_3-\frac{2}{3k^3},\dots)}{\tau_{BKP}(x_1,x_3,\dots)}$$

$$= \frac{\tau_{KP}(x_1-\frac{1}{k},-\frac{1}{2k^2},x_3-\frac{1}{3k^3},\dots)}{\tau_{KP}(x_1,0,x_3,0,\dots)} .$$

Hence, setting

$$g(x) = \log \tau_{BKP}(x_1,x_3,x_5,\dots)$$

and

$$f(x,t) = \log \tau_{KP}(x_1,-\frac{t^2}{2},x_3,-\frac{t^4}{4},x_5,\dots)$$

with $x = (x_1,x_3,x_5,\dots)$, we have

$$f(x,0) = \log \tau_{KP}(x_1,0,x_3,0,x_5,\dots)$$

and

$$g(x_1-2t, x_3-\frac{2t^2}{3}, \ldots)-g(x)$$

$$= f(x_1-t, x_3-\frac{t^3}{3}, \ldots, t)-f(x,0).$$

Replacing x_j by $x_j+\frac{t^j}{j}$ we obtain

$$g(x_1-t, x_3-\frac{t^2}{3}, \ldots)-g(x_1+t, x_3+\frac{t^2}{3}, \ldots)$$

$$= f(x,t)-f(x_1+t, x_3+\frac{t^3}{3}, \ldots, 0).$$

If we change t to $-t$, we obtain

$$g(x_1+t, x_3+\frac{t^2}{3}, \ldots)-g(x_1-t, x_3-\frac{t^2}{3}, \ldots)$$

$$= f(x,t)-f(x_1-t, x_3-\frac{t^3}{3}, \ldots, 0).$$

Thus we have

$$2(g(x_1+t, x_3+\frac{t^3}{3}, \ldots)-g(x_1-t, x_3-\frac{t^3}{3}, \ldots))$$

$$= f(x_1+t, x_3+\frac{t^3}{3}, \ldots, 0)-f(x_1-t, x_3-\frac{t^3}{3}, \ldots, 0),$$

which implies $f(x,0) = 2g(x)+\text{const}$.

Vertex operators and soliton solutions.

We shall introduce the operator

$$X_B(p,q) = e^{\tilde{\xi}(x,p)+\tilde{\xi}(x,q)}e^{-2\tilde{\xi}(\tilde{\partial},p^{-1})-2\tilde{\xi}(\tilde{\partial},q^{-1})}$$

75

where $\tilde{\partial} = (\frac{\partial}{\partial x_1}, \frac{1}{3}\frac{\partial}{\partial x_3}, \frac{1}{5}\frac{\partial}{\partial x_5}, \ldots)$. Then $X_B(p,q)$ acts infinitesimally on the space of the τ functions of the BKP hierarchy. The N soliton solution is given by

$$\tau_B(x;a_1,p_1,q_1,\ldots,a_N,p_N,q_N)$$

$$= e^{\sum a_j X_B(p_j,q_j)} \cdot 1$$

$$= 1+\sum a_j e^{\xi_j} + \sum c_{jk} a_j a_k e^{\xi_j+\xi_k} + \ldots$$

$$= \sum_{\substack{0\le r\le N \\ 1\le j_1<\ldots<j_r\le N}} \prod_{\nu=1}^{r} a_{j_\nu} \prod_{1\le\nu<\mu\le r} c_{j_\nu j_\mu} e^{\sum_{\nu=1}^{r} \xi_{j_\nu}},$$

where $\xi_j = \tilde{\xi}(x,p_j)+\tilde{\xi}(x,q_j)$ and

$$c_{jk} = \frac{(p_j-p_k)(p_j-q_k)(q_j-p_k)(q_j-q_k)}{(p_j+p_k)(p_j+q_k)(q_j+p_k)(q_j+q_k)}.$$

1.10. Quasi-periodic solution. The construction of quasi-periodic solutions to the KP equation by starting from a curve is due to Krichever [15]. For this class of solutions, the wave function is called Baker-Akhiezer's function, and the τ function is nothing other than Riemann's theta function associated with the curve. Here we exploit the bilinear identity to study quasi-periodic solutions of the KP and BKP hierarchies. For the KP hierarchy, this approach is identical with that of Cherednik [16].

Let $\bar{C}$ be a compact curve, and let ∞ be a non-singular point of $\bar{C}$. Set $C = \bar{C}-\{\infty\}$. We denote by $\mathcal{O}(C)$ the space of regular functions on C. Now we fix a local coordinate k^{-1} around ∞ such that $k^{-1} = 0$ at ∞. From (1.3.5), $f(L)_+ = \sum_{n\geq 0} a_n B_n$ for $f(k) = \sum a_n k^n$. Sato remarked that Baker-Akhiezer's function is characterized by

$$f(L)_+ w(x,k) = f(k)w(x,k) \tag{1.10.1}$$

for any $f \in \mathcal{O}(C)$ along with (1.3.1). This condition is interpreted in terms of the τ function:

$$\left(\sum_{n>0} a_n \partial_n\right)\tau(x) - \left(\sum_{n<0} n a_n x_{-n}\right)\tau(x) = c(f)\tau(x) \tag{1.10.2}$$

for any $f = \sum a_n k^n \in \mathcal{O}(C)$ where $c(f)$ is a constant depending on f.

Let us recall Krichever's construction. Let g be the genus of $\overline{C}$ and choose g points $P_1,\ldots,P_g$ in the general position. We take a meromoprhic function $w(x,k)$ defined on C with at most simple poles at P_j $(j = 1,\ldots,g)$, satisfying

$$w(x,k) = \hat{w}(x,k)e^{\xi(x,k)} \quad \text{as} \quad k \to \infty. \tag{1.10.3}$$

Here $\hat{w}(x,k)$ is a holomorphic funciton defined on a neighborhood of ∞. We normalize it so that $\hat{w}(x,\infty) = 1$. Such a wave function is unique.

Now, for $f \in \mathcal{O}(C)$, $f(L)_+ w(x,k) - f(k)w(x,k)$ has the same property with $w(x,k)$ except for the fact that $e^{-\xi(x,k)} \times (f(L)_+ w(x,k) - f(k)w(x,k))$ has a zero at $k = \infty$. The uniqueness of w then implies (1.10.1).

Let us take g points $P_1^*,\ldots,P_g^*$ so that $P_1^*+\ldots+P_g^*+P_1+\ldots+P_g-2\infty$ is linearly equivalent to the canonical divisor, and let us take a one-form θ such that

$$(\theta) = P_1+\ldots+P_g+P_1^*+\ldots+P_g^*-2\infty.$$

Now, let $w^*(x,k)$ be a meromorphic function defined on C which has at most simple poles at $P_1^*,\ldots,P_g^*$ and $e^{\xi(x,k)}w^*(x,k)$ is holomorphic at ∞. Then, by the residue theorem

$$\int w(x,k)w^*(x',k)\theta = 0$$

where the integral is performed along a contour around ∞. Thus, if we write $\theta = \varphi(k)dk$, then $w(x,k)$ is a solution to the KP hierarchy and const. $\varphi(k)w^*(x,k)$ is its adjoint wave function. The corresponding τ function is written by using the ϑ function of $\overline{C}$.

The solution to the BKP hierarchy can be constructed in a similar way by imposing additional conditions. We start from a curve with an involution ι with two fixed points. We take one of two fixed points as ∞ and denote the other one by 0. We take P_j $(j = 1,\ldots,g)$ such that $P_1+\ldots+P_g+\iota(P_1)+\ldots+\iota(P_g)-\infty-0$ is linearly equivalent to the canonical divisor.

Let $w(x,P)$ be a function constructed in the same way as in the KP case. Let θ be the meromorphic one-form such that $(\theta) = P_1+\ldots+P_g+\iota(P_1)+\ldots+\iota(P_g)-\infty-0$ and with the

78

residues -1 at ∞ and 1 at 0. Then $\iota\theta = \theta$. We choose a local coordinate k^{-1} at ∞ such that $\theta = dk/2\pi ik$ and $k\circ\iota = -k$. Then $w(x,k)$ is a wave function for the BKP hierarchy. In fact, $\mathrm{Res}_{P=\infty} w(x,P)w(x,\iota(P))\theta = -1$ implies that $\mathrm{Res}_{P=0} w(x,P)w(x,\iota(P))\theta = 1$, and hence $w(x,0) = \pm 1$. Therefore we have the following bilinear identity characteristic to the BKP hierarchy.

$$-\mathrm{Res}_{k=\infty} w(x,k)w(x',-k)\frac{dk}{2\pi ik}$$

$$= \mathrm{Res}_{P=0} w(x,P)w(x',\iota(P))\theta$$

$$= 1.$$

The corresponding τ function is the theta function on the Prym variety. The relation between the τ function of the KP hierarchy and that of the BKP hierarchy stated in 1.9 corresponds to the knwon fact that the theta function of the Prym variety is the square root of the restriction of the theta function on the Jacobian variety. (See, for example [17]). For the detail we refer the reader to [1].

§2. The operator Theory for the KP and the BKP hierarchy.

Here we reveal the intimate relation between the KP hierarchy and the infinite dimensional Lie algebra $\mathfrak{gl}(\infty)$ using the language of free fermion operators (cf. [18]).

In §2.1 we define charged free fermion operators. Qudratic operators form an infinite dimensional Lie algebra. This Lie algebra, which we denote by $\mathfrak{g}(V,V^*)$, and the corresponding group, which we denote by $G(V,V^*)$, are the center of our consideration.

In §2.2 we construct a representation of the operator algebra (the Fock representation) in an abstract manner. The representation module has a cyclic vector called the vacuum. The $G(V,V^*)$-orbit of the vacuum is shown to be an infinite dimensional Grassmann manifold. The vacuum expectation value defined there gives us a clue to obtain explicit formulas.

In §2.3 $x = (x_1,x_2,x_3,\ldots)$ flows are introduced and τ functions are defined. The latter are parametrized by the $G(V,V^*)$-orbit of the vacuum.

In §2.4 the character polynomials are shown to be τ functions.

In §2.5 and §2.6 an explicit realization of the Fock representation is given by using vertex operators acting on $\mathbb{C}[x_1,x_2,x_3,\ldots]$. The N soliton is also naturally obtained in terms of vertex operators.

In §2.7 we give a simple proof of the bilinear identity for the wave function. This identity is equivalent to Hirota's bilinear equations for τ functions. A simultaneous linearization (the Zakharov-Shabat problem) of the KP hierarchy is also given in Hirota's form.

In §2.8 we define $\mathfrak{gl}(\infty)$ as an extension of $\mathfrak{g}(V,V^*)$. This is a preparation for §3.

The BKP theory is briefly described in §2.9. The Lie algebra we are now concerned with is $\mathfrak{go}(\infty)$ and the corresponding operator theory is based on neutral free fermion operators.

§2.1. The operator algebra. Consider a non-commutative algebra $\mathbf{A}$ generated by ψ_n and ψ_n^* $(n \in \mathbb{Z})$ which satisfy the defining relations

$$[\psi_m,\psi_n]_+ (\overset{=}{\text{def}}\ \psi_m\psi_n+\psi_n\psi_m) = 0,$$
$$[\psi_m^*,\psi_n^*]_+ = 0,\ [\psi_m,\psi_n^*]_+ = \delta_{mn}. \tag{2.1.1}$$

It is called the operator algebra, and its element is called an operator. In particular, the generators ψ_n and ψ_n^* $(n \in \mathbb{Z})$ are called free fermion operators. We set

$$V = \bigoplus_{n\in\mathbb{Z}} \mathbb{C}\psi_n,\ V^* = \bigoplus_{n\in\mathbb{Z}} \mathbb{C}\psi_n^*,\ W = V^*\oplus V. \tag{2.1.2}$$

By the pairing $\langle\psi_m,\psi_n^*\rangle = \delta_{mn}$, V and V^* are dual and ψ_n $(n \in \mathbb{Z})$ and ψ_n^* $(n \in \mathbb{Z})$ constitute dual basis.

Quadratic operators $\psi_m\psi_n^*$ $(m,n \in \mathbb{Z})$ satisfy the following Lie bracket relations:

$$[\psi_m\psi_n^*,\psi_{m'}\psi_{n'}^*] = \delta_{nm'}\psi_m\psi_{n'}^*-\delta_{mn'}\psi_{m'}\psi_n^*. \tag{2.1.3}$$

Thus $\psi_m\psi_n^*$ $(m,n \in \mathbb{Z})$ together with 1 span an infinite dimensional Lie algebra. We denote it by $\mathfrak{g}(V,V^*)$. (We reserve $\mathfrak{gl}(\infty)$ for an extension of $\mathfrak{g}(V,V^*)$ discussed in §2.8.) The corresponding group is characterized as follows.

$$G(V,V^*) = \{g \in \mathbf{A} \mid \exists g^{-1},\ gVg^{-1} = V,\ gV^*g^{-1} = V^*\}. \tag{2.1.4}$$

An operator g in $G(V,V^*)$ induces linear transformations in $GL(V)$ and $GL(V^*)$ separately through $g \cdot g^{-1}$. Namely, it satisfy commutation relations with free fermion operators of the following form.

$$g\psi_n = \sum_{m\in\mathbb{Z}} \psi_m g a_{mn},\ \psi_n^* g = \sum_{m\in\mathbb{Z}} g\psi_m^* a_{nm}. \tag{2.1.5}$$

Example 1.

$$g = 1-(\psi_\mu^*+\psi_\nu^*)(\psi_\mu+\psi_\nu),\ g^{-1} = -1+(\psi_\mu+\psi_\nu)(\psi_\mu^*+\psi_\nu^*),$$

$$a_{mn} = \delta_{mn}-(\delta_{\mu m}+\delta_{\nu m})(\delta_{\mu n}+\delta_{\nu n}). \qquad (2.1.6)$$

§2.2. The Fock representation. Consider an $\mathbb{A}$ left module with a cyclic vector $|vac\rangle$ satisfying

$$\psi_n|vac\rangle = 0 \quad (n < 0),\ \psi_n^*|vac\rangle = 0 \quad (n \geq 0). \qquad (2.2.1)$$

This module $\mathbb{A}|vac\rangle$ is called the Fock space and the representation of $\mathbb{A}$ on it is called the Fock representation.

We also consider an $\mathbb{A}$ right module with a cyclic vector $\langle vac|$ satisfying

$$\langle vac|\psi_n = 0 \quad (n \geq 0),\ \langle vac|\psi_n^* = 0 \quad (n < 0). \qquad (2.2.2)$$

From (2.2.1) and (2.2.2) a $\mathbb{C}$-bilinear pairing

$$\begin{array}{ccc} \langle vac|\mathbb{A} \underset{\mathbb{A}}{\otimes} \mathbb{A}|vac\rangle & \longrightarrow & \mathbb{C} \\ \cup & & \cup \\ \langle vac|a_1 \otimes a_2|vac\rangle & \longmapsto & \langle vac|a_1a_2|vac\rangle \end{array} \qquad (2.2.3)$$

is uniquely determined up to constant multiple. We normalize it so that

$$\langle vac|1|vac\rangle = 1. \qquad (2.2.4)$$

The quantity $\langle vac|a|vac\rangle$ $(a \in \mathbb{A})$ is called the vacuum expectation value of a.

Example 2. For $w_1,\dots,w_n \in V$, $w_1^*,\dots,w_n^* \in V^*$ and $g \in G(V,V^*)$,

$$\frac{\langle vac|gw_1^*\dots w_n^*w_n\dots w_1|vac\rangle}{\langle vac|g|vac\rangle} = \det\left(\frac{\langle vac|gw_j^*w_k|vac\rangle}{\langle vac|g|vac\rangle}\right)_{j,k=1,\dots,n}$$

(2.2.5)

In particular,

$$\langle vac|w_1^*\dots w_n^* w_n\dots w_1|vac\rangle = \det(\langle vac|w_j^* w_k|vac\rangle)_{j,k=1,\dots,n}. \tag{2.2.6}$$

We shall see below that τ functions of the KP hierarchy is parametrized by the $G(V,V^*)$-orbit of $|vac\rangle$. This orbit modulo constant multiple can be identified with an <u>infinite dimensional Grassmann manifold</u>: A vector $v = g|vac\rangle$ in $G(V,V^*)|vac\rangle$ determines a subspace $\mathbf{V}(v)$ in V,

$$\mathbf{V}(v) = \{\phi \in V \mid \phi v = 0\}. \tag{2.2.7}$$

In particular,

$$\mathbf{V}(|vac\rangle) = \bigoplus_{n<0} \mathbb{C}\psi_n. \tag{2.2.8}$$

Thus $G(V,V^*)|vac\rangle / GL(1)$ is identified with the Grassmann manifold $GL(V)\mathbf{V}(|vac\rangle)$.

Example 3. For $\mu < 0 \leq \nu$, we set

$$\begin{aligned} v &= (1-(\psi_\mu^*+\psi_\nu^*)(\psi_\mu+\psi_\nu))|vac\rangle \\ &= -\psi_\mu^*\psi_\nu|vac\rangle. \end{aligned} \tag{2.2.9}$$

Then we have

$$\mathbf{V}(v) = \Big(\bigoplus_{\substack{n<0\\ n\neq\mu}} \mathbb{C}\psi_n\Big) \oplus \mathbb{C}\psi_\nu. \tag{2.2.10}$$

§2.3. <u>The τ functions</u>. We introduce an infinite set of variables

$$x = (x_1, x_2, x_3, \dots) \tag{2.3.1}$$

and set

$$H(x) = \sum_{\ell=1}^{\infty} \sum_{n\in\mathbb{Z}} x_\ell \psi_n \psi_{n+\ell}^*. \tag{2.3.2}$$

The latter is called the Hamiltonian of the KP hierarchy. From (2.2.1) we have

$$H(x)|vac\rangle = 0. \tag{2.3.3}$$

Note that $\langle vac|H(x) \neq 0$. The x-evolution of an operator a is defined to be

$$a(x) = e^{H(x)} a\, e^{-H(x)}. \tag{2.3.4}$$

The Hamiltonian $H(x)$ belongs to the formal completion of $\mathfrak{g}(V,V^*)$, and $e^{H(x)}$ belongs to that of $G(V,V^*)$. We denote by Λ the following infinite matrix

$$\Lambda = (\delta_{m+1,n})_{m,n\in\mathbf{Z}} = \begin{pmatrix} \ddots & & & & \\ & 0 & 1 & & \\ & & 0 & 1 & \\ & & & 0 & 1 \\ & & & & \ddots \end{pmatrix}. \tag{2.3.5}$$

Using (2.1.1), we see that the action of $H(x)$ on V through $[H(x),\cdot]$ is represented by

$$\sum_{\ell=1}^{\infty} x_\ell \Lambda^\ell = \begin{pmatrix} \ddots & \ddots & \ddots & \ddots & & & \\ & 0 & x_1 & x_2 & x_3 & \cdots & \\ & & 0 & x_1 & x_2 & x_3 & \cdots \\ & & & 0 & x_1 & x_2 & x_3 & \cdots \\ & & & & \ddots & \ddots & \ddots & \ddots \end{pmatrix}. \tag{2.3.6}$$

Define polynomials $p_\ell(x)$ by

$$e^{\sum_{\ell=1}^{\infty} x_\ell k^\ell} = \sum_{\ell=0}^{\infty} p_\ell(x) k^\ell. \tag{2.3.7}$$

The action of $e^{H(x)}$ on V is represented by

$$e^{\sum_{\ell=1}^{\infty} x_\ell k^\ell} = \begin{pmatrix} \cdot & \cdot & \cdot & & & & & \\ & \cdot & \cdot & \cdot & & & & \\ & & 1 & p_1(x) & p_2(x) & p_3(x) & \cdot & \cdot \\ & & & 1 & p_1(x) & p_2(x) & p_3(x) & \cdot & \cdot \\ & & & & 1 & p_1(x) & p_2(x) & p_3(x) & \cdot & \cdot \\ & & & & & \cdot & \cdot & \cdot \\ & & & & & & \cdot & \cdot & \cdot \end{pmatrix} \tag{2.3.8}$$

The action on V^* is contra-gredient. To sum up, we have (see (2.3.4))

$$\begin{aligned} \psi_n(x) &= \sum_\ell \psi_{n-\ell} p_\ell(x), \\ \psi_n^*(x) &= \sum_\ell \psi_{n+\ell}^* p_\ell(-x). \end{aligned} \tag{2.3.9}$$

For a vector $v = g|vac\rangle$ in $G(V,V^*)|vac\rangle$, we define

$$\tau(x;v) = \langle vac|g(x)|vac\rangle. \tag{2.3.10}$$

From (2.3.3) and (2.3.4) we have

$$\tau(x;\ v) = \langle vac|e^{H(x)}g|vac\rangle. \tag{2.3.11}$$

Thus $\tau(x;v)$ is actually determined by v.

Example 4. (See (2.2.9).)

$$\begin{aligned} \tau(x;\psi_m^*\psi_n|vac\rangle) &= \langle vac|\psi_m^*(x)\psi_n(x)|vac\rangle, \\ &= \sum_{\ell\geq 0} p_{\ell-m}(-x)p_{n-\ell}(x). \end{aligned} \tag{2.3.12}$$

The following table shows $-\tau(x;\psi_m^*\psi_n|vac\rangle)$.

	n=0	n=1	n=2	n=3
m=-1	x_1	$x_2+\frac{x_1^2}{2}$	$x_3+x_1x_2+\frac{x_1^3}{6}$	$x_4+x_1x_3+\frac{x_2^2}{2}+\frac{x_1^2x_2}{2}+\frac{x_1^4}{24}$
m=-2	$x_2-\frac{x_1^2}{2}$	$x_3-\frac{x_1^3}{3}$	$x_4+\frac{x_2^2}{2}-\frac{x_1^2x_2}{2}-\frac{x_1^4}{8}$	
m=-3	$x_3-x_1x_2+\frac{x_1^3}{6}$	$x_4-\frac{x_2^2}{2}-\frac{x_1^2x_2}{2}+\frac{x_1^4}{8}$		
m=-4	$x_4-x_1x_3-\frac{x_2^2}{2}+\frac{x_1^2x_2}{2}-\frac{x_1^4}{24}$			

(2.3.13)

Exploiting Theorem 1.5.3 in [19], the expectation value in (2.3.11) can be calculated. Recall the commutation relation (2.1.5). We set $(d_{mn}) = (a_{mn})^{-1}$. Denote by $*$ the anti-automorphism of $\mathbf{A}$ such that $w^* = w$ for $w \in W$. Then $nr(g) = gg^* = g^*g$ is a constant. We define a constant c_g by

$$c_g^2 = nr(g)\,\frac{\det(d_{mn})_{m,n\geq 0}}{\det(a_{mn})_{m,n<0}}. \qquad (2.3.14)$$

Then

$$\tau(x;v) = c_g \det\Big(\sum_{m\in\mathbb{Z}} p_{m-n}(x)a_{mn'}\Big)_{n,n'<0} \qquad (2.3.15)$$

This expression for the τ function of the KP hierarchy is obtained by Sato [2].

§2.4. The character polynomials. Take integers $m_1,\ldots,m_k$ and $n_1,\ldots,n_k$ such that $m_1<\ldots<m_k<0\leq n_k<\ldots<n_1$, and consider the following subspace of V.

$$\Big(\bigoplus_{\substack{n<0\\ n\neq m_1,\ldots,m_k}} \mathbb{C}\psi_n\Big) \oplus \mathbb{C}\psi_{n_1} \oplus \ldots \oplus \mathbb{C}\psi_{n_k}. \qquad (2.4.1)$$

The corresponding vector (see (2.2.7)) in $G(V,V^*)|vac\rangle$ is given by

$$\psi^*_{m_1}\cdots\psi^*_{m_k}\psi_{n_k}\cdots\psi_{n_1}|\text{vac}>. \tag{2.4.2}$$

By using (2.2.6), the τ function for it is given by

$$\begin{aligned}
&<\text{vac}|e^{H(x)}\psi^*_{m_1}\cdots\psi^*_{m_k}\psi_{n_k}\cdots\psi_{n_1}|\text{vac}> \\
&= <\text{vac}|\psi^*_{m_1}(x)\ldots\psi^*_{m_k}(x)\psi_{n_k}(x)\ldots\psi_{n_1}(x)|\text{vac}> \\
&= \det(<\text{vac}|\psi^*_m(x)\psi_n(x)|\text{vac}>)_{\substack{m=m_1,\ldots,m_k \\ n=n_1,\ldots,n_k}}.
\end{aligned} \tag{2.4.3}$$

The last expression is related to <u>the character polynomials</u> [20]. Let $Y = (f_1,f_2,\ldots,f_s)$ be a Young diagram.

$$Y = \tag{2.4.4}$$

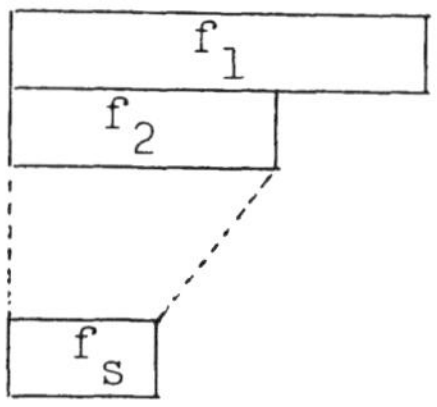

Choose a sufficiently large integer N. For an element g of $GL(N,\mathbb{C})$, we set

$$x_\ell = \frac{1}{\ell}\,\text{trace}\, g^\ell. \tag{2.4.5}$$

The character corresponding to Y is a polynomial in x_1, $x_2,\ldots$ independent of the choice of large N. We denote it by $\chi_Y(x)$.

The character polynomial for the ℓ-th symmetric tensor (i.e. $Y = (\ell)$) coincides with $p_\ell(x)$. We denote by $q_\ell(x)$ the character polynomials for the ℓ-th skew-symmetric tensor (i.e. $Y = (\underbrace{1,1,\ldots,1}_{\ell})$). Then

$$q_\ell(x) = (-)^\ell p_\ell(-x). \tag{2.4.6}$$

For convenience, we set $p_\ell(x) = q_\ell(x) = 0$ for $\ell < 0$.

We also denote by $\chi_{mn}(x)$ $(m < 0 \leq n)$ that for $Y = (n+1, \underbrace{1,1,\ldots,1}_{-m+1})$. Then

$$\chi_{mn}(x) = (-)^m \sum_{\ell \geq 0} p_{\ell-m}(-x)p_{n-\ell}(x),$$
$$= (-)^{m+1} \sum_{\ell \leq -1} p_{\ell-m}(-x)p_{n-\ell}(x). \tag{2.4.7}$$

Example 5. (See (2.3.13).)

$$\chi_{\square}(x) = x_1,\ \chi_{\square\square}(x) = x_2 + \frac{x_1^2}{2},\ \chi_{\begin{smallmatrix}\square\\\square\end{smallmatrix}}(x) = -x_2 + \frac{x_1^2}{2},$$
$$\chi_{\square\square\square}(x) = x_3+x_1x_2 + \frac{x_1^3}{6},\ \chi_{\begin{smallmatrix}\square\square\\\square\end{smallmatrix}}(x) = -x_3 + \frac{x_1^3}{3},$$
$$\chi_{\begin{smallmatrix}\square\\\square\\\square\end{smallmatrix}}(x) = x_3-x_1x_2 + \frac{x_1^3}{6}. \tag{2.4.8}$$

In general, let Y be of the following form.

$Y =$ 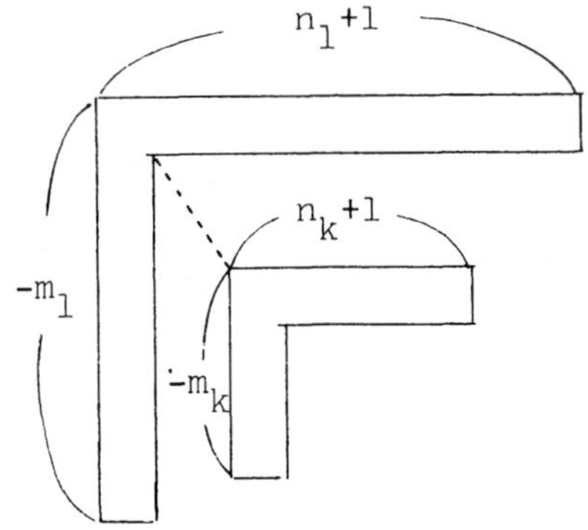

(2.4.9)

Then we have

$$\chi_Y(x) = \det(\chi_{mn}(x))_{\substack{m=m_1,\ldots,m_k \\ n=n_1,\ldots,n_k}}. \tag{2.4.10}$$

From (2.3.12), (2.4.3), (2.4.7) and (2.4.10), we have

$$\chi_Y(x) = (-)^{m_1+\ldots+m_k} \langle vac|e^{H(x)}\psi^*_{m_1}\cdots\psi^*_{m_k}\psi_{n_k}\cdots\psi_{n_1}|vac\rangle. \tag{2.4.11}$$

We have shown that the character polynomial $\chi_Y(x)$ corresponding to a Young diagram Y is written as $\tau(x;v)$ for an appropriate choice of v in $G(V,V^*)|vac\rangle$. In the following sections we show that $\tau(x;v)$ solves the Hirota equations of the KP hierarchy. In conclusion, the character polynomials are τ functions for the KP hierarchy. This is originally due to Mikio and Yasuko Sato.

§2.5. Realization of the Fock space. We define a degree (we call it the charge) of operators by setting

$$\deg \psi_n = 1, \ \deg \psi_n^* = -1, \qquad (2.5.1)$$

and denote by $\mathbb{A}(m)$ the subspace of charge m operators.

The set of vectors $\psi_{m_1}^* \ldots \psi_{m_k}^* \psi_{n_k} \ldots \psi_{n_1} |vac\rangle$ with $m_1 < \ldots < m_k < 0 \le n_k < \ldots < n_1$ constitutes a basis of $\mathbb{A}(0)|vac\rangle$, while the character polynomials constitute a basis of $\mathbb{C}[x_1, x_2, x_3, \ldots]$. Thus we have an isomorphism.

$$\begin{array}{ccc} \mathbb{A}(0)|vac\rangle & \longrightarrow & \mathbb{C}[x_1,x_2,x_3,\ldots] \\ \cup & & \cup \\ a|vac\rangle & \longmapsto & \langle vac|e^{H(x)}a|vac\rangle \end{array} \qquad (2.5.2)$$

In this section, we extend this isomorphism to the entire Fock space $\mathbb{A}|vac\rangle$.

Let us define a state of charge n:

$$\langle n| = \begin{cases} \langle vac|\psi_{-1} \ \ldots \ \psi_{-n} & (n < 0). \\ \langle vac| & (n = 0) \\ \langle vac|\psi_0^* \ \ldots \ \psi_{n-1}^* & (n > 0) \end{cases}$$

Introducing an auxiliary parameter u, we have an isomorp

$$\iota: \begin{array}{ccc} \mathbb{A}|vac\rangle & \longrightarrow & \mathbb{C}[x_1,x_2,x_3,\ldots,u,u^{-1}]. \\ \cup & & \cup \\ a|vac\rangle & \longmapsto & \sum_{m\in\mathbb{Z}} \langle m|e^{H(x)}a|vac\rangle u^m \end{array} \qquad (2.5.3)$$

In fact, the following formulas which generalize (2.4.11) are available: For a Young diagram Y of the form

89

$Y =$ 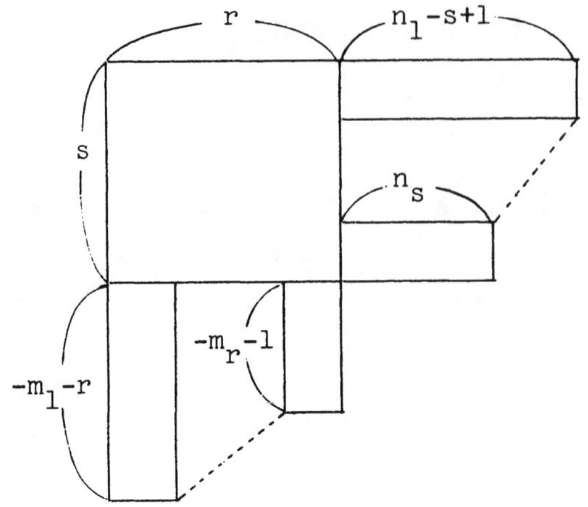

, (2.5.4)

we have

$$\chi_Y(x) = (-)^{m_1+\ldots+m_r+(s-r)(s-r-1)/2}$$
$$\times \langle s-r|e^{H(x)}\psi^*_{m_1}\cdots\psi^*_{m_r}\psi_{n_s}\cdots\psi_{n_1}|\mathrm{vac}\rangle. \tag{2.5.5}$$

If $s \geq r$ (2.5.5) gives the formula

$$\chi_Y(x) = \det\begin{pmatrix} p_{n_1-s+r+1}(x) & \ldots & p_{n_s-s+r+1}(x) \\ \vdots & & \vdots \\ p_{n_1}(x) & \ldots & p_{n_s}(x) \\ \chi_{m_1n_1}(x) & \ldots & \chi_{m_1n_s}(x) \\ \vdots & & \vdots \\ \chi_{m_rn_1}(x) & \ldots & \chi_{m_rn_s}(x) \end{pmatrix}. \tag{2.5.6}$$

If $s \leq r$ (2.5.5) gives the formula

$$\chi_Y(x) = \det\begin{pmatrix} q_{-m_1-r+s}(x) & \ldots & q_{-m_1-1}(x)\chi_{m_1n_1}(x) & \ldots & \chi_{m_1n_s}(x) \\ \vdots & & & & \\ q_{-m_r-r+s}(x) & \ldots & q_{-m_r-1}(x)\chi_{m_rn_1}(x) & \ldots & \chi_{m_rn_s}(x) \end{pmatrix}. \tag{2.5.7}$$

§2.6. Realization of the Fock representation. In this section we shall see that the action of $\mathbf{A}$ on $\mathbb{C}[x_1,x_2,x_3,\ldots,u,u^{-1}]$ through the isomorphism (2.5.3) is realized by linear differential operators of infinite order.

For the purpose, we introduce another basis of free fermion operators.

$$\psi(k) = \sum_{n\in\mathbf{Z}} \psi_n k^n, \quad \psi^*(k) = \sum_{n\in\mathbf{Z}} \psi_n^* k^{-n}. \tag{2.6.1}$$

The x-evolution is diagonalized by this choice.

$$e^{H(x)}\psi(k)e^{-H(x)} = e^{\xi(x,k)}\psi(k), \quad e^{H(x)}\psi^*(k)e^{-H(x)} = e^{-\xi(x,k)}\psi^*(k). \tag{2.6.2}$$

We set

$$\tilde{\partial} = \left(\frac{\partial}{\partial x_1}, \frac{1}{2}\frac{\partial}{\partial x_2}, \frac{1}{3}\frac{\partial}{\partial x_3}, \ldots\right). \tag{2.6.3}$$

The following differential operators of infinite order are called vertex operators.

$$X(k) = e^{\xi(x,k)}e^{-\xi(\tilde{\partial},k^{-1})}, \quad X^*(k) = e^{-\xi(x,k)}e^{\xi(\tilde{\partial},k^{-1})}. \tag{2.6.4}$$

A calculation shows the following.

$$\begin{aligned} \langle m|e^{H(x)}\psi(k) &= k^{m-1}X(k)\langle m-1|e^{H(x)} \\ \langle m|e^{H(x)}\psi^*(k) &= k^{-m}X^*(k)\langle m+1|e^{H(x)} \end{aligned} \tag{2.6.5}$$

We define operators $S(k)$ and $S^*(k)$ acting on $f(x,u) \in \mathbb{C}[x_1,x_2,x_3,\ldots,u,u^{-1}]$ by

$$\begin{aligned} (S(k)f)(x,u) &= uf(x,ku), \\ (S^*(k)f)(x,u) &= ku^{-1}f(x,k^{-1}u). \end{aligned} \tag{2.6.6}$$

91

By using (2.6.5), the action of $\psi(k)$ and $\psi^*(k)$ on the Fock space $\mathbf{A}|vac\rangle \cong \mathbb{C}[x_1,x_2,x_3,\ldots,u,u^{-1}]$ is interpreted as follows.

$$\psi(k) = X(k)S(k),\ \psi^*(k) = X^*(k)S^*(k). \tag{2.6.7}$$

Exploiting (2.6.7) we show that the N soliton solution for the KP hierarchy is given by (2.3.10) with

$$\begin{aligned} g &= e^{\sum_{j=1}^{N} a_j \frac{p_j-q_j}{q_j}\psi(p_j)\psi^*(q_j)}, \\ &= \prod_{j=1}^{N} (1+a_j\frac{p_j-q_j}{q_j}\psi(p_j)\psi^*(q_j)). \end{aligned} \tag{2.6.8}$$

We set

$$X(p,q) = e^{\xi(x,p)-\xi(x,q)}e^{-\xi(\tilde{\partial},p^{-1})+\xi(\tilde{\partial},q^{-1})}. \tag{2.6.9}$$

Then we have

$$X(p)X^*(q) = \frac{1}{1-q/p}X(p,q). \tag{2.6.10}$$

For $a \in \mathbf{A}(0)$, (2.6.6) and (2.6.7) show that (see (2.5.3))

$$\begin{aligned} \iota(a|vac\rangle) &= \langle vac|e^{H(x)}a|vac\rangle, \\ \iota(\psi^*(q)a|vac\rangle) &= qu^{-1}X^*(q)\langle vac|e^{H(x)}a|vac\rangle, \\ \iota(\psi(p)\psi^*(q)a|vac\rangle) &= \frac{q}{p}X(p)X^*(q)\langle vac|e^{H(x)}a|vac\rangle, \\ &= \frac{q}{p-q}X(p,q)\langle vac|e^{H(x)}a|vac\rangle. \end{aligned} \tag{2.6.11}$$

Thus using (2.6.8) we have the N soliton as follows.

$$\begin{aligned} &\langle vac|e^{H(x)}g|vac\rangle \\ &= \prod_{j=1}^{N} (1+a_jX(p_j,q_j))\cdot 1, \end{aligned}$$

$$= \sum_{n=0}^{N} \sum_{1\le j_1<\dots<j_n\le N} a_{j_1}\dots a_{j_n}$$

$$\times \prod_{\substack{j,j'=j_1,\dots,j_n \\ j<j'}} \frac{(p_j-p_{j'})(q_j-q_{j'})}{(p_j-q_{j'})(q_j-p_{j'})} e^{\sum_{j=j_1,\dots,j_n} (\xi(x,p_j)-\xi(x,q_j))} .$$

§2.7. <u>The bilinear equations</u>. For given τ function $\tau(x)$, the wave functions $w(x;k)$ and $w^*(x;k)$ are defined by

$$w(x;k) = X(k)\tau(x)/\tau(x),$$

$$w^*(x;k) = X^*(k)\tau(x)/\tau(x). \tag{2.7.1}$$

Let us show the bilinear identity

$$\oint \frac{dk}{2\pi i} w(x;k)w^*(x';k) = 0, \tag{2.7.2}$$

by taking $\tau(x) = \tau(x;v)$ with $v = g|vac> \in G(V,V^*)|vac>$.

From (2.6.5) and (2.7.1) we have

$$w(x;k) = <1|e^{H(x)}\psi(k)g|vac>/\tau(x),$$

$$w^*(x;k) = <-1|e^{H(x)}\psi^*(k)g|vac>/k\tau(x). \tag{2.7.3}$$

Hence we have

$$\oint \frac{dk}{2\pi i} w(x;k)w^*(x';k)$$

$$= \oint \frac{dk}{2\pi ik} \frac{<1|e^{H(x)}\psi(k)g|vac>}{\tau(x)} \cdot \frac{<-1|e^{H(x')}\psi^*(k)g|vac>}{\tau(x')},$$

$$= \sum_{n\in\mathbf{Z}} \frac{<1|e^{H(x)}\psi_n g|vac>}{\tau(x)} \cdot \frac{<-1|e^{H(x')}\psi_n^* g|vac>}{\tau(x')}$$

(2.7.4)

By using (2.1.5) the last expression is rewritten as

$$\sum_{n\in\mathbf{Z}} - \frac{\langle 1|e^{H(x)}g\psi_n|vac\rangle}{\tau(x)} \cdot \frac{\langle -1|e^{H(x')}g\psi_n^*|vac\rangle}{\tau(x')} \qquad (2.7.5)$$

Since either ψ_n or ψ_n^* annihilates $|vac\rangle$, this is identically zero. Thus we have proved (2.7.2).

The bilinear identity (2.7.2) is equivalently rewritten as Hirota's bilinear equations for the KP hierarchy.

$$\sum_{j=0}^{\infty} p_j(-2y)p_{j+1}(\tilde{D}_x)e^{\sum_{\ell=1}^{\infty} y_\ell D_{x_\ell}}\tau(x)\cdot\tau(x) = 0 \qquad (2.7.6)$$

Similarly, we obtain

$$\int \frac{dk'}{2\pi ik'}\langle 2|e^{H(x)}\psi(k')\psi(k)g|vac\rangle\langle -1|e^{H(x')}\psi^*(k')g|vac\rangle = 0, \qquad (2.7.7)$$

which is equivalent to

$$\sum_{j=0}^{\infty} p_j(-2y)p_{j+2}(\tilde{D}_x)e^{\sum_{\ell=1}^{\infty} y_\ell D_{x_\ell}}(w(x;k)\tau(x))\cdot\tau(x) = 0. \qquad (2.7.8)$$

This is nothing but linear equations for the KP hierarchy.

From (2.7.1) $w(x,k)$ admits an expansion at $k = \infty$ of the form

$$w(x,k) = (1+\sum_{n=-\infty}^{-1} f_n k^n)e^{\xi(x,k)}. \qquad (2.7.9)$$

We set $f_0 = 1$. From (2.7.3) it also admits another expansion of the form

$$w(x,k) = \sum_{m\in\mathbb{Z}} \langle 1|e^{H(x)}\psi_m g|vac\rangle k^m. \qquad (2.7.10)$$

From (2.1.5) and (2.2.1), the coefficients satisfy the following linear relations

$$\sum_{m\in\mathbb{Z}} \langle 1|e^{H(x)}\psi_m g|vac\rangle a_{mn'} = 0 \qquad (n' < 0), \qquad (2.7.11)$$

while by using (2.7.9) we have

$$\langle 1|e^{H(x)}\psi_m g|vac\rangle = \sum_{n=-\infty}^{0} f_n p_{m-n}(x). \qquad (2.7.12)$$

Thus, we obtain a linear equation for $(f_n)_{n<0}$:

$$\sum_{\substack{m\in\mathbf{Z}\\ n<0}} f_n p_{m-n}(x) a_{mn'} + \sum_{m\in\mathbf{Z}} p_m(x) a_{mn'} = 0 \quad (n' < 0). \qquad (2.7.13)$$

Hence the determinant expression (2.3.15) for the τ function is naturally connected with the existence of the wave function satisfying (2.7.9) and (2.7.11).

§2.8. <u>The Lie algebra $\mathfrak{gl}(\infty)$.</u> In this section we extend the Lie algebra $\mathfrak{g}(V,V^*)$. Denote by E_{mn} the following matrix of infinite size.

$$E_{mn} = (\delta_{jm}\delta_{kn})_{j,k\in\mathbf{Z}}. \qquad (2.8.1)$$

We define $\tilde{\mathfrak{gl}}(\infty)$ to be the following infinite dimensional algebra

$$\{\sum_{m,n} c_{mn}E_{mn} | c_{mn} = 0 \quad \text{for} \quad |m-n| >> 0\}. \qquad (2.8.2)$$

The action of $:\psi_m\psi_n^*:$ on V through $[:\psi_m\psi_n^*:,\cdot]$ is as follows.

$$[:\psi_m\psi_n^*:,\psi_k] = \sum_j \psi_j\delta_{jm}\delta_{kn}. \qquad (2.8.3)$$

Hence we have the following identification.

$$\tilde{\mathfrak{gl}}(\infty) = \{\sum_{m,n} c_{mn}:\psi_m\psi_n^*:|c_{mn} = 0 \quad \text{for} \quad |m-n| >> 0\}. \qquad (2.8.4)$$

The reason why we take $:\psi_m\psi_n^*:$ instead of $\psi_m\psi_n^*$ is to avoid infinity of the vacuum expectation value of an element in $\tilde{\mathfrak{gl}}(\infty)$. The effect will be clear in (2.8.6)

We define

$$Y_+(n) = \begin{cases} 1 & n \geq 0. \\ 0 & n < 0 \end{cases} \qquad (2.8.5)$$

Then we have

$$[:\psi_m\psi_n^*:, :\psi_{m'}\psi_{n'}^*:]$$

$$= \delta_{nm'}:\psi_m\psi_{n'}^*: - \delta_{mn'}:\psi_{m'}\psi_n^*: + \delta_{nm'}\delta_{mn'}(Y_+(n)-Y_+(m)). \qquad (2.8.6)$$

We use (2.8.6) as the definition of the following central extension of $\tilde{gl}(\infty)$.

$$gl(\infty) = \tilde{gl}(\infty) \oplus \mathbb{C}1. \qquad (2.8.7)$$

Since $\psi_m\psi_n^* = :\psi_m\psi_n^*: + \langle\psi_m\psi_n^*\rangle$, $gl(V,V^*)$ is a subalgebra of $gl(\infty)$.

Now let us consider the representation of $gl(\infty)$ on $\mathbf{A}(0)|vac\rangle$. Note that

$$\langle\psi(p)\psi^*(q)\rangle = \frac{q}{p-q}. \qquad (2.8.8)$$

From (2.6.11) and (2.8.8), the action of $:\psi(p)\psi^*(q):$ on $\mathbf{A}(0)|vac\rangle$ is given by

$$\frac{q}{p-q}(X(p,q)-1). \qquad (2.8.9)$$

In (2.8.9) there is no singularity at $p = q$. We define a differential operator Z_{mn} by

$$\frac{q}{p-q}(X(p,q)-1) = \sum_{m,n} Z_{mn}p^m q^{-n}. \qquad (2.8.10)$$

Let us count the homogeneous degree of elements in $\mathbb{C}[x] = \mathbb{C}[x_1,x_2,x_3,\ldots]$ in such a way that

$$\deg x_\ell = \ell. \qquad (2.8.11)$$

We set

$$\mathbb{C}[x]_r = \{f \in \mathbb{C}[x] \mid \deg f \leq r\}. \tag{2.8.12}$$

Then we see that

$$Z_{mn}\mathbb{C}[x]_r \subset \mathbb{C}[x]_{r+m-n},$$

$$Z_{mn}|_{\mathbb{C}[x]_r} = 0 \quad \text{unless} \quad m \geq -r,\ n < r \quad \text{and} \quad m-n > -r. \tag{2.8.13}$$

Hence, if $c_{mn} = 0$ for $|m-n| \gg 0$, the operator $\sum_{m,n} c_{mn}Z_{mn}$ actually acts on $\mathbb{C}[x]$. Thus identifying $:\psi_m\psi_n^*:$ with Z_{mn}, we construct a representation of $\mathfrak{gl}(\infty)$ on $\mathbb{C}[x]$ in terms of differential operators.

We note that by setting $p = q$ in (2.8.10) we have the following formulas.

$$\ell x_\ell = \sum_{n\in\mathbf{Z}} Z_{n\ n-\ell},$$

$$0 = \sum_{n\in\mathbf{Z}} Z_{nn},$$

$$\frac{\partial}{\partial x_\ell} = \sum_{n\in\mathbf{Z}} Z_{n\ n+\ell}. \tag{2.8.14}$$

Similarly we have

$$\sum_{\ell=1}^{\infty} \ell x_\ell \frac{\partial}{\partial x_\ell} = -\sum_{n\in\mathbf{Z}} nZ_{nn}. \tag{2.8.15}$$

§2.9. <u>The BKP hierarchy</u>. We denote by **B** a non-commutative algebra generated by ϕ_n $(n \in \mathbf{Z})$ satisfying

$$[\phi_m, \phi_n]_+ = (-)^m\delta_{m,-n}. \tag{2.9.1}$$

We set $W_B = \bigoplus_{n\in\mathbf{Z}} \mathbb{C}\phi_n$ and equip it with an inner product $\langle\phi_m, \phi_n\rangle = (-)^m\delta_{m,-n}$. The algebra **B** is called the Clifford algebra on W_B. There is a unique isomorphism of **B** which

97

negates ϕ_n $(n \in \mathbb{Z})$. We denote by $\mathbb{B}^+$ (resp. $\mathbb{B}^-$) the subspace which is invariant (negated) by this isomorphism.

Quadratic elements $\phi_m\phi_n$ $(m,n \in \mathbb{Z})$ span an infinite dimensional Lie algebra, which we denote by $\mathfrak{g}(W_B)$. The corresponding group is given by

$$G^+(W_B) = \{g \in \mathbb{B}^+ | \exists g^{-1},\ gW_Bg^{-1} = W_B\}. \tag{2.9.2}$$

An element g of $G^+(W_B)$ induces an orthogonal transformation of W_B.

We consider a left $\mathbb{B}$ module with a cyclic vector $|vac\rangle$ satisfying

$$\phi_n|vac\rangle = 0 \qquad (n < 0), \tag{2.9.3}$$

and also a right $\mathbb{B}$ module with a cyclic vector $\langle vac|$ satisfying

$$\langle vac|\phi_n = 0 \qquad (n > 0). \tag{2.9.4}$$

By setting $\langle vac|1|vac\rangle = 1$ and $\langle vac|\phi_0|vac\rangle = 0$, the vacuum expectation value is determined.

We set

$$L_0 = \{\phi \in W_B | \phi|vac\rangle = 0\} = \bigoplus_{n<0} \mathbb{C}\phi_n, \tag{2.9.5}$$

and

$$I = \{T \in SO(W_B) | TL_0 = L_0\}. \tag{2.9.6}$$

The homogeneous space $SO(W_B)/I$ is called the orthogonal Grassmann manifold. We have an isomorphism

$$G^+(W_B)|vac\rangle/GL(1) \cong SO(W_B)/I. \tag{2.9.7}$$

We introduce an infinite set of variables $x = (x_1, x_3, x_5, \ldots)$ and set

$$H_B(x) = \frac{1}{2} \sum_{\ell:\mathrm{odd}} \sum_{n\in\mathbb{Z}} (-)^{n+1} x_\ell \phi_n \phi_{-n-\ell}. \qquad (2.9.8)$$

We define polynomials $\tilde{p}_\ell(x)$ by

$$e^{\tilde{\xi}(x,k)} = \sum_{\ell=0}^{\infty} \tilde{p}_\ell(x) k^\ell. \qquad (2.9.9)$$

The x-evolution of ϕ_n is given by

$$\phi_n(x) = e^{H_B(x)} \phi_n e^{-H_B(x)} = \sum_{\ell=0}^{\infty} \phi_{n-\ell} \tilde{p}_\ell(x). \qquad (2.9.10)$$

For a vector $v = g|vac\rangle \in G^+(W_B)|vac\rangle$, we define

$$\tau(x;v) = \langle vac|e^{H_B(x)} g|vac\rangle. \qquad (2.9.11)$$

For $m \leq -1$ and $n \geq 0$ we set

$$\tilde{\chi}_{m,n}(x) = (-)^{m+1} \sum_{\ell \leq -1} \tilde{p}_{\ell-m}(-x) \tilde{p}_{n-\ell}(x),$$

$$= (-)^m \sum_{\ell \geq 0} \tilde{p}_{\ell-m}(-x) \tilde{p}_{n-\ell}(x). \qquad (2.9.12)$$

Then we have for $m > n \geq 0$

$$\langle vac|e^{H_B(x)} \phi_m \phi_n|vac\rangle$$

$$= \langle vac|\phi_m(x)\phi_n(x)|vac\rangle,$$

$$= \begin{cases} \frac{1}{2} \tilde{p}_m(x) & (n = 0). \\ \frac{1}{2}(\tilde{\chi}_{-m,n}(x) - \tilde{\chi}_{-n,m}(x)) & (n > 0) \end{cases} \qquad (2.9.13)$$

In general, the vector $\phi_{m_1} \cdots \phi_{m_{2k}}|vac\rangle$ $(m_1 > \cdots > m_{2k} \geq 0)$ corresponds to the space

$$\left(\bigoplus_{\substack{n<0 \\ n \neq -m_1, \ldots, -m_{2k}}} \mathbb{C}\phi_n \right) \oplus \left(\bigoplus_{\substack{m=m_1, \ldots, m_{2k} \\ m \neq 0}} \mathbb{C}\phi_m \right) \qquad (2.9.14)$$

in $SO(W_B)/I$, and the τ function is given by

$$\langle vac|e^{H_B(x)}\phi_{m_1}\cdots\phi_{m_{2k}}|vac\rangle$$

$$= \text{Pfaffian}(\langle vac|\phi_m(x)\phi_{m'}(x)|vac\rangle)_{m,m'=m_1,\ldots,m_{2k}} \tag{2.9.15}$$

We have the following isomorphism

$$\begin{array}{llll} & \mathbb{B}|vac\rangle & \longrightarrow & \mathbb{C}[x_1,x_3,x_5,\ldots]\oplus\mathbb{C}[x_1,x_3,x_5,\ldots]\phi_0 \\ \iota: & \quad\psi & & \qquad\qquad\psi \\ & a|vac\rangle & \longmapsto & \langle vac|e^{H_B(x)}a|vac\rangle\oplus\langle vac|\phi_0 e^{H_B(x)}a|vac\rangle\phi_0 \end{array} \tag{2.9.16}$$

We note that $\phi_0^2 = \frac{1}{2}$ and that ϕ_0 acts on $\mathbb{C}[x_1,x_3,x_5,\ldots] \oplus \mathbb{C}[x_1,x_3,x_5,\ldots]\phi_0$ by multiplication.

We set

$$\phi(k) = \sum_{n\in\mathbb{Z}}\phi_n k^n, \tag{2.9.17}$$

$$X_B(k) = e^{\tilde{\xi}(x,k)}e^{-2\tilde{\xi}(\tilde{\partial},k^{-1})} = \sum_{n\in\mathbb{Z}} X_{Bn}k^n,$$

$$(\tilde{\partial} = (\frac{\partial}{\partial x_1}, \frac{1}{3}\frac{\partial}{\partial x_3}, \frac{1}{5}\frac{\partial}{\partial x_5},\ldots)). \tag{2.9.18}$$

They are related to each other by

$$X_B(k)\langle vac|e^{H_B(x)} = 2\langle vac|\phi_0 e^{H_B(x)}\phi(k),$$

$$X_B(k)\langle vac|\phi_0 e^{H_B(x)} = \langle vac|e^{H_B(x)}\phi(k). \tag{2.9.19}$$

Thus the action of ϕ_n on $\mathbb{C}[x_1,x_3,x_5,\ldots]\oplus\mathbb{C}[x_1,x_3,x_5,\ldots]\phi_0$ through the isomorphism (2.9.16) is given by

$$\phi_n = \phi_0 X_{Bn}. \tag{2.9.20}$$

We define

100

$$X_B(p,q) = e^{\tilde{\xi}(x,p)+\tilde{\xi}(x,q)} e^{-2(\tilde{\xi}(\tilde{\partial},p^{-1})+\tilde{\xi}(\tilde{\partial},q^{-1}))}. \tag{2.9.21}$$

Then the action of $\phi(p)\phi(q)$ on $\mathbb{C}[x_1,x_3,x_5,\ldots]$ is given by $\frac{1}{2}\frac{p-q}{p+q}X_B(p,q)$. The N soliton solution for the BKP hierarchy is given by

$$\langle vac|e^{H_B(x)} e^{\sum_{j=1}^{N} a_j \frac{p_j+q_j}{p_j-q_j}\phi(p_j)\phi(q_j)} |vac\rangle$$

$$= \sum_{n=0}^{N} \sum_{1\le j_1<\ldots<j_n\le N} a_{j_1}\ldots a_{j_n}$$

$$\times \prod_{\substack{i,j'=j_1,\ldots,j_n\\ j<j'}} \frac{(p_j-p_{j'})(p_j-q_{j'})(q_j-p_{j'})(q_j-q_{j'})}{(p_j+p_{j'})(p_j+q_{j'})(q_j+p_{j'})(q_j+q_{j'})} e^{\sum_{j=j_1,\ldots,j_n} (\tilde{\xi}(x,p_j)+\tilde{\xi}(x,q_j))}. \tag{2.9.22}$$

From (2.9.19) the wave function is given by

$$w(x,k)=2\langle vac|\phi_0 e^{H_B(x)} \phi(k)g|vac\rangle/\langle vac|e^{H_B(x)} g|vac\rangle. \tag{2.9.23}$$

Exploiting this expression we can easily show the bilinear identity

$$\oint\frac{dk}{2\pi ik}\, w(x,k)w(x',-k) = 1. \tag{2.9.24}$$

We define

$$\mathcal{O}(\infty) = \{ \sum_{m,n} c_{mn}:\phi_m\phi_n: | c_{mn} = 0 \quad \text{for} \quad |m-n| >> 0\}. \tag{2.9.25}$$

We can identify $:\phi_m\phi_{-n}:$ with

$$(-)^n E_{mn} - (-)^m E_{-n-m}. \tag{2.9.26}$$

Note that

$$[:\phi_m\phi_{-n}:,\ :\phi_{m'}\phi_{-n'}:]$$

$$= (-)^n\delta_{nm'}:\phi_m\phi_{-n'}:-(-)^n\delta_{n,-n'}:\phi_m\phi_{m'}:$$

$$+ (-)^m\delta_{m,-m'}:\phi_{-n'}\phi_{-n}:-(-)^m\delta_{mm'}:\phi_{m'}\phi_{-n}:$$

$$+ (-)^{m+n}(\delta_{nm'}\delta_{mn'}-\delta_{m,-m'}\delta_{n-n'})(Y_B(n)-Y_B(m))$$

where $Y_B(n) = \begin{cases} 1 & n > 0. \\ \frac{1}{2} & n = 0 \\ 0 & n < 0 \end{cases}$ (2.9.27)

Using (2.9.27) we define the central extension

$$\mathfrak{go}(\infty) = \mathfrak{o}(\infty) \oplus \mathbb{C}1. \tag{2.9.28}$$

The representation of $\mathfrak{go}(\infty)$ on $\mathbb{C}[x_1,x_3,x_5,\ldots]$ is given by

$$:\phi(p)\phi(q): \leftrightarrow \frac{1}{2}\cdot\frac{p-q}{p+q}(X_B(p,q)-1). \tag{2.9.29}$$

§3. Reductions of the KP and the BKP hierarchies, Euclidean Lie algebras and the number of Hirota's bilinear equations

Our purpose in this section is two-fold: i) to describe how sub-holonomic hierarchies of soliton equations, such as the KdV, the Bonssinesq and the Sawada-Kotera hierarchies [21], are obtained from the KP or the BKP hierarchies (the reduction), and ii) to count the number of Hirota's equations for the KP, the BKP and these sub-holonomic hierarchies. Here we describe i) in terms of transformation groups. In the course, Euclidean Lie algebras appear as subalgebras of $\mathfrak{gl}(\infty)$ or $\mathfrak{go}(\infty)$ and constitute infinitesimal transformation groups for these sub-holonomic hierarchies. Then the character formula of Kac-Moody Lie algebras affords us a systematic way of studying ii).

3.1 In the previous section, we have seen that the totality of τ-functions of the KP(resp. the BKP) hierarchy is the orbit space of 1 under the representation of $\mathfrak{gl}(\infty)$ (resp. $\mathfrak{go}(\infty)$) on $\mathbb{C}[x_1, x_2, x_3, \ldots]$(resp. $\mathbb{C}[x_1, x_3, x_5, \ldots]$) through the isomorphism ι induced from the Fock representation (cf. (2.5.3), (2.6.11)).

Let ℓ be a positive integer and consider the following subalgebras of $\mathfrak{gl}(\infty)$ and $\mathfrak{go}(\infty)$:

$$\mathfrak{gl}(\infty)_\ell = \{X \in \mathfrak{gl}(\infty) \mid [X, \Lambda_{\ell j}] = 0,\ j \in \mathbf{Z}\}$$

$$\mathfrak{go}(\infty)_\ell = \{X \in \mathfrak{go}(\infty) \quad [X, \Lambda_{\ell j}] = 0,\ j \in \mathbf{Z};\ \text{odd}\}$$

where $\Lambda_j = \Sigma_{n\in\mathbf{Z}} E_{n,n+j}$. Noting the relation

$\partial/\partial x_j = \Lambda_j$, $jx_j = \Lambda_{-j}$ (for KP), $= 2\Lambda_{-j}$ (for BKP), we find that $\mathfrak{gl}(\infty)_\ell$ (resp. $\mathfrak{go}(\infty)_\ell$) acts on $\mathbb{C}[x_i;\ i \not\equiv 0\ (\mathrm{mod}.\ell)]$

(resp. $\mathbb{C}[x_i, i : \text{odd} \not\equiv 0 \ (\text{mod}, \ell)]$).

We define the space of τ-functions of the ℓ-reduced KP ($(KP)_\ell$) and the ℓ-reduced BKP ($(BKP)_\ell$) hierarchies to be the orbit space of 1 under the representation of $\mathfrak{gl}(\infty)_\ell$ and $\mathfrak{go}(\infty)_\ell$ obtained by restricting the above representations of $\mathfrak{gl}(\infty)$ and $\mathfrak{go}(\infty)$ respectively. We note that τ-functions of the $(KP)_\ell$ and the $(BKP)_\ell$ (ℓ:odd) hierarchies do not depend on $x_{\ell j}$ $j>0$ and $x_{\ell j}$ j : odd >0 respectively.

In terms of linear problems, the $(KP)_\ell$ hierarchy is described in the following manner. Using the relation between τ-functions and wave functions $w(x,k)$ and the fact that τ-functions of the $(KP)_\ell$ hierarchy do not depend on $x_{\ell j}$ $j > 0$, we find that wave functions for the $(KP)_\ell$ hierarchy satisfy

$$(\partial/\partial x_{\ell j})w = k^{\ell j}w, \quad j > 0.$$

Namely, the linear problem of the $(KP)_\ell$ hierarchy is obtained from the linear problem of the KP hierarchy (1.3.1)by imposing the additional constraint

$$L^\ell = \text{a differential operator} = B_\ell$$

on the micro-differential operator L (1.3.1).In other words, the $(KP)_\ell$ hierarchy is the integrability condition of the linear problem

$$(\partial/\partial x_j)w = k^j w, \quad j \equiv 0 \ (\text{mod. } \ell) \ j > 0$$

$$(\partial/\partial x_k)w = (L^k)_+ w, \quad k \not\equiv 0 \ (\text{mod. } \ell) \ \ k \geq 1.$$

The same description applies for the odd-reductions of the BKP bierarchy. However, to describe the even reductions of the

BKP hierarchy in terms of linear problems is rather subtler, since for the BKP hierarchy the even time variables x_{even} are excluded from the outset.

Through the above description of reduced hierarchies, we see that the KdV, the Boussinesq and the Sawada-Kotera hierarchies are obtained as the $(KP)_2$, the $(KP)_3$ and the $(BKP)_3$ hierarchies respectively.

Recalling that the actions of $\mathfrak{gl}(\infty)$ and $\mathfrak{go}(\infty)$ are realized by vertex operators $Z(p, q)$ and $Z_B(p, q)$ and noting the conditions defining $\mathfrak{gl}(\infty)_\ell$ and $\mathfrak{go}(\infty)_\ell$, we know that the actions of $\mathfrak{gl}(\infty)_\ell$ and $\mathfrak{go}(\infty)_\ell$ are realized by the specialized vertex operators

$$Z(p, \omega p) = \frac{\omega}{1-\omega}[\exp(\Sigma_{j\geq 1}(1-\omega^j)p^j x_j)\exp(\Sigma_{j\geq 1}(1-\omega^{-j})\frac{1}{j}\frac{\partial}{\partial x_j})-1]$$

$$= \Sigma Z_j(\omega)p^j \qquad (\omega^\ell = 1, \omega \neq 1)$$

$$Z_B(p, -\omega p) = \frac{1}{2}\frac{1+\omega}{1-\omega}[\exp(\Sigma_{j:odd\geq 1}(1-\omega^j)p^j x_j)\exp(-2\Sigma_{j:odd\geq 1}$$

$$(1-\omega^{-j})\frac{1}{j}\frac{\partial}{\partial x_j})-1]$$

$$= \Sigma Z_{B,j}(\omega)p^j, \qquad (\omega^\ell = 1, \omega \neq 1).$$

Corresponding to the sepecialization of vertex operators, we obtain N-soliton solutions of these reduced hierarchies from those of original hierarchies by specializing the choice of parameters p_j, q_j as $p_j^\ell = q_j^\ell$ or $p_j^\ell = (-q_j)^\ell$.

Now let us see how these subholonomic hierarchies are described by Hirota's bilinear equations. In view of the definition of reduced τ-functions, we see that τ-functions of the $(KP)_\ell$ or the $(BKP)_\ell$ (ℓ:odd) hierarchy satisfy bilinear equations of the original hierarchies in which the substitution $D_{\ell j} = 0 \quad j > 0$ or $D_{\ell j} = 0 \quad j : \text{odd} > 0$ are made and that τ-functions of the $(BKP)_\ell$ (ℓ = even) hierarchy satisfy the bilinear equations of the BKP hierarchy. Also we can show that reduced τ-funcitons satisfy new bilinear equations

$$\Sigma_{i\geq 0}p_i(2u)p_{i+\ell+1}(-\widetilde{D})e^{<u,D>}\tau_{(KP)_\ell}\cdot\tau_{(KP)_\ell} = 0$$

$$\Sigma_{i>0}p_i(2u)p_{i+\ell}(-2\widehat{D})e^{<u,D>}\tau_{(BKP)_\ell}\cdot\tau_{(BKP)_\ell} = 0$$

where $<u, D> = \Sigma_{j=1}^{\infty}u_jD_j$ (for KP), $= \Sigma_{j:\text{odd}\geq 1}u_jD_j$ (for BKP) and u_i are auxiliary parameters. The first few bilinear equations for these reduced hierarchies are given in [1].

In this way, τ-functions of the KP or the BKP or the $(KP)_\ell$ or the $(BKP)_\ell$ hierarchy satisfy many bilinear equations. Then, the the following question naturally arises: how many are they?

More precisely, we formulate this problem as follows. For a hierarchy S of soliton equations whose bilinear equations have the form $P(D)\tau\cdot\tau = 0$, we put

$$H_S(m) = \{P \in \mathbb{C}[D_i, i \in I_S] \,|\, P(D)\tau_S\cdot\tau_S = 0, \quad \text{for any } \tau_S, \deg P = m\},$$

Here I_S denotes the set of independent variables attached to S and the degree of D_j is counted to be j. Then our problem is to count the dimension

106

$$h_S(m) = \dim H_S(m).$$

This problem was proposed by M. and Y. Sato [13]. They calculated $h_S(m)$ for S = the KdV, the modified KdV and the nonlinear Schrödinger hierarchies (for the latter two, we must modify the definition of $H_S(m)$). For the KdV hierarchy their result states

$$h_{KdV}(m) = p(m;\text{ positive odd integer})$$

$$-p(m;\text{ distinct positive integer}).$$

Here, for example, p(m; positive odd integer) signifies the number of partitions of m into positive integers. We also use this notation in the sequel. Further they conjectured for the KP and the Sawada-Kotera hierarchies the following:

$$h_{KP}(m) = p(m-1;\text{ positive integer}) \tag{3.1.1}$$

$$h_{\text{Sawada-Kotera}}(m) = p(m;\text{ positive integer} \equiv \pm 1 \pmod{6}) \tag{3.1.2}$$

$$-p(m:\text{ positive integer} \equiv \pm 2 \pmod{10}).$$

In §3.3, we explain how $h_S(m)$ can be calculated by using the character formula of Kac-Moody Lie algebras

<u>3.2</u> Let us study the structure of $\mathfrak{gl}(\infty)_\ell$ and $\mathfrak{go}(\infty)_\ell$ more closely. As a result, Euclidean Lie algebras will appear.

Let $X = A \oplus Cz$, $A = a_{ij}E_{ij}$ be an element of $\mathfrak{gl}(\infty)_\ell$, then the definition of $\mathfrak{gl}(\infty)_\ell$ implies the relation

$$a_{i+\ell,j+\ell} = a_{i,j}.$$

This allows us to identify A with an $\ell\times\ell$ matrix of Laurent polynomials

$$A(t) = \Sigma_{n\in\mathbb{Z}} A_n t^n , \quad A_n = (a_{i,j+n\ell})_{0\le i,j\le \ell-1}. \tag{3.2.1}$$

If we identify $\mathbb{C}^\infty$ on which $\mathfrak{gl}(\infty)$ acts with $\mathbb{C}[k, k^{-1}]$ via $E_{ij}k^{-\nu} = \delta_{j\nu}k^{-i}$, Λ_n is multiplication by k^n. The definition of $\mathfrak{gl}(\infty)_\ell$ implies that the matrix A commute with the multiplication by $t = k^\ell$ and consequently belongs to $\mathrm{End}_{\mathbb{C}[t,t^{-1}]}\mathbb{C}[k,k^{-1}) \cong \mathfrak{gl}(\ell, \mathbb{C}[t,t^{-1}])$. Above inentification realizes this view point. With this identification, the bracket of $\mathfrak{gl}(\infty)_\ell$ is given by

$$[A(t) \oplus cz, A'(t) \oplus c'z] = [A(t),A'(t)] \oplus c(A(t), A'(t))z.$$

where the bracket in the right hand side is the usual one and

$$c(A(t), A'(t)) = \mathrm{Res}\ \mathrm{tr}\ \frac{dA(t)}{dt} A'(t) = \Sigma n \mathrm{tr} A_n A'_n .$$

Further for $A(t) \oplus cz \in \mathfrak{gl}(\infty)_\ell$, the definition of $\mathfrak{gl}(\infty)_\ell$ and the above bilinear form c implies $A(t) \in \mathfrak{sl}(\ell, \mathbb{C}[t,t^{-1}])$.

Thus we have

$$\mathfrak{gl}(\infty)_\ell = \mathfrak{sl}(\ell, \mathbb{C}[t,t^{-1}]) \oplus \mathbb{C}z.$$

The right hand side is a realization of Enclidean Lie algebra $A^{(1)}_{\ell-1}$.

Analogously an element $X = A \oplus cz$, $A = \Sigma a_{ij}E_{ij}$ of $\mathfrak{go}(\infty)_\ell$

can be identified with a $\ell\times\ell$ matrix of Laurent polynomial by (3.2.1). We note that $\mathfrak{o}(\infty)$ can be regarded as the orthogonal Lie algebra on $\mathbb{C}[k,k^{-1}]$ with respect to the inner product

$$(f,\ g) = \Sigma(-)^{i}\xi_{i}\eta_{-i} = \mathrm{Res}_{k=0}f(k)g(-k)dk/k.$$

$$f(k) = \Sigma\xi_{i}k^{-i},\ g(k) = \Sigma\eta_{i}k^{-i}.$$

For $f,\ g \in \mathbb{C}[k,\ k^{-1}]$, we put

$$\begin{aligned}<f,\ g>_{\ell}(t) &= \Sigma_{j\in\mathbf{Z}}(\Lambda_{-\ j},\ f,\ g)t^{j}\\ &= \Sigma_{j\in\mathbf{Z}}\int k^{-\ell j}f(k)f(-k)dk/k\cdot t^{j} \in \mathbb{C}[t,\ t^{-1}].\end{aligned}$$

Then for even ℓ, $<f,\ g>_{\ell}(t)$ is symmetric and for odd ℓ, we have

$$<g,\ f>_{\ell}(t) = <f,\ g>_{\ell}(-t).$$

Since elements of $\mathfrak{go}(\infty)_{\ell}$ preserves the above bilinear form, $\mathfrak{go}(\infty)_{\ell}$ is realized as a Lie algebra over Laurent polynomials as

$$\mathfrak{go}(\infty)_{\ell} = \begin{cases} \{A(t)\in\mathfrak{sl}(\ell,\mathbb{C}[t,t^{-1}]\,|\,J_{\ell}(t)A(t) + {}^{t}A(-t)J_{\ell}(t) = 0\}\oplus\mathbb{C}z & \ell:\text{odd}\\ \{A(t)\in\mathfrak{sl}(\ell,\mathbb{C}[t,t^{-1}]\,|\,J_{\ell}(t)A(t) + {}^{t}A(t)J_{\ell}(t) = 0\}\oplus\mathbb{C}z & \ell:\text{even}.\end{cases}$$

where

$$J_{\ell}(t) = (<k^{-i},\ k^{-j}>(t))_{0\le i,j\le\ell-1} = \left[\begin{array}{c|ccc} 1 & & & \\ \hline & & & (-)^{\ell-1}t^{-1}\\ & & \cdot^{\cdot^{\cdot}} & \\ & -t^{-1} & t^{-1} & \end{array}\right].$$

The rule of central extension is given by the skew-symmetric bilinear form

$$\tilde{c}_B(A(t), A'(t)) = \frac{1}{2} \text{Res } \text{tr}\frac{dA(t)}{dt}A'(t).$$

These turn out to be realizations of Euclidean Lie algebras $A^{(2)}_{\ell-1}$ and $D^{(2)}_{\ell/2}$.

Here we recall the definition of Kac-Moody Lie algebras and their standard representations. Euclidean Lie algebras form a subclass of Kac-Moody Lie algebras.

Let $C = (c_{ij})_{0\leq 1, j\leq n}$ be an $(n+1)\times(n+1)$ integral matrix with the following properties : $c_{ii} = 2$, $c_{ij} \leq 0$ for $i \neq j$ and $c_{ij} = 0$ implies $c_{ji} = 0$. A matrix with these properties is called the (generalized)Cartan matrix. Then a Kac-Moody Lie algebra with a Cartan matrix C is defined to be a complex Lie algebra with $3(n+1)$ generators e_i, f_i, h_i $(0\leq i\leq n)$ and defining relations

$$[e_i, f_j] = \delta_{ij}h_i, \quad [h_i, h_j] = 0$$

$$[h_i, e_j] = c_{ij}e_j, \quad [h_i, f_1] = -c_{ij}f_j$$

$$(\text{ad } e_i)^{1-c_{ij}}e_j = \underbrace{[e_i,[\ldots[e_i, e_j]]}_{1-c_{ij} \text{ times}} = 0$$

$$(\text{ad } f_i)^{1-c_{ij}}f_j = 0.$$

Though in general for the definition of a Kac-Moody Lie algebra one more step is needed, for Euclidean Lie algebras the above relations are sufficient. Cartan matrices defining $A^{(1)}_n$, $A^{(2)}_{2n}$ and $D^{(1)}_n$ are given in Table 1 of [1]VII and the identifications of $\mathfrak{gl}(\infty)_\ell$ and $\mathfrak{go}(\infty)$ with $A_\ell^{(1)}$, $A_{\ell-1}^{(2)}$ and $D_{\ell/2}^{(2)}$ throught the generators are given in Table 2 of [1]VII. Standard representations of Kac-Moody Lie algebras are defined as follcws Let $\mathfrak{g}$ be a Kac-Moody Lie algebra and let λ be a linear form on $\mathfrak{h} = \bigoplus_{i=0}^{n} \mathbb{C}\, h_i \in \mathfrak{g}$ such that $\lambda(h_i) \in \mathbb{Z}_+$ $i = 0,1,..,n$. An irreducible $\mathfrak{g}$-module V is called the standard module with the highest weight λ, if there exists a nonzero vector $v_\lambda \in V$ satisfying

$$hv_\lambda = \lambda(h)v_\lambda \text{ for any } h \in \mathfrak{h}$$

$$e_i v_\lambda = 0 \quad i = 0,1,...,n.$$

The vector v_λ is unique up to constant multiple, and is called the highest weight vector. The basic representation is the one corresponding to the linear form Λ

$$\Lambda(h_0) = L,\ \Lambda(h_1)=0\ ,\ ...,\ \Lambda(h_n) = 0.$$

From the viewpoint of representation of Kac-Moody Lie algebras, the homogeneous components of specialized vertex operator $Z_j(\omega)$ or $Z_{B_j}(\omega)$ gives a linear representation of $A_{\ell-1}^{(1)}$, $A_{\ell-1}^{(2)}$, $D_{\ell/2}^{(2)}$ on the polynomial ring of infinitaly many variable $\mathbb{C}[x_j]$, $j \not\equiv \ell$ (mod. ℓ) $j > 0$ $j \not\equiv 0$ (mod. ℓ) j : odd > 0 and j : odd > 0 respectively together with 1, x_j, $\partial/\partial x_j$.
This representation turns out to be the basic representation with the highest weight vector 1 and is identical with the one constructed by Kac-Kazhdan-Lepowsky-Wilson [10]. This means, in purticular, the totality of reduced (polynomial) τ-functions is the orbit space of the highest weight vector 1 under the basic representation of $A_{\ell-1}^{(1)}$ or $A_{\ell-1}^{(2)}$ or $D_{\ell/2}^{(2)}$.

3.3 Now we count $h_S(m)$ for S = KP, BKP, $(KP)_\ell$ and $(BKP)_\ell$. Hereafter we denote by $\mathfrak{g}_S$ the coresponding Euclidean Lie algebra.

Let us see how the representation of $\mathfrak{g}_S$ relates to the calculation of $h_S(m)$. We put $H_S = \oplus_{m\in \mathbb{Z}_+} H_S(m)$. In view of the definition of the Hirota bilinear operator

$$P(D_x)f\cdot g = P(\partial_y)f(x+y)g(x-y)|_{y=0},$$

we write the product $\tau_S(x+y)\tau_S(x-y)$ as

$$\tau_S(x+y)\tau_S(x-y) = \Sigma F_i(x;\tau_S)G_i(y;\tau_S)$$

where in the right hand side we take the shortest representation. Then the linear hull of $G_i(y,\tau_S)$'s is uniquely determined, i.e. not depend on the choice of G_i's. We denote this space by G. Therefore if we introduce a pairing of $\mathbb{C}[\partial_{y_i}, i\in I_S]$ and $\mathbb{C}[y_i; i\in I_S]$ by

$$\langle a, b\rangle = a(\partial_y)b(y)|_{y=0}, \quad a \in \mathbb{C}[\partial_{y_i}; i\in I_S], \ b \in \mathbb{C}[y_i, i\in I_S],$$

we have

$$H_S = G^{\perp}.$$

By counting the degree of ∂_i to be i, we introduce the gradu- ation in G, $G = \oplus_{m\in \mathbb{Z}_+} G_m$. Then we have

$$h_S(m) = \dim\{P \in \mathbb{C}[\partial_i, i\in I_i] \deg P = m\} - \dim G_m$$

$$= p(m, \text{integers in } I_S) - \dim G_m.$$

Thus our problem is reduced to count dim G_m.

Next we show that the space G can be described in term of representations of $\mathcal{G}_S$. Since the totality of τ-functions of the hierarchy S is the orbit space of 1 under the basic representation of $\mathcal{G}_S$ on $\mathbb{C}[x_i, i\in I_S]$, the irreducible representation space $V(2\Lambda)$ with the highest weight 2Λ is the space

$$\text{linear hull of } \tau_S(x^{(1)})\tau_S(x^{(2)})$$

where $x^{(1)}$ and $x^{(2)}$ are two copies of independent variables attached to S.

In the space $V(2\Lambda)$ we consider the subspace

$$\Omega = \{v \in V(2\Lambda) | (\partial/\partial x_i^{(1)} + \partial/\partial x_i^{(2)})v = 0, \ \forall i\in I_S\}$$

$$= \{v \in V(2\Lambda) | \ v\text{: function of } y\}$$

where we have set

$$2x_i = x_i^{(1)} + x_i^{(2)}, \qquad 2y_i = x_i^{(1)} - x_i^{(2)} ,$$

Let s be the principal subalgebra of $\mathcal{G}_S$, which in the basic representation of $\mathcal{G}_S$ on $V(\Lambda) = \mathbb{C}[x_i, i\in I_S]$ corresponds to the infinite dimensional Heisenberg subalgebra spanned be 1, x_i, $\partial/\partial x_i$, $i\in I_S$.

Then following the argument of Lepowsky-Wilson [22], we have as s-modules

$$V(2\Lambda) = V(\Lambda)\otimes_{\mathbb{C}}\Omega.$$

113

On the other hand by recalling the definition of G and by using the complete reducibility of s-modules, we have

$$G = \Omega.$$

In this way, the space of our concern is described in terms of representations of $\mathfrak{g}_S$.

We can introduce a natural gradation (the principal gradation) in a standard module of $\mathfrak{g}_S$. In our case, the principal degree of an element $v \in \Omega$ is given by reversing the degree of v regarded as an element of G.

By the may, the principally specialized character is nothing but the generating funciton of dimensions of principally graded subspaces of standard modules, such as Ω. Also a method of calculating it is known. Therefore by appling that formula, we can calculate $\dim G_m$ and consequently $h_S(m)$. For details, we refer to [1]. Below we list the results.

For the $(KP)_{n+1}$ hierarchy, we have

$$h_{(KP)_{n+1}}(m) = p(m; \text{positive integer} \not\equiv 0 \ (\text{mod. } n+1))$$

$$-p(m; \text{positive integer} \not\equiv 0,\ 1 \ (\text{mod. } n+3)).$$

In particular, for the $(KP)_2$ (= the KdV) hierarchy, we have

$$h_{KdV} = p(m; \text{positive odd integer})$$

$$- p(m; \text{integer} \equiv 2 \ (\text{mod. } 4)).$$

By virtue of a theorem of Euler the second term of the right hand side is equal to

p(m; distinct positive even integer).

Hence the result of M. and Y. Sato is recovered.

For the $(BKP)_{2n+1}$ and the $(BKP)_{2n+2}$ hierarchies, we have

$$h_{(BKP)_{2n+1}}(m) = p(m;\ \text{positive odd integer} \not\equiv 0 \pmod{2n+1})$$
$$- p(m;\ \text{positive even integer} \not\equiv 2(n+1), 2(n+2) \pmod{4n+6})$$
$$h_{(BKP)_{2n+2}}(m) = p(m;\ \text{positive odd integer})$$
$$- p(m;\ \text{positive even integer} \not\equiv 0 \pmod{(2n+2)})$$

For the $(BKP)_3$ (= the Sawada-Kotera) hierarchy, this proves the conjecture of M. and Y. Sato (3.1.2).

The dimensions of $H_{KP}(m)$ and $H_{BKP}(m)$ are calculated in the following way

$$h_{KP}(m) = \lim_{n\to\infty} h_{(KP)_n}(m)$$

$$h_{BKP}(m) = \lim_{n\to\infty} h_{(BKP)_n}(m).$$

By using the above results, we have

$$h_{KP}(m) = p(m;\ \text{positive integer}) - p(m;\ \text{integer} \geq 2)$$

$$= p(m-1;\ \text{positive integer})$$

which proves the conjecture of M. and Y. Sato (3.1.1).
Also we have

$$h_{BKP} = p(m, \text{positive odd integer}) - p(m, \text{positive even integer}).$$

References

[1] M. Kashiwara and T. Miwa, Proc. Japan Acad. 57 A (1981) 342.
E. Date, M. Kashiwara and T. Miwa, ibid. 57 A (1981) 387.
E. Date, M. Jimbo, M. Kashiwara and T. Miwa, J. Phys. Soc. Jpn. 50 (1981) 3806, 3813.
————, Physica 4 D (1982) 343; Publ. RIMS, Kyoto Univ. 18 (1982) 1077, 1111.

[2] M. Sato, Soliton equations as dynamical systems on infinite dimensional Grassmann manifolds, RIMS Kokyuroku 439 (1981) 30.

[3] K. Ueno, The infinite dimensional Lie algebras acting on SU(n) chiral field and Riemann-Hilbert problem, RIMS preprint 374 (1981).
K. Ueno and Y. Nakamura, Transformation theory for anti-self-dual equations in four-dimensional Euclidean space and Riemann-Hilbert problem, RIMS preprint 375 (1981).

[4] L. Dolan, Phys. Rev. Lett. 47 (1981) 1371.

[5] A. G. Reiman and M. A. Semenov-Tjan-Šanskii, Soviet Math. Dokl. 21 (1980) 630.

[6] V. G. Drinfeld and V. V. Sokolov, Dokl. Akad. Nauk USSR, 258 (1981) 11.

[7] M. Adler and P. van Moerbeke, Advances in Math. 38 (1980) 267, 318.

[8] H. Yosida, in this volume.

[9] R. Hirota, Direct method of finding exact solutions of nonlinear evolution equations, Springer Lecture Notes in Math. 515 (1976) 40.

[10] V. G. Kac, D. A. Kazhdan, J. Lepowsky and R. L. Wilson, Advances in Math. 42 (1981) 83.

[11] J. Lepowsky and R. L. Wilson, Commun. math. Phys. 62 (1978) 43.

[12] E. Date, M. Jimbo, M. Kashiwara and T. Miwa, J. Phys. A 16 (1983) 221.

[13] M. Sato and Y. Sato(Mori), On Hirota's bilinear equations I, II, RIMS Kokyuroku 388 (1980) 183, 414 (1981) 181.

[14] J. Satsuma, J. Phys. Soc. Jpn. 40 (1976) 286.

[15] I. M. Krichever, Russ. Math. Surveys 32 (1977) 185.

[16] I. V. Cherednik, Funct. Anal. and Its Appl. 12 (1978) 195.

[17] J. D. Fay, Theta functions on Rieman Surfaces, Springer Lecture Notes in Math. 352 (1973).

[18] M. Jimbo, T. Miwa, M. Sato and Y. Mori, Holonomic quantum fields, in Springer Lecture Notes in Phys. 116 (1979) 119.
M. Sato, T. Miwa and M. Jimbo, Aspects of holonomic quantum fields, in Springer Lecture Notes in Phys. 126 (1979) 429.

[19] M. Sato, T. Miwa and M. Jimbo, Publ. RIMS, Kyoto Univ. 14 (1978) 223.

[20] D. E. Littlewood, The theory of group characters and matrix representations of groups, Oxford, 1958.

[21] K. Sawada and T. Kotera, Prog. Theo. Phys. 51 (1974) 1355.

[22] J. Lepowsky and R. L. Wilson, A Lie theoretic interpretation and proof of the Rogers-Ramanujan identities, preprint (1981).

[23] E. Date, M. Jimbo and T. Miwa, J. Phys. Soc. Jpn. (1982) 4116, 4125; M. Jimbo and T. Miwa, Lett. Math. Phys. 6 (1982) 463.

(Received in February, 1982)

JGP - Vol. 5, n. 3, 1989

Quantization of symplectic orbits of compact Lie groups by means of the functional integral

A. ALEKSEEV, L. FADDEEV, S. SHATASHVILI

Leningrad Branch of V.A. Steklov Mathematical Institute,
Fontanka 27, 191011, Leningrad, USSR

Dedicated to I.M. Gelfand
on his 75th birthday

Abstract. *The functional integral for the quantization of the coadjoint orbits of the unitary and orthogonal groups is given by means of an explicit construction of the corresponding «Darboux» variables.*

INTRODUCTION

In this paper we describe a method for the quantization of the Hamiltonian action of compact group on its coadjoint orbit. In the spirit of the method of orbits [1, 2] this quantization is supposed to give the corresponding representation of the group. Our method is based on the functional integral form of the matrix elements of the representation.

The simplest case of the group $SO(3)$ was considered before [3]. We shall present it in the introduction to illustrate the underlying ideas of our general approach. In the main text we shall consider unitary and orthogonal groups. Symplectic and exceptional cases can be treated in analogous way and we plan to present the explicit formulas in a separate publication.

The main technical problem in the generalization of the $SO(3)$ example to the higher rank case consists in finding the «Darboux variables» for the canonical symplectic

---

Key-Words: Quantization, compact Lie groups, functional integral.
***1980 MSC:** 81 E 10, 22 5 05.*

form on the orbit. We describe the derivation of such variables for unitary and orthogonal cases in §1 and §2, correspondingly. In §3 we use them to evaluate the functional integral by reducing it to finite-dimensional one.

The condition of quantization of the parameters of the orbit [1, 2] is used in a natural way in our formalism. Indeed, it allows to satisfy the requirement of single-valuedness of the expression $\exp\{i \times \text{action}\}$.

We prove that the representation obtained by our method coincides with that associated with the orbit by an explicit evaluation of its character. Of course the matrix elements and the character are calculated for the action of a chosen Cartan subgroup. This subgroup is associated in a natural way with the choice of the Darboux variables.

Now let us consider the illustrative case of the group $SO(3)$. Its orbits are realized as spheres S^2 parametrized by their radius. It is convenient to use for the beginning coordinates $x^1, x^2, x^3 : \sum_i (x^i)^2 = m^2$. The symplectic form Ω is given by

(1) $$\Omega = \frac{1}{2m^2}\epsilon^{abc}x^a \mathrm{d}\, x^b \mathrm{d}\, x^c.$$

Functions x^a correspond to the generators of $SO(3)$ with the Poisson brackets

$$\{x^a, x^b\} = \epsilon^{abc}x^c.$$

Quantization condition for orbits leads to the requirement that m is an integer or a half-integer. The Darboux variables are given by the spherical coordinates

$$x^1 = m \sin\, \theta \cos\, \varphi,\ x^2 = m \sin\, \theta \cos\, \varphi,\ x^1 = m \sin\, \cos\, \theta,$$

so that

$$\Omega = m \mathrm{d}\, \varphi\, \mathrm{d}\, \sin\, \theta.$$

The 1-form $\omega = -\mathrm{d}^{-1}\Omega$, necessary to define the action, is not single-valued, because Ω is not an exact 2-form. If we choose

$$\omega = (\gamma + m \cos\, \theta)\mathrm{d}\, \varphi$$

then the action of the path

$$\varphi = \varphi(t),\ \theta = \theta(t), \qquad 0 \leq t \leq T,$$

is given by

(2) $$S_0 = m\int_0^T \cos\, \theta \mathrm{d}\, \varphi + \gamma \int_0^T \mathrm{d}\, \varphi.$$

For infinitesimal closed contours around the singular points $\theta = 0, \pi$ this action is given by $2\pi(\gamma \pm m)$ so does do not contribute to the $\exp\{i \times$ action for integer values of $m \pm \gamma$. Thus we can choose $\gamma = 1/2$ if m is a halfinteger and $\gamma = 0$ if m is an integer.

The expression for the matrix element of the operator $\exp\{-ix_3 T\}$ will be taken in the form

(3) $$G(\varphi', \varphi'' = \langle \varphi'' | e^{-ix^3 T} | \varphi' \rangle = \int \prod_t d\eta(t) d\varphi(t) \exp\{iS\},$$

where $\eta = \cos\theta$,

(4) $$S = \int_0^T (m\eta\dot{\varphi} - h) dt + \gamma \int_0^T \dot{\varphi} dt = S_0 - \int_0^T h dt$$

and $h = x^3 = m \cos\theta$. Here the trajectories $\eta(t), \varphi(t)$ (over which we sum the function $\exp\{i \times$ action$\}$) are subject to the condition

$$\varphi(0) = \varphi', \quad \varphi(T) = \varphi'' + 2\pi n$$

for any integer n.

Alternativly, this matrix element can be rewritten as a sum

(5) $$G(\varphi', \varphi'') = \sum_n G_0(\varphi', \varphi'' + 2\pi n),$$

where the matrix element $G_{01}(\varphi', \varphi'')$ is given by the same functional integral but with fixed boundary values $\varphi(0) = \varphi', \varphi(T) = \varphi''$. The variables $\eta(t)$ have values in the interval $-1 \leq \eta \leq 1$, whereas $\varphi(t)$ runs over the whole real axis $-\infty < \varphi(t) < \infty$.

The following comments to this definition are appropriate:

1. Expression (3) essentially coincides with the general formula for the functional integral on the phase space with the chosen «coordinate-like» variables [see formula 4]. The role of coordinates is played by φ in our case. The difference with the usual case of linear phase space in that our phase space is compact.

2. The phase space S^2 is replaced by the strip $-1 \leq \eta \leq 1, -\infty < \varphi < \infty$ which is a covering of the cylinder $-1 \leq \eta \leq 1, 0 \leq \varphi < 2\pi$ i.e. of the sphere with two deleted points, which correspond to the singularities of our coordinate system. It is a common wisdom in quantum mechanics that one has to use the covering space in the functional integral.

3. The summation in (5) is an averaging over the corresponding discrete fundamental group, restoring the periodicity of the matrix element $< \varphi'' | \exp\{-ix^3 T\} | \varphi' >$.

After these comments we shall calculate the path integral, using its definition by means of the finite dimensional approximation. We have

$$G_0(\varphi', \varphi'') =$$
$$= \lim (\frac{1}{2\pi})^N \int_k \prod d\eta_k d\varphi_k e^{i\sum_{k=1}^{N}\{m\eta_k(\varphi_k - \varphi_{k-1}) - m\eta_k \frac{T}{N} + \gamma(\varphi_k - \varphi_{k-1})\}},$$

(6) $$\varphi_0 = \varphi', \ \varphi_N = \varphi'',$$

where the number of integrations over η exceeds by one that φ. Integrals over φ give δ-functions which can be integrated over $\eta - s$. As a result we obtain

(7)
$$G_0(\varphi', \varphi'') =$$
$$= \frac{1}{2\pi} \int_{-1}^{1} d\eta e^{im\eta(\varphi'', \varphi' - T) + i\gamma(\varphi'', \varphi')} =$$
$$= \frac{1}{2\pi} \left[\frac{e^{im(\varphi'' - \varphi' - T)}}{m(\varphi'' - \varphi' - T)} - \frac{e^{im(\varphi'' - \varphi' - T)}}{m(\varphi'' - \varphi' - T)} \right]$$
$$e^{i\gamma(\varphi'' - \varphi')}.$$

We can say that the whole functional integral is given by the contribution of the «semi-classical trajectories» $\eta(t) = \eta$.

To perform the averaging (5) we are to choose the regularization for the infinite sum

$$\sum_{-\infty}^{\infty} \frac{1}{2\pi n + \varphi'' - \varphi' - T}.$$

We shall use the following regularizations

(8) $$S^{\pm} = \lim_{\epsilon \to 0} \sum \frac{e^{\pm i\epsilon 2\pi n}}{2\pi n + \alpha} = \frac{e^{\pm \frac{1}{2} i\alpha}}{\sin \frac{\alpha}{2}}$$

and associate S^+ with the first term in (7) and S^- with the second. As a result we shall obtain

(9)
$$G(\varphi', \varphi'') =$$
$$= \frac{1}{2\pi} e^{i\gamma(\varphi'', \varphi')} \frac{\sin(m + \frac{1}{2})(\varphi'', \varphi' - T)}{\sin \frac{1}{2}(\varphi'', \varphi' - T)} =$$
$$= \frac{1}{2\pi} e^{i\gamma(\varphi'', \varphi')} \sum_{m_z = -m}^{m} e^{im_z(\varphi'', \varphi' - T)}.$$

In particular, for $T = 0$ (zero Hamiltonian) we get

$$\langle\varphi'' \mid \varphi'\rangle = \tag{10}$$
$$= \frac{1}{2\pi} e^{i\gamma(\varphi'',\varphi')} \sum_{-m}^{m} e^{im_s(\varphi'',\varphi')},$$

which is a projector onto the states $|k> = e^{ik\varphi}, -m+\gamma \leq k \leq m+\gamma$ defining the basis for the finite dimensional representation of spin m. For the character we obtain

$$\chi(T) = \int_0^{2\pi} d\varphi G(\varphi,\varphi) = \frac{\sin(m+\frac{1}{2})T}{\sin\frac{1}{2}T}, \tag{11}$$

which is of course the character of the spin m representation. The role of the regularization (8) becomes clear if we calculate (5) interchanging the sum and the integral over η. Indeed we can use the formula

$$\sum_n e^{2\pi ni\sigma} = \sum_k \delta(\sigma - k),$$

where the sum in the RHS is over all integer k, to perform the η integral. However the zeros of δ-function $\delta(\eta-1)$ and $\delta(\eta+1)$ correspond to the end points in the η integral. Our regularization corresponds to taking their contribution with coefficient 1. In other words, the η-integral is understood as the limit

$$\int_{-1}^{1} d\eta(\ldots) = \lim_{\epsilon\to 0}\int_{-1-\epsilon}^{1+\epsilon} d\eta(\ldots)$$

The tricks developed on this example will be used in the main text to define an appropriated functional integral for the higher rank case.

Let us note that an alternative method of evaluating the correlation functions, corresponding to the action $\int\omega$ exists. It is based on the Ward identities and BJL method (for the case of any compact group) [5]. The quantization of spin is closly related to the problem of construction of the propagator of relativistic spinning particle in the first quantized formalism [6]. The integral of type (3) can be used to solve this problem [5].

The action (2) is the simplest example of the Wess-Zumino action. It is tempting to say that our method could be used in the case of infinite dimensional group, for example, for the group diff S^1. The corresponding Wess-Zumino action is connected with the quantum gravity in $1+1$ dimension [7]. Also, the quantization of the theories with the Wess-Zumino action is connected with selfconsistent treatment of anomalous theories [8]. We shall consider these problems in separate publications.

This paper is dedicated to the 75th birthday of professor Gelfand. We hope that the new interpretation of the Gelfand-Zetlin basis will be interesting to him.

1. DARBOUX VARIABLES FOR THE UNITARY GROUP

In this section we shall reduce the symplectic 2-form on the coadjoint orbit $SU(n)$ to the manifestly canonical form, convenient for the functional integral.

In what follows we shall use a natural identification of coadjoint orbits with the adjoint ones. The point of the orbit has the form

$$M = gM_0\, g^+, \tag{12}$$

where $g \in SU(n)$ and M_0 is a fixed antihermitian traceless matrix. The symplectic form Ω can be written in the form

$$\Omega = \frac{1}{2} tr MYY, \tag{13}$$

where the 1-form Y with the values in the Lie algebra satisfies the equation

$$\mathrm{d}\, M = [Y, M] \tag{14}$$

and tr is the Killing-Cartan form. It is clear from (12) that $Y = \mathrm{d}\, g \cdot g^{-1}$ solves this equation; different choice of solution leads to the same expression for Ω.

We shall find explicit coordinates in which Ω has the Darboux form. Let $e^1, \ldots, e^n$ be an orthogonal set of eigenvectors of matrix M_0 with eigenvalues $2im_1, \ldots, 2im_n$ $m_1 \geq \ldots \geq m_n$ and

$$a^1 = ge^i. \tag{15}$$

Formulae (13) and (15) imply that Ω can be written as

$$\begin{aligned} &\Omega = -\mathrm{d}\, \alpha, \\ &\alpha = \frac{i}{2} \sum_1^n m_k[(a^k, \mathrm{d}\, a^k) - (\mathrm{d}\, a^k, a^k)] \end{aligned} \tag{16}$$

where we use the complex scalar product.

If the complex variables a_i^k were independent, the form Ω would be given in explicitly canonical form in the linear phase of real dimension $2n^2$ which we shall call auxiliary space in the future. However, the variables a_i^k are subject to the orthonormality constraint of

$$\varphi_{ij} \equiv (a^i, a^j) - \delta^{ij} = 0 \tag{17}$$

This means that (16) is still not of the desired form and one must make a reduction with respect to these constraints [4].

The constraints (17) are naturally divided into two sets of the first and second class. The first set consists of the constraints

$$\varphi_i \equiv \varphi_{ii} = (a^i, a^i) - 1 = 0 \tag{18}$$

Indeed, their Poisson brackets on the auxiliary space vanish. Moreover, the constraints φ_i commute with the coordinate functions M_{ij} defining the orbit. This means that

$$\{\varphi_i, f(M)\} = 0$$

for any function on the orbit.

The constraints $\varphi_{ij}, i \neq j$ are of the second class in the generic case of distinct eigenvalues m_i. Indeed we have

$$\{\varphi_{ij}, \varphi_{kl}\}|_{\varphi=0} = i\delta^{il}\delta^{jk}\left(\frac{1}{m_i} - \frac{1}{m_j}\right) \tag{19}$$

and this matrix is nondegenerate in the subspace $i \neq j, k \neq l$. Thus the dimension of the physical phase space, i.e. the dimension of the auxiliary space minus twice the number of 1-st class constraints minus the number of 2-nd class constraints is given by

$$2n^2 - 2n - 2\frac{n^2 - n}{2} = n^2 - n.$$

This number coincides with the dimension of the maximal (nondegenerate) orbit given by the dimension of the group minus its rank. This means that the Hamiltonian reduction of the form (16) with respect constraints (17) leads to the Kirillov form on the orbit. The explicit reduction is realized by introduction of supplementary conditions in quantity equal to that of 1-st constraints, so that the matrix of all Poisson brackets of constraints and supplementary conditions is nondegenerate. The conditions

$$x^i \equiv a^i_n - \bar{a}^i_n = 0, \tag{20}$$

where a^i_n is the last component of a^i in a fixed basis, serve this purpose at a generic point. Indeed, we have the following Poisson brackets on the auxiliary space

$$\{x^i, x^j\} = 0, \{\varphi_{ij}, x^k\} = i\frac{\delta^{ik}}{m_i}a^j_n + i\frac{\delta^{jk}}{m_j}\bar{a}^j_n. \tag{21}$$

The last matrix in nondegenerate on the subspace $i \neq j$ (if $\mathrm{Re}\, a^j_n \neq 0$) which is sufficient for our statement.

Now, the form we are looking for is that induced by Ω on the submanifold $\varphi_{ij} = x^k = 0$. Our aim is to find explicit canonical (angle-action) variables for it. Making use of $x^i = 0$ we can rewrite α in the form

$$\alpha = \frac{1}{2}\sum m_k[(a^k_\perp, \mathrm{d}\, a^k_\perp) - (\mathrm{d}\, a^k_\perp, a^k_\perp)], \tag{22}$$

where $a^k_\perp$ are projections of a^k onto the subspace orthogonal to the last basic vector. In can be shown that this projection can be written as

$$a^k_\perp = \sum_{p=1}^{n-1} C^{kp} e'_p \tag{23}$$

with some moving orthogonal frame e'_p in this subspace and real coefficients C^{kp}, constituting a rectangular matrix C. The last condition leads to cancellation of terms with $\mathrm{d}\, C^{kp}$ in α which takes the form

$$\alpha = \frac{i}{2}\sum_{k=1}^{n}\sum_{p=1}^{n-1}\sum_{q=1}^{n-1}{}' m^k C^{kp} C^{kq}[(e'_p, \mathrm{d}\, e'_q) - \mathrm{h.c.}]. \tag{24}$$

The matrix C satisfies the normalization condition

$$(a^i_\perp, a^j_\perp) = \sum C^{ip} C^{jp} = \delta^{ij} - a^i_n a^j_n; \tag{25}$$

along with the condition $\sum_{i=1}^{n}(a^i_n)^2 = 1$ it implies that the eigenvalues of the matrix M' with matrix elements

$$M'_{pq} = \sum C^{ip} m_i C^{jq} \tag{26}$$

vary in the domain (see [9])

$$m_i \geq m'_i \geq m_{i+1} \tag{27}$$

Due to its reality it can be diagonalized by a real orthogonal transformation which can be absorbed into the choice of matrix C This means that the form α now takes the form

$$\alpha = \frac{i}{2}\sum_{1=i}^{k-1} m'_i\, [(e'_i, \mathrm{d}\, e'_i) - \mathrm{h.c.}], \tag{28}$$

where e'_i are $h-1$ orthogonal vectors. Parametrizing e'_i as follows

$$e'_i = e^{-i\varphi'_i} a^{i\prime}, \quad Im a^{i\prime}_n = 0,$$

we get

$$\alpha = \sum_1^{n-1} m_i' d\varphi_i' + \frac{i}{2}\sum_1^{n-1} m_i'[(a_i', d a_i') - \text{h.c.}]. \tag{29}$$

Now the second term in RHS of (29) looks exactly as (16) with the only change $n \to n-1$. We can repeat our procedure once more and so on. Finally, we get the expression we were looking for

$$\alpha = \sum_1^{n-1} m_i' d\varphi_i' + \frac{i}{2}\sum_1^{n-2} m_i'' + \ldots + m_1^{(n-1)} d\varphi_1^{(n-1)}. \tag{30}$$

Here φ are angle variables, $0 \le \varphi < 2\pi$, and the domain of $m_i^{(k)}$ is determined by our recurrent procedure of descent

$$m_i^{(k-1)} \ge m_i^{(k)} \ge m_{i+1}^{(k-1)}. \tag{31}$$

It is certainly worth mentioning that these coordinates constitute the classical analog of the Gelfand-Zetlin table with parameters of orbit $m_1, \ldots, m_n$ giving the first line. In a different context they were introduced in [10], however their connection with the symplectic structure on the orbits was not discussed.

In the course of deriving the final result (30) by means of the hamiltonian reduction we referred several times to the generic situation. In particular, matrix in (21) is singular when $\operatorname{Re} a_n^i = 0$ and reduction is to be continued in this case. This means that the coordinates $(m_i^{(k)}, \varphi_i^{(k)})$ which over the manifold $\prod^{\frac{n^2-n}{2}} \times T^{\frac{n^2-n}{2}}$, where $\prod$ is polyhedron (31) and T-torus, are coordinates on the orbit with some «singular» set being deleted. The action which we introduce on the orbit in §3 is unsensible to this deletion mod 2π.

2. DARBOUX VARIABLES FOR THE ORTHOGONAL GROUP

The Lie algebra $SO(n)$ is generated by real antisymmetric matrices $M_{ij} = -M_{ji}$. The point of the orbit is given by

$$M = g M_0 g^T, \tag{32}$$

where $g \in SO(n)$ and M_0 can be taken in the form

$$M_{PF} = \begin{pmatrix} 0\, m_1 & & 0 \\ -m_1 0 & & 0 \\ & \ddots & \\ 0 & & 0\, m_r \\ 0 & & -m_r 0 \end{pmatrix}, \qquad n = 2r, \tag{33}$$

or

$$M_{PF} = \begin{pmatrix} 0\, m_1 & & 0 \\ -m_1 0 & & 0 \\ & \ddots & \\ 0 & & 0\, m_r \\ 0 & & -m_r 0 \\ & & 0 \end{pmatrix}, \qquad \begin{matrix} n = 2r + 1, \\ \\ m_l \geq m_2 \ldots \geq m_r . \end{matrix}$$

We see that the orbit is parametrized by the numbers $m_1, \ldots, m_r$ (where $r = [n/2]$ is the rank of the algebra) which can be taken positive if $n = 2r + 1$, and all but one positive if $n = 2r$.

Let $f_1, \ldots, f_{2r}$ be a set of orthonormal vectors such that

$$M_{PF} f_{2k-1} = -m_k f_{2k},$$
$$M_{PF} f_{2k} = -m_k f_{2k-1}.$$

Then using the notation

$$g f_{2k-1} = g_k, \; g f_{2k} = p_k,$$

we have $\Omega = -d\alpha$, where

$$\alpha = \frac{1}{2} \sum_{1}^{r} m_k [(p^k, d q^k) - (q^k, d p^k)] \tag{34}$$

and now our scalar product is real.

The real variables (p, q) span the auxiliary space with dimension $2rn$, and the form α must be obtained from (34) by hamiltonian reduction with respect to constraints

$$\begin{aligned} \varphi^{(1)}_{kl} &= (p^k, p^l) - \delta^{kl} = 0, \\ \varphi^{(2)}_{kl} &= (q^k, q^l) - \delta^{kl} = 0, \\ \varphi^{(3)}_{kl} &= (p^k, q^l) - \delta^{kl} = 0. \end{aligned} \tag{35}$$

The constraints

$$\varphi_i \equiv (p^i, p^i) + (q^i, q^i) - 2 \tag{36}$$

are of the first class, all others belong to the second class in a generic point $m_k \neq m_l, k \neq l$ with nonzero Poisson brackets

$$\begin{aligned} \{\varphi^{(3)}_{kl}, \varphi^{(1)}_{st}\}\,|_{\varphi=0} &= -\delta^{ls}\delta^{kt}\frac{1}{m_e} - \delta^{ls}\delta^{kt}\frac{1}{m_e}, \; s \geq t \\ \{\varphi^{(3)}_{kl}, \varphi^{(1)}_{st}\}\,|_{\varphi=0} &= -\delta^{ks}\delta^{lt}\frac{1}{m_k} - \delta^{kt}\delta^{ls}\frac{1}{m_k}. \end{aligned} \tag{37}$$

The number of physical degrees of freedom in the generic case of distinct m_i is given by

$$2rn - 2r - 2\frac{r^2 - r}{2} - r^2 - r = 2r(n - r - 1),$$

which coincides with the dimension of a nondegenerate orbit given by $\dim G - \operatorname{rank} G$.

The supplementary conditions needed for the reduction will be taken in the form

(38) $$\chi^i \equiv p_n^i = 0.$$

Their Poisson brackets with the 1-class constraints

(39) $$\{\chi^i, \varphi_k\} = \delta^{ik} q_n^i$$

are nonzero for a generic point $q_n^i \neq 0$.

To perform the reduction let us first consider the case $n = 2r + 1$. Let $q_\perp^i$ be the projection of q^i onto the subspace orthogonal to the last basic vector; $q_\perp^i$ form an r-dimensional subspace in the auxiliary space and can be expanded as

(40) $$q_\perp^i = \sum c^{ik} e_k$$

with respect to some moving orthogonal frame e_k in this subspace; c^{ik} are real coefficients constituting an $r \times r$ matrix C. As in the unitary case differentials of this matrix C are absent in α due to respect to the condition $(p^l, e_k) = 0 = (p^l, q_\perp^k)$. Hence α takes the form

(41) $$\alpha = \frac{1}{2} \sum m_i c^{ik} [(p^i, \mathrm{d} e_k) - (e_k, \mathrm{d} p_i)].$$

The matrix $d_{ik} = m_i c^{ik}$ which appears in (41) can be written as

(42) $$\mathrm{d} = u^T m' V,$$

where u and v are orthogonal matrices and m' is diagonal: $m' = \operatorname{diag}(m', \ldots, m'_r)$. If we denote

(43) $$p^{i'} = up^i, \quad q^{i'} = vq^i,$$

then

(44) $$\alpha = \frac{1}{2} \sum m'_i [(p^{i'}, \mathrm{d} q^{i'}) - (q^{i'}, \mathrm{d} p^{i'})].$$

and here we can suppose that $m'_1 \geq m'_2 \geq \ldots m'_r$.

It is easy to prove that m'_i satisfy the condition

$$(45) \qquad \begin{aligned} m_i &\geq m'_i \geq m_{i+1}, \\ m_r &\geq m'_r \geq -m_r. \end{aligned}$$

To prove this let us consider a symmetric positive matrix $\mathcal{D} = \mathrm{d}\mathrm{d}^T$ with eigenvalues $m_i'^2$. It follows from the definition of C that

$$(46) \qquad \sum c^{ik} c^{jk} = \delta^{ij} - q^i_n q^j_n,$$

where q^i_n is the last component of q^i in the fixed basis. Thus

$$(47) \qquad \mathcal{D} = M^2 - \Delta,$$

where M^2 is diagonal with eigenvalues m_i^2 and Δ is a positive matrix of rank 1 with matrix elements

$$(48) \qquad \Delta_{ij} = m_i q^i_n m_j q^j_n.$$

This information is sufficient to state that [9]

$$(49) \qquad m_i^2 \geq m_i'^2 \geq m_{i+1}^2; \quad m_{r+1} = 0.$$

Generically we can choose such orthogonal matrices u and v that all eigenvalues m'_i but one will be positive. Thus we obtain the condition (45).

Set of vectors $p^{i'}, q^{i'}$ form an orthogonal moving basis in the $2r$-dimensional subspace of the n-dimensional vector space. We parametrize these vectors as follows

$$(50) \qquad \begin{aligned} p^{i'} &= \hat{p}_i \cos \varphi'_i + \hat{q}_i \sin \varphi'_i, \\ q^{i'} &= \hat{q}_i \sin \varphi'_i + \hat{q}_i \cos \varphi'_i, \end{aligned}$$

where $\hat{p}^i, \hat{q}^i$ is some new orthonormal basis in a plane $p^{i'}, q^{i'}$ with the condition: $\hat{p}^i_{n-1} = 0$. Here $\hat{p}^i_{n-1} = 0$ is the last component of $\hat{p}^i$ in a fixed basis. In the new variables α takes the form

$$(51) \qquad \begin{aligned} \alpha &= \sum_{i=1}^{r} m'_i \mathrm{d}\varphi'_i \\ &+ \frac{1}{2} \sum_{i=1}^{r} m'_i [\left(\hat{p}^{i'}, \mathrm{d}\hat{q}^i \right) - \left(\hat{q}^{i'}, \mathrm{d}\hat{p}'^i \right)]. \end{aligned}$$

The second term in RHS of (51) looks as (34) with the change $n = 2r + 1 \mapsto n = 2r$, where n is dimension of vector space.

Now we briefly discuss the case $n = 2r$:

$$\alpha' = \sum_1^r m'_i\,(\hat{p}^i, d\,\hat{q}^i) = \sum_1^r m'_i\,(\hat{p}_i, d\,\hat{q}^i_\perp), \tag{52}$$

here $\hat{q}^i_\perp$ as above is the projection of $\hat{q}^i$ onto the subspace orthogonal to the last basic vector in the $2r$-dimensional vector space. But now $\hat{q}^i_\perp$ form an $(r-1)$-dimensional subspace in this space spanned by orthogonal moving vectors $\hat{e}_1, \ldots, \hat{e}^{(r-1)}$ and therefore

$$\hat{q}^i_\perp = \sum_{k=1}^{r-1} \hat{c}^{ik} \hat{e}_k. \tag{53}$$

The coefficient $\hat{C}^{ik}$ form an $r \times (r-1)$ rectangular matrix $\hat{C}$. In the $\hat{p}, \hat{e}$ variables we have

$$\alpha' = \sum m'_i\,\hat{c}^{ik}(\hat{p}^i, d\,\hat{e}_k). \tag{54}$$

Now the matrix $\hat{d} : \hat{d}_{ik} = m'_i\,\hat{c}^{ik}$ can be written in the form

$$\hat{d} = \hat{u}^T m'' \hat{v},$$

with $\hat{u}$ and $\hat{v}$ being orthogonal matrices of dimension r and $(r-1)$ respectively, and m'' a diagonal matrix with eigenvalues $(m''_1, \ldots, m''_{r-1}, 0)$. Put

$$\hat{q}^{k\prime} = \hat{v}\hat{e}_k, \quad \hat{p}^{k\prime} = \hat{u}\hat{p}^k. \tag{55}$$

Then α' takes the form

$$\alpha' = \sum_1^r m''_i\,(\hat{p}^{i\prime}, d\,\hat{q}^{i\prime}), \tag{56}$$

with $\hat{p}'_r$ omitted.

It can be shown as in the previous case that the eigenvalues of the positive matrix $\mathcal{D} = \hat{d}\,\hat{d}^T$ satisfy conditions

$$m_i'^{/2} \geq m_i''^{/2} \geq m_{i+1}'^{/2} \tag{57}$$

and now all m_i'' can be chosen positive:

$$m_i' \geq m_i'' \geq m_{i+1}' . \tag{58}$$

Thus after the second step of our descent procedure we obtain the expression

$$\alpha = \sum_{i=1}^{r} m_i' \mathrm{d}\varphi_i' + \sum_{1}^{r-1} m_i'' \mathrm{d}\varphi_i'' + \sum_{1}^{r-1} m_i''(p_i'', \varphi_i''), \tag{59}$$

and now the last term looks exactly as in (34) with the only change $r \to r-1$. We can repeat our procedure once more and so on. Then we get the expression we are looking for:

$$\alpha = \sum_{1}^{r} m_i' \mathrm{d}\varphi_i' + \sum_{1}^{r-1} m_i'' \mathrm{d}\varphi_i'' + \ldots + m_1^{(2r-1)} \mathrm{d}\varphi_1^{(2r-1)} \tag{60}$$

for the case $n = 2r + 1$, and

$$\alpha = \sum_{1}^{r} m_i' \mathrm{d}\varphi_i' + \sum_{1}^{r} m_i'' \mathrm{d}\varphi_i'' + \ldots + m_1^{(2r-2)} \mathrm{d}\varphi_1^{(2r-2)}$$

if $n = 2r$. Here $\varphi_i^{(k)}$ are angle variables $0 \leq \varphi_i^{(k)} < 2\pi$ and the domain of $m_i^{(k)}$ is determined by our recurrent descent procedure

$$m_i^{(k-1)} \geq m_i^{(k)} \geq m_{i+1}^{(k-1)}, \tag{61}$$

but for the last elements, in each line which satisfy

$$m_{i_0}^{(k-1)} \geq m_{i_0}^{(k)} \geq |m_{i_0+1}^{(k-1)}| \qquad \text{if } n-k = 2p+1$$
$$m_{i_0}^{(k-1)} \geq m_{i_0}^{(k)} \geq -m_{i_0}^{(k-1)} \qquad \text{if } n-k = 2p.$$

This is the classical analog of the Gelfand-Zetlin table for the $SO(n)$ algebra.

As in §2, the form (60) defined on the polyhedron (61) times the torus coincides with the Kirillov form on the orbit with some «singular» submanifold deleted.

3. CALCULATION OF THE FUNCTIONAL INTEGRAL

After the introduction of convenient canonical coordinates on the orbit the functional integral for the higher rank can be treated exactly as that in the introduction for the $SO(3)$ case. We introduce the main functional integral in the form

$$G(\{\varphi'\},\{\varphi''\}) \int_{i,k,t} \prod \mathrm{d}\, m_i^{(k)} \mathrm{d}\varphi_i^{(k)} \; e^{i\int \left[m_i^{(k)}(t) + \gamma_i^{(k)})\dot{\varphi}_i^{(k)} - h(M(t))\right] \mathrm{d}t} \tag{62}$$

where the hamiltonian $h(M)$ is a function on the orbit and $\gamma_i^{(k)}$ are some constants; we integrate over the phase space trajectories $m_i^{(k)}(t), \varphi_i^{(k)}(t)$ lying within the boundaries (31), (61) for $m(t)$ and $-\infty < \varphi(t) < \infty; 0 < t < T$. The boundary conditions are given by

$$\varphi_i^{(k)}(0) = \varphi_i^{(k)\prime}, \varphi_i^{(k)}(T) = \varphi_i^{(k)\prime\prime} (\text{mod } 2\pi). \tag{63}$$

Alternatively, we can fix the boundary conditions completely and average over the φ'' shifted by any integer multiples of 2π.

The condition of single-valuedness of the action leads to intergral orbits which correspond to the cases when the parameters m_i are all integers or half integers. In the former case the action is single-valued for $\gamma_i^{(k)} = 0$. In the latter case we are to take all $\gamma_i^{(k)} = 1/2$.

It is clear that the functional integral (62) can be reduced to the finite-dimensional one if the hamiltonian is any function of $m_i^{(k)} = 0$. In particular, the generators belonging to some Cartan subgroup are specific linear combinations of m-variables. For the $SU(n)$ case the explicit formula looks as follows [11]

$$P^{(k)}(m) = \sum_i m_i^{(k)} = 0$$

The functional integral for fixed boundary condition is reduced to

$$G(\{\varphi'\}, \{\varphi''\}) \int \mathrm{d}m \, e^{i \sum_{i,k} \left[m_i^{(k)} (\varphi_i^{(k)\prime\prime} - \varphi_i^{(k)\prime}) - h(m)T + \gamma_i^{(k)} (\varphi_i^{(k)\prime\prime} - \varphi_i^{(k)\prime}) \right]} \tag{64}$$

and the averaging over final φ-variables gives

$$\begin{aligned} G(\{\varphi'\}, \{\varphi''\}) &= \sum_{\{n_i^{(k)}\}} G_0(\{\varphi'\}, \{\varphi'' + 2\pi n\}) = \\ &= \int \mathrm{d}m \, e^{i \sum_{i,k} \left[m_i^{(k)} + \gamma_i^{(k)})(\varphi_i^{(k)\prime\prime} - \varphi_i^{(k)\prime}) - h(m)T \right]} \delta(\alpha, m), \end{aligned} \tag{65}$$

where $\delta(\alpha, m)$ is the periodic δ-function supported at all integer points for $\alpha_i^{(k)} = 0$ and at all half-integer points for $\alpha_i^{(k)} = 1/2$ This δ-function in the integral allows to perform the last integration over m. However, we are to remember the regularization, analogous to that in the introduction (8), namely integration over $m_i^{(k)}$ is to be taken over the domain

$$m_i^{(k+1)} + \epsilon \geq m_i^{(k)} \geq m_i^{(k-1)} - \epsilon \tag{66}$$

with the limit $\epsilon \to 0$.

As a result we shall get an expression of the type

$$G(\{\varphi'\},\{\varphi''\}) = \sum_{m} e^{i\left(\sum\left[m_i^{(k)}+(\varphi_i^{(k)\prime\prime}-\varphi_i^{(k)\prime})-h(m)T\right)\right]} \tag{67}$$

where the sum is taken over all quantized Gelfand-Zetlin tables with fixed $m_1, \ldots, m_z$, the latter evidently playing the role of the highest weight of the representation corresponding to the orbit.

REFERENCES

[1] A. KIRILLOV: *Elements of the representation theory*, Nauka, Moscow, 1972 (in Russian).

[2] A. A. KIRILLOV: *Characters of unitary representations of Lie groups*, Funkt. anal. i ego prilozh., 2, 2, 40-55, 1968 (in Russian).

[3] L. D. FADDEEV, 1965, unpublished.
H. B. NIELSEN, D. ROHRLICH: *A Path integral to quantize Spin*, Nucl. Phys., B299, 471-483, 1988.

[4] A. A. SLAVNOV, L. D. FADDEEV: *Introduction to quantum theory of gauge fields*, Nauka, Mo scow, 1978 (in Russian).

[5] A. YU ALEKSEEV, S. L. SHATASHVILI, *Propagator for the relativistic spinning particle via functional integral over trajectories*, Mod. Phys. Lett. A, Vol. 3, N. 16 (1988), 1151-1159.

[6] A. M. POLYAKOV, *Fermi-Bose transmutations induced by gauge fields*, Mod. Phys. Lett. A, **3**, 325-334, 1988.

[7] A. M. POLYAKOV: *Quantum gravity in two dimensions*, Mod. Phys. Lett. A, **11**, 893-898, 1987.

[8] L. D. FADDEEV, S. L. SHATASHVILI: *Realization of the Schwinger term in the Gauss law and possibility of correct quantization of theory with anomalies*, Phys. Lett. **167B**, 225-228, 1986.

[9] F. R. GANTMAKHER: *Theory of matrices*, Nauka, Moscow, 1966 (in Russian).

[10] I. M. GELFAND, M. A. NAIMARK: *Unitary representation of classical groups*, Proceedings of Steklov Institite, **36**, Moscow, 1950.

[11] D. P. ZHELOBENKO: *Compact groups and their representations*, Nauka, Moscow, 1970 (in Russian).

Manuscript received: September 20, 1988

Nuclear Physics B327 (1989) 399–414
North-Holland, Amsterdam

COHERENT-STATE PATH INTEGRALS FOR LOOP GROUPS AND NON-ABELIAN BOSONISATION

Michael STONE

Department of Theoretical Physics, Schuster Laboratory, University of Manchester, Manchester M13 9PL, UK
and
Loomis Laboratory of Physics, University of Illinois, 1110 W. Green Street, Urbana, IL 61801, USA*

Received 28 April 1989

After a discussion of coherent states for the loop group LU(N), we use them to write down a bosonic path integral which describes the level-one representations of LU(N). The construction uses the description of the representations in terms of free right-going Weyl fermions and so provides one with an explicit geometrical interpretation for a Fermi–Bose equivalence. The bosonic system is easily seen to have a single Kac–Moody algebra as its Poisson brackets and coincides with the non-abelian chiral boson model that has been introduced by Sonnenschein et al. [Nucl. Phys. B301 (1988) 346; B309 (1988) 752].

1. Introduction

Two-dimensional field theories possess remarkable equivalences between fermions and bosons. These equivalences were originally discovered for abelian symmetries [1, 2] but a partial non-abelian bosonisation was observed soon after [3]: here the fermions transform under a non-abelian group but the boson fields are restricted to an abelian maximal torus in the non-abelian group, the rest of the symmetries being hidden in vertex operator constructions [4]. It was only much later that Fermi–Bose equivalences were extended to bosonic fields transforming under the full non-abelian group [5, 6]. A general algebraic framework for these relations is provided by the representation theory of Kac–Moody algebras (for a review, see ref. [7]) but the condition for a representation to be constructed out of free fermions: that the fermions transform in the same way as the tangents to a point in a symmetric space under its isotropy group, has a pronounced geometric flavour.

It has been known for some time [8, 9] that a natural geometric approach to the bosonic description of fermionic systems is provided by coherent state path integrals. Coherent states, and coherent state path integrals, have a long history [10] and

* Permanent address.

0550-3213/89/$03.50 © Elsevier Science Publishers B.V.
(North-Holland Physics Publishing Division)

find applications in a number of areas of quantum physics. However, while Grassman valued coherent states are invariably used in driving path integrals for Fermi systems, most of the applications to bosonic quantum field theories have used Feyman's route to the path integral. Recently Weigman [11, 12] has pointed out that there is a large body of mathematics literature where coherent states, although not necessarily under that name, are used to discuss the dynamics [13] and representations [14] of ordinary Lie groups. He has pointed out that the same techniques apply to infinite dimensional groups such as those associated with Kac–Moody and Virasoro algebras, and lead naturally to coherent state path integrals for two-dimensional quantum fields and for string theory [12]. Similar ideas motivate recent work by Witten [15].

The constructions refs. [8, 9] work in any number of space-time dimensions, but in more than two dimensions they replace a few fermionic Grassman variables by a large number of commuting variables and this limits their utility. It seems that only in two dimensions, where the rich and well-explored structure of loop groups [16] and their algebras give compact formulae, do simple equivalences exist.

Since boson fields normally come with both left and right propagating modes, bosonisation formulae are most easily obtained for Dirac fermions. For some applications one can obtain an equivalence for Weyl fermions by coupling external probes to only one of the chiral components of the Dirac fermion. Unfortunately gravitational interactions will automatically couple to both components and both will contribute to the conformal anomaly, so, for applications to the heterotic string for example, we would need to bosonise chiral fermions directly. An action for non-abelian chiral bosons has recently been proposed [17] and the consequent equation of motion has only right-moving solutions and admits only one Kac–Moody algebra [18]. This makes it a candidate for a bosonic equivalent for a chiral fermion.

In this paper we illustrate some of these ideas, and begin a geometric study of bosonisation, by deriving the coherent state path integral for the current algebra of a system of right-moving free fermions using the methods of refs. [8, 9]. In this case the chiral fermions seem to be simpler than Dirac fermions and using the special properties of two-dimensional loop groups we will be able to simplify the general expressions from these papers and show that the resultant two-dimensional bosonic field theory coincides with the proposed chiral boson action of ref. [17]. We also demonstrate the geometric origin of the single Kac–Moody current algebra found in ref. [18]*.

Sect. 2 provides us with a review of the coherent state path integral for finite dimensional compact Lie groups and discusses the geometry of the Poisson brackets. Sect. 3 defines the infinite dimensional analogue of the group coherent states and exhibits some of their properties that have no finite dimensional analogue. Sect.

* A brief version of these results has appeared in ref. [27].

4 obtains some necessary expectation values and explains the origin of the curvature, or Berry phase, of the vacuum. Sect. 5 uses these results to write down the path integral and shows that the boson field theory has the correct Poisson brackets.

2. Coherent state path integrals for Lie groups

In this section we will review the basic idea of the coherent state path integral for compact simple groups and its connection with classical dynamics [11–13, 19]. Suppose $g \to D(g)$ is an irreducible unitary representation of a compact simple group G and $|0\rangle$ is any state in this representation. We define a collection of coherent states, labelled by elements $g \in \mathrm{G}$, by

$$|g\rangle = D(g)|0\rangle . \tag{2.1}$$

Because of the irreducibility, Schur's lemma shows that these states satisfy an overcompleteness relation

$$1 = \text{const.} \times \int \mathrm{d}[g] |g\rangle\langle g| \tag{2.2}$$

where $\mathrm{d}[g]$ is the Haar measure on G.

We can use these states to give a path integral representation for the thermodynamic partition function

$$\mathscr{Z} = \mathrm{Tr}(\mathrm{e}^{-\beta \hat{H}}), \tag{2.3}$$

where the trace is taken only over the states in the representation $D(g)$. We obtain the path integral by repeatedly inserting the resolution of the identity, eq. (2.2), into the trace. Ignoring any subtleties in the calculus, we obtain the formal path integral

$$\mathscr{Z} = \int \mathrm{d}[g] \exp\left[- \int_0^\beta (\langle g|\dot{g}\rangle + \langle g|\hat{H}|g\rangle)\, \mathrm{d}t \right]. \tag{2.4}$$

Here $\mathrm{d}[g]$ is the path measure made from the Haar measure at each step and the integral is over all periodic paths in G taking "time" β.

Because of the casual use of ordinary calculus ideas, i.e. regarding $|g(t+\delta t)\rangle - |g(t)\rangle$ to be $0(\delta t)$, we should be careful in interpreting this expression, but for the moment we will pretend that we can take the exponent in eq. (2.4) literally as an action for the classical dynamics of the system.

This classical action is first order in time derivatives and in Minkowski space seems to be an example of the form

$$S = \int_{\gamma = \partial \Gamma} \mathrm{d}t \left(H(x) - a(x)_\mu \dot{x}^\mu \right) = \int_\gamma H \,\mathrm{d}t - \int_\Gamma \omega , \tag{2.5}$$

where

$$\omega = \tfrac{1}{2}\omega_{\mu\nu}\, \mathrm{d}x^{\mu} \wedge \mathrm{d}x^{\nu}, \qquad \omega_{\mu\nu} = \partial_{\mu}a_{\nu} - \partial_{\nu}a_{\mu}. \tag{2.6}$$

The variation principle produces the classical equations of motion associated with the action (2.5) as

$$\partial_{\mu}H = \omega_{\mu\nu}\dot{x}^{\nu} \tag{2.7}$$

and Poisson brackets can be defined by

$$\{f, g\} = \omega^{\mu\nu}\partial_{\mu}f\partial_{\nu}g, \tag{2.8}$$

where $\omega^{\mu\nu}$ is the matrix inverse to $\omega_{\mu\nu}$.

This formalism does not apply immediately to the action in eq. (2.4). The matrix ω occurring there turns out not to have an inverse. Indeed there are no unique solutions to the equations of motion arising from eq. (2.4). This is because of a "gauge invariance" hidden in the action: In addition to being overcomplete, the physically distinct states are also overdescribed by the label g. The elements in G that leave $|0\rangle$ fixed up to a phase

$$D(g)|0\rangle = \mathrm{e}^{i\varphi}|0\rangle \tag{2.9}$$

form a subgroup H, which we will call the isotropy group [20] of $|0\rangle$. Choosing, for example, the state $|0\rangle$ to be a non-degenerate weight vector in the representation D leads to H being the maximal torus, T, generated by the Cartan sub-algebra; the phase φ is the weight of the element that exponentiates to $g \in \mathrm{T}$. The rays representing the physical states of the system are in one-to-one correspondence with elements of the quotient space, G/H and the path integral can also be regarded as being over G/H. That $\langle g|\hat{H}|g\rangle$ is independent of the phase φ is clear but it is also true that the integral $\oint\langle g|\mathrm{d}g\rangle$, around any contractable closed loop in G, is independent of choice of representatives in the cosets gH since it can be written as $\int\langle \mathrm{d}g|\mathrm{d}g\rangle$ over any surface spanned by the loop. The integral over H therefore factors out the overall path integral. If we were to take open paths from $g = \textit{identity}$ to any other $g \in \mathrm{G/H}$, the integral would depend on the choice of representative at the end point, but this would be a gauge choice for the wave functions which are thus sections of a line-bundle over G/H [20, 21].

Part of the exponent, $\int\langle g|\dot{g}\rangle\,\mathrm{d}t = \oint\langle g|\mathrm{d}g\rangle$, is purely imaginary, even in euclidean space, and should be recognisable as the Berry phase [22] which would result if the states $|g\rangle$ were carried adiabatically around the path in the coset space G/H. Of course there is nothing adiabatic about the variations in the path integral and Berry's phase is here playing its other role as a natural connexion on the hermitian line bundle associated with the principal bundle $\pi: \mathrm{G} \to \mathrm{G/H}$. The two-form

$\omega = \langle dg|dg \rangle$ plays a double role as the curvature of the line bundle and as the symplectic form defining the classical hamiltonian dynamics [23, 24].

Bearing in mind these slight complications let us derive the equations of motion from (2.4). In this section we will work in euclidean space so the reader should bear in mind that some factors of i would be different in real time.

We have

$$S = -\int (\langle 0|g^{-1}\dot{g} + g^{-1}\hat{H}g|0\rangle)\,dt, \tag{2.10}$$

and the variational principle reads, in terms of $\eta = g^{-1}\delta g$,

$$0 = \delta S = \int \langle 0|\left[\eta, (g^{-1}\hat{H}g + g^{-1}\dot{g})\right]|0\rangle\,dt. \tag{2.11}$$

From this we find that

$$0 = g^{-1}\hat{H}g + g^{-1}\dot{g} + \lambda(t) \tag{2.12}$$

where $\lambda(t)$ is any, in general t-dependent, element of the Lie algebra of H. The solutions of the equation of motion are therefore

$$g(t) = e^{-\hat{H}t}h(t), \tag{2.13}$$

which are only unique as elements of G/H and not as elements of G.

We now want to define the Poisson bracket by its basic property

$$\frac{df(t)}{dt} = \{H, f\}. \tag{2.14}$$

Because of the ambiguity of the solutions of the equations of motion, we can only make sense of this definition if we restrict ourselves to functions on G/H i.e. to functions that obey $f(g) = f(gh)$ for $h \in$ H. If λ_i are a set of generators for the Lie algebra of G obeying $[\lambda_i, \lambda_j] = f_{ij}^k\lambda_k$, we can make functions J_i with the required property by taking $J_i(g) = \langle g|\lambda_i|g\rangle$. From eq. (2.12) we find

$$\frac{dJ_i(g)}{dt} = -\langle 0|\left[g^{-1}\dot{g}, g^{-1}J_i g\right]|0\rangle = \langle g|\left[\hat{H}, \lambda_i\right]|g\rangle, \tag{2.15}$$

and taking $\hat{H} = \lambda_j$ we find

$$\{J_i(g), J_j(g)\} = \langle g|\left[\lambda_i, \lambda_j\right]|g\rangle = f_{ij}^k J_k(g). \tag{2.16}$$

So the algebra of the *classical* Poisson brackets coincides with algebra of the quantum mechanical operators (In Minkowski space there is of course a factor of i between them). The assignment of a function $J_i(g)$ to a generator of a group action on the phase space of a mechanical system, where it then generates the symmetry via Hamilton's equations, is known as the *moment map*. It is described extensively in ref. [23]. The general construction being described here is the inverse of the method of geometric quantization [24].

The problems with the operator ordering and the need for non-naive calculus become apparent when one considers what in the path integral expression determines the representation $D(g)$. The only place where any of the properties of the starting representation appear is in the Berry phase and its curvature ω. This curvature however depends only on the Lie algebra weight of the cyclic state $|0\rangle$ and this is not unique to the representation. Clearly something in the way we regularise the path integral must also play a role. If we take as $|0\rangle$ to be a highest weight vector in the representation then this does determine the representation uniquely. In this case the resulting coherent states, which have smallest spread and are in some sense closest to the classical limit, lead to the wave functions being *holomorphic* sections of the line bundle over G/H [19, 20, 23, 24]. Further, in this case only, the phase space is a Kähler manifold [23]. We conjecture that taking $|0\rangle$ as a greatest weight and regularising in a way that preserves the Kähler structure will be enough to determine the path integral and the representation – but more work is needed on this issue.

3. Coherent states for loop groups

We want to apply the simple formalism of the previous section to two-dimensional quantum field theories. The groups to use for the construction of the coherent states are the loop groups that occur naturally as gauge groups in 1 + 1 dimensions. An element of the loop group LG associated with the compact simple group G is simply a mapping

$$g(x) : \mathrm{S}^1 \to \mathrm{G} \tag{3.1}$$

with multiplication being taken pointwise i.e. $(g_1 g_2)(x) = g_1(x) g_2(x)$. Despite the seeming triviality of the definition, all the representations of these groups that are both non-trivial and physical (in the sense that they are unitary and have vacua with energy bounded below) are projective representations. They correspond to greatest weight representations of the affine Lie algebras reviewed in ref. [7]. For the present we wish to proceed *more geometrico* and will draw extensively on the ideas and constructions to be found in ref. [16] which is recommended reading although we will try to make the exposition self-contained.

The present work will be restricted to the group LU(N) which is the gauge group for a set of N free right-moving Fermi fields with standard anti-commutation relations

$$\{\psi_i^\dagger(x), \psi_j(x')\} = \delta_{ij}\delta(x-x'). \tag{3.2}$$

We will take the simplest possible hamiltonian:

$$H = \int \mathrm{d}x \sum_{i=1}^{N} \left(\psi_i^\dagger(-i\partial x)\psi_i - \mu\psi_i^\dagger\psi_i\right). \tag{3.3}$$

For convenience with factors of 2π in intermediate calculations, the fermions will live on a line of length 2π with periodic boundary conditions (the difference between Ramond and Neveu–Schwartz fermions being taken care of by the chemical potential μ). This does not affect any of the factors of 2π in the final formulae.

The second-quantized system is constructed on a vacuum state made by filling the Dirac sea up to $E=\mu$. The levels filled correspond to the states $|n\rangle$ of the first quantized system. These are eigenstates of the first quantized hamiltonian

$$(-i\partial_x)|n\rangle = n|n\rangle, \qquad \langle x|n\rangle = (1/\sqrt{2\pi})\mathrm{e}^{inx}. \tag{3.4}$$

This vacuum state is a highest weight vector and will be our cyclic state $|0\rangle$. We construct the system of coherent states $|g(x)\rangle$, by replacing each first-quantized state in the sea by its image under the loop group. We will denote the image by $|n, g(x)\rangle$, where

$$\langle x|n, g(x)\rangle = g(x)\mathrm{e}^{inx}. \tag{3.5}$$

If we arrange for the Fermi level $E=\mu$ to fall between two energy levels (Neveu–Schwartz fermions), the vacuum is non-degenerate and is left fixed, up to a (formally infinite) phase, by those elements in $LU(N)$ which are independent of x, since these only mix states within a given energy level. The isotropy group is therefore isomorphic to the group $U(N)$ itself, and the system of coherent states forms a line bundle over $LU(N)/U(N)$. Because, for any given coset in $LU(N)/U(N)$, we can always choose a representative having the property $g(0)=$ *identity*, we can identify the quotient space $LU(N)/U(N)$ with the group of based loops $\Omega U(N) = \{g(x)\colon S^1 \to U(N) | g(0) = \textit{identity}\}$ but one should beware that this is a topological identification only: $U(N)$ is *not* a normal subgroup of $LU(N)$ and so $LU(N)/U(N)$ is not a group, in contrast to $\Omega U(N)$ which is.

If the Fermi surface falls exactly on an energy level (Ramond fermions), the vacuum will form a representation of $U(N)$ depending on exactly how many of the N states at $E=0$ are actually filled. The isotropy group will then be smaller than $U(N)$. If we are only interested in taking the thermodynamic limit, which is usually the case in physical application outside string theory, the existence of a few extra particles somewhere in the universe should be of no account and the physics will be locally identical with the previous case. For string applications this degeneracy must be dealt with separately. Since our interests are ultimately with large systems we will not do this here.

The first sign of the profound difference between these infinite dimensional groups and their finite dimensional counterparts is seen in the spectral flow that they can produce. Unlike $\mathrm{U}(N)$, the group $\mathrm{LU}(N)$ falls into disconnected parts: those $g(x)$ whose determinant winds n times as x traverses $(0, 2\pi]$ cannot be continuously deformed into $g(x)$ with other winding numbers. If we form the infinite matrix $U(g)$ with entries

$$U_{nm} = U_{n-m} = \langle n|g|m\rangle = \frac{1}{2\pi}\int_0^{2\pi} \mathrm{e}^{-(n-m)x} g(x)\,\mathrm{d}x \tag{3.6}$$

and partition it in accordance with the decomposition of the one-particle Hilbert space into positive and negative energy subspaces, $\mathscr{H} = \mathscr{H}^+ \oplus \mathscr{H}^-$ as

$$U(g) = \begin{matrix} \mathscr{H}^- \\ \mathscr{H}^- \end{matrix} \overset{\begin{matrix}\mathscr{H}^+ & \mathscr{H}^-\end{matrix}}{\begin{pmatrix} U_{++} & U_{+-} \\ U_{-+} & U_{--} \end{pmatrix}}, \tag{3.7}$$

then modest smoothness requirements on $g(x)$ make the off-diagonal matrices, $U_{\pm\mp}$, Hilbert–Schmidt, i.e. $\mathrm{Tr}(U^\dagger_{\pm\mp} U_{\pm\mp})$ converges because the Fourier components fall off rapidly as $n - m \to \infty$. The matrices on the diagonal, $U_{\pm\pm}$, are then Fredholm, i.e. have finite dimensional kernel and co-kernel. For example, if we take $g = \mathrm{e}^{ix} \in \mathrm{U}(1)$ which winds once, then the matrix U_{nm} has 1's immediately above the main diagonal and zeros elsewhere. Then $\mathrm{Tr}(U^\dagger_{+-} U_{+-}) = 1$, the kernel of U_{--} is one-dimensional while the image of $\mathscr{H}^+$ under U_{++} fails to fill out $\mathscr{H}^+$ by just one state i.e. the dimension of $\mathrm{coker}(U_{++})$ is 1. It is easy to show, from the unitarity of U, that

$$\mathrm{index}\, U_{++} = \dim\ker U_{++}(g) - \dim\mathrm{coker}\, U_{++}(g) = -\Gamma(\det g(x)), \tag{3.8}$$

where $\Gamma(f)$ is the winding number of the complex number f around the origin in the Argand diagram. Also

$$\mathrm{index}\, U_{++} = -\mathrm{index}\, U_{--}. \tag{3.9}$$

As a result of acting by one of the disconnected parts of the group, a number of levels equal to the winding number have crossed form below the Fermi surface to above. Thus the disconnected parts of $\mathrm{LU}(N)$ act to change the fermion number, the charge of the vacuum under the U(1) part of U(N), by

$$\Delta Q=\frac{1}{2\pi i}\int_0^{2\pi}\partial_x(\log \operatorname{Det}(g(x))\,\mathrm{d}x=\int \mathrm{d}x\,\frac{1}{2\pi i}\operatorname{Tr}\left(g^{-1}\partial_x g(x)\right). \quad (3.10)$$

This means that the irreducible representation of LG built on the vacuum will contain states of various different fermions numbers.

4. Expectation values and the vacuum curvature

By writing down a path integral for the loop group coherent states we will have obtained a bosonic path integral equivalent to the original fermionic theory, or at least that part of it which is contained in the irreducible representation built on the vacuum by the action of the currents. We need to compute the ingredients used in eq. (2.4). In particular we need the expectation of the hamiltonian $\langle g|\hat{H}|g\rangle$, the Berry connexion, $\langle g|\dot{g}\rangle$, and, for the computation of the Poisson brackets, the expectation values of the currents that generate the Lie algebra of the loop group, $\psi^\dagger\lambda_i\psi$.

Expectations, such as $\langle g|\psi^\dagger\hat{O}\psi|g\rangle$, of the second quantized form of one-particle operator $\hat{O}$, can be evaluated, after normal ordering so that they vanish for $g=identity$, as

$$\langle g|\psi^\dagger\hat{O}\psi|g\rangle=-\tfrac{1}{2}\sum_n \operatorname{sgn}(E_n)\left(\langle n,g|\hat{O}|n,g\rangle-\langle n|\hat{O}|n\rangle\right), \quad (4.1)$$

or, in terms of the matrix

$$J=\begin{matrix} & \mathscr{H}^+ & \mathscr{H}^- \\ \mathscr{H}^+ & +1 & 0 \\ \mathscr{H}^- & 0 & -1\end{matrix}, \quad (4.2)$$

as

$$\langle g|\psi^\dagger\hat{O}\psi|g\rangle=-\tfrac{1}{2}\operatorname{Tr}\left(\hat{O}(gJg^{-1}-J)\right). \quad (4.3)$$

The best way to proceed is to take a variation

$$\delta(\langle g|\psi^\dagger\hat{O}\psi|g\rangle)=-\tfrac{1}{2}\operatorname{Tr}\left(\hat{O}\left[\xi,g^{-1}Jg\right]\right),\qquad \xi=\delta g g^{-1},$$

$$=-\tfrac{1}{2}\operatorname{Tr}\left(g^{-1}\hat{O}g[\eta,J]\right),\qquad \eta=g^{-1}\delta g. \quad (4.4)$$

Now

$$[\eta,J]=\begin{pmatrix}0 & -2\eta_{+-}\\ 2\eta_{-+} & 0\end{pmatrix} \quad (4.5)$$

where the matrix η is given, in the energy basis, by

$$\langle n|\eta(x)|m\rangle = (1/2\pi)\int \mathrm{d}x\, \mathrm{e}^{-i(n-m)x}\,\eta(x) = \eta_{n-m}. \tag{4.6}$$

We evaluate the traces using

$$\begin{pmatrix} \hat{O}_{++} & \hat{O}_{+-} \\ \hat{O}_{-+} & \hat{O}_{--} \end{pmatrix}\begin{pmatrix} 0 & \eta_{+-} \\ -\eta_{-+} & 0 \end{pmatrix} = \begin{pmatrix} -\hat{O}_{+-}\eta_{-+} & \hat{O}_{++}\eta_{+-} \\ -\hat{O}_{--}\eta_{-+} & \hat{O}_{-+}\eta_{+-} \end{pmatrix}. \tag{4.7}$$

Let us begin with $\langle g|\psi^\dagger\lambda_i\psi|g\rangle$ where

$$\langle n|g^{-1}\hat{O}g|m\rangle = (1/2\pi)g^{-1}\lambda_i g\, \mathrm{e}^{-i(n-m)x} \tag{4.8}$$

and we find

$$\begin{aligned}\delta\langle g|\psi^\dagger\lambda_i\psi|g\rangle &= \frac{1}{2\pi}\mathrm{Tr}\left(g^{-1}(x)\lambda_i g(x)\sum_{n=-\infty}^{\infty} n\,\mathrm{e}^{inx}\eta_n\right)\\ &= \frac{1}{2\pi}\mathrm{Tr}(g^{-1}\lambda_i g(-i\partial_x\eta(x))\\ &= \frac{1}{2\pi i}\mathrm{Tr}(g^{-1}\lambda_i g\,\partial_x(g^{-1}\delta g)) = \delta\left(\frac{1}{2\pi i}\mathrm{Tr}(\lambda_i\,\partial_x g g^{-1})\right).\end{aligned} \tag{4.9}$$

Since $\langle J_i\rangle = 0$ at $g = \mathit{identity}$ we have

$$\langle g|J_i|g\rangle = (1/2\pi i)\mathrm{Tr}(\lambda_i\,\partial_x g g^{-1}). \tag{4.10}$$

For the hamiltonian we have

$$\langle n|g^{-1}\hat{O}g|m\rangle = (g^{-1}(-i\partial_x)g)_{n-m} + m\delta_{nm}. \tag{4.11}$$

We find that

$$\langle g|\psi^\dagger(-i\partial_x)\psi|g\rangle = (1/4\pi)\mathrm{Tr}(g^{-1}\partial_x g)^2 \tag{4.12}$$

plus a part coming from $m\delta_{nm}$ which needs regularising. It depends on the spectral asymmetry of the one particle states and we will set it to zero. It is strictly only zero for Neveu–Schwartz fermions.

To compute the curvature of the line bundle we note that the decomposition $\mathscr{H}=\mathscr{H}^+\oplus\mathscr{H}^-$ defines an element of a grassmanian manifold $\mathrm{Gr}(\mathscr{H})$. The images of the subspace $\mathscr{H}^-$ under $\mathrm{LU}(N)$ form a submanifold of $\mathrm{Gr}(\mathscr{H})$. The process of filling the Dirac sea is the same as taking exterior powers of the orthonormal vectors that span $\mathscr{H}^-$ until we reach the highest exterior power, $\Lambda^{(\max)}(\mathscr{H}^-)$, which is the full Dirac sea. The submanifold is essentially the space of Slater determinant wave functions.

Up to a phase we can identify the subspace $g\mathscr{H}^- \in \mathrm{Gr}(\mathscr{H})$ with any element in $\Lambda^{(\max)}(g\mathscr{H}^-)$ and the process of forgetting the phase is the projection that defines the bundle. The curvature of this vacuum line bundle turns out to be the first Chern character of the pull-back of the curvature of the tautological bundle on $\mathrm{Gr}(\mathscr{H})$, i.e. the bundle whose fibre above $g\mathscr{H}$ is $g\mathscr{H}$ itself. To understand this statement and make it plausible consider first a finite dimensional grassmanian $\mathrm{Gr}_k(\mathscr{V}_n)$ consisting of k dimensional sub-spaces $\mathscr{W}_k$ of the $n = k + l$ dimensional vector space, $\mathscr{V}_n$. If the k-dimensional subspaces $\mathscr{W}_k$ are spanned by $u_1, u_2, \ldots, u_k$ then, up to the coefficient, these are in one-to-one correspondence with exterior powers of the form $u_1 \wedge u_2 \cdots \wedge u_k \in \Lambda^{(k)}(\mathscr{V}_n)$. Let $\mathscr{D}^k$ be the manifold of decomposable elements in $\Lambda^k(\mathscr{V}_n)$, i.e. those that can be written as a product of one forms, then the projection $\pi: \mathscr{D}^k \to \mathrm{Gr}_k$ defines a line bundle on Gr_k. Its curvature is found as follows: For each element of Gr_k let u_i be an orthonormal basis for $\mathscr{V}_n$ adapted to the decomposition $\mathscr{V}_n = \mathscr{W}_k \oplus \mathscr{W}_k^{\perp}$, i.e. u_i, $i = 1, k$ span $\mathscr{W}_k$ and u_i, $i = k+1, n$ span $\mathscr{W}_k^{\perp}$. Define a connexion by the $n \times n$ matrix valued one-form on the grassmanian manifold given by $\omega_{ij} = \langle u_i | \mathrm{d}u_j \rangle \in \Lambda^1(\mathrm{Gr}_k)$. It is a flat connexion, i.e. it obeys

$$\Omega = \mathrm{d}\omega + \omega \wedge \omega = 0 . \tag{4.13}$$

Partition ω:

$$\omega = \begin{array}{c} \\ \mathscr{W}_k^{\perp} \\ \mathscr{W}_k \end{array} \begin{array}{c} \begin{array}{cc} \mathscr{W}_k^{\perp} & \mathscr{W}_k \end{array} \\ \begin{pmatrix} \omega_{++} & \omega_{+-} \\ \omega_{-+} & \omega_{--} \end{pmatrix} \end{array} . \tag{4.14}$$

Then the curvature form Ω_{--} on the k dimensional vector bundle given by projecting back onto $\mathscr{W}_k$ is given by

$$\Omega_{--} = \delta\omega_{--} + \omega_{--} \wedge \omega_{--} = -\omega_{-+} \wedge \omega_{+-} \tag{4.15}$$

and it is not flat.

This construction is exactly how parallel transport is defined on a sphere: At each point on the surface we take a basis for $\mathbb{R}^3$ with two vectors tangent to the sphere surface and one normal to it. If we move a 3-vector parallel to itself in $\mathbb{R}^3$ the connexion is flat, but if we constantly project back into the tangent plane we get the usual definition of parallel transport in the two-dimensional tangent space and this is not flat.

We want the curvature on the line bundle $\pi: \mathscr{D}^k \to \mathrm{Gr}_k$. The curvature comes from the Berry connection A^- on the exterior powers corresponding to the relevant grassmanian element:

$$A^{-}=\langle u_1\wedge u_2\cdots\wedge u_k|\mathrm{d}(u_1\wedge u_2\cdots\wedge u_k)\rangle. \tag{4.16}$$

The inner product on $\Lambda(\mathscr{W}_k)$ is the usual one induced on the exterior powers of $\mathscr{W}_k$ from the product on $\mathscr{W}_k$. Evaluating eq. (4.16) is straightforward and we find that the curvature of the line bundle is given by the first Chern character of the connexion on the bundle whose fibre is $\mathscr{W}_k$

$$\mathrm{d}A^{-}=F^{-}=\mathrm{Tr}(\Omega_{--})=-\mathrm{Tr}(\omega_{-+}\wedge\omega_{+-}). \tag{4.17}$$

We can similarly define F^{+} from the Chern character on the bundle whose fibre is $\mathscr{W}_k^{\perp}$ and find that $F^{+}+F^{-}=0$ in accordance with the theorem that Chern characters behave additively under Whitney sums.

In the infinite dimensional case the matrix elements ω_{nm} are given by $\langle n|g^{-1}\mathrm{d}g|m\rangle$ and because ω_{nm} goes to zero rapidly as $n-m\to\infty$, the sums in eq. (4.17) are convergent and the same calculation gives the curvature of the space of infinite Slater determinants. We find the vacuum curvature to be

$$\Omega(\delta_1 g,\delta_2 g)=(1/2\pi)\int\mathrm{Tr}\big(g^{-1}\delta_1 g\,\partial_x(g^{-1}\delta_2 g)\big). \tag{4.18}$$

5. The path integral

We now have all the ingredients for our path integral. Putting them together, we find the action to be

$$S=\frac{1}{4\pi}\int\mathrm{d}x\,\mathrm{d}t\,\mathrm{Tr}(g^{-1}\partial_x g)^2+i\frac{1}{2\pi}\int\mathrm{d}x\,\mathrm{d}t\,\mathrm{d}\tau\,\mathrm{Tr}\big(g^{-1}\partial_t g\,\partial_x(g^{-1}\partial_\tau g)\big). \tag{5.1}$$

The extra variable τ is the one normally required for the global definition of Wess–Zumino terms. At the moment this S is apparently *not* that found in refs. [17, 18] but be rewriting it slightly we will find it is the same. We have to note that our two-form Ω is not quite the one found by integrating the conventional Wess–Zumino three-form,

$$\mathrm{WZ}=(1/12\pi)\int\mathrm{Tr}\big((g^{-1}\delta g)^3\big), \tag{5.2}$$

over S^1 to make a two-form, but it is *cohomologous* to it. However changing the symplectic form defining a dynamical system, even by a co-boundary, changes the equations of motion so we must keep track of the difference. The conventional

Wess–Zumino term can be written as

$$\mathrm{WZ} = (1/4\pi)\int \mathrm{Tr}\left(\left[g^{-1}\partial_x g, g^{-1}\partial_t g\right] g^{-1}\partial_\tau g\right) \mathrm{d}x\,\mathrm{d}t\,\mathrm{d}\tau \tag{5.3}$$

and after some integration by parts as

$$\mathrm{WZ} = -\frac{1}{2\pi}\int \mathrm{Tr}\left(\left(g^{-1}\partial_t g\right)\partial_x\left(g^{-1}\partial_\tau g\right)\right)\mathrm{d}x\,\mathrm{d}t\,\mathrm{d}\tau - \frac{1}{4\pi}\int \mathrm{Tr}\left(g^{-1}\partial_x g g^{-1}\partial_t g\right)\mathrm{d}x\,\mathrm{d}t \tag{5.4}$$

exhibiting it as our curvature term Ω together with a part, written in terms of x and t only. This latter expression can be lumped together with the energy as part of a conventional action.

Using this we rewrite S as

$$S = \frac{1}{4\pi}\int \mathrm{Tr}\left(\left(g^{-1}\partial_x g\right)g^{-1}\left(\partial_x - i\partial_t\right)g\right)\mathrm{d}x\,\mathrm{d}t - \frac{i}{12\pi}\int \left(g^{-1}\mathrm{d}g\right)^3 \tag{5.5}$$

which is the same as refs. [17, 18].

The equation of motion (it seems more natural in this section to work in Minkowski space) is

$$\partial_x\left(g^{-1}\partial_x g + g^{-1}\partial_t g\right) = 0, \tag{5.6}$$

which tells us that

$$g^{-1}\partial_x g + g^{-1}\partial_t g + T(t) = 0, \tag{5.7}$$

where $T(t)$ is any x independent element of the Lie algebra. Thus the solutions are

$$g(x,t) = g(x-t)h(t) \tag{5.8}$$

describing right-going waves of $\mathrm{LU}(N)/\mathrm{U}(N)$. The arbitrary, x independent, factor of $h(t)$ occurs because of the gauge choice involved in the projection onto the coset, just as in eq. (2.13). This equation of motion is also discussed in ref. [16].

We now derive the Poisson brackets corresponding to the quantum operators $\psi^\dagger\lambda_i\psi$. We have already computed the required expectation values.

$$J_i(g) = \langle g|J_i|g\rangle = (1/2\pi i)\mathrm{Tr}\left(\lambda_i\,\partial_x g g^{-1}\right) \tag{5.9}$$

and we see that they are independent of the gauge choice as required. Since the

Poisson brackets will involve distributions it is convenient to take the functional

$$F(\phi) = \int dx\, \mathrm{Tr}\big(\lambda_i (\partial_x g) g^{-1}\big)\phi \tag{5.10}$$

to drive the motion. Using F as the hamiltonian, we find the equation of motion

$$\partial_x\big((1/2\pi) g^{-1}\dot{g} + g^{-1}\lambda_i g\phi\big) = 0, \tag{5.11}$$

or

$$\dot{g}(x) = -2\pi\lambda_i g\phi(x) + gT(t). \tag{5.12}$$

From this we compute

$$\frac{d}{dt}\mathrm{Tr}\big(\lambda_j\, \partial_x g g^{-1}\big) = \mathrm{Tr}\big(g^{-1}\lambda_j g\, \partial_x(g^{-1}\dot{g})\big)$$

$$= -2\pi\, \mathrm{Tr}\big(\lambda_i\lambda_j\big)\partial_x\phi + 2\pi\, \mathrm{Tr}\big(\big[\lambda_i, \lambda_j\big]\, \partial_x g g^{-1}\big)\phi(x). \tag{5.13}$$

The dependence on $T(t)$ drops out, as it must, and we find the Poisson bracket

$$\{J_i(x), J_j(y)\} = i f_{ij}^k J_k(x)\delta(x-y) - (1/2\pi)\mathrm{Tr}\big(\lambda_i\lambda_j\big)\partial_x\delta(x-y), \tag{5.14}$$

which shows, as expected, that the original fermionic operators generate a Kac–Moody algebra. The boson equivalent of $\psi^\dagger\lambda_i\psi$ is therefore

$$(1/2\pi i)Tr\big(\lambda_i\, \partial_x g g^{-1}\big).$$

6. Conclusion and discussion

In this paper we have shown that a formal use of the coherent state path integral for a loop group provides a natural explanation for the equivalence between free right-moving fermions and a chiral boson field taking values in LU(N)/U(N). We see that all the results for the equations of motion and classical Poisson brackets are analagous to those for finite dimensional compact simple groups.

In evaluating the curvature and the required expectation values we have had to use some ideas from the differential geometry of infinite grassmanian manifolds. This may give the results obtained a spurious air of sophistication but, although *these* manipulations can be made rigorous [16], no argument based on path integral methods can be regarded as a proof of anything without much more work: all the loop group constructions depend on smoothness assumptions about $g(x,t)$ and these are not satisfied by typical field configurations in a path integral. Further even the finite dimensional coherent state path integrals need care in their use because of operator ordering ambiguities. Despite these problems, the coherent state argument shows *why* there is a Fermi–Bose equivalence and why it takes the form it does. A rigorous argument for the equivalence can be constructed on the lines of ref. [5] by observing the essential uniqueness of the level-1 representations of the Kac–Moody algebra.

It is worth asking whether there is a natural geometric origin for the non-interaction theorem alluded to in the first paragraph of this paper (see ref. [7] for a discussion). Work on this topic is in progress.

An obvious question to ask is whether this approach to non-abelian bosonisation will work for the Dirac fermions. One has to be careful because the bosonic path integral for the Dirac fermions is not quite a product of decoupled left and right-handed boson fields: the requirement of vector gauge invariance couples the two modes together. Discussions of the relevant geometry of the coupled Kac–Moody algebras is given in ref. [25]. The appropriate coherent states to use may be analogues of those introduced for a top (whose configuration space is $SU(2) \otimes SU(2)/SU(2)$) in ref. [26]. The Witten non-abelian bosonisation also has the feature of being manifestly Lorentz invariant, a property that is quite cryptic in the present formalism.

I would like to thank Fedele Lizzi for valuable conversations and for showing me ref. [18], thus persuading me that my ideas on the subject might be of interest. I would also like to thank Alan McKane and Hugh Luckock for help while I was organizing my thoughts. I am grateful to the Physics Department of the University of Manchester for hospitality. This work was supported by SERC grant GR/E/91301 and by NSF-DMR-84-15063.

References

[1] S. Coleman, Phys. Rev. D11 (1975) 2088
[2] S. Mandelstam, Phys. Rev. D11 (1975) 3026
[3] M. Halpern, Phys. Rev. D12 (1975) 1684
[4] J. Lepowsky, S. Mandelstam and I. M. Singer, ed. Vertex operators in mathematics and physics (Springer, New York, 1985)
[5] E. Witten, Commun. Math. Phys. 92 (1984) 455
[6] A.M. Polyakov and P.B. Wiegman, Phys. Lett. B131 (1983) 121
[7] P. Goddard and D. Olive, Int. J. Modern Physics A1 (1986) 303
[8] J-P. Blaizot and H. Orland, Phys. Rev. C24 (1981) 1740
[9] H. Kuratsuji and T. Suzuki, Supp. Prog. Theor. Phys. 74 & 75 (1983) 209
[10] J.R. Klauder and B.-S. Skagerstam, ed., Coherent States (World Scientific, Singapore, 1985)p. 48
[11] P.B. Wiegman, Phys. Rev. Lett. 60 (1988) 821
[12] P.B. Wiegman, MIT preprint submitted to Phys. Lett. A
[13] J.M. Souriau, Structure des systems dynamics (Dunod, Paris, 1970)
[14] A.A. Kirillov, Elements of the theory of representations (Springer, Berlin, 1976)
[15] E. Witten, Commun. Math. Phys. 114 (1988) 1
[16] A. Pressley and G. Segal, Loop groups (Clarendon, Oxford, 1986) p. 48

[17] J. Sonnenschein, Nucl. Phys. B309 (1988) 752;
Y. Frishman and J. Sonnenschein, Nucl. Phys. B301 (1988) 346
[18] P. Salomonson, B.-S. Skagerstam and A. Stern, Phys. Rev. Lett. 62 (1989) 1817
[19] M. Stone, Nucl. Phys. B314 (1989) 557
[20] E. Onofri, J. Math. Phys. 16 (1975) 1087
[21] A.M. Perelomov, Commun. Math. Phys. 26 (1972) 222
[22] M.V. Berry, Proc. Roy. Soc. London A392 (1984) 45;
B. Simon, Phys. Rev. Lett. 51 (1983) 2167
[23] V. Guillemin and S. Sternberg, Symplectic techniques in physics (Cambridge Univ. Press, Cambridge, 1984)
[24] N. Woodhouse, Geometric quantization (Clarendon, Oxford, 1980)
[25] G. Felder, K. Gawedzki and A. Kupiainen, Commun. Math. Phys. 117 (1988) 127;
A. Abouelsaood and D. Gepner, Phys. Lett. B176 (1986) 380;
D. Gepner and E. Witten, Nucl. Phys. B278 (1986) 483
[26] D. Janssen, Sov. J. Nucl. Phys. 25 (1977) 479
[27] M. Stone, Phys. Rev. Lett. 63 (1989) 731